BIOLOGY

Science for Life

THIRD EDITION

Colleen Belk

University of Minnesota–Duluth

Virginia Borden Maier

St. John Fisher College

Benjamin Cummings

San Francisco Boston New York
Cape Town Hong Kong London Madrid Mexico City
Montreal Munich Paris Singapore Sydney Tokyo Toronto

Vice President, Editor-in-Chief: Beth Wilbur
Senior Acquisitions Editor: Star MacKenzie
Executive Director of Development: Deborah Gale
Development Editor: Cathy Murphy
Project Editor: Dusty Friedman
Art Editor: Elisheva Marcus
Assistant Editor: Erin Mann
Senior Media Producer: Jonathan Ballard
Executive Marketing Manager: Lauren Harp
Senior Marketing Manager: Jay Jenkins
VP, Director of Media Strategy: Stacy Treco
Associate Director of Production: Erin Gregg
Managing Editor: Michael Early
Senior Production Supervisor: Shannon Tozier
Production Service: Pre-Press PMG
Illustrations: Imagineering Media Services, Inc.
Text and Cover Design: Derek Bacchus
Manufacturing Buyer: Michael Penne
Director, Image Resource Center: Melinda Patelli
Manager, Rights and Permissions: Zina Arabia
Manager, Visual Research: Elaine Soares
Image Permission Coordinator: Cynthia Vincenti
Photo Research: Yvonne Gerin
Cover Printer: Phoenix Color
Printer and Binder: Quebecor World, Dubuque
Cover Image: Shutterstock Images; Sergey Popov V/Shutterstock Images; M.I. Walker/Science Photo Library/Photo Researchers, Inc; CreativEye99/Shutterstock Images

Library of Congress Cataloging-in-Publication Data

Belk, Colleen.
Biology : science for life / Colleen Belk, Virginia Borden Maier. – 3rd ed.
p. cm.
ISBN-13: 978-0-321-55959-3
ISBN-10: 0-321-55959-2
1. Biology. I. Borden, Virginia. II. Title.
QH307.2.B43 2010b
570–dc22

2008041338

ISBN 10-digit: 0-321-55959-2; 13-digit: 978-0-321-55959-3 (Student Edition)
ISBN 10-digit: 0-321-59553-X; 13-digit: 978-0-321-59553-9 (Professional Copy)
ISBN 10-digit: 0-321-59593-9; 13-digit: 978-0-321-59593-5 (Books a la Carte Edition)

Benjamin Cummings
is an imprint of

www.pearsonhighered.com

1 2 3 4 5 6 7 8 9 10—QWD—12 11 10 09

ABOUT THE AUTHORS

Colleen Belk and **Virginia Borden Maier** collaborated on teaching biology to nonmajors for over a decade together at the University of Minnesota–Duluth. This collaboration has continued through Virginia's move to St. John Fisher College in Rochester, New York, and has been enhanced by their differing but complementary areas of expertise. In addition to the nonmajors course, Colleen Belk teaches general biology for majors, genetics, cell biology, and molecular biology courses. Virginia Borden Maier teaches general biology for majors, evolutionary biology, plant biology, ecology, and conservation biology courses.

After several somewhat painful attempts at teaching all of biology in a single semester, the two authors came to the conclusion that they needed to find a better way. They realized that their students were more engaged when they understood how biology directly affected their lives. Colleen and Virginia began to structure their lectures around stories they knew would interest students. When they began letting the story drive the science, they immediately noticed a difference in student interest, energy, and willingness to work harder at learning biology. Not only has this approach increased student understanding, but it has also increased the authors' enjoyment in teaching the course—presenting students with fascinating stories infused with biological concepts is simply a lot more fun. This approach served to invigorate their teaching. Knowing that their students are learning the biology that they will need now and in the future gives the authors a deep and abiding satisfaction.

Preface

To the Student

As you worked your way through high school or otherwise worked to prepare yourself for college, you were probably unaware that an information explosion was taking place in the field of biology. This explosion, brought on by advances in biotechnology and communicated by faster, more powerful computers, has allowed scientists to gather data more quickly and disseminate data to colleagues in the global scientific community with the click of a mouse. Every discipline of biology has benefited from these advances, and today's scientists collectively know more than any individual could ever hope to understand.

Paradoxically, as it becomes more and more difficult to synthesize huge amounts of information from disparate disciplines within the broad field of biology, it becomes more vital that we do so. The very same technologies that led to the information boom, coupled with expanding human populations, present us with complex ethical questions. These questions include whether it is acceptable to clone humans, when human life begins and ends, who owns living organisms, what our responsibilities toward endangered species are, and many more. No amount of knowledge alone will provide satisfactory answers to these questions. Addressing these kinds of questions requires the development of a scientific literacy that surpasses the rote memorization of facts. To make decisions that are individually, socially, and ecologically responsible, you must not only understand some fundamental principles of biology but also be able to use this knowledge as a tool to help you analyze ethical and moral issues involving biology.

To help you understand biology and apply your knowledge to an ever-expanding suite of issues, we have structured each chapter of *Biology: Science for Life* around a compelling story in which biology plays an integral role. Through the story you not only will learn the relevant biological principles but also will see how science can be used to help answer complex questions. As you learn to apply the strategies modeled by the text, you will begin developing your critical thinking skills.

By the time you finish this book, you should have a clear understanding of many important biological principles. You will also be able to critically evaluate which information is most reliable instead of simply accepting all the information you hear or read about. Even though you may not be planning to be a practicing biologist, well-developed critical thinking skills will enable you to make decisions that affect your own life, such as whether to take nutritional supplements, and decisions that affect the lives of others, such as whether to believe the DNA evidence presented to you as a juror in a criminal case.

It is our sincere hope that understanding how biology applies to important personal, social, and ecological issues will convince you to stay informed about such issues. On the job, in your community, at the doctor's office, in the voting booth, and at home reading the paper or surfing the web, your knowledge of the basic biology underlying so many of the challenges that we as individuals and as a society face will enable you to make well-informed decisions for your home, your nation, and your world.

To the Instructor

By now you are probably all too aware that teaching nonmajors students is very different from teaching biology majors. You know that most of these students will never take another formal biology course; therefore, your course may be the last chance for these students to see the relevance of science in their everyday lives and the last chance to appreciate how biology is woven throughout the fabric of their lives. You recognize the importance of engaging these students because you know that these students will one day be voting on issues of scientific importance, holding positions of power in the community, serving on juries, and making health care decisions for themselves and their families. You know that your students' lives will be enhanced if they have a thorough grounding in basic biological principles and scientific literacy. To help your efforts toward this end, this text is structured around several themes.

Themes Throughout *Biology: Science for Life*

The Story Drives the Science. We have found that students are much more likely to be engaged in the learning process when the textbook and lectures capitalize on their natural curiosity. This text accomplishes this by using a story to drive the science in every chapter. Students get caught up in the story and become interested in learning the biology so they can see how the story is resolved. This approach allows us to cover the key areas of biology, including basic chemistry, the unity and diversity of life, cell structure and function, classical and molecular genetics, evolution, and ecology, in a manner that makes students want to learn. Not only do students want to learn, but also this approach allows students both to connect the science to their everyday lives and to integrate the principles and concepts for later application to other situations. This approach will give you flexibility in teaching and will support you in developing students' critical thinking skills.

The Process of Science. This book also uses another novel approach in the way that the process of science is modeled. The first chapter is dedicated to the scientific method and hypothesis testing, and each subsequent chapter weaves the scientific method and hypothesis testing throughout the story. The development of students' critical thinking skills is thus reinforced for the duration of the course. Students will see that the application of the scientific method is often the best way to answer questions raised in the story. This practice not only allows students to develop their critical thinking skills but also, as they begin to think like scientists, helps them understand why and how scientists do what they do.

Integration of Evolution. Another aspect of *Biology: Science for Life* that sets it apart from many other texts is the manner in which evolutionary principles are integrated throughout the text. The role of evolutionary processes is highlighted in every chapter, even when the chapter is not specifically focused on an evolutionary question. For example, when discussing infectious diseases, the evolution of antibiotic-resistant strains of bacteria is addressed. The physiology unit includes an essay on evolution in each chapter. These essays illustrate the importance of natural selection in the development of various organs and organ systems across a wide range of organisms. With evolution serving as an overarching theme, students are better able to see that all of life is connected through this process.

Pedagogical Elements

Open the book and flip through a few pages and you will see some of the most inviting, lively, and informative illustrations you have ever seen in a biology text. The illustrations are inviting because they have a warm, hand-drawn quality that is clean and uncluttered. Most important, the illustrations are informative not only because they were carefully crafted to enhance concepts in the text but also because they employ techniques like the "pointer" that help draw the students' attention to the important part of the figure (see p. 184). Likewise, tables are more than just tools for organizing information; they are illustrated to provide attractive, easy references for the student. We hope that the welcoming nature of the art and tables in this text will encourage nonmajors to explore ideas and concepts instead of being overwhelmed before they even get started.

In addition to lively illustrations of conventional biology concepts, this text also uses analogies to help students understand difficult topics. For example, the process of translation is likened to baking a cake (see p. 201). These analogies and illustrations are peppered throughout the text.

Students can reinforce and assess what they are learning in the classroom by reading the chapter, studying the figures, and answering the end-of-chapter questions. We have written these questions in every format likely to be used by an instructor during an exam so that students have practice in answering many different types of questions. We have also included "Connecting the Science" questions that would be appropriate for essay exams, class discussions, or use as topics for term papers.

Improvements in the Third Edition

The positive feedback garnered by previous editions assured us that presenting science alongside a story works for students and instructors alike. With this edition, we focused on improving flexibility for instructors, helping students assess their understanding, and providing students with opportunities to be better consumers of popular media.

- In an effort to accommodate the wide range of teaching and learning styles of those using our book, we rearranged some content (and increased our coverage of some topics) into sections titled **A Closer Look.** Preceding each **A Closer Look** section is a general overview of the topic sufficient to provide nonmajors students with a basic understanding. **A Closer Look** then provides details that add depth to this understanding. This feature provides instructors with teaching flexibility; for example, some instructors are satisfied with a general overview of cellular respiration, including the reactants and products, and similar basics of metabolism. Others want to teach electron transport and oxidative phosphorylation. **A Closer Look** facilitates both approaches. Preceeding this section is the general overview. The **Closer Look** section itself is differentiated from the rest of the text, making it easy to include or omit from students' assigned reading.

- Each chapter contains several **Stop and Stretch** questions designed to provide rest stops where students can assess whether they have understood the material they just completed reading. These questions require students to synthesize material they have read and apply their knowledge in a new situation. Many of these questions are designed to familiarize students with data analysis and help them develop skill at interpreting and understanding scientific data. Answers to these questions are provided in the appendix.
- Another feature that aids in student learning is **Visualize This,** questions in figure captions that test a student's understanding of the concepts contained in the art. These questions are designed to encourage students to spend extra time with complex figures and develop a more sophisticated understanding of the concepts they present. **Visualize This** questions are included with three or more figures in each chapter and answers to these questions can be found in the appendix.
- Another new feature that we are very excited about is an end-of-chapter feature titled **Savvy Reader.** We developed this feature in response to the desire of so many of the professors teaching this course to help students learn to critically evaluate science presented in the media. We all want our students to become better, more informed consumers of science. Each **Savvy Reader** includes an article or advertisement taken from any of a variety of media sources. After reading the excerpt, the student then answers a series of questions written with the goal of helping them critically evaluate the scientific information presented in the article. Students learn to evaluate claims made in the popular press about issues ranging from the benefits of particular health care products to whether our religious preferences are encoded in our genes.

Supplements

A group of talented and dedicated biology educators teamed up with us to build a set of resources that equip nonmajors with the tools to achieve scientific literacy that will allow them to make informed decisions about the biological issues that affect them daily. The student resources offer several ways of reviewing and reinforcing the concepts and facts covered in this textbook. We provide instructors with a test bank, Instructor Guide, and the Instructor Resource DVD which includes an updated and expanded suite of lecture presentation materials, and a valuable source of ideas for educators to enrich their instruction efforts. Available in print and media formats, the *Biology: Science for Life with Physiology* resources are easy to navigate and support a variety of learning and teaching styles.

We believe you will find that the design and format of this text and its supplements will help you meet the challenge of helping students both succeed in your course and develop science skills—for life.

Supplement Authors

Study Guide
Catherine Podeszwa

Instructor Guide
Patricia Johnson, *Palm Beach Community College*

Test Bank/Web Quizzes
Deborah Cato, *Wheaton College*
Gregory S. Pryor, *Francis Marion University*
Troy Rohn, *Boise State University*
Carol St. Angelo, *Hofstra University*

Study Card
Patricia Johnson, *Palm Beach Community College*

PowerPoint Lectures
Steve McCommas, *Souther Illinois University*
James Hutcheon, *Georgia Southern University*

Supplements Contributors
Deborah Cato, *Wheaton College*
William H. Coleman, *University of Hartford*
Cynthia Ghent, *Towson University*
Diane Melroy, *University of North Carolina, Wilmington*
Anthony Palombella, *Longwood University*
Gregory S. Pryor, *Francis Marion University*
Troy Rohn, *Boise State University*
Carol St. Angelo, *Hofstra University*

Supplements Reviewers
James S. Backer, *Daytona Beach Community College*
David Belt, *Penn Valley Community College*
Barbara Blonder, *Flagler College*
Kimberly Cline-Brown, *University of Northern Iowa*
Judy Dacus, *Cedar Valley College*
Lisa Delissio, *Salem State College*
Chris Farrell, *Trevecca Nazarene University*
Stewart Frankel, *University of Hartford*
Anthony Ippolito, *DePaul University*
Michael A. Kotarski, *Niagara University*
Michelle Mabry, *Davis and Elkins College*
Matthew J. Maurer, *University of Virginia's College at Wise*
Wilma Robertson, *Boise State University*
Joanne Russell, *Manchester Community College*
Carol St. Angelo, *Hofstra University*
Joyce Tamashiro, *University of Puget Sound*
Michael Troyan, *Pennsylvania State University*

* * *

As with previous editions, the overall goal of the text remains providing a thorough overview of the essentials of biological science while trying to avoid overloading students with information. We worked closely with instructors using prior editions, as well as other reviewers, to pinpoint essential content to include in this edition while staying true to the book's philosophy of learning science in a story format. The development of the third edition has truly been a collaborative process among ourselves, the students and instructors who used prior editions, and our many thoughtful reviewers. We look forward to learning about your experience with *Biology: Science for Life, 3rd edition.*

> "*Because science, told as a story, can intrigue and inform the non-scientific minds among us, it has the potential to bridge the two cultures into which civilization is split—the sciences and the humanities. For educators, stories are an exciting way to draw young minds into the scientific culture.*"
>
> —E.O. Wilson

Compelling stories make biology more accessible and understandable

Captivating narrative helps demystify biology topics

Every chapter introduces biology topics through a story, helping to explain important biological processes using concrete examples and applications.

Real-life Examples

Each chapter story draws from real-life examples, making the reading more engaging and accessible.

Chapter 5 frames the discussion of DNA Synthesis, Mitosis, and Meiosis around the interesting story of a college student who receives a cancer diagnosis.

Nicole's early college career was similar to that of most students. She enjoyed her independence and the wide variety of courses required for her double major in biology and psychology. She worried about her grades and finding ways to balance her course work with her social life. She also tried to find time for lifting weights in the school's athletic center and snowboarding at a local ski hill. Some weekends, to take a break from school, she would ride the bus home to see her family.

Managing to get schoolwork done, see friends and family, and still have time left to exercise had been difficult, but possible, for Nicole during her first two years at school. That changed drastically during her third year of school.

One morning in October of her junior year, Nicole began having severe pains in her abdomen. The first time this happened, she was just beginning an experiment in her cell biology laboratory. Hunched over and sweating, she barely managed to make it through her two-hour biology lab. Over the next few days, the pain intensified so much that she was unable to walk from her apartment to her classes without stopping several times to rest.

Later that week, as she was preparing to leave for class, the pain was so severe that she had to lie d... ...allway of her apartment. When her ro... ...got home a few

The story is revisited throughout its chapter. In the example below, the story narrative provides an opportunity for students to learn about cellular reproduction as they follow the cancer patient's experience.

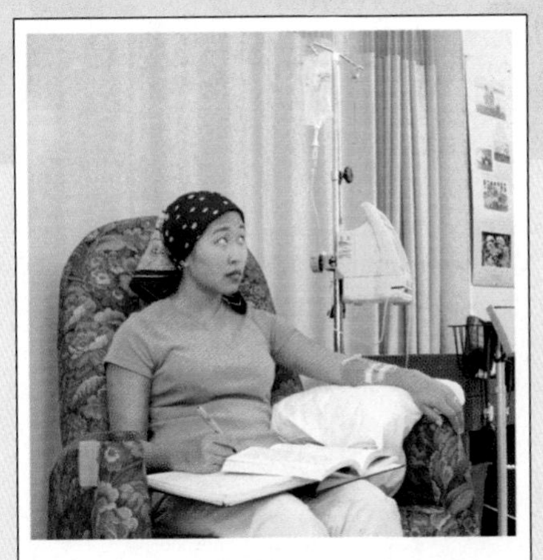

Figure 5.17 **Chemotherapy.** Many chemotherapeutic agents, such as Nicole's Taxol, are administered through an intravenous (IV) needle.

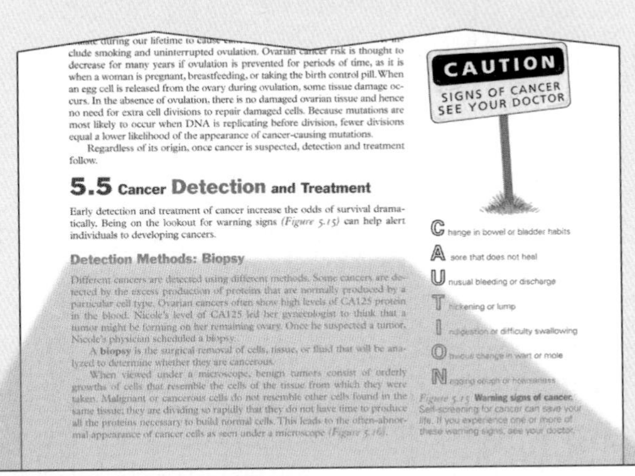

...during our lifetime to cause... include smoking and uninterrupted ovulation. Ovarian cancer risk is thought to decrease for many years if ovulation is prevented for periods of time, as it is when a woman is pregnant, breastfeeding, or taking the birth control pill. When an egg cell is released from the ovary during ovulation, some tissue damage occurs. In the absence of ovulation, there is no damaged ovarian tissue and hence no need for extra cell divisions to repair damaged cells. Because mutations are most likely to occur when DNA is replicating before division, fewer divisions equal a lower likelihood of the appearance of cancer-causing mutations.

Regardless of its origin, once cancer is suspected, detection and treatment follow.

5.5 Cancer Detection and Treatment

Early detection and treatment of cancer increase the odds of survival dramatically. Being on the lookout for warning signs (*Figure 5.15*) can help alert individuals to developing cancers.

Detection Methods: Biopsy

Different cancers are detected using different methods. Some cancers are detected by the excess production of proteins that are normally produced by a particular cell type. Ovarian cancers often show high levels of CA125 protein in the blood. Nicole's level of CA125 led her gynecologist to think that a tumor might be forming on her remaining ovary. Once he suspected a tumor, Nicole's physician scheduled a biopsy.

A **biopsy** is the surgical removal of cells, tissue, or fluid that will be analyzed to determine whether they are cancerous.

When viewed under a microscope, benign tumors consist of orderly growths of cells that resemble the cells of the tissue from which they were taken. Malignant or cancerous cells do not resemble other cells found in the same tissue; they are dividing so rapidly that they do not have time to produce all the proteins necessary to build normal cells. This leads to the often-abnormal appearance of cancer cells as seen under a microscope (*Figure 5.16*).

Figure 5.15 **Warning signs of cancer.** Self-screening for cancer can save your life. If you experience one or more of these warning signs, see your doctor.

Detection Methods: Biopsy

Different cancers are detected using different methods. Some cancers are detected by the excess production of proteins that are normally produced by a particular cell type. Ovarian cancers often show high levels of CA125 protein in the blood. Nicole's level of CA125 led her gynecologist to think that a tumor might be forming on her remaining ovary. Once he suspected a tumor, Nicole's physician scheduled a biopsy.

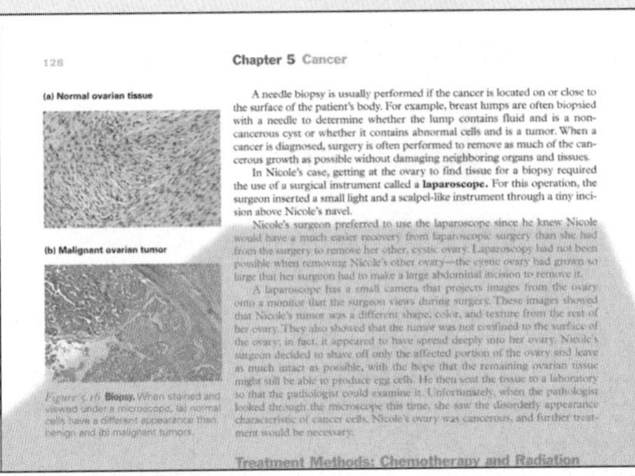

128 **Chapter 5** Cancer

(a) Normal ovarian tissue

(b) Malignant ovarian tumor

Figure 5.16 **Biopsy.** When stained and viewed under a microscope, (a) normal cells have a different appearance than benign and (b) malignant tumors.

A needle biopsy is usually performed if the cancer is located on or close to the surface of the patient's body. For example, breast lumps are often biopsied with a needle to determine whether the lump contains fluid and is a noncancerous cyst or whether it contains abnormal cells and is a tumor. When a cancer is diagnosed, surgery is often performed to remove as much of the cancerous growth as possible without damaging neighboring organs and tissues.

In Nicole's case, getting at the ovary to find tissue for a biopsy required the use of a surgical instrument called a **laparoscope.** For this operation, the surgeon inserted a small light and a scalpel-like instrument through a tiny incision above Nicole's navel.

Nicole's surgeon preferred to use the laparoscope since he knew Nicole would have a much easier recovery from laparoscopic surgery than she had from the surgery to remove her other, cystic ovary. Laparoscopy had not been possible when removing Nicole's other ovary—the cystic ovary had grown so large that her surgeon had to make a large abdominal incision to remove it.

A laparoscope has a small camera that projects images from the ovary onto a monitor that the surgeon views during surgery. These images showed that Nicole's tumor was a different shape, color, and texture from the rest of her ovary. They also showed that the tumor was not confined to the surface of the ovary; in fact, it appeared to have spread deeply into her ovary. Nicole's surgeon decided to shave off only the affected portion of the ovary and leave as much intact as possible, with the hope that the remaining ovarian tissue might still be able to produce egg cells. He then sent the tissue to a laboratory so that the pathologist could examine it. Unfortunately, when the pathologist looked through the microscope this time, she saw the disorderly appearance characteristic of cancer cells. Nicole's ovary was cancerous, and further treatment would be necessary.

Treatment Methods: Chemotherapy and Radiation

Nicole's surgeon preferred to use the laparoscope since he knew Nicole would have a much easier recovery from laparoscopic surgery than she had from the surgery to remove her other, cystic ovary. Laparoscopy had not been possible when removing Nicole's other ovary—the cystic ovary had grown so large that her surgeon had to make a large abdominal incision to remove it.

A laparoscope has a small camera that projects images from the ovary onto a monitor that the surgeon views during surgery. These images showed that Nicole's tumor was a different shape, color, and texture from the rest of her ovary. They also showed that the tumor was not confined to the surface of the ovary; in fact, it appeared to have spread deeply into her ovary. Nicole's surgeon decided to shave off only the affected portion of the ovary and leave as much intact as possible, with the hope that the remaining ovarian tissue might still be able to produce egg cells. He then sent the tissue to a laboratory so that the pathologist could examine it. Unfortunately, when the pathologist looked through the microscope this time, she saw the disorderly appearance characteristic of cancer cells. Nicole's ovary was cancerous, and further treatment would be necessary.

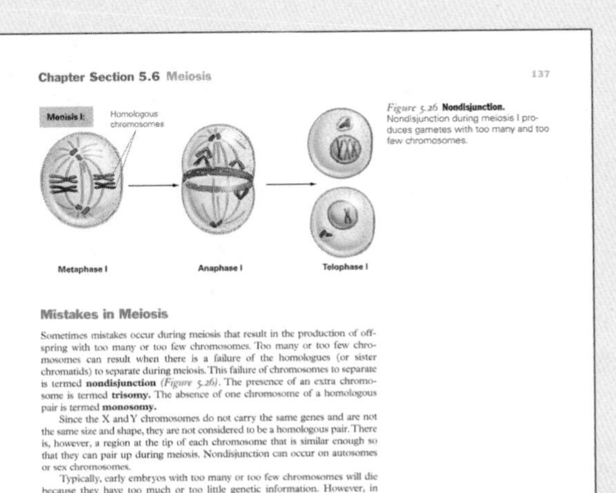

Chapter Section 5.6 Meiosis 137

Figure 5.26 **Nondisjunction.** Nondisjunction during meiosis I produces gametes with too many and too few chromosomes.

Mistakes in Meiosis

Sometimes mistakes occur during meiosis that result in the production of offspring with too many or too few chromosomes. Too many or too few chromosomes can result when there is a failure of the homologues (or sister chromatids) to separate during meiosis. This failure of chromosomes to separate is termed **nondisjunction** (*Figure 5.26*). The presence of an extra chromosome is termed **trisomy.** The absence of one chromosome of a homologous pair is termed **monosomy.**

Since the X and Y chromosomes do not carry the same genes and are not the same size and shape, they are not considered to be a homologous pair. There is, however, a region at the tip of each chromosome that is similar enough so that they can pair up during meiosis. Nondisjunction can occur on autosomes or sex chromosomes.

Typically, early embryos with too many or too few chromosomes will die because they have too much or too little genetic information. However, in some situations, such as when the extra or missing chromosome is very small (such as in chromosomes 21, 13, and 18) or contains very little genetic information (such as the Y chromosome or an X chromosome that will later be inactivated), the embryo can survive. *Table 5.2* lists some chromosomal anomalies in humans that are compatible with life.

From the previous discussions, you have learned that cells undergo mitosis for growth and repair and meiosis to produce gametes. *Figure 5.27* compares the significant features of mitosis and meiosis.

It is now possible to revisit the question of whether Nicole will pass on cancer-causing genes to any children she may have. Because Nicole developed cancer at such a young age, it seems likely that she may have inherited at least one mutant cell-cycle control gene; thus, she may or may not pass that gene on. If Nicole has both a normal and a mutant version of a cell-cycle control gene, then she will be able to make gametes with and without the mutant allele. Therefore, she could pass on the mutant allele if a gamete containing that allele is involved in fertilization. We have also seen that it takes many "hits" or mutations for a cancer to develop. Therefore, even if Nicole does pass on one or a few mutant versions of cell-cycle control genes to a child, environmental conditions will dictate whether enough other mutations will accumulate to allow a cancer to develop.

Mutations caused by environmental exposures are not passed from parents to children unless the mutation happens to occur in a cell of the gonads that will be used to produce a gamete. Nicole's cancer occurred in the ovary,

It is now possible to revisit the question of whether Nicole will pass on cancer-causing genes to any children she may have. Because Nicole developed cancer at such a young age, it seems likely that she may have inherited at least one mutant cell-cycle control gene; thus, she may or may not pass that gene on. If Nicole has both a normal and a mutant version of a cell-cycle control gene, then she will be able to make gametes with and without the mutant allele. Therefore, she could pass on the mutant allele if a gamete containing that allele is involved in fertilization. We have also seen that it takes many "hits" or mutations for a cancer to develop. Therefore, even if Nicole does pass on one or a few mutant versions of cell-cycle control genes to a child, environmental conditions will dictate whether enough other mutations will accumulate to allow a cancer to develop.

A completely re-designed art program gives the book a unified, fresh look

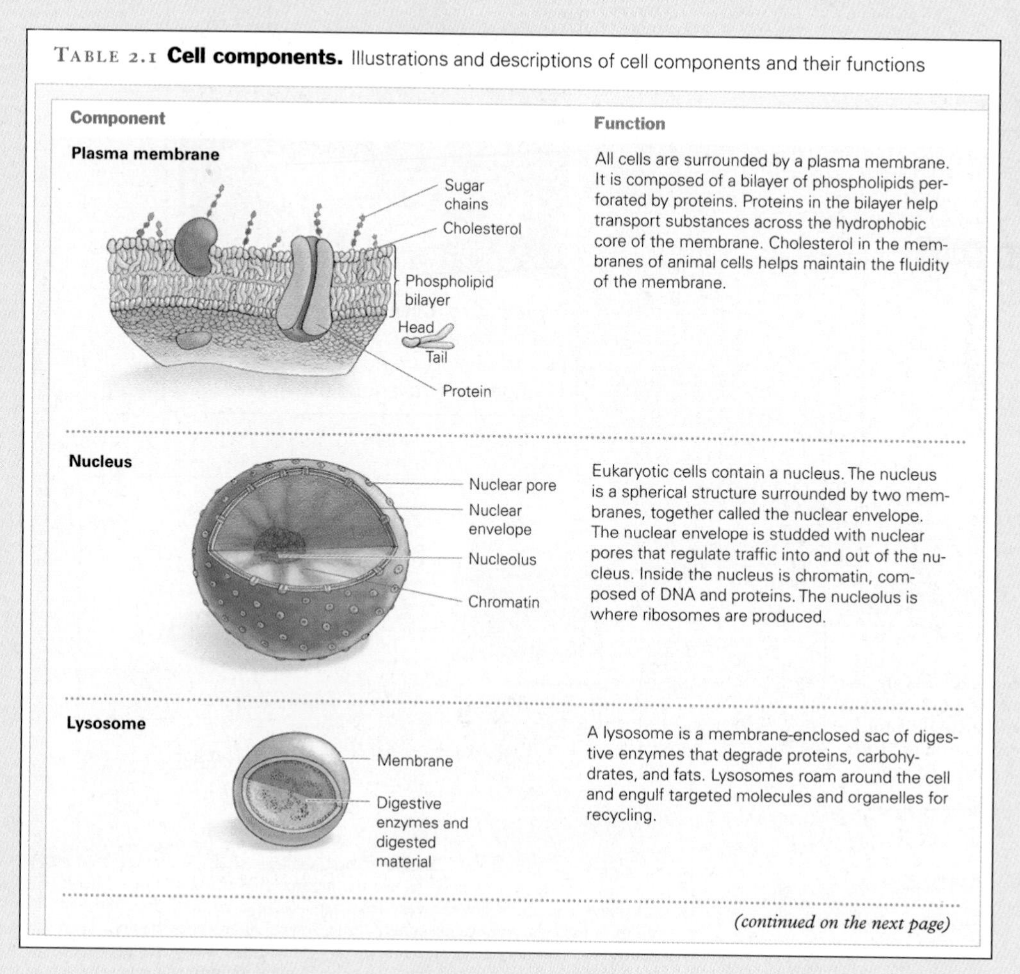

TABLE 2.1 **Cell components.** Illustrations and descriptions of cell components and their functions

Component	Function
Plasma membrane (Sugar chains; Cholesterol; Phospholipid bilayer; Head; Tail; Protein)	All cells are surrounded by a plasma membrane. It is composed of a bilayer of phospholipids perforated by proteins. Proteins in the bilayer help transport substances across the hydrophobic core of the membrane. Cholesterol in the membranes of animal cells helps maintain the fluidity of the membrane.
Nucleus (Nuclear pore; Nuclear envelope; Nucleolus; Chromatin)	Eukaryotic cells contain a nucleus. The nucleus is a spherical structure surrounded by two membranes, together called the nuclear envelope. The nuclear envelope is studded with nuclear pores that regulate traffic into and out of the nucleus. Inside the nucleus is chromatin, composed of DNA and proteins. The nucleolus is where ribosomes are produced.
Lysosome (Membrane; Digestive enzymes and digested material)	A lysosome is a membrane-enclosed sac of digestive enzymes that degrade proteins, carbohydrates, and fats. Lysosomes roam around the cell and engulf targeted molecules and organelles for recycling.

(continued on the next page)

Illustrated Tables

Unique comprehensive tables organize information in one place and provide easy visual references for students.

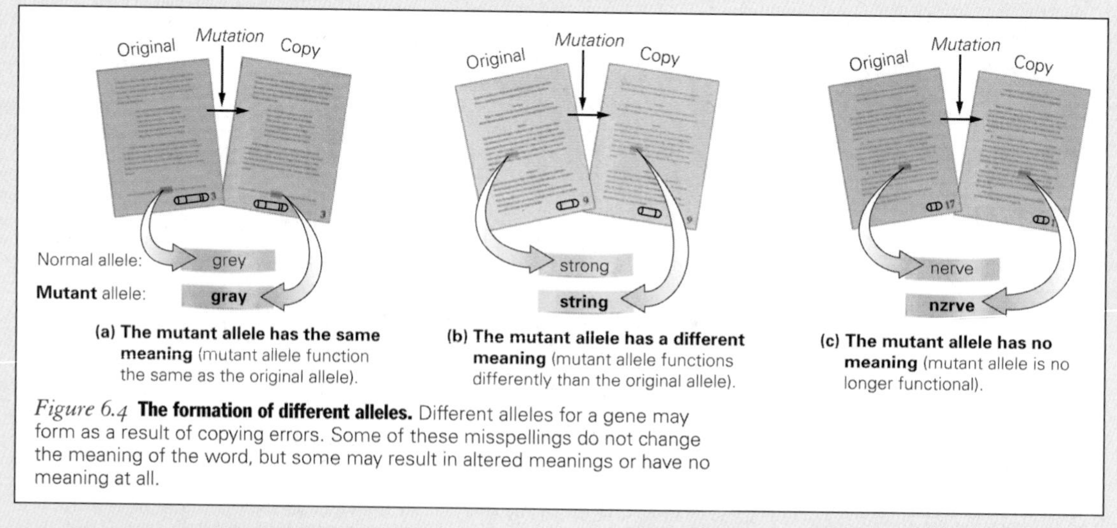

(a) The mutant allele has the same meaning (mutant allele function the same as the original allele).

(b) The mutant allele has a different meaning (mutant allele functions differently than the original allele).

(c) The mutant allele has no meaning (mutant allele is no longer functional).

Figure 6.4 **The formation of different alleles.** Different alleles for a gene may form as a result of copying errors. Some of these misspellings do not change the meaning of the word, but some may result in altered meanings or have no meaning at all.

Visual Analogies

Coordinating with the authors' use of analogies in their writing, the art program uses *visual analogies* to help students better understand complex processes. In Figure 6.4, genes are illustrated as words in the pages of an instruction manual.

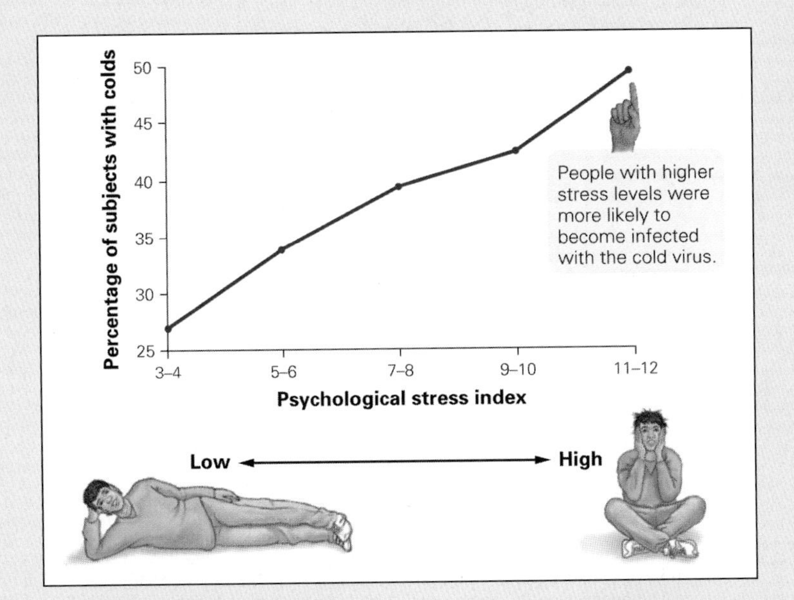

Informative Illustrations

Select illustrations use a **"pointer hand"** to draw students' attention to the important part of the figure, emulating an instructor's hand at the whiteboard.

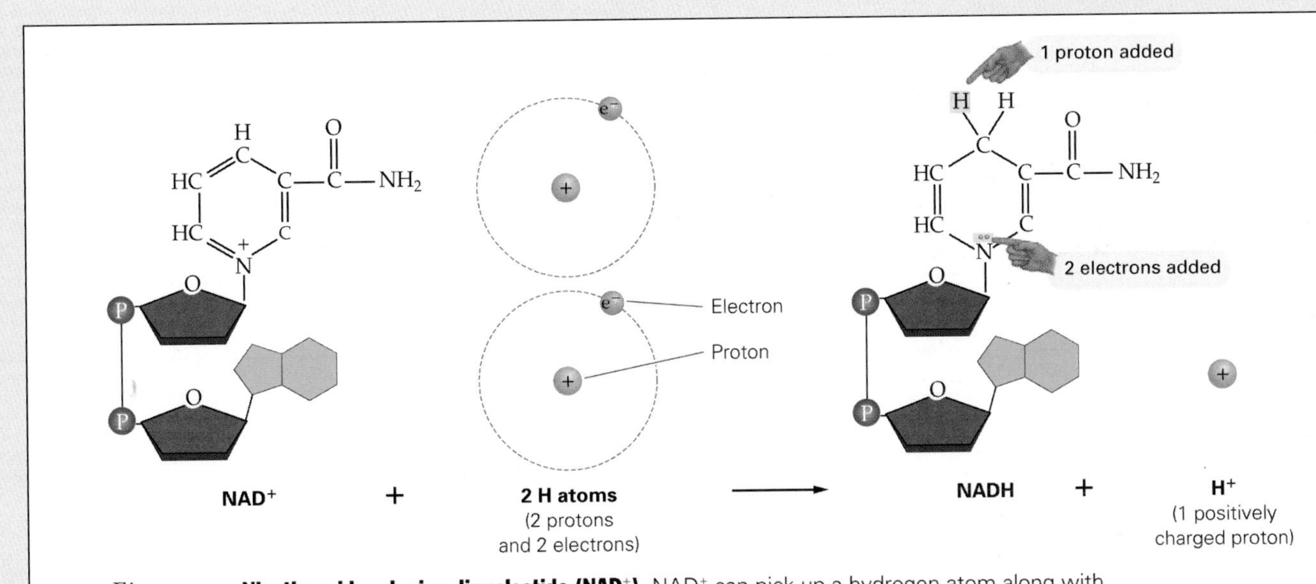

Figure 4.20 **Nicotinamide adenine dinucleotide (NAD^+).** NAD^+ can pick up a hydrogen atom along with its electrons. Hydrogen atoms are composed of 1 negatively charged electron that circles around the 1 positively charged proton. When NAD^+ encounters 2 hydrogen atoms (from food), it utilizes each hydrogen atom's electron and only 1 proton, thus releasing 1 proton. **Visualize This:** Why does the number of double bonds in the nitrogenous base change when a proton is added to NAD^+ to produce NADH?

NEW! Visualize This

These questions within select figure legends encourage students to delve into a figure's content to reinforce their understanding. Suggested answers are provided in the "Answers" section at the back of the book.

New to the Third Edition

Flexible Overview *and* A Closer Look *Sections*

Selected topics are now set off in sections titled **Overview** and **A Closer Look.**

This delineation allows instructors to assign an **Overview** of a complex biological topic such as membrane transport...

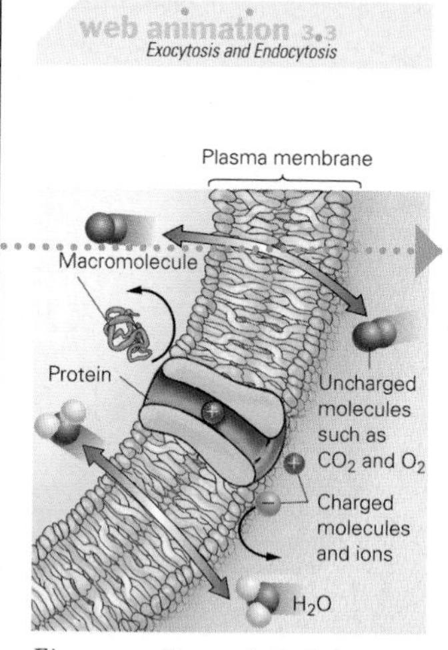

Figure 3.10 **Transport of substances across membranes.** The ability of a substance to cross a membrane is, in part, a function of its size and charge.

3.3 Transport Across Membranes

Building-block molecules must cross the plasma membrane to gain access to the inside of the cell, where they can be used to synthesize cell components or be metabolized to provide energy for the cell. The chemistry of the membrane facilitates the transport of some substances and prevents the transport of others.

Overview: Membrane Transport

Recall from Chapter 2 that the plasma membrane that surrounds cells is composed of a phospholipid bilayer. The interior of the bilayer is hydrophobic. Hydrophobic substances can dissolve in the membrane and pass through it more easily than hydrophilic ones. In this sense, the membrane of the cell is differentially permeable to the transport of molecules, allowing some to pass and blocking others from passing.

Substances that can cross the membrane will do so until the concentration is equal on both sides of the membrane. Carbon dioxide, water, and oxygen move freely across the membrane. Larger molecules and charged molecules and ions cannot cross the lipid bilayer on their own. If these substances need to be moved across the membrane they must move through proteins embedded in the membrane. Proteins in the membrane can serve as channels to allow such molecules to cross until their concentration is equal on both sides of the membrane. Proteins in the membrane can also move substances across the membrane past the point of their chemical equilibrium when more of a substance is required on one side of the membrane than the other *(Figure 3.10)*.

...or instructors can expand the reading assignment to include **A Closer Look,** which provides greater detail on processes such as passive transport, active transport, exocytosis, and endocytosis.

A Closer Look: Membrane Transport

As you learned in the preceding section, the manner in which a substance is moved across a membrane depends on its particular chemistry and concentration in the cell. Each type of transport has its own characteristics and energy requirements.

Passive Transport: Diffusion, Facilitated Diffusion, and Osmosis. All molecules contain energy that makes them vibrate and bounce against each other, scattering around like billiard balls during a game of pool. In fact, molecules will bounce against each other until they are spread out over all

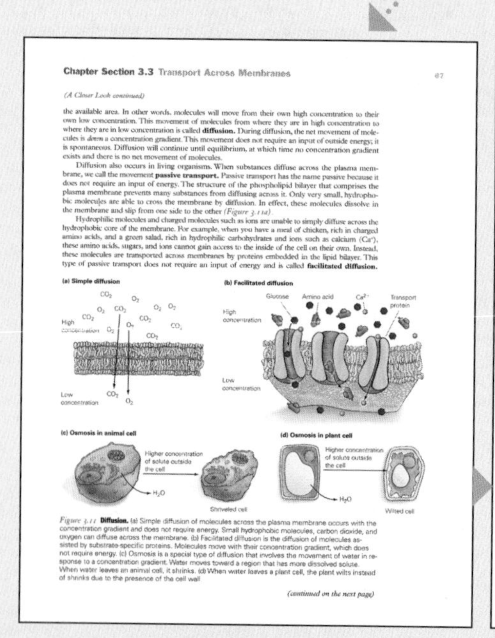
Chapter Section 3.3 Transport Across Membranes

(A Closer Look continued)

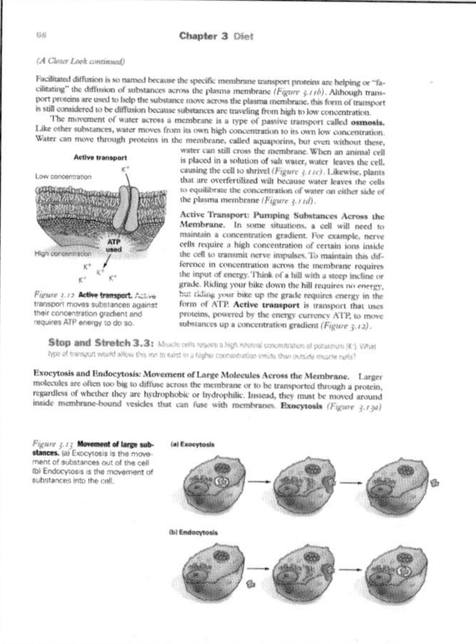
Chapter 3 Diet

(A Closer Look continued)

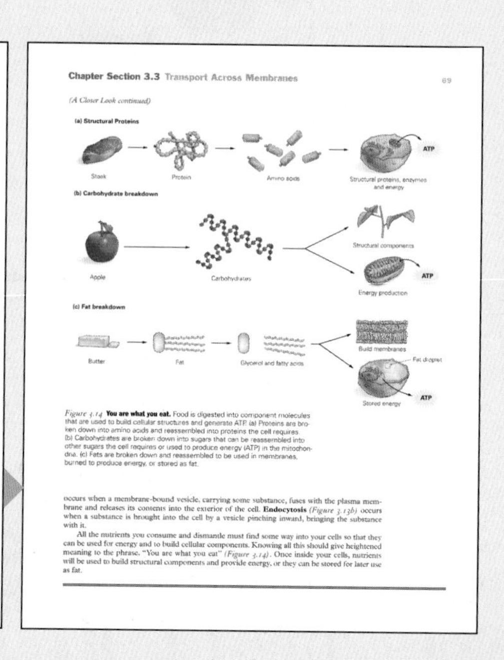
Chapter Section 3.3 Transport Across Membranes

(A Closer Look continued)

Stop and Stretch

These questions help students pace their reading, think critically, and apply the concepts they have just read. Suggested answers are provided in the "Answers" section at the back of the book.

Stop and Stretch 3.1: Many cultures have characteristic meals that include protein from two different sources. For example, beans and rice, eggs and toast, and cereal and milk are common food pairings. Why might this practice of pairing different foods have evolved?

(a) Lentils are high in lysine and low in valine. (b) Rice is low in lysine and high in valine.

The side groups of lysine and valine are different.

Lysine Valine Lysine Valine

Figure 3.2 **Essential amino acids.** Essential amino acids, such as lysine and valine, cannot be synthesized by the body and must be obtained from the diet. Some proteins are high in one amino acid but low in another. Eating a wide variety of proteins thus ensures that all the necessary amino acids are available for growth, development, and maintenance.

essential amino acids your body needs. Proteins obtained by eating meat are more likely to be complete than are those obtained by eating plants; plant proteins can often be missing one or more essential amino acids (*Figure 3.2*).

In the past, some nutritionists believed that vegetarians might be at risk for deficiencies in certain amino acids. However, scientific studies have shown that there is little cause for concern. If a vegetarian's diet is rich in a wide variety of plant-based foods, the body will have little trouble obtaining all the amino acids it needs to build proteins.

Stop and Stretch 3.1: Many cultures have characteristic meals that include protein from two different sources. For example, beans and rice, eggs and toast, and cereal and milk are common food pairings. Why might this practice of pairing different foods have evolved?

Even though you can obtain all the proteins you need by eating a variety of plants, Americans tend to eat a lot of meat as well. In addition to being rich in proteins, meat tends to be rich in fat.

Fats as Nutrients. The body uses fat, a type of lipid, as a source of energy. Gram for gram, fat contains a little more than twice as much energy as carbohydrates or protein. This energy is stored in the carbon, oxygen, and hydrogen bonds of the fat molecule.

Foods that are rich in fat include meat, milk, cheese, vegetable oils, and nuts. Muscle is often surrounded by stored fat, but some animals store fat throughout muscle, leading to the marbled appearance of some red meat. Other animals—chickens, for example—store fat on the surface of the muscle, making it easy to remove for cooking (*Figure 3.3*). Most mammals, including humans, store fat just below the skin to help cushion and protect vital organs, to insulate the body from cold weather, and to store energy in case of famine. Some scientists believe that prehistoric humans often faced times of famine and may have evolved to crave fat.

Recall from Chapter 2 that fats consist of a glycerol molecule with hydrogen and carbon-rich fatty acid tails attached. Your body can synthesize most of the fatty acids it requires. Those that cannot be synthesized are called

Figure 3.3 **Fat storage.** Fat can be intertwined with muscle tissue, as seen in this marbled piece of beef (a), or it can lie on the surface, as seen on this chicken leg (b).

Savvy Reader

These boxes feature a short excerpt drawn from a variety of sources (newspaper, magazine, website) followed by questions to help students interpret and evaluate scientific information and data found in everyday media. Suggested answers are provided in the "Answers" section at the back of the book.

Savvy Reader Do Scientists Doubt Evolution?

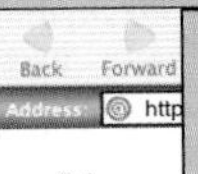

Savvy Reader Global Warming Killing the Polar Bears?

Savvy Reader Oxygizer Improves Performance?

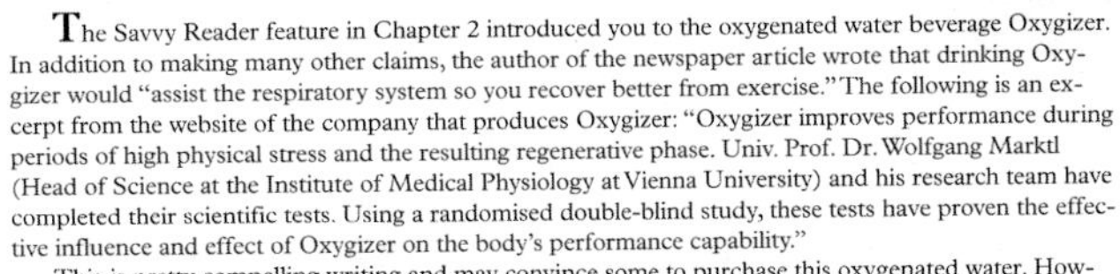

The Savvy Reader feature in Chapter 2 introduced you to the oxygenated water beverage Oxygizer. In addition to making many other claims, the author of the newspaper article wrote that drinking Oxygizer would "assist the respiratory system so you recover better from exercise." The following is an excerpt from the website of the company that produces Oxygizer: "Oxygizer improves performance during periods of high physical stress and the resulting regenerative phase. Univ. Prof. Dr. Wolfgang Marktl (Head of Science at the Institute of Medical Physiology at Vienna University) and his research team have completed their scientific tests. Using a randomised double-blind study, these tests have proven the effective influence and effect of Oxygizer on the body's performance capability."

This is pretty compelling writing and may convince some to purchase this oxygenated water. However, let's also look at an excerpt from the actual scientific study performed by Dr. Marktl and published in the *International Journal of Sports Medicine* in March 2006. "Results showed no significant influence on aerobic parameters or lactate metabolism, neither at submaximal nor at maximal levels. We conclude that the consumption of oxygenated water does not enhance aerobic performance."

1. Does it appear that the author of the newspaper article read the actual study or the promotional material only?
2. How are claims made in the newspaper and on websites different from claims made by authors of articles published in scientific journals?
3. The Oxygizer website also includes some data (http://www.oxygizer.com/default.aspx?lngId=2) that seem to support their claims. Private companies can hire their own scientists to perform studies that often have results that differ from those of government and university-sponsored scientists. Would you be more skeptical of results produced by scientists hired by the company whose product they are testing or scientists who work for the government or a University?
4. Carefully consider the following two sentences from the Oxygizer website: "Univ. Prof. Dr. Wolfgang Marktl (Head of Science at the Institute of Medical Physiology at Vienna University) and his research team have completed their scientific tests. Using a randomised double-blind study, these tests have proven the effective influence and effect of Oxygizer on the body's performance capability." Each of these sentences, read separately, is true. Dr. Marktl and his team did complete their tests, and the Oxygizer scientists did produce data showing increased performance capability. However, placed adjacent to each other, these sentences seem to be indicating that Dr. Marktl's university-sponsored research came up with results that were actually produced by the Oxygizer scientists. Do you think this is a willful attempt to deceive potential customers? Most people don't have time to do such a thorough analysis of every newspaper article they read. This is why it is helpful to develop a general level of skepticism about most product claims.

NEW! mybiology saves time for students and instructors

mybiology is the central place for students to access the online materials needed to succeed in non-majors biology courses. Following a new four-step learning path, students can make the most of their study time by accessing high quality, book-specific practice materials before exams.

www.mybiology.com

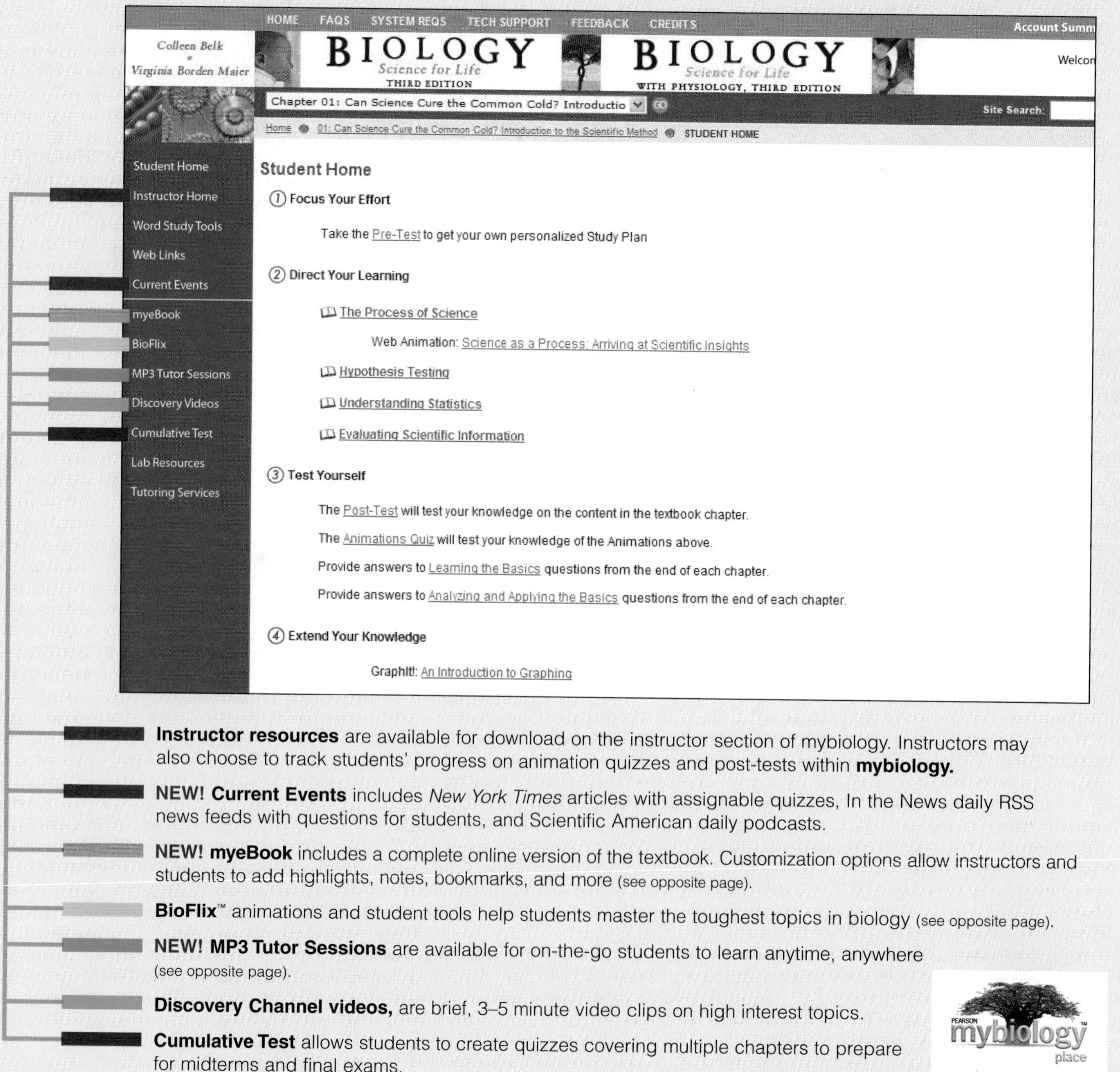

- **Instructor resources** are available for download on the instructor section of mybiology. Instructors may also choose to track students' progress on animation quizzes and post-tests within **mybiology.**
- **NEW! Current Events** includes *New York Times* articles with assignable quizzes, In the News daily RSS news feeds with questions for students, and Scientific American daily podcasts.
- **NEW! myeBook** includes a complete online version of the textbook. Customization options allow instructors and students to add highlights, notes, bookmarks, and more (see opposite page).
- **BioFlix**™ animations and student tools help students master the toughest topics in biology (see opposite page).
- **NEW! MP3 Tutor Sessions** are available for on-the-go students to learn anytime, anywhere (see opposite page).
- **Discovery Channel videos,** are brief, 3–5 minute video clips on high interest topics.
- **Cumulative Test** allows students to create quizzes covering multiple chapters to prepare for midterms and final exams.

BioFlix™

NEW! BioFlix animations and student tools invigorate classroom lectures and cover the most difficult biology topics, including cellular respiration, photosynthesis, and how neurons work. BioFlix include 3-D, movie-quality animations, labeled slide shows, carefully constructed student tutorials, study sheets, and quizzes.

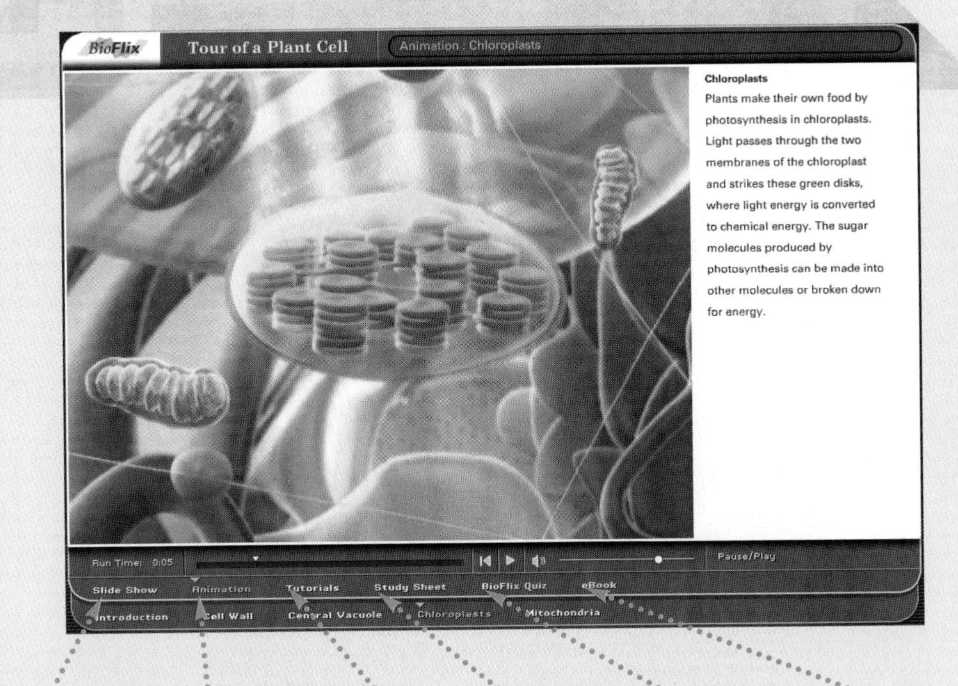

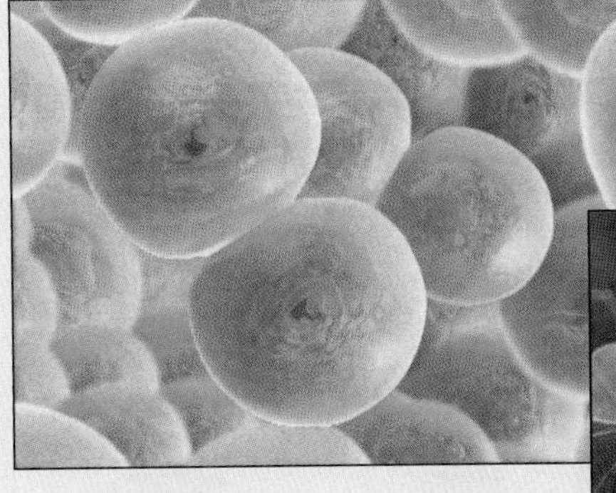

NEW! MP3 Tutor Sessions are audio tutorials—virtual office hours—for students to take advantage of anytime, anywhere. MP3 Tutor Sessions can be downloaded onto a computer and transferred to an iPod or any digital-media device that plays MP3 files.

From **mybiology,** MP3 Tutor Sessions are tied, chapter-by-chapter, to the textbook and contain the following sections:

- Big Ideas: The "need to know" for each chapter
- Practice Questions: A spot check for the big ideas tells students if they should keep studying
- Key Terms: Audio "flashcards" help review key concepts and vocabulary
- Rapid Review: A quick drill session for studying right before the test

NEW! myeBook gives students access to the text wherever they have access to the Internet. myeBook provides the full text, including art and figures with a zoom feature that can be used to enlarge them for better viewing. Within myeBook students are also able to pop-up definitions and terms to help with vocabulary and the reading of the material. Students can also personalize myeBook using the notes and highlight features at the top of each page.

Dynamic resources help you teach biology

REVISED! *Instructor Resource DVD*

The Instructor Resource DVD package combines all of the instructor media in one convenient location, organized chapter-by-chapter. **978-0-321-58736-7 • 0-321-58736-7**

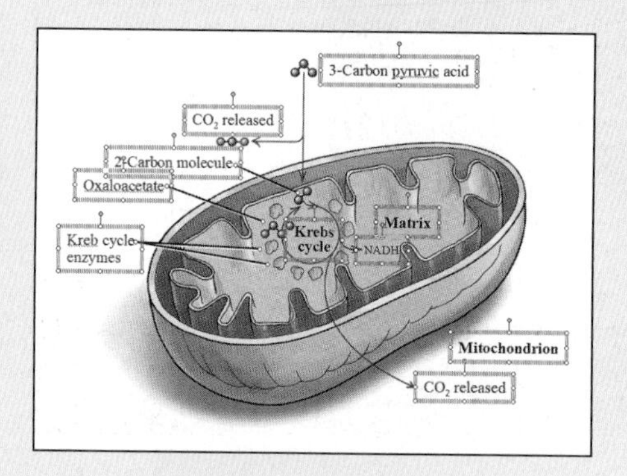

Customizable PowerPoint® lectures outline the major concepts presented in each chapter and include key figures to get the points across.

NEW! Author teaching videos show Virginia Borden Maier explaining how she successfully uses a narrative approach in her class. She demonstrates an active learning exercise using a "Mr. Potato Head" to explain genetic concepts, and more. These videos can also be downloaded from the instructor area of **mybiology.**

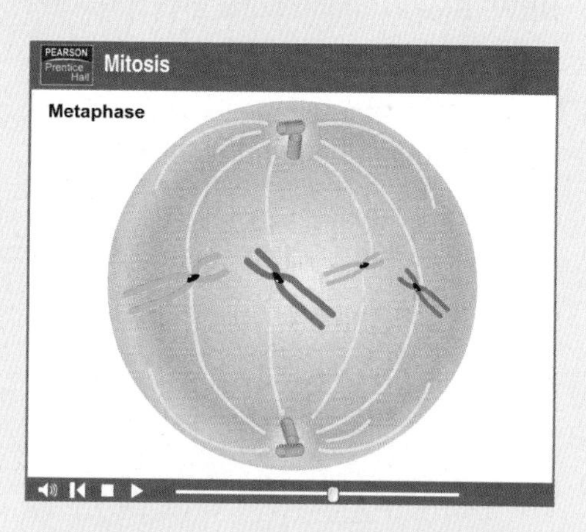

Animations are book-specific animations that illustrate key concepts in a chapter and can be used through PowerPoint for lecture presentations.

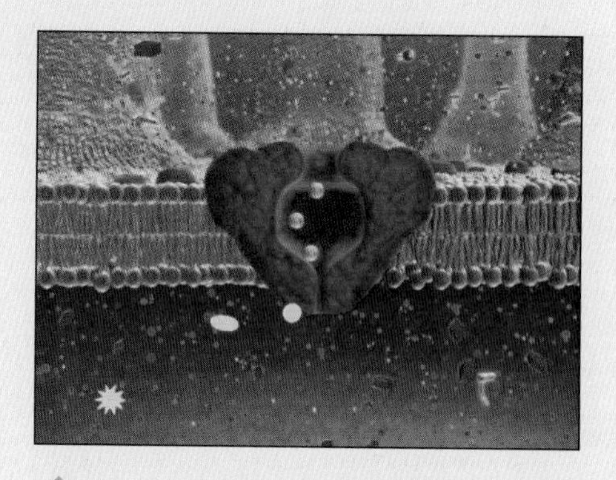

NEW! BioFlix™ animations invigorate classroom lectures with 3-D, movie-quality graphics. Each animation is typically three minutes long.

BioFlix™

Discovery Channel videos are brief 3-5 minute video clips covering topics from fighting cancer to antibiotic resistance and introduced species.

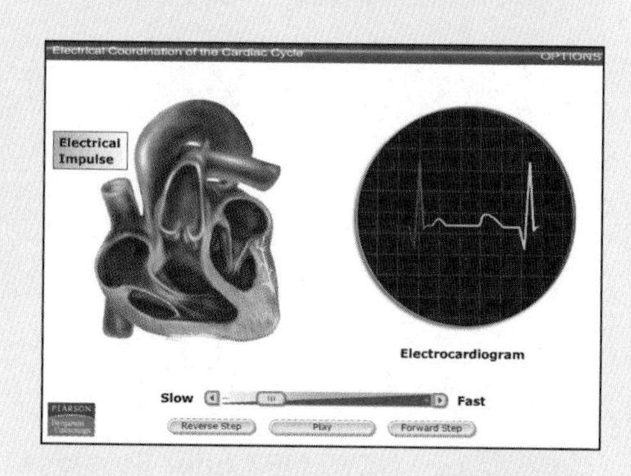

Extra animations are also provided for instructors to step through in PowerPoint.

All of the art, tables, and photos are available as JPEGs and in PowerPoint formats. Stepped-art and PowerPoint art files allow instructors to customize labels to best fit their teaching needs.

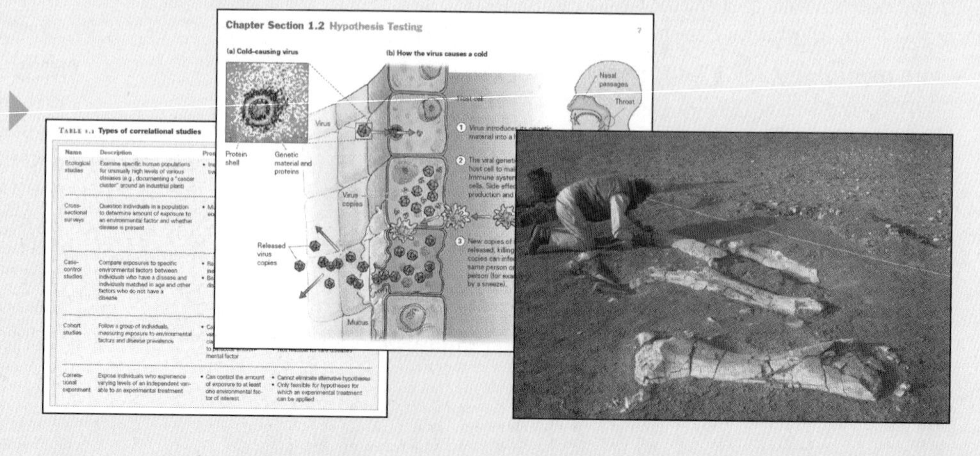

Expanded Test Bank questions are available as downloadable Microsoft® Word files and in TestGen® format.

Scientific American *Current Issues in Biology*

Scientific American *Current Issues in Biology* gives your students the best of both worlds—accessible, dynamic, relevant articles from Scientific American magazine that present key issues in biology, paired with the authority, reliability, and clarity of Pearson's non-majors biology texts. Articles include questions to help students check their comprehension and make connections to science and society.

Volume 1: 978-0-8053-7507-7 • 0-8053-7507-4

Table of Contents

Volume 2: 978-0-8053-7108-6 • 0-8053-7108-7

Table of Contents

Volume 3: 978-0-8053-7527-5 • 0-8053-7527-9

Table of Contents

Volume 4: 978-0-8053-3566-8 • 0-8053-3566-8

Table of Contents

Volume 5: 978-0-321-54187-1 • 0-321-54187-1

Table of Contents

Instructor and Student Supplements

For Instructors

Instructor Resource DVD
978-0-321-58736-7 I 0-321-58736-7
The Instructor Resource DVD includes four DVDs with a full suite of JPEG and PowerPoint® options; more than 150 instructor animations including dynamic BioFlix™; additional lecture and course support items; author teaching videos; the Test Bank in TestGen® software and Microsoft® Word format; and Discovery Videos.

Instructor Guide
978-0-321-58739-8 I 0-321-58739-1
This printed guide includes new Teacher and Student Objectives, expanded Lecture Outlines, and Lecture Activities and Handouts. It is also available in Microsoft Word format on the Instructor Resource Center and on the Instructor Resource DVD.

Printed Test Bank
978-0-321-58740-4 I 0-321-58740-5
All of the exam questions in the Test Bank have been peer reviewed, providing questions that set the standard for quality and accuracy. This edition includes new art-based questions and more questions per chapter. The Test Bank is also available in TestGen software, Course Management Systems, and Microsoft Word format on the Instructor Resource DVD.

mybiology
mybiology is the central place for all beginning biology students to access the online materials needed to succeed in non-majors biology courses. Students can make the most of their study time by accessing high quality, book-specific practice materials before exams. Following a four-step learning path, students can identify what they don't know and focus on activities tailored to their learning style.
www.mybiology.com

Transparency Acetates
978-0-321-58735-0 I 0-321-58735-9
These 200 four-color acetates include key illustrations from the text. They are enhanced for classroom presentation with enlarged labels and increased color saturation for better viewing by the students.

Course Management Options

CourseCompass™
This nationally-hosted, dynamic, interactive online course management system is powered by Blackboard, the leading platform for Internet-based learning tools. This easy-to-use and customizable program enables professors to tailor content to meet individual course needs! Every CourseCompass course includes pre-loaded content such as testing and assessment question pools.
www.pearsonhighered.com/coursecompass

CourseCompass with myeBook
978-0-321-58752-7 I 0-321-58752-9
This contains all the content of CourseCompass in addition to myeBook content.

WebCT Premium
This course management system contains pre-loaded content such as testing and assessment question pools.
www.pearsonhighered.com/webct

Blackboard Premium
This course management system contains pre-loaded content such as testing and assessment question pools.
www.pearsonhighered.com/blackboard

For Students

Study Guide
978-0-321-58743-5 I 0-321-58743-X
The Study Guide includes a concise summary of each chapter section, new key term flashcards, and an expanded self test with answers and feedback, which helps students determine if they understand the chapter coverage.

mybiology
www.mybiology.com
See description above.

Study Card
978-0-321-58738-1 I 0-321-58738-3
This laminated quick reference Study Card is developed specifically for your text and can be value packaged with the book at no additional charge. The Study Card provides a brief, chapter-by-chapter review, including key figures, to help students review the most important topics in biology.

Acknowledgments

Reviewers

Each chapter of this book was thoroughly reviewed several times as it moved through the development process. Reviewers were chosen on the basis of their demonstrated talent and dedication in the classroom. Many of these reviewers were already trying various approaches to actively engage students in lectures and to raise the scientific literacy and critical thinking skills among their students. Their passion for teaching and commitment to their students were evident throughout this process. These devoted individuals scrupulously checked each chapter for scientific accuracy, readability, and coverage level. In addition to general reviewers, we also had a team of expert reviewers evaluate individual chapters to ensure that the content was accurate and that all the necessary concepts were included.

All of these reviewers provided thoughtful, insightful feedback, which improved the text significantly. Their efforts reflect their deep commitment to teaching nonmajors and improving the scientific literacy of all students. We are very thankful for their contributions.

We express sincere gratitude to the expert reviewers who worked so carefully with the authors in reviewing manuscript to ensure the scientific accuracy of the text and art.

Janet Bester-Meredith, *Seattle Pacific University*
Barry Beutler, *College of Eastern Utah*
Bruno Borsari, *Winona State University*
Robert S. Boyd, *Auburn University*
Charles Cottingham, *Frederick Community College*
Deborah Dardis, *Southeastern Louisiana University*
Stephen Ebbs, *Southern Illinois University*
Ana Escandon, *Los Angeles Harbor College*
Marirose Ethington, *Genessee Community College*
Bruce Griffis, *Kentucky State University*
Michael Groesbeck, *Brigham Young University, Idaho*
Laurie Host, *Harford Community College*
Judy Kaufman, *Monroe Community College*
Brenda Leady, *University of Toledo*
Lee Likins, *University of Missouri, Kansas City*
Geri Mayer, *Florida Atlantic University*
Diane Melroy, *University of North Carolina, Wilmington*
Lori Nicholas, *New York University*
Erin O'Brien, *Dixie College*
Murray Paton Pendarvis, *Southeastern Louisiana University*
Indiren Pillay, *Culver-Stockton College*
Wilma Robertson, *Boise State University*
Mary Severinghaus, *Parkland College*
Bryan Spohn Florida, *Community College at Jacksonville, Kent Campus*
Glena Temple, *Viterbo*
Janet Vigna, *Grand Valley State*
John Zook, *Ohio University*

Prior Edition Reviewers

Daryl Adams, *Minnesota State University, Mankato*
Karen Aguirre, *Clarkson University*
Susan Aronica, *Canisius College*
Mary Ashley, *University of Chicago*
James S. Backer, *Daytona Beach Community College*
Ellen Baker, *Santa Monica College*
Gail F. Baker, *LaGuardia Community College*
Neil R. Baker, *The Ohio State University*
Andrew Baldwin, *Mesa Community College*
Thomas Balgooyen, *San Jose State University*
Tamatha R. Barbeau, *Francis Marion University*
Sarah Barlow, *Middle Tennessee State University*
Andrew M. Barton, *University of Maine, Farmington*
Vernon Bauer, *Francis Marion University*
Paul Beardsley, *Idaho State University*
Donna Becker, *Northern Michigan University*
Steve Berg, *Winona State University*
Carl T. Bergstrom, *University of Washington*
Donna H. Bivans, *Pitt Community College*
Lesley Blair, *Oregon State University*
John Blamire, *City University of New York, Brooklyn College*
Barbara Blonder, *Flagler College*
Susan Bornstein-Forst, *Marian College*
Bruno Borsari, *Winona State University*
James Botsford, *New Mexico State University*
Robert S. Boyd, *Auburn University*
Bryan Brendley, *Gannon University*
Eric Brenner, *New York University*
Peggy Brickman, *University of Georgia*
Carol Britson, *University of Mississippi*
Carole Browne, *Wake Forest University*
Neil Buckley, *State University of New York, Plattsburgh*
Stephanie Burdett, *Brigham Young University*
Warren Burggren, *University of North Texas*
Nancy Butler, *Kutztown University*
Suzanne Butler, *Miami-Dade Community College*
Wilbert Butler, *Tallahassee Community College*
David Byres, *Florida Community College at Jacksonville*
Tom Campbell, *Pierce College, Los Angeles*
Deborah Cato, *Wheaton College*
Peter Chabora, *Queens College*
Bruce Chase, *University of Nebraska, Omaha*

Thomas F. Chubb, *Villanova University*
Gregory Clark, *University of Texas, Austin*
Kimberly Cline-Brown, *University of Northern Iowa*
Mary Colavito, *Santa Monica College*
William H. Coleman, *University of Hartford*
William F. Collins III, *Stony Brook University*
Walter Conley, *State University of New York, Potsdam*
Jerry L. Cook, *Sam Houston State University*
Melanie Cook, *Tyler Junior College*
Scott Cooper, *University of Wisconsin, La Crosse*
Erica Corbett, *Southeastern Oklahoma State University*
George Cornwall, *University of Colorado*
James B. Courtright, *Marquette University*
Angela Cunningham, *Baylor University*
Garry Davies, *University of Alaska, Anchorage*
Miriam del Campo, *Miami-Dade Community College*
Judith D'Aleo, *Plymouth State University*
Juville Dario-Becker, *Central Virginia Community College*
Garry Davies, *University of Alaska, Anchorage*
Edward A. DeGrauw, *Portland Community College*
Heather DeHart, *Western Kentucky University*
Veronique Delesalle, *Gettysburg College*
Lisa Delissio, *Salem State College*
Beth De Stasio, *Lawrence University*
Elizabeth Desy, *Southwest Minnesota State University*
Donald Deters, *Bowling Green State University*
Gregg Dieringer, *Northwest Missouri State*
Diane Dixon, *Southeastern Oklahoma State University*
Christopher Dobson, *Grand Valley State University*
Cecile Dolan, *New Hampshire Community Technical College, Manchester*
Matthew Douglas, *Grand Rapids Community College*
Lee C. Drickamer, *Northern Arizona University*
Susan Dunford, *University of Cincinnati*
Douglas Eder, Southern Illinois University, Edwardsville
Patrick J. Enderle, *East Carolina University*
William Epperly, *Robert Morris College*
Dan Eshel, *City University of New York, Brooklyn College*
Deborah Fahey, *Wheaton College*
Richard Firenze, *Broome Community College*
Lynn Firestone, *Brigham Young University*
Brandon L. Foster, *Wake Technical Community College*
Richard A. Fralick, *Plymouth State University*
Barbara Frank, *Idaho State University*
Lori Frear, *Wake Technical Community College*
David Froelich, *Austin Community College*
Suzanne Frucht, *Northwest Missouri State University*
Edward Gabriel, *Lycoming College*
Anne Galbraith, *University of Wisconsin, La Crosse*
Patrick Galliart, *North Iowa Area Community College*
Wendy Garrison, *University of Mississippi*
Alexandros Georgakilas, *East Carolina University*
Robert George, *University of North Carolina, Wilmington*
Tammy Gillespie, *Eastern Arizona College*
Sharon Gilman, *Coastal Carolina University*
Mac F. Given, *Neumann College*
Bruce Goldman, *University of Connecticut, Storrs*
Andrew Goliszek, *North Carolina Agricultural and Technical State University*
Beatriz Gonzalez, *Sante Fe Community College*
Eugene Goodman, *University of Wisconsin, Parkside*
Lara Gossage, *Hutchinson Community College*
Tamar Goulet, *University of Mississippi*
Becky Graham, *University of West Alabama*
John Green, *Nicholls State University*
Robert S. Greene, *Niagara University*
Tony J. Greenfield, *Southwest Minnesota State University*
Mark Grobner, *California State University, Stanislaus*
Stanley Guffey, *University of Tennessee*
Mark Hammer, *Wayne State University*
Blanche Haning, *North Carolina State University*
Robert Harms, *St. Louis Community College*
Craig M. Hart, *Louisiana State University*
Patricia Hauslein, *St. Cloud State University*
Stephen Hedman, *University of Minnesota, Duluth*
Bethany Henderson-Dean, *University of Findlay*
Julie Hens, *University of Maryland University College*
Peter Heywood, *Brown University*
Julia Hinton, *McNeese State University*
Phyllis C. Hirsh, *East Los Angeles College*
Elizabeth Hodgson, *York College of Pennsylvania*
Leland Holland, *Pasco-Hernando Community College*
Jane Horlings, *Saddleback Community College*
Margaret Horton, *University of North Carolina, Greensboro*
David Howard, *University of Wisconsin, La Crosse*
Michael Hudecki, *State University of New York, Buffalo*
Michael E. S. Hudspeth, *Northern Illinois University*
Laura Huenneke, *New Mexico State University*
Pamela D. Huggins, *Fairmont State University*
Sue Hum-Musser, *Western Illinois University*
Carol Hurney, *James Madison University*
James Hutcheon, *Georgia Southern University*
Carl Johansson, *Fresno City College*
Thomas Jordan, *Pima Community College*
Jann Joseph, *Grand Valley State University*
Mary K. Kananen, *Penn State University, Altoona*
Arnold Karpoff, *University of Louisville*
Judy Kaufman, *Monroe Community College*
Michael Keas, *Oklahoma Baptist University*
Judith Kelly, *Henry Ford Community College*
Karen Kendall-Fite, *Columbia State Community College*
Andrew Keth, *Clarion University*
David Kirby, *American University*
Stacey Kiser, *Lane Community College*
Dennis Kitz, *Southern Illinois University, Edwardsville*
Carl Kloock, *California State, Bakersfield*
Jennifer Knapp, *Nashville State Technical Community College*

Loren Knapp, *University of South Carolina*
Michael A. Kotarski, *Niagara University*
Michelle LaBonte, *Framingham State College*
Phyllis Laine, *Xavier University*
Dale Lambert, *Tarrant County College*
Tom Langen, *Clarkson University*
Lynn Larsen, *Portland Community College*
Mark Lavery, *Oregon State University*
Brenda Leady, *University of Toledo*
Mary Lehman, *Longwood University*
Doug Levey, *University of Florida*
Abigail Littlefield, *Landmark College*
Andrew D. Lloyd, *Delaware State University*
Jayson Lloyd, *College of Southern Idaho*
Judy Lonsdale, *Boise State University*
Kate Lormand, *Arapahoe Community College*
Paul Lurquin, *Washington State University*
Kimberly Lyle-Ippolito, *Anderson University*
Douglas Lyng, *Indiana University/Purdue University*
Michelle Mabry, *Davis and Elkins College*
Stephen E. MacAvoy, *American University*
Molly MacLean, *University of Maine*
Charles Mallery, *University of Miami*
Ken Marr Green, *River Community College*
Kathleen Marrs, *Indiana University/Purdue University*
T. D. Maze, *Lander University*
Steve McCommas, *Southern Illinois University, Edwardsville*
Colleen McNamara, *Albuquerque Technical Vocational Institute*
Mary McNamara, *Albuquerque Technical Vocational Institute*
John McWilliams, *Oklahoma Baptist University*
Susan T. Meiers, *Western Illinois University*
Diane Melroy, *University of North Carolina, Wilmington*
Joseph Mendelson, *Utah State University*
Paige A. Mettler-Cherry, *Lindenwood University*
Debra Meuler, *Cardinal Stritch University*
James E. Mickle, *North Carolina State University*
Craig Milgrim, *Grossmont College*
Hugh Miller, *East Tennessee State University*
Jennifer Miskowski, *University of Wisconsin, La Crosse*
Ali Mohamed, *Virginia State University*
Stephen Molnar, *Washington University*
James Mone, *Millersville University*
Daniela Monk, *Washington State University*
David Mork, *Yakima Valley Community College*
Bertram Murray, *Rutgers University*
Ken Nadler, *Michigan State University*
John J. Natalini, *Quincy University*
Alissa A. Neill, *University of Rhode Island*
Dawn Nelson, *Community College of Southern Nevada*
Joseph Newhouse, *California University of Pennsylvania*
Jeffrey Newman, *Lycoming College*
David L.G. Noakes, *University of Guelph*
Shawn Nordell, *St. Louis University*
Tonye E. Numbere, *University of Missouri, Rolla*
Igor Oksov, *Union County College*
Kevin Padian, *University of California, Berkeley*
Arnas Palaima, *University of Mississippi*
Anthony Palombella, *Longwood University*
Marilee Benore Parsons, *University of Michigan, Dearborn*
Steven L. Peck, *Brigham Young University*
Javier Penalosa, *Buffalo State College*
Rhoda Perozzi, *Virginia Commonwealth University*
John Peters, *College of Charleston*
Patricia Phelps, *Austin Community College*
Polly Phillips, *Florida International University*
Francis J. Pitocchelli, *Saint Anselm College*
Roberta L. Pohlman, *Wright State University*
Calvin Porter, *Xavier University*
Linda Potts, *University of North Carolina, Wilmington*
Robert Pozos, *San Diego State University*
Marion Preest, *The Claremont Colleges*
Gregory Pryor, *Francis Marion University*
Rongsun Pu, *Kean University*
Narayanan Rajendran, *Kentucky State University*
Anne E. Reilly, *Florida Atlantic University*
Michael H. Renfroe, *James Madison University*
Laura Rhoads, *State University of New York, Potsdam*
Gwynne S. Rife, *University of Findlay*
Todd Rimkus, *Marymount University*
Laurel Roberts, *University of Pittsburgh*
Wilma Robertson, *Boise State University*
William E. Rogers, *Texas A&M University*
Troy Rohn, *Boise State University*
Deborah Ross, *Indiana University/Purdue University*
Christel Rowe, *Hibbing Community College*
Joanne Russell, *Manchester Community College*
Michael Rutledge, *Middle Tennessee State University*
Wendy Ryan, *Kutztown University*
Christopher Sacchi, *Kutztown University*
Kim Sadler, *Middle Tennessee State University*
Brian Sailer, *Albuquerque Technical Vocational Institute*
Jasmine Saros, *University of Wisconsin, La Crosse*
Ken Saville, *Albion College*
Louis Scala, *Kutztown University*
Daniel C. Scheirer, *Northeastern University*
Beverly Schieltz, *Wright State University*
Nancy Schmidt, *Pima Community College*
Robert Schoch, *Boston University*
Julie Schroer, *Bismarck State College*
Fayla Schwartz, *Everett Community College*
Steven Scott, *Merritt College*
Gray Scrimgeour, *University of Toronto*
Roger Seeber, *West Liberty State College*
Allison Shearer, *Grossmont College*
Robert Shetlar, *Georgia Southern University*
Cara Shillington, *Eastern Michigan University*
Beatrice Sirakaya, *Pennsylvania State University*
Cynthia Sirna, *Gadsden State Community College*
Thomas Sluss, *Fort Lewis College*
Brian Smith Black, *Hills State University*
Douglas Smith, *Clarion University of Pennsylvania*
Mark Smith, *Chaffey College*
Gregory Smutzer, *Temple University*
Sally Sommers, *Smith Boston University*
Anna Bess Sorin, *University of Memphis*
Carol St. Angelo, *Hofstra University*
Amanda Starnes, *Emory University*
Susan L. Steen, *Idaho State University*
Timothy Stewart, *Longwood College*
Shawn Stover, *Davis and Elkins College*
Bradley J. Swanson, *Central Michigan University*
Joyce Tamashiro, *University of Puget Sound*
Jeffrey Taylor, *Slippery Rock University*
Martha Taylor, *Cornell University*
Tania Thalkar, *Clarion University of Pennsylvania*
Alice Templet, *Nicholls State University*
Jeffrey Thomas, *University of California, Los Angeles*

Janis Thompson, *Lorain County Community College*
Nina Thumser, *California University of Pennsylvania*
Alana Tibbets, *Southern Illinois University, Edwardsville*
Martin Tracey, *Florida International University*
Jeffrey Travis, *State University of New York, Albany*
Robert Turgeon, *Cornell University*
Michael Tveten, *Pima Community College, Northwest Campus*
James Urban, *Kansas State University*
Brandi Van Roo, *Framingham State College*
John Vaughan, *St. Petersburg Junior College*
Martin Vaughan, *Indiana State University*
Mark Venable, *Appalachian State University*
Paul Verrell, *Washington State University*
Tanya Vickers, *University of Utah*
Janet Vigna, *Grand Valley State University*
Sean Walker, *California State University, Fullerton*
Don Waller, *University of Wisconsin, Madison*
Tracy Ware, *Salem State College*
Jennifer Warner, *University of North Carolina, Charlotte*
Lisa Weasel, *Portland State University*
Carol Weaver, *Union University*
Frances Weaver, *Widener University*
Elizabeth Welnhofer, *Canisius College*
Marcia Wendeln, *Wright State University*
Shauna Weyrauch, *Ohio State University, Newark*
Wayne Whaley, *Utah Valley State College*
Howard Whiteman, *Murray State University*
Vernon Wiersema, *Houston Community College*
Gerald Wilcox, *Potomac State College*
Peter J. Wilkin, *Purdue University North Central*
Robert R. Wise, *University of Wisconsin, Oshkosh*
Michelle Withers, *Louisiana State University*
Art Woods, *University of Texas, Austin*
Elton Woodward, *Daytona Beach Community College*
Kenneth Wunch, *Sam Houston State University*
Donna Young, *University of Winnipeg*
Michelle L. Zjhra, *Georgia Southern University*
Michelle Zurawski, *Moraine Valley Community College*

Student Focus Group Participants

California State University, Fullerton:
Danielle Bruening
Leslie Buena
Andrés Carrillo
Victor Galvan
Jessica Ginger
Sarah Harpst
Robin Keber
Ryan Roberts
Melissa Romero
Erin Seale
Nathan Tran
Tracy Valentovich
Sean Vogt

Fullerton College:
Michael Baker
Mahetzi Hernandez
Heidi McMorris
Daniel Minn
Sam Myers
David Omut
Jonathan Pistorino
James W. Pura
Samantha Ramirez
Tiffany Speed
Tristan Terry

The Book Team

We are indebted to our editor Star MacKenzie. Star is a delight to work with—insightful, funny, kind, and generous with her time. Star is truly committed to producing an excellent book that meets the needs of students and instructors. We feel fortunate to have had her guidance throughout the project.

We also owe a great deal of thanks to our development editor, Cathy Murphy. Cathy read every word we wrote from a student's perspective and helped us effectively address issues raised by users and reviewers. Her ideas and organizational skills helped make this a better book.

Art Development Editor Ellie Marcus was responsible for taking our ideas about art and transferring them to the page. She combed through each and every figure in an attempt to find ways to improve the design and layout to facilitate student understanding of difficult concepts. In addition to being a talented artist, she has been a pleasure to work with.

And finally, thanks to Dusty Friedman our project editor whose attention to detail and organizational skills helped keep this project on track and on time.

This book is dedicated to our families, friends, and colleagues who have supported us over the years. Having loving families, great friends, and a supportive work environment enabled us to make this heartfelt contribution to nonmajors biology education.

Colleen Belk and
Virginia Borden Maier

Contents

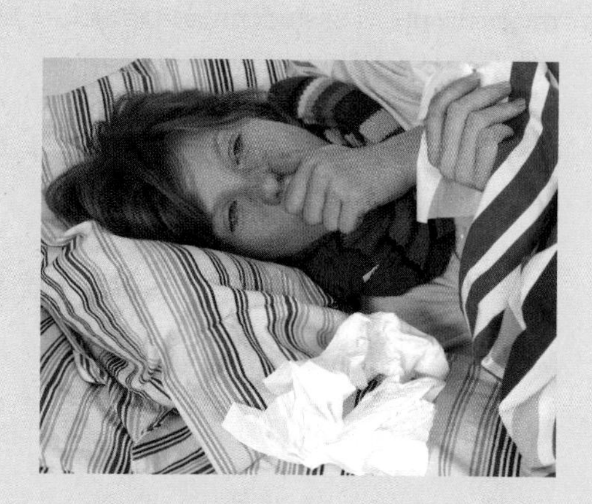

Chapter 3

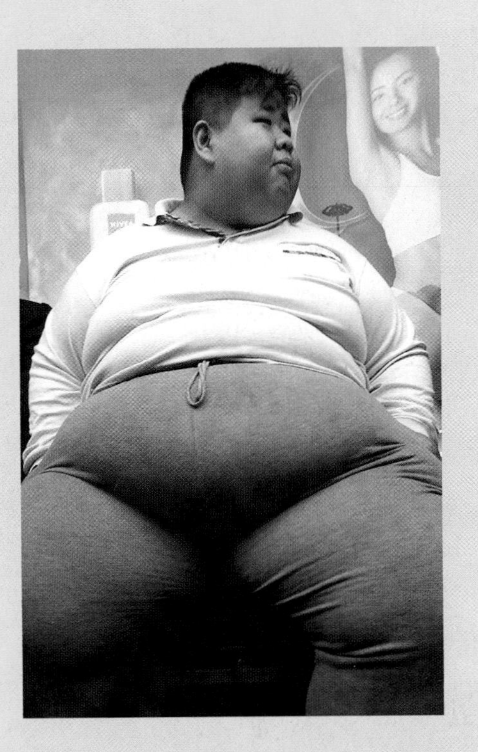

Diet

Life in the Greenhouse

Unit Two
Genetics

Chapter 5

Cancer

Contents

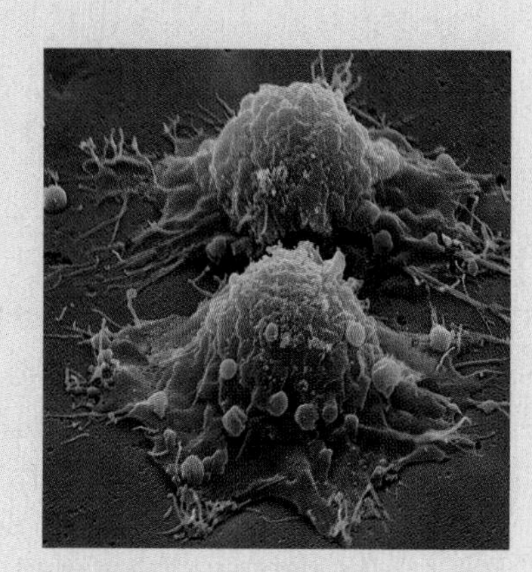

Chapter 6

Are You Only As Smart As Your Genes?

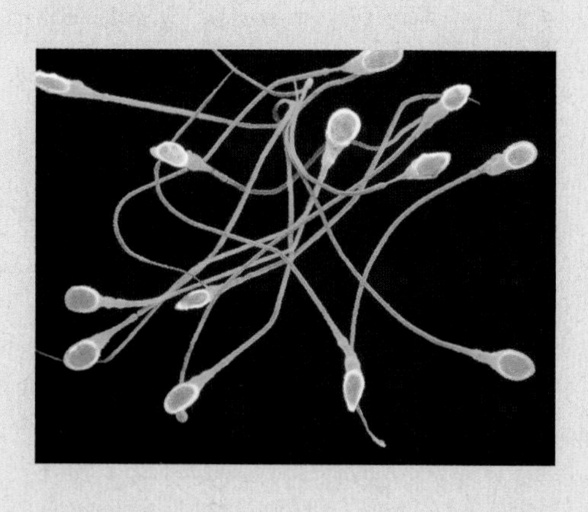

STOP NOW
BEFORE IT'S
TOO LATE!

Unit Three
Evolution

Unit Four
Ecology

CHAPTER 1

Can Science Cure the Common Cold?

Introduction to the Scientific Method

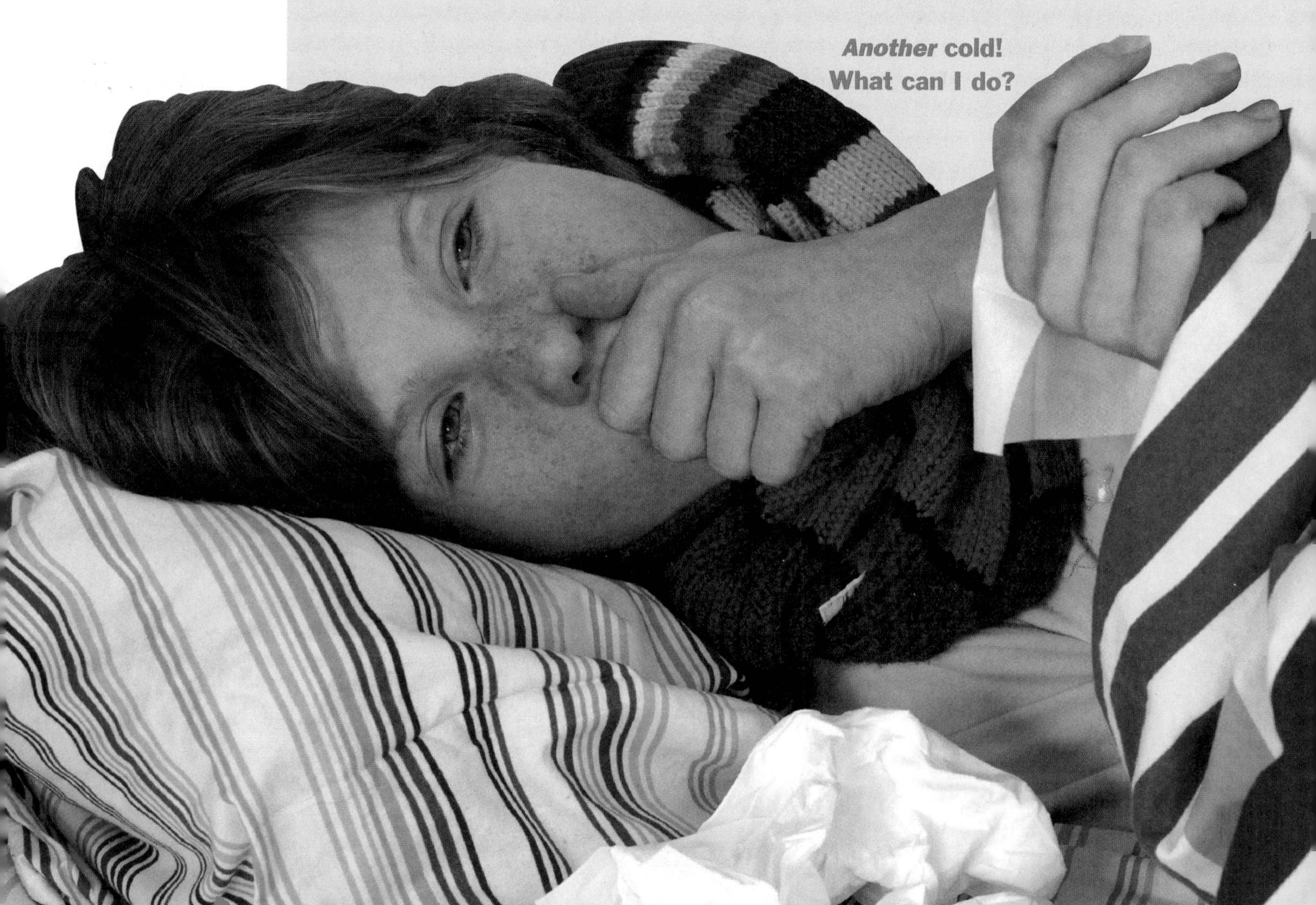

Take massive doses of Vitamin C?

We have all been there—you just recover from one bad head cold, and on a morning soon after you notice that scratchy feeling in your throat that signals a new one is about to begin. It is always at the worst time, too, when you have an important exam coming up, a term paper due, and a packed social calendar. Why are you sick yet again? What can you do about it?

If you ask your friends and relatives, you will hear the usual advice on how to prevent and treat colds: Take massive doses of vitamin C. Suck on zinc lozenges. Drink plenty of echinacea tea. Meditate. Get more rest. Exercise vigorously every day. Put that hat on when you go outside! You are left with an overwhelming list of options, some of which are contradictory or even counter to common sense. If you keep up with health news, you may be even more confused. One website reports that a popular over-the-counter cold treatment is effective, while a local TV news story details the risks of using this remedy and highlights its ineffectiveness. How do you decide what to do?

Drink echinacea tea?

Faced with this bewildering situation, most people follow the advice that makes the most sense to them, and if they find they still feel terrible, they try another remedy. Testing ideas and discarding ones that don't work is a kind of "everyday science." However, this technique has its limitations—for example, even if you feel better after trying a new cold treatment, you can't know if your recovery occurred because the treatment was effective or because the cold was ending anyway.

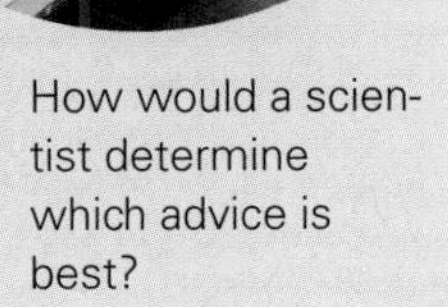

How would a scientist determine which advice is best?

What professional scientists do is a more refined version of this everyday science, using strategies that help eliminate other possible explanations for a result. And while some fields of science may use unintelligible words or complicated and expensive equipment, the basic process for testing ideas is universal to all areas of science. An understanding of this process can help you evaluate information about many issues that may concern and intrigue you—from health issues, to global warming, to the origin of life and the universe—with more confidence. In this chapter, we introduce you to the powerful process scientists use by asking the question we've considered here: Is there a cure for the common cold?

web animation 1.1 *Science as a Process: Arriving at Scientific Insights*

1.1 The Process of Science

The term *science* can refer to a body of knowledge—for example, the science of **biology** is the study of living organisms. Your impression of science may be that it requires near-perfect recall of specific sets of facts about the world. In reality, this goal is impossible and unnecessary—we do have reference books, after all. The real action in science is not in memorizing what is already known about the world but in using the process of science to discover something new and unknown.

This process—making observations of the world, proposing ideas about how something works, testing those ideas, and discarding (or modifying) our ideas in response to the results of a test—is the essence of the **scientific method.** The scientific method allows us to solve problems and answer questions efficiently and effectively. Can we use the scientific method to solve the complicated problem of preventing and treating colds?

The Nature of Hypotheses

The statements our friends and family make about which actions will help us remain healthy (for example, the advice to wear a hat) are in some part based on the advice giver's understanding of how our bodies resist colds. Ideas about "how things work" are called **hypotheses.** Or, more formally, a hypothesis is a proposed explanation for one or more **observations.**

Hypotheses in biology come from knowledge about how the body and other biological systems work, experiences in similar situations, our understanding of other scientific research, and logical reasoning; they are also shaped by our creative mind *(Figure 1.1)*. When your mom tells you to dress warmly in order to avoid colds, she is basing her advice on the following hypothesis: Becoming chilled makes you more susceptible to illness.

The hallmark of science is that hypotheses are subject to rigorous testing. Therefore, scientific hypotheses must be **testable**—it must be possible to evaluate a hypothesis through observations of the measurable universe. Not all hypotheses are testable. For instance, the statement that "colds are generated by disturbances in psychic energy" is not a scientific hypothesis because psychic energy cannot be seen or measured—it does not have a material nature and therefore cannot be put to a test.

In addition, hypotheses that require the intervention of a supernatural force cannot be tested scientifically. If something is **supernatural,** it is not constrained by any laws of nature, and its behavior cannot be predicted using our current understanding of the natural world.

A scientific hypothesis must also be **falsifiable;** that is, an observation or set of observations could potentially prove it false. The hypothesis that exposure

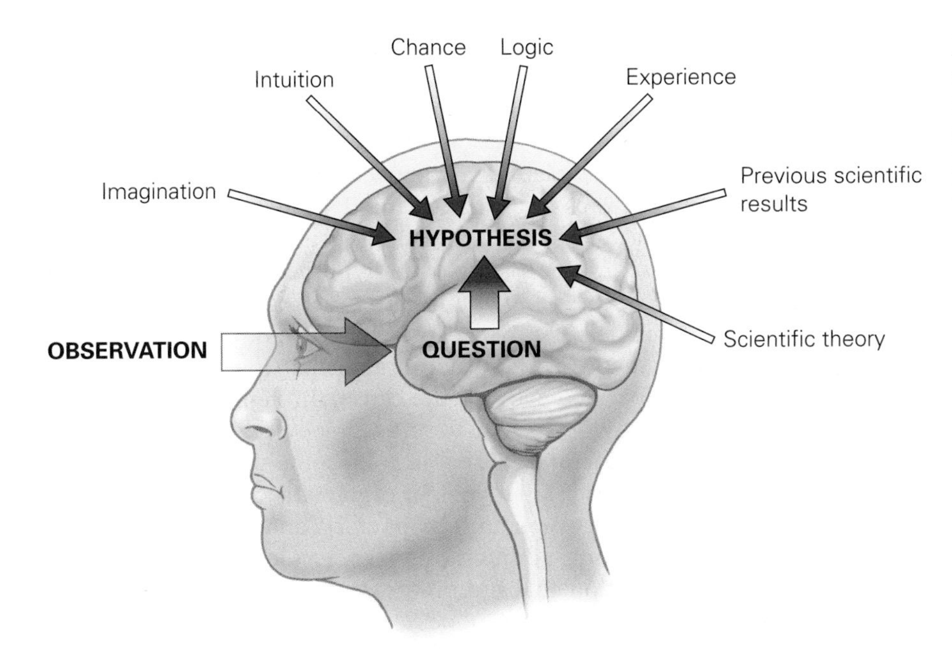

Figure 1.1 **Hypothesis generation.** All of us generate hypotheses. Many different factors, both logical and creative, influence the development of a hypothesis. Scientific hypotheses are both testable and falsifiable.

to cold temperatures increases your susceptibility to colds is falsifiable; we can imagine an observation that would cause us to reject this hypothesis (for instance, the observation that people exposed to cold temperatures do not catch more colds than people protected from chills). Of course, not all hypotheses are proved false, but it is essential in science that incorrect ideas are discarded, and that can occur only if it is possible to prove those ideas false. Lack of falsifiability is another reason supernatural hypotheses cannot be scientific. Because a supernatural force can cause any possible result of a test, hypotheses that rely on supernatural forces cannot be falsified.

Finally, statements that are value judgments, such as, "It is wrong to cheat on an exam," are not scientific because different people have different ideas about right and wrong. It is impossible to falsify these types of statements. To find answers to questions of morality, ethics, or justice, we turn to other methods of gaining understanding—such as philosophy and religion.

Scientific Theories

Most hypotheses fit into a larger picture of scientific understanding. We can see this relationship when examining how research upended a commonly held belief about diet and health—that chronic stomach and intestinal inflammation is caused by eating too much spicy food. This belief directed the standard medical practice for ulcer treatment. Patients were prescribed drugs that reduced stomach acid levels and advised to avoid eating acidic or highly spiced foods. These treatments were rarely successful, and ulcers were considered chronic, possibly lifelong problems.

In 1982, Australian scientists Robin Warren and Barry Marshall discovered that a bacterium, later named *Helicobacter pylori*, was present in nearly all samples of ulcer tissue that they examined *(Figure 1.2)*. From this observation, Warren and Marshall reasoned that *H. pylori* infection was the cause of most ulcers. Barry Marshall even tested this hypothesis on himself by consuming live *H. pylori* and subsequently suffering from acute stomach pain.

Warren and Marshall's colleagues were at first unconvinced that ulcers could have such a simple cause. Today, the hypothesis that *H. pylori* infection is responsible for most ulcers is accepted as fact. The primary reasons why this is the case? (1) No reasonable alternative hypotheses about the causes of

(a) *Helicobacter pylori*

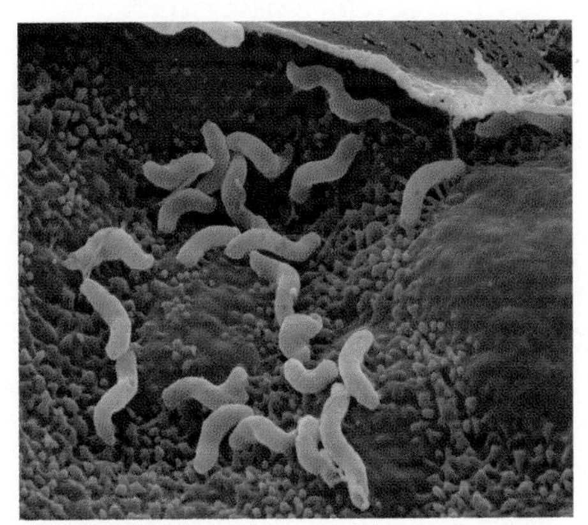

(b) Robin Warren and Barry Marshall, 2005 Nobel Prize winners in Medicine

Figure 1.2 **A scientific breakthrough.** (a) *Helicobacter pylori* (b) Robin Warren and Barry Marshall won the 2005 Nobel Prize in Medicine for their discovery of the link between *H. pylori* and ulcers.

ulcers (for instance, consumption of highly spiced foods) has been consistently supported by hypothesis tests; and (2) the hypothesis has not been rejected after carefully designed experiments demonstrated that eliminating *H. pylori* cured most ulcers.

The third reason that the relationship between *H. pylori* and ulcers is considered fact is that it conforms to a well-accepted scientific principle, namely, the germ theory of disease. A **scientific theory** is an explanation of a set of related observations based on well-supported hypotheses from several different, independent lines of research. *Germ theory* took shape from the accumulated observations of biologists such as Louis Pasteur and Robert Koch. The basic premise of germ theory is that microorganisms (that is, organisms too small to be seen with the naked eye) are the cause, through infection, of some or all human diseases.

Pasteur observed that bacteria cause milk to become sour. From this observation, he reasoned that these same types of organisms could injure humans. Koch demonstrated a link between anthrax bacteria and a specific set of fatal symptoms in mice, providing additional evidence. Germ theory is further supported by the observation that antibiotic treatment that targets particular microorganisms can cure certain illnesses—as is the case with bacteria-caused ulcers.

In everyday speech, the word *theory* is synonymous with "untested hypothesis based on little information." In contrast, scientists use the term when referring to well-supported ideas of how the natural world works. The supporting foundation of all scientific theories is multiple hypothesis tests.

The Logic of Hypothesis Tests

One very common hypothesis about cold prevention is that taking vitamin C supplements keeps you healthy. This hypothesis is very appealing, especially given the following generally known facts:

1. Fruits and vegetables contain a lot of vitamin C.
2. People with diets rich in fruits and vegetables are generally healthier than people who skimp on these food items.
3. Vitamin C is known to be an anti-inflammatory agent, reducing throat and nose irritation.

With these facts in mind, we can state the following falsifiable hypothesis:

> Consuming vitamin C decreases the risk of catching a cold.

This hypothesis makes sense given the statements just listed and the experiences of the many people who insist that vitamin C keeps them healthy. The process used to construct this hypothesis is called **inductive reasoning**—combining a series of specific observations (here, statements 1–3) to discern a general principle. Inductive reasoning is an essential tool for understanding the world. However, a word of caution is in order: Just because an inductively deduced hypothesis makes sense does not mean that it is true. The following example demonstrates this point.

Consider the ancient hypothesis that the sun revolves around Earth. This hypothesis was sensible based on the observation that the sun rose in the east every morning, traveled across the sky, and set in the west every night. For almost all of history, this hypothesis was considered to be a "fact" by nearly all of Western society. To most people, the hypothesis made perfect sense since the common religious belief was that Earth was the center of the universe and surrounded by the heavens. It wasn't until the early seventeenth century that this hypothesis was falsified as the result of Galileo Galilei's observations of Venus. Galileo's work helped to confirm the more modern hypothesis, proposed by Nicolaus Copernicus, that Earth revolves around the sun.

So, even though the hypothesis about vitamin C is sensible, it needs to be tested. Hypothesis testing is based on **deductive reasoning** or deduction. Deduction involves using a general principle to predict an expected observation. This **prediction** concerns the outcome of an action, test, or investigation. In other words, the prediction is the result we expect from a hypothesis test.

Deductive reasoning takes the form of "if/then" statements. That is, if our general principle is correct, then we expect to observe a specific outcome. A prediction based on the vitamin C hypothesis could be:

If vitamin C decreases the risk of catching a cold, *then* people who take vitamin C supplements with their regular diets will experience fewer colds than will people who do not take supplements.

Stop and Stretch 1.1: Consider the following scenario: Homer, your coworker, is a notorious doughnut lover who has a "nose" for free food. You walk into the break room one morning to discover a box from the doughnut shop that is completely empty. According to this information, what most likely happened to the doughnuts? Is this an inductive or deductive hypothesis?

Deductive reasoning, with its resulting predictions, is a powerful method for testing hypotheses. However, the structure of such a statement means that hypotheses can be clearly rejected if untrue, but impossible to prove if true (*Figure 1.3*). This shortcoming is illustrated using the if/then statement concerning vitamin C and colds.

Consider the possible outcomes of a comparison between people who supplement with vitamin C and those who do not. People who take vitamin C supplements may suffer through more colds than people who do not; they may have the same number of colds as the people who do not supplement, or supplementers may in fact experience fewer colds. What does each of these results tell us about the hypothesis?

If people who take vitamin C have more colds or the same number of colds as those who do not supplement, then the hypothesis that vitamin C provides protection against colds can be rejected. But what if people who supplement with vitamin C do experience fewer colds? If this is the case, then we can only say that the hypothesis has been supported and not disproven.

Why is it impossible to say that the hypothesis that vitamin C prevents colds is true? Because there are **alternative hypotheses** that explain why people with different vitamin-taking habits are different in their cold susceptibility. In other words, demonstrating the truth of the *then* portion of a deductive statement does not prove that the *if* portion is true.

Stop and Stretch 1.2: Consider this if/then statement: If your coworker Homer ate all the chocolate doughnuts, then there won't be any doughnuts left in the box that was placed in the break room 30 minutes ago. Imagine you find that the doughnuts are all gone—did you just prove that Homer ate them? Why or why not?

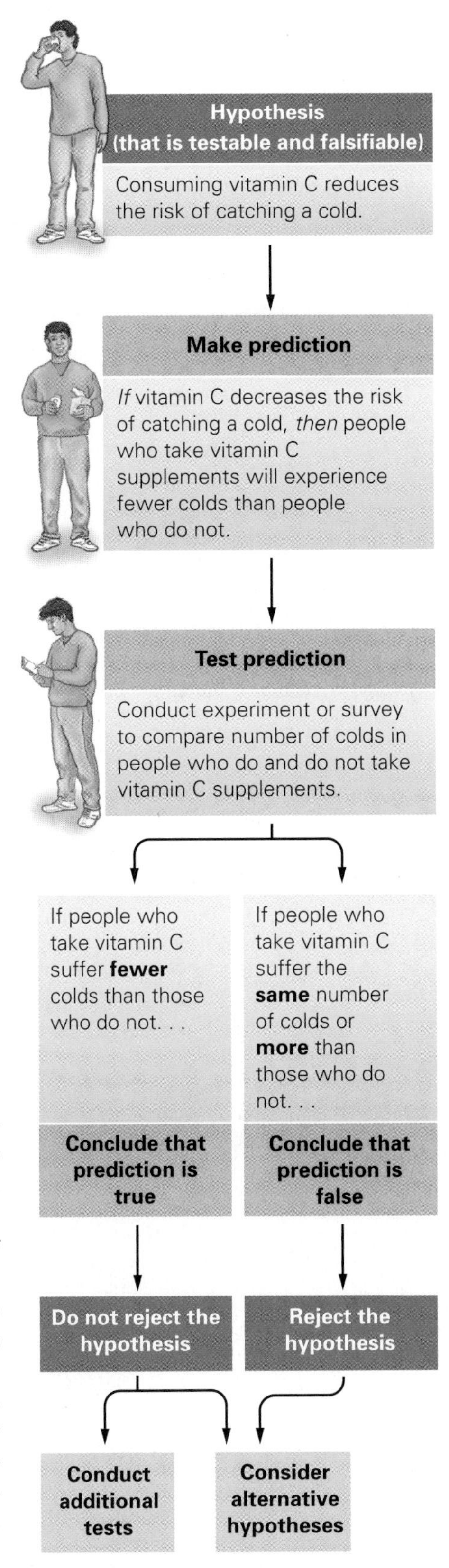

Figure 1.3 **The scientific method.** Tests of hypotheses follow a logical path. This flowchart illustrates the process of deduction as practiced by scientists. **Visualize This:** According to this flowchart, scientists should consider alternative hypotheses even if their hypothesis is supported by their research. Explain why this is the case.

Consider the alternative hypothesis that frequent exercise reduces susceptibility to catching a cold. And, also suppose that people who take vitamin C supplements are more likely to engage in regular exercise. If both of these hypotheses are true, then the prediction that vitamin C supplementers experience fewer colds than people who do not supplement would be true but not because the original hypothesis (vitamin C reduces the risk of colds) is true. Instead, people who take vitamin C supplements experience fewer colds because they are also more likely to exercise, and it is exercise that reduces cold susceptibility.

A hypothesis that seems to be true because it has not been rejected by an initial test may be rejected later because of a different test. This is what happened to the hypothesis that vitamin C consumption reduces susceptibility to colds. The argument for the power of vitamin C was popularized in 1970 by Nobel Prize–winning chemist Linus Pauling. Pauling based his assertion—that large doses of vitamin C reduce the incidence of colds by as much as 45%—on the results of a few studies that had been published between the 1930s and 1970s. However, repeated, careful tests of this hypothesis have since failed to support it. In many of the studies Pauling cited, it appears that alternative hypotheses explain the difference in cold incidence between vitamin C supplementers and nonsupplementers. Today, most health scientists agree that the hypothesis that vitamin C prevents colds has been convincingly falsified.

The example of the vitamin C hypothesis also highlights a challenge of communicating scientific information. You can see why the belief that vitamin C prevents colds is so widespread. If you don't know that scientific knowledge relies on rejecting incorrect ideas, a book by a Noble Prize–winning scientist may seem like the last word on the benefits of vitamin C. It took many years of careful research to show that this "last word" was, in fact, wrong.

1.2 Hypothesis Testing

The previous discussion may seem discouraging: How can scientists determine the truth of any hypothesis when there is always a chance that the hypothesis could be falsified? Even if one of the hypotheses about cold prevention is supported, does the difficulty of eliminating alternative hypotheses mean that we will never know which approach is truly best? The answer is yes—and no.

Hypotheses cannot be proven absolutely true; it is always possible that the true cause of a phenomenon may be found in a hypothesis that has not yet been tested. However, in a practical sense, a hypothesis can be proven beyond a reasonable doubt. That is, when one hypothesis has not been disproven through repeated testing and all reasonable alternative hypotheses have been eliminated, scientists accept that the well-supported hypothesis is, in a practical sense, true. "Truth" in science can therefore be defined as what we know and understand based on all currently available information. But scientists always leave open the possibility that what seems true now may someday be proven false.

One of the most effective ways to test many hypotheses is through rigorous scientific experiments. Experimentation has enabled scientists to prove beyond a reasonable doubt that the common cold is caused by a virus. A virus has a very simple structure—it typically contains a short strand of genetic material and a few proteins encased in a relatively tough protein shell and sometimes surrounded by a membrane. A virus must enter, or infect, a cell to reproduce. Of the over 200 types of viruses that are known to cause the common cold, most infect the cells in our noses and throats. The sneezing, coughing, congestion, and sore throat of a cold appear to result from the body's immune response to a viral invasion *(Figure 1.4)*.

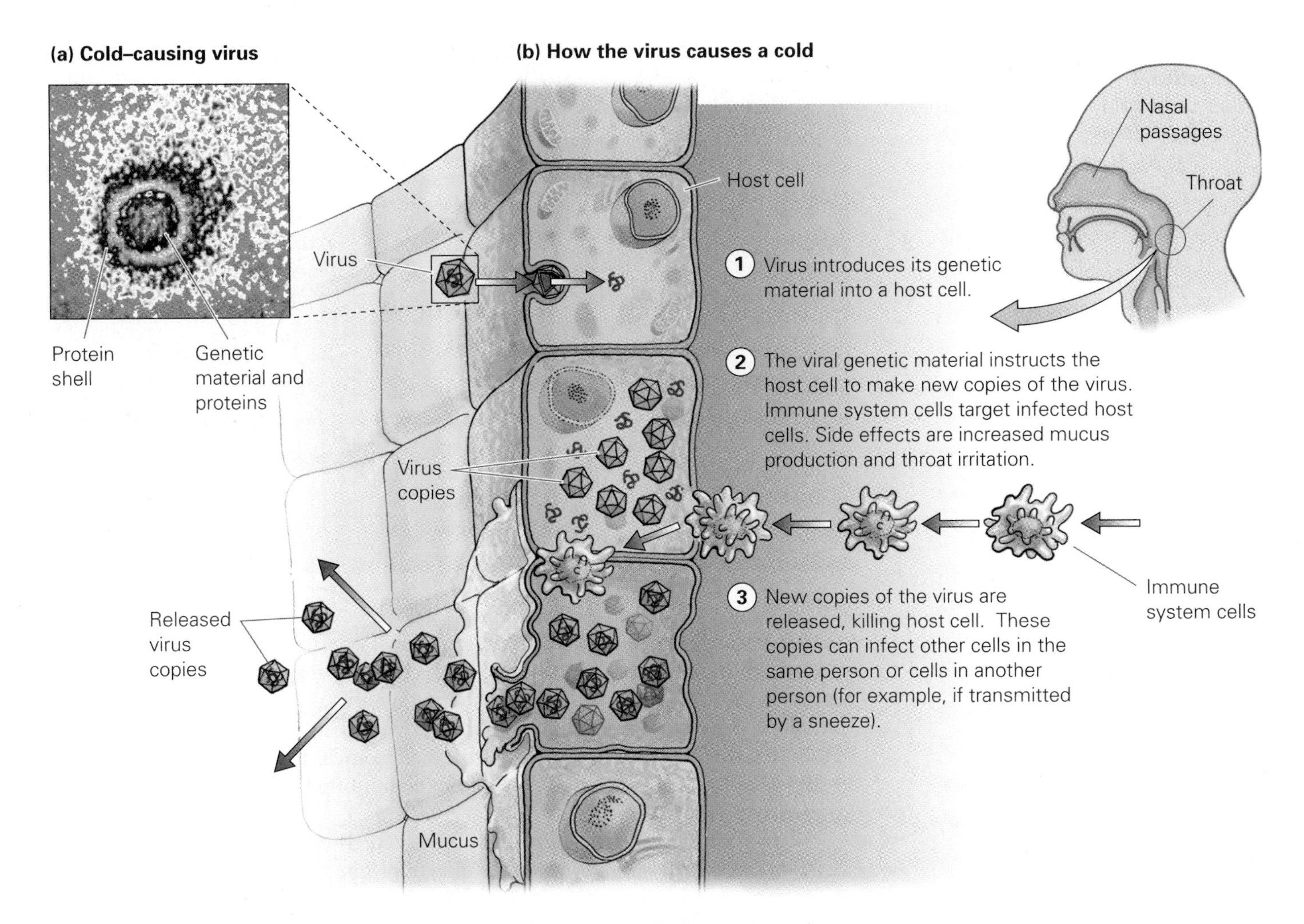

Figure 1.4 **A cold-causing virus.** (a) An electron microscope image of a typical rhinovirus, one of the many types of viruses that cause the common cold. (b) A rhinovirus causes illness by invading nose and throat cells and using them as "factories" to make virus copies. Cold symptoms result from immune system attempts to eliminate the virus. **Visualize This:** Find two points in this process where intervention by drugs or other treatment would disrupt it and lead to reduced cold symptoms.

As you may know, if we survive a virus infection, we are unlikely to experience a recurrence of the disease the virus causes. For example, it is extremely rare to suffer from chicken pox twice since one exposure to the chicken pox virus (either through infection or vaccination) usually provides lifelong immunity to future infection. The sheer number of cold viruses makes immunity to the common cold—and the development of a vaccine to prevent it—impossible. Scientists thus focus their experimental research about common colds on methods of prevention and treatment.

The Experimental Method

Experiments are sets of actions or observations designed to test specific hypotheses. Generally, an experiment allows a scientist to control the conditions that may affect the subject of study. Manipulating the environment allows a scientist to eliminate some alternative hypotheses that may explain the result.

Experimentation in science is analogous to what a mechanic does when diagnosing a car problem. There are many reasons why a car engine won't

Figure 1.5 **Testing hypotheses through observation.** The fossil record provides a source of data to test hypotheses about evolutionary history.

start. If a mechanic starts by tinkering with all possible fixes before restarting the car, the mechanic would not know which component had caused the problem and would have an unhappy customer who was charged for a lot of parts and labor. Instead, a mechanic would start with testing the battery for power; if it is fine, then the starter motor is the next logical problem. Likewise, a scientist systematically attempts to eliminate hypotheses that might explain a particular phenomenon.

Not all scientific hypotheses can be tested through experimentation. For instance, hypotheses about how life on Earth originated or the cause of dinosaur extinction are usually not testable in this way. These hypotheses are instead tested using careful observation of the natural world. For instance, the examination of fossils and other geological evidence allows scientists to test hypotheses regarding the extinction of the dinosaurs *(Figure 1.5)*.

The information collected by scientists during hypothesis testing is known as **data.** The data are collected on the **variables** of the test, that is, any factor that can change in value under different conditions. In an experimental test, scientists manipulate an **independent variable** (one whose value can be freely changed) to measure the effect on a **dependent variable.** The dependent variable may or may not be influenced by changes in the independent variable, but it cannot be systematically changed by the researchers. For example, to measure the effect of vitamin C on cold prevention, scientists can vary individuals' vitamin C intake (the independent variable) and measure their susceptibility to illness on exposure to a cold virus (the dependent variable).

Data obtained from well-designed experiments should allow researchers to convincingly reject or support a hypothesis. This is more likely to occur if the experiment is controlled.

Controlled Experiments

Control has a very specific meaning in science. A **control** for an experiment is a subject who is similar to an experimental subject except that the control is not exposed to the experimental treatment. If the control and experimental groups differ at the end of a well-designed test, then the difference is likely due to the experimental treatment.

An extract of *Echinacea purpurea* (a common North American prairie plant) in the form of echinacea tea has been touted as a treatment to reduce the likelihood as well as the severity and duration of colds *(Figure 1.6)*. A recent scientific experiment on the efficacy of *Echinacea* involved asking individuals suffering from colds to rate the effectiveness of a tea in relieving their symptoms. In this study, people who used Echinacea tea felt that it was 33% more effective. The "33% more effective" is in comparison to the rated effectiveness of a tea that did not contain *Echinacea* extract—that is, the results from the control group.

Figure 1.6 **Echinacea purpurea, an American coneflower.** Extracts from the leaves and roots of this plant are among the most popular herbal remedies sold in the United States.

Control groups are designed to eliminate as many alternative hypotheses as possible for the results of experiment. The first step is to select a pool of subjects in such a way that it eliminates differences between groups in a variety of independent variables, for instance, participants' ages, diets, stress levels, and likelihood of visiting a health care provider.

One effective way to minimize differences between groups is the **random assignment** of individuals. For example, a researcher might put all of the volunteers' names in a hat, draw out half, and designate these people as the experimental group and the remainder as the control group. As a result, there is unlikely to be a systematic difference between the experimental and control groups—each group should be a rough cross section of the population in the study. In the echinacea tea experiment just described, members of both the experimental and control groups were female employees of a nursing home who sought relief from their colds at their employer's clinic. The volunteers were randomly assigned into either the experimental or control group as they came into the clinic.

Stop and Stretch 1.3: In the echinacea tea experiment, a nonrandom assignment scheme might have put the first 25 visitors to the clinic in the control group and the next 25 in the experimental group. Imagine that in this version of the experiment, the experimental group did recover in fewer days than the control group. Describe an alternative hypothesis related to the nonrandom assignment of subjects that has not been eliminated and that could explain the results.

The second step in designing a good control is to treat all subjects identically during the course of the experiment. In this study, all participants received the same information about the supposed benefits of echinacea tea, and during the course of the experiment, all participants were given tea to drink five to six times daily until their symptoms subsided. However, individuals in the control group received "sham tea" that did not contain *Echinacea* extract. Treating all participants the same ensures that no factor related to the interaction between subject and researcher influences the results.

The sham tea in this experiment would be equivalent to the sugar pills that are given to control subjects during drug trials. Like other intentionally ineffective medical treatments, sham tea and sugar pills are called **placebos.** Employing a placebo generates only one consistent difference between individuals in the two groups—in this case, the type of tea they consumed.

In the echinacea tea study, the data indicated that cold severity was lower in the experimental group compared to those who received placebo. Because their study utilized controls, the researchers can be confident that the groups differed because of the effect of *Echinacea.* By reducing the likelihood that alternative hypotheses could explain their results, the researchers could strongly infer that they were measuring a real, positive effect of echinacea tea on colds *(Figure 1.7)*.

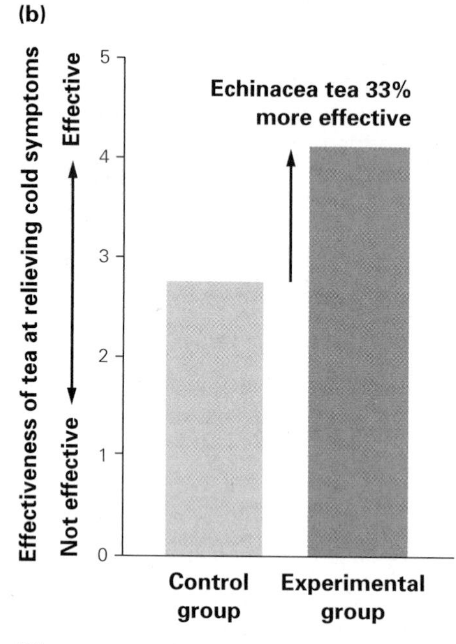

Figure 1.7 **A controlled experiment.** (a) In a controlled experiment testing echinacea tea as a treatment for colds, all subjects were treated identically except for the type of tea they consumed. (b) The results of the experiment indicated that echinacea tea was 33% more effective than the placebo.

The study described here supports the hypothesis that echinacea tea reduces the severity of colds. However, it is extremely rare that a single experiment will cause the scientific community to accept a hypothesis beyond a reasonable doubt. Dozens of studies, each using different experimental designs and many using extracts from different parts of the plant, have investigated the effect of *Echinacea* on common colds and other illnesses. Some of these studies have shown a positive effect, but others have shown none. In the medical community as a whole, the jury is still out regarding the effectiveness of this popular herb as a cold treatment. Only through continued controlled tests of the hypothesis will we discern an accurate answer to the question, Is *Echinacea* an effective cold treatment?

Minimizing Bias in Experimental Design

Scientists and human research subjects may have strong opinions about the truth of a particular hypothesis even before it is tested. These opinions may cause participants to unfairly influence, or **bias,** the results of an experiment.

One potential source of bias is subject expectation. Individual experimental subjects may consciously or unconsciously model the behavior they feel the researcher expects from them. For example, an individual who knew she was receiving echinacea tea may have felt confident that she would recover more quickly. This might cause her to underreport her cold symptoms. This potential problem is avoided by designing a **blind experiment,** in which individual subjects are not aware of exactly what they are predicted to experience. In experiments on drug treatments, this means not telling participants whether they are receiving the drug or a placebo.

Another source of bias arises when a researcher makes consistent errors in the measurement and evaluation of results. This phenomenon is called observer bias. In the echinacea tea experiment, observer bias could take various forms. Expecting a particular outcome might lead a scientist to give slightly different instructions about which symptoms constituted a cold to subjects who received echinacea tea. Or, if the researcher expected people who drank echinacea tea to experience fewer colds, she might make small errors in the measurement of cold severity that influenced the final result.

To avoid the problem of experimenter bias, the data collectors themselves should be "blind." Ideally, the scientist, doctor, or technician applying the treatment does not know which group (experimental or control) any given subject is part of until after all data have been collected and analyzed *(Figure 1.8)*. Blinding the data collector ensures that the data are **objective,** or in other words, without bias.

We call experiments **double blind** when both the research subjects and the technicians performing the measurements are unaware of either the hypothesis or whether a subject is in the control or experimental group. Double-blind experiments nearly eliminate the effects of human bias on results. When both researcher and subject have few expectations about the outcome, the results obtained from an experiment are more credible.

Using Correlation to Test Hypotheses

Double-blind, placebo-controlled, randomized experiments represent the gold standard for medical research. However, well-controlled experiments can be difficult to perform when humans are the experimental subjects. The requirement that both experimental and control groups be treated nearly identically means that some people receive no treatment. In the case of healthy volunteers with head colds, the placebo treatment of sham tea did not hurt those who received it.

However, placebo treatments are impractical or unethical in many cases. For instance, imagine testing the effectiveness of a birth control drug using a

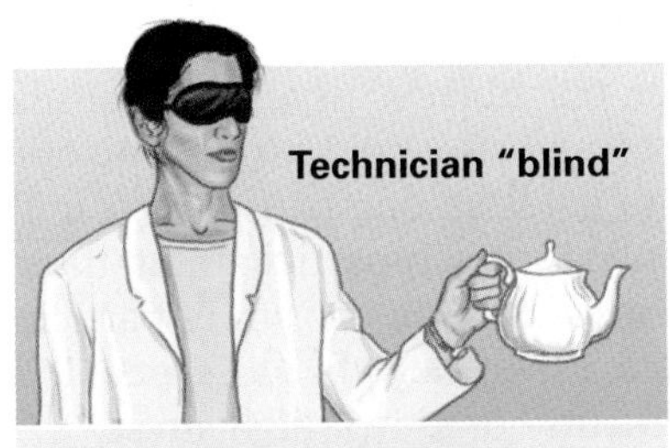

	Technician "blind"	Subject "blind"
What they know	• **Limited knowledge** of experimental hypothesis • **No knowledge** of which group participants belong to	• **Limited knowledge** of experimental hypothesis • **No knowledge** of which group he or she belongs to
How they behave	• **No difference** in instructions to participants • **No difference** in treatment of participants • **No difference** in data collection	• **Unbiased** reporting of symptoms or effects of treatment

Figure 1.8 **Double-blind experiments.** Double-blind experiments result in data that are more objective.

controlled experiment. This would require asking women to take a pill that may or may not prevent pregnancy while not using any other form of birth control!

Experiments on Model Organisms. Scientists can use **model organisms** when testing hypotheses that would raise ethical or practical problems when tested on people. In the case of research on human health and disease, model organisms are typically other mammals. Mammals are especially useful as model organisms in medical research because they are closely related to us evolutionarily. Like us, they have hair and give birth to live young, and thus they also share with us similarities in anatomy and physiology.

The vast majority of animals used in biomedical research are rodents such as rats, mice, and guinea pigs, although some areas of research require animals that are more similar to humans in size, such as dogs or pigs, or share a closer evolutionary relationship, such as chimpanzees (*Figure 1.9*).

The use of model organisms allows experimental testing on potential drugs and other therapies before these methods are employed on people. Research on model organisms has contributed to a better understanding of nearly every serious human health threat, including cancer, heart disease, Alzheimer's disease, and AIDS (acquired immunodeficiency syndrome). However, ethical concerns about the use of animals in research persist and can complicate such studies. In addition, the results of animal studies are not always directly applicable to humans—despite a shared evolutionary history, animals still can have important differences from humans in physiology. Testing hypotheses about human health in human beings still provides the clearest answer to these questions.

Looking for Relationships Between Factors. Scientists can also test hypotheses using correlations when controlled experiments on humans are difficult or impossible to perform. A **correlation** is a relationship between two variables.

Suggestions about using meditation to reduce susceptibility to colds are based on a correlation between high levels of psychological stress and increased

(a) Rat

(b) Dog

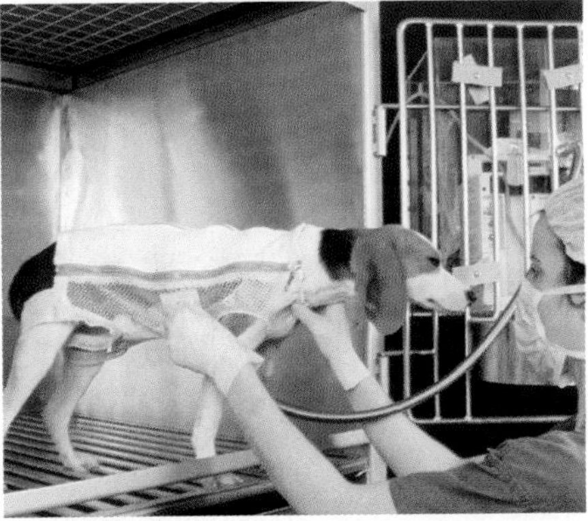

(c) Chimpanzee

Figure 1.9 **Model organisms in science.** (a) The classic "lab rat" is easy to raise and care for and has little genetic diversity to confound the results of an experiment. (b) Dogs have cardiovascular systems that are larger than rats' and similar to those of humans. (c) Chimpanzees are humans' closest biological relatives.

susceptibility to cold virus infections *(Figure 1.10)*. This correlation was generated by researchers who collected data on subjects' psychological stress levels before giving them nasal drops that contained a cold virus. Doctors later reported on the incidence and severity of colds among participants in the study. Note that while the cold virus was applied to each participant in the study, the researchers had no influence on the stress level of the study participants—in other words, this was not a controlled experiment because people were not randomly assigned to different "treatments" of low or high stress.

Stop and Stretch 1.4: Even though the researchers did not manipulate the independent variable in this study, there are still an independent variable and a dependent variable. What are they?

Let's examine the results presented in Figure 1.10. The horizontal axis of the graph, or **x-axis,** illustrates the independent variable. This variable is stress, and the graph ranks subjects along a scale of stress level—from low stress on the left edge of the scale to high stress on the right. The vertical axis of the graph, the **y-axis,** is the dependent variable—the percentage of study participants who developed colds as reported by their doctors. Each point on the graph represents a group of individuals and tells us what percentage of people in each stress category had clinical colds.

The line connecting the 5 points on the graph illustrates a correlation—the relationship between stress level and susceptibility to cold virus infections. Because the line rises to the right, it illustrates a positive correlation. These data tell us that people who have higher stress levels were more likely to come down with colds. But does this relationship mean that high stress causes increased cold susceptibility?

To conclude that stress causes illness, we need the same assurances that are given by a controlled experiment. In other words, we must assume that the individuals measured for the correlation are similar in every way except for their stress levels. Is this a good assumption? Not necessarily. Most correlations cannot control for alternative hypotheses. People who feel more stressed may have poorer diets because they feel time limited and rely on fast food more often. In addition, people who feel highly stressed may be in situations where they are exposed to more cold viruses. These differences among people who differ in stress level may also influence their cold susceptibility *(Figure 1.11)*. Therefore, even with a strong correlational relationship between the two factors, we cannot strongly infer that stress causes decreased resistance to colds.

Researchers who use correlational studies can eliminate a number of alternative hypotheses by closely examining their subjects. For example, this study on stress and cold susceptibility collected data from subjects on age, weight, sex, education, and their exposure to infected individuals. None of these factors differed consistently among low-stress and high-stress groups. While this analysis increases the strength of the inference that high stress levels truly do increase susceptibility to colds, people with high-stress lifestyles still may have important differences from those with low-stress

Figure 1.10 **Correlation between stress level and illness.** The graph indicates that people reporting higher levels of stress became infected after exposure to a cold virus more often than did people who reported low levels of stress. **Visualize This:** This graph groups people with similar, but not identical, stress index measures. Why might this have been necessary? If people with stress indices 3 and 4 have the same susceptibility to colds, does this call into question the correlation?

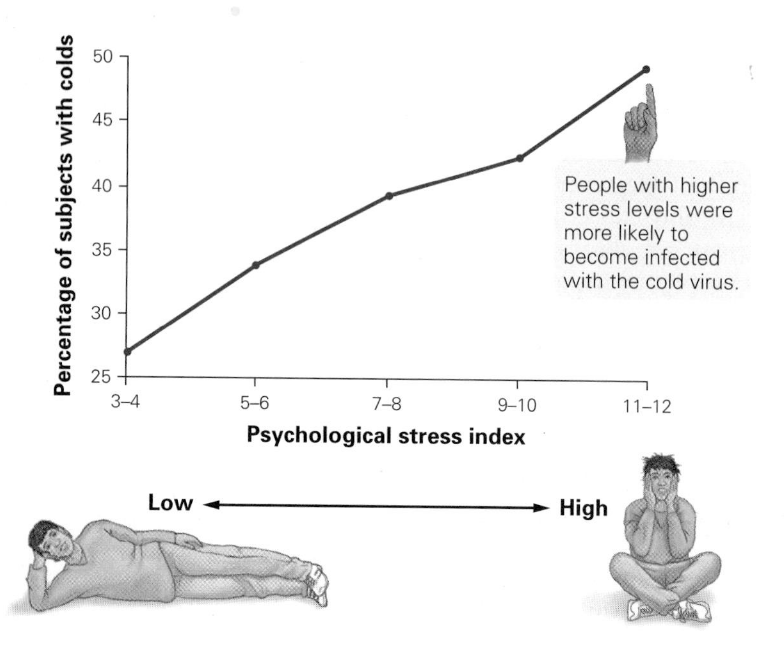

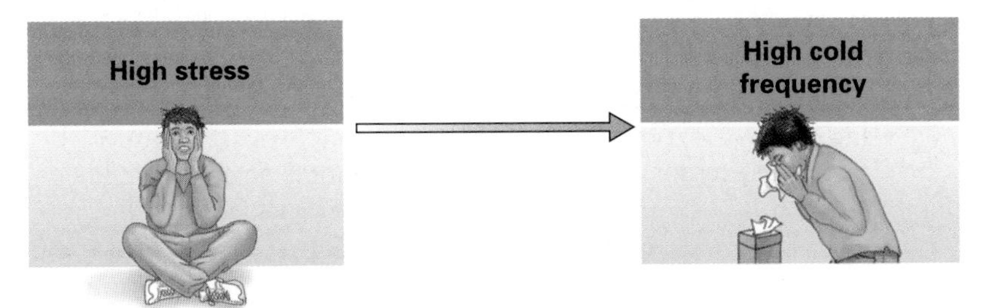

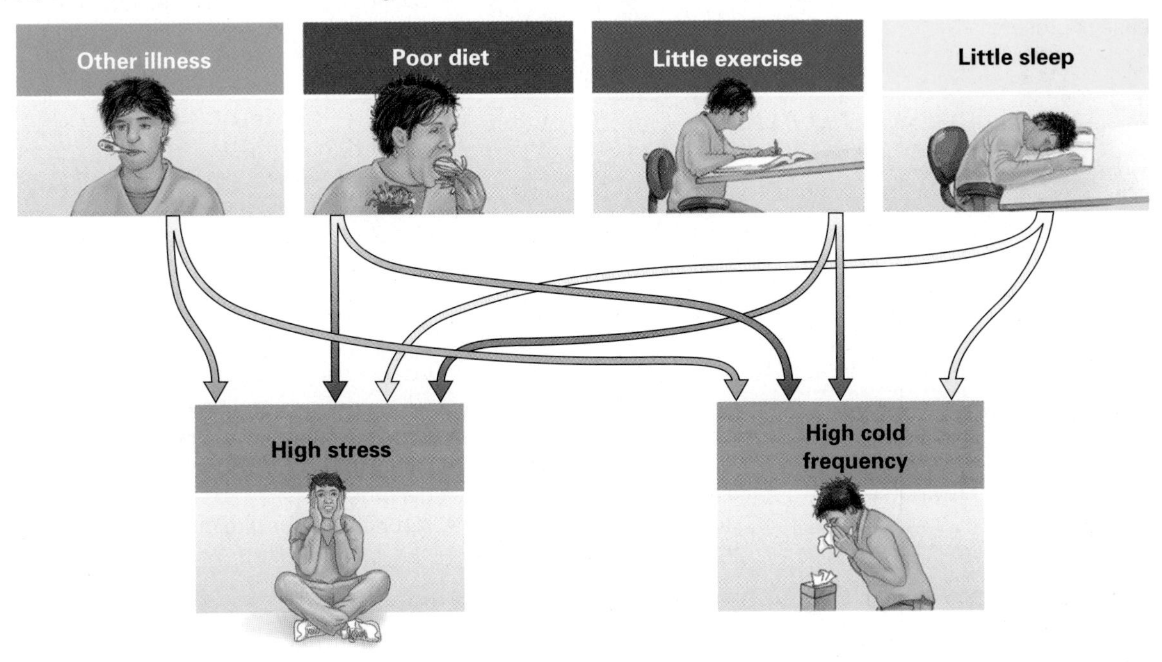

Figure 1.11 **Correlation does not signify causation.** A correlation typically cannot eliminate all alternative hypotheses.

lifestyles. It is possible that one of those differences is the real cause of disparities in cold frequency.

As you can see, it is difficult to demonstrate a cause-and-effect relationship between two factors simply by showing a correlation between them. In other words, correlation does not equal causation. For example, a commonly understood correlation exists between exposure to cold air and epidemics of the common cold. It is true that as outdoor temperatures drop, the incidence of colds increases. But numerous controlled experiments indicate that chilling does not increase susceptibility to colds. Instead, cold outdoor temperatures mean increased close contact with other people (and their viruses). Despite the correlation, cold air does not cause colds—exposure to viruses does.

Correlational studies are the main tool of epidemiology, the study of the distribution and causes of diseases. One commonly used epidemiology technique is a cross-sectional survey. In this type of survey, many individuals are both tested for the presence of a particular condition and asked about their exposure to various factors. The limitations of cross-sectional surveys include the effect of subject bias and poor recall by survey participants, in addition to all of the problems associated with interpreting correlations. *Table 1.1* provides an overview of the variety of correlational strategies employed to study the links between our environment and our health.

TABLE 1.1 **Types of correlational studies**

Name	Description	Pros	Cons
Ecological studies	Examine specific human populations for unusually high levels of various diseases (e.g., documenting a "cancer cluster" around an industrial plant)	• Inexpensive and relatively easy to do	• Unsure whether exposure to environmental factor is actually related to onset of the disease
Cross-sectional surveys	Question individuals in a population to determine amount of exposure to an environmental factor and whether disease is present	• More specific than ecological study	• Expensive • Subjects may not know exposure levels • Cannot control for other factors that may be different among individuals in survey • Cannot be used for rare diseases
Case-control studies	Compare exposures to specific environmental factors between individuals who have a disease and individuals matched in age and other factors who do not have a disease	• Relatively fast and inexpensive • Best method for rare diseases	• Does not measure absolute risk of disease as a result of exposure • Difficult to select appropriate controls to eliminate alternative hypotheses • Examines just one disease possibly associated with an environmental factor
Cohort studies	Follow a group of individuals, measuring exposure to environmental factors and disease prevalence	• Can determine risk of various diseases associated with exposure to particular environmental factor	• Expensive and time consuming • Difficult to control for alternative hypotheses • Not feasible for rare diseases
Correlational experiment	Expose individuals who experience varying levels of an independent variable to an experimental treatment	• Can control the amount of exposure to at least one environmental factor of interest	• Cannot eliminate alternative hypotheses • Only feasible for hypotheses for which an experimental treatment can be applied

1.3 Understanding Statistics

During a review of scientific literature on cold prevention and treatment, you may come across statements about the "significance" of the effects of different cold-reducing measures. For instance, one report may state that factor A reduced cold severity but that the results of the study were "not significant." Another study may state that factor B caused a "significant reduction" in illness. We might then assume that this statement means factor B will help us feel better, whereas factor A will have little effect. This is an incorrect assumption because in scientific studies, *significance* is defined a bit differently from its usual sense. To evaluate the scientific use of this term, we need a basic understanding of statistics.

Overview: What Statistical Tests Can Tell Us

We often use the term *statistics* to refer to a summary of accumulated information—for instance, a baseball player's success at hitting is summarized by his batting average, the total number of hits he made divided by the number of opportunities he had to hit. The science of **statistics** is a bit different; it is a specialized branch of mathematics used to evaluate and compare data.

An experimental test utilizes a small subgroup, or **sample,** of a population. Statistical methods can summarize data from the sample—for instance, we can describe the average, also known as the **mean,** length of colds experienced by experimental and control groups. **Statistical tests** can then be used to extend the results from a sample to the entire population.

When scientists conduct an experiment, they hypothesize that there is a true, underlying effect of their experimental treatment on the entire population. An experiment on a sample of a population can only estimate this true effect because a sample is always an imperfect "snapshot" of an entire population.

Consider a hypothesis that the average hair length of women in a college class in 1963 was shorter than the average hair length at the same college today. To test this hypothesis, we could compare a sample of snapshots from college yearbooks. If hairstyles were very similar among the women in a snapshot, you could reasonably assume that the average hair length in the college class is close to the average length in the snapshot. However, what if you see that women in a snapshot have a variety of hairstyles, from crew cuts to long braids? In this case, it is difficult to determine whether the average hair length in the snapshot is at all close to the average for the class. With so much variation, the snapshot could, by chance, contain a surprisingly high frequency of women with very long hair, causing the average length in the sample to be much longer than the average length for the entire class *(Figure 1.12)*.

A statistical test calculates the likelihood, given the number of individuals sampled and the variation within samples, that the difference between two samples reflects a real, underlying difference between the populations from which these samples were drawn. A **statistically significant** result is one that is very unlikely to be due to chance differences between the experimental and control groups, so thus likely represents a true difference between the groups.

In the experiment with the echinacea tea, statistical tests indicated that the 33 percent reduction in cold severity observed by the researchers was statistically significant. In other words, there is a low probability that the difference between the two groups is due to chance. If the experiment is properly designed, a statistically significant result allows researcher to infer that the treatment had an effect.

(a) Average hair length in this snapshot is shorter...

(b) ...than average hair length in this snapshot.

So, is hair today longer than in 1963?

- Little variability
- High probability of reflecting average of all women in the class

- High variability
- Low probability of reflecting average of all women in the class

Figure 1.12 **The role of statistics.** Statistical tests calculate the variability within groups to determine the probability that groups differ only by chance.

A Closer Look: Statistics

We can explore the role that statistical tests play more closely by evaluating another study on cold treatments. This study examined the efficacy of lozenges containing zinc on reducing cold severity.

Some forms of zinc can block common cold viruses from invading cells in the nasal cavity. This observation led to the hypothesis that consuming zinc at the start of a cold could decrease the severity of cold symptoms by reducing the number of cells that become infected. Researchers at the Cleveland Clinic tested this hypothesis using a sample of 100 of their employees who volunteered for a study within 24 hours of developing cold symptoms. The researchers randomly assigned subjects to control or experimental groups. Members of the experimental group received lozenges containing zinc, while members of the control group received placebo lozenges. Members of both groups received the same instructions for using the lozenges and were asked to rate their symptoms until they had recovered. The experiment was double blind.

When the data from the zinc lozenge experiment were summarized, the statistics indicated that the mean recovery time was more than 3 days shorter in the zinc group than in the placebo group *(Figure 1.13)*. On the surface, this result appears to support the hypothesis. However, recall the example of the snapshot of women's hair length. A statistical test is necessary because of the effect of chance.

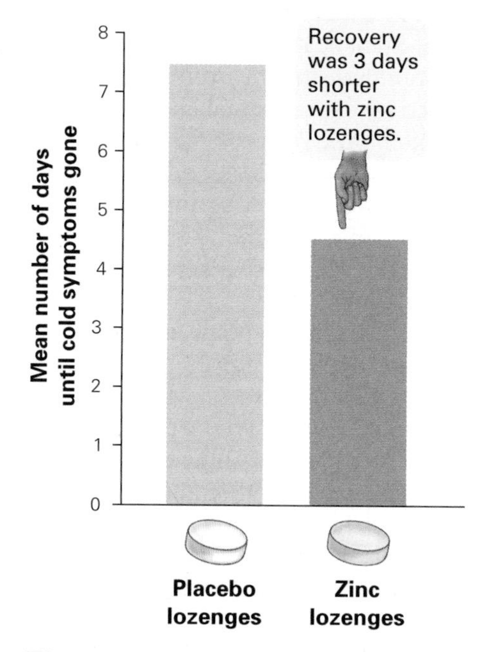

Figure 1.13 **Zinc lozenges reduce the duration of colds.** Individuals taking zinc lozenges had colds lasting about 4½ days as opposed to approximately 7½ days for individuals taking placebo.

The Problem of Sampling Error. The effect of chance on experimental results is known as **sampling error**—more specifically, sampling error is the difference between a sample and the population from which it was drawn. Similarly, in any experiment, individuals in the experimental group will differ from individuals in the control group in random ways. Even if there is *no* true effect of an experimental treatment, the data from the experimental group will never be identical to the data from the control group.

For example, we know that people differ in their ability to recover from a cold infection. If we give zinc lozenges to one volunteer and placebo lozenges to another, it is likely that the two volunteers will have colds of different lengths. But even if the zinc-taker had a shorter cold than the placebo-taker, you would probably say that the test did not tell us much about our hypothesis—the zinc-taker might just have had a less-severe cold for other reasons.

Now imagine that we had 5 volunteers in each group and saw a difference, or that the difference was only 1 day instead of 3 days. How would we determine if the lozenges had an effect? Statistical tests allow researchers to look at their data and determine how likely it is—that is, the **probability**—that the result is due to sampling error. In this case, the statistical test distinguished between two possibilities for why the experimental group had shorter colds: Either the difference was due to the effectiveness of zinc as a cold treatment or it was due to a chance difference from the control group. The results indicated that there was a low probability, less than 1 in 10,000 (0.01%), that the experimental and control groups were so different simply by chance. In other words, the result is **statistically significant.**

A statistical measure of the amount of variability in a sample is often expressed as the **standard error.** The standard error is used to generate the **confidence interval**—the range of values that has a high probability (usually 95%) of containing the true population mean *(Figure 1.14)*. Put simply, although the average of a sample is unlikely to be exactly the average of the population, the average plus the standard error represents the highest likely value for the population average, and the average minus the standard error represents the lowest likely average value. The confidence interval provides a way to express how much sampling error is influencing the results. A smaller confidence interval indicates that sampling error is likely to be small, and results with small confidence intervals are more likely to be statistically significant if the hypothesis is true.

(A Closer Look continued)

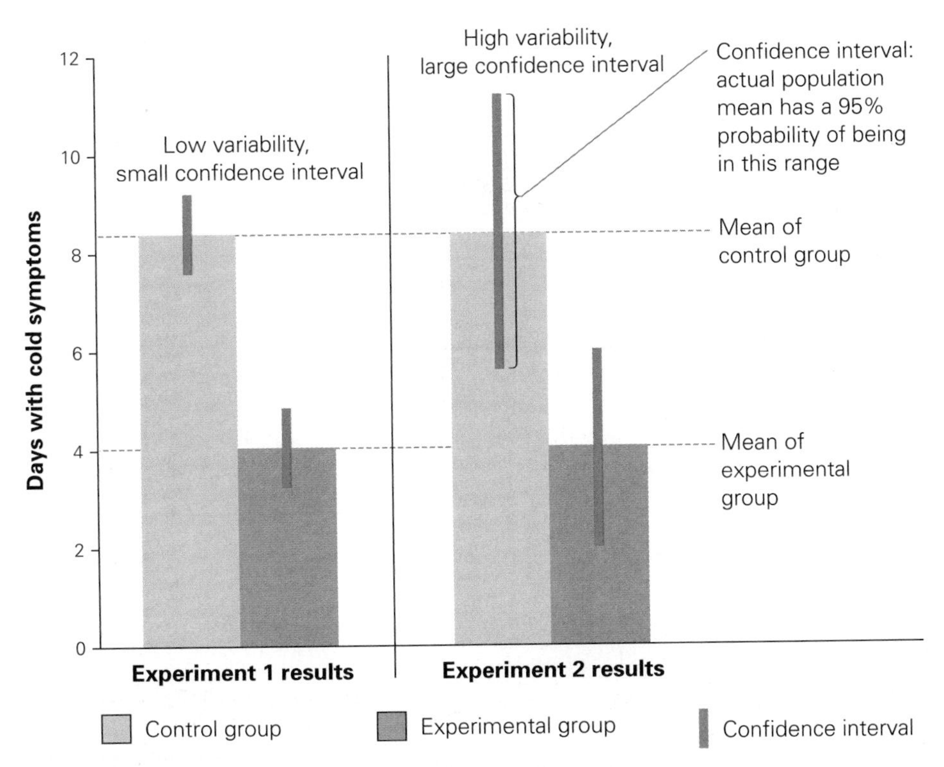

Figure 1.14 **Confidence interval.** The red lines on these bar graphs represent the confidence interval for each sample mean. A more variable sample has a larger confidence interval than a less-variable sample. Even though the means are the same for both experimental and control groups in these two sets of bars, only the data summarized at left illustrate a significant difference because of the greater variability in the samples illustrated on the right side.

Stop and Stretch 1.5: Opinion polling before a recent election indicated that candidate A was favored by 47% of likely voters, and candidate B was favored by 51% of likely voters. The standard error was 3%. Why did reporters refer to this poll as a "statistical tie"?

Factors That Influence Statistical Significance. One characteristic of experiments that influences sampling error is **sample size**—the number of individuals in the experimental and control groups. If a treatment has no effect, a small sample size could return results that appear significantly different because of an unusually large and consistent sampling error. This was the case with the vitamin C hypothesis described earlier. Subsequent tests with larger sample sizes, encompassing a wider variety of individuals with different underlying susceptibilities to colds, allowed scientists to reject the hypothesis that vitamin C prevents colds.

Conversely, if the effect of a treatment is real but the sample size of the experiment is small, a single experiment may not allow researchers to determine convincingly that their hypothesis has support. Several of the experiments that tested the efficacy of *Echinacea* extract demonstrate this phenomenon. For example, one experiment performed at a Wisconsin clinic with 48 participants indicated that echinacea tea drinkers were 30% less likely to experience any cold symptoms after virus exposure as compared to individuals who received a placebo tea. However, the small sample size of the study meant that this result was not statistically significant.

The more participants there are in a study, the more likely it is that researchers will see a true effect of an experimental treatment even if it is very small. For example, a study of over 21,000 male smokers over 6 years in Finland demonstrated that men who took vitamin E supplements had 5% fewer colds than the men who did not take these supplements. In this case, the large sample size allowed researchers to see that vitamin E has a real, but relatively tiny, effect on cold incidence. In other words, this statistically significant result has little real-world *practical* significance; a 5% effect shouldn't convince people to begin supplementing with vitamin E to prevent colds. The relationship among hypotheses, experimental tests, sample size, and statistical and practical significance is summarized in *Figure 1.15*.

(continued on the next page)

(A Closer Look continued)

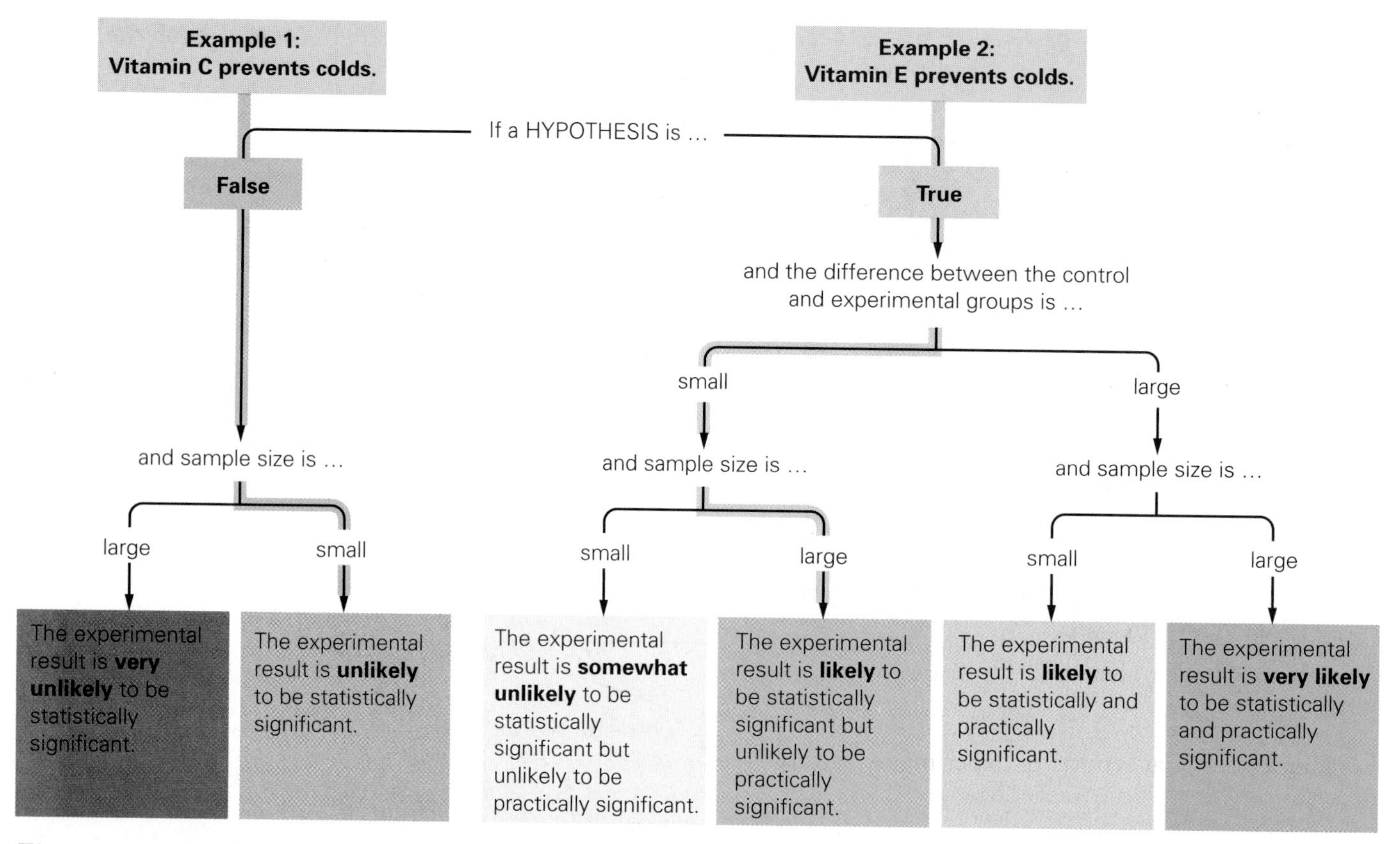

Figure 1.15 **Factors that influence statistical significance.** This flowchart summarizes the relationship between the true effect of a treatment and the sample size of an experiment on the likelihood of obtaining statistical significance. **Visualize This:** The Nurse's Health Survey, which has followed nearly 80,000 women for over 20 years, has found that a diet low in refined carbohydrates such as white flour and sugar but relatively high in vegetable fat and protein cuts heart disease risk by 30% relative to women with a high refined-carbohydrate diet. How does this result map out on the flowchart here?

There is one final caveat to this discussion. A statistically significant result is typically defined as one that has a *probability* of 5% or less of being due to chance alone. But probability is not certainty. If all scientific research uses this same standard, as many as 1 in every 20 statistically significant results (that is, 5% of the total) is actually reporting an effect that is *not real.* In other words, some statistically significant results are "false positive," representing a surprisingly large difference between experimental and control groups that occurred only as a result of sampling error. This potential error explains why one supportive experiment is not enough to convince all scientists that a particular hypothesis is accurate.

Even with a statistical test indicating that the result had a probability of less than 0.01% of occurring by chance, we should begin to feel assured that taking zinc lozenges will reduce the duration of colds only after additional hypothesis tests that give similar results. In fact, scientists continue to test this hypothesis, and there is still no consensus about the effectiveness of zinc as a cold treatment.

What Statistical Tests Cannot Tell Us

All statistical tests operate with the assumption that the experiment was designed and carried out correctly. In other words, a statistical test evaluates the chance of sampling error, not observer error, and a statistically significant result is not the last word on an experimentally tested hypothesis. An examination of the experiment itself is required.

In the test of the effectiveness of zinc lozenges, the experimental design minimized the likelihood that alternative hypotheses could explain the results by being double blind, randomized, and controlled. Given such a well-designed experiment, this statistically significant result allows researchers to strongly infer that consuming zinc lozenges reduces the duration of colds.

As we saw with the experiment on vitamin E intake and cold prevention, statistical significance is also not equivalent to *significance* as we usually define the term, that is, as "meaningful or important." Unfortunately, experimental results reported in the news often use the term *significant* without clarifying this important distinction. Understanding that problem, as well as other misleading aspects of how science can be presented, will enable you to better use scientific information.

1.4 Evaluating Scientific Information

The previous sections should help you see why definitive scientific answers to our questions are slow in coming. However, a well-designed experiment can certainly allow us to approach the truth.

Primary Sources

Looking critically at reports of experiments can help us make well-informed decisions about actions to take. Most of the research on cold prevention and treatment is first published as **primary sources,** written by the researchers themselves and reviewed within the scientific community *(Figure 1.16)*. The process of **peer review,** in which other scientists critique the results and conclusions of an experiment before it is published, helps increase confidence in scientific information. Peer-reviewed research articles in journals such as *Science*, *Nature*, the *Journal of the American Medical Association*, and hundreds of others represent the first and most reliable sources of current scientific knowledge.

However, evaluating the hundreds of scientific papers that are published weekly is a task no one of us can perform. Even if we focused only on a particular field of interest, the technical jargon used in many scientific papers may be a significant barrier to the nonexpert public.

Instead of reading the primary literature, most of us receive our scientific information from **secondary sources** such as books, news reports, and advertisements. How can we evaluate information in this context? The following sections provide strategies for doing so, and a recurring Savvy Reader feature in this textbook will help you practice these evaluation skills.

Information from Anecdotes

Information about dietary supplements such as echinacea tea and zinc lozenges is often in the form of **anecdotal evidence**—meaning that the advice is based on one individual's personal experience. A friend's enthusiastic plug for vitamin 22C, because she felt it helped her, is an example of a testimonial—a common form of anecdote. Advertisements that use a celebrity to pitch a product "because it worked for them" are classic forms of testimonials.

You should be cautious about basing decisions on anecdotal evidence, which is not at all equivalent to well-designed scientific research. For example, many of us have heard anecdotes along the lines of the grandpa who was a pack-a-day smoker and lived to the age of 94. However, hundreds of

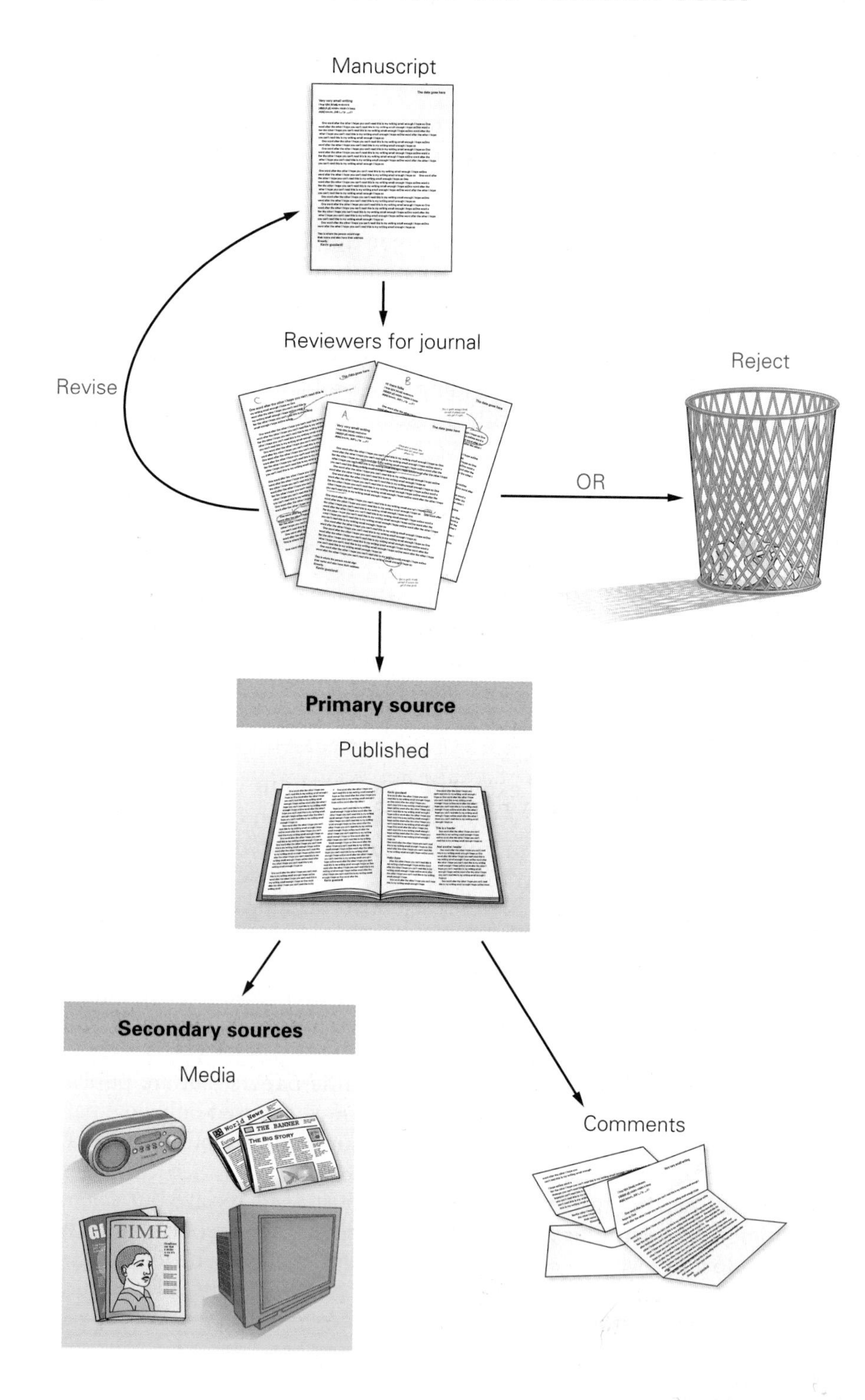

Figure 1.16 **Primary sources: publishing scientific results.** Most scientific journals require papers to go through stringent review before publication.

studies have demonstrated the clear link between cigarette smoking and premature death. Although anecdotes may indicate that a product or treatment has merit, only well-designed tests of the hypothesis can determine its safety and efficacy.

Stop and Stretch 1.6: One of the challenges of presenting science is that anecdotes make a much bigger impression on people than statistical summaries of data. Why do you think this is the case?

Science in the News

Popular news sources provide a steady stream of science information. However, stories about research results often do not contain information about the adequacy of controls, the number of subjects, the experimental design, or the source of the scientist's funding. How can anyone evaluate the quality of research that supports dramatic headlines about risks to human health or environmental catastrophes?

First, you must consider the source of media reports. Certainly news organizations will be more reliable reporters of fact than will entertainment tabloids, and news organizations with science writers should be better reporters of the substance of a study than those without. Television talk shows, which need to fill airtime, regularly have guests who promote a particular health claim. Too often these guests may be presenting information that is based on anecdotes or an incomplete summary of the primary literature.

Paid advertisements are a legitimate means of disseminating information. However, claims in advertising should be carefully evaluated. Advertisements of over-the-counter and prescription drugs must conform to rigorous government standards regarding the truth of their claims. However, lower standards apply to advertisements for herbal supplements, many health food products, and diet plans. Be sure to examine the fine print because advertisers often are required to clarify the statements made in their ads.

Another commonly used source for health information is the Internet. As you know, anyone can post information on this great resource. Typing in "common cold prevention" on a standard web search engine will return thousands of web pages—from highly respected academic and government sources to small companies trying to sell their products or individuals who have strong, sometimes completely unsupported, ideas about cures. Often it can be difficult to determine the reliability of a well-designed website, and even well-used sites such as Wikipedia may contain erroneous or misleading information. Here are some things to consider when using the web as a resource for health information:

1. Choose sites maintained by reputable medical establishments, such as the National Institutes of Health (NIH) or the Mayo Clinic.
2. It costs money to maintain a website. Consider whether the site seems to be promoting a product or agenda. Advertisements for a specific product should alert you to possible bias.
3. Check the date when the website was last updated and see whether the page has been updated since its original posting. Science and medicine are disciplines that must frequently incorporate new data into hypotheses. A reliable website will be updated often.
4. Determine whether unsubstantiated claims are being made. Look for references and be suspicious of any studies that are not from peer-reviewed journals.

Understanding Science from Secondary Sources

Once you are satisfied that a media source is relatively reliable, you can examine the scientific claim that it presents. Use your understanding of the process of science and of experimental design to evaluate the story and the science. Does the story about the claim present the results of a scientific study, or is it built around an untested hypothesis? Were the results obtained using the scientific method, with tests designed to reject false hypotheses? Is the story confusing correlation with causation? Does it seem that the information is applicable to nonlaboratory situations, or is it based on results from preliminary or animal studies?

Look for clues about how well the reporters did their homework. Scientists usually discuss the limitations of their research in their papers. Are these cautions noted in an article or television piece? If not, the reporter may be overemphasizing the applicability of the results.

Then, note if the scientific discovery itself is controversial. That is, does it reject a hypothesis that has long been supported? Does it concern a subject that is controversial (like the origin of racial differences or homosexuality)? Might it lead to a change in social policy? In these cases, be extremely cautious. New and unexpected research results must be evaluated in light of other scientific evidence and understanding. Reports that lack comments from other experts may miss problems with a study or fail to place it in the context of other research. The Savvy Reader feature in this chapter provides a checklist of questions to ask and answer as you evaluate news reports.

Finally, the news media generally highlight only those science stories that editors find newsworthy. As we have seen, scientific understanding accumulates relatively slowly, with many tests of the same hypothesis finally leading to an accurate understanding of a natural phenomenon. News organizations are also more likely to report a study that supports a hypothesis rather than one that gives less-supportive results, even if both types of studies exist.

In addition, even the most respected media sources may not be as thorough as readers would like. For example, a recent review published in the *New England Journal of Medicine* evaluated the news media's coverage of new medications. Of 207 randomly selected news stories, only 40% that cited experts with financial ties to a drug told readers about this relationship. This potential conflict of interest calls into question the expert's objectivity. Another 40% of the news stories did not provide basic statistics about the drugs' benefits. Most of the news reports also failed to distinguish between absolute benefits (how many people were helped by the drug) and relative benefits (how many people were helped by the drug relative to other therapies for the condition). The journal's review is a vivid reminder that we need to be cautious when reading or viewing news reports on scientific topics.

Even after following all of these guidelines, you will still find well-researched news reports on several scientific studies that seem to give conflicting and confusing results. As you now know, such confusion is the nature of the scientific process—early in our search for understanding, many hypotheses are proposed and discussed; some are tested and rejected immediately, and some are supported by one experiment but later rejected by more thorough experiments. It is only by clearly understanding the process and pitfalls of scientific research that you can distinguish "what we know" from "what we don't know."

1.5 Is There a Cure for the Common Cold?

So where does our discussion leave us? Will we ever find the best way to prevent a cold or reduce its effects? In the United States, over 1 billion cases of the common cold are reported per year, costing billions of dollars in medical visits, treatment, and lost workdays. Consequently, there is an enormous effort to find effective protection from the different viruses that cause colds.

A search of medical publication databases indicates that every year, nearly 100 scholarly articles regarding the biology, treatment, and consequences of common cold infection are published. This research has led to several important discoveries about the structure and biochemistry of common cold viruses, how they enter cells, and how the body reacts to these infections.

Despite all of the research and the emergence of some promising possibilities, the best prevention method known for common colds is still the old standby—keep your hands clean. Numerous studies have indicated that rates of common cold infection are 20% to 30% lower in populations who employ effective hand-washing procedures. Cold viruses can survive on surfaces for many hours; if you pick them up on your hands from a surface and transfer them to your mouth, eyes, or nose, you may inoculate yourself with a 7-day sniffle *(Figure 1.17)*.

Of course, not everyone gets sick when exposed to a cold virus. The reason one person has more colds than another might not be due to a difference in personal hygiene. The correlation that showed a relationship between stress and cold susceptibility appears to have some merit. Research indicates that among people exposed to viruses, the likelihood of ending up with an infection increases with high levels of psychological stress—something that many college students clearly experience.

Research also indicates that vitamin C intake, diet quality, exposure to cold temperatures, and exercise frequency appear to have no effect on cold susceptibility, although along with echinacea tea and zinc lozenges, there is some evidence that vitamin C may reduce cold symptoms a bit. Surprisingly, even though medical research has led to the elimination of killer viruses such as smallpox and polio, scientists are still a long way from "curing" the common cold.

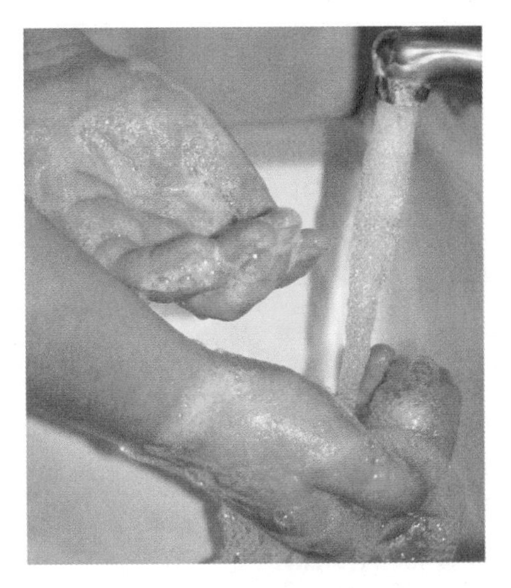

Figure 1.17 **Preventing colds.** The best defense against infection is effective hand washing. Lather with a liquid soap for 20 seconds, then rinse under running water for 10 seconds. Alcohol-based hand gels (without water) containing at least 60% ethanol are less effective but will work in a pinch.

Savvy Reader A Toolkit for Evaluating Science in the News

The following checklist can be used as a guide to evaluating science information in the news. Although any issue that raises a red flag should cause you to be cautious about the conclusions of the story, no one issue should cause you to reject the story (and its associated research) out of hand. However, in general, the fewer red flags raised by a story, the more reliable the report of the significance of the scientific study. Use Table 1.2 to evaluate this extract from an article by Terry Gupta in the Nashua (New Hampshire) *Telegraph*:

BY TERRY GUPTA | THE NASHUA TELEGRAPH (NEW HAMPSHIRE) | OCTOBER 26, 2007

[Paula] Fortier, a licensed massage therapist, shared her special brew. To boiled water and tea, she adds cayenne pepper, honey and fresh squeezed lemon. Fortier's enthusiasm may be more contagious than the flu when she describes this heated, "unbeatable" combination.

"Your throat opens up," she said. "You're less irritated and you sweat out some unwanted toxins." She varies the exact measures of the ingredients according to tolerance and taste.

While on the topic of hot water, one of Fortier's favorite remedies for "an achy body that feels sore all over is a hot bath in a steeping tub with 2 cups of cider vinegar for 15 minutes. This draws off toxins, reduces inflammation and leaves you feeling better."

(continued on the next page)

Savvy Reader (continued)

TABLE 1.2 **A guide for evaluating science in the news.** *For each question, check the appropriate box.*

Question	Possible answers: Preferred answer		Possible answers: Raises a red flag	
1. What is the basis for the story?	Hypothesis test	○	Untested assertion *No data to support claims in the article.*	○
2. What is the affiliation of the scientist?	Independent (university or government agency)	○	Employed by an industry or advocacy group *Data and conclusions could be biased.*	○
3. What is the funding source for the study?	Government or nonpartisan foundation (without bias)	○	Industry group or other partisan source (with bias) *Data and conclusions could be biased.*	○
4. If the hypothesis test is a correlation: Did the researchers attempt to eliminate reasonable alternative hypotheses?	Yes	○	No *Correlation does not equal causation. One hypothesis test provides poor support if alternatives are not examined.*	○
If the hypothesis test is an experiment: Is the experimental treatment the only difference between the control group and the experimental group?	Yes	○	No *An experiment provides poor support if alternatives are not examined.*	○
5. Was the sample of individuals in the experiment a good cross section of the population?	Yes	○	No *Results may not be applicable to the entire population.*	○
6. Was the data collected from a relatively large number of people?	Yes	○	No *Study is prone to sampling error.*	○
7. Were participants blind to the group they belonged to and/or to the "expected outcome" of the study?	Yes	○	No *Subject expectation can influence results.*	○
8. Were data collectors and/or analysts blinded to the group membership of participants in the study?	Yes	○	No *Observer bias can influence results.*	○
9. Did the news reporter put the study in the context of other research on the same subject?	Yes	○	No *Cannot determine if these results are unusual or fit into a broader pattern of results.*	○
10. Did the news story contain commentary from other independent scientists?	Yes	○	No *Cannot determine if these results are unusual or if the study is considered questionable by others in the field.*	○
11. Did the reporter list the limitations of the study or studies on which he or she is reporting ?	Yes	○	No *Reporter may not be reading study critically and could be overstating the applicability of the results.*	○

CHAPTER Review

For study help, animations, and more quiz questions go to www.mybiology.com.

Summary

1.1 The Process of Science

- Science is a process of testing hypotheses—statements about how the natural world works. Scientific hypotheses must be testable and falsifiable (p. 2).
- A scientific theory is an explanation of a set of related observations based on well-supported hypotheses from several different, independent lines of research (p. 3).
- Hypotheses are tested via the process of deductive reasoning, which allows researchers to make specific predictions about expected observations (pp. 4–5).
- Absolutely proving hypotheses is impossible. However, well-designed scientific experiments can allow researchers to strongly infer that their hypothesis is correct (p. 5).

web animation 1.1
Science as a Process: Arriving at Scientific Insights

1.2 Hypothesis Testing

- Controlled experiments test hypotheses about the effect of experimental treatments by comparing a randomly assigned experimental group with a control group. Controls are individuals who are treated identically to the experimental group except for application of the treatment (pp. 8–9).
- Bias in scientific results can be minimized with double-blind experiments that keep subjects and data collectors unaware of which individuals belong in the control or experimental group (p. 10).
- If performing controlled experiments on humans is considered unethical, scientists sometimes use correlations. The correlation of structure and function between humans and model organisms, such as other mammals, allows experimental tests of these hypotheses (p. 11).
- If experiments are impossible, scientists try to discern associations between two factors. A correlation study can describe a relationship between two factors, but it does not strongly imply that one factor causes the other (pp. 12–14).

1.3 Understanding Statistics

- Statistics help scientists evaluate the results of their experiments by determining if results appear to reflect the true effect of an experimental treatment on a sample of a population (p. 15).
- A statistical test indicates the role that chance plays in the experimental results; this is called sampling error. A statistically significant result is one that is very unlikely to be due to chance differences between the experimental and control groups (pp. 16–17).
- Even when an experimental result is highly significant, hypotheses are tested multiple times before scientists come to consensus on the true effect of a treatment (p. 18).

1.4 Evaluating Scientific Information

- Primary sources of information are experimental results published in professional journals and peer reviewed by other scientists before publication (p. 20).
  - Most people get their scientific information from secondary sources such as the news media. The ability to evaluate science from these sources is an important skill (p. 20).
- Anecdotal evidence is an unreliable means of evaluating information, and media sources are of variable quality; distinguishing between news stories and advertisements is important when evaluating

Summary (continued)

the reliability of information. The Internet is a rich source of information, but users should look for clues to a particular website's credibility (pp. 20–21).

- Stories about science should be carefully evaluated for information on the actual study performed, the universality of the claims made by the researchers, and other studies on the same subject. Sometimes confusing stories about scientific information are a reflection of controversy within the scientific field itself (p. 22).

Roots to Remember

The following roots of words come mainly from Latin and Greek and will help you to decipher terms:

bio- means life.
deduc- means to reason out, working from facts.
induc- means to rely on reason to derive principles (and also to cause to happen).
hypo- means under, below, or basis.
-ology means the study of or branch of knowledge about.

Learning the Basics

Answers to all *Learning the Basics, Analyzing and Applying the Basics,* and *Stop and Stretch* questions can be found in Appendix A.

1. What is the value of a placebo in an experimental design?

2. Which of the following is an example of inductive reasoning? **A.** All cows eat grass; **B.** My cow eats grass and my neighbor's cow eats grass; therefore all cows probably eat grass; **C.** If all cows eat grass, when I examine a random sample of all the cows in Minnesota, I will find that they all eat grass; **D.** Cows may or may not eat grass, depending on what type of farm they live on.

3. A scientific hypothesis is ______________.
A. an opinion; **B.** a proposed explanation for an observation; **C.** a fact; **D.** easily proved true; **E.** an idea proposed by a scientist.

4. How is a scientific theory different from a scientific hypothesis?
A. It is based on weaker evidence; **B.** It has not been proved true; **C.** It is not falsifiable; **D.** It can explain a large number of observations; **E.** It must be proposed by a professional scientist.

5. Which of the following is a prediction of the hypothesis: Eating chicken noodle soup is an effective treatment for colds?
A. People who eat chicken noodle soup have shorter colds than do people who do not eat chicken noodle soup; **B.** People who do not eat chicken noodle soup experience unusually long and severe colds; **C.** Cold viruses cannot live in chicken noodle soup; **D.** People who eat chicken noodle soup feel healthier than do people who do not eat chicken noodle soup; **E.** Consuming chicken noodle soup causes people to sneeze.

6. If I perform a hypothesis test in which I demon-strate that the prediction I made above is true, I have ______________.
A. proved the hypothesis; **B.** supported the hypothesis; **C.** not falsified the hypothesis; **D.** B and C are correct; **E.** A, B, and C are correct

7. Control subjects in an experiment ______________.
A. should be similar in most ways to the experimental subjects; **B.** should not know whether they are in the control or experimental group; **C.** should have essentially the same interactions with the researchers as the experimental subjects; **D.** help eliminate alternative hypotheses that could explain experimental results; **E.** all of the above

8. An experiment in which neither the participants in the experiment nor the technicians collecting the data know which individuals are in the experimental group and which ones are in the control group is known as ______________.
A. controlled; **B.** biased; **C.** double blind; **D.** falsifiable; **E.** unpredictable

9. A relationship between two factors, for instance between outside temperature and the number of people with active colds in a population, is known as a(n) ______________.
A. significant result; **B.** correlation; **C.** hypothesis; **D.** alternative hypothesis; **E.** experimental test

10. A primary source of scientific results is ______________.
A. the news media; **B.** anecdotes from others; **C.** articles in peer-reviewed journals; **D.** the Internet; **E.** all of the above

Analyzing and Applying the Basics

1. There is a strong correlation between obesity and the occurrence of a disease known as type 2 diabetes—that is, obese individuals have a higher instance of diabetes than nonobese individuals do. Does this mean that obesity causes diabetes? Explain.

2. In an experiment examining vitamin C as a cold treatment, students with cold symptoms who visited the campus medical center either received vitamin C or were treated with over-the-counter drugs. Students then reported on the length and severity of their colds. Both the students and the clinic health providers knew which treatment students were receiving. This study indicated that vitamin C significantly reduced the length and severity of colds. Which factors make this result somewhat unreliable?

3. Brain-derived neurotrophic factor (BDNF) is a substance produced in the brain that helps nerve cells (neurons) to grow and survive. BDNF also increases the connectivity of neurons and improves learning and mental function. A 2002 study that examined the effects of intense wheel-running on rats and mice found a positive correlation between BDNF levels and running distance. What could you conclude from this result?

Connecting the Science

1. Much of the research on common cold prevention and treatment is performed by scientists employed or funded by drug companies. Often these companies do not allow scientists to publish the results of their research for fear that competitors at other drug companies will use this research to develop a new drug before they do. Should our society allow scientific research to be owned and controlled by private companies?

2. Should society restrict the kinds of research performed by government-funded scientists? For example, many people believe that research performed on tissues from human fetuses should be restricted; these people believe that such research would justify abortion. If most Americans feel this way, should the government avoid funding this research? Are there any risks associated with *not* funding research with public money?

CHAPTER 2

Are We Alone in the Universe?

Water, Biochemistry, and Cells

Does life exist on Mars?

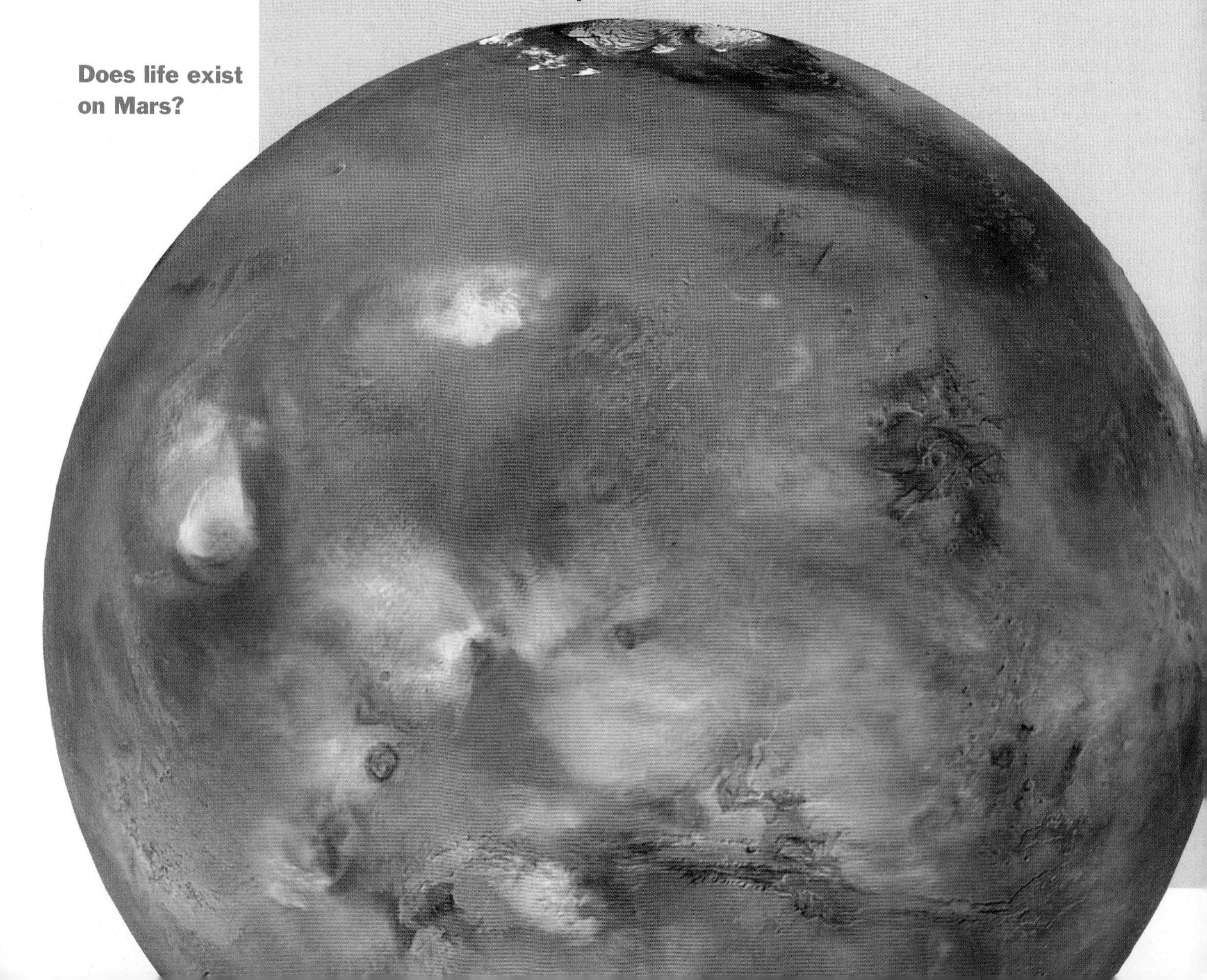

Popular images aside, we know of no other intelligent life in our solar system.

Is there life on other planets? Scientists have found persuasive, if not conclusive, evidence of life on Mars. When NASA scientist David McKay first proposed that this cold, dry, rather harsh planet could harbor life, people were astounded.

The evidence of life found by Dr. McKay and his team did not in any way resemble the cartoon images often used to depict Martians. Instead, what these scientists found was evidence of life in a 3.6-billion-year-old, potato-size rock. They believe that the rock had been ejected from the surface of Mars around 15 million years ago and had traveled through space for nearly that entire time. Ultimately the rock crashed to Earth, landing in Antarctica about 13,000 years ago and remaining there until discovered by scientists in 1984. This meteorite, drably named ALH84001, appeared to contain the same features that scientists use to demonstrate the existence of life in 3.6-billion-year-old Earth rocks—there were fossils, various minerals that are characteristic of life, and evidence of complex chemicals typically produced by living organisms.

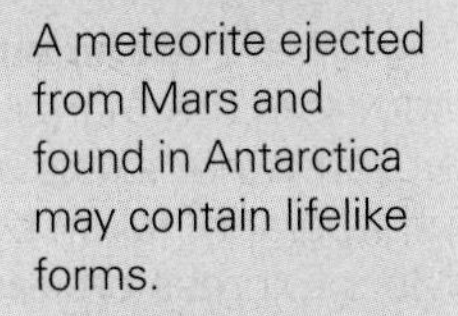

A meteorite ejected from Mars and found in Antarctica may contain lifelike forms.

While many scientists debate the assertion that this rock provides evidence of life on Mars, the announcement served to inject new energy into Mars exploration. Since then, multiple robotic rovers and mapping satellites have been sent to the planet, and the U.S. government announced an initiative to send astronauts to the red planet by the 2020s. While there are many reasons to explore Mars, the question that remains most intriguing—and is a significant focus of several of these missions—is whether life ever existed there.

The fascination about potential Martian life speaks to a fundamental question that many humans share: Are the creatures on Earth the only living organisms in the universe? Our galaxy is filled with countless stars and planets, and the universe teems with galaxies. Even if we find no convincing evidence of life on Mars, there are a seemingly infinite number of places to look for other living beings. In this chapter, we discuss the characteristics and requirements of life and examine techniques that scientists use to search the universe for other living creatures.

Will the Mars rover find evidence of life outside Earth?

web animation 2.1
Chemistry and Water

web animation 2.2
Nucleic Acids

2.1 What Does Life Require?

Because the galaxy likely contains billions of planets, scientists looking for life elsewhere seek to identify the range of conditions under which they would expect life to arise. What is it that scientists look for when identifying a planet (or moon) as a candidate for hosting life?

A Definition of Life

Figure 2.1 **Homeostasis.** Black-capped chickadees can maintain a core body temperature of 108°F (~42°C) during the day, even when the air temperature is well below zero.

In science-fiction movies, alien life-forms are often obviously alive and even somewhat familiar looking. But in reality, living organisms may be truly alien; that is, they may look nothing like organisms we are familiar with on Earth. So how would we determine whether an entity found on another planet was actually alive?

Surprisingly, biologists do not have a simple definition for a "living organism." A list of the attributes found in most earthly life-forms includes growth, movement, reproduction, response to external environmental stimuli, and **metabolism** (all of the chemical processes that occur in cells, including the breakdown of substances to produce energy, the synthesis of substances necessary for life, and the excretion of wastes generated by these processes). However, this definition could apply to things that no one considers to be living. For example, fire can grow, consume energy, give off waste, move, reproduce by sending off sparks, and change in response to environmental conditions. And some organisms that are clearly living do not conform to this definition. Male mules grow, metabolize, move, and respond to stimuli, but they are sterile (unable to reproduce).

If we examine more closely the characteristics of living organisms on Earth, we will see that all organisms contain a common set of biological molecules, are composed of cells, and can maintain **homeostasis,** that is, a roughly constant internal environment despite an ever-changing external environment *(Figure 2.1)*. The ability to maintain homeostasis requires complex feedback mechanisms between multiple sensory and physiological systems and is possible only in living organisms. In addition, populations of living organisms can evolve, that is, change in average physical characteristics over time. If we search the universe for planets that could support life similar to that found on Earth—and thus organisms that we would clearly identify as "living"—the list of planetary requirements becomes more stringent. In particular, an Earth-like planet should have abundant liquid **water** available.

Figure 2.2 **Water on Mars.** (a) This image from the Mars rover indicates that frozen water exists on Mars. (b) This photograph, taken by the European Mars Express orbiter, shows a channel on Mars that may have been formed by running water.

(a) Frozen water

(b) Running water

The Properties of Water

Water is a requirement for life. Although Mars does not currently appear to have any liquid water, ice is found at its poles *(Figure 2.2a)*, and features of its surface indicate that it once contained salty seas and flowing water *(Figure 2.2b)*. The presence of liquid water on Mars would fulfill an essential prerequisite for the appearance of life. But why is water such an important feature?

Water is made up of two elements: hydrogen and oxygen. **Elements** are the fundamental forms of matter and are composed of atoms that cannot be broken down by normal physical means such as boiling.

Atoms are the smallest units that have the properties of any given element. Ninety-two natural elemental atoms have been described by chemists, and several more have been created in laboratories. Hydrogen, oxygen, and calcium are examples of elements commonly found in living organisms. Each element has a one- or two-letter symbol: H for hydrogen, O for oxygen, and Ca for calcium, for example.

Atoms are composed of subatomic particles called **protons, neutrons,** and **electrons.** Protons have a positive electric charge; these particles and the uncharged neutrons make up the **nucleus** of an atom. All atoms of a particular element have the same number of protons, giving the element its **atomic number.** The negatively charged electrons are found outside the nucleus in an "electron cloud." Electrons are attracted to the positively charged nucleus *(Figure 2.3)*. A *neutral atom* has equal numbers of protons and electrons. Electrically charged **ions** do not have an equal number of protons and electrons. In this case, the atom is not neutral and is instead charged.

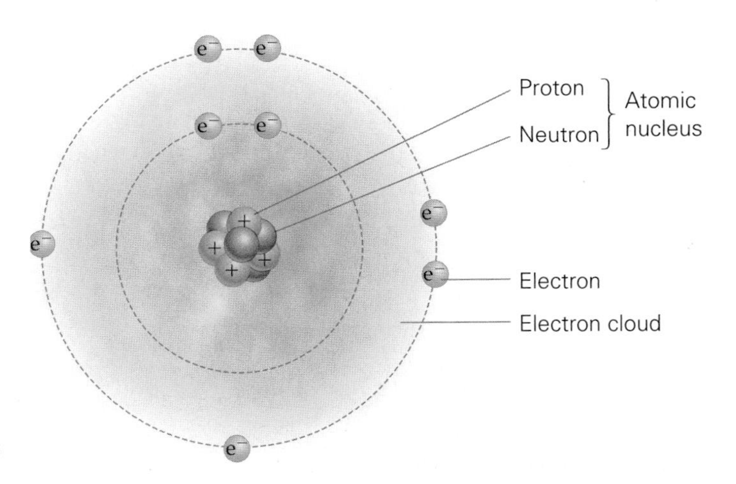

Figure 2.3 **Atomic structure.** An oxygen atom contains a nucleus made up of 8 protons and 8 neutrons. Orbiting electrons surround the nucleus. Although the number of particles within each atom differs, all atoms have the same basic structure.

The Structure of Water. The chemical formula for water is H_2O, indicating that it contains two hydrogen atoms for every one oxygen. Water, like other molecules, consists of two or more atoms joined by chemical bonds. A molecule can be composed of the same or different atoms. For example, a molecule of oxygen consists of two oxygen atoms joined to each other, while a molecule of carbon dioxide consists of carbon and oxygen atoms.

Water Is a Good Solvent. Water has the ability to dissolve a wide variety of substances. A substance that dissolves when mixed with another substance is called a solute. When a solute is dissolved in a liquid, such as water, the liquid is called a solvent. Once dissolved, components of a particular solute can pass freely throughout the water, making a chemical mixture or solution.

Water is a good solvent because it is **polar,** meaning that different regions, or poles, of the molecule have different charges. The polarity arises because oxygen is more attractive to electrons, that is, it is more **electronegative,** than most other atoms, including hydrogen. As a result of oxygen's electronegativity, electrons in a water molecule spend more time near the nucleus of the oxygen atom than near the nuclei of the hydrogen atoms. With more negatively charged electrons near it, the oxygen in water carries a partial negative charge, symbolized by the Greek letter delta, δ^-. The hydrogen atoms thus have a partial positive charge, symbolized by δ^+ *(Figure 2.4)*. When atoms of a molecule carry no partial charge, they are said to be **nonpolar.**

Water Facilitates Chemical Reactions. Because it is such a powerful solvent, water can facilitate **chemical reactions,** which are changes in the chemical composition of substances. Solutes in a mixture, called **reactants,** can come in contact with each other, permitting the modification of chemical bonds that occur during a reaction. The molecules formed as a result of a chemical reaction are known as **products.**

Water molecules tend to orient themselves so that the hydrogen atom (with its partial positive charge) of one molecule is near the oxygen atom (with its partial negative charge) of another molecule *(Figure 2.5a)*. The weak attraction between hydrogen atoms and oxygen atoms in adjacent molecules forms a **hydrogen bond.** Hydrogen bonding is a type of weak chemical bond

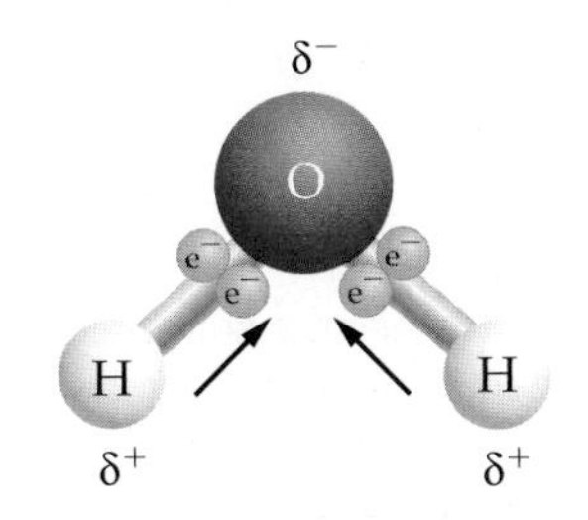

Figure 2.4 **Polarity in a water molecule.** Water is a polar molecule. Its atoms do not share electrons equally. **Visualize This:** Toward which atom are the electrons of a water molecule pulled?

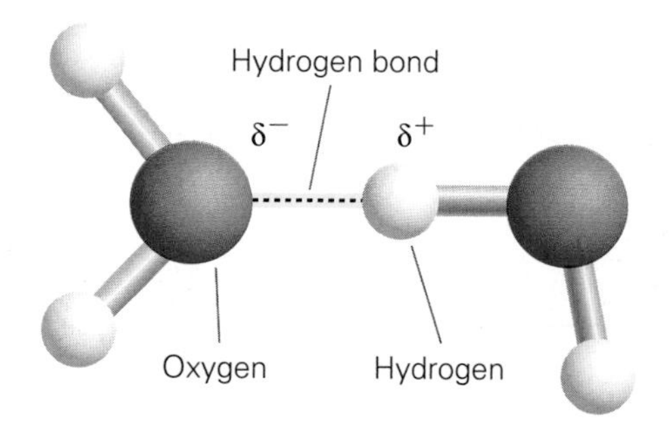

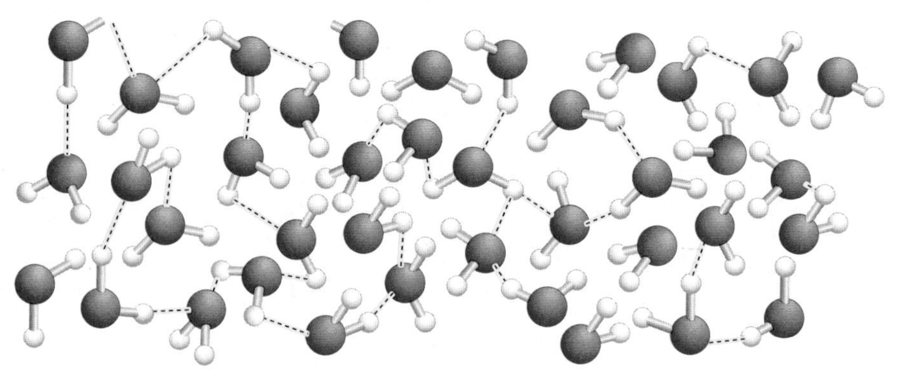

Figure 2.5 **Hydrogen bonding.** Hydrogen bonding can occur when there is a weak attraction between the hydrogen and oxygen atoms between (a) two or (b) many different water molecules.

that forms when a partially positive hydrogen atom is attracted to a partially negative atom. Hydrogen bonds can be intramolecular, involving different regions within the same molecule, or they can be intermolecular, between different molecules, as is the case in hydrogen bonding between different water molecules. *Figure 2.5b* shows the hydrogen bonding that occurs between water molecules in liquid form.

Water Is Cohesive. The tendency of like molecules to stick together is called **cohesion.** Cohesion is much stronger in water than in most liquids as a result of hydrogen bonding and is an important property of many biological systems. For instance, many plants depend on cohesion to help transport a continuous column of water from the roots to the leaves.

When heat energy is added to water, its initial effect is to disrupt the hydrogen bonding among water molecules. Therefore, this heat energy can be absorbed without changing the temperature of water. Only after the hydrogen bonds have been broken can added heat increase the temperature. In other words, the initial input of energy is absorbed.

The flowing water that was once found on Mars is now only in the form of ice. Until scientists can land on Mars and collect ice samples for analysis, its actual composition is a matter of conjecture. However, images taken by a NASA rover have led scientists to believe that some of the rocks on Mars were probably produced from deposits at the bottom of a body of salt water. Salt water on Mars is likely to be the same as salt water on Earth, a solution of water and sodium chloride. We know from surveying Earth's oceans that salt water is hospitable to millions of different life-forms. In fact, most hypotheses about the origin of life on Earth presume that our ancestors first arose in the salty oceans.

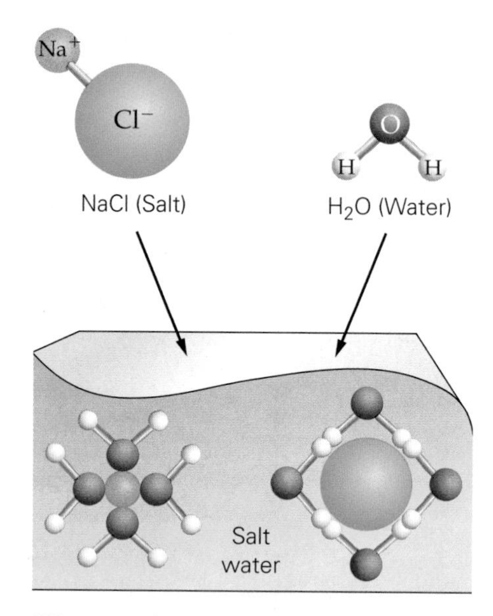

Figure 2.6 **Water as a solvent.** When salt is placed in water, the negatively charged regions of the water molecules surround the positively charged sodium ion, and the positively charged regions of the water molecules surround the negatively charged chlorine ion, breaking the bond holding sodium and chloride together and dissolving the salt.

The ability of water to dissolve substances such as sodium chloride is a direct result of its polarity. Each molecule of sodium chloride is composed of one sodium ion (Na^+) and one chloride ion (Cl^-). In the case of sodium chloride, the negative pole of water molecules will be attracted to a positively charged sodium ion and separate it from a negatively charged chloride ion *(Figure 2.6)*. Water can also dissolve other polar molecules, such as alcohol, in a similar manner. Polar molecules are called **hydrophilic** ("water loving") because of their ability to dissolve in water.

Salts are produced by the reaction of an **acid** (a substance that donates H^+ ions to a solution) with a **base** (a substance that accepts H^+ ions). Water can break apart or dissociate into H^+ and OH^- ions. The **pH** scale is a measure of the relative amounts of these ions in a solution. The more acidic a solution is, the higher the H^+ concentration is relative to the OH^- ions. Hydrogen ion concentration is inversely related to pH, so the higher the H^+ concentration, the lower the pH. Basic solutions have fewer H^+ ions relative to OH^- ions and thus a higher pH *(Figure 2.7)*. These ions can react with other

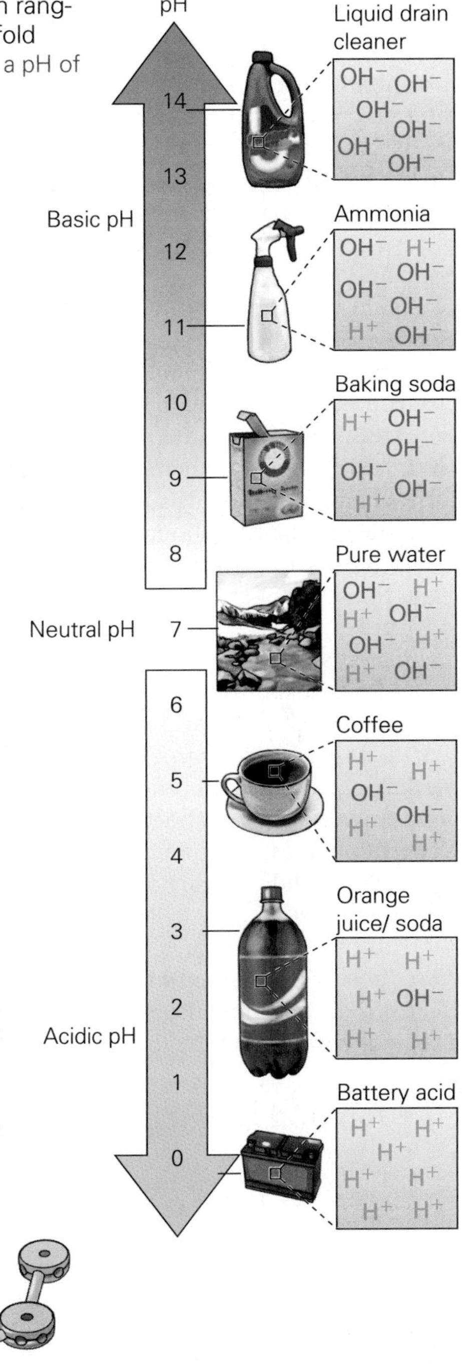

Figure 2.7 **The pH scale.** The pH scale is a measure of hydrogen ion concentration ranging from 0 (most acidic) to 14 (most basic). Each pH unit actually represents a 10-fold (10×) difference in the concentration of H^+ ions. **Visualize This:** A substance with a pH of 5 would have how many times more H^+ ions than a substance with a pH of 7?

charged molecules and help to bring them into the water solution. At any given time, a small percentage of water molecules in a pure solution will be dissociated. There are equal numbers of these ions in pure water, and so it is neutral, which on the pH scale is 7. The pH of most cells is very close to 7.

Nonpolar molecules, such as oil, do not contain charged atoms and are referred to as **hydrophobic** ("water fearing") because they do not easily mix with water.

When a European Space Agency probe landed on Titan, one of Saturn's moons and a place where the chemical composition of the atmosphere may be similar to that found on early Earth, the photos transmitted by the probe indicated that liquid was present on the surface of this bitterly cold place. At atmospheric temperatures of approximately −292°F (−180°C) the liquid is obviously not water; instead, it is most likely a mixture of ethane and methane, both nonpolar molecules. As a result, oceans on Titan are much poorer solvents than are oceans on Earth, and conditions in these oceans are probably not suitable for the evolution of life.

Organic Chemistry

The Martian meteorite ALH84001 had one characteristic that provided some evidence that the rock once contained living organisms—the presence of complex molecules containing the element carbon.

All life on Earth is based on the chemistry of the element carbon. The branch of chemistry that is concerned with complex carbon-containing molecules is called **organic chemistry.**

Overview: Chemical Bonds. Chemical bonds between atoms and molecules involve attractions that help stabilize various configurations. In general, this involves the sharing or transfer of electrons. One of the most important elements in biology is carbon, which is often involved in chemical bonding because of its ability to make bonds with up to four other elements. Like a Tinkertoy™ connector, carbon has multiple sites for connections that allow carbon-containing molecules to take an almost infinite variety of shapes *(Figure 2.8)*.

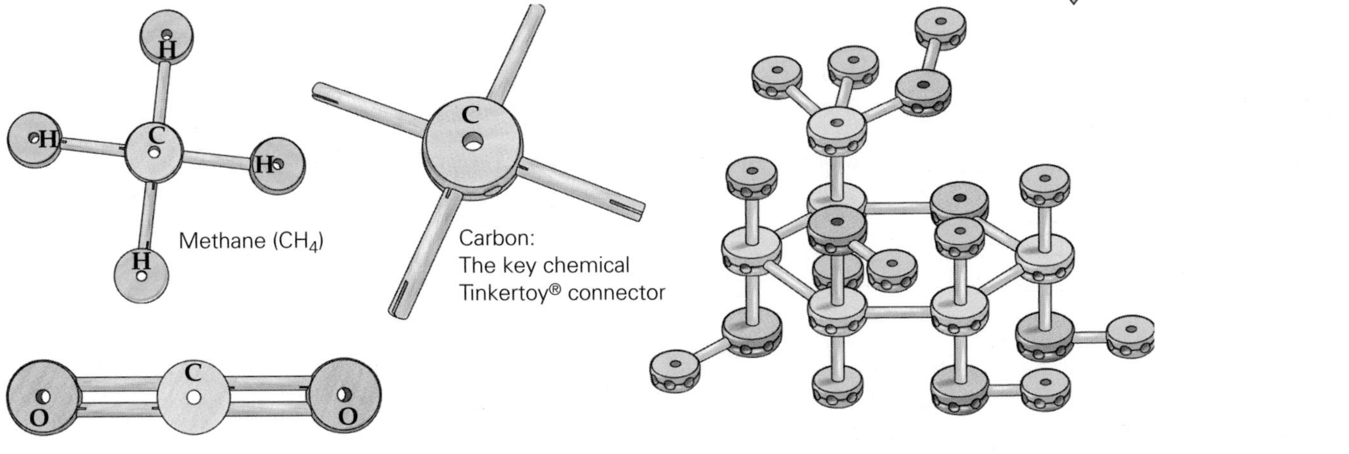

Figure 2.8 **Carbon, the chemical Tinkertoy™ connector.** Because carbon forms four covalent bonds at a time, carbon-containing compounds can have diverse shapes.

A Closer Look: Chemical Bonds

The ability of elements to make chemical bonds depends on the atom's electron configuration. The electrons in the electron cloud that surrounds the atom's nucleus have different energy levels based on their distance from the nucleus. The first energy level, or **electron shell,** is closest to the nucleus, and the electrons located there have the lowest energy. The second energy level is a little farther away, and the electrons located in the second shell have a little more energy. The third energy level is even farther away, and its electrons have even more energy, and so on.

Each energy level can hold a specific maximum number of electrons. The first shell holds 2 electrons, and the second and third shells each hold a maximum of 8. Electrons fill the lowest energy shell before advancing to fill a higher energy-level shell. For example, hydrogen with its 1 electron needs only one more electron to fill its first shell.

Atoms with the same number of electrons in their outermost energy shell, called the **valence shell,** exhibit similar chemical behaviors. When the valence shell is full, the atom will not normally form chemical bonds with other atoms. Atoms whose valence shells are not full of electrons often combine via chemical bonds.

Atoms with 4 or 5 electrons in the outermost valence shell tend to share electrons to complete their valence shells. When atoms share electrons, a type of bond called a **covalent bond** is formed.

Carbon, with its 4 valence electrons is said to be tetravalent *(Figure 2.9a)*. In other words, it can form up to four bonds. Carbon can form 4 single bonds, 2 double bonds, 1 double bond and 2 single bonds, and so on, depending on the number of electrons needed by the atom that is its partner. *Figure 2.9b* shows carbon covalently bonded to 4 hydrogens to produce methane, an organic compound that is common in the atmosphere of Titan. Covalent bonds are symbolized by a short line indicating a shared pair of electrons *(Figure 2.10a)*. When an element such as carbon enters into bonds involving two *pairs* of shared electrons, this is called a double bond. A carbon-to-carbon double bond is symbolized by two horizontal lines *(Figure 2.10b)*.

Atoms with 1, 2, or 3 electrons in their valence shell tend to lose electrons and therefore become positively charged ions, while atoms with 6 or 7 electrons in the valence shell tend to gain electrons and become negatively charged ions. Positively and negatively charged ions associate into a type of bond called the ionic bond.

Stop and Stretch 2.1: What does not make sense (chemically) about the structure below?

H—C=C—H

Ionic bonds form between charged atoms attracted to each other by similar, opposite charges. For example, the sodium atom forms an ionic bond with a chlorine atom to produce table salt (sodium chloride) when the sodium atom gives up an electron and the chlorine atom gains one

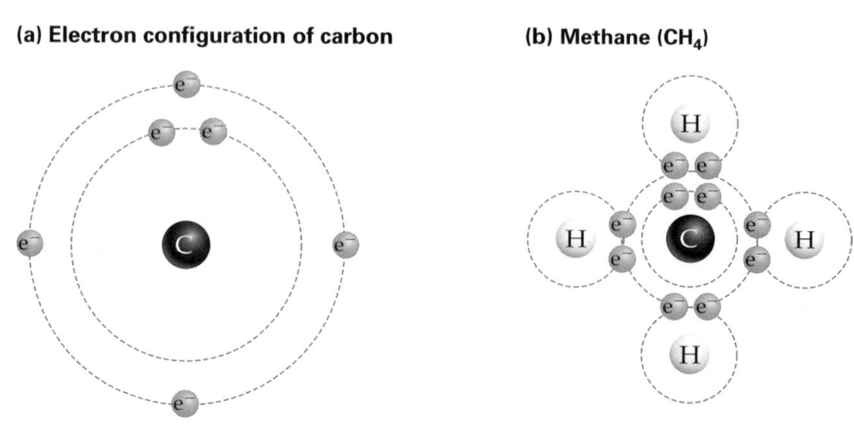

Figure 2.9 **Covalent bonding.** (a) Carbon has 4 unpaired electrons in its valence shell and is thus able to make up to 4 chemical bonds. (b) Methane consists of carbon covalently bonded to 4 hydrogens.

(A Closer Look continued)

(Figure 2.11). More than 2 atoms can be involved in an ionic bond. For instance, calcium will react with 2 chlorine atoms to produce calcium chloride ($CaCl_2$). This is because calcium has 2 electrons in its valence shell—when it loses these it has 2 more protons than neutrons, giving it a double-positive charge. Each chlorine atom, with 7 electrons, picks only 1 more electron to have a stable outer shell and a single negative charge. Thus, 2 chlorine ions will be attracted to a single calcium ion.

Ionic bonds are about as strong as covalent bonds. They can be more easily disrupted, however, when mixed with certain liquids containing electrical charges. Water is one liquid that causes ions in molecules to dissociate or fall apart.

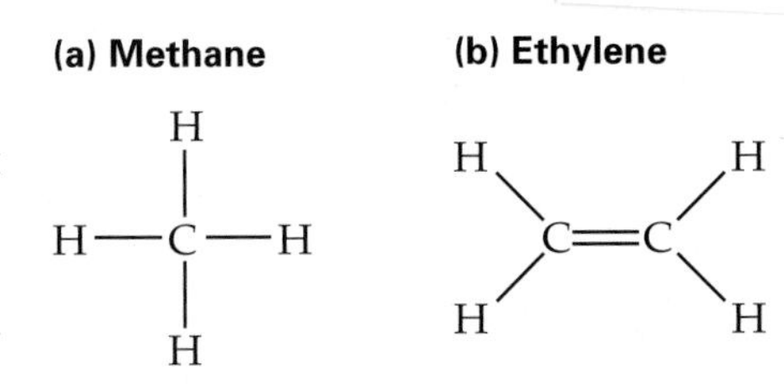

Figure 2.10 **Single and double bonds.** (a) Covalent bonds are symbolized by a short line indicating a shared pair of electrons. (b) Double covalent bonds involve two pairs of shared electrons, symbolized by two horizontal lines.

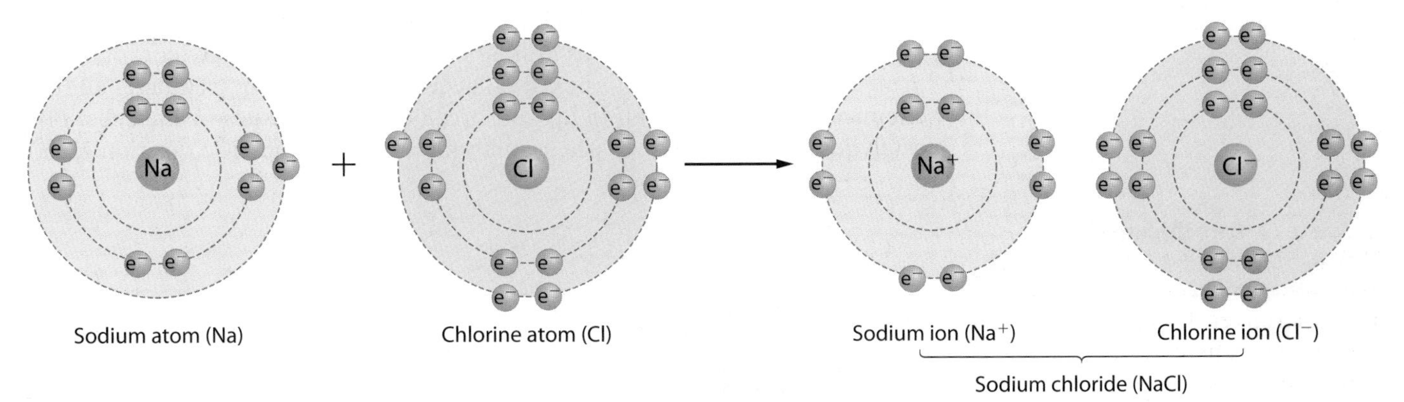

Figure 2.11 **Ionic bonding.** Ionic bonds form when electrons are transferred between charged atoms.

The simple organic molecules found in the Martian meteorite that appear to have formed on Mars are carbonates, molecules containing carbon and oxygen, and **hydrocarbons,** made up of chains and rings of carbon and hydrogen. Carbonates and hydrocarbons can form under certain natural conditions even without the presence of life. However, the meteorite lacked convincing evidence of **macromolecules**—organic molecules that are known to be produced only by living organisms.

Structure and Function of Macromolecules

The macromolecules present in living organisms are carbohydrates, proteins, lipids, and nucleic acids. To date, every living Earth organism, whether bacteria, plant, or animal, has been found to contain these same macromolecules.

Carbohydrates. Sugars, or **carbohydrates,** provide the major source of energy for daily activities. Carbohydrates also play important structural roles in cells. The simplest carbohydrates are composed of carbon, hydrogen, and oxygen in the ratio (CH_2O). For example, the carbohydrate glucose is symbolized as $6(CH_2O)$ or $C_6H_{12}O_6$. Glucose is a simple sugar, or monosaccharide, that consists of a single ring-shape structure. Disaccharides are two rings joined together. Table sugar, called sucrose, is a disaccharide composed of glucose and fructose, a sugar found in fruits.

Joining many individual subunits, or monomers, together produces polymers. Polymers of sugar monomers are called **polysaccharides** *(Figure 2.12)*. Plants use tough polysaccharides in their cell walls as a sort of structural skeleton. The polysaccharide cellulose, found in plant cell walls, is the most abundant carbohydrate on Earth. The external skeletons of insects, spiders, and lobsters are composed of the polysaccharide chitin, and the cell walls that surround bacterial cells are rich in structural polysaccharides.

According to David McKay and his colleagues, the particular set of hydrocarbons found in the Martian meteorite is identical to the set formed when carbohydrates in certain bacteria on Earth break down. These trace remains of possible Martian carbohydrates are an important piece of evidence that scientists use to argue that Mars once harbored Earth-like life. Evidence of the presence of proteins on the meteorite is less convincing.

Proteins. Living organisms require **proteins** for a wide variety of processes. Proteins are important structural components of cells; they are integral to the structure of cell membranes and make up half the dry weight of most cells. Some cells, such as animal muscle cells, are largely composed of proteins. Proteins called **enzymes** accelerate and help regulate all the chemical reactions that build up and break down molecules inside cells. The catalytic power of enzymes (their ability to drastically increase reaction rates)

Figure 2.12 **Carbohydrates.** Monosaccharides, which tend to form ring structures in aqueous solutions, are individual sugar molecules. Disaccharides are 2 monosaccharides joined together, and polysaccharides are long chains of sugars joined together. The monosaccharide glucose and the disaccharide sucrose are important sources of energy, and cellulose plays a structural role in plant cell walls.

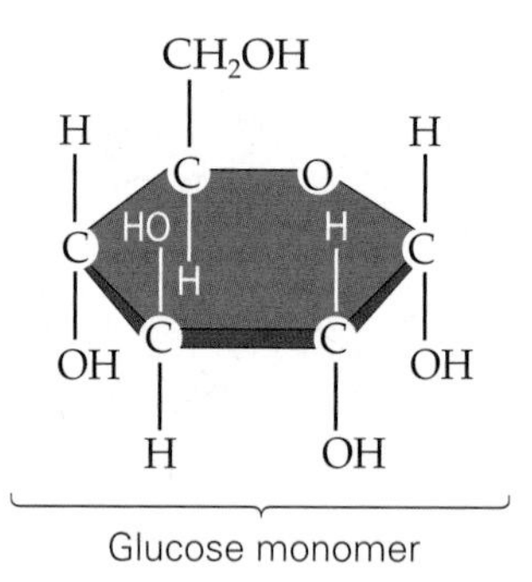

Glucose

Fructose

Sucrose—a disaccharide

Cellulose—a polysaccharide

allows metabolism to occur under normal cellular conditions. Proteins can also serve as channels through which substances are brought into cells, and they can function as hormones that send chemical messages throughout an organism's body.

Proteins are large molecules made of monomer subunits called **amino acids.** There are 20 commonly occurring amino acids. Like carbohydrates, amino acids are made of carbons, hydrogens, and oxygens; these form the amino acid's carboxyl group. In addition, amino acids have nitrogen as part of an amino ($-NH_2^+$) group along with various side groups. Side groups are chemical groups that give amino acids different chemical properties *(Figure 2.13a)*.

Polymers of amino acids can be joined together in various sequences called polypeptides. The chemical bond joining adjacent amino acids is a **peptide bond.** *Figure 2.13b* shows three amino acids—valine, alanine, and phenylalanine—joined by peptide bonds. Precisely folded polypeptides produce specific proteins in much the same manner that children can use differently shaped beads to produce a wide variety of structures *(Figure 2.13c)*. Each amino acid side group has unique chemical properties, including being polar or nonpolar. Since each protein is composed of a particular sequence of amino acids, each protein has a unique shape and therefore specialized chemical properties.

Scientists have found no evidence of proteins in the Martian meteorite, although one group of investigators did report the presence of tiny amounts of three amino acids within the rock. However, it may be the case that these amino acids are contaminants; that is, they are present in the meteor because the meteor has been on protein-rich Earth for several thousand years. In addition, some amino acids are known to form under conditions where life is not present, so the presence of amino acids is not necessarily evidence of life.

Figure 2.13 **Amino acids, peptide bonds, and proteins.** (a) All amino acids have the same backbone but different side groups. (b) Amino acids are joined together by peptide bonds. Long chains of these are called polypeptides. (c) Polypeptide chains fold upon themselves to produce proteins.

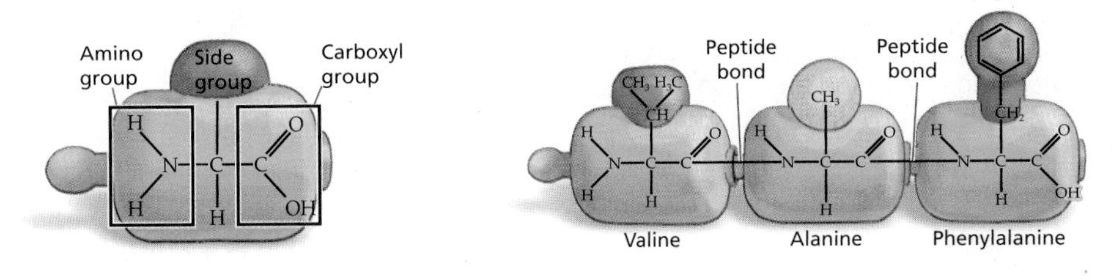

(c) Protein

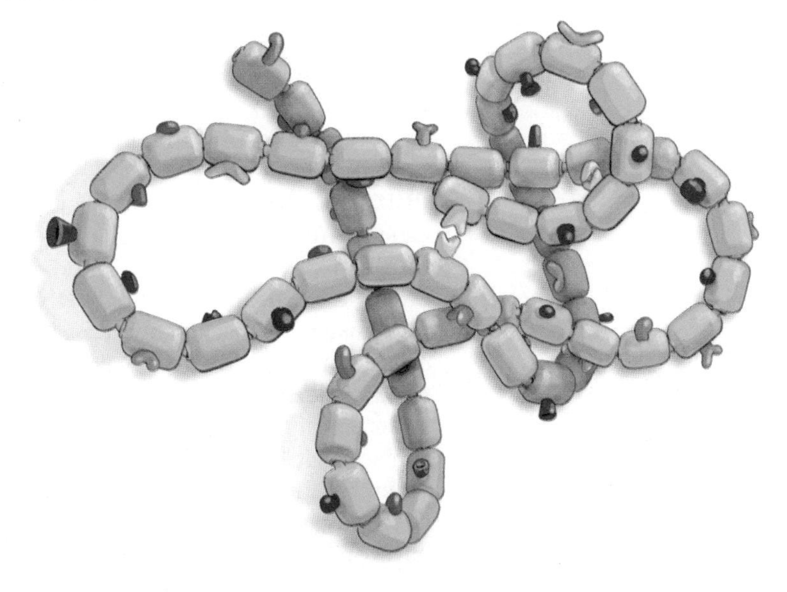

Lipids.

One type of organic molecule, abundant in living organisms, that has not been found in the Martian meteorite is lipids. **Lipids** are partially or entirely hydrophobic organic molecules made primarily of hydrocarbons. Important lipids include fats, steroids, and phospholipids.

Fat.

The structure of a **fat** is that of a 3-carbon glycerol molecule with up to 3 long hydrocarbon chains attached to it *(Figure 2.14a)*. Like the hydrocarbons present in gasoline, these can be burned to produce energy. The long hydrocarbon chains are called **fatty acid** tails of the fat. Fats are hydrophobic and function in energy storage within living organisms.

Steroids.

Steroids are composed of 4 fused carbon-containing rings. Cholesterol *(Figure 2.14b)* is one steroid that is probably familiar; its primary function in animal cells (plant

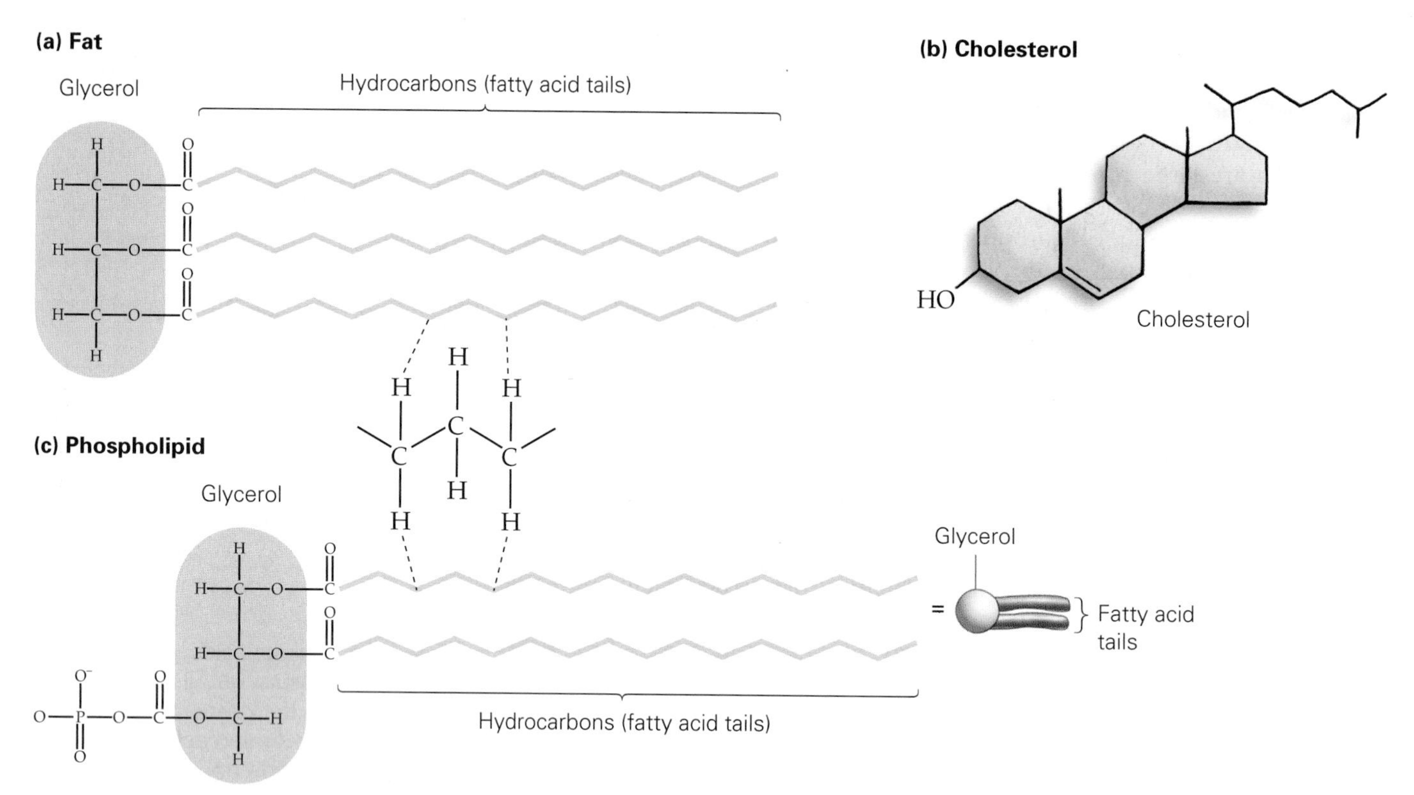

Figure 2.14 **Three types of lipids.** (a) Fats are composed of a glycerol molecule with 3 hydrocarbon-rich fatty acid tails attached. (b) Cholesterol is a steroid common in animal cell membranes. (c) Phospholipids are composed of a glycerol backbone with 2 fatty acids attached and 1 phosphate head group. The cartoon drawing to the right shows how phospholipids are often depicted.

cells do not contain cholesterol) is to help maintain the fluidity of membranes. Other steroids include the sex hormones testosterone, estrogen, and progesterone, which are produced by the sex organs and have effects throughout the body.

Phospholipids. **Phospholipids** are similar to fats except that each glycerol molecule is attached to 2 fatty acid tails (not 3, as you would find in a dietary fat). The third bond in a phospholipid is to a phosphate head group. The phosphate head group is hydrophilic, and the two tails are hydrophobic *(Figure 2.14c)*. Phospholipids often have an additional head group, attached to the phosphate, that also confers unique chemical properties on the individual phospholipid. Phospholipids are important constituents of the membranes that surround cells and that designate compartments within cells.

Even if the Martian meteorite contained unambiguous traces of carbohydrates, proteins, and lipids, the source of these molecules would not clearly be living organisms without a mechanism for passing information about their traits to the next generation. The hereditary, or genetic, information common to all life on Earth is in the form of nucleic acids.

Nucleic Acids. Nucleic acids are composed of long strings of monomers called **nucleotides.** A nucleotide is made up of a sugar, a phosphate, and a nitrogen-containing base. There are two classes of nucleic acids in living organisms. **Ribonucleic acid (RNA)** plays a key role in helping cells synthesize proteins and is discussed in detail in later chapters. The nucleic acid that serves as the primary storage of genetic information in nearly all living organisms is **deoxyribonucleic acid (DNA).** *Figure 2.15* shows the three-dimensional structure of a DNA molecule and zooms inward to the chemical structure. You can see that DNA is composed of two curving strands that wind around each other to form a double helix. The sugar in DNA is the 5-carbon sugar deoxyribose. The nitrogen-containing bases, or **nitrogenous bases,** of DNA have one of four different chemical structures, each with a different name:

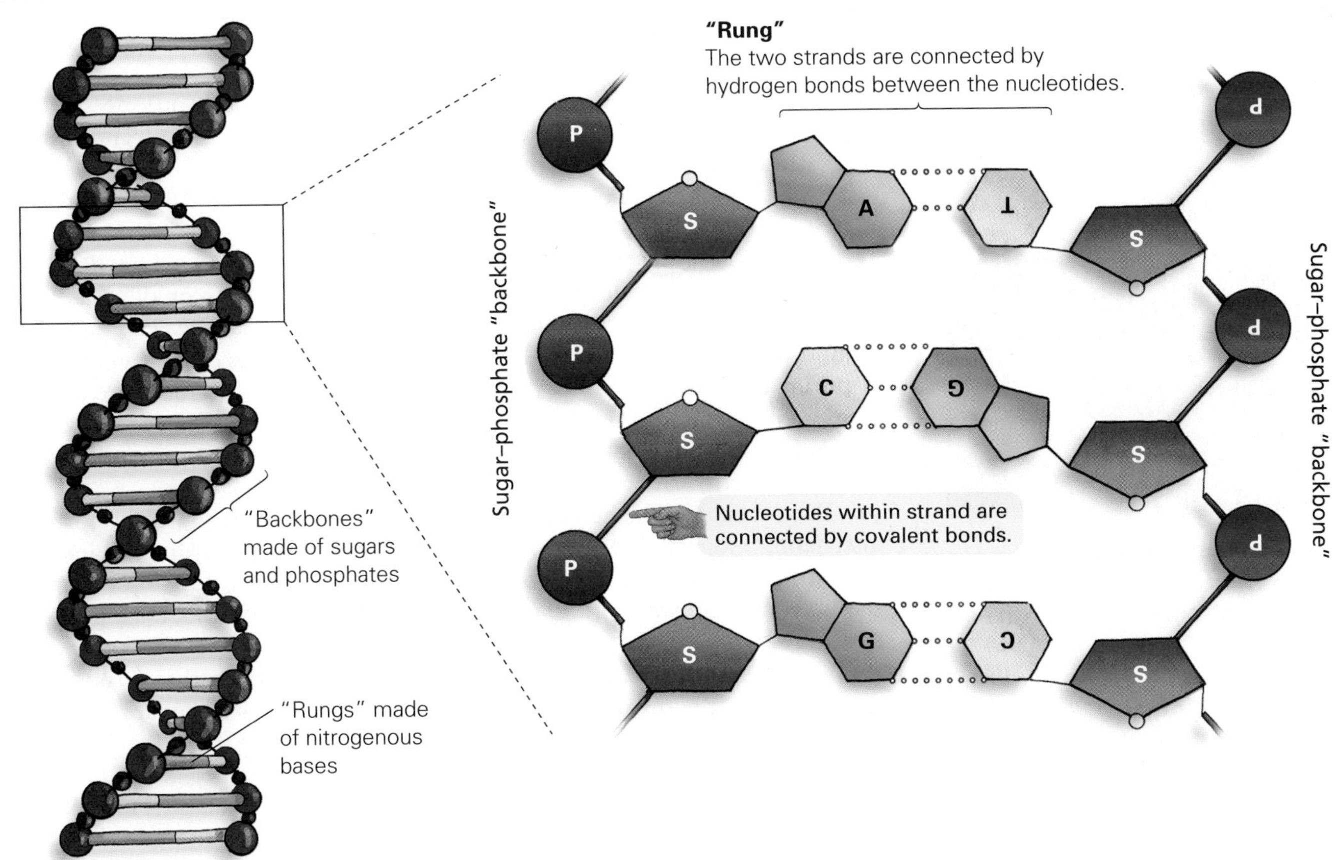

(c) Each nucleotide is composed of a phosphate, a sugar, and a nitrogenous base.

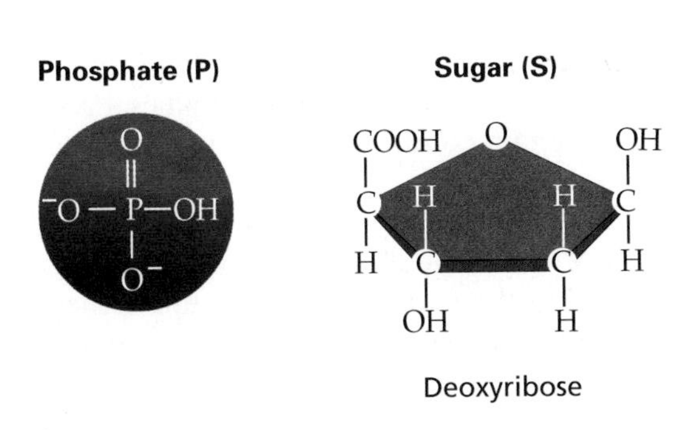

Nitrogenous bases

Purines		Pyrimidines
Adenine (A)	A always pairs with T (see part b)	Thymine (T)
Guanine (G)	G always pairs with C (see part b)	Cytosine (C)

Figure 2.15 **DNA structure.** (a) DNA is a double-helical structure composed of sugars, phosphates, and nitrogenous bases. **Visualize This:** Point out the sugar-phosphate backbone and the nitrogenous bases in this image. Why are there two colors and sizes of nitrogenous bases? (b) Each strand of the helix is composed of repeating units of sugars and phosphates, making the sugar-phosphate backbone, and of nitrogenous bases. (c) A phosphate, a sugar, and a nitrogenous base comprise the structure of a nucleotide. Adenine and guanine are purines, which have a double-ring structure; cytosine and thymine are pyrimidines, which have a single-ring structure.

adenine (A), guanine (G), thymine (T), and **cytosine (C).** Nucleotides are joined to each other along the length of the helix by covalent bonds.

Nitrogenous bases form hydrogen bonds with each other across the width of the helix. On a DNA molecule, an adenine (A) on one strand always pairs with a thymine (T) on the opposite strand. Likewise, guanine (G) always pairs with cytosine (C). The term **complementary** is used to describe these pairings. For example, A is complementary to T, and C is

Figure 2.16 **The DNA model.** American James Watson (left) and Englishman Francis Crick are shown with the three-dimensional model of DNA they devised while working at the University of Cambridge in England.

complementary to G. Therefore, the order of nucleotides on one strand of the DNA helix predicts the order of nucleotides on the other strand. Thus, if one strand of the DNA molecule is composed of nucleotides AACGATCCG, then we know that the order of nucleotides on the other strand is TTGCTAGGC.

As a result of this **base-pairing rule** (A pairs with T; G pairs with C), the width of the DNA helix is uniform. There are no bulges or dimples in the structure of the DNA helix because A and G, called **purines,** are structures composed of two rings; C and T are single-ring structures called **pyrimidines.** A purine always pairs with a pyrimidine and vice versa, so there are always 3 rings across the width of the helix. A-to-T base pairs have 2 hydrogen bonds holding them together. G-to-C pairs have 3 hydrogen bonds holding them together.

Each strand of the helix thus consists of a series of sugars and phosphates alternating along the length of the helix, the **sugar-phosphate backbone.** The strands of the helix align so that the nucleotides face "up" on one side of the helix and "down" on the other side of the helix. For this reason, the two strands of the helix are said to be antiparallel.

The overall structure of a DNA molecule can be likened to a rope ladder that is twisted, with the sides of the ladder composed of sugars and phosphates (the sugar-phosphate backbone) and the rungs of the ladder composed of the nitrogenous-base sequences A, C, G, and T. The structure of DNA was determined by a group of scientists in the 1950s, most notably James Watson and Francis Crick *(Figure 2.16)*.

How Might Macromolecules on Other Planets Differ? Many scientists argue that the fundamental constituents described here—car-bohydrates, proteins, lipids, and nucleic acids—will be essentially similar wherever life is found. They will readily admit that the finer details are very likely to differ, however. For example, all proteins known on Earth contain only 20 different amino acids, despite an infinite number of possibilities. Presumably, proteins on other planets could contain completely different amino acids and many more than 20.

Not all scientists agree with this position, which they call "carbon chauvinism." Carbon is not the only chemical Tinkertoy connector; other elements, including silicon, can also make connections with four other atoms. Silicon is also relatively abundant in the universe and could theoretically form the backbone of an alternative organic chemistry. The basic constituents of silicon-based life may be very different from the chemical building blocks of life on Earth.

Even if all life in the universe is based on carbon chemistry, it is very unlikely that the suite of organisms found on another planet will look much like life on our planet. However, understanding the history of life on Earth also provides insight into the possible nature of life elsewhere in the universe.

BioFlix™ Tour of an Animal Cell

BioFlix™ Tour of a Plant Cell

2.2 Life on Earth

One of the most dramatic features of the Martian meteorite is the presence of fossils that look remarkably like the tiniest living organisms known from Earth. The largest of these fossils is less than 1/100th of the diameter of a human hair, and most are about 1/1000th of the diameter of a human hair—small enough that it would take about 1000 laid end to end to span the dot at the end of this sentence. Some are egg shaped, while others are tubular. These fossils appear similar to the simplest and most ancient of known organisms and are the strongest piece of evidence supporting the hypothesis that Mars once was home to living organisms.

Prokaryotic and Eukaryotic Cells

David McKay and his colleagues argue that the fossil structures in the Martian meteorite are the remains of tiny cells. A **cell** is the fundamental structural unit of life on Earth, separated from its environment by a membrane and sometimes an external wall. Bacteria are composed of single cells, which perform all of the activities required for life. More complex organisms can be composed of trillions of cells working together and do not have any cells that could survive and reproduce independently.

All cells can be placed into one of two categories, prokaryotic or eukaryotic, based on the presence or absence of certain cellular structures. Bacteria are **prokaryotic** cells. Prokaryotes do not have a **nucleus,** a separate membrane-bound compartment that contains genetic material in the form of DNA. They also do not contain any membrane-bound internal structures. Prokaryotic cells are much smaller than eukaryotic cells *(Figure 2.17a)*, and according to the fossil record, they pre-date eukaryotic cells. The fossils in the Martian meteorite resemble modern prokaryotic cells known as nanobacteria. However, bacterial cells also have a **cell wall** that helps them maintain their shape *(Figure 2.17b)*. The fossils in the Martian meteorite show no evidence of similar walls, which does not support the hypothesis that the fossils are relics of once-living organisms.

Eukaryotic cells have a nucleus and other internal structures with specialized functions, called **organelles,** that are surrounded by membranes. Eukaryotic organisms include single-celled organisms such as amoebas and yeast as well as multicellular plants, fungi, and animals, to name a few. As you will learn in Chapter 12, scientists believe that the first prokaryotic cells

(a) Different sizes: eukaryotic vs. prokaryotic cells

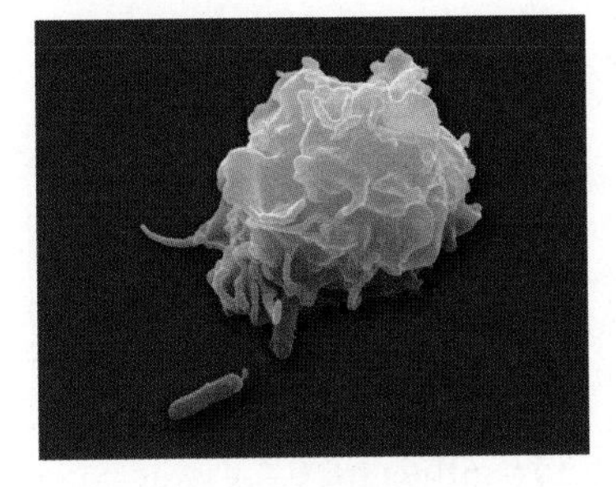

Figure 2.17 **Prokaryotic and eukaryotic cells.** (a) Prokaryotic cells are typically about 1/10 the diameter of a eukaryotic cell as evidenced by the size of the two bacterial cells and a white blood cell. (b) Prokaryotic cells are structurally less complex than eukaryotic cells.

(b) Prokaryotic cell features

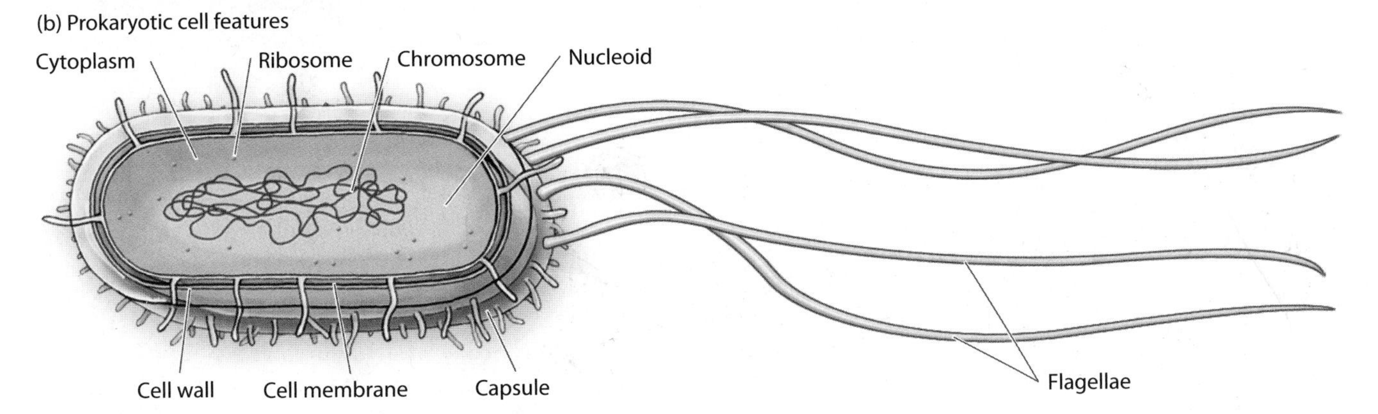

appeared on Earth over 3.5 billion years ago, and that the first eukaryotes appeared about 1.7 billion years later.

Many scientists dispute David McKay's interpretation of the tubular structures in ALH84001. In fact, similar structures can be formed in the absence of life by certain minerals under extremes of heat and pressure. If the Martian fossils are indeed cells, they likely contained features found inside earthly cells, some of which should be visible in the fossils.

Cell Structure

Each living cell can be considered a veritable factory working to break down nutrients and to recycle its components. We start from the outside of the cell and examine the structure and function of various cell components as we work our way into the cell *(Table 2.1)*.

TABLE 2.1 **Cell components.** Illustrations and descriptions of cell components and their functions

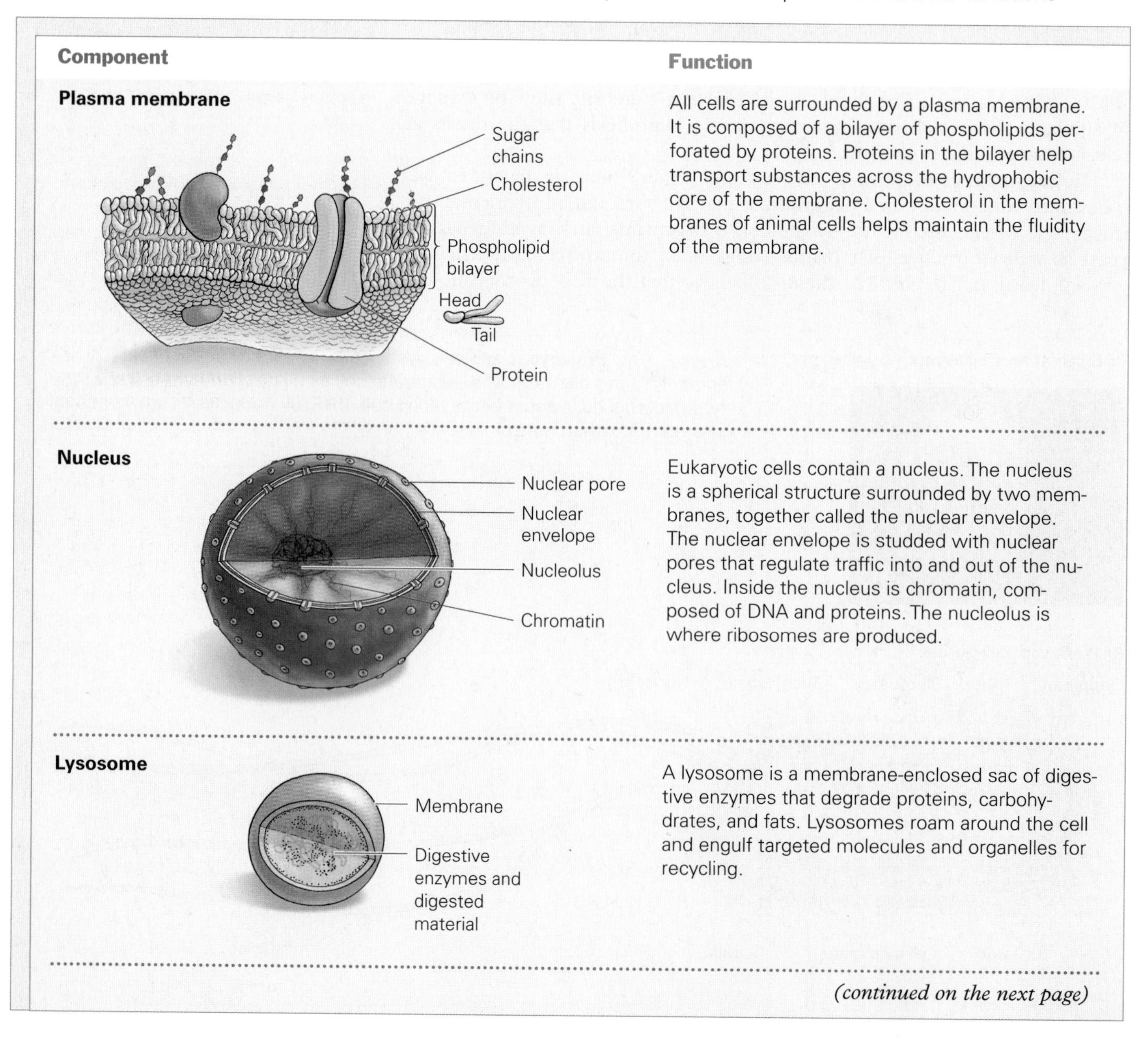

Component	Function
Plasma membrane	All cells are surrounded by a plasma membrane. It is composed of a bilayer of phospholipids perforated by proteins. Proteins in the bilayer help transport substances across the hydrophobic core of the membrane. Cholesterol in the membranes of animal cells helps maintain the fluidity of the membrane.
Nucleus	Eukaryotic cells contain a nucleus. The nucleus is a spherical structure surrounded by two membranes, together called the nuclear envelope. The nuclear envelope is studded with nuclear pores that regulate traffic into and out of the nucleus. Inside the nucleus is chromatin, composed of DNA and proteins. The nucleolus is where ribosomes are produced.
Lysosome	A lysosome is a membrane-enclosed sac of digestive enzymes that degrade proteins, carbohydrates, and fats. Lysosomes roam around the cell and engulf targeted molecules and organelles for recycling.

(continued on the next page)

TABLE 2.1 **Cell components.** Illustrations and descriptions of cell components and their functions. *(Continued)*

Component	Function
Chloroplast 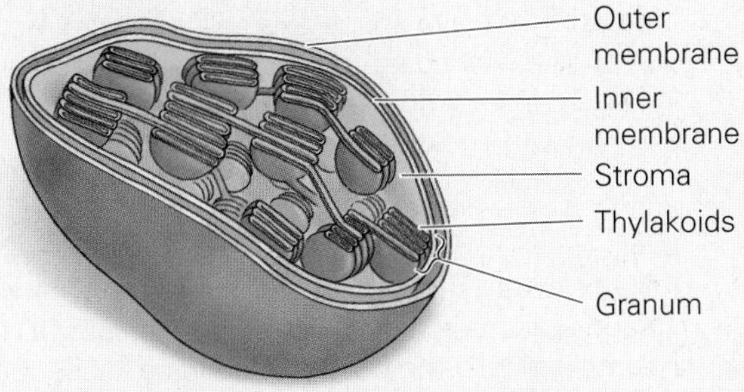 Outer membrane Inner membrane Stroma Thylakoids Granum	An important organelle present in plant cells, the chloroplast uses the sun's energy to convert carbon dioxide and water into sugars. Each chloroplast has an outer membrane, an inner membrane, a liquid interior called the stroma, and a network of membranous sacs called thylakoids that stack on one another to form structures called grana (singular: granum). Chloroplasts also contain pigment molecules that give green parts of plants their color.
Ribosomes 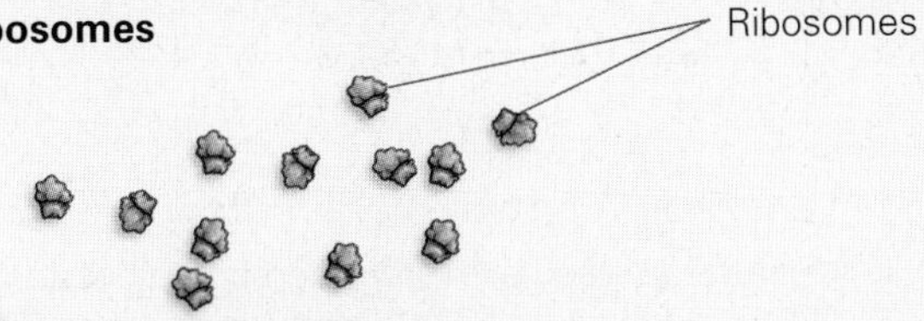 Ribosomes	Ribosomes are found in eukaryotic and prokaryotic cells. Ribosomes are built in the nucleolus and shipped out of the nucleus through nuclear pores to the cytoplasm, where they are used as workbenches for protein synthesis. They can be found floating in the cytoplasm or tethered to the ER.
Endoplasmic reticulum (ER) 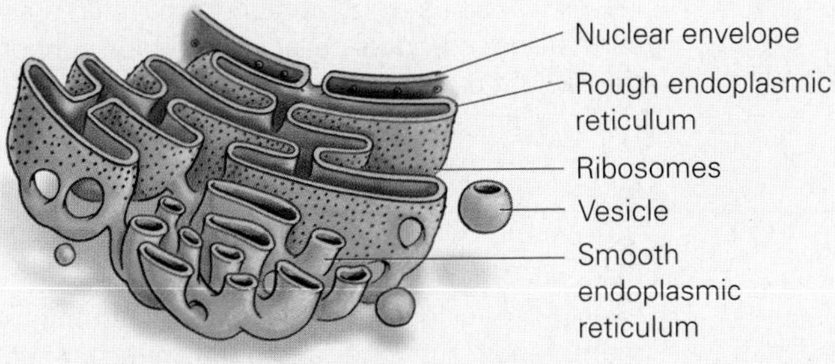 Nuclear envelope Rough endoplasmic reticulum Ribosomes Vesicle Smooth endoplasmic reticulum	The ER is a large network of membranes that begins at the nuclear envelope and extends into the cytoplasm of a eukaryotic cell. ER with ribosomes attached is called rough ER. Proteins synthesized on rough ER will be secreted from the cell or will become part of the plasma membrane. ER without ribosomes attached is called smooth ER. The function of the smooth ER depends on cell type but includes tasks such as detoxifying harmful substances and synthesizing lipids. Vesicles are pinched-off pieces of membrane that transport substances to the Golgi apparatus or plasma membrane.
Golgi apparatus 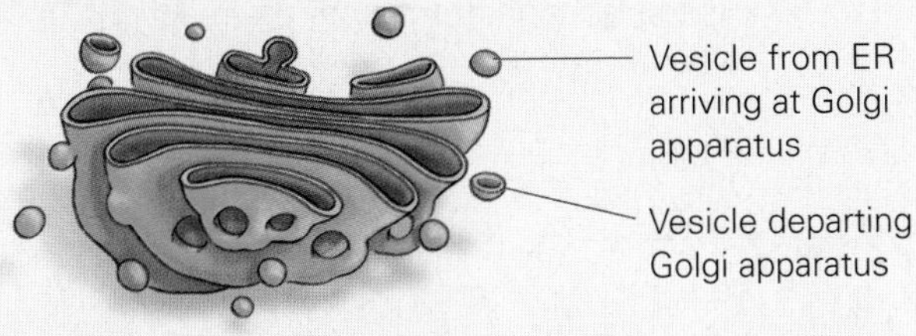 Vesicle from ER arriving at Golgi apparatus Vesicle departing Golgi apparatus	The Golgi apparatus is a stack of membranous sacs. Vesicles from the ER fuse with the Golgi apparatus and empty their protein contents. The proteins are then modified, sorted, and sent to the correct destination in new transport vesicles that bud off from the sacs.
Centrioles 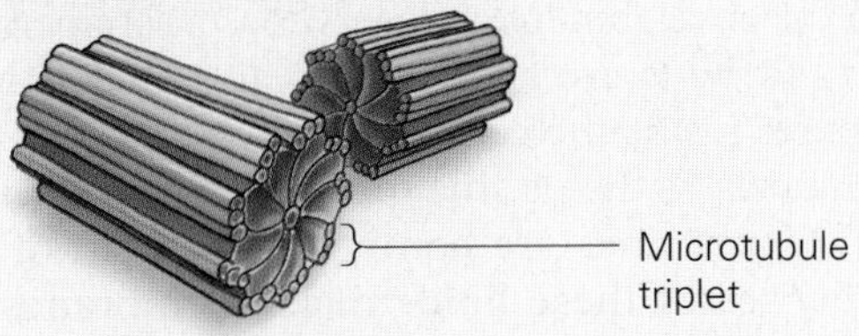 Microtubule triplet	Centrioles are barrel-shaped rings composed of microtubules that help move chromosomes around when a cell divides. Centrioles are involved in microtubule formation during cell division and the formation of cilia and flagella.

(continued on the next page)

TABLE 2.1 **Cell components.** Illustrations and descriptions of cell components and their functions. *(Continued)*

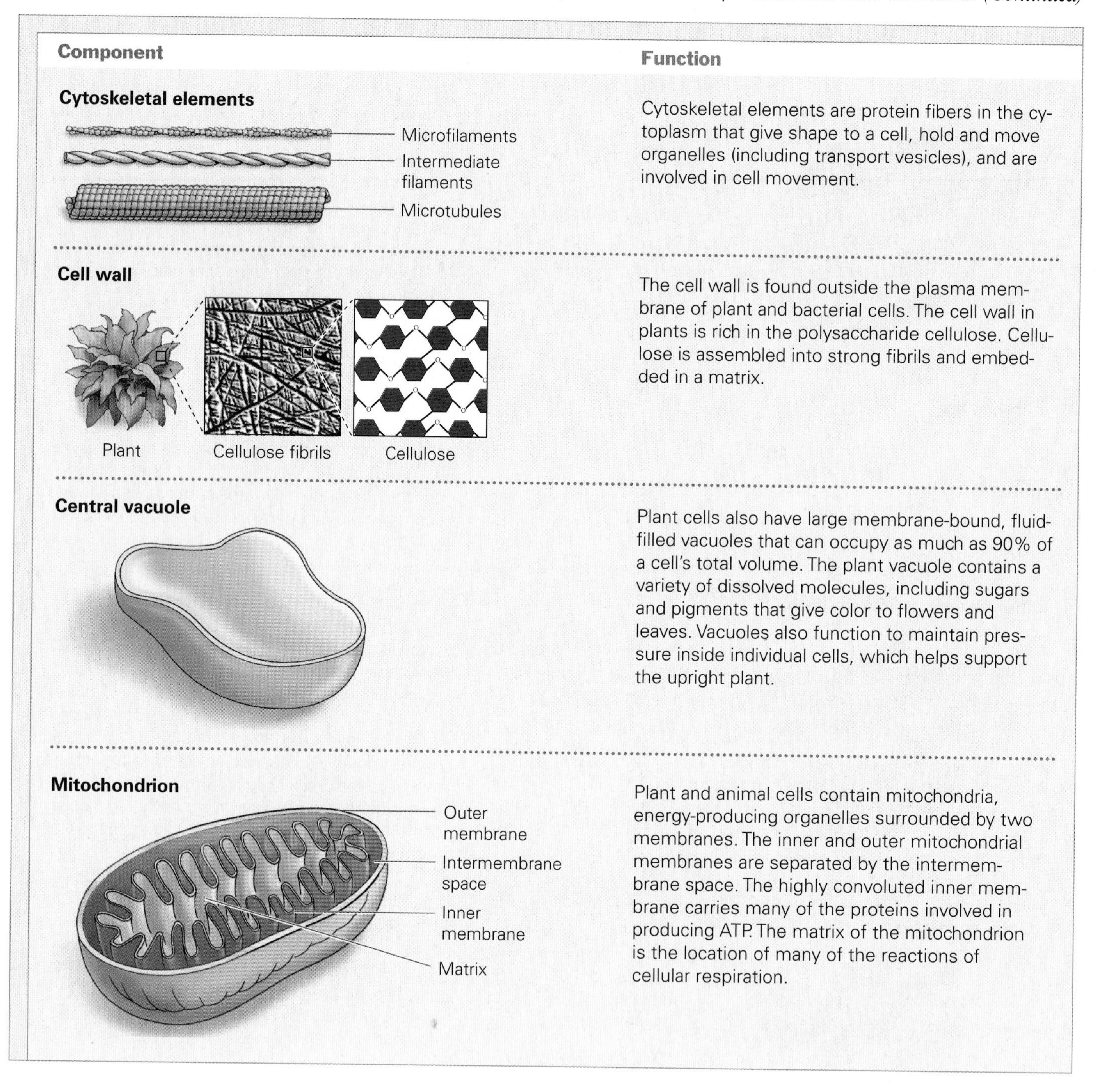

Component	Function
Cytoskeletal elements (Microfilaments; Intermediate filaments; Microtubules)	Cytoskeletal elements are protein fibers in the cytoplasm that give shape to a cell, hold and move organelles (including transport vesicles), and are involved in cell movement.
Cell wall (Plant; Cellulose fibrils; Cellulose)	The cell wall is found outside the plasma membrane of plant and bacterial cells. The cell wall in plants is rich in the polysaccharide cellulose. Cellulose is assembled into strong fibrils and embedded in a matrix.
Central vacuole	Plant cells also have large membrane-bound, fluid-filled vacuoles that can occupy as much as 90% of a cell's total volume. The plant vacuole contains a variety of dissolved molecules, including sugars and pigments that give color to flowers and leaves. Vacuoles also function to maintain pressure inside individual cells, which helps support the upright plant.
Mitochondrion (Outer membrane; Intermembrane space; Inner membrane; Matrix)	Plant and animal cells contain mitochondria, energy-producing organelles surrounded by two membranes. The inner and outer mitochondrial membranes are separated by the intermembrane space. The highly convoluted inner membrane carries many of the proteins involved in producing ATP. The matrix of the mitochondrion is the location of many of the reactions of cellular respiration.

Plasma Membrane. All cells are enclosed by a structure called a **plasma membrane.** The plasma membrane defines the outer boundary of each cell, isolates the cell's contents from the environment, and serves as a semipermeable barrier that determines which nutrients are allowed into and out of the cell. Membranes that enclose structures inside the cell are usually referred to as cell membranes, while the outer boundary is the plasma membrane.

Internal and external cell membranes are composed, in part, of phospholipids. The chemical properties of these lipids make membranes flexible and self-sealing. When phospholipid molecules are placed in a watery solution, such as in a cell, they orient themselves so that their hydrophilic heads are exposed to the water and their hydrophobic tails are away from the water. They cluster into a form called a **phospholipid bilayer,** in which the tails of the phospholipids interact with themselves and exclude water, while the heads maximize their exposure to the surrounding water both inside and outside the membrane. The bilayer of phospholipids is stuffed with proteins that carry out enzymatic functions, serve as receptors, and help transport substances.

A Fluid Mosaic of Lipids and Proteins. All of the lipids and most of the proteins in the plasma membrane are free to bob about, sliding from one location in the membrane to another. Because lipids and proteins move about laterally within the membrane, the membrane is a **fluid mosaic** of lipids and proteins. The membrane is fluid since the composition of any one location on the membrane can change. In the same manner that a patchwork quilt is a mosaic (different fabrics making up the whole quilt), so, too, is the membrane a mosaic with different regions of membrane being composed of different types of phospholipids and proteins.

Cell membranes are **semipermeable** in the sense that they allow some substances to cross and prevent others from crossing. This characteristic allows cells to maintain a different internal composition from the surrounding solution.

Nucleus. All eukaryotic cells contain a nucleus surrounded by a double nuclear membrane, which houses the DNA. Inside the nucleus is the nucleolus, which is where ribosomes are assembled.

Cytosol. Between the nucleus and the plasma membrane lies the cytosol, a watery matrix containing water, salts, and many of the enzymes required for cellular reactions. The cytosol houses the subcellular structures called organelles. The term **cytoplasm** includes the cytosol and organelles.

Organelles. Organelles are to cells as organs are to the body. Each organelle performs a specific job required by the cell, and all organelles work together to keep an individual cell healthy and to produce the raw materials that the cell needs to survive. Each organelle is enclosed in its own lipid bilayer. Some organelles are involved in metabolism. For example, organelles called **mitochondria** help the cells convert food energy into a form usable by cells, called ATP (adenosine triphosphate), while **chloroplasts** in plant cells use energy from sunlight to make sugars. **Lysosomes** help break down substances that are ingested before they are sent to the mitochondria. Other organelles are involved in producing proteins. The endoplasmic reticulum (ER) is an extensive membranous organelle that can be studded with ribosomes (**rough endoplasmic reticulum**) and involved in protein synthesis or tubular in shape and involved in lipid synthesis (**smooth endoplasmic reticulum**). Proteins that are assembled on the membranes of the rough ER can be modified and sorted in a membranous structure called the **Golgi apparatus.**

There are other important subcellular structures that are not considered organelles because they are not bounded by membranes. **Ribosomes** are workbenches where proteins are assembled. **Centrioles** are involved in moving genetic material around when a cell divides, and many fibers that compose the cytoskeleton help maintain the cell shape. Some organelles and subcellular structures are found in certain cell types only. For instance, in addition to having chloroplasts and a cell wall, the plant cell also has a **vacuole** to store water, sugars, and pigments. Table 2.1 describes the

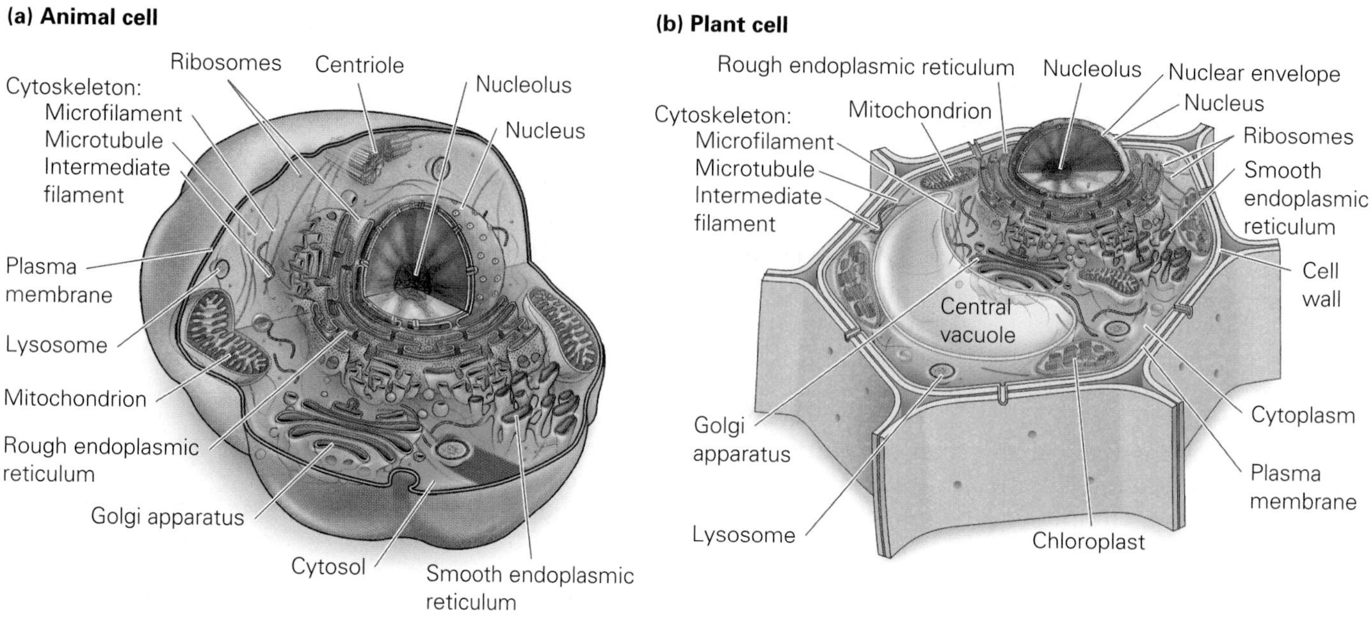

Figure 2.18 **Animal and plant cells.** These drawing of a generalized (a) animal cell and (b) plant cell show the locations and sizes of organelles and other structures. **Visualize This:** What three structures are present in plant but not animal cells?

structures and functions of most cellular organelles in greater detail. *Figure 2.18* shows an animal cell and a plant cell complete with their complement of organelles.

Stop and Stretch 2.2: Would you expect prokaryotic cells to contain ribosomes? Why or why not?

The Tree of Life and Evolutionary Theory

Biologists disagree about the total number of different **species,** or types of living organisms, that are present on Earth today. This uncertainty stems from lack of knowledge. Although scientists likely have identified most of the larger organisms—such as land plants, mammals, birds, reptiles, and fish—millions of species of insects, fungi, bacteria, and other microscopic organisms remain unknown to science. Amazingly, credible estimates of the number of species on Earth range from 5 million to 100 million. Given this level of uncertainty, most biologists think that the likeliest number is near 10 million.

Theory of Evolution. While the diversity of living organisms is tremendous, there exist remarkable similarities among all known species. All have the same basic biochemistry, including carbohydrates, lipids, proteins, and nucleic acids. All consist of cells surrounded by a plasma membrane. All eukaryotic organisms (including fungi, animals, and plants) contain nearly the same suite of cellular organelles. The best explanation for the shared characteristics of all species, what biologists refer to as "the unity of life," is that all living organisms share a common ancestor that arose on Earth nearly 4 billion years ago. The divergence and differences among modern species arose as a result of changes in the characteristics of populations, both in response to environmental change (a process called *natural selection*) and due to chance. These ideas underlie the entire science of biology and are known as the **theory of evolution.**

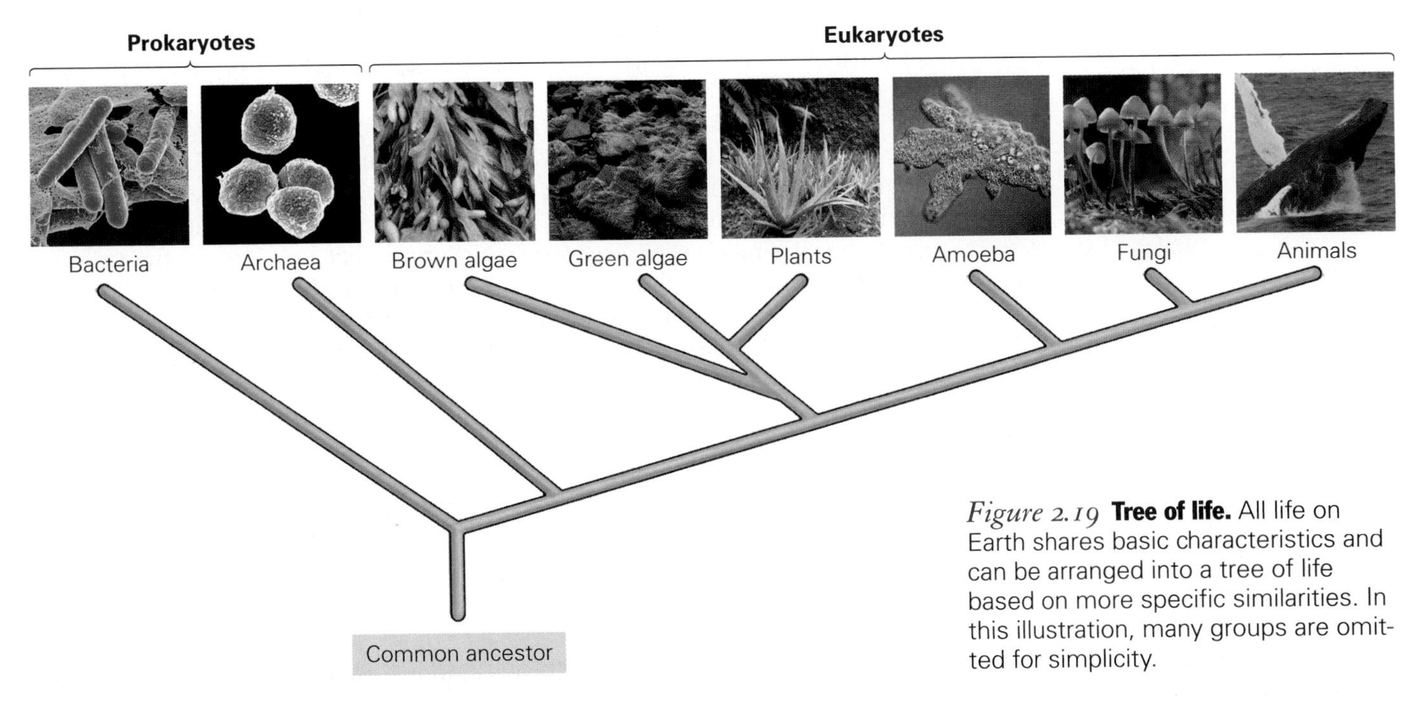

Figure 2.19 **Tree of life.** All life on Earth shares basic characteristics and can be arranged into a tree of life based on more specific similarities. In this illustration, many groups are omitted for simplicity.

A major process in the diversification of life from a single ancestor was natural selection. The theory of natural selection is discussed in detail in Chapter 10, but the basic principle is simple: Individual organisms vary from each other, and some of these variations increase their chances of survival and reproduction. A genetic trait that increases survival and reproduction should become more prevalent over time. In contrast, less-successful variants should eventually be lost from the population.

The common ancestor can be thought of as the starting place for life on Earth, and the continual divergence among species and groups of species can be thought of as life's branching. Modern organisms can therefore be arranged on a tree of life that reflects their basic unity and relationships. According to current understanding, living organisms can be grouped into three large groups: two that are prokaryotic and one containing all eukaryotes. Eukaryotes can be further grouped into several categories made up primarily of free-living, single-celled organisms (such as amoebas and algae) and the three major multicellular groups—plants, fungi, and animals *(Figure 2.19)*. Chapter 12 provides a deeper exploration of the diversity of life on Earth.

Because evolutionary change results from chance events and environmental changes (including the appearance of other species), the group of species present on Earth today represents only one set of an infinite number of possibilities. In other words, life on other planets need not look identical to life on Earth. For example, instead of the common body form found in animals, called bilateral symmetry, by which bodies can be visually divided into two mirror-image halves, life on other planets could be primarily radially symmetric and thus look very different *(Figure 2.20)*. In fact, it is possible that life on other planets might not even be based on carbon. Scientists have no examples of what living organisms would look like on a planet where the organic molecules were based on silicon or bathed in liquid methane.

Figure 2.20 **Diversity of body form.** Not all animals are bilaterally symmetric like us, with two eyes, two ears, two arms, two legs, an obvious head and "tail," and one central axis. A sea star is radially symmetric—it can be divided into two equal halves in any direction. **Visualize This:** Is an earthworm bilaterally symmetric?

Life in the Universe. Do other living organisms exist in the universe? Given the universe's sheer size and complexity, most scientists who study this question think that the existence of life on other planets is nearly certain.

While the evidence of life in the Martian meteorite is unconvincing to many scientists, we may find out in our lifetimes that life exists, or once existed, on our planetary neighbor.

What about the existence of intelligent life that could communicate with us? Some scientists argue that as a result of natural selection, the evolution of intelligence is inevitable wherever life arises. Others point to the history of life on Earth—consisting of at least 2.5 billion years, during which all life was made up of single-celled organisms—to argue that most life in the universe must be "simple and dumb." It is clear from our explorations of the solar system that none of the sun's other planets host intelligent life. The nearest sun-like stars that could host an Earth-like planet, Alpha Centauri A and B, are over 4 light years away—nearly 40 trillion miles. With current technologies, it would take nearly 50,000 years to reach the Alpha Centauri stars, and there is certainly no guarantee that intelligent life would be found on any planets that circle them. For all practical purposes, at this time in human history, we are still alone in the universe.

Savvy Reader Detox Drinks

A CLEAR WINNER IN THE FEEL-GOOD STAKES | *BY CAROLINE STACEY* | *THE INDEPENDENT* (LONDON) | JANUARY 3, 2004

So you thought water was just a drink? Think again. It's a lifestyle choice. We can all safely drink our litre or more a day straight from the tap. But where's the cachet or the profit in that? It's almost as free as air. And wonderful and hydrating though tap water is, the latest bottled waters offer so much more—to make you sportier, healthier, and less hungover.

With Oxygizer you pay for air and water together. It's oxygenated, but not fizzy. Bottled in the Tyrolean mountains by a company based in Innsbruck, Austria, it describes itself as "a sip of fresh air." Already big in the Middle East—where water's a more precious commodity than it is here—it has been launched in Europe and now in the UK.

Oxygizer doesn't just slake a thirst, it provides the body with extra oxygen too. A litre contains 150 mg of oxygen, around 25 times more than what's in a litre of tap water. This apparently helps remove toxins and ensures a stronger immune system, as well as assisting the respiratory system so you recover better from exercise. Some claim detox benefits, it helps hangovers, and even enhances flavours to make food taste better.

1. List the claims made by this article. Is there enough information presented in this article to back up the claims made?
2. Use the appropriate questions in the checklist provided in Chapter 1, Table 1.2, to evaluate this newspaper article. What types of information are missing from this article?
3. Is any data presented to substantiate the claim that oxygenated water improves health?

CHAPTER Review

For study help, animations, and more quiz questions go to www.mybiology.com.

Summary

2.1 What Does Life Require?

- Living organisms must be able to grow, metabolize substances, reproduce, and respond to external stimuli (p. 30).
- Living organisms contain a common set of biological molecules, are composed of cells, and can maintain homeostasis and evolve (p. 30).
- Water is a good solvent in part because of its polarity (p. 31).
- Hydrogen bonding occurs when a weak attraction develops between hydrogen and other atoms (pp. 31–32).
- The polarity of water also facilitates the dissolving of salts. Salts are produced by the reaction of an acid with a base (p. 32).
- The pH scale is a measure of the relative percentages of and ions in a solution and ranges from 0 (acidic or rich in ions) to 14 (basic or rich in ions) (pp. 32–33).
- Chemical bonding depends on an element's electron configuration. Electrons closer to the nucleus have less energy than those that are farther away from the nucleus. The first energy level can hold 2 electrons. The next 2 levels each hold 8 electrons. Atoms that have space in their valence shell form chemical bonds (pp. 33–34).
- Covalent bonds form when atoms share electrons. These tend to be strong bonds (pp. 34–35).
- Ionic bonds form between positively and negatively charged ions. These tend to be weak bonds (p. 35).
- Life on Earth is based on the chemistry of the element carbon, which can make bonds with up to four other elements (p. 35).
- Carbohydrates function in energy storage and play structural roles. They can be single-unit monosaccharides or multiple-unit polysaccharides with sugar monomers arranged in different orders (p. 36).
- Proteins play structural, enzymatic, and transport roles in cells. They are composed of amino acid monomers arranged in different orders (p. 37).
- Lipids are partially or entirely hydrophobic and come in three different forms. Fats are composed of glycerol and three fatty acids. Fats store energy. Phospholipids are composed of glycerol, 2 fatty acids, and a phosphate group. They are important structural components of cell membranes. Steroids are composed of 4 fused rings. Cholesterol is a steroid found in some animal cell membranes and helps maintain fluidity. Other steroids function as hormones (pp. 37–38).
- Nucleic acids are polymers of nucleotides, each of which is composed of a sugar, a phosphate, and a nitrogen-containing base (pp. 38–40).

web animation 2.1 *Chemistry and Water*

web animation 2.2 *Nucleic Acids*

2.2 Life on Earth

- There are two main categories of cells: Those with nuclei and membrane-bound organelles are eukaryotes; those lacking a nucleus and membrane-bound organelles are prokaryotes (pp. 41–42).
- The plasma membrane that surrounds cells is a semipermeable boundary composed of a phospholipid bilayer that has embedded proteins and cholesterol (p. 45).
- Lipids and proteins can move about the membrane. This fluidity of the membrane allows changes in the protein and lipid composition (p. 45).
- Some organisms, such as plants and bacterial cells, have a cell wall outside the plasma membrane that helps protect these cells and maintain their shape (p. 44).
- Subcellular organelles and structures perform many different functions within the cell. Mitochondria and chloroplasts are involved in energy conversions. Lysosomes are involved in breakdown of macromolecules. Ribosomes serve as sites for protein synthesis. Proteins can be synthesized on ribosomes attached to rough endoplasmic reticulum.

Summary (continued)

Smooth endoplasmic reticulum synthesizes lipids. The Golgi apparatus sorts proteins and sends them to their cellular destination. Centrioles help cells divide. The plant cell vacuole stores water and other substances (pp. 42, 45–46).

- There may be nearly 10 million unique life-forms on Earth. Despite all of this diversity, all life on Earth shares the same organic chemistry, genetic material, and basic cellular structures (p. 46).
- The similarities among living organisms on Earth provide support for the theory of evolution, which states that all life on Earth derives from a common ancestor. The process of evolutionary change since the origin of that ancestor led to the modern relationships among organisms, called the tree of life (pp. 46–47).

BioFlix™ Tour of a Plant Cell

BioFlix™ Tour of an Animal Cell

web animation 2.3 *A Comparison of Prokaryotic and Eukaryotic Cells*

Roots to Remember

The following roots of words come mainly from Latin and Greek and will help you to decipher terms:

cyto- and **-cyte** mean cell or a kind of cell.
hydro- means water.
-mer means subunit.
macro- means large.
micro- means small.
mono- means one.
-philic means to love.
-phobic means to fear.
-plasm means a fluid.
poly- means many.

Learning the Basics

1. List the four biological molecules commonly found in living organisms.

2. Describe the structure and function of the subcellular organelles.

3. Describe the structure of the plasma membrane.

4. Water ______________.
A. is a good solute; **B.** dissociates into H^+ and OH^- ions; **C.** serves as an enzyme; **D.** makes strong covalent bonds with other molecules; **E.** has an acidic pH

5. Electrons ______________.
A. are negatively charged; **B.** along with neutrons comprise the nucleus; **C.** are attracted to the negatively charged nucleus; **D.** located closest to the nucleus have the most energy; **E.** all of the above are true

6. Which of the following terms is least like the others?
A. monosaccharide; **B.** phospholipid; **C.** fat; **D.** steroid; **E.** lipid

7. Different proteins are composed of different sequences of ______________.
A. sugars; **B.** glycerols; **C.** fats; **D.** amino acids

8. Proteins may function as ______________.
A. the genetic material; **B.** cholesterol molecules; **C.** fat reserves; **D.** enzymes; **E.** all of the above.

9. A fat molecule consists of ______________.
A. carbohydrates and proteins; **B.** complex carbohydrates only; **C.** saturated oxygen atoms; **D.** glycerol and fatty acids

10. Eukaryotic cells differ from prokaryotic cells in that ______________.
A. only eukaryotic cells contain DNA; **B.** only eukaryotic cells have a plasma membrane; **C.** only eukaryotic cells are considered to be alive; **D.** only eukaryotic cells have a nucleus; **E.** only eukaryotic cells are found on Earth

Analyzing and Applying the Basics

1. A virus is made up of a protein coat surrounding a small segment of genetic material (either DNA or RNA) and a few proteins. Some viruses are also enveloped in membranes derived from the virus's host cell. Viruses cannot reproduce without taking over the genetic "machinery" of their host cell. Based on this description and biologists' definition of life, should a virus be considered a living organism?

2. Any molecule containing oxygen can be polar. The structure of methanol (CH_3OH) is drawn in *Figure 2.21*. Which part of this molecule will have a partial negative charge, and which will have a partial positive charge?

3. Some scientists have argued that silicon (Si) could also be an appropriate basis for organic chemistry because it is abundant and can form bonds with many other atoms. Carbon contains 6 electrons, and silicon contains 14. Recalling that the lowest electron shell contains 2 electrons, and the next 2 shells can contain a maximum of 8, how many "spaces" does silicon have in its valence shell? How does this compare to carbon?

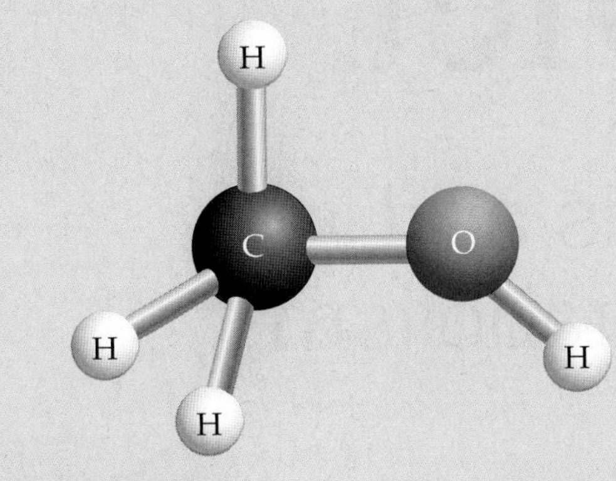

Figure 2.21 **Methanol.**

Connecting the Science

1. Water's characteristic as an excellent solvent means that many human-created chemicals (including some that are quite toxic) can be found in water bodies around the globe. How would our use and manufacture of toxic chemicals be different if most of these chemicals could not be dissolved and diluted in water but instead accumulated where they were produced and used?

2. Do you believe that humans should expend considerable energy and resources looking for life, even intelligent life, elsewhere in the universe? Why or why not?

CHAPTER 3

Diet

Cells and Metabolism

In spite of all the attention paid to diet and exercise, obesity is on the rise.

Most college students are, for the first time, making all their own choices about food.

The average college student spends a lot of time thinking about his or her body and ways to make it more attractive. While people come in all shapes and sizes, attractiveness tends to be more narrowly defined by the images of men and women we see in the popular media. Nearly all media images equate attractiveness and desirability with a very limited range of body types.

For men, the ideal includes a tall, broad-shouldered, muscular physique with so little body fat that every muscle is visible. The standards for female beauty are equally unforgiving and include small hips; long, thin limbs; large breasts; and no body fat.

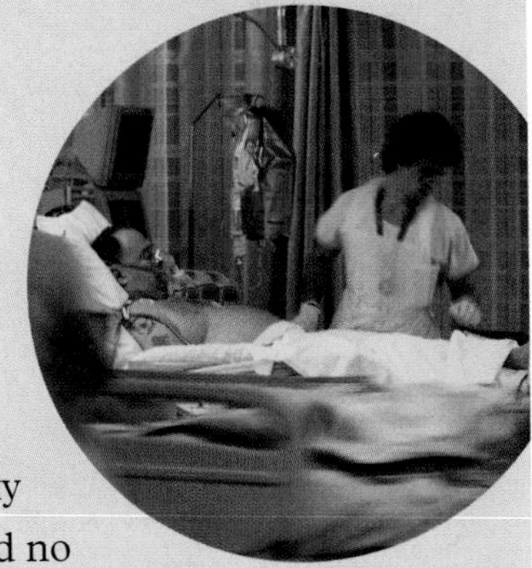

Making unhealthy choices now can lead to future health problems associated with obesity

Into this milieu steps the average college student—worried about appearance, trying to find time to study, exercise, and socialize—now making all of his or her own decisions about food, often on a limited budget.

Making choices that are good for long-term health is not easy. Typical meal-plan choices at college dining centers are often greasy and fat laden—and available in unlimited portions. Making healthful choices is further complicated by the presence of campus snack shops, vending machines, and conveniently located fast-food restaurants that offer time-pressed students easily accessible, inexpensive foods containing little nutrition.

Coupling tremendous pressure to be thin with a glut of readily available unhealthful foods can lead students to establish poor eating habits that persist far beyond college life. In many cases, these conflicting pressures may cause students to develop eating habits that result in lifelong battles with obesity and eating disorders.

. . . or anorexia.

Learning about how much and what kinds of foods to eat, along with what's in the food we eat, from nutrients to empty calories, will help you make good decisions about eating that will help you maintain a body weight that is healthful for you.

3.1 Nutrients

The food that organisms ingest provides building-block molecules that can be broken down and used as raw materials for growth, for maintenance and repair, and as a source of energy. Another name for the substances in food that provide structural materials or energy is **nutrients**.

Macronutrients

Nutrients that are required in large amounts are called **macronutrients.** These include water, carbohydrates, proteins, and fats.

Water and Nutrition. Most animals can survive for several weeks with no nutrition other than water. However, survival without water is limited to just a few days. Besides helping the body disperse other nutrients, water helps dissolve and eliminate the waste products of digestion. Water helps to maintain blood pressure and is involved in virtually all cellular activities.

A decrease below the body's required water level, called **dehydration,** can lead to muscle cramps, fatigue, headaches, dizziness, nausea, confusion, and increased heart rate. Severe dehydration can result in hallucinations, heat stroke, and death. The body, including the brain, relies on water in the circulatory system to deliver nutrients as well as some oxygen. When water levels are low, the delivery system becomes less effective.

In addition, evaporation of water from the skin (sweating) helps maintain body temperature. When water is low and sweating decreases, the body temperature can rise to a harmful level.

Every day, humans lose about 3 liters of water as sweat, in urine, and in feces. To avoid dehydration, we must replace this water. We can replace some of it by consuming food that contains water. A typical adult obtains about 1.5 liters of water per day from food consumption, leaving a deficit of about 1.5 liters that must be replaced.

In addition to obtaining a healthy dose of water every day, people must consume foods that contain carbohydrates, proteins, and fats. In Chapter 2, we explored the structure of these macromolecules, and now we focus on how they function in the body.

Carbohydrates as Nutrients. Foods such as bread, cereal, rice, and pasta, as well as fruits and vegetables, are rich in sugars called carbohydrates. Carbohydrates are the major source of energy for cells. Energy is stored in the chemical bonds between the carbons, hydrogens, and oxygens that comprise carbohydrate molecules. Carbohydrates can exist as single-unit monomers or can be bonded to each other to produce longer-chain polysaccharide polymers.

The single-unit simple sugars are digested and enter the bloodstream quickly after ingestion. Sugars found in milk, juice, honey, and most refined foods are simple sugars. Fructose, the sugar found in corn syrup, is shown in *Figure 3.1a*.

When multisubunit sugars are composed of many different branching chains of sugar monomers, they are called **complex carbohydrates.** Complex carbohydrates are found in vegetables, breads, legumes, and pasta.

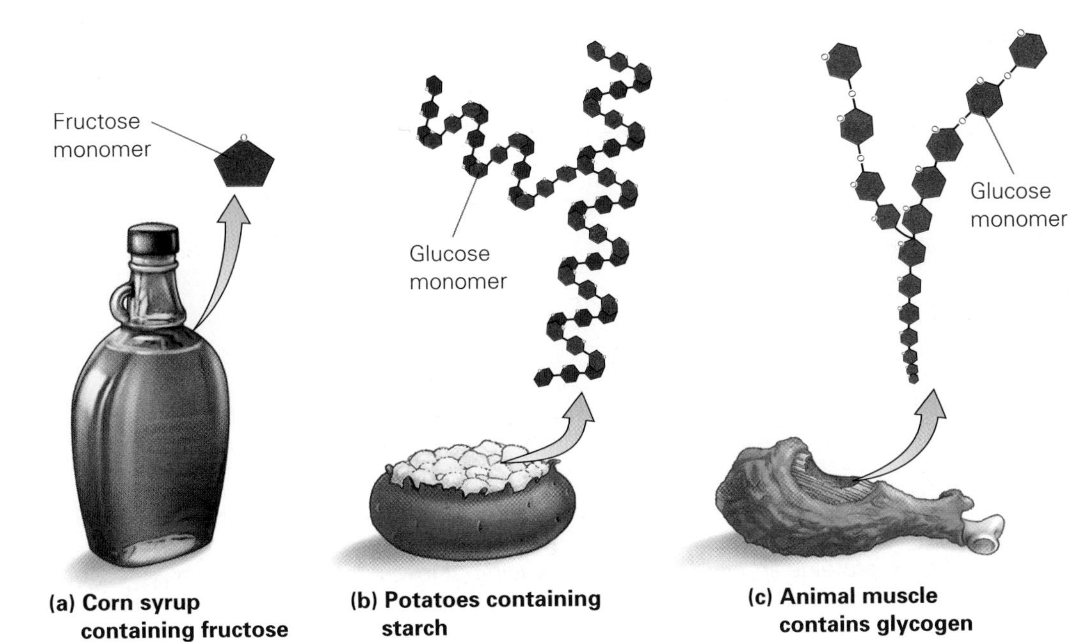

Figure 3.1 **Dietary carbohydrates.** (a) The simple sugar fructose is found in corn syrup. Complex carbohydrates include (b) starch in potatoes and other vegetables and (c) glycogen found in animal muscle.

When more sugar is present than needed, it can be stored for later use. Plants, such as potatoes, store their excess carbohydrates as the complex carbohydrate starch *(Figure 3.1b)*. Animals store their excess carbohydrates as the complex carbohydrate glycogen in muscles and the liver *(Figure 3.1c)*. Both starch and glycogen are polymers of glucose.

The body digests complex carbohydrates more slowly than it does simpler sugars because complex carbohydrates have more chemical bonds to break. Endurance athletes will load up on complex carbohydrates for several days before a race to increase the amount of easily accessible energy they can draw on during competition.

Nutritionists agree that most of the carbohydrates in a healthful diet should be in the form of complex carbohydrates, and that we should consume only minimal amounts of refined and processed sugars. When you consume complex carbohydrates in fruits, vegetables, and grains, you are also consuming many vitamins and minerals as well as fiber.

Dietary fiber, also called *roughage,* is composed mainly of those complex carbohydrates that humans cannot digest. For this reason, dietary fiber is passed into the large intestine, where some of it is digested by bacteria, and the remainder gives bulk to the feces. Whole grains, beans, and many fruits and vegetables are good sources of dietary fiber.

Although fiber is not a nutrient because it is not absorbed by the body, it is still an important part of a healthful diet. Fiber helps maintain healthy cholesterol levels. Fiber may also decrease your risk of various cancers.

Proteins as Nutrients. Protein-rich foods include beef, poultry, fish, beans, eggs, nuts, and dairy products such as milk, yogurt, and cheese.

Chapter 2 acquainted you with the idea that proteins are composed of amino acids, and that amino acids differ from each other based on their side groups. Amino acids are bonded to each other in an infinite variety of combinations to produce a diverse array of proteins with many different functions.

Your body is able to synthesize many of the commonly occurring amino acids. Those your body cannot synthesize are called **essential amino acids** and must be supplied by the foods you eat. Complete proteins contain all the

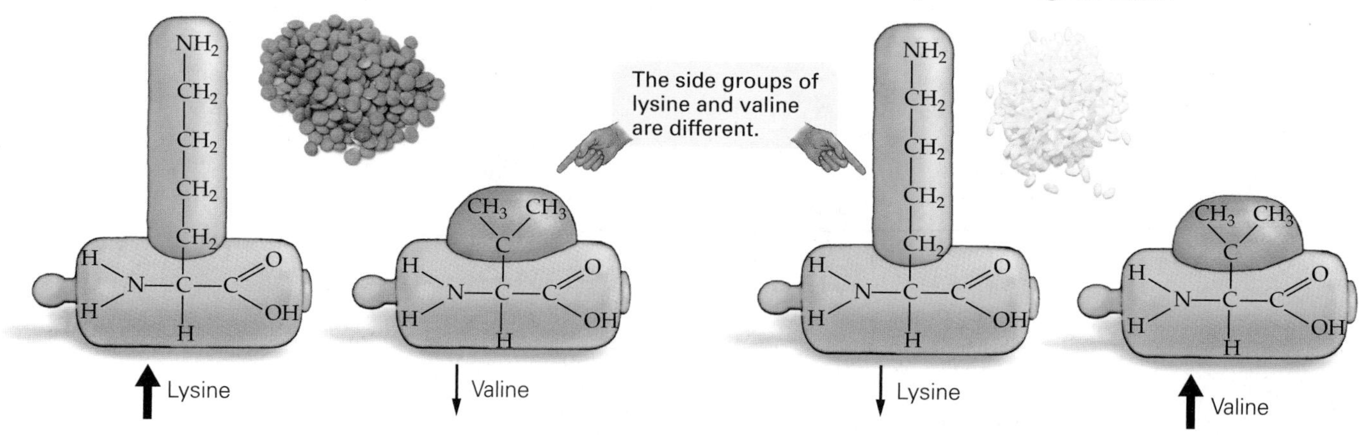

Figure 3.2 **Essential amino acids.** Essential amino acids, such as lysine and valine, cannot be synthesized by the body and must be obtained from the diet. Some proteins are high in one amino acid but low in another. Eating a wide variety of proteins thus ensures that all the necessary amino acids are available for growth, development, and maintenance.

essential amino acids your body needs. Proteins obtained by eating meat are more likely to be complete than are those obtained by eating plants; plant proteins can often be missing one or more essential amino acids *(Figure 3.2)*.

In the past, some nutritionists believed that vegetarians might be at risk for deficiencies in certain amino acids. However, scientific studies have shown that there is little cause for concern. If a vegetarian's diet is rich in a wide variety of plant-based foods, the body will have little trouble obtaining all the amino acids it needs to build proteins.

Stop and Stretch 3.1: Many cultures have characteristic meals that include protein from two different sources. For example, beans and rice, eggs and toast, and cereal and milk are common food pairings. Why might this practice of pairing different foods have evolved?

Even though you can obtain all the proteins you need by eating a variety of plants, Americans tend to eat a lot of meat as well. In addition to being rich in proteins, meat tends to be rich in fat.

Fats as Nutrients. The body uses fat, a type of lipid, as a source of energy. Gram for gram, fat contains a little more than twice as much energy as carbohydrates or protein. This energy is stored in the carbon, oxygen, and hydrogen bonds of the fat molecule.

Foods that are rich in fat include meat, milk, cheese, vegetable oils, and nuts. Muscle is often surrounded by stored fat, but some animals store fat throughout muscle, leading to the marbled appearance of some red meat. Other animals—chickens, for example—store fat on the surface of the muscle, making it easy to remove for cooking *(Figure 3.3)*. Most mammals, including humans, store fat just below the skin to help cushion and protect vital organs, to insulate the body from cold weather, and to store energy in case of famine. Some scientists believe that prehistoric humans often faced times of famine and may have evolved to crave fat.

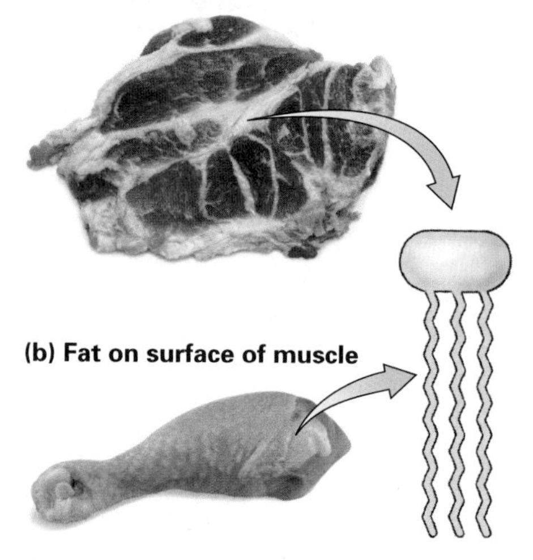

Figure 3.3 **Fat storage.** Fat can be intertwined with muscle tissue, as seen in this marbled piece of beef (a), or it can lie on the surface, as seen on this chicken leg (b).

Recall from Chapter 2 that fats consist of a glycerol molecule with hydrogen and carbon-rich fatty acid tails attached. Your body can synthesize most of the fatty acids it requires. Those that cannot be synthesized are called

essential fatty acids. Like essential amino acids, essential fatty acids must be obtained from the diet. Omega-3 and omega-6 fatty acids are essential fatty acids that can be obtained by eating fish.

The fatty acid tails of a fat molecule can differ in the number and placement of double bonds *(Figure 3.4a)*. When the carbons of a fatty acid are bound to as many hydrogens as possible, the fat is said to be a **saturated fat** (saturated in hydrogens). When there are carbon-to-carbon double bonds, the fat is not saturated in hydrogens, and it is therefore an **unsaturated fat** *(Figure 3.4b)*. The more double bonds, the higher the degree of unsaturation. When it contains many unsaturated carbons, the fat is referred to as **polyunsaturated.** The double bonds in unsaturated fats make the structures kink instead of lying flat. This form prevents the adjacent fat molecules from packing tightly together, so unsaturated fat tends to be liquid at room temperature. Cooking oil is an example of an unsaturated fat. Unsaturated fats are more likely to come from plant sources, while the fats found in animals are typically saturated. Saturated fats, with their absence of carbon-to-carbon double bonds, pack tightly together to make a solid structure. This is why saturated fats, such as butter, are solid at room temperature.

Commercial food manufacturers sometimes add hydrogen atoms to unsaturated fats by combining hydrogen gas with vegetable oils under pressure. This process, called **hydrogenation,** increases the fat's level of saturation; it retards spoilage and solidifies liquid oils, thereby making food seem less greasy and extending their shelf life. Margarine is vegetable oil that has undergone hydrogenation.

When hydrogen atoms are on the same side of the carbon-to-carbon double bond, they are said to be in the *cis configuration*—naturally occurring

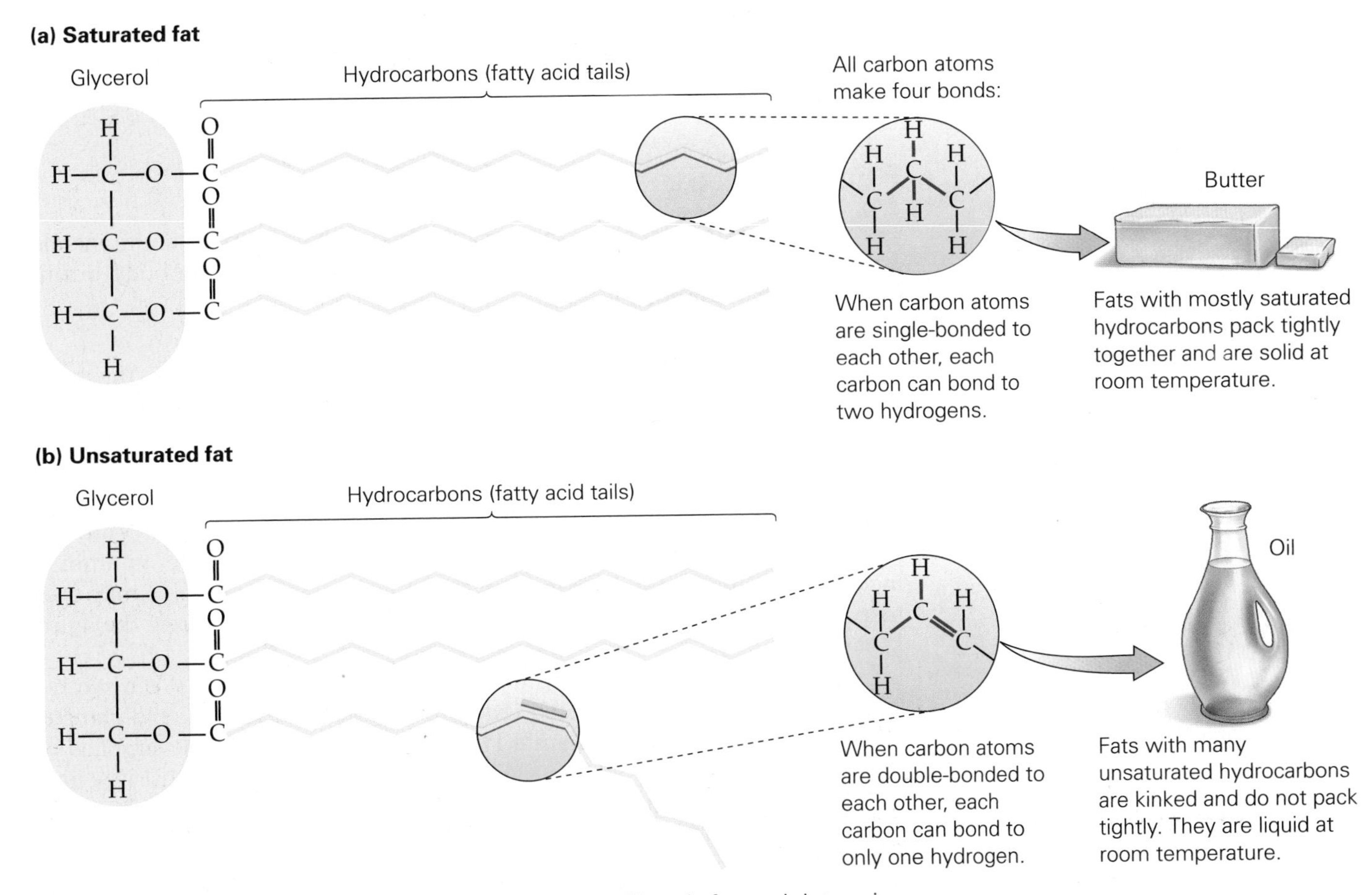

Figure 3.4 **Saturated and unsaturated fats.** The types of bonds formed determine whether a fat will be (a) solid or (b) liquid at room temperature.

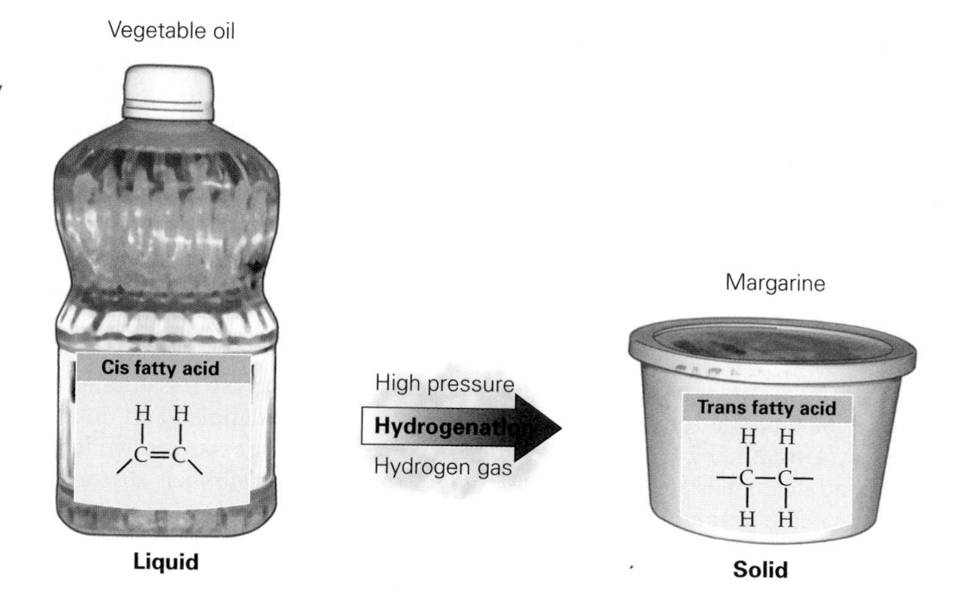

Figure 3.5 **Hydrogenation.** Adding hydrogen gas to vegetable oil forces the addition of hydrogen atoms to the fatty acid chains, some of which are not incorporated in the normally occurring cis configuration, but instead are incorporated in the trans configuration. The addition of hydrogens means that there are fewer double bonds; hence liquids can be solidified by this process. **Visualize This:** Count the number of hydrogens in the vegetable oil and in the margarine. Did the number of hydrogens increase or decrease during hydrogenation?

unsaturated fats have their hydrogen atoms in *cis* configuration. When hydrogen atoms are on opposite sides of the double bond, they are said to be in the *trans* configuration *(Figure 3.5)*. **Trans fats** are straight-chain saturated fats with fatty acids in a *trans* form. However, in contrast to most fats, trans fats are not required or beneficial.

While definitive studies are currently under way, the potential health risks of consuming foods rich in trans-fatty acids, common in fast foods, include increased risk of clogged arteries, heart disease, and diabetes. Because fat contains more stored energy per gram than carbohydrate and protein do, and because excess fat intake is associated with several diseases, nutritionists recommend that you limit the amount of fat in your diet.

Micronutrients

Nutrients that are essential in minute amounts, such as vitamins and minerals, are called **micronutrients.** They are neither destroyed by the body during use nor burned for energy.

Vitamins. Vitamins are organic substances, most of which the body cannot synthesize. Most vitamins function as **coenzymes,** or molecules that help enzymes, and thus speed up the body's chemical reactions. When a vitamin is not present in sufficient quantities, deficiencies can affect every cell in the body because many different enzymes, all requiring the same vitamin, are involved in numerous different bodily functions. Vitamins also help with the absorption of other nutrients; for example, vitamin C increases the absorption of iron from the intestine. Some vitamins may even help protect the body against cancer and heart disease and slow the aging process.

Vitamin D, also called calcitriol, is the only vitamin your cells can synthesize. Because sunlight is required for synthesis, people living in cold climates can develop deficiencies in vitamin D. All other vitamins must be supplied by the foods you eat. Many vitamins are water soluble, so boiling causes them to leach out into the water—this is why fresh vegetables are more nutritious than cooked ones. Steaming vegetables or using the vitamin-rich broth of canned vegetables when making soup helps preserve the vitamin content. Because the body does not store them, water-soluble vitamins are more likely than fat-soluble vitamins to be the source of dietary deficiencies. Vitamins A, D, E, and K are fat soluble and build up in stored fat; allowing an excess of these vitamins to accumulate in the body can be toxic. *Table 3.1* lists some vitamins and their roles in the body.

TABLE 3.1 **Vitamins.** Water-soluble and fat soluble vitamins.

Water-Soluble Vitamins

- Small organic molecules
- Will dissolve in water
- Cannot be synthesized by body
- Supplements packaged as pressed tablets
- Excesses usually not a problem because water-soluble vitamins are excreted in urine, not stored

Vitamin	Sources	Functions	Effects of Deficiency
Thiamin (B1)	Pork, whole grains, leafy green vegetables	Required component of many enzymes	Water retention and heart failure
Riboflavin (B2)	Milk, whole grains, leafy green vegetables	Required component of many enzymes	Skin lesions
Folic acid	Dark green vegetables, nuts, legumes (dried beans, peas, and lentils), whole grains	Required component of many enzymes	Neural tube defects, anemia, and gastrointestinal problems
B12	Chicken, fish, red meat, dairy	Required component of many enzymes	Anemia and impaired nerve function
B6	Red meat, poultry, fish, spinach, potatoes, and tomatoes	Required component of many enzymes	Anemia, nerve disorders, and muscular disorders
Pantothenic acid	Meat, vegetables, grains	Required component of many enzymes	Fatigue, numbness, headaches, and nausea
Biotin	Legumes, egg yolk	Required component of many enzymes	Dermatitis, sore tongue, and anemia
C	Citrus fruits, strawberries, tomatoes, broccoli, cabbage, green pepper	Collagen synthesis; improves iron absorption	Scurvy and poor wound healing
Niacin (B3)	Nuts, leafy green vegetables, potatoes	Required component of many enzymes	Skin and nervous system damage

TABLE 3.1 **Vitamins.** Water-soluble and fat soluble vitamins. *(Continued)*

Fat-Soluble Vitamins

- Small organic molecules
- Will not dissolve in water
- Cannot be synthesized by body (except vitamin D)
- Supplements packaged as oily gel caps
- Excesses can cause problems since fat-soluble vitamins are not excreted readily

Vitamin	Sources	Functions	Effects of Deficiency	Effects of Excess
A	Leafy green and yellow vegetables, liver, egg yolk	Component of eye pigment	Night blindness, scaly skin, skin sores, and blindness	Drowsiness, headache, hair loss, abdominal pain, and bone pain
D	Milk, egg yolk	Helps calcium be absorbed and increases bone growth	Bone deformities	Kidney damage, diarrhea, and vomiting
E	Dark green vegetables, nuts, legumes, whole grains	Required component of many enzymes	Neural tube defects, anemia, and gastrointestinal problems	Fatigue, weakness, nausea, headache, blurred vision, and diarrhea
K	Leafy green vegetables, cabbage, cauliflower	Helps blood clot	Bruising, abnormal clotting, and severe bleeding	Liver damage and anemia

Minerals. Minerals are substances that do not contain carbon but are essential for many cell functions. Because they lack carbon, minerals are said to be inorganic. They are important for proper fluid balance, in muscle contraction and conduction of nerve impulses, and for building bones and teeth. Calcium, chloride, magnesium, phosphorus, potassium, sodium, and sulfur are all minerals. Like some vitamins, minerals are water soluble and can leach out into the water during boiling. Also like vitamins, minerals are not synthesized in the body and must be supplied through your diet. *Table 3.2* lists the various functions of minerals that your body requires and describes what happens when there is a deficiency or an excess in certain minerals.

Processed Versus Whole Foods

Food that has undergone extensive processing has been stripped of much of its nutritive value. For example, refining flour removes the nutrient-rich, outer parts of the grain (called *bran*) along with the nutrient-rich, inner germ portion during processing, resulting in the loss of many vitamins and minerals and much of the fiber. It is best to limit your consumption of processed foods in general. Likewise, sweets (highly processed, sugar-rich foods) should occupy a very small portion of your diet because they provide no real nutrition, just calories. This is why sweets are often referred to as "empty" calories.

Foods that have not been stripped of their nutrition by processing are called whole foods. Eating a wide variety of whole foods such as fruits, vegetables, and grains gives you a much better chance of achieving a healthful diet

TABLE 3.2 **Minerals.** The minerals we require and their roles in the body.

Minerals

- Will dissolve in water
- Inorganic elements (do not contain carbon)
- Cannot be synthesized by body
- Supplements packaged as pressed tablets

Mineral	Sources	Functions	Effects of Deficiency	Effects of Excess
Calcium	Milk, cheese, dark green vegetables, legumes	Bone strength, blood clotting	Stunted growth, osteoporosis	Kidney stones
Chloride	Table salt, processed foods	Formation of stomach acid	Muscle cramps, reduced appetite, poor growth	High blood pressure
Magnesium	Whole grains, leafy green vegetables, legumes, dairy, nuts	Required component of many enzymes	Muscle cramps	Neurological disturbances
Phosphorus	Dairy, red meat, poultry, grains	Bone and tooth formation	Weakness, bone damage	Impaired ability to absorb nutrients
Potassium	Meats, fruits, vegetables, whole grains	Water balance, muscle function	Muscle weakness	Muscle weakness, paralysis, and heart failure
Sodium	Table salt, processed foods	Water balance, nerve function	Muscle cramps, reduced appetite	High blood pressure
Sulfur	Meat, legumes, milk, eggs	Components of many proteins	None known	None known

than eating highly refined, fatty foods that are low in complex carbohydrates and vitamins—also known as junk food.

In addition to containing vitamins and minerals, many whole foods contain molecules called **antioxidants** that are thought to play a role in the prevention of many diseases, including cancer. Biologists are currently investigating antioxidants to see whether these substances can slow the aging process. Antioxidants protect cells from damage caused by highly reactive molecules that are generated by normal cell processes. These highly reactive molecules, called free radicals, have an incomplete electron shell, which makes them more chemically reactive than molecules with complete electron shells. Antioxidants are abundant in fruits and vegetables, nuts, grains, and some meats. *Table 3.3* describes food sources of common antioxidants.

3.2 Enzymes and Metabolism

web animation 3.1
Enzymes

In addition to eating a well-balanced diet rich in unprocessed foods, it is important for people to eat the right amount of food. All food, whether carbohydrate, protein, or fat, can be turned into fat when too much is consumed. In

Table 3.3 Antioxidants. Antioxidants are being investigated for their disease-preventing abilities.

Antioxidants	• Present in whole foods • Protect cells from damage caused by free radicals • Thought to have a role in disease prevention
Antioxidant	**Source**
Beta-carotene	Foods rich in beta-carotene are orange in color; they include carrots, cantaloupe, squash, mangoes, pumpkin, and apricots. Beta-carotene is also found in some leafy green vegetables such as collard greens, kale, and spinach.
Flavenoid	Cocoa and dark chocolate contain flavenoids.
Lutein	Lutein, which is known to help keep eyes healthy, is also found in leafy green vegetables such as collard greens, kale, and spinach.
Lycopene	Lycopene is a powerful antioxidant found in watermelon, papaya, apricots, guava, and tomatoes.
Selenium	Selenium is a mineral (not an antioxidant) that serves as a coenzyme for many antioxidant enzymes, thereby increasing their effectiveness. Rice, wheat, meats, bread, and Brazil nuts are major sources of dietary selenium.
Vitamin A	Foods rich in vitamin A include sweet potatoes, liver, milk, carrots, egg yolks, and mozzarella cheese.
Vitamin C	Foods rich in vitamin C include most fruits, vegetables, and meats.
Vitamin E	Vitamin E is found in almonds, many cooking oils, mangoes, broccoli, and nuts.

this manner, energy stored in the chemical bonds of food is converted into fat and stored for later use.

The amount of fat that a given individual will store depends partly on how quickly or slowly he or she breaks down food molecules into their component parts. **Metabolism** is a general term used to describe all of the chemical reactions occurring in the body.

Enzymes

All metabolic reactions are regulated by proteins called **enzymes** that speed up, or **catalyze,** the rate of reactions. Enzymes help your body break down the foods you ingest and liberate the energy stored in their chemical bonds. Enzymes are usually named for the reaction they catalyze and end in the suffix *-ase.* For example, sucrase is the enzyme that breaks down the table sugar sucrose.

To break chemical bonds, molecules must absorb energy from their surroundings, often by absorbing heat. This is why heating a chemical reaction will speed up the reaction. Heating cells to an excessively high temperature can damage or kill them. Enzymes help catalyze the body's chemical reactions without requiring heat for the reactants to break their chemical bonds. Therefore, by decreasing the energy required to start the reaction, enzymes allow chemical reactions to occur more quickly.

Activation Energy. The energy required to start the metabolic reaction serves as a barrier to catalysis and is called the **activation energy** *(Figure 3.6)*. If not for the activation energy barrier, all of the chemical reactions in cells would occur relentlessly, whether the products of the reactions were needed or not. Because most metabolic reactions need to surpass the activation energy barrier before proceeding, they can be regulated by the presence or absence of enzymes. In other words, a given chemical reaction will occur only if the proper enzyme is available. How do enzymes decrease the activation energy barrier?

Induced Fit. The chemicals that are metabolized by an enzyme-catalyzed reaction are called the enzyme's **substrate.** Enzymes decrease activation energy by binding to their substrate and placing stress on its chemical bonds, decreasing the amount of initial energy required to break the bonds. The region of the enzyme where the substrate binds is called the enzyme's **active site.** Each active site has its own shape and chemistry. When the substrate binds to the active site, the enzyme changes shape slightly to envelop the substrate. This shape change by the enzyme in response to substrate binding is called **induced fit** because the substrate induces the enzyme to change shape to conform to the substrate's contours. When the enzyme changes shape, it places stress on the chemical bonds of the substrate, making them easier to break. In this manner, the enzyme helps convert the substrate to a reaction product and then resumes its original shape so that it can perform the reaction again *(Figure 3.7)*.

Different enzymes catalyze different reactions by a property called **specificity.** The specificity of an enzyme is the result of its shape and the shape of its active site. Different enzymes have unique shapes because they are composed of amino acids in varying sequences. The 20 amino acids, each with its own unique side group, are arranged in distinct orders for each enzyme, producing enzymes of all shapes and sizes, each with an active site that can bind with its particular substrate. Although an infinite variety of enzymes could be produced, it is quite often the case that different organisms will utilize very similar enzymes, likely due to their evolution from a common ancestor.

(a) No enzyme present

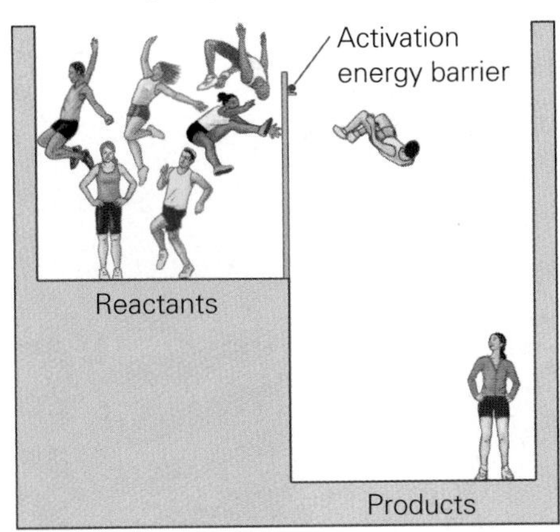

(b) Enzyme present

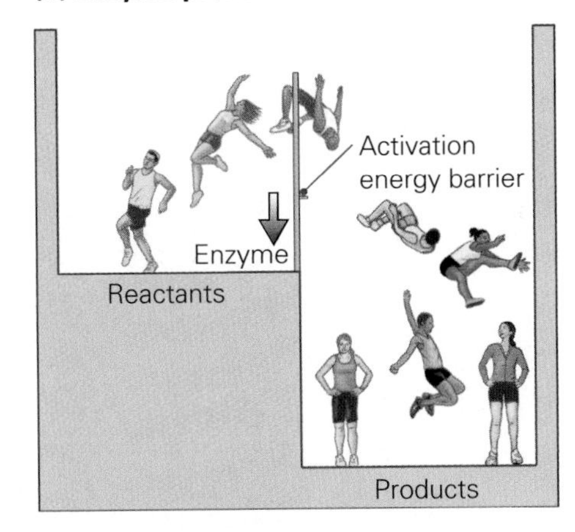

Figure 3.6 **Activation energy.** (a) Few high jumpers can make it across the bar when the bar is high. Likewise, to break bonds in a reactant and form products, an activation energy barrier must be surmounted. (b) Lowering the high jump bar allows more jumpers to clear the bar. Enzymes lower the activation energy barrier in cells.

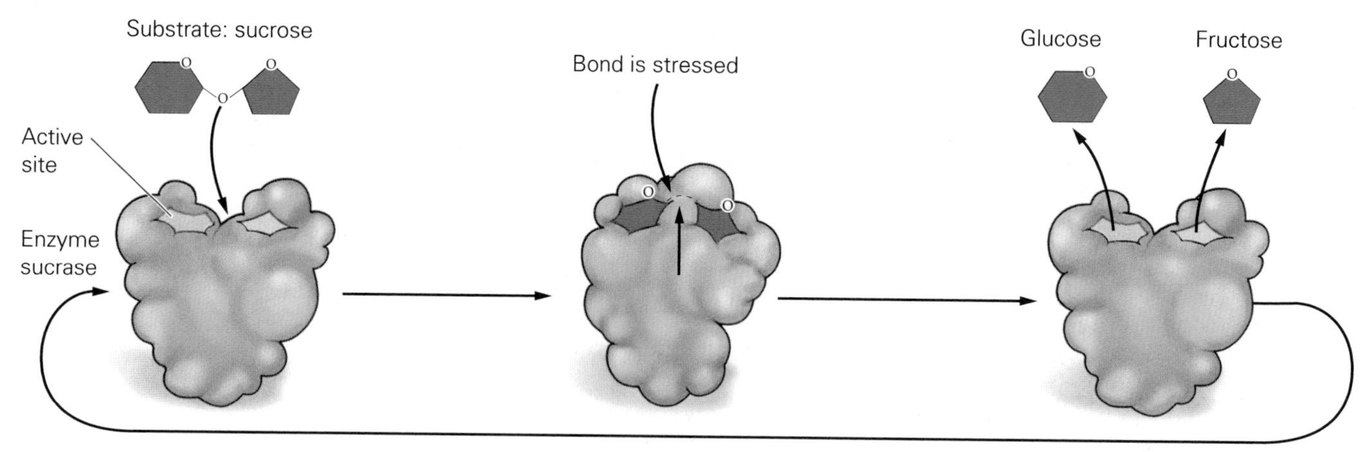

Figure 3.7 **Enzymes.** The enzyme sucrase is cleaving (splitting) the disaccharide sucrose into its monosaccharide subunits, fructose and glucose. **Visualize This:** Is the enzyme itself changed by the process of catalysis?

Lactose intolerance is a common dietary problem caused by an enzyme deficiency. People with lactose intolerance are unable to digest large amounts of lactose (the most common sugar in milk). This condition results from a shortage of the enzyme **lactase,** typically produced by the cells of the small intestine. Lactase breaks down lactose into simpler forms (the monomers glucose and galactose) that can then be absorbed into the bloodstream. When there is not enough lactase available to digest lactose, bacteria in the intestine metabolize the sugar, producing lactic acid. The buildup of lactic acid causes bloating, gas, cramps, and diarrhea.

Most babies can digest milk sugars efficiently, but as they age, their ability to produce this enzyme declines. Scientists hypothesize that in the evolutionary past, the production of this enzyme after weaning was unnecessary, and most adults were probably lactose intolerant. However, with the addition of dairy products to the human diet, selective pressure now favors continuing to produce this enzyme into adulthood. Lactose intolerance can be treated by taking dietary supplements that contain the lactase enzyme.

Enzymes do more than just the break down milk sugars. They mediate all of the metabolic reactions occurring in an organism's cells. Enzymes even affect the rate of a particular individual's metabolism. This means that two similar size individuals might need to consume different amounts of food to meet their daily energy requirements.

Calories and Metabolic Rate

Energy is measured in units called calories. A **calorie** is the amount of energy required to raise the temperature of 1 gram of water by 1 degree Celsius (1°C). In scientific literature, energy is usually reported in kilocalories, and 1 kilocalorie equals 1000 calories of energy. However, in physiology—and on nutritional labels—the prefix *kilo-* is dropped, and a kilocalorie is referred to as a **Calorie** (with a capital *C*). Calories are consumed to supply the body with energy to do work, which includes maintaining body temperature.

Balancing energy intake versus energy output means eating the correct amount of food to maintain health. When foods are eaten, they are broken down into their component subunits. The energy stored in the chemical bonds of food can be used to make a form of energy that the cell can use. When the supply of Calories is greater than the demand, the excess Calories can be stored by the body as fat.

The speed and efficiency of many different enzymes will lead to an overall increase or decrease in the rate at which a person can break down food. Thus, when you say that your metabolism is slow or fast, you are actually referring to the speed at which enzymes catalyze chemical reactions in your body.

A person's **metabolic rate** is a measure of his or her energy use. This rate changes according to the person's activity level. For example, we require less energy when asleep than we do when exercising. The **basal metabolic rate** represents the resting energy use of an awake, alert person. The average basal metabolic rate is 70 Calories per hour, or 1680 Calories per day. However, this rate varies widely among individuals because many factors influence each person's basal metabolic rate: exercise habits, body weight, sex, age, and genetics. Overall nutritional status can also affect metabolism because an enzyme that is missing its vitamin coenzyme may not be able to perform at its optimal rate.

Exercise requires energy, which allows you to consume more Calories without having to store them. As for body weight, a heavy person utilizes more Calories during exercise than a thin person does. *Figure 3.8* shows the number of Calories used per hour for individuals of different body weights. Males require more Calories per day than females do because testosterone, a hormone produced in larger quantities by males, increases the rate at which fat breaks down. Men also have more muscle than women, which requires more energy to maintain than fat does.

Age and genetics also play a role in metabolic rate. Two people of the same size and sex, who consume the same number of Calories and exercise the same amount, will not necessarily store the same amount of fat. The rate at which the foods you eat are metabolized slows as you age, and some people are simply born with lower basal metabolic rates. To obtain a rough measure of how many Calories you should consume per day, multiply the weight you wish to maintain by 11 and add the number of Calories you burn during exercise.

The properties of metabolic enzymes, like those of all proteins, are determined by the genes that encode them. Genes that influence a person's rate of fat storage and utilization are passed from parents to children. All of these variables help explain why some people seem to eat and eat and never gain an ounce, while others struggle with their weight for their entire lives.

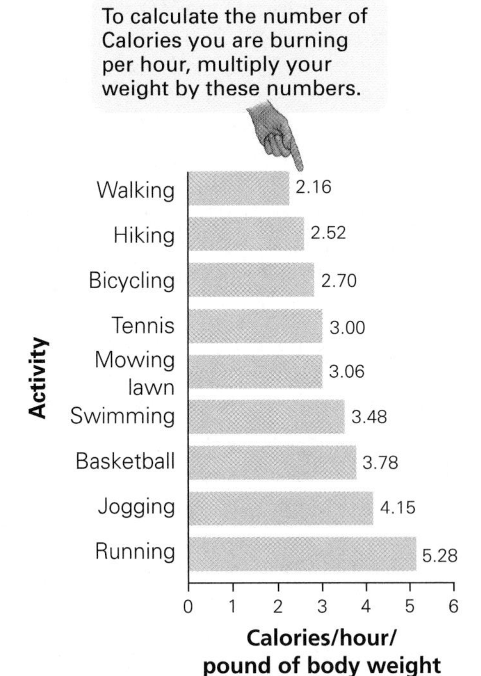

Figure 3.8 **Energy expenditures for various activities.** This bar graph can help you determine how many Calories you burn during certain activities. **Visualize This:** How many Calories would a 160-pound person burn in 30 minutes of swimming?

Stop and Stretch 3.2: Losing 1 pound of fat requires you to burn 3500 more Calories than you consume. If a person is trying to lose 1 pound per week and he decreases his caloric intake by 300 Calories per day, how many more Calories must he burn each day by exercising to reach his goal?

To be fully metabolized, food is broken down by the digestive system and then transported to individual cells via the bloodstream *(Figure 3.9)*. Inside the mitochondria of cells, food energy can be converted into chemical energy in the form of **adenosine triphosphate** or **ATP.** ATP and energy generation are covered fully in Chapter 4.

Once nutrients arrive at cells, they must traverse the membrane that surrounds cells, the plasma membrane.

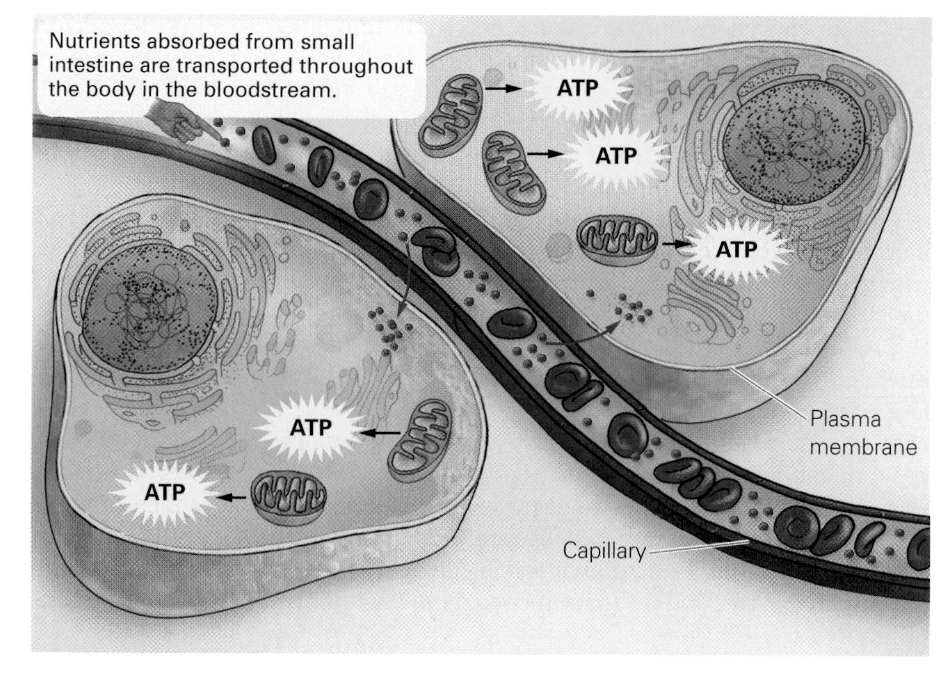

Figure 3.9 **Nutrients move from the bloodstream to cells.** Substances absorbed from the small intestine into the bloodstream cross the plasma membrane of individual cells. Once inside a cell, the food can be broken down further to release the energy stored in its chemical bonds. This energy is then used to produce ATP to power cellular activities.

web animation 3.2
Passive and Active Transport

web animation 3.3
Exocytosis and Endocytosis

3.3 Transport Across Membranes

Building-block molecules must cross the plasma membrane to gain access to the inside of the cell, where they can be used to synthesize cell components or be metabolized to provide energy for the cell. The chemistry of the membrane facilitates the transport of some substances and prevents the transport of others.

Overview: Membrane Transport

Recall from Chapter 2 that the plasma membrane that surrounds cells is composed of a phospholipid bilayer. The interior of the bilayer is hydrophobic. Hydrophobic substances can dissolve in the membrane and pass through it more easily than hydrophilic ones. In this sense, the membrane of the cell is differentially permeable to the transport of molecules, allowing some to pass and blocking others from passing.

Substances that can cross the membrane will do so until the concentration is equal on both sides of the membrane. Carbon dioxide, water, and oxygen move freely across the membrane. Larger molecules and charged molecules and ions cannot cross the lipid bilayer on their own. If these substances need to be moved across the membrane they must move through proteins embedded in the membrane. Proteins in the membrane can serve as channels to allow such molecules to cross until their concentration is equal on both sides of the membrane. Proteins in the membrane can also move substances across the membrane past the point of their chemical equilibrium when more of a substance is required on one side of the membrane than the other *(Figure 3.10)*.

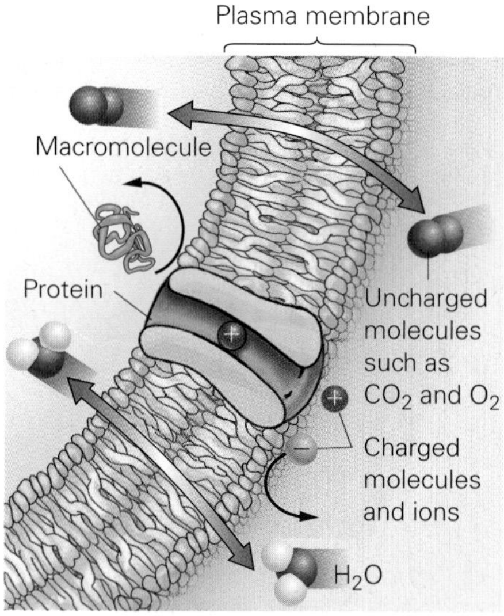

Plasma membrane
Macromolecule
Protein
Uncharged molecules such as CO_2 and O_2
Charged molecules and ions
H_2O

Figure 3.10 **Transport of substances across membranes.** The ability of a substance to cross a membrane is, in part, a function of its size and charge.

A Closer Look: Membrane Transport

As you learned in the preceding section, the manner in which a substance is moved across a membrane depends on its particular chemistry and concentration in the cell. Each type of transport has its own characteristics and energy requirements.

Passive Transport: Diffusion, Facilitated Diffusion, and Osmosis. All molecules contain energy that makes them vibrate and bounce against each other, scattering around like billiard balls during a game of pool. In fact, molecules will bounce against each other until they are spread out over all

(A Closer Look continued)

the available area. In other words, molecules will move from their own high concentration to their own low concentration. This movement of molecules from where they are in high concentration to where they are in low concentration is called **diffusion.** During diffusion, the net movement of molecules is *down* a concentration gradient. This movement does not require an input of outside energy; it is spontaneous. Diffusion will continue until equilibrium, at which time no concentration gradient exists and there is no net movement of molecules.

Diffusion also occurs in living organisms. When substances diffuse across the plasma membrane, we call the movement **passive transport.** Passive transport has the name passive because it does not require an input of energy. The structure of the phospholipid bilayer that comprises the plasma membrane prevents many substances from diffusing across it. Only very small, hydrophobic molecules are able to cross the membrane by diffusion. In effect, these molecules dissolve in the membrane and slip from one side to the other *(Figure 3.11a)*.

Hydrophilic molecules and charged molecules such as ions are unable to simply diffuse across the hydrophobic core of the membrane. For example, when you have a meal of chicken, rich in charged amino acids, and a green salad, rich in hydrophilic carbohydrates and ions such as calcium (Ca^{+}), these amino acids, sugars, and ions cannot gain access to the inside of the cell on their own. Instead, these molecules are transported across membranes by proteins embedded in the lipid bilayer. This type of passive transport does not require an input of energy and is called **facilitated diffusion.**

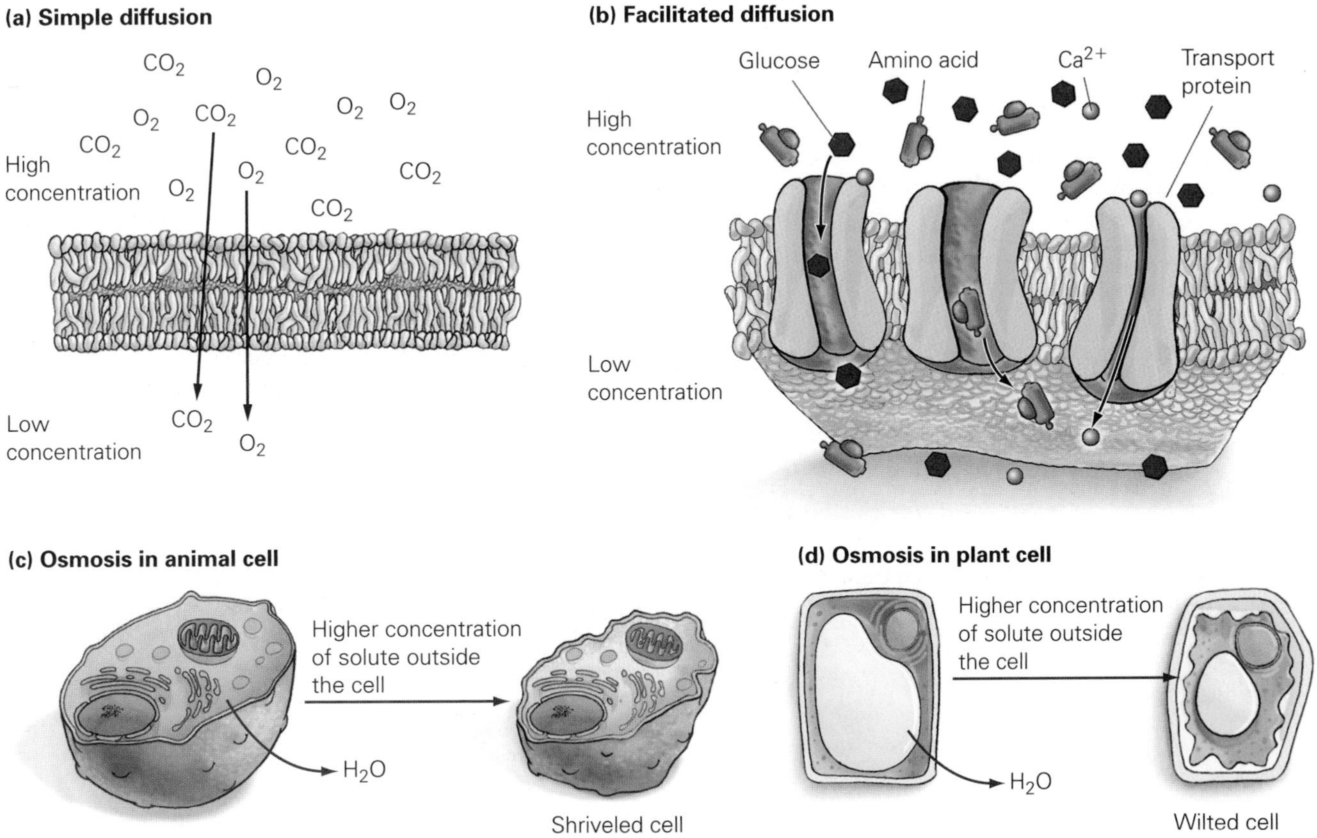

Figure 3.11 **Diffusion.** (a) Simple diffusion of molecules across the plasma membrane occurs with the concentration gradient and does not require energy. Small hydrophobic molecules, carbon dioxide, and oxygen can diffuse across the membrane. (b) Facilitated diffusion is the diffusion of molecules assisted by substrate-specific proteins. Molecules move with their concentration gradient, which does not require energy. (c) Osmosis is a special type of diffusion that involves the movement of water in response to a concentration gradient. Water moves toward a region that has more dissolved solute. When water leaves an animal cell, it shrinks. (d) When water leaves a plant cell, the plant wilts instead of shrinks due to the presence of the cell wall.

(continued on the next page)

(A Closer Look continued)

Facilitated diffusion is so named because the specific membrane transport proteins are helping or "facilitating" the diffusion of substances across the plasma membrane *(Figure 3.11b)*. Although transport proteins are used to help the substance move across the plasma membrane, this form of transport is still considered to be diffusion because substances are traveling from high to low concentration.

The movement of water across a membrane is a type of passive transport called **osmosis.** Like other substances, water moves from its own high concentration to its own low concentration. Water can move through proteins in the membrane, called aquaporins, but even without these, water can still cross the membrane. When an animal cell is placed in a solution of salt water, water leaves the cell, causing the cell to shrivel *(Figure 3.11c)*. Likewise, plants that are overfertilized wilt because water leaves the cells to equilibrate the concentration of water on either side of the plasma membrane *(Figure 3.11d)*.

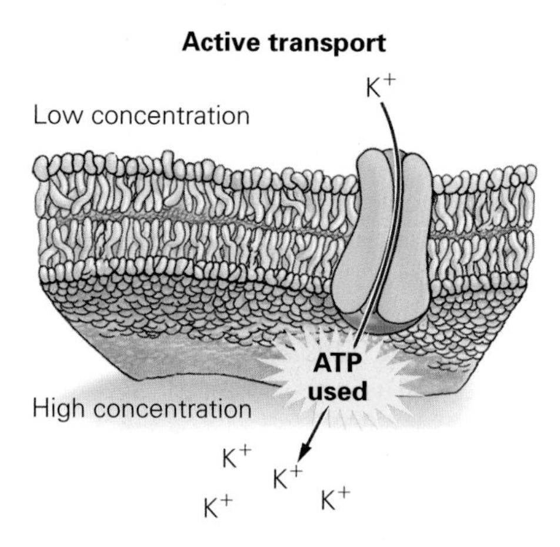

Figure 3.12 **Active transport.** Active transport moves substances against their concentration gradient and requires ATP energy to do so.

Active Transport: Pumping Substances Across the Membrane. In some situations, a cell will need to maintain a concentration gradient. For example, nerve cells require a high concentration of certain ions inside the cell to transmit nerve impulses. To maintain this difference in concentration across the membrane requires the input of energy. Think of a hill with a steep incline or grade. Riding your bike down the hill requires no energy, but riding your bike up the grade requires energy in the form of ATP. **Active transport** is transport that uses proteins, powered by the energy currency ATP, to move substances up a concentration gradient *(Figure 3.12)*.

Stop and Stretch 3.3: Muscle cells require a high internal concentration of potassium (K^+). What type of transport would allow this ion to exist in a higher concentration inside than outside muscle cells?

Exocytosis and Endocytosis: Movement of Large Molecules Across the Membrane. Larger molecules are often too big to diffuse across the membrane or to be transported through a protein, regardless of whether they are hydrophobic or hydrophilic. Instead, they must be moved around inside membrane-bound vesicles that can fuse with membranes. **Exocytosis** *(Figure 3.13a)*

Figure 3.13 **Movement of large substances.** (a) Exocytosis is the movement of substances out of the cell. (b) Endocytosis is the movement of substances into the cell.

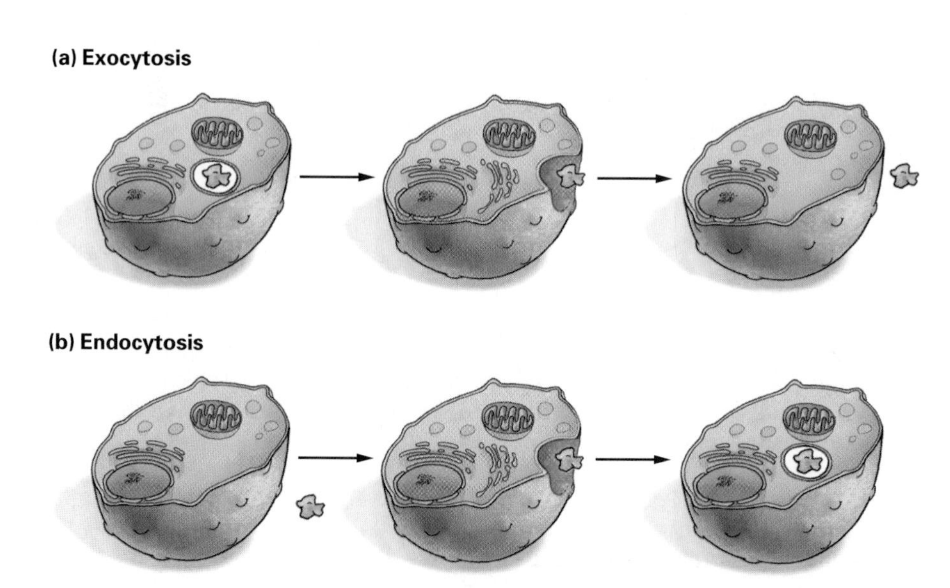

(A Closer Look continued)

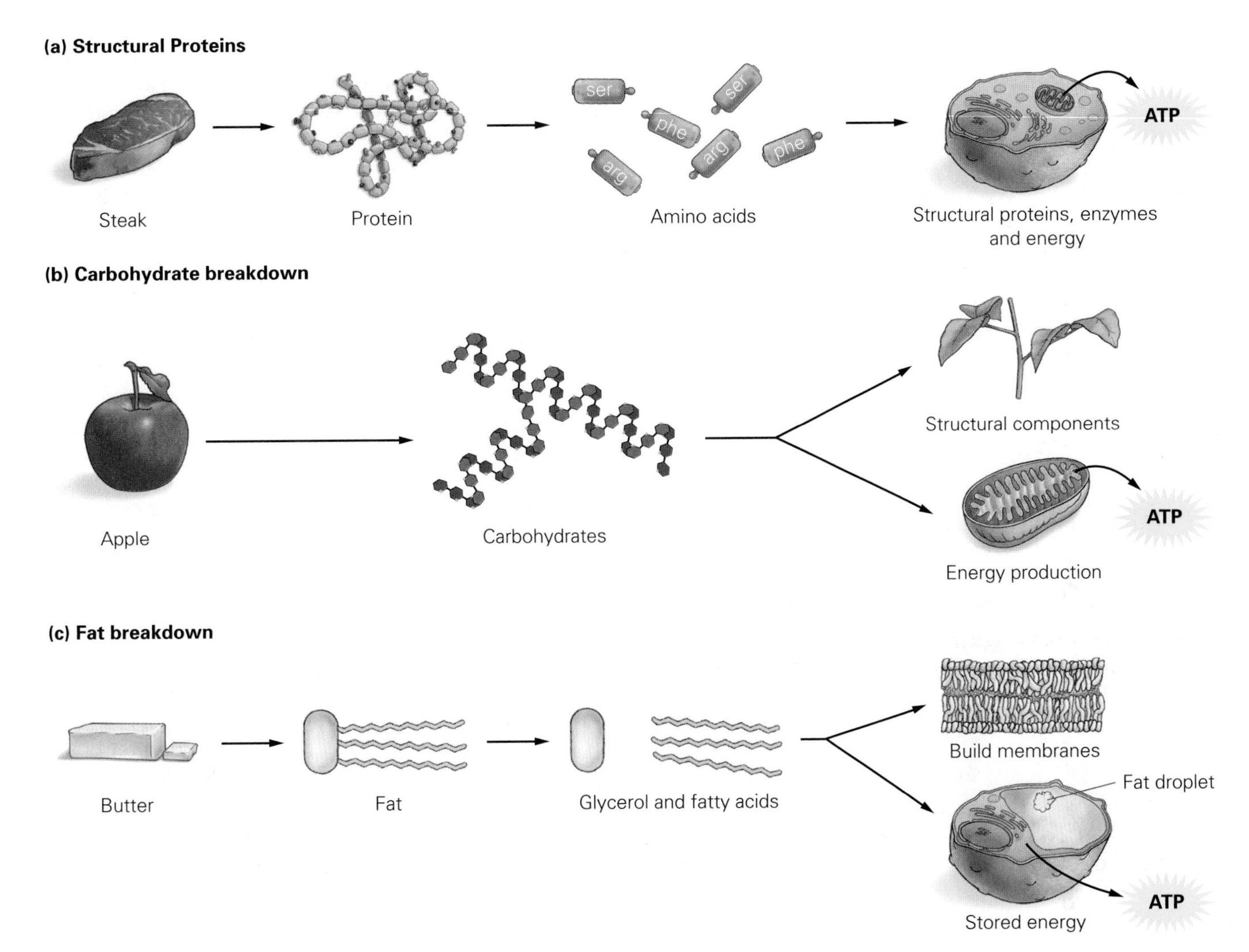

Figure 3.14 **You are what you eat.** Food is digested into component molecules that are used to build cellular structures and generate ATP. (a) Proteins are broken down into amino acids and reassembled into proteins the cell requires. (b) Carbohydrates are broken down into sugars that can be reassembled into other sugars the cell requires or used to produce energy (ATP) in the mitochondria. (c) Fats are broken down and reassembled to be used in membranes, burned to produce energy, or stored as fat.

occurs when a membrane-bound vesicle, carrying some substance, fuses with the plasma membrane and releases its contents into the exterior of the cell. **Endocytosis** *(Figure 3.13b)* occurs when a substance is brought into the cell by a vesicle pinching inward, bringing the substance with it.

All the nutrients you consume and dismantle must find some way into your cells so that they can be used for energy and to build cellular components. Knowing all this should give heightened meaning to the phrase, "You are what you eat" *(Figure 3.14)*. Once inside your cells, nutrients will be used to build structural components and provide energy, or they can be stored for later use as fat.

3.4 Body Fat and Health

Figure 3.15 **The perception of beauty.** (a) GI Joe has become more muscular over time. (b) Female sex symbols have become thinner over time.

A clear understanding of how much body fat is healthful is hard to come by because cultural and biological definitions of the term overweight differ markedly. Cultural definitions of overweight have changed over the years. Men and women who were considered to be of normal weight in the past might not be seen as meeting today's standards. In the United States, the evolution of this trend has been paralleled by changes in children's action figures and dolls and in celebrities and movie stars over the last several decades. You need only compare the physiques of action figures from the 1960s and 1970s to the physiques seen on today's action figures to see how the standards have changed for males *(Figure 3.15a)*. Standards for women have also changed. The 1960 sex symbol and movie star Marilyn Monroe was much curvier than more recent stars, such as Nicole Richie *(Figure 3.15b)*.

The next time you read the newspaper, take note of advertisements for diets promoting diet products. It is often the case that "before" pictures show individuals of healthful weights, and "after" pictures show people who are too thin. Indeed, the average woman in the United States weighs 140 pounds and wears a size 12. The average model weighs 103 pounds and wears a size 4. With these distorted messages about body fat, it is difficult for average people to know how much body fat is right for them.

Evaluating How Much Body Fat Is Healthful

A person's sex, along with other factors, determines his or her ideal amount of body fat. Women need more body fat than men do to maintain their fertility. On average, healthy women have 22% body fat, and healthy men have 14%. To maintain essential body functions, women need at least 12% body fat but not more than 32%; for men, the range is between 3% and 29%. This difference between females and males, a so-called sex difference, exists because women store more fat on their breasts, hips, and thighs than men do. A difference in muscle mass leads to increased energy use by males because muscles use more energy than does fat.

Women also have an 8% thicker layer of tissue called the dermis under the outer epidermal layer of the skin as compared to men. This means that in a woman and a man of similar strength and body fat, the woman's muscles would look smoother and less defined than the man's muscles would.

A person's frame size also influences body fat—larger-boned people carry more fat. In addition, body fat tends to increase with age.

Unfortunately, it is a bit tricky to determine what any individual's ideal body weight should be. In the past, you simply weighed yourself and compared your weight to a chart showing a range of acceptable weights for given heights. The weight ranges on these tables were associated with the weights of a group of people who bought life insurance in the 1950s and whose health was monitored until they died. The problem with using these tables is that the subjects may not have been representative of the whole population. As you learned in Chapter 1, generalizing results seen in one group to another group can lead to erroneous conclusions. People who had the money to buy life insurance tended to have the other benefits of money as well, including easier access to health care, better nutrition, *and* lower body weight. Their longer lives may have had more to do with better health care and nutrition than with their weight.

To deal with some of the ambiguities associated with the insurance company's weight tables, a new measure of weight and health risk, the **body mass index (BMI),** has been developed. BMI is a calculation that uses both height and weight to determine a value that correlates an estimate of body fat with the risk of illness and death *(Table 3.4)*.

TABLE 3.4 **Body mass index (BMI).** A chart based on height and weight correlations.

Height	Weight											
4'10"	91	96	100	105	110	115	119	124	129	134	138	143
4'11"	94	99	104	109	114	119	124	128	133	138	143	148
5'0"	97	102	107	112	118	123	128	133	138	143	148	153
5'1"	100	106	111	116	122	127	132	137	143	148	153	158
5'2"	103	109	115	120	126	131	136	142	148	153	158	164
5'3"	107	113	118	124	130	135	141	146	152	158	163	169
5'4"	110	116	122	128	134	140	145	151	157	163	169	174
5'5"	114	120	126	132	138	144	150	156	162	168	174	180
5'6"	117	124	130	136	142	148	155	161	167	173	179	186
5'7"	121	127	134	140	146	153	159	166	172	178	185	191
5'8"	125	131	138	144	151	158	164	171	177	184	190	197
5'9"	129	135	142	149	155	162	169	176	183	189	196	203
5'10"	132	139	146	153	160	167	174	181	188	195	202	207
5'11"	136	143	150	157	165	172	179	186	193	200	208	215
6'0"	140	147	154	162	169	177	184	191	198	206	213	221
6'1"	144	151	159	166	174	182	189	197	205	212	219	227
6'2"	148	155	163	171	179	186	194	202	210	218	225	233
6'3"	151	160	168	176	184	192	200	208	216	224	232	240
6'4"	156	164	172	180	189	197	205	213	221	230	238	246
BMI	**19**	**20**	**21**	**22**	**23**	**24**	**25**	**26**	**27**	**28**	**29**	**30**

16 20 25 30

Anorexic | Underweight and possibly anorexic | Healthy | Overweight | Obese

Although the BMI measurement is a better approximation of ideal weight than are the insurance charts of the past, it is not perfect; BMI still does not account for differences in frame size, gender, or muscle mass. In fact, studies show that as many as one in four people may be misclassified by BMI tables because this measurement provides no means to distinguish between lean muscle mass and body fat. For example, an athlete with a lot of muscle will weigh more than a similar size person with a lot of fat because muscle is heavier than fat.

If your BMI falls within the healthy range (BMI of 20–25), you probably have no reason to worry about health risks from excess weight. If your BMI is high, you may be at increased risk for diseases associated with obesity.

Obesity

Close to one in three Americans has a BMI of 30 or greater and is therefore considered to be obese. This crisis in **obesity** is the result you would expect when constant access to cheap, high-fat, energy-dense, unhealthful food is combined with lack of exercise. This relationship is clearly illustrated by the case of the Pima Indians.

Several hundred years ago, the ancestral population of Pima Indians split into two tribes. One branch moved to Arizona and adopted the American diet and lifestyle; the typical Pima of Arizona gets as much exercise as the average

American and, like most Americans, eats a high-fat, low-fiber diet. Unfortunately, the health consequences for these people are more severe than they are for most other Americans—close to 60% of the Arizona Pima are obese and diabetic. In contrast, the Pima of Mexico maintained their ancestral farming life; their diet is rich in fruits, vegetables, and fiber. The Pima of Mexico also engage in physical labor for close to 22 hours per week and are on average 60 pounds lighter than their Arizona relatives. Consequently, diabetes is virtually unheard of in this group.

This example illustrates the impact of lifestyle on health: The Pima of Arizona share many genes with their Mexican relatives but have far less healthful lives due to their diet and lack of exercise. The example also shows that genes influence body weight because the Pima of Arizona have higher rates of obesity and diabetes than those of other Americans whose lifestyle they share.

Whether obesity is the result of genetics, diet, or lack of exercise, the health risks associated with obesity are the same. As your weight increases, so do your risks of diabetes, hypertension, heart disease, stroke, and joint problems.

Diabetes. **Diabetes** is a disorder of carbohydrate metabolism characterized by the impaired ability of the body to produce or respond to insulin. **Insulin** is a hormone secreted by beta cells, which are located within clusters of cells in the pancreas. Insulin's role in the body is to trigger cells to take up glucose so that they can convert the sugar into energy. People with diabetes are unable to metabolize glucose; as a result, the level of glucose in the blood rises *(Figure 3.16)*.

There are two forms of the disease. Type 1, insulin-dependent diabetes mellitus (IDDM) usually arises in childhood. People with IDDM cannot produce insulin because their immune systems mistakenly destroy their own beta cells. When the body is no longer able to produce insulin, daily injections of the hormone are required. Type 1 diabetes is not correlated with obesity.

Type 2, **non-insulin-dependent diabetes mellitus** (NIDDM), usually occurs after 40 years of age and is more common in the obese. NIDDM arises

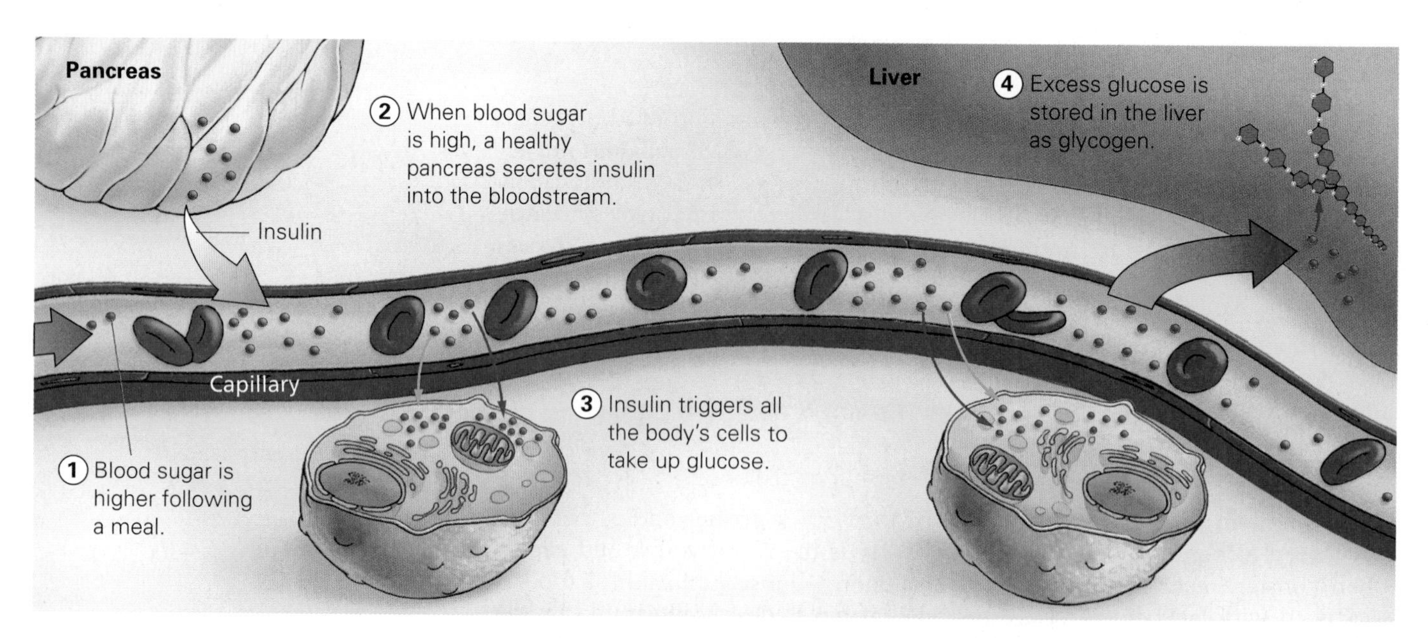

Figure 3.16 **Blood glucose metabolism and diabetes.** Insulin serves as a key to unlock cells so that glucose can enter. A diabetic person either does not produce enough insulin or does not respond properly to the insulin produced by the body. As a result, glucose stays in the blood instead of fueling cellular activities.

either from decreased pancreatic secretion of insulin or from reduced responsiveness to secreted insulin in target cells. People with NIDDM are able to control blood glucose levels through diet and exercise and, if necessary, by insulin injections.

Hypertension. **Hypertension,** or high blood pressure, places increased stress on the circulatory system and causes the heart to work too hard. Compared to a person with normal blood pressure, a hypertensive person is six times more likely to have a heart attack.

Blood pressure is the force, originated by the pumping action of the heart, exerted by the blood against the walls of the blood vessels. Blood vessels expand and contract in response to this force. Blood pressure is reported as two numbers: The higher number, called the **systolic blood pressure,** represents the pressure exerted by the blood against the walls of the blood vessels as the heart contracts; the lower number, called the **diastolic blood pressure,** is the pressure that exists between contractions of the heart when the heart is relaxing. Normal blood pressure is around 120 over 80 (symbolized as 120/80). Blood pressure is considered to be high when it is persistently above 140/90.

Problematic weight gain is typically the result of increases in the amount of fatty tissue versus increases in muscle mass. Fat, like all tissues, relies on oxygen and other nutrients from food to produce energy. As the amount of fat on your body increases, so does the demand for these substances. Therefore, the amount of blood required to carry oxygen and nutrients also increases. Increased blood volume means that the heart has to work harder to keep the blood moving through the vessels, thus placing more pressure on blood vessel walls and leading to increased heart rate and blood pressure.

Heart Attack, Stroke, and Cholesterol. A **heart attack** occurs when there is a sudden interruption in the supply of blood to the heart caused by the blockage of a vessel supplying the heart. A **stroke** is a sudden loss of brain function that results when blood vessels supplying the brain are blocked or ruptured. Heart attack and stroke are more likely in obese people because the elevated blood pressure caused by obesity also damages the lining of blood vessels and increases the likelihood that cholesterol will be deposited there. Cholesterol-lined vessels are said to be *atherosclerotic.*

Because lipids like cholesterol are not soluble in aqueous solutions, cholesterol is carried throughout the body, attached to proteins in structures called lipoproteins. **Low-density lipoproteins (LDLs)** have a high proportion of cholesterol (in other words, they are low in protein). LDLs distribute both the cholesterol synthesized by the liver and the cholesterol derived from diet throughout the body. LDLs are also important for carrying cholesterol to cells, where it is used to help make plasma membranes and hormones. **High-density lipoproteins (HDL)** contain more protein than cholesterol. HDLs scavenge excess cholesterol from the body and return it to the liver, where it is used to make bile. The cholesterol-rich bile is then released into the small intestine, and from there much of it exits the body in the feces. The LDL/HDL ratio is an index of the rate at which cholesterol is leaving body cells and returning to the liver.

Your physician can measure your cholesterol level by determining the amounts of LDL and HDL in your blood. If your total cholesterol level is over 200 or your LDL level is above 100 or so, then your physician may recommend that you decrease the amount of cholesterol and saturated fat in your diet. This may mean eating more plant-based foods and less meat since plants do not have cholesterol, as well as reducing the amount of saturated fats in your diet. Saturated fat is thought to raise cholesterol levels by stimulating the liver to step up its production of LDLs and slowing the rate at which LDLs are cleared from the blood.

Stop and Stretch 3.4: A friend of yours has his cholesterol level checked and tells you that he is really relieved because his cholesterol is 195. What other factors should your friend be considering before he decides his cholesterol level gives him no cause to be concerned?

Cholesterol is not all bad; in fact, some cholesterol is necessary—it is present in cell membranes to help maintain their fluidity, and it is the building block for steroid hormones such as estrogen and testosterone *(Figure 3.17)*. You do, however, synthesize enough cholesterol so that you do not need to obtain much from your diet.

For some people, those with a genetic predisposition to high cholesterol, controlling cholesterol levels through diet is difficult because dietary cholesterol makes up only a fraction of the body's total cholesterol. People with high cholesterol who do not respond to dietary changes may have inherited genes that increase the liver's production of cholesterol. These people may require prescription medications to control their cholesterol levels.

Cholesterol-laden, atherosclerotic vessels increase your risk of heart disease and stroke. Fat deposits narrow your heart's arteries, so less blood can flow to your heart. Diminished blood flow to your heart can cause chest pain, or angina. A complete blockage can lead to a heart attack. Lack of blood flow to the heart during a heart attack can cause the oxygen-starved heart tissue to die, leading to irreversible heart damage.

The same buildup of fatty deposits also occurs in the arteries of the brain. If a blood clot forms in a narrowed artery in the brain, it can completely block blood flow to an area of the brain, resulting in a stroke. If oxygen-starved brain tissue dies, permanent brain damage can result.

Anorexia and Bulimia

Eating disorders that make you underweight cause health problems that are as severe as those caused by too much weight *(Table 3.5)*. **Anorexia,** or self-starvation, is rampant on college campuses. Estimates suggest that 1 in 5 college women and 1 in 20 college men restrict their intake of Calories so severely that they are essentially starving themselves to death. Others allow themselves to eat—sometimes very large amounts of food (called binge eating)—but prevent the nutrients from being turned into fat by purging themselves, often by vomiting. Binge eating followed by purging is called **bulimia.**

Anorexia has serious long-term health consequences. Anorexia can starve heart muscles to the point that altered rhythms develop. Blood flow is reduced,

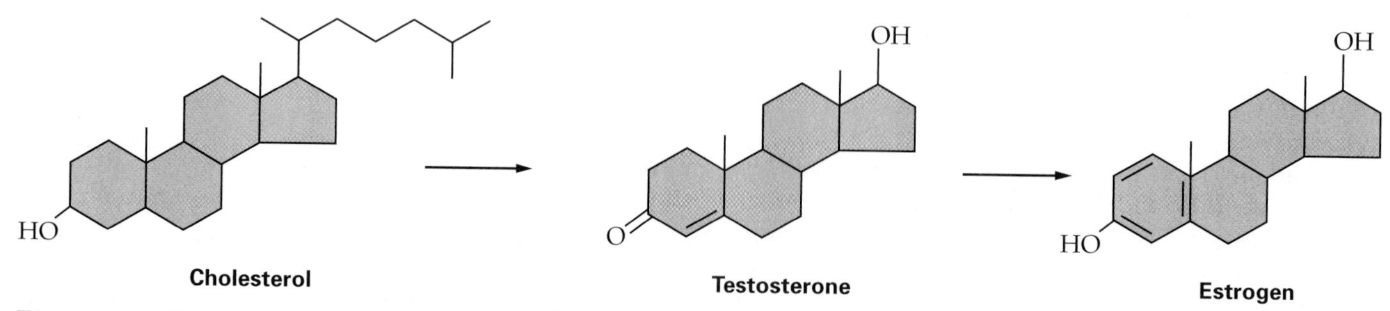

Figure 3.17 **Cholesterol is a building-block molecule.** Cholesterol can serve as a precursor molecule from which both testosterone and estrogen can be manufactured.

TABLE 3.5 **Obesity and anorexia or bulimia.** Health problems result from being either overweight or underweight.

Health Problems Resulting from Obesity	Health Problems Resulting from Anorexia and Bulimia
• Adult-onset diabetes • Hypertension (high blood pressure) • Heart attack • Stroke • Joint problems	• Altered heart rhythms • Amenorrhea (cessation of menstruation) • Osteoporosis (weakened bones) • Ruptured stomach • Dental/gum problems • Dehydration

and blood pressure drops so much that the little nourishment present cannot get to the cells. The lack of body fat accompanying anorexia can also lead to the cessation of menstruation, a condition known as amenorrhea. Amenorrhea occurs when a protein called leptin, which is secreted by fat cells, signals the brain that there is not enough body fat to support a pregnancy. Hormones (such as estrogen) that regulate menstruation are blocked, and menstruation ceases. Amenorrhea can be permanent and causes sterility in a substantial percentage of anorexics.

The damage done by the lack of estrogen is not limited to the reproductive system; bones are affected as well. Estrogen secreted by the ovaries during the menstrual cycle acts on bone cells to help them maintain their strength and size. Anorexics reduce the development of dense bone and put themselves at a much higher risk of breaking their weakened bones, in a condition called **osteoporosis.**

Besides experiencing the same health problems that anorexics face, bulimics can rupture their stomachs through forced vomiting. They often have dental and gum problems caused by stomach acid being forced into their mouths during vomiting, and they can become fatally dehydrated.

Achieving Ideal Weight

As you have seen, the health problems associated with obesity, anorexia, and bulimia are severe. To avoid these problems it is best to focus more on fitness and healthy eating and less on body weight.

As the newly designed USDA Food Guide Pyramid attempts to illustrate *(Figure 3.18)*, working slowly toward being fit and eating healthfully rather than trying the latest fad diet are more realistic and attainable ways to achieve the positive health outcomes that we all desire. In fact, fitness may be more important than body weight in terms of health. Studies show that fit but overweight people have better health outcomes than unfit slender people. In other words, lack of fitness is associated with higher health risks than excess body weight. Therefore, it makes more sense to focus on eating right and exercising than it does to focus on the number on the scale.

Figure 3.18 **USDA Food Guide Pyramid.** The newly designed pyramid stresses the importance of physical activity. The size of each triangle represents the relative proportion of your diet that should be composed of each food group. **Visualize This:** What does the unlabeled yellow triangle represent? Why is this triangle smaller than the others?

Savvy Reader Oxygizer Improves Performance?

The Savvy Reader feature in Chapter 2 introduced you to the oxygenated water beverage Oxygizer. In addition to making many other claims, the author of the newspaper article wrote that drinking Oxygizer would "assist the respiratory system so you recover better from exercise." The following is an excerpt from the website of the company that produces Oxygizer: "Oxygizer improves performance during periods of high physical stress and the resulting regenerative phase. Univ. Prof. Dr. Wolfgang Marktl (Head of Science at the Institute of Medical Physiology at Vienna University) and his research team have completed their scientific tests. Using a randomised double-blind study, these tests have proven the effective influence and effect of Oxygizer on the body's performance capability."

This is pretty compelling writing and may convince some to purchase this oxygenated water. However, let's also look at an excerpt from the actual scientific study performed by Dr. Marktl and published in the *International Journal of Sports Medicine* in March 2006. "Results showed no significant influence on aerobic parameters or lactate metabolism, neither at submaximal nor at maximal levels. We conclude that the consumption of oxygenated water does not enhance aerobic performance."

1. Does it appear that the author of the newspaper article read the actual study or the promotional material only?
2. How are claims made in the newspaper and on websites different from claims made by authors of articles published in scientific journals?
3. The Oxygizer website also includes some data (http://www.oxygizer.com/default.aspx?lngId=2) that seem to support their claims. Private companies can hire their own scientists to perform studies that often have results that differ from those of government and university-sponsored scientists. Would you be more skeptical of results produced by scientists hired by the company whose product they are testing or scientists who work for the government or a University?
4. Carefully consider the following two sentences from the Oxygizer website: "Univ. Prof. Dr. Wolfgang Marktl (Head of Science at the Institute of Medical Physiology at Vienna University) and his research team have completed their scientific tests. Using a randomised double-blind study, these tests have proven the effective influence and effect of Oxygizer on the body's performance capability." Each of these sentences, read separately, is true. Dr. Marktl and his team did complete their tests, and the Oxygizer scientists did produce data showing increased performance capability. However, placed adjacent to each other, these sentences seem to be indicating that Dr. Marktl's university-sponsored research came up with results that were actually produced by the Oxygizer scientists. Do you think this is a willful attempt to deceive potential customers? Most people don't have time to do such a thorough analysis of every newspaper article they read. This is why it is helpful to develop a general level of skepticism about most product claims.

CHAPTER Review

For study help, animations, and more quiz questions go to www.mybiology.com.

Summary

3.1 Nutrients

- Nutrients provide structural units and energy for cells (p. 54).
- Water is an important dietary constituent that helps dissolve and eliminate wastes and maintain blood pressure and body temperature (p. 54).
- Macronutrients are required in large amounts for proper growth and development. Macronutrients include carbohydrates, proteins, and fats. All of these molecules are composed of subunits that can be broken down for use by the cell (pp. 54–58).
- Micronutrients are dietary substances required in minute amounts for proper growth and development; they include vitamins and minerals (p. 58).
- Vitamins are organic substances, most of which the body cannot synthesize. Many vitamins serve as coenzymes to help enzymes function properly (pp. 58–60).
- Minerals are inorganic substances essential for many cell functions (p. 60).
- Processing foods decreases their nutritive value (pp. 60–61).

3.2 Enzymes and Metabolism

- Metabolic reactions include all the chemical reactions that occur in cells to build up or break down macromolecules (p. 63).
- Metabolism is governed by enzymes. Enzymes are proteins that catalyze specific cellular reactions, first by binding the substrate to the enzyme's active site. This binding causes the enzyme to change shape (induced fit), placing stress on the bonds of the substrate and thereby lowering the activation energy (pp. 63–64).
- Energy is measured in units called Calories (pp. 64–65).
- An individual's metabolic rate is affected by many factors, including age, sex, exercise level, body weight, and genetics (p. 65).
- The energy stored in the chemical bonds of food can be released by metabolic reactions and stored in the bonds of ATP. Cells use ATP to power energy-requiring processes (pp. 65–66)

web animation 3.1 *Enzymes*

3.3 Transport Across Membranes

- To gain access to cells, nutrients move across the plasma membrane, which functions as a semipermeable barrier that allows some substances to pass and prevents others from crossing (p. 66).
- The plasma membrane is composed of two layers of phospholipids, in which are embedded proteins and cholesterol (p. 66).
- Passive transport mechanisms include simple diffusion and facilitated diffusion (diffusion through proteins). Passive transport always moves substances with their concentration gradient and does not require energy (pp. 66–67).
  - Osmosis, the diffusion of water across a membrane, can involve the movement of water through protein pores in the membrane (pp. 67–68).
- Active transport is an energy-requiring process that requires proteins in cell membranes to move substances against their concentration gradients (pp. 67–68).
- Larger molecules move into and out of cells enclosed in membrane-bound vesicles (p. 68).

web animation 3.2 *Passive and Active Transport*

web animation 3.3 *Exocytosis and Endocytosis*

Summary (continued)

3.4 Body Fat and Health

- Determining ideal body weight is difficult with conventional methods (pp. 70–71).
- Obesity is associated with many health problems, including hypertension, heart attack and stroke, diabetes, and joint problems (pp. 70–74).
- Anorexia and bulimia are very common on college campuses and result in serious long-term health problems (pp. 74–75).

Roots to Remember

The following roots of words come mainly from Latin and Greek and will help you to decipher terms:

-ase is a common suffix in names of enzymes.
endo- means inside.
exo- means outside.
iso- means the same or equal.
lipo- refers to fat or lipid.
osmo- refers to water.
osteo- refers to bone.

Learning the Basics

1. Define the term *metabolism.*

2. List three common cellular substances that can pass through cell membranes unaided.

3. Macronutrients ______________.
A. include carbohydrates and vitamins; B. should comprise a small percentage of a healthful diet; C. are essential in minute amounts to help enzymes function; D. include carbohydrates, fats, and proteins; E. are synthesized by cells and not necessary to obtain from the diet

4. The function of low-density lipoproteins (LDLs) is to ______________.
A. break down proteins; B. digest starch; C. transport cholesterol from the liver; D. carry carbohydrates into the urine

5. Which of the following is a *false* statement regarding enzymes?
A. Enzymes are proteins that speed up metabolic reactions; B. Enzymes have specific substrates; C. Enzymes supply ATP to their substrates; D. An enzyme may be used many times.

6. Enzymes speed up chemical reactions by ______________.
A. heating cells; B. binding to substrates and placing stress on their bonds; C. changing the shape of the cell; D. supplying energy to the substrate

7. A substance moving across a membrane against a concentration gradient is moving by ______________.
A. passive transport; B. osmosis; C. facilitated diffusion; D. active transport; E. diffusion

8. A cell that is placed in salty seawater will ______________.
A. take sodium and chloride ions in by diffusion; B. move water out of the cell by active transport; C. use facilitated diffusion to break apart the sodium and chloride ions; D. lose water to the outside of the cell via osmosis

9. Which of the following forms of membrane transport require specific membrane proteins?
A. diffusion; B. exocytosis; C. facilitated diffusion; D. active transport; E. facilitated diffusion and active transport

10. Water crosses cell membranes .
A. by active transport; B. through protein channels called aquaporins; C. against its concentration gradient; D. in plant cells but not in animal cells

Analyzing and Applying the Basics

1. A friend of yours does not want to eat meat, so instead she consumes protein powders that she buys at a nutrition store. What would be the disadvantages of this practice?

2. Two people with very similar diets and similar exercise levels have very different amounts of body fat. Why might this be the case?

3. What would you say to a friend who believes that he is fat even though his BMI places him in the "normal" range? How about a friend who qualifies as obese on a BMI chart but who exercises regularly and eats a well-balanced diet?

Connecting the Science

1. Why do you think that anorexia and bulimia are more common among women than men?

2. The New York City Board of Health recently adopted the nation's first major municipal ban on the use of trans fats in restaurant cooking. Do you think state health regulators should also ban trans fats in commercially prepared/ packaged food like cookies and potato chips?

CHAPTER 4

Life in the Greenhouse

Photosynthesis, Cellular Respiration, and Global Warming

Sea levels are rising, threatening the survival of low-lying coastal cities like New Orleans.

Global warming is causing glaciers to melt, raising sea levels.

Melting ice endangers the habitat of polar bears.

In August 2005, Hurricane Katrina came onshore along the Gulf Coast of the United States. Much of the low-lying coast and the city of New Orleans, Louisiana, were in Katrina's path. Shortly after the hurricane, water levels rose and levees broke, swamping the city and stranding many residents. This disaster, many say, could have been averted with a stronger levee system and better planning. However, New Orleans lies below sea level. Its future, like that of low-lying islands and coastal communities around the world, is threatened by global warming.

Global warming has many effects, including higher temperatures, rising sea levels, and more intense storms. Sea levels have risen by 10 to 20 cm (4 to 8 inches) in the twentieth century. Worldwide rain and snowfall over land has increased as well. Many scientists predict that as water temperatures increase, storm intensity will increase as well, resulting in more and more hurricanes and catastrophic flooding akin to what we saw in New Orleans in 2005.

This is happening in part because ice is melting and falling into the ocean. In Antarctica, rising temperatures have led to the collapse of massive ice shelves. In recent years, two massive chunks of ice, each about the size of Rhode Island, have fallen into the ocean. The Greenland ice sheet is becoming thinner at its margins every year.

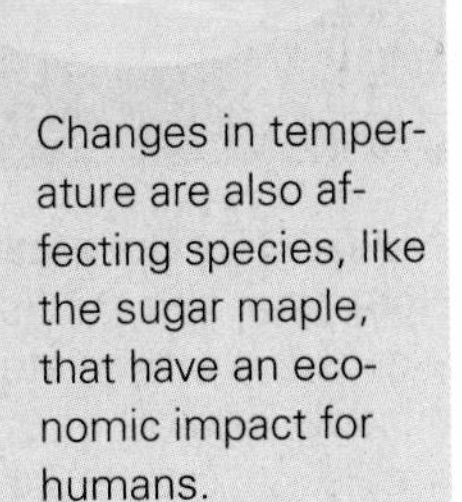

Changes in temperature are also affecting species, like the sugar maple, that have an economic impact for humans.

Glaciers are melting worldwide. The glaciers in Glacier National Park, located in the northwest corner of Montana, decrease in size and number every year. As the glaciers shrink, they take with them natural habitat set aside for protection in this national park. Some of the park's glaciers have already shrunk to half their original size, and the total number of glaciers has decreased from approximately 150 in 1850 to around 35 today.

Like ice masses all over the world, these glaciers are slowly succumbing to warmer temperatures. According to the U.S. National Climate Data Center, the entire planet has warmed by 0.25°C (0.5°F) each decade during the twentieth century. If this trend continues, scientists predict that by the year 2030 not a single glacier will be left in Glacier National Park.

A review published in the journal *Nature* described various species that have been affected by climate change. Many of these species are temperature sensitive, and they must move closer to the poles or to higher elevations to find regions with the proper climate. Arctic foxes are retreating northward and being replaced by the less-cold-hardy red fox. Edith's checkerspot butterfly is now found 124 m higher in elevation and 92 km north of its range in 1900, and a wide variety of corals have experienced a dramatic increase in the frequency and extent of damage resulting from increased ocean temperatures. Shrinking polar ice is reducing the habitat of polar bears, which may soon become an endangered species.

Plant species are affected also. New England risks losing its profitable maple syrup industry along with its leaf-watching tourists as the cool-weather-adapted sugar maple population declines in a warming climate. Turning the maples' sugar into syrup requires nighttime temperatures that are below freezing and daytime temperatures in the mid-40s. Warmer temperatures overall have led to tapping seasons that start earlier, end sooner, and produce syrup of a lesser quality. A report by the U.S. Office of Science and Technology Policy indicated that the ideal range of the sugar maple is now close to 300 miles north of New England.

While the public seems to believe that there is debate among scientists and government-appointed panels about global warming, that is not really the case. Scientists who publish in peer-reviewed journals, the Intergovernmental Panel on Climate Control (IPCC), National Academy of Sciences, and the American Association for the Advancement of Science (AAAS) all agree that climate is warming and that most of the warming observed recently is attributable to human activities.

Why is the climate changing, and what can we do to help mitigate the problems the future is likely to bring?

4.1 The Greenhouse Effect

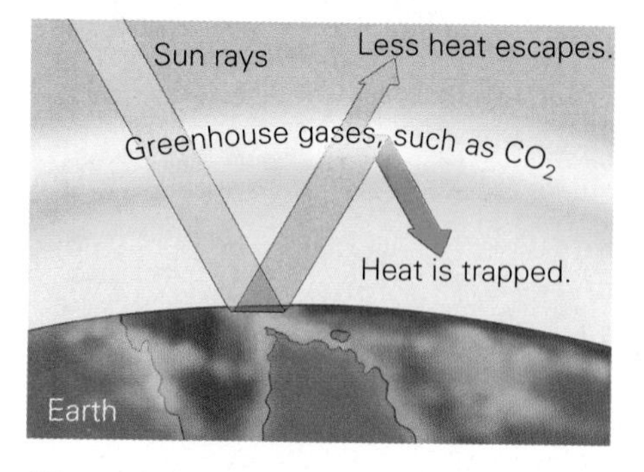

Figure 4.1 **The greenhouse effect.** Heat from the sun is trapped in the atmosphere by water vapor, carbon dioxide, and other greenhouse gases. **Visualize This:** What effect would increased levels of carbon dioxide have on the greenhouse effect?

Global warming is the progressive increase of Earth's average temperature that has been occurring over the past century. Scientists agree that global warming is caused by recent increases in the concentrations of particular atmospheric gases, including methane, nitrous oxide, water vapor, and carbon dioxide. These increases are on top of the natural increases that have been occurring since the last ice age. Because increases in carbon dioxide seem to be the major source of problems related to global warming, we focus mainly on that gas for the rest of this discussion.

The presence of carbon dioxide in the atmosphere leads to a phenomenon called the **greenhouse effect.** The greenhouse effect works like this: Warmth from the sun heats Earth's surface, which then radiates the heat energy absorbed outward. Most of this heat is radiated back into space, but some of the heat is retained in the atmosphere. The retention of heat is facilitated by carbon dioxide molecules, which act like a blanket to trap the heat radiated by Earth's surface *(Figure 4.1)*. When you sleep under a blanket at night, your body heat is trapped and helps keep you warm. When the levels of greenhouse gases in the atmosphere increase, the effect is similar to sleeping under too many blankets—the temperature increases. The trapping of this warmth radiating from Earth is known as the greenhouse effect.

This is not exactly how panes of glass in a greenhouse function, that is, by allowing radiation from the sun to penetrate into the greenhouse and then trapping the heat that radiates from the warmed-up surfaces inside the greenhouse. But the overall effect is the same—the air temperature increases.

The greenhouse effect is not in itself a dangerous or unhealthy phenomenon. If Earth's atmosphere did not have some greenhouse gases, too much heat would be lost to space, and Earth would be too cold to support life. It is the excess warming due to more and more carbon dioxide accumulating in the atmosphere as a result of coal, oil, and natural gas burning that is causing problems.

Stop and Stretch 4.1: The moon's average daytime temperature is 107°C (225°F), and its average nighttime temperature is −153°C (−243°F). But the moon and Earth are the same distance from the sun. Think about the carbon dioxide blanket on Earth to come up with a hypothesis for why temperatures on the moon might fluctuate so dramatically.

In the absence of excess greenhouse gases, water vapor and carbon dioxide work together to keep temperatures on Earth hospitable for life.

Water, Heat, and Temperature

Bodies of water absorb heat and help maintain stable temperatures on Earth. You have probably noticed that when you heat water on a stove, the metal pot becomes hot before the water. This is because water heats more slowly than metal and has a stronger resistance to temperature changes than most substances.

Heat and temperature are measures of energy. **Heat** is the total amount of energy associated with the movement of atoms and molecules in a substance. **Temperature** is a measure of the intensity of heat—for example, how fast the molecules in the substance are moving. When you are swimming in a cool lake your body has a higher temperature than the water but the lake contains more heat because of its large volume of water.

Water molecules are attracted to each other, resulting in the formation of hydrogen bonds between neighboring molecules. When water is heated, the heat energy disrupts the hydrogen bonds. Water remains in the liquid form because not all the hydrogen bonds are broken at any one time. Only after the hydrogen bonds have been broken can heat cause individual water molecules to move faster, thus increasing the temperature. In other words, the initial input of heat used to break hydrogen bonds between water molecules does not immediately raise the temperature of water; instead, it breaks hydrogen bonds. Therefore, water can absorb and store a large amount of heat while warming up only a few degrees in temperature. When water cools, hydrogen bonds reform between adjacent molecules, releasing heat into the atmosphere. Water can release a large amount of heat into the surroundings while not decreasing the temperature of the body of water very much *(Figure 4.2)*.

Water's high heat-absorbing capacity has important effects on Earth's climate. The vast amount of water contained in Earth's oceans and lakes moderates temperatures by storing huge amounts of heat radiated by the sun and giving off heat that warms the air during cooler times. Therefore, the balance between releasing and maintaining heat energy is vital to the maintenance of

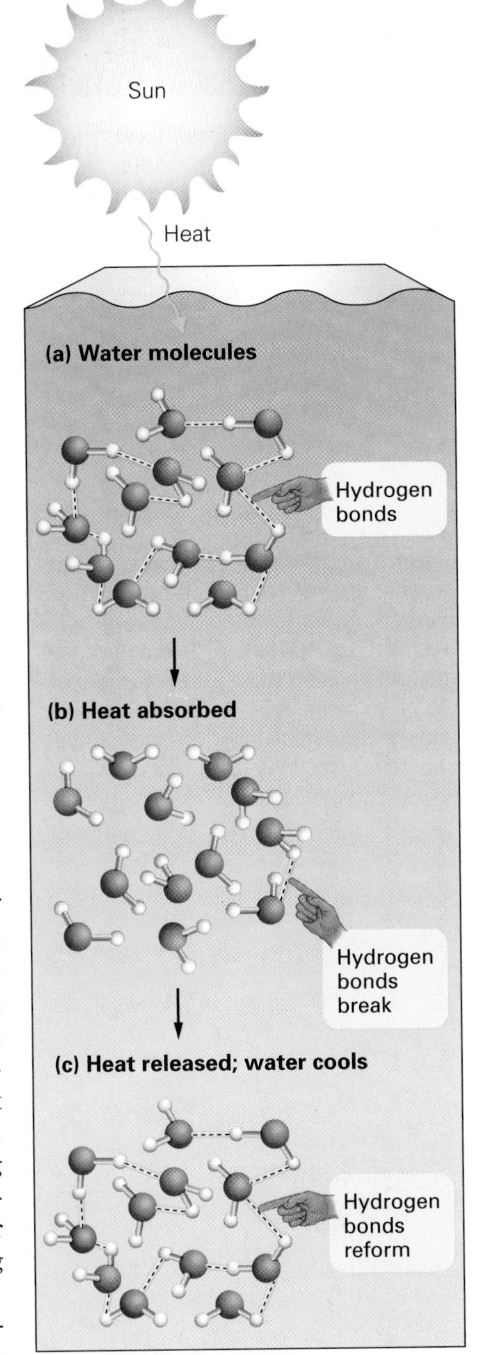

Figure 4.2 **Hydrogen bonding in water.** Hydrogen bonds break as they absorb heat and reform as water releases heat.

climate conditions on Earth. This balance can be disrupted when increasing levels of carbon dioxide cause more heat to be trapped. Carbon dioxide can come from many different sources.

Figure 4.4 **Burning fossil fuels.** The burning of fossil fuels by industrial plants and automobiles adds more carbon dioxide to the environment.

4.2 The Flow of Carbon

Many of the atoms found in complex molecules of living organisms are broken down into simpler molecules and recycled for use in different capacities. Carbon dioxide (CO_2) is no different. The carbon dioxide you exhale is released into the atmosphere, where it can absorb heat, diffuse into the oceans to be absorbed by plants and animals that live there, or be absorbed by forests and soil. Volcanic eruptions return carbon dioxide trapped within Earth's surface to the atmosphere. As you can see in *Figure 4.3*, carbon cycles back and forth between living organisms, the atmosphere, bodies of water, and soil.

The ocean has served as Earth's largest carbon dioxide and heat reservoir, but oceanic and atmospheric scientists are very concerned about the ocean's ability to absorb carbon dioxide at the rate that it is being emitted into the atmosphere. This is because human activities have rapidly increased the rate of carbon release into the atmosphere, largely by burning fossil fuels *(Figure 4.4)*.

Fossil fuels are the buried remains of ancient plants and microorganisms that have been transformed by heat and pressure into coal, oil, and natural gas. These fuels are rich in carbon because plants remove carbon from the atmosphere during photosynthesis; consequently, plant structures are rich in organic carbon. Dead plant materials that are buried before they decompose, and thus before their carbon is released in the form of carbon dioxide, can produce fossil fuels. Humans combust this stored organic carbon to produce energy. The plants that made up the majority of fossil fuels lived from 290 to 362 million years ago, during a geological period called the Carboniferous Period.

Burning these fossil fuels to generate electricity, power our cars, and heat our homes releases carbon dioxide into the atmosphere. Increases in carbon dioxide are well documented by direct measurements of the atmosphere over the past 50 years *(Figure 4.5)*.

Scientists can also directly measure the amount of carbon dioxide that was present in the atmosphere in the past by examining cores of ice sheets that have existed for thousands of years. This is because snow near the surface of ice

Figure 4.3 **The flow of carbon.** Living organisms and volcanoes produce CO_2 Forests, oceans, and soil absorb CO_2 from the air. **Visualize This:** Based on the observation that carbon dioxide levels have increased in the last 100 years, and that volcanic activity has remained constant, which would you predict releases more carbon dioxide into the air: volcanic or human activity?

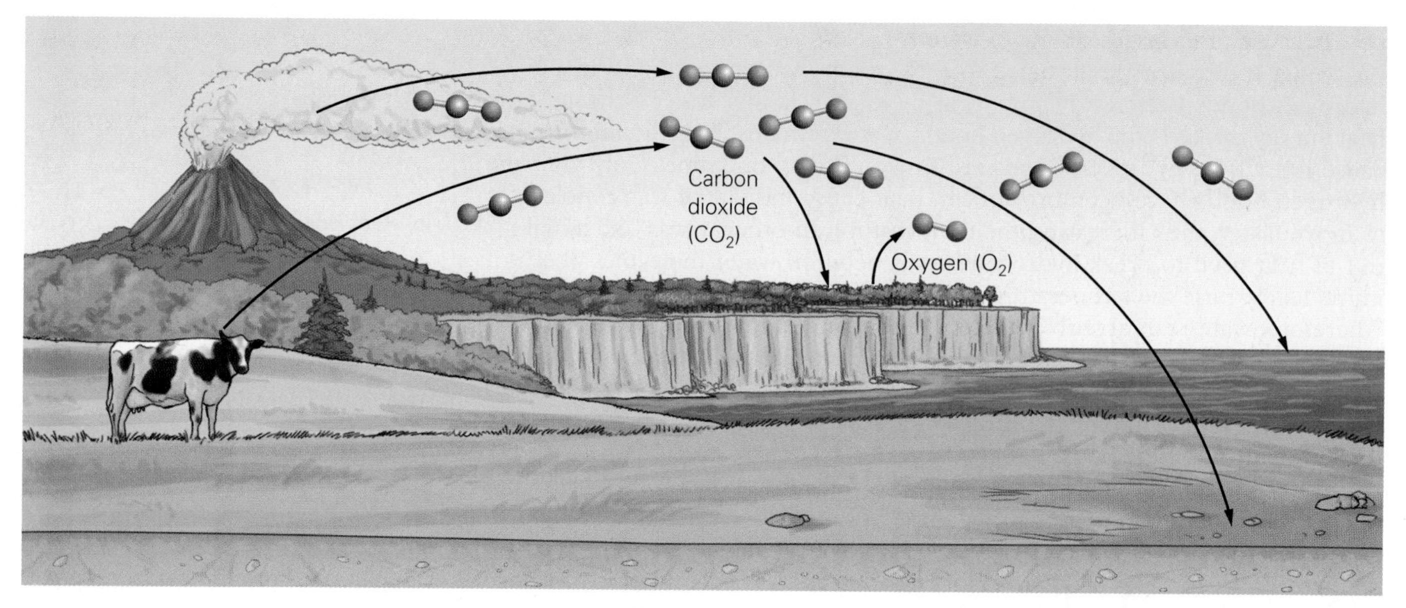

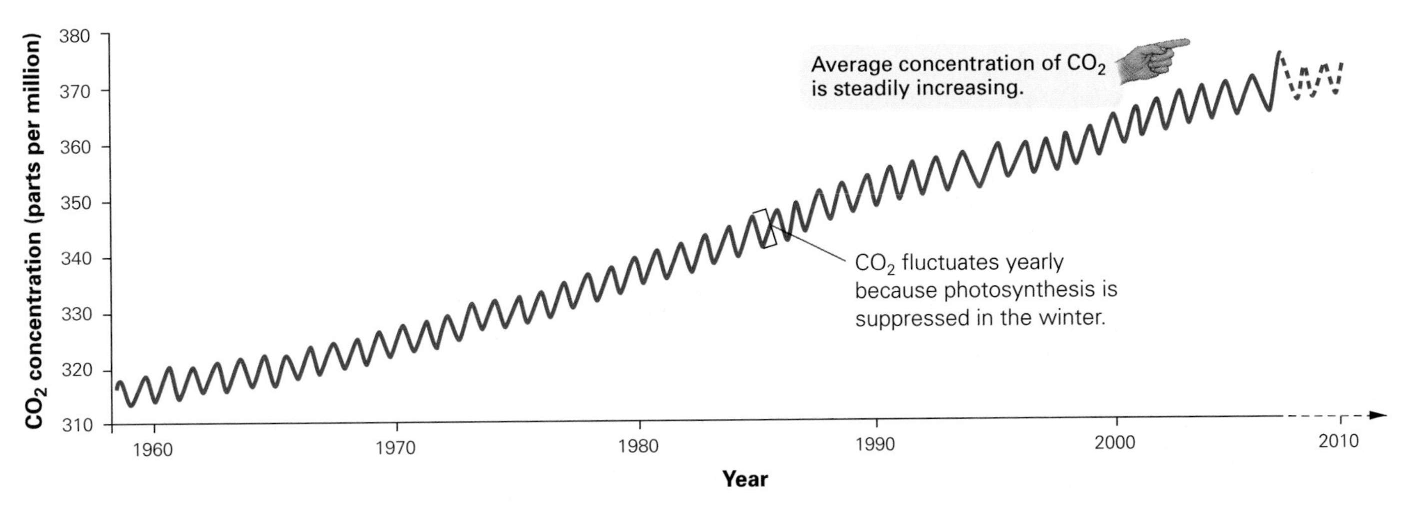

Figure 4.5 **Increases in atmospheric carbon dioxide.** Carbon dioxide levels are increasing. **Visualize This:** Compare the rate of increase from 1960 to 1965 with the rate of increase from 2000 to 2005. Did carbon dioxide rates increase more rapidly in the early 1960s or the early 2000s?

traps air. As more snow accumulates, the underlying snow is compressed into ice that contains air bubbles. Cores can be removed from long-lived ice sheets and analyzed to determine the concentration of carbon dioxide trapped in air bubbles. These bubbles are actual samples of the atmosphere from up to hundreds of thousands of years ago *(Figure 4.6)*. In addition, certain characteristics of gases trapped in the bubbles of ice cores can provide indirect information about temperatures at the time the bubbles formed. Ice core data from Antarctica *(Figure 4.7)* indicate that the concentration of carbon dioxide in the atmosphere is much higher now than at any time in the past 400,000 years and that increased levels of carbon dioxide are correlated with increased temperatures.

Although Earth has gone through temperature cycles many times in the past, human activities are inflating the rate of increase, and these increases may persist for thousands of years. It is already clear that when carbon dioxide levels increase, so do temperatures *(Figure 4.8)*.

Reducing the biological, economic, and social losses caused by global warming will require not only slowing the rate of warming but also mediating the effects of increasing temperatures that are inevitable given current atmospheric carbon dioxide levels.

Figure 4.6 **Ice core.** By analyzing ice cores, scientists can measure the concentration of carbon dioxide that was present in early atmospheres.

Stop and Stretch 4.2: Even if humans stopped releasing carbon dioxide from fossil fuels today, high levels of the gas would persist in the atmosphere for hundreds of years. Given the movement of carbon within the carbon cycle, explain why it will take so long for CO_2 levels to decline.

Predictions for future conditions are based on climate models, which take into account many different factors (solar radiation, ocean currents, precipitation, and cloud cover, for example). One challenge in developing these models is understanding climate feedbacks. Negative feedbacks are feedbacks that negate change. Higher temperatures leading to more cloud production and reduction of warming is a negative-feedback mechanism. Positive feedbacks promote change. For example, melting polar ice decreases the reflectiveness of the ocean, which can make warming more dramatic.

Another example of positive feedback occurs when temperatures increase. Chemical reactions, including those taking place within living organisms,

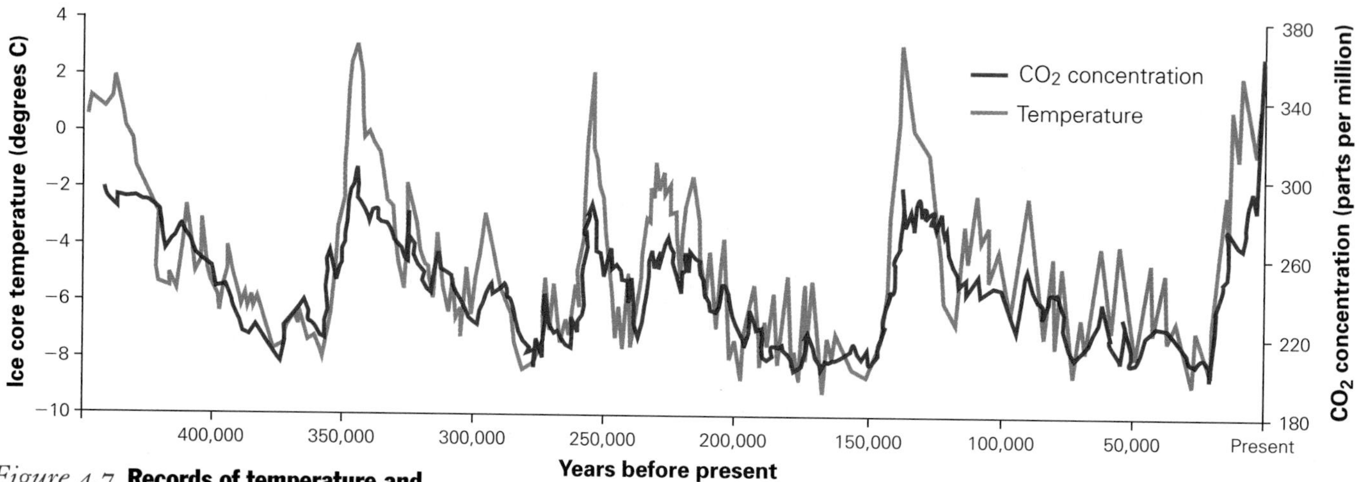

Figure 4.7 **Records of temperature and atmospheric carbon dioxide concentration from Antarctic ice cores.** These data indicate that increases in carbon dioxide levels are correlated with higher temperatures.

generally happen more rapidly at higher temperatures. One reaction in particular, cellular respiration, produces carbon dioxide as a by-product. Therefore, when metabolism increases, the production of greenhouse gases increases.

4.3 Cellular Respiration

Increasing temperatures can change an organism's energy needs and can affect how rapidly it grows, develops, and reproduces. For some organisms, increasing energy needs associated with higher temperatures can cause them to be outcompeted for resources by other organisms, ultimately requiring species to move toward the poles or higher elevations. For other organisms, higher temperatures allow them to go through their life cycles more rapidly, leading to increased populations. In both cases, the process that causes these effects is cellular respiration.

The main function of cellular respiration is to convert the energy stored in chemical bonds of food into energy cells can use. Energy is stored in the electrons of chemical bonds, and when bonds are broken in a three-step process, **adenosine triphosphate,** or **ATP,** is produced. ATP can supply energy to cells because it stores energy obtained from the movement of electrons that originated in food into its own bonds. Before trying to understand cellular respiration, it is important to have a better understanding of this chemical.

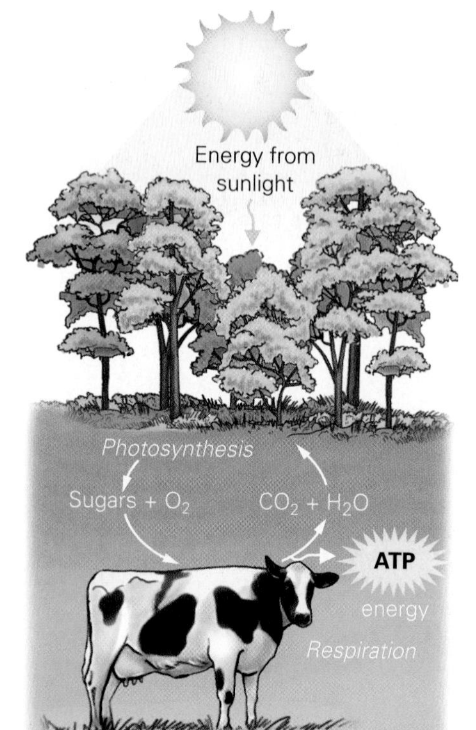

Figure 4.8 **The flow of chemicals and energy.** Energy enters biological systems in the form of sunlight which drives the synthesis of sugars. Sugars are broken down to produce energy in cells.

Structure and Function of ATP

Structurally, ATP is a nucleotide triphosphate. It contains the nitrogenous base adenine, a sugar, and three phosphates *(Figure 4.9)*. Each phosphate in the series of three is negatively charged. These negative charges repel each other, which contributes to the stored energy in this molecule.

Removal of the terminal phosphate group of ATP releases energy that can be used to perform cellular work. After removal of a phosphate group ATP becomes adenosine diphosphate (ADP). In this manner, ATP behaves much like a coiled spring. To think about this, imagine loading a dart gun. Pushing the dart into the gun requires energy from your arm muscles, and the energy you exert will be stored in the coiled spring inside the dart gun *(Figure 4.10)*. When you shoot the dart gun, the energy is released from the gun and used to perform some work—in this case, sending a dart through the air.

The phosphate group that is removed from ATP can be transferred to another molecule. Thus, ATP can energize other compounds through **phosphorylation,** which means that it transfers a phosphate to another molecule. You can think of the donated phosphate as a little bag of energy. When a mol-

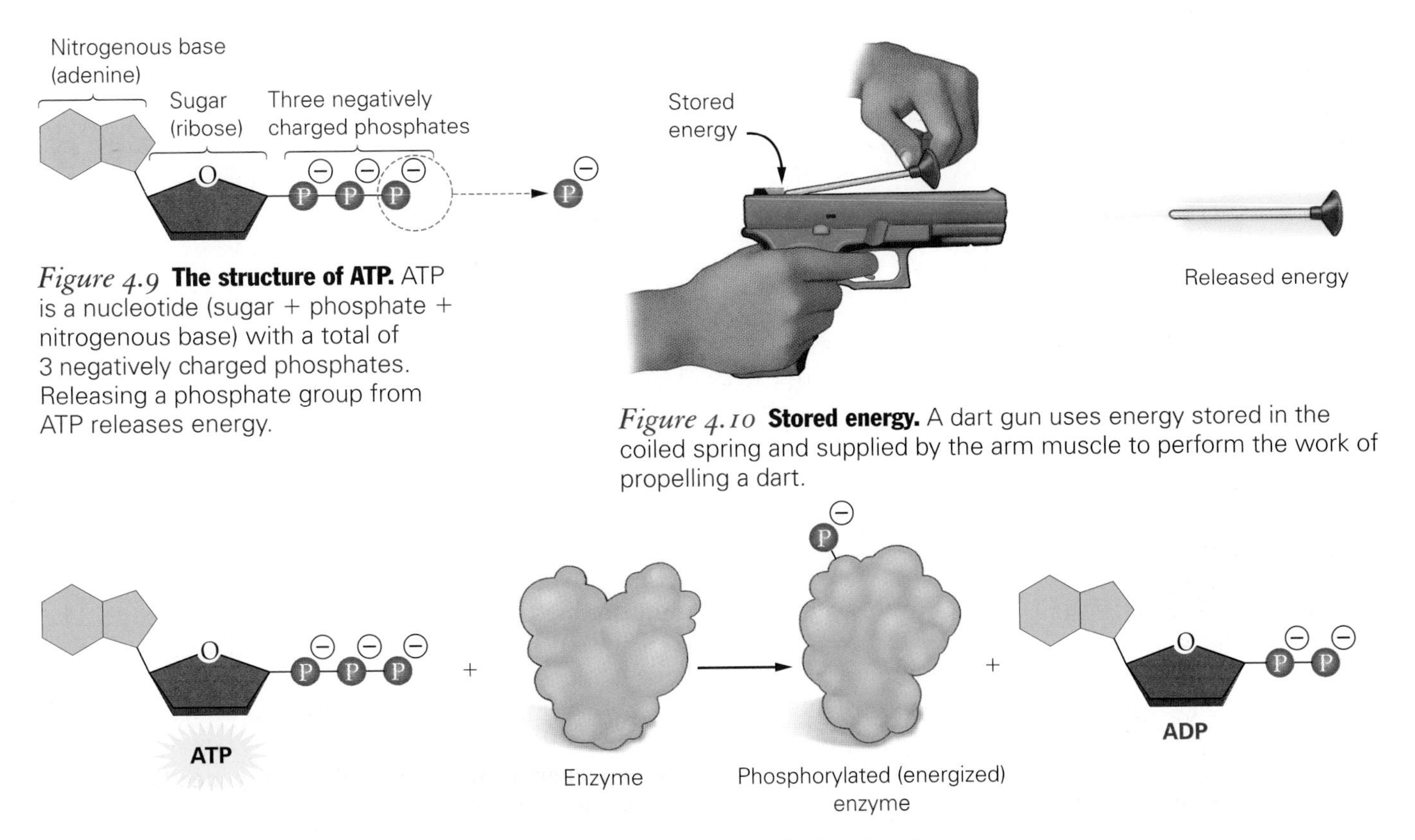

Figure 4.9 **The structure of ATP.** ATP is a nucleotide (sugar + phosphate + nitrogenous base) with a total of 3 negatively charged phosphates. Releasing a phosphate group from ATP releases energy.

Figure 4.10 **Stored energy.** A dart gun uses energy stored in the coiled spring and supplied by the arm muscle to perform the work of propelling a dart.

Figure 4.11 **Phosphorylation.** The terminal phosphate group of an ATP molecule can be transferred to another molecule, in this case an enzyme, to energize it. When ATP loses a phosphate, it becomes ADP.

ecule, say an enzyme, needs energy, the phosphate group is transferred from ATP to the enzyme, and the enzyme now has the energy it needs to perform its job *(Figure 4.11)*. The energy released by the removal of the outermost phosphate of ATP can be used to help cells perform many different kinds of work. ATP helps power mechanical work such as the movement of proteins in muscles, transport work such as the movement of substances across membranes during active transport, and chemical work such as the making of complex molecules from simpler ones *(Figure 4.12)*.

Cells are continuously using ATP. Exhausting the supply of ATP means that more ATP must be regenerated. ATP is synthesized by adding back a phosphate group to ADP during the process of cellular respiration *(Figure 4.13)*.

Figure 4.12 **ATP and cellular work.** ATP powers (a) mechanical work, such as the moving of the flagella of this single-celled green algae of the **Chlamydomonas species;** (b) transport work, such as the active transport of a substance across a membrane from its own low to high concentration; and (c) chemical work, such as the enzymatic conversion of substrates to a product.

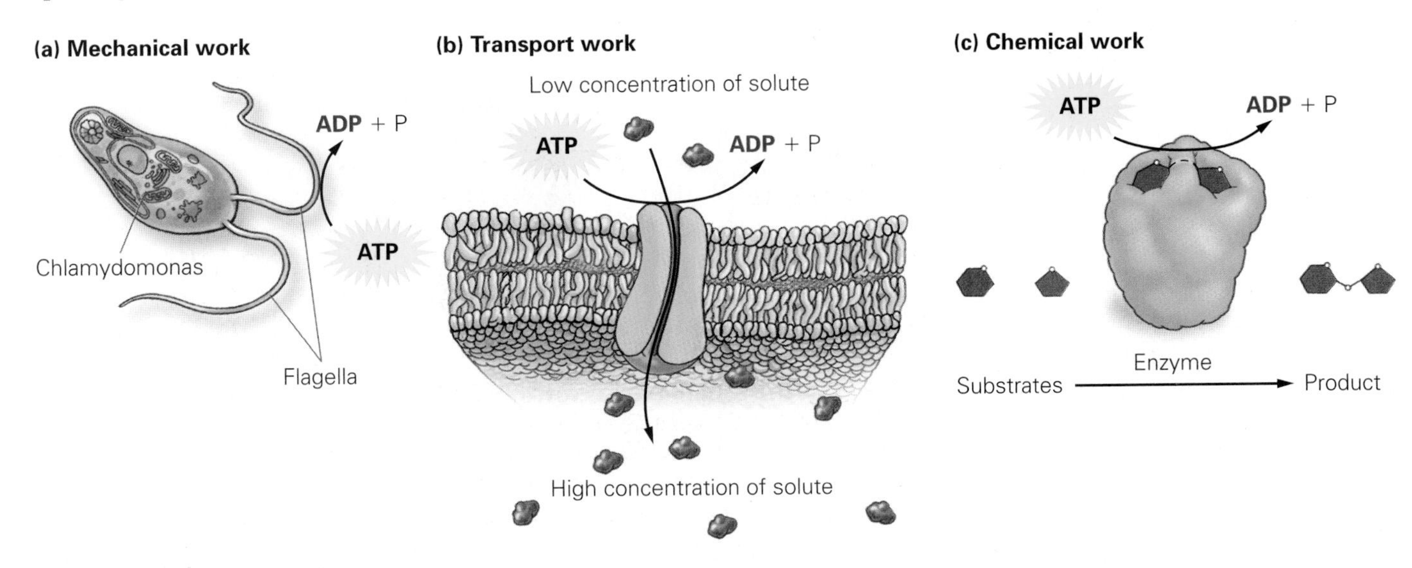

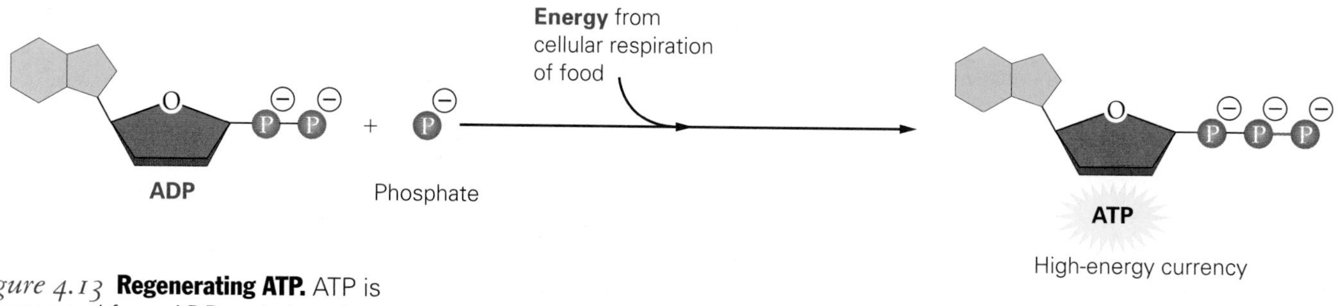

Energy from cellular respiration of food

ADP

Phosphate

ATP

High-energy currency

Figure 4.13 **Regenerating ATP.** ATP is regenerated from ADP and phosphate during the process of cellular respiration.

During this process, cells produce carbon dioxide and use oxygen to produce water. Because some of the steps in cellular respiration require oxygen, they are said to be **aerobic** reactions, and this type of cellular respiration is called **aerobic respiration.**

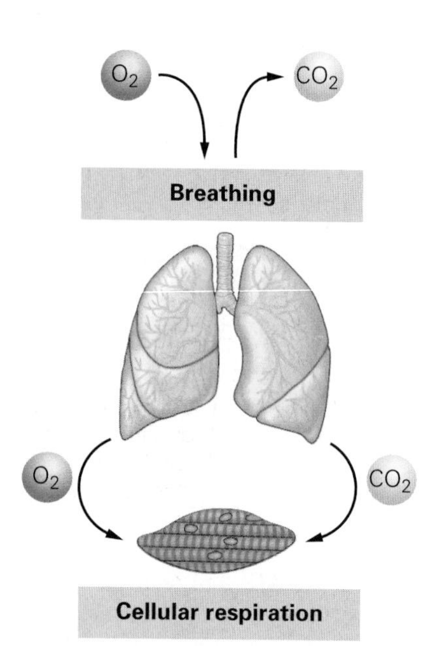

Figure 4.14 **Breathing and cellular respiration.** (a) When you inhale, you bring oxygen from the atmosphere into your lungs. This oxygen is delivered through the bloodstream to tissues that use it to drive cellular respiration. (b) The carbon dioxide produced by cellular respiration is released from cells and diffuses into the blood and to the lungs. Carbon dioxide is released from the lungs when you exhale.

An Overview of Cellular Respiration

The word respiration can also be used to describe breathing. When we breathe, we take oxygen in through our lungs and expel carbon dioxide. The oxygen we breathe in is delivered to cells, which undergo cellular respiration and release carbon dioxide *(Figure 4.14)*.

Most foods can be broken down to produce ATP as they are routed through this process. Carbohydrate metabolism begins at the beginning of the pathway, while proteins and fats enter at later points.

The equation for carbohydrate breakdown is

$$C_6H_{12}O_6 + 6\,O_2 \longrightarrow 6\,CO_2 + 6\,H_2O$$

$$\text{Glucose} + \text{Oxygen} \longrightarrow \text{Carbon dioxide} + \text{Water}$$

Glucose is an energy-rich sugar, but the products of its digestion—carbon dioxide and water—are energy poor. So where does the energy go? The energy released during the conversion of glucose to carbon dioxide and water is used to synthesize ATP. Many of the chemical reactions in this process occur in the mitochondria, organelles that are found in both plant and animal cells, through a series of complex reactions that break apart the glucose molecule. In doing so, the carbons and oxygens that make up the original glucose molecule are released from the cell as carbon dioxide. Hydrogens from glucose combine with oxygen to produce water *(Figure 4.15)*. Gaining an appreciation for *how* this happens requires a more in-depth look.

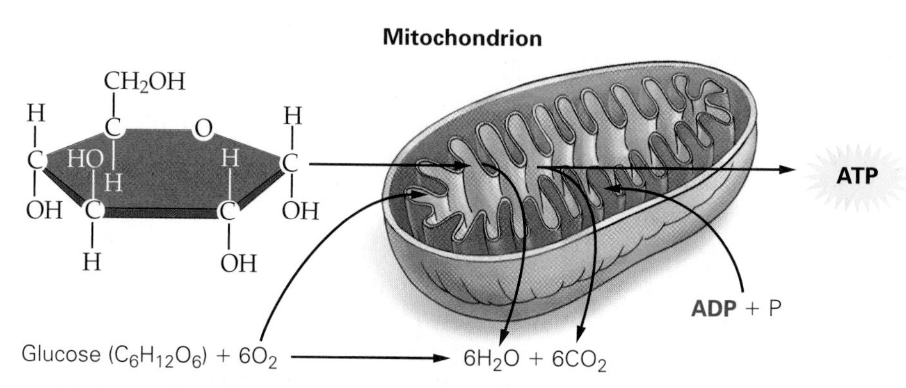

Figure 4.15 **Overview of cellular respiration.** The breakdown of glucose by cellular respiration requires oxygen and ADP plus phosphate. The energy stored in the bonds of glucose is harvested to produce ATP (from ADP and P), releasing carbon dioxide and water.

A Closer Look: Glycolysis, the Citric Acid Cycle, Electron Transport, and ATP Synthesis

Cellular respiration occurs in three steps: glycolysis, the citric acid cycle, and the electron transport chain *(Figure 4.16)*.

Step 1: Glycolysis. To harvest energy from glucose, the 6-carbon glucose molecule is first broken down into two 3-carbon **pyruvic acid** molecules *(Figure 4.17)*. This part of the process of cellular respiration actually occurs outside of any organelle, in the fluid cytosol. Glycolysis does not require oxygen but does produce 2 molecules of ATP. After glycolysis, the pyruvic acid is decarboxylated (loses a carbon dioxide molecule), and the 2-carbon fragment that is left is further metabolized inside the mitochondria.

Mitochondria are surrounded by an inner and an outer membrane. The space between the two membranes is called the **intermembrane space.** The semifluid medium inside the mitochondrion is called the matrix *(Figure 4.18)*.

Once inside the mitochondrion, the energy stored in the bonds of pyruvic acid is converted into the energy stored in the bonds of ATP. The first step of this conversion is called the citric acid cycle.

Step 2: The citric acid cycle. The citric acid cycle removes electrons from the carbon-containing compounds it receives from glycolysis because these electrons can be used later to generate ATP. The citric acid cycle is a series of reactions catalyzed by eight different enzymes, located in the matrix of each mitochondrion. This cycle breaks down the remains of a carbohydrate, harvesting its electrons and releasing carbon dioxide into the atmosphere *(Figure 4.19)*. These reactions are a cycle because every trip around the pathway regenerates the first reactant in the cycle. Therefore, the first reactant in the cycle, a 4-carbon molecule called oxaloacetate (OAA), is always available to react with carbohydrate fragments entering the citric acid cycle.

In addition to removing carbon dioxide, the citric acid cycle removes electrons for use in producing ATP during the final step of cellular respiration. These electrons do not simply float around in a cell; this would damage the cell. Instead, they are carried

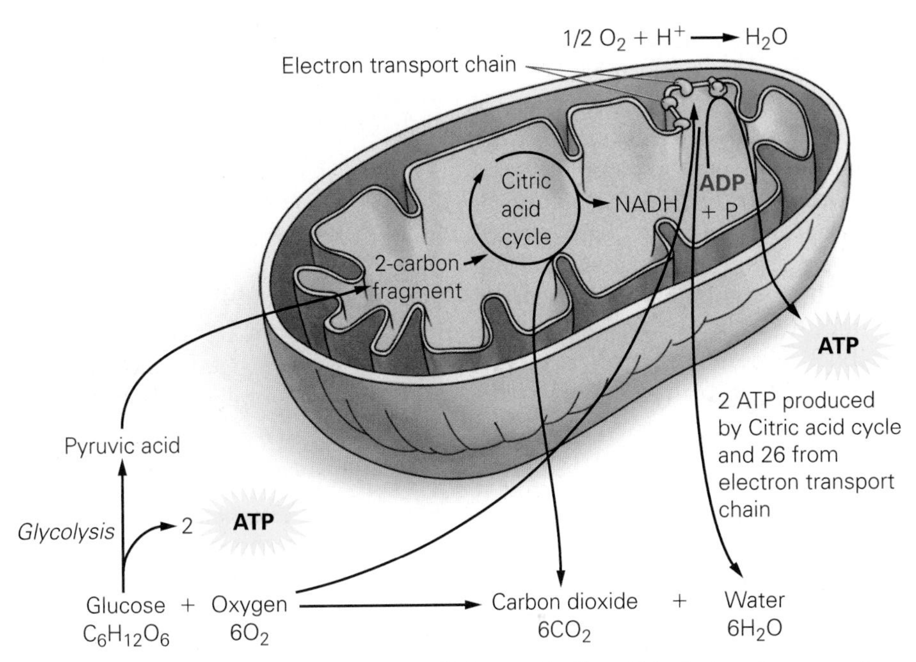

Figure 4.16 **A more detailed look at cellular respiration.** This figure diagrams the inputs and outputs of cellular respiration.

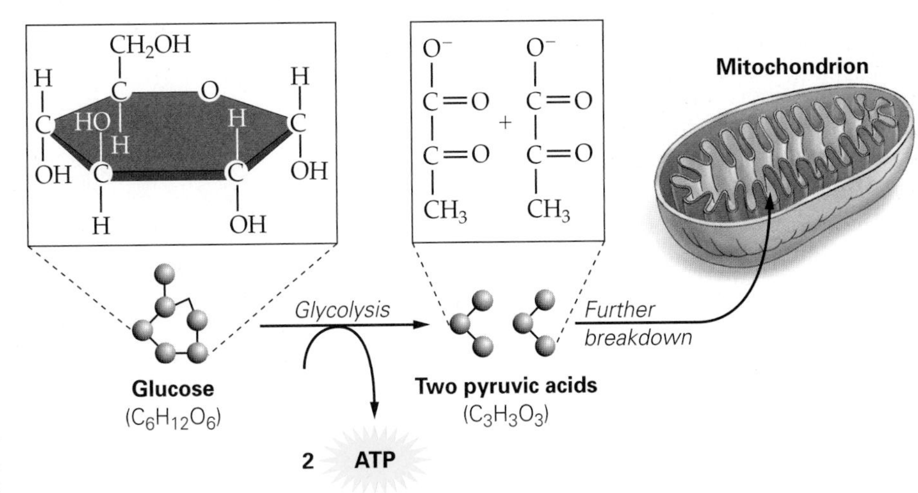

Figure 4.17 **Glycolysis.** Glycolysis is the enzymatic conversion of glucose into two pyruvic acid molecules. The pyruvic acid molecules are further broken down in the mitochondrion. Two ATP are made during glycolysis. **Visualize This:** Why is glycolysis an apt name for this process?

(continued on the next page)

(A Closer Look continued)

by molecules called *electron carriers*. One of the electron carriers utilized by cellular respiration is a chemical called **nicotinamide adenine dinucleotide (NAD^+).** NAD^+ picks up 2 hydrogen atoms (along with their electrons) and releases 1 positively charged hydrogen ion (H^+), becoming **NADH** *(Figure 4.20)*.

NADH serves as a sort of taxicab for electrons. The empty taxicab (NAD^+) picks up electrons. The full taxicab (NADH) carries electrons to their destination, where they are dropped off, and the empty taxicab returns for more electrons. NADH deposits its electrons at the electron transport chain *(Figure 4.21)*.

Step 3: Electron transport and ATP synthesis. The electron transport chain is a series of proteins embedded in the inner mitochondrial membrane that functions as a sort of conveyer belt for electrons, moving them from one protein to another. The electrons move toward the bottom of the electron transport chain toward the matrix of the mitochondrion, where they combine with oxygen to produce water.

Each time an electron is picked up by a protein or handed off to another protein, the protein moving it changes shape. This shape change facilitates the movement of hydrogen ions (H^+) from the matrix of the mitochondrion to the intermembrane space. So, while the proteins in the electron transport chain are moving electrons down the electron transport chain toward oxygen, they are also moving H^+ ions across the inner mitochondrial membrane and into the intermembrane space. This decreases the concentration of H^+ ions in the matrix and increases their concentration within the intermembrane space. As you learned in Chapter 3, whenever a concentration gradient of a molecule exists, molecules will diffuse from an area of high concentration to an area of low concentration. Since charged ions cannot diffuse across the hydrophobic core of the membrane, they escape through a protein channel in the membrane called **ATP synthase.** This enzyme uses the energy generated by the rushing H^+ ions to synthesize 26 ATP from ADP and phosphate in the same manner that water rushing through a mechanical turbine can be used to generate electricity *(Figure 4.22)*.

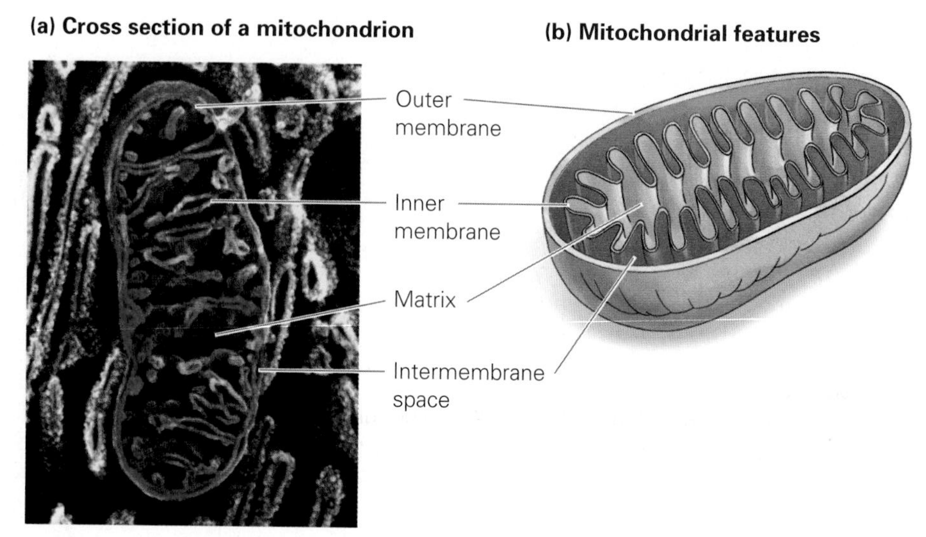

Figure 4.18 **Mitochondria.** The colorized cross section (a) and drawing (b) of a mitochondrion show the structures of the mitochondrion involved in cellular respiration.

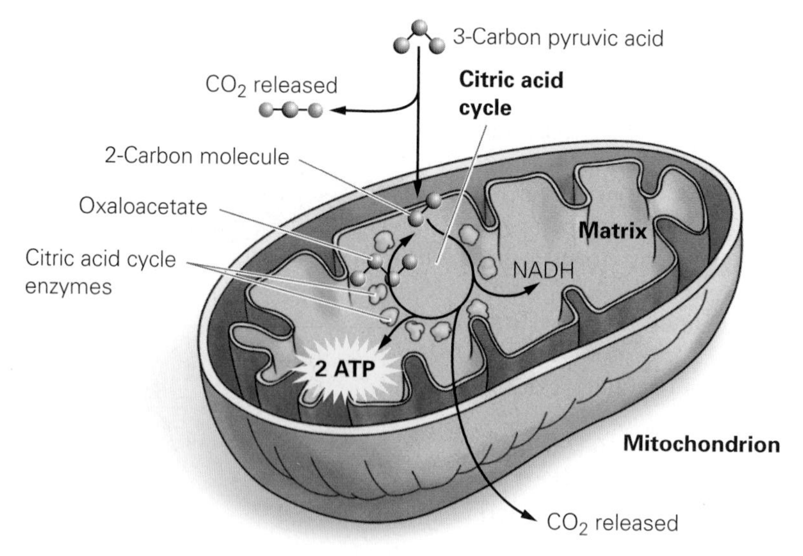

Figure 4.19 **The citric acid cycle.** The 3-carbon pyruvic acid molecules generated by glycolysis are decarboxylated, leaving a 2-carbon molecule that enters the citric acid cycle within the mitochondrial matrix. The 2-carbon fragment reacts with a 4-carbon OAA molecule and proceeds through a stepwise series of reactions that results in the production of more carbon dioxide and regenerates OAA. NADH and two ATP are also produced.

(A Closer Look continued)

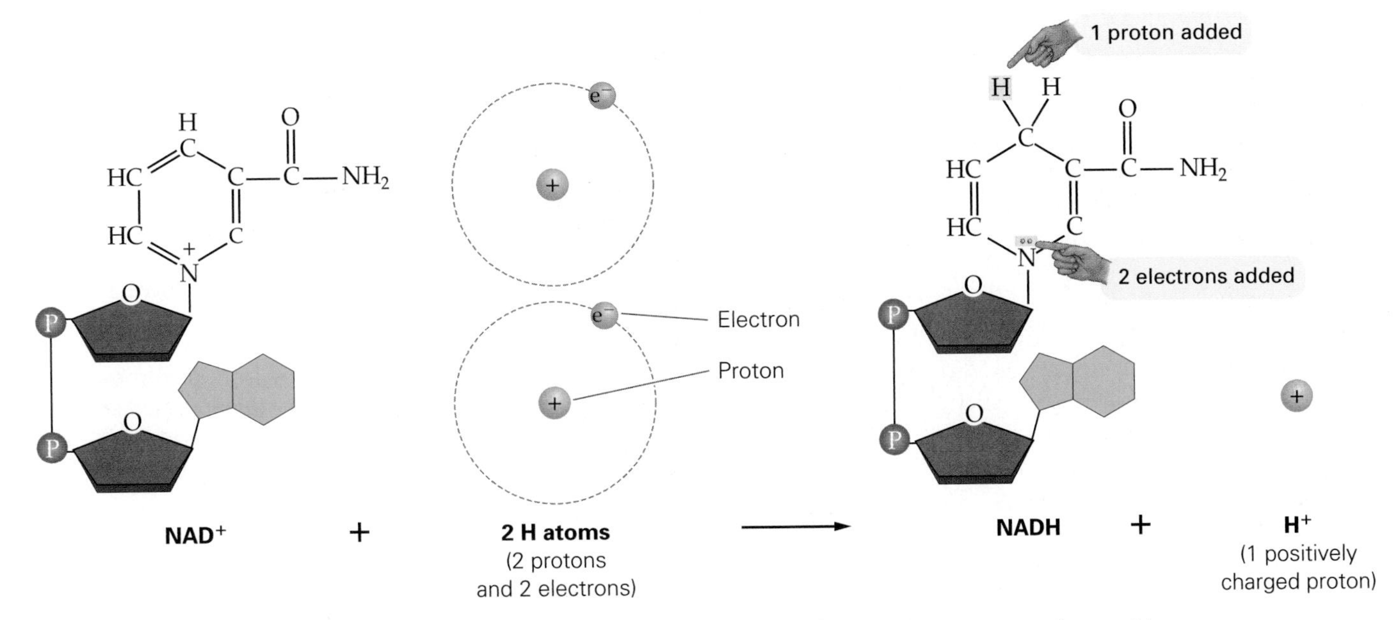

Figure 4.20 **Nicotinamide adenine dinucleotide (NAD^+).** NAD^+ can pick up a hydrogen atom along with its electrons. Hydrogen atoms are composed of 1 negatively charged electron that circles around the 1 positively charged proton. When NAD^+ encounters 2 hydrogen atoms (from food), it utilizes each hydrogen atom's electron and only 1 proton, thus releasing 1 proton. **Visualize This:** Why does the number of double bonds in the nitrogenous base change when a proton is added to NAD^+ to produce NADH?

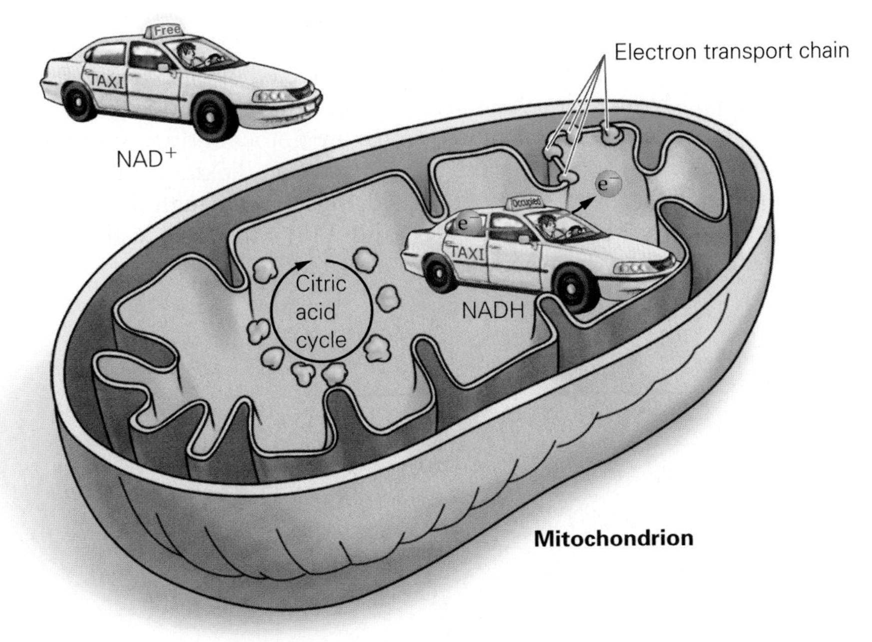

Figure 4.21 **Electron carriers.** NADH serves as an electron carrier, bringing electrons removed from the original glucose molecule to the electron transport chain. After dropping off its electrons, the electron carrier can be loaded up again and bring more electrons to the electron transport chain.

Overall, the two pyruvic acids produced by the breakdown of glucose during glycolysis are converted into carbon dioxide and water. Carbon dioxide is produced when it is removed from the pyruvic acid molecules during the citric acid cycle, and water is formed when oxygen combines with hydrogen ions at the bottom of the electron transport chain (see Figure 4.16).

(continued on the next page)

(A Closer Look continued)

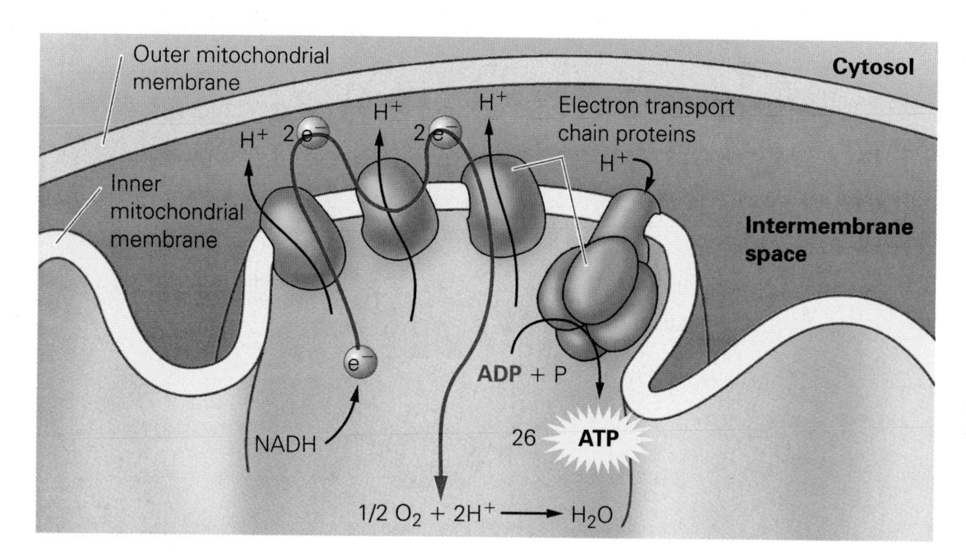

Figure 4.22 **The electron transport chain of the inner mitochondrial membrane.** NADH brings electrons from the citric acid cycle to the electron transport chain. As electrons move through the proteins of the electron transport chain, hydrogen ions are pumped into the intermembrane space. Hydrogen ions flow back through an ATP synthase protein, which converts ADP to ATP. In this manner energy from electrons added to the electron transport chain is used to produce ATP. **Visualize This:** What is meant by ½ oxygen in this figure?

Metabolism of Other Nutrients

Proteins and fats are broken down and their subunits merge with the carbohydrate breakdown pathway. *Figure 4.23* shows the points of entry for proteins and fats. Protein is broken down into component amino acids, which are then used to synthesize new proteins. Most organisms can also break down proteins to supply energy. However, this process takes place only when fats or carbohydrates are unavailable. In humans and other animals, the first step in producing energy from the amino acids of a protein is to remove the nitrogen-containing amino group of the amino acid. Amino groups are then converted to a compound called urea, which is excreted in the urine. The carbon, oxygen, and hydrogen remaining after the amino group is removed undergo further breakdown and eventually enter the mitochondria, where they are fed through the citric acid cycle and produce carbon dioxide, water, and ATP. The subunits of fats (glycerol and fatty acids) also go through the citric acid cycle and produce carbon dioxide, water, and ATP. Most cells will break down fat only when carbohydrate supplies are depleted.

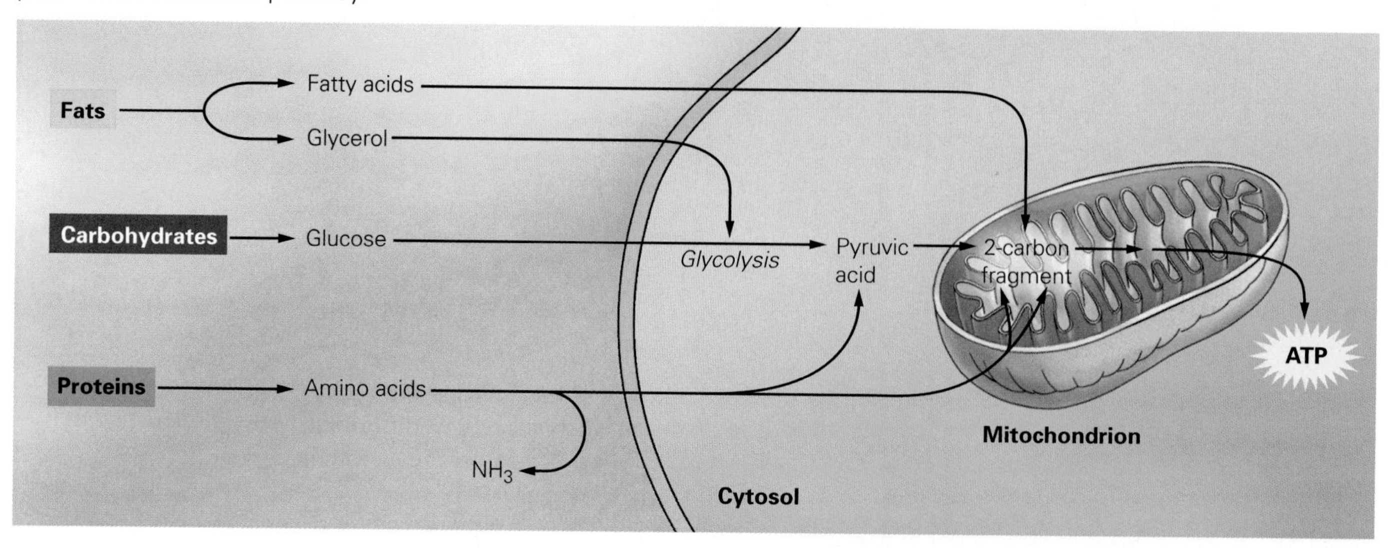

Figure 4.23 **Metabolism of other macromolecules.** Carbohydrates, proteins and fats can all undergo cellular respiration; they just feed into different parts of the metabolic pathway.

Metabolism Without Oxygen: Anaerobic Respiration and Fermentation

Aerobic cellular respiration is one way for organisms to generate energy. It is also possible for cells to generate energy in the absence of oxygen, by a process called **anaerobic respiration.**

Muscle cells normally produce ATP by aerobic respiration. However, oxygen supplies diminish with intense exercise. When muscle cells run low on oxygen, they must get most of their ATP from glycolysis, the only step in the cellular respiration process that does not require oxygen. When glycolysis happens without aerobic respiration, cells can run low on NAD^+ because it is converted to NADH during glycolysis. When this happens, cells use a process called **fermentation** to regenerate NAD^+. No usable energy is produced by fermentation; fermentation simply recycles NAD^+.

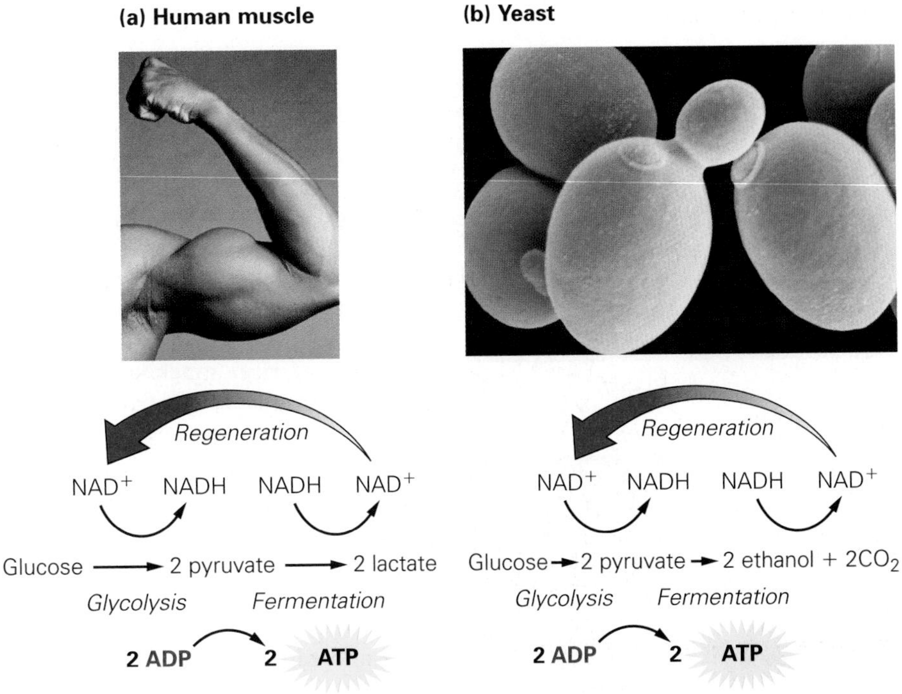

Fermentation cannot, however, be used for very long because one of the by-products of this reaction leads to the buildup of a compound called lactic acid. Lactic acid is produced by the actions of the electron acceptor NADH, which has no place to dump its electrons during fermentation since there is no electron transport chain and no oxygen to accept the electrons. Instead, NADH deposits its electrons by giving them to the pyruvic acid produced by glycolysis *(Figure 4.24a)*. Lactic acid is transported to the liver, where liver cells use oxygen to convert it back to pyruvic acid.

This requirement for oxygen, to convert lactic acid to pyruvic acid, explains why you continue to breathe heavily even after you have stopped working out. Your body needs to supply oxygen to your liver for this conversion, sometimes referred to as "paying back your oxygen debt." The accumulation of lactic acid also explains the phenomenon called "hitting the wall." Anyone who has ever felt as though their legs were turning to wood while running or biking knows this feeling. When your muscles are producing lactic acid by fermentation for a long time, the oxygen debt becomes too large, and muscles shut down until the rate of oxygen supply outpaces the rate of oxygen utilization.

(a) Human muscle

Regeneration

NAD^+ NADH NADH NAD^+

Glucose → 2 pyruvate → 2 lactate

Glycolysis *Fermentation*

2 ADP → 2 ATP

(b) Yeast

Regeneration

NAD^+ NADH NADH NAD^+

Glucose → 2 pyruvate → 2 ethanol + $2CO_2$

Glycolysis *Fermentation*

2 ADP → 2 ATP

Figure 4.24 **Metabolism without oxygen.** Glycolysis can be followed by (a) lactate fermentation to regenerate NAD^+. This pathway also produces 2 ATP during glycolysis. Glycolysis followed by (b) alcohol fermentation also regenerates NAD^+ and produces 2 ATP.

Stop and Stretch 4.3: Aerobic exercise (such as running, swimming, and biking) strengthens the heart, allowing it to pump more blood per beat. Given the effects of anaerobic respiration on the body, how does aerobic exercise increase stamina?

Some fungi and bacteria also produce lactic acid during fermentation. Certain microbes placed in an anaerobic environment transform the sugars in milk into yogurt, sour cream, and cheese. It is the lactic acid present in these dairy products that gives them their sharp or sour flavor. Yeast in an anaerobic environment produces ethyl alcohol instead of lactic acid. Ethyl alcohol is formed when carbon dioxide is removed from pyruvic acid *(Figure 4.24b)*.

The yeast used to help make beer and wine converts sugars present in grains (beer) or grapes (wine) into ethyl alcohol and carbon dioxide. Carbon dioxide, produced by baker's yeast, helps bread to rise.

The ability to generate energy by performing metabolic reactions can be affected by rising temperatures, leading to some devastating effects.

Global Warming and Cellular Respiration

Alaska's Kenai Peninsula is experiencing firsthand some effects of global warming on cellular respiration. Increases in temperature have helped to speed up the life cycle of the spruce bark beetle *(Figure 4.25a)*. These beetles, about the size of a grain of rice, attack spruce trees by boring through the outer bark to the phloem, a thin layer directly beneath the outer bark that transports food manufactured by photosynthesis from the foliage down to the roots. Once inside the phloem, the beetle feeds and lays eggs. The resulting damage and blockage of the phloem prevents nutrient transport to the roots, and the tree dies *(Figure 4.25b)*. Over the past decade, the spruce population has suffered huge losses—close to 4 million acres of trees on southeastern Alaska's Kenai Peninsula; nearly all of the spruce trees there have been killed by infestations of these bark beetles.

Populations of spruce bark beetles are normally kept in check by cool summers and bitterly cold winters. Cooler summers help control the number of beetles because they cannot fly when temperatures are below 60°F. This limits the beetles' ability to colonize other trees. Cold winters can kill beetles and their larvae. The warmer temperatures not only fail to kill beetles in the winter but also speed up this insect's rate of reproduction. Typically, it takes a spruce bark beetle 2 years to develop from an egg to an adult, but the warmer summers and winters have allowed the beetle to develop into an adult and lay new eggs during just one summer. More beetles mean more destruction to forests.

(a) Spruce bark beetle

(b) Spruce tree

Figure 4.25 **Spruce bark beetle.** The spruce bark beetle (a) kills spruce trees (b) by blocking water and nutrient flow.

(a) Forest fire—burning quickly releases heat and light energy.

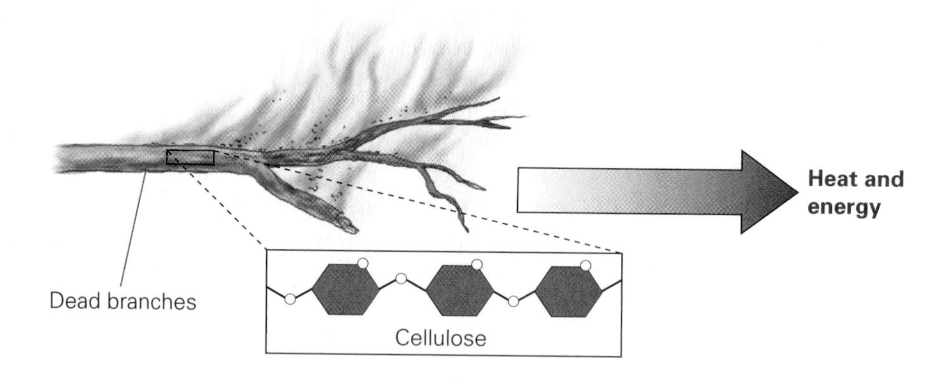

(b) Cellular respiration—ATP energy release is slow, controlled.

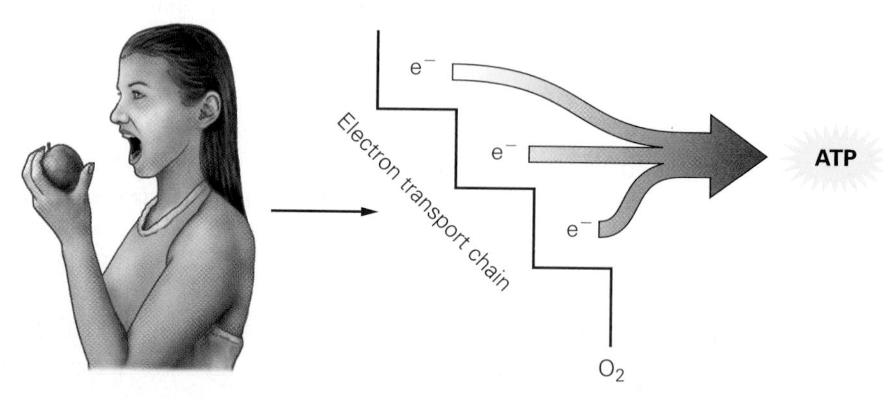

Figure 4.26 **Burning carbohydrates releases energy.** (a) Plant cells have rigid cell walls composed of cellulose, a polymer of glucose. When carbohydrate burning is uncontrolled, as in a forest fire, energy is released as heat and light. (b) Cellular respiration is a controlled burn. Carbohydrates that are eaten have electrons removed during cellular respiration. As these electrons move through the electron transport chain, they release energy that is used to drive the synthesis of ATP.

The accelerated life cycle of spruce bark beetles can be credited, in part, to the speeding up of cellular respiration. The enzymes that catalyze the reactions of cellular respiration, like all enzymes, are affected by temperature. Warmer temperatures typically speed up the rate at which enzymes catalyze reactions unless the temperature gets too hot, in which case the enzyme loses its characteristic shape and can no longer perform its job. In a sense, warmer temperatures make the citric acid cycle spin faster, producing more energy, which allows the beetles to grow and reproduce more quickly as well as to fly earlier in the year and thus disperse to a greater number of trees.

As the beetles do their damage, trees drop their dried-out dead needles and limbs on the ground, providing fuel for forest fires. Forest fires release even more carbon dioxide into the atmosphere as the carbohydrate that comprises much of the plant tissue, such as cellulose that makes up the cell wall, is burned. Cellular respiration is a controlled burn of carbohydrates by which the energy released from breaking the bonds of sugars is used to make ATP. Combustion by fire releases all the stored energy without harvesting any for ATP production (*Figure 4.26*).

You have learned that increases in carbon dioxide levels are causing global warming and that increasing temperatures can cause organisms to undergo cellular respiration at a faster rate, releasing even more CO_2 into the environment. Can the additional CO_2 be removed from the` atmosphere by increased photosynthesis?

Figure 4.27 **Overview of photosynthesis.** Sunlight drives the synthesis of glucose from carbon dioxide and water.

4.4 Photosynthesis

Plants and other photosynthetic organisms remove carbon dioxide from the atmosphere and use it to make sugars and other macromolecules by the process of photosynthesis. In the process, they release oxygen into the atmosphere.

BioFlix™ Photosynthesis

An Overview of Photosynthesis

The equation summarizing photosynthesis is as follows:

$$6\ CO_2 + 6\ H_2O + \text{Light energy} \longrightarrow C_6H_{12}O_6 + 6\ O_2$$

$$\text{Carbon dioxide} + \text{Water} + \text{Light energy} \longrightarrow \text{Glucose} + \text{Oxygen}$$

The sun is the ultimate source of energy for living organisms. Plants transform energy from the sun into chemical energy through the process of photosynthesis by using light energy to rearrange the atoms present in carbon dioxide and water into energy-rich carbohydrates, producing oxygen as a waste product (*Figure 4.27*). In land plants, carbon dioxide enters, and oxygen gas is released through adjustable microscopic structures called **stomata** that are located on the surface of the leaf (*Figure 4.28*).

Plants use the carbohydrates that they produce by photosynthesis to grow and supply energy to their cells. They, along with the organisms that eat them, liberate the energy stored in the chemical bonds of sugars by undergoing the process of cellular respiration. Both plants and animals perform cellular respiration, but animals cannot perform photosynthesis.

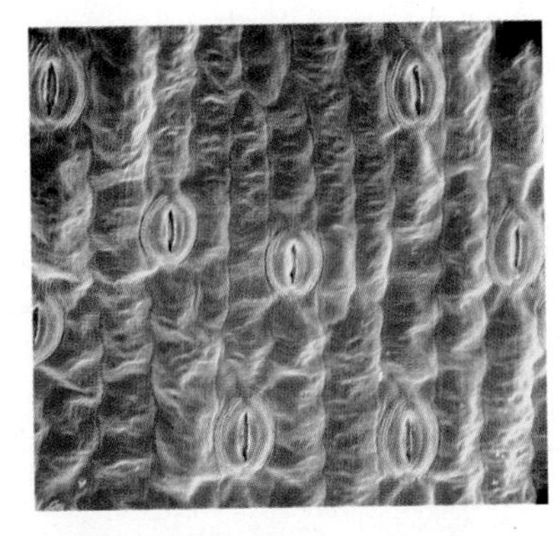

Figure 4.28 **Stomata.** Stomata are adjustable microscopic pores found on the surface of leaves that allow for gas exchange.

Stop and Stretch 4.4: In what ways are photosynthesis and cellular respiration related?

A Closer Look: The Light Reactions and the Calvin Cycle

Green tissues in plants contain specialized organelles that serve as the sites of photosynthesis. You learned in Chapter 2 that **chloroplasts** are surrounded by two membranes. The inner and outer membranes together are called the chloroplast envelope. The chloroplast envelope encloses a compartment filled with **stroma,** the thick fluid that houses some of the enzymes of photosynthesis. Suspended in the stroma are disk-like membranous structures called **thylakoids.** When thylakoids are stacked on top of each other, like pancakes, the stacks are called **grana** *(Figure 4.29)*. The large amount of membrane inside the chloroplast provides more surface area upon which some of the reactions of photosynthesis can occur.

On the surface of the thylakoid membrane are millions of pigment molecules, called **chlorophyll,** that absorb energy from the sun. It is the chlorophyll molecule that gives leaves and other plant structures their green color. Like all pigments, chlorophyll absorbs light. Light is made up of rays with different colors, or levels of energy, and each energy level has a different wavelength—to the human eye, shorter and middle wavelengths appear violet to green, and longer wavelengths appear yellow to red. Different organisms can perceive different wavelengths of light. For example, bees can see ultraviolet light, which is invisible to humans. Differences in wavelength visibility help bees see colors and patterns in floral structures as an aid to direct them to the sexual organs of the plant.

Chlorophyll looks green to human eyes because it absorbs the shorter (blue) and longer (red) wavelengths of visible light and reflects the middle (green) range of wavelengths. Leaves on deciduous trees change color in the fall because the abundant chlorophyll breaks down, making visible other less-abundant pigments present in the leaf that reflect red, orange, and yellow wavelengths.

Photosynthesis can be broken down into two steps. The first or "photo" step harvests energy from the sun during a series of reactions called the **light reactions,** which occur when there is sunlight. The second or "synthesis" step, called the **Calvin cycle,** uses the harvested energy to synthesize sugars in either the presence or the absence of sunlight. For this reason, the Calvin cycle is also sometimes referred to as the light-independent reactions.

The Light Reactions. When a pigment such as chlorophyll absorbs sunlight, electrons associated with the pigment become excited or increase their energy level. In effect, the light energy is transferred to the chlorophyll and becomes chemical energy. For most pigments, the molecule remains excited for a very brief amount of time before the chemical energy is lost as heat. This is why a surface that looks black (that is, one composed of a pigment that absorbs all visible light wavelengths) heats up quickly in comparison to a surface that looks white (one that absorbs no visible light wavelengths). Inside a chloroplast, however, the chemical energy of the excited chlorophyll molecules is not allowed to be released as heat; instead, the energy is captured.

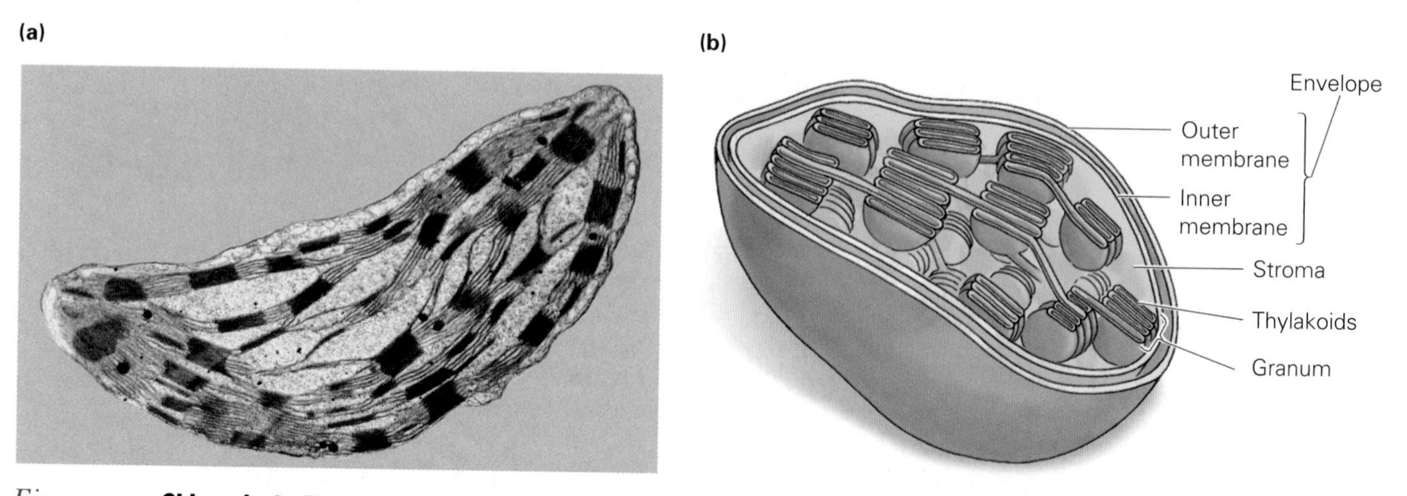

Figure 4.29 **Chloroplasts.** The cross section (a) and drawing (b) of a chloroplast show the structures involved in photosynthesis.

(A Closer Look continued)

When sunlight strikes the chlorophyll molecule and electrons are excited, they move to a higher energy level. The electrons are then transferred to other molecules in an electron transport chain located in the thylakoid membrane. As the electrons are handed down the electron transport chain, some ATP is produced. Some of the proteins in the electron transport chain not only move electrons to a lower energy level, they also pump protons into the interior of the thylakoid, setting up a gradient in protons. The protons then rush through an ATP synthase enzyme located in the thylakoid membrane and produce ATP in the same way that mitochondria make ATP. The newly synthesized ATP is released into the stroma, where it can be used by the enzymes of the Calvin cycle to produce sugars and other organic molecules.

During the light reactions, oxygen is produced when water (H_2O) is "split" into $2H^+$ ions and a single oxygen atom (O). Two oxygen atoms combine to produce O_2 which is released from the chloroplast. Since the hydrogen atom usually contains a single proton, around which orbits a single electron, the splitting of water to produce two H^+ ions also releases 2 electrons. These 2 electrons are transferred back to the chlorophyll molecule to replace those passed along the electron transport chain.

At the end of the electron transport chain, electrons are transferred to the electron carrier for plants, which is nicotinamide adenine dinucleotide phosphate, or NADP. Just like the NAD^+ involved in cellular respiration, $NADP^+$ functions as an electron taxicab. The difference between NAD^+ and $NADP^+$ is simply the presence of an extra phosphate group. The $NADP^+$ used during photosynthesis picks up 2 electrons and 1 H^+ ion to become NADPH. NADPH ferries electrons to the stroma, where the enzymes of the Calvin cycle will use the electrons in assembling sugars *(Figure 4.30)*. Thus, the light reactions produce ATP, a source of electrons for the synthesis step, and release oxygen as a by-product.

Calvin Cycle. The Calvin cycle is a series of enzymes located in the stroma that uses the ATP and NADPH produced during photosynthesis to convert carbon dioxide (CO_2) into sugars (CH_2O).

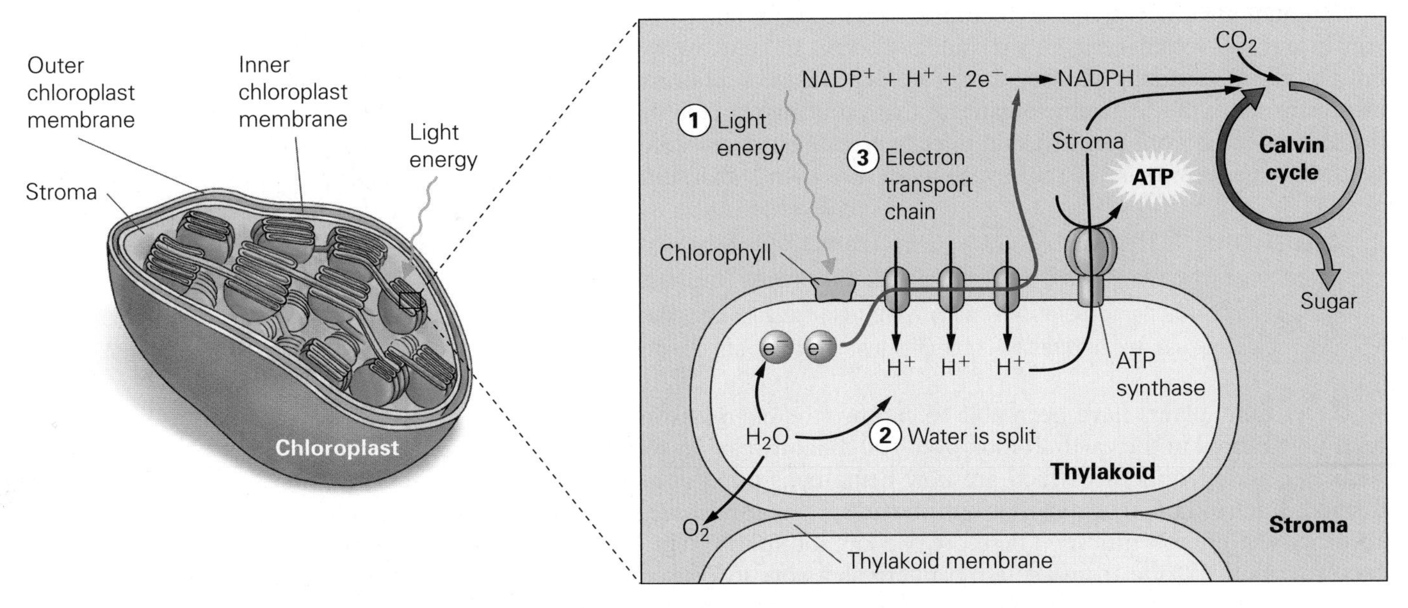

Figure 4.30 **The light reactions of photosynthesis.** During the light reactions of photosynthesis (1) sunlight strikes chlorophyll molecules located in the thylakoid membrane, exciting electrons that then move to a higher energy level. (2) Water is split. Electrons removed from water are used to replace those lost from chlorophyll. The remaining hydrogen ions stay in the thylakoid and oxygen is released. (3) The energy of the excited electrons is harvested in a stepwise manner as the electrons are handed down an electron transport chain, producing ATP and NADPH. ATP and NADPH are produced in the stroma, where they will be available to the enzymes of the Calvin cycle.

(continued on the next page)

(A Closer Look continued)

CH_2O is the general formula for sugars. For example, glucose is $C_6H_{12}O_6$ or $6(CH_2O)$. A quick comparison of the composition of these molecules makes it obvious that converting CO_2 into CH_2O requires the incorporation of hydrogen atoms and their associated electrons. Hydrogen atoms are removed from NADPH to produce sugars, thereby regenerating $NADP^+$ *(Figure 4.31)*.

During the Calvin cycle, carbon dioxide from the environment reacts with a 5-carbon molecule that is generated by the Calvin cycle and called ribulose bisphosphate or RuBP. The enzyme that performs this reaction is ribulose bisphosphate carboxylase oxygenase, or **rubisco.** This reaction produces an unstable 6-carbon molecule that immediately breaks down into two 3-carbon molecules called three-phosphoglyceric acid, or 3-PGA, which is rearranged to produce glyceraldehyde three-phosphate or G3P, a 3-carbon sugar that the cell uses to produce glucose and other compounds. RuBP is regenerated, completing the cycle.

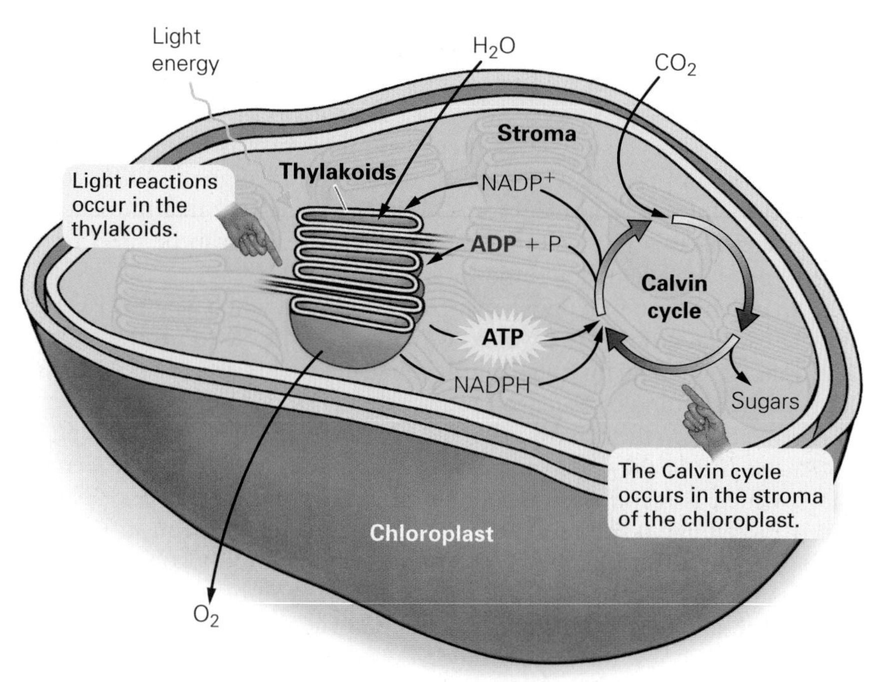

Figure 4.31 **The Calvin cycle.** Carbon dioxide enters the Calvin cycle. The energy of ATP is used to add hydrogens and electrons from NADPH to CO_2 to produce sugars. **Visualize This:** Would an absence of sunlight affect the Calvin cycle?

Without this reaction in the chloroplasts—when solar energy is transformed into the chemical energy in glucose—cellular respiration could not occur, and glucose would not be used to synthesize ATP. For this reason, virtually all living things are dependent on photosynthesis for food, even meat-eating organisms, since they consume organisms that eat plants. In fact, the only two items in the human diet that do not come from plants (either directly or indirectly) are water and salt. Plants and other photosynthetic organisms make all the oxygen in the atmosphere that humans require for respiration.

Stop and Stretch 4.5: Plants can make all of the macromolecules they need to survive from simple, inorganic components. They accumulate carbon from the atmosphere and hydrogen and oxygen from water. The remaining elements they require are from the soil. Consider the constituents of proteins and nucleic acids and list two or three soil nutrients that you think are most important to plant growth.

Over time, plants have been able to capture the energy from light to form carbon-rich fossil fuels now buried in the earth. It took over 100 million years to form these nonrenewable resources such as coal, oil, and gas, which are now being consumed at a much faster rate than they were formed. The result is that more carbon dioxide is being released into the atmosphere than can be absorbed via photosynthesis. Therefore, we cannot simply rely on photosynthesis to remove the excess carbon dioxide from the atmosphere as a way to prevent global warming. In fact, rising temperatures can actually slow photosynthesis in certain types of plants.

C_3, C_4, and CAM Plants. You learned earlier that plants can bring carbon dioxide into leaves through openings called stomata and that the carbon dioxide brought in is used to produce sugars during the Calvin cycle. Stomatal openings are located on the surface of a leaf and are surrounded by two kidney-bean-shaped cells called **guard cells.** When the guard cells are compressed against each other, the stomata are closed, thus restricting the flow of gases into or out of the plant. When the guard cells change shape to create an opening between them, the stomata are open, and carbon dioxide and oxygen gases can be exchanged. In addition to the exchange of gases, water can move out of the plant through the stomatal opening via a process called **transpiration** *(Figure 4.32)*. The

(A Closer Look continued)

transpired water is replaced when water from the soil is brought into the plant, bringing with it minerals that the plant needs to synthesize many compounds.

In most plant species, the Calvin cycle produces 3-carbon sugars, which are converted into the sugars that are either stored as food for the plant or transported to growing leaves, roots, and reproductive structures. Plants that produce the 3-carbon molecule are called **C_3 plants.** C_3 plants are the most abundant type of plant on Earth and include agriculturally important species such as soybeans, wheat, and rice.

Rising temperatures can reduce the rate of photosynthesis because on hot, dry days plants close their stomata to reduce the rate of water lost to transpiration. Closing stomatal openings prevents carbon dioxide from entering the plant, and the rate of photosynthesis declines.

Closing the stomatal openings to prevent water loss causes another series of reactions to occur, called **photorespiration.** During photorespiration, the first enzyme in the Calvin cycle uses oxygen instead of carbon dioxide as its substrate. While most enzymes display a high degree of specificity for their particular substrate, some enzymes have additional substrates to which they have lesser affinities. The enzyme that catalyzes the first step of the Calvin cycle, rubisco, is one such enzyme. The most common protein on Earth, rubisco is the also most common protein in leafy tissue. Carbon dioxide is its preferred substrate, but when carbon dioxide is low, rubisco will also allow oxygen into its active site. That is, it behaves as an oxygenase. When oxygen is high, such as when photosynthesis is occurring but the stomata are closed, oxygen will be used as the substrate of rubisco, and the plant will undergo photorespiration. During photorespiration, the rubisco enzyme catalyzes the incorporation of oxygen into a compound called glycolate. Glycolate cannot be used for food or for the synthesis of structural components. In fact, it must be destroyed by the plant since high levels of glycolate inhibit photosynthesis. The breakdown of glycolate requires ATP and releases carbon dioxide that had been previously incorporated into sugars during the Calvin cycle.

You might wonder why it is that plant cells perform this wasteful process. It has to do with the evolution of photosynthesis on early Earth. Photosynthesis evolved when the atmosphere was largely devoid of oxygen. Under these conditions, the enzyme's affinity for oxygen did not present the problems it does in today's oxygen-rich atmosphere. All descendants of the first photosynthesizers inherited the instructions for producing the same rubisco enzyme, and so modern plants are "stuck" with this somewhat inefficient system. In dry environments, natural selection should favor plants that can minimize photorespiration despite having closed stomata much of the time. Two mechanisms for reducing photorespiration are known as C_4 and CAM photosynthesis.

C_4 plants, like all plants, conserve water during hot, sunny periods by closing their stomata. However, these plants carry an additional enzyme (compared to C_3 plants) that allows them to avoid photorespiration and continue to make sugars even though carbon dioxide levels within the plants are low during these periods. This enzyme is present in cells closest to the stomata and has a much higher affinity for carbon dioxide than does rubisco. It is able to procure carbon dioxide even when the stomata are almost closed. The carbon dioxide is combined with a 3-carbon acceptor molecule to produce a 4-carbon compound, hence the name C_4 plants. The 4-carbon compound then migrates to cells deeper within the leaf, where the carbon dioxide is released and produces a locally high concentration of carbon dioxide that enables rubisco to function as a carboxylase in the Calvin cycle. The 3-carbon acceptor molecule returns to the cells nearest the stomata. Corn and sugar cane are C_4 plants that can keep making sugars even though their stomata are almost closed. However, C_4 photosynthesis carries a cost—C_3 plants require 3 ATP molecules to convert 1 molecule of carbon dioxide into sugar, but C_4 plants require 5 molecules of ATP. The

(continued on the next page)

(a) Open

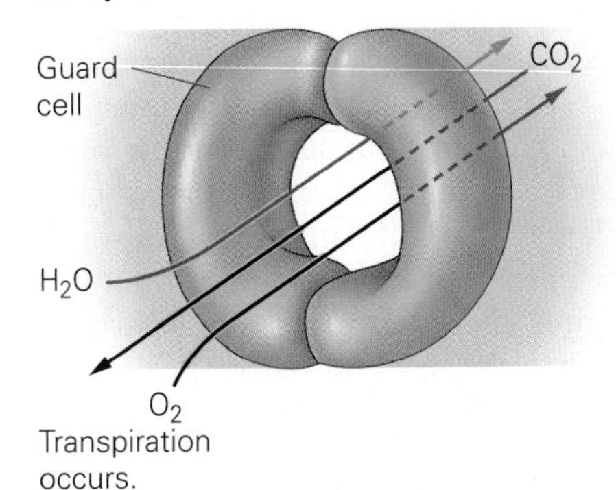

Transpiration occurs.

(b) Closed

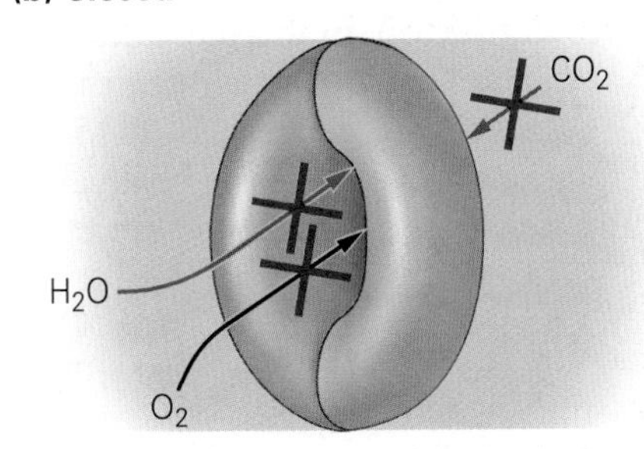

Transpiration does not occur.

Figure 4.32 **Gas exchange and water loss.** (a) When the stomata are open, carbon dioxide and oxygen gases can be exchanged. Water can be lost through a process called transpiration. (b) When the guard cells change shape to block the opening, gas exchange and transpiration do not occur.

(A Closer Look continued)

enzymes used in C_4 photosynthesis are also more sensitive to cold temperatures than are the enzymes of the Calvin cycle. Thus, C_3 plants have an advantage in certain environmental situations (cool and shady), and C_4 plants have an advantage in others (hot and sunny).

One other photosynthetic adaptation involves **CAM plants.** CAM stands for crassulacean acid metabolism, named for the plant family Crassulaceae, in which this mechanism was first discovered. Members of the Crassulaceae include the jade plant and other succulent (water-storing) plants. A CAM plant conserves water by opening its stomata at night only. The carbon dioxide that comes in during the night cannot immediately be used for photosynthesis because that process requires energy from the sun. During the night, the carbon is stored as an acid that is broken down during the day and releases carbon dioxide while the stomata are closed. This carbon dioxide can then be used for photosynthesis when sunlight becomes available. Therefore, carbon dioxide can enter at night, be stored as an acid, and then be used by the Calvin cycle during the subsequent day, even if the stomata are closed during the day to conserve water. Growth is limited in CAM plants because the amount of carbon dioxide stored in acid during the night is limited; the plants use it all up early in the day and cannot perform any more photosynthesis. *Table 4.1* summarizes C_3, C_4, and CAM plant strategies.

Scientists are concerned that increasing temperatures may favor plants with these water-saving adaptations and change the relative percentages of C_3, C_4, and CAM plants in existence. This change could negatively affect some agricultural crops and could change the relative percentages of native

TABLE 4.1 **C_3, C_4, and CAM plant photosynthesis.** Plants have evolved adaptations that prevent water loss.

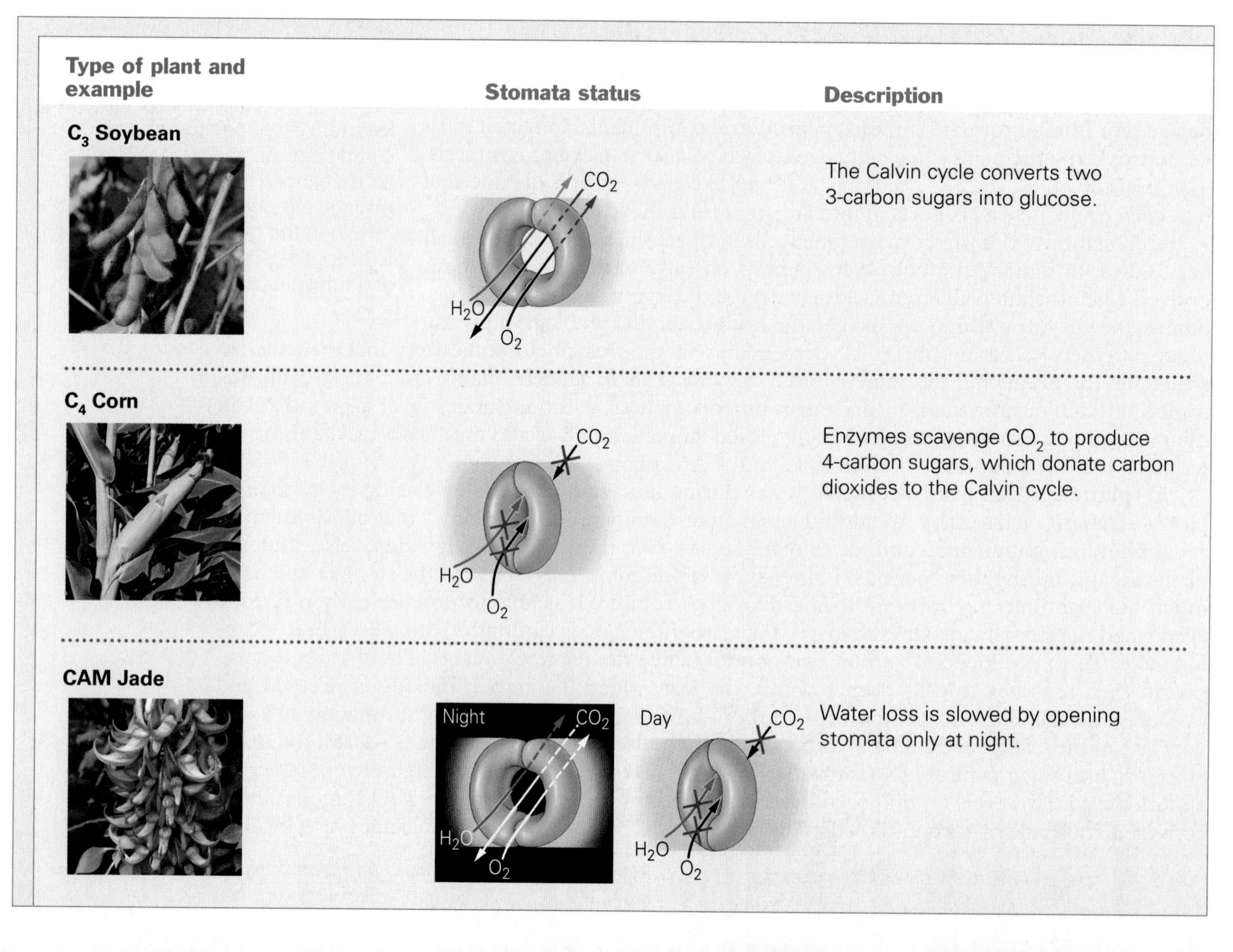

Type of plant and example	Stomata status	Description
C_3 Soybean		The Calvin cycle converts two 3-carbon sugars into glucose.
C_4 Corn		Enzymes scavenge CO_2 to produce 4-carbon sugars, which donate carbon dioxides to the Calvin cycle.
CAM Jade		Water loss is slowed by opening stomata only at night.

(A Closer Look continued)

plant species in a given region since certain adaptations gain relative advantages as the climate changes. Like the sugar maple, those species that cannot migrate to more appropriate environments or cannot compete with other plants in a changing environment may become extinct. As plant communities change and lose species, many of the animals that depend on these communities may suffer.

Global Warming and Photosynthesis

Photosynthesis cannot keep up with the amount of carbon dioxide that is currently being released. While humans are adding more carbon dioxide to the environment, we are also removing organisms that take up our excess carbon dioxide. The process that occurs when forests are cleared for logging, farming, and ever-expanding human settlements is called **deforestation.** Deforestation contributes directly to the increase in carbon dioxide within the atmosphere. Current estimates are that up to 25% of the carbon dioxide introduced into the atmosphere originates from the cutting and burning of forests in the tropics. The loss of trees as a result of deforestation also indirectly contributes to rising carbon dioxide levels when these forests are replaced by pasture or cropland. Because net rates of photosynthesis (as measured by grams of carbon dioxide removed from the atmosphere per acre per year) in grasslands or agricultural fields are 30% to 60% less than rates in forests, the loss of forests significantly decreases the removal of carbon dioxide from the atmosphere.

Replanting trees in deforested areas may help increase the rate at which carbon dioxide is removed from the atmosphere because young trees have a faster net photosynthetic rate than older trees. This is because older trees have lots of nonphotosynthetic, woody tissues that use the products of photosynthesis and that seedlings have yet to develop. In other words, young trees can put more of their organic carbon into storage as wood, while older trees use more of the carbon simply to maintain themselves. However, when these trees are logged, their roots and branches are left behind to decompose. In addition, most of the wood that is harvested is turned into paper, which will decompose after a few years, or fuel, which will be burned. Decomposition and burning result in the release into the environment of carbon compounds that were once part of the trees. Therefore, even though replanting after deforestation helps remove some of the carbon dioxide from the atmosphere, it does not result in a return to previous levels.

4.5 Decreasing the Effects of Global Warming

The devastation seen in New Orleans is paralleled by environmental damage seen worldwide. Many other countries are affected by decisions made by the people of the United States. Home to only 4% of the world's population, the United States produces close to one-fourth of the carbon dioxide emitted by fossil fuel burning. The per capita emissions rate of carbon dioxide for Americans is twice that of the Japanese or Germans, three times that of the global average, four times that of Swedes, and 20 times that of the average Indian.

Most of the emissions for an individual country come from industry, followed by transportation and then by commercial, residential, and agricultural emissions. All of us can work to reduce our personal contribution to global warming by decreasing residential and transport emissions. Most residential emissions are from energy used to heat and cool homes and to power electrical appliances. Transportation emissions are affected by the fuel economy of cars and the number of miles traveled. *Table 4.2* describes many ways that you can decrease your greenhouse gas

TABLE 4.2 **Decreasing your greenhouse gas emissions.** Here are some ideas that you can use to help slow the rate of global warming.

Action	Annual decrease in carbon dioxide production
Drive an energy-efficient vehicle. SUVs average 16 miles per gallon, while smaller cars average 25 miles per gallon.	 13,000 pounds
Carpool 2 days per week.	 1,590 pounds
Recycle glass bottles, aluminum cans, plastic, newspapers, and cardboard.	 850 pounds
Walk 10 miles per week instead of driving.	 590 pounds
Buy high-efficiency appliances.	 400 pounds per appliance
Buy food and other products with reusable or recyclable packaging, or reduced packaging, to save the energy required to manufacture new containers.	 230 pounds
Use a push mower instead of a power mower.	 80 pounds
Plant shade trees around your home to decrease energy consumption and to remove carbon dioxide by photosynthesis.	 50 pounds

emissions and indicates the number of pounds of carbon dioxide that each action would save annually. These reductions may seem trivial in comparison to the scope of the problem, but when they are multiplied by the almost 300 million people in the United States, the savings become significant.

Having an effect on industrial, commercial, and agricultural sectors is difficult for any one individual. Instead, this will take leadership from the policy makers who are committed to reducing emissions. To do so requires that our leaders, and all of us, understand that even though the causes, implications, and solutions of global warming may be open to debate, the fact that it is occurring at an unprecedented rate is not.

Savvy Reader Global Warming Killing the Polar Bears?

THE WALL STREET JOURNAL | DECEMBER 14, 2005 | BY JIM CARLTON

It may be the latest evidence of global warming: Polar bears are drowning.

Scientists for the first time have documented multiple deaths of polar bears off Alaska, where they likely drowned after swimming long distances in the ocean amid the melting of the Arctic ice shelf. The bears spend most of their time hunting and raising their young on ice floes.

In a quarter-century of aerial surveys of the Alaskan coastline before 2004, researchers from the U.S. Minerals Management Service said they typically spotted a lone polar bear swimming in the ocean far from ice about once every two years. Polar-bear drownings were so rare that they have never been documented in the surveys.

But in September 2004, when the polar ice cap had retreated a record 160 miles north of the northern coast of Alaska, researchers counted 10 polar bears swimming as far as 60 miles offshore. Polar bears can swim long distances but have evolved to mainly swim between sheets of ice, scientists say.

The researchers returned to the vicinity a few days after a fierce storm and found four dead bears floating in the water. "Extrapolation of survey data suggests that on the order of 40 bears may have been swimming and that many of those probably drowned as a result of rough seas caused by high winds," the researchers say in a report set to be released today.

In addition to documenting polar-bear deaths, the Minerals Management Service researchers, Chuck Monnett, Jeffrey Gleason and Lisa Rotterman, also found a striking shift in the bears' habits. From 1979 to 1991, 87% of the bears spotted were found mostly on sea ice. From 1992 to 2004, the percentage dropped to 33%. Most of the remaining bears have been found either in the ocean or on beaches, congregating around carcasses of whales butchered by hunters. In the past, polar bears were rarely seen at such kill sites, because they spent their time hunting their favorite meal—seals—on sea ice.

1. What is the hypothesis of this excerpted article?
2. Did the author explore other hypotheses about the deaths of the polar bears, even if only to refute them?
3. What do you think that the scientists should do next to continue testing this hypothesis?

CHAPTER Review

For study help, animations, and more quiz questions go to www.mybiology.com.

Summary

4.1 The Greenhouse Effect

- The planet is warming. This warming will lead to a rise in ocean levels, changes in weather patterns, and the disruption of biological communities (pp. 81–82).
- Greenhouse gases, particularly carbon dioxide, increase the amount of heat retained in Earth's atmosphere, which then leads to increased temperatures (pp. 82–83).
- Water can absorb large amounts of heat without undergoing rapid or drastic changes in temperature because heat must first be used to break hydrogen bonds between adjacent water molecules. A high heat-absorbing capacity is a characteristic of water (p. 83).

4.2 The Carbon Cycle

- Carbon dioxide cycles between animals, plants, soil, oceans, and the atmosphere (p. 84).
- Carbon dioxide levels in the atmosphere are increasing. This increase is caused by human activities such as the burning of fossil fuels and is leading to global warming (pp. 84–86).

4.3 Cellular Respiration

- Cellular respiration converts the energy stored in chemical bonds of food into ATP (p. 86).
- ATP is a nucleotide triphosphate. The nucleotide found in ATP contains a sugar and the nitrogenous base, adenine (pp. 86–87).
- Breaking the terminal phosphate bond of ATP releases energy that can be used to perform cellular work and produces ADP plus a phosphate (pp. 86–87).
- ATP is generated in most organisms by the process of cellular respiration, which consumes carbohydrates and releases water and carbon dioxide as waste products (pp. 87–88).
- Cellular respiration begins in the cytosol, where a 6-carbon sugar is broken down into two 3-carbon pyruvic acid molecules during the anaerobic process of glycolysis (p. 89).
- The pyruvic acid molecules then move across the two mitochondrial membranes, where they are decarboxylated. The remaining 2-carbon fragment then moves into the matrix of the mitochondrion, where the citric acid cycle strips them of carbon dioxide and electrons (pp. 89–90).
- The electrons are carried by electron carriers to the inner mitochondrial membrane; there they are added to a series of proteins called the electron transport chain. At the bottom of the electron transport chain, electronegative oxygen pulls the electrons toward itself. As the electrons move down the electron transport chain, the energy that they release is used to drive protons (H^+) into the intermembrane space. Once there, the protons rush through the enzyme ATP

synthase and produce ATP from ADP and phosphate (pp. 90–92).

- When electrons reach the oxygen at the bottom of the electron transport chain, they combine with the oxygen and hydrogen ions to produce water (p. 92).

web animation 4.1
Glucose Metabolism

BioFlix™ **Cellular Respiration**

4.4 Photosynthesis

- Photosynthesis utilizes carbon dioxide from the atmosphere to make sugars and other substances (p. 95).
- During photosynthesis, energy from sunlight is used to rearrange the atoms of carbon dioxide and water to produce sugars and oxygen (p. 95).
- Photosynthesis occurs in chloroplasts. Sunlight strikes the chlorophyll molecule within chloroplasts, boosting electrons to a higher energy level. These excited electrons are dropped down an electron transport chain located in the thylakoid membrane, and ATP is made (pp. 96–97).
- Electrons are also passed to electron carriers that transport electrons to the Calvin cycle. The electrons that are lost become replaced by electrons acquired during the splitting of water, and oxygen is released. The Calvin cycle utilizes the products of the light reactions (ATP and the electron carrier NADPH) to incorporate carbon dioxide into sugars (pp. 97–98).
- Photosynthesis removes carbon dioxide from the air, potentially reducing the risk of global warming. However, humans are also deforesting Earth's land surface, reducing the global rate of photosynthesis (p. 98).
- Stomata on a plant's surface not only allow in carbon dioxide for photosynthesis but also allow water to escape from the plant. Guard cells surrounding the stomata can change shape to close the stomata and restrict water loss. However, when stomata are closed, carbon dioxide declines in the plant, and the energy-wasting process of photorespiration may occur. C_4 and CAM plants have evolved to perform photosynthesis while reducing the risk of photorespiration in dry conditions (pp. 98–101).

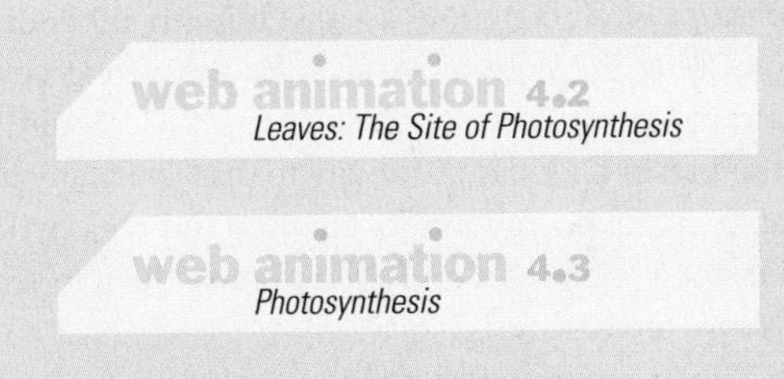

BioFlix™ **Photosynthesis**

4.5 Decreasing the Effects of Global Warming

- The effects of global warming will not be as severe if humans can reduce carbon dioxide emissions (pp. 101–103).

Roots to Remember

The following roots of words come mainly from Latin and Greek and will help you decipher terms:

an- means absence of, as in anaerobic.
glyco- means sugar.
photo- means light.

Learning the Basics

1. What are the reactants and products of cellular respiration and photosynthesis?

2. Carbon dioxide functions as a greenhouse gas by ______.

A. interfering with water's ability to absorb heat; **B.** increasing the random molecular motions of oxygen; **C.** allowing radiation from the sun to reach Earth and absorbing the reradiated heat; **D.** splitting into carbon and oxygen and increasing the rate of cellular respiration

3. Water has a high heat-absorbing capacity because ______.

A. the sun's rays penetrate to the bottom of bodies of water, mainly heating the bottom surface; **B.** the strong covalent bonds that hold individual water molecules together require large inputs of heat to break; **C.** it has the ability to dissolve many heat-resistant solutes; **D.** initial energy inputs are used to break hydrogen bonds between water molecules and then to raise the temperature; **E.** all of the above are true

4. Cellular respiration involves ______.

A. the aerobic metabolism of sugars in the mitochondria by a process called glycolysis; **B.** an electron transport chain that releases carbon dioxide; **C.** the synthesis of ATP, which is driven by the rushing of protons through an ATP synthase; **D.** electron carriers that bring electrons to the citric acid cycle; **E.** the production of water during the citric acid cycle

5. The electron transport chain ______.

A. is located in the matrix of the mitochondrion; **B.** has the electronegative carbon dioxide at its base; **C.** is a series of nucleotides located in the inner mitochondrial membrane; **D.** is a series of enzymes located in the intermembrane space; **E.** moves electrons from protein to protein and moves protons from the matrix into the intermembrane space

6. Which of the following **does not** occur during the light reactions of photosynthesis?

A. Oxygen is split, releasing water; **B.** Electrons from chlorophyll are added to an electron transport chain; **C.** An electron transport chain drives the synthesis of ATP for use by the Calvin cycle; **D.** NADPH is produced and will carry electrons to the Calvin cycle; **E.** Oxygen is produced when water is split.

7. Which of the following is a **false** statement about photosynthesis?

A. During the Calvin cycle, electrons and ATP from the light reactions are combined with atmospheric carbon dioxide to produce sugars; **B.** The enzymes of the Calvin cycle are located in the chloroplast stroma; **C.** Oxygen produced during the Calvin cycle is released into the atmosphere; **D.** Sunlight drives photosynthesis by boosting electrons found in chlorophyll to a higher energy level; **E.** Electrons released when sunlight strikes chlorophyll are replaced by electrons from water.

8. Select the **true** statement regarding metabolism in plant and animal cells.

A. Plant and animal cells both perform photosynthesis and aerobic respiration; **B.** Animal cells perform aerobic respiration only, and plant cells perform photosynthesis only; **C.** Plant cells perform aerobic respiration only, and animal cells perform photosynthesis only; **D.** Plant cells perform aerobic respiration and photosynthesis, and animal cells perform aerobic respiration only.

9. Which human activity generates the most carbon dioxide?

A. driving; **B.** cooking; **C.** bathing; **D.** using aerosol sprays.

10. True or false: The products of cellular respiration are the reactants in photosynthesis.

Analyzing and Applying the Basics

1. Are sugars the only macromolecules that can be broken down to produce ATP? If not, how are other nutrients metabolized?

2. How do the different strategies employed by C_3, C_4, and CAM plants help them adapt to their environments?

3. Describe the sites of aerobic respiration and photosynthesis. What organelles are involved? Where in each organelle do the different steps of each process occur?

Connecting the Science

1. What can individuals do to slow the effects of global warming?

2. Do you think it is okay for the individuals of one country to produce more greenhouse gases than do individuals of other countries? Why or why not?

CHAPTER 5

Cancer

DNA Synthesis, Mitosis, and Meiosis

Cancer cells divide when they should not.

Nicole got sick during her junior year of college.

She had to undergo some procedures to see if she had cancer.

Nicole's early college career was similar to that of most students. She enjoyed her independence and the wide variety of courses required for her double major in biology and psychology. She worried about her grades and finding ways to balance her course work with her social life. She also tried to find time for lifting weights in the school's athletic center and snowboarding at a local ski hill. Some weekends, to take a break from school, she would ride the bus home to see her family.

Managing to get schoolwork done, see friends and family, and still have time left to exercise had been difficult, but possible, for Nicole during her first two years at school. That changed drastically during her third year of school.

One morning in October of her junior year, Nicole began having severe pains in her abdomen. The first time this happened, she was just beginning an experiment in her cell biology laboratory. Hunched over and sweating, she barely managed to make it through her two-hour biology lab. Over the next few days, the pain intensified so much that she was unable to walk from her apartment to her classes without stopping several times to rest.

She wants to understand why she got cancer.

Later that week, as she was preparing to leave for class, the pain was so severe that she had to lie down in the hallway of her apartment. When her roommate got home a few minutes later, she took Nicole to the student health center for an emergency visit. The physician at the health center first determined that Nicole's appendix had not burst and then made an appointment for Nicole to see a gynecologist the next day.

After hearing Nicole's symptoms, the gynecologist pressed on her abdomen and felt what he thought was a mass on her right ovary. He used a noninvasive procedure called ultrasound to try to get an image of her ovary. This procedure requires the use of high-frequency sound waves. These waves, which cannot be heard by humans, bounce off tissues and produce a pattern of echoes that can be used to create a picture called a sonogram. Healthy tissues, fluid-filled cysts, and tumors all look different on a sonogram *(Figure 5.1)*.

Nicole's sonogram convinced her gynecologist that she had a large growth on her ovary. He told her that he suspected this growth was a *cyst*, or fluid-filled sac. Her gynecologist told her that cysts often go away without treatment, but this one seemed to be quite large and should be surgically removed.

Even though the idea of having an operation was scary for Nicole, she was relieved to know that the pain would stop. Her gynecologist also assured her that she had nothing to worry about because cysts are not cancerous. A week after the abdominal pain began, Nicole's gynecologist removed the cyst and her completely engulfed right ovary through an incision just below her navel. He then sent the cystic ovary to a physician who specializes in determining whether tissues are cancerous. This physician, called a *pathologist*, determined that Nicole's doctor had been right—there was no sign of cancer.

(a) Healthy ovary

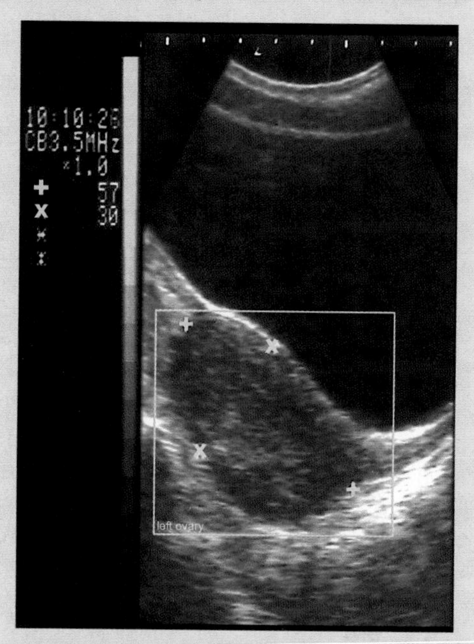

(b) Cancerous ovarian tissue

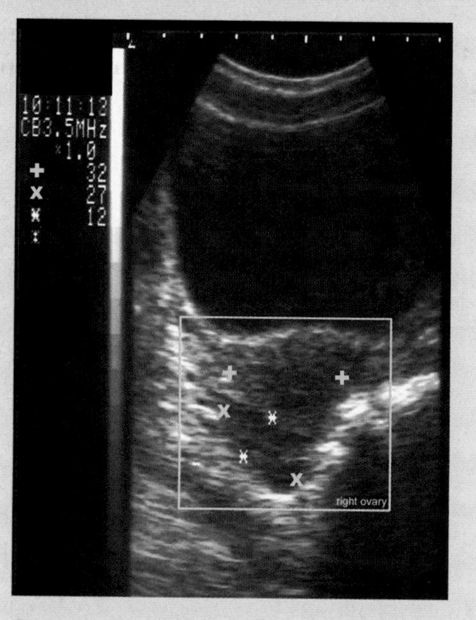

Figure 5.1 **Normal and cancerous tissues.** Physicians can sometimes use ultrasound to determine whether ovarian tissues are (a) normal or (b) contain tumors.

After the operation, Nicole's gynecologist assured her that the remaining ovary would compensate for the missing ovary by ovulating (producing an egg cell) every month. He added that he would have to monitor her remaining ovary carefully to make sure that it did not become cystic, or even worse, cancerous. She could not afford to lose another ovary if she wanted to have children someday.

Monitoring her remaining ovary involved monthly visits to her gynecologist's office, where Nicole had her blood drawn and analyzed. The blood was tested for the level of a protein called CA125, which is produced by ovarian cells. Higher-than-normal CA125 levels usually indicate that the ovarian cells have increased in size or number and are thus associated with the presence of an ovarian tumor.

Nicole went to her scheduled checkups for five months after the original surgery. The day after her March checkup, Nicole received a message from her doctor asking that she come to see him the next day. Because she needed to study for an upcoming exam, Nicole tried to push aside her concerns about the appointment. By the time she arrived at her gynecologist's office, she had convinced herself that nothing serious could be wrong. She thought a mistake had probably been made and that he just wanted to perform another blood test.

The minute her gynecologist entered the exam room, Nicole could tell by his demeanor that something was wrong. As he started speaking to her, she began to feel very anxious—when he said that she might have a tumor on her remaining ovary, she could not believe her ears. When she heard the words *cancer* and *biopsy,* Nicole felt as though she was being pulled underwater. She could see that her doctor was still talking, but she could not hear or understand him. She felt too nauseous to think clearly, so she excused herself from the exam room, took the bus home, and immediately called her parents.

After speaking with her parents, Nicole realized that she had many questions to ask her doctor. She did not understand how it was possible for such a young woman to have lost one ovary to a cyst and then possibly to have a tumor on the other ovary. She wondered how this tumor would be treated and what her prognosis would be. Before seeing her gynecologist again, Nicole decided to do some research in order to make a list of questions for her doctor.

5.1 What Is Cancer?

Cancer is a disease that begins when a single cell replicates itself although it should not. **Cell division** is the process a cell undergoes to make copies of itself. This process is normally regulated so that a cell divides only when more cells are required and when conditions are favorable for division. A cancerous cell is a rebellious cell that divides without being given the go-ahead.

Tumors Can Be Cancerous

Unregulated cell division leads to a pileup of cells that form a lump or **tumor.** A tumor is a mass of cells that has no apparent function in the body. Tumors that stay in one place and do not affect surrounding structures are said to be **benign.** Some benign tumors remain harmless; others become cancerous. Tumors that invade surrounding tissues are **malignant** or cancerous. The cells of a malignant tumor can break away and start new cancers at distant locations through a process called **metastasis** *(Figure 5.2)*.

Cancer cells can travel virtually anywhere in the body via the lymphatic and circulatory systems. The **lymphatic system** collects fluids lost from microscopic blood vessels called **capillaries. Lymph nodes** are structures that filter the lost fluids, or lymph. When a cancer patient is undergoing surgery, the surgeon will often remove a few lymph nodes to see if any cancer cells are in the nodes. If cancer cells appear in the nodes, then some cells have left the original tumor and are moving through the bloodstream. If this has happened, cancerous cells likely have metastasized to other locations in the body.

When cancer cells metastasize, they can gain access not only to the **circulatory system,** which includes blood vessels to transport the blood, but

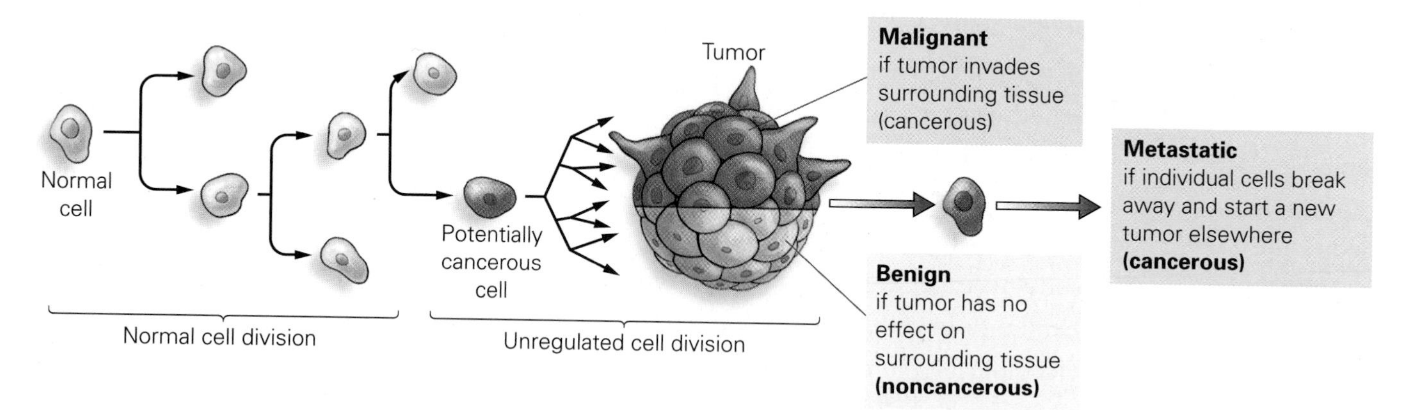

Figure 5.2 **What is cancer?** A tumor is a clump of cells with no function. Tumors may remain benign, or they can invade surrounding tissues and become malignant. Tumor cells may move, or metastasize, to other locations in the body. Malignant and metastatic tumors are cancerous.

also to the heart, which pumps the blood. Once inside a blood vessel, cancer cells can drift virtually anywhere in the body.

Cancer cells differ from normal cells in three ways: (1) They divide when they should not; (2) they invade surrounding tissues; and (3) they can move to other locations in the body. All tissues that undergo cell division are susceptible to becoming cancerous. However, there are ways to increase or decrease the probability of getting cancer.

Risk Factors for Cancer

Certain exposures and behaviors, called **risk factors,** increase a person's risk of obtaining a disease. General risk factors for virtually all cancers include tobacco use, a high-fat and low-fiber diet, lack of exercise, obesity, excess alcohol consumption, and increasing age. *Table 5.1* outlines other risk factors that are linked to particular cancers.

Tobacco Use. The use of tobacco of any type, whether cigarettes, cigars, pipes, or chewing tobacco, increases your risk of many cancers. While smoking is the cause of 90% of all lung cancers, it is also the cause of about one-third of all cancer deaths. Cigar smokers have increased rates of lung, larynx, esophagus, and mouth cancers. Chewing tobacco increases the risk of cancers of the mouth, gums, and cheeks. People who do not smoke but who are exposed to secondhand smoke have increased lung cancer rates.

Tobacco smoke contains more than 20 known cancer-causing substances called **carcinogens.** For a substance to be considered carcinogenic, exposure to the substance must be correlated with an increased risk of cancer. Examples of carcinogens include cigarette smoke, radiation, ultraviolet light, asbestos, and some viruses.

The carcinogens that are inhaled during smoking come into contact with cells deep inside the lungs. Chemicals present in cigarettes and cigarette smoke have been shown to increase cell division, inhibit a cell's ability to repair damaged DNA, and prevent cells from dying when they should.

Stop and Stretch 5.1: In addition to dividing uncontrollably, cancer cells also fail to undergo a type of programmed cell death called apoptosis, during which a cell uses specialized chemicals to kill itself. Why do you think it is useful that our own cells can "commit suicide" in certain situations?

Chemicals in cigarette smoke also disrupt the transport of substances across cell membranes and alter many of the enzyme reactions that occur within cells. They have also been shown to increase the generation of *free*

TABLE 5.1 **Cancer risk.** Risk factors and detection methods for particular cancers are given.

Cancer Location	Risk Factors	Detection	Comments
Ovary Oviduct Ovary 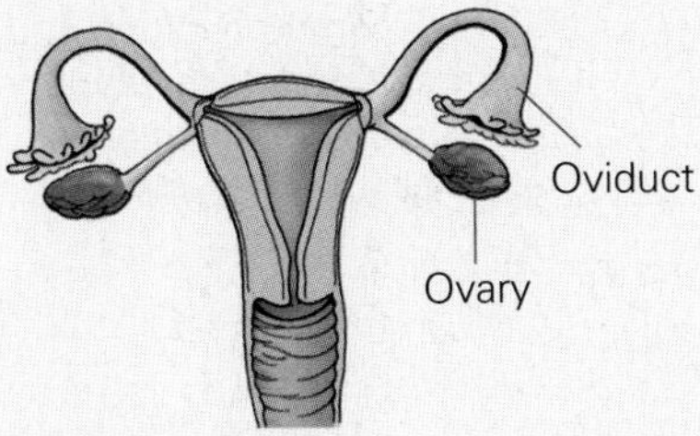	• Smoking • Mutation to *BRCA2* gene • Advanced age • Oral contraceptive use and pregnancy decrease risk	• Blood test for elevated CA125 level • Gynecological exam	• Fifth leading cause of death among women in the United States
Breast 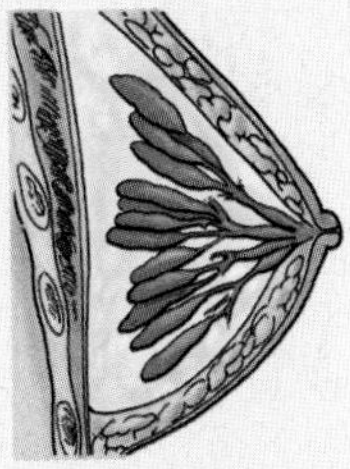	• Smoking • Mutation to *BRCA1* gene • High-fat, low-fiber diet • Use of oral contraceptives may slightly increase risk	• Monthly self-exams, look and feel for lumps or changes in contour • Mammogram	• Only 5% of breast cancers are due to *BRCA1* mutations. • Second-highest cause of cancer-related deaths • 1% of breast cancer occurs in males
Cervix 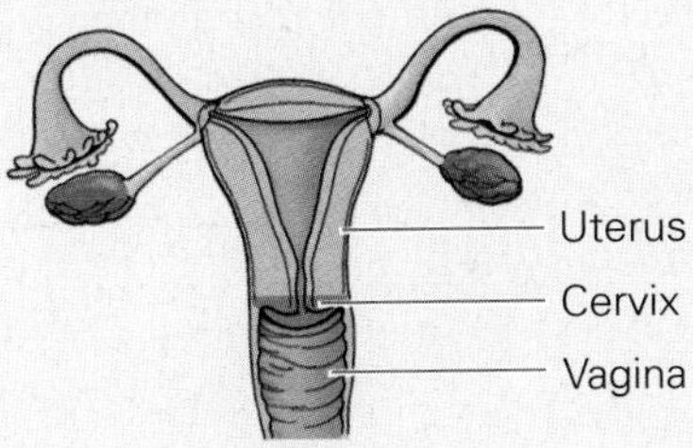 Uterus Cervix Vagina	• Smoking • Exposure to sexually transmitted human papilloma virus (HPV)	• Annual Pap smear tests for the presence of precancerous cells	• Precancerous cells can be removed by laser surgery or cryotherapy (freezing) before they become cancerous
Skin 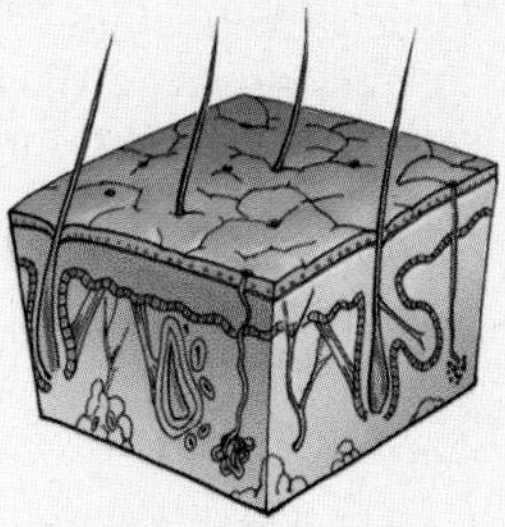	• Smoking • Fair skin • Exposure to ultraviolet light from the sun or tanning beds	• Monthly self-exams, look for growths that change in size or shape	• Skin cancer is the most common of all cancers; usually curable if caught early
Blood (leukemia) 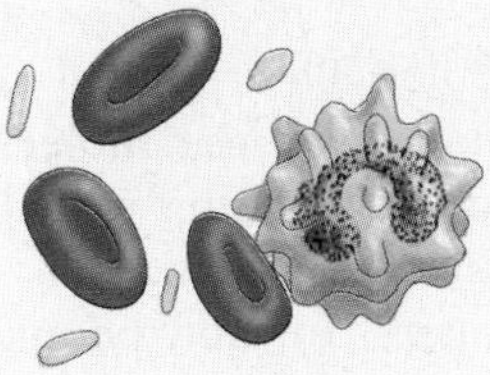	• Exposure to high-energy radiation such as that produced by atomic bomb explosions in Japan during World War II	• A sample of blood is examined under a microscope	• Cancerous white blood cells cannot fight infection efficiently; people with leukemia often succumb to infections

(continued on the next page)

TABLE 5.1 Cancer risk. Risk factors and detection methods for particular cancers are given. *(Continued)*

Cancer Location	Risk Factors	Detection	Comments
Lung	• Smoking • Exposure to second-hand smoke • Asbestos inhalation	• X-ray	• Lung cancer is the most common cause of death from cancer, and the best prevention is to quit, or never start, smoking
Colon and rectum Small intestine Colon	• Smoking • Polyps in the colon • Advanced age • High-fat, low-fiber diet	• Change in bowel habits • Colonoscopy is an examination of the rectum and colon using a lighted instrument	• Benign buds called polyps can grow in the colon; removal prevents them from mutating and becoming cancerous
Prostate Bladder Prostate Rectum	• Smoking • Advanced age • High-fat, low-fiber diet	• Blood test for elevated level of prostate-specific antigen (PSA) • Physical exam by physician, via rectum	• More common in African American men than Asian, white, or Native American men
Testicle Testicle Scrotum	• Abnormal testicular development	• Monthly self-exam, inspect for lumps and changes in contour	• Testicular cancer accounts for only 1% of all cancers in men but is the most common form of cancer found in males between the ages of 15 and 35

radicals, which remove electrons from other molecules. The removal of electrons from DNA or other molecules causes damage to these molecules—damage that, over time, may lead to cancer. Cigarette smoking provides so many different opportunities for DNA damage and cell damage that tumor formation and metastasis are quite likely for smokers. In fact, people who smoke cigarettes increase their odds of developing almost every cancer.

A High-Fat, Low-Fiber Diet. Cancer risk may also be influenced by diet. The American Cancer Society recommends eating at least 5 servings of fruits and vegetables every day as well as 6 servings of food from other plant sources, such as breads, cereals, grains, rice, pasta, or beans. Plant foods are low in fat and high in fiber. A diet high in fat (greater than 15% of all calories obtained from fats) and low in fiber (fewer than 30 g per day) is associated with increased risk of cancer. Fruits and vegetables are also rich in *antioxidants.* These substances help to neutralize the electrical charge on free radicals and thereby prevent the free radicals from taking electrons from other molecules, including DNA. There is some evidence that antioxidants may help prevent certain cancers by minimizing the number of free radicals that may damage the DNA in our cells.

Lack of Exercise. Regular exercise decreases the risk of most cancers, partly because exercise keeps the immune system functioning effectively. The immune system helps destroy cancer cells when it can recognize them as foreign to the host body. Unfortunately, since cancer cells are actually your own body's cells run amok, the immune system cannot always differentiate between normal cells and cancer cells.

Obesity. Exercise also helps prevent obesity, which is associated with increased risk for many cancers, including cancers of the breast, uterus, ovary, colon, gallbladder, and prostate. The abundance of fatty tissue has been hypothesized to increase the odds of hormone-sensitive cancers such as breast, uterine, ovarian, and prostate cancer.

Excess Alcohol Consumption. Drinking alcohol is associated with increased risk of some types of cancer. Men who want to decrease their cancer risk should have no more than two alcoholic drinks a day, and women one or none. People who both drink and smoke increase their odds of cancer in a multiplicative rather than additive manner. In other words, if one type of cancer occurs in 10% of smokers and in 2% of drinkers, someone who smokes and drinks multiplies his chances to a rate that is closer to 20% than 12%. The risk factor percentages are multiplied rather than added.

Increasing Age. As you age, your immune system weakens, and its ability to distinguish between cancer cells and normal cells decreases. This weakening is part of the reason many cancers are far more likely in elderly people. Additional factors that help explain the higher cancer risk with increasing age include cumulative damage. If we are all exposed to carcinogens during our lifetime, then the longer we are alive, the greater the probability that some of those carcinogens will mutate genes involved in regulating the cell cycle. Also, because multiple mutations are necessary for a cancer to develop, it often takes many years to progress from the initial mutation to a tumor and then to full-blown cancer. Scientists estimate that most cancers large enough to be detected have been growing for at least five years and are composed of close to one billion cells.

Nicole's cancer affected ovarian tissue. Why might ovarian cells be more likely to become cancerous than some other types of cells? Cells that divide frequently are more prone to cancer than those that don't divide often. When an egg cell is released from the ovary during ovulation, the tissue of the ovary becomes perforated. Cells near the perforation site undergo cell division to heal the damaged surface of the ovary. For Nicole, these cell divisions may have become uncontrolled, leading to the growth of a tumor.

web animation 5.1
The Structure of DNA

5.2 Passing Genes and Chromosomes to Daughter Cells

(a) Amoeba

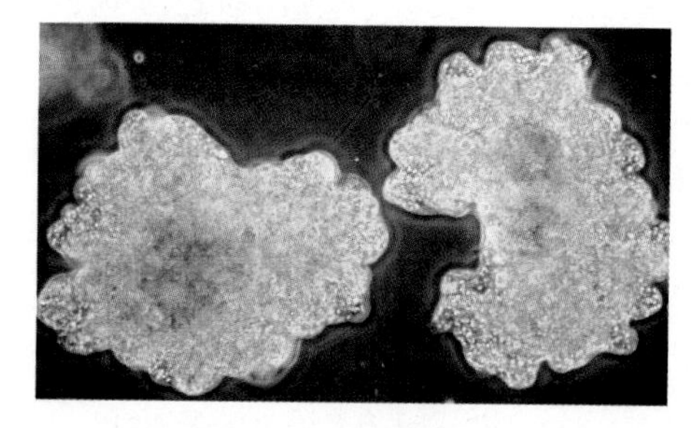

(b) English ivy

Figure 5.3 **Asexual reproduction.** (a) This single-celled amoeba divides by copying its DNA and producing offspring that are genetically identical to the original, parent amoeba. (b) Some multicellular organisms, such as this English ivy plant, can reproduce asexually from cuttings.

Most cell division does not lead to cancer. Cell division produces new cells to heal wounds, replace damaged cells, and help organisms grow and reproduce themselves. Each of us begins life as a single fertilized egg cell that undergoes millions of rounds of cell division to produce all the cells that comprise the tissues and organs of our bodies.

Some organisms reproduce by producing exact copies of themselves via cell division. Reproduction of this type, called **asexual reproduction,** does not require genetic input from two parents and results in offspring that are genetically identical to the original parent cell. Single-celled organisms, such as bacteria and amoeba, reproduce in this manner *(Figure 5.3a)*. Some multicellular organisms can reproduce asexually also. For example, most plants can grow from clippings of the stem, leaves, or roots and thereby reproduce asexually *(Figure 5.3b)*. Organisms whose reproduction requires genetic information from two parents undergo **sexual reproduction.** Humans reproduce sexually when sperm and egg cells combine their genetic information at fertilization.

Genes and Chromosomes

Whether reproducing sexually or asexually, all dividing cells must first make a copy of their genetic material, the **DNA (deoxyribonucleic acid).** The DNA carries the instructions, called **genes,** for building all of the proteins that cells require. The DNA in the nucleus is wrapped around proteins to produce structures called **chromosomes.** Chromosomes can carry hundreds of genes along their length. Different organisms have different numbers of chromosomes in their cells. For example, dogs have 78 chromosomes in each cell; humans have 46, and dandelions have 24.

Chromosomes are in an uncondensed, string-like form prior to cell division *(Figure 5.4a)*. Before cell division occurs, the DNA in each chromosome is condensed (compressed) in a short, linear form *(Figure 5.4b)*. Condensed linear chromosomes are easier to maneuver during cell division and are less likely to become tangled or broken than the uncondensed and string-like structures are. When a chromosome is replicated, a copy is produced that carries the same genes. The copied chromosomes are called **sister chromatids,** and each sister chromatid is composed of one DNA molecule. Sister chromatids are attached to each other at a region toward the middle of the replicated chromosome, called the **centromere.**

You learned in Chapter 2 that the DNA molecule itself is double stranded and can be likened to a twisted rope ladder. The backbone or "handrails" of each strand are composed of alternating sugar and phosphate groups. Across the width or "rungs" of the DNA helix are the nitrogenous bases, paired together via hydrogen bonds such that adenine (A) makes a base pair with thymine (T), and guanine (G) makes a base pair with cytosine (C).

You also learned that two of the people credited with determining DNA structure are James Watson and Francis Crick. Watson and Crick reported their hypothesis about the structure of the DNA molecule in a 1953 paper for the journal *Nature*. Although they did not go so far as to propose a detailed model for how the DNA molecule was replicated, they did say, "It has not escaped our notice that the specific pairing we have postulated immediately suggests a copying mechanism for the genetic material." The copying mechanism that Watson and Crick referred to is also called *DNA replication.*

DNA Replication

During the process of **DNA replication** that precedes cell division, the double-stranded DNA molecule is copied, first by splitting the molecule in half up the middle of the helix. New nucleotides are added to each side of the original parent molecule, maintaining the A-to-T and G-to-C base pairings. This process results in two daughter DNA molecules, each composed of one strand of parental nucleotides and one newly synthesized strand *(Figure 5.5a)*. Because each newly formed DNA molecule consists of one-half conserved parental DNA and one-half new daughter DNA, this method of DNA replication is referred to as semiconservative replication.

Replicating the DNA requires an enzyme that assists in DNA synthesis. This enzyme, called **DNA polymerase,** moves along the length of the unwound helix and helps bind incoming nucleotides to each other on the newly forming daughter strand *(Figure 5.5b)*. When free nucleotides floating in the nucleus have an affinity for each other (A for T and G for C), they bind to each other across the width of the helix. Nucleotides that bind to each other are said to be *complementary* to each other.

The DNA polymerase enzyme catalyzes the formation of the covalent bond between nucleotides along the length of the helix. The paired nitrogenous bases are joined across the width of the backbone by hydrogen bonding, and the DNA polymerase advances along the parental DNA strand to the next unpaired nucleotide. When an entire chromosome has been replicated, the newly synthesized copies are identical to each other. They are attached at the centromere as sister chromatids *(Figure 5.6)*.

(a) Uncondensed DNA

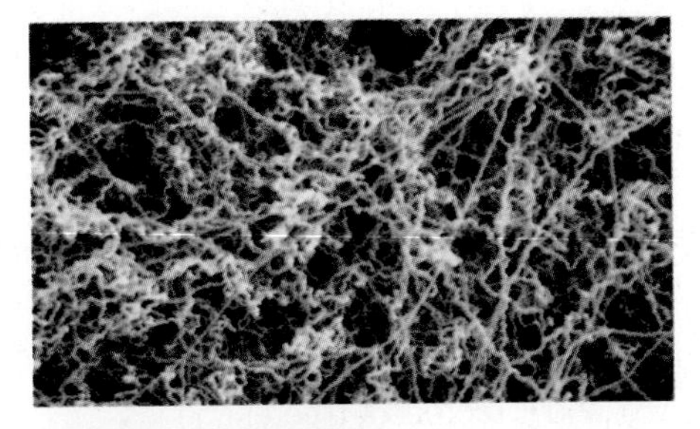

(b) DNA condensed into chromosomes

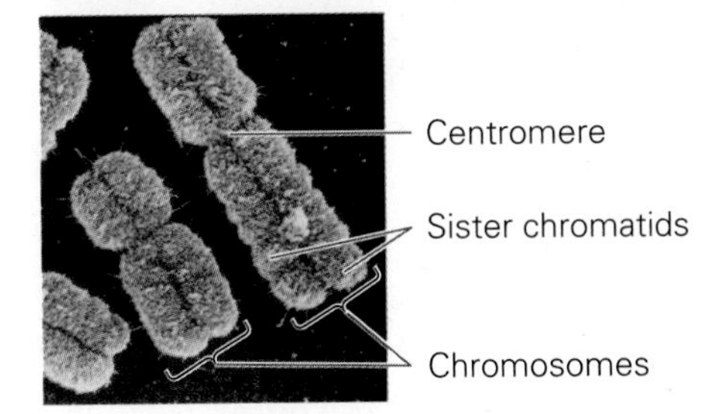

Figure 5.4 **DNA condenses during cell division.** (a) DNA in its replicated but uncondensed form prior to cell division. (b) During cell division, each copy of DNA is wrapped neatly around many small proteins, forming the condensed structure of a chromosome. After DNA replication, two identical sister chromatids are produced and joined to each other at the centromere.

(a) DNA replication

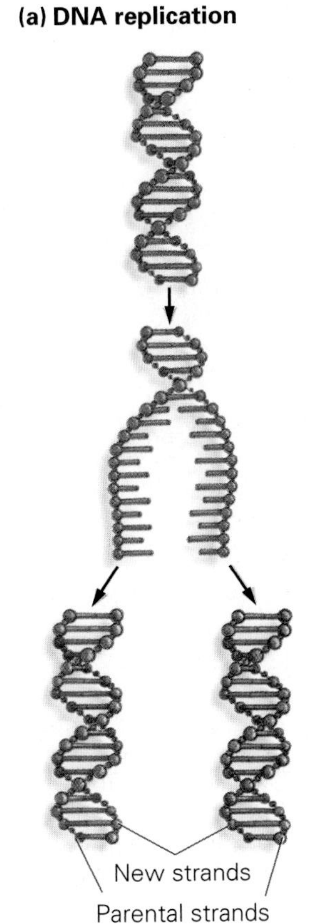

(b) The DNA polymerase enzyme facilitates replication.

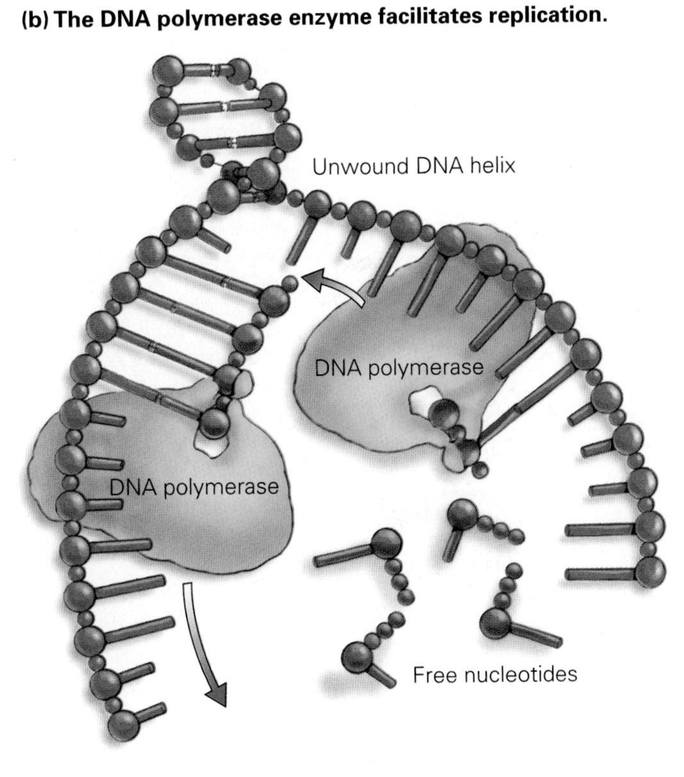

Figure 5.5 **DNA replication.** (a) DNA replication results in the production of two identical daughter DNA molecules from one parent molecule. Each daughter DNA molecule contains half of the parental DNA and half of the newly synthesized DNA. **Visualize This:** Assume another round of replication were to occur with the incoming nucleotides still being purple in color. How many total DNA molecules would be produced, and what proportion of each DNA molecule would be purple? (b) The DNA polymerase enzyme moves along the unwound helix, tying together adjacent nucleotides on the newly forming daughter DNA strand. Free nucleotides have three phosphate groups, two of which are hydrolyzed to provide energy for this reaction before the nucleotide is added to the growing chain.

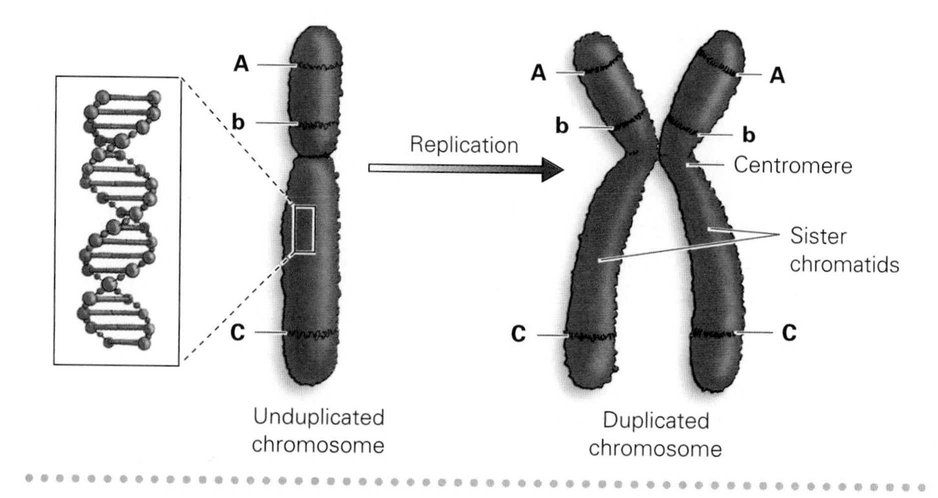

Figure 5.6 **Unreplicated and replicated chromosomes.** An unreplicated chromosome is composed of one double-stranded DNA molecule. A replicated chromosome is X-shaped and composed of two identical double-stranded DNA molecules. Each DNA molecule of the duplicated chromosome is a copy of the original chromosome and is called a sister chromatid. In this illustration, the letters A, b, and C represent different genes along the length of the chromosome.

Stop and Stretch 5.2: DNA replication is not entirely analogous to photocopying. Explain how it differs.

After a cell's DNA has been replicated, the cell is ready to divide into daughter cells. Mitosis is one way in which this division occurs.

web animation 5.2 *Mitosis*

web animation 5.3 *The Cell Cycle*

***BioFlix*™ Mitosis**

5.3 The Cell Cycle and Mitosis

Mitosis is an asexual division that produces daughter cells that are exact copies of the parent cell. Mitosis is part of the **cell cycle** or life cycle of the cell.

For cells that divide by mitosis, the cell cycle includes three steps: (1) *interphase*, when the DNA replicates; (2) *mitosis*, when the copied chromosomes move into the daughter nuclei; and (3) *cytokinesis*, when the cytoplasm of the parent cell splits *(Figure 5.7a)*. As you will see, interphase and mitosis are further divided into steps as well.

(a) Copying and partitioning DNA

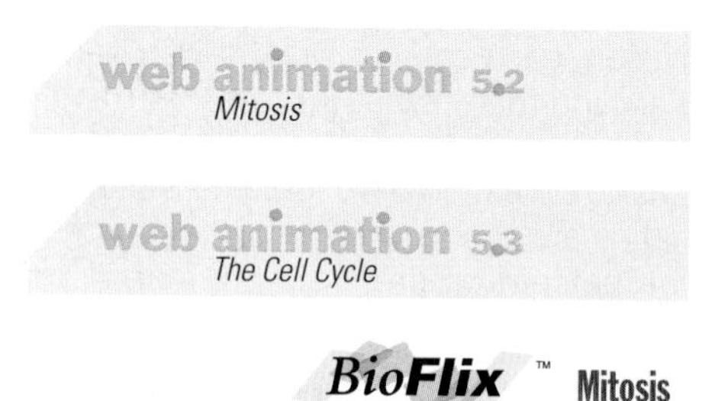

(b) Steps in the cell cycle

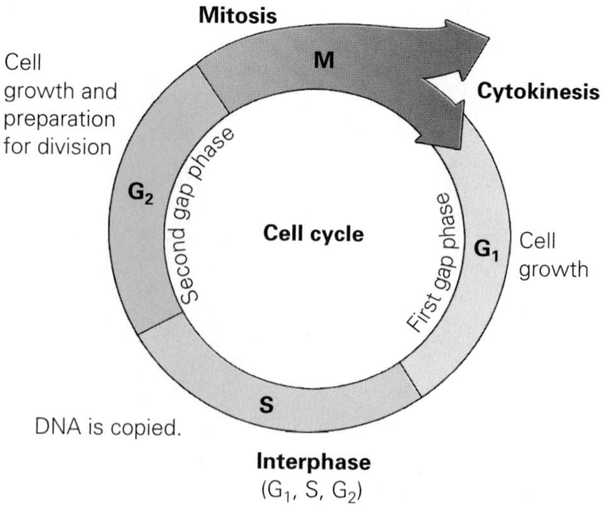

Figure 5.7 **The cell cycle.** (a) During interphase, the DNA is copied. Separation of the DNA into two daughter cells occurs during mitosis. Cytokinesis is the division of the cytoplasm, creating two cells. (b) During interphase, there are two stages when the cell grows in preparation for cell division, G_1 and G_2. During the S stage of interphase, the DNA replicates. The chromosomes are separated and two daughter cells are formed during the M phase.

Interphase

A normal cell spends most of its time in **interphase** *(Figure 5.7b)*. During this phase of the cell cycle, the cell performs its typical functions and produces the proteins required for the cell to do its particular job. For example, during interphase, a muscle cell would be producing proteins required for muscle contraction, and a blood cell would be producing proteins required to transport oxygen. Different cell types spend varying amounts of time in interphase. Cells that frequently divide, like skin cells, spend less time in interphase than do those that seldom divide, such as some nerve cells. A cell that will divide also begins preparations for division during interphase. Interphase can be separated into three phases: G_1, S, and G_2.

During the G_1 (first gap or growth) phase, most of the cell's organelles duplicate. Consequently, the cell grows larger during this phase. During the S (synthesis) phase, the DNA in the chromosomes replicates. During the G_2 (second gap) phase of the cell cycle, proteins are synthesized that will help drive mitosis to completion. The cell continues to grow and prepare for the division of chromosomes that will take place during mitosis.

Mitosis

The movement of chromosomes into new cells occurs during **mitosis.** Mitosis takes place in all cells with a nucleus, although some of the specifics of cell division differ among kingdoms. Whether these phases occur in a plant or an animal, the outcome of mitosis and the next phase, cytokinesis, is the same: the production of genetically identical daughter cells. To achieve this outcome, the sister chromatids of a replicated chromosome are pulled apart, and one copy of each is placed into each newly forming nucleus. Mitosis is accomplished during four stages: prophase, metaphase, anaphase, and telophase.

During **prophase,** the replicated chromosomes condense, allowing them to move around in the cell without becoming entangled. Protein structures called **microtubules** also form and grow, ultimately radiating out from opposite ends, or **poles,** of the dividing cell. The growth of microtubules helps the cell to expand. Motor proteins attached to microtubules also help pull the chromosomes around during cell division. The membrane that surrounds the nucleus, called the **nuclear envelope,** breaks down so that the microtubules can gain access to the replicated chromosomes. At the poles of each dividing animal cell, structures called **centrioles** physically anchor one end of each forming microtubule. Plant cells do not contain centrioles, but microtubules in these cells do remain anchored at a pole.

During **metaphase,** the replicated chromosomes are aligned across the middle, or equator, of each cell. To do this the microtubules, which are attached to each chromosome at the centromere, line up the chromosomes in single file across the middle of the cell.

During **anaphase,** the centromere splits, and the motor proteins pull each sister chromatid of a chromosome to opposite poles of the cell.

In the last stage of mitosis, **telophase,** the nuclear envelopes re-form around the newly produced daughter nuclei, and the chromosomes revert to their uncondensed form. Cytokinesis divides the cytoplasm, and daughter cells are produced. *Figure 5.8* summarizes the cell cycle in animal cells. The four stages of mitosis are nearly identical in plant cells.

Stop and Stretch 5.3: How does the similarity in the process of mitosis between animals and plants support the idea that all organisms share a common ancestor?

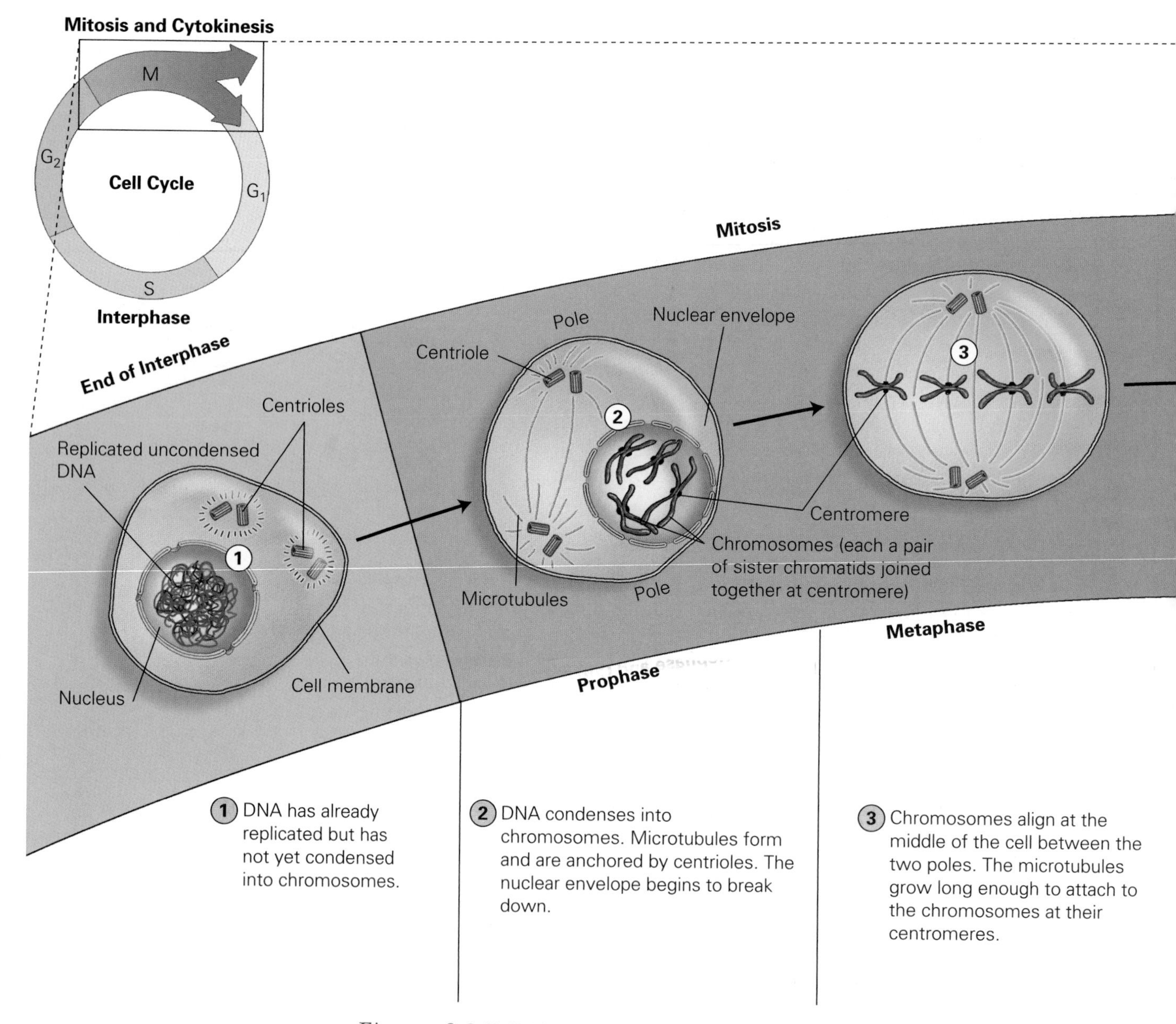

Figure 5.8 **Cell division in animal cells.** This diagram illustrates how cell division proceeds from interphase through mitosis and cytokinesis.

Cytokinesis

Cytokinesis means "cellular movement." Cytokinesis in plant cells requires that cells build a new **cell wall,** an inflexible structure surrounding the plant cells. During telophase of mitosis in a plant cell, membrane-bound vesicles from the Golgi apparatus deliver the materials required for building the cell wall to the center of the cell. The materials include a tough, fibrous carbohydrate called **cellulose** as well as some proteins. The membranes surrounding the vesicles gather in the center of the cell to form a structure called a **cell plate.** The cell plate and cell wall grow across the width of the cell and form a barrier that eventually separates the products of mitosis into two daughter cells *(Figure 5.9a)*.

Because animal cells do not have a rigid cell wall, they have evolved a different method for separating the products of mitosis into daughter cells.

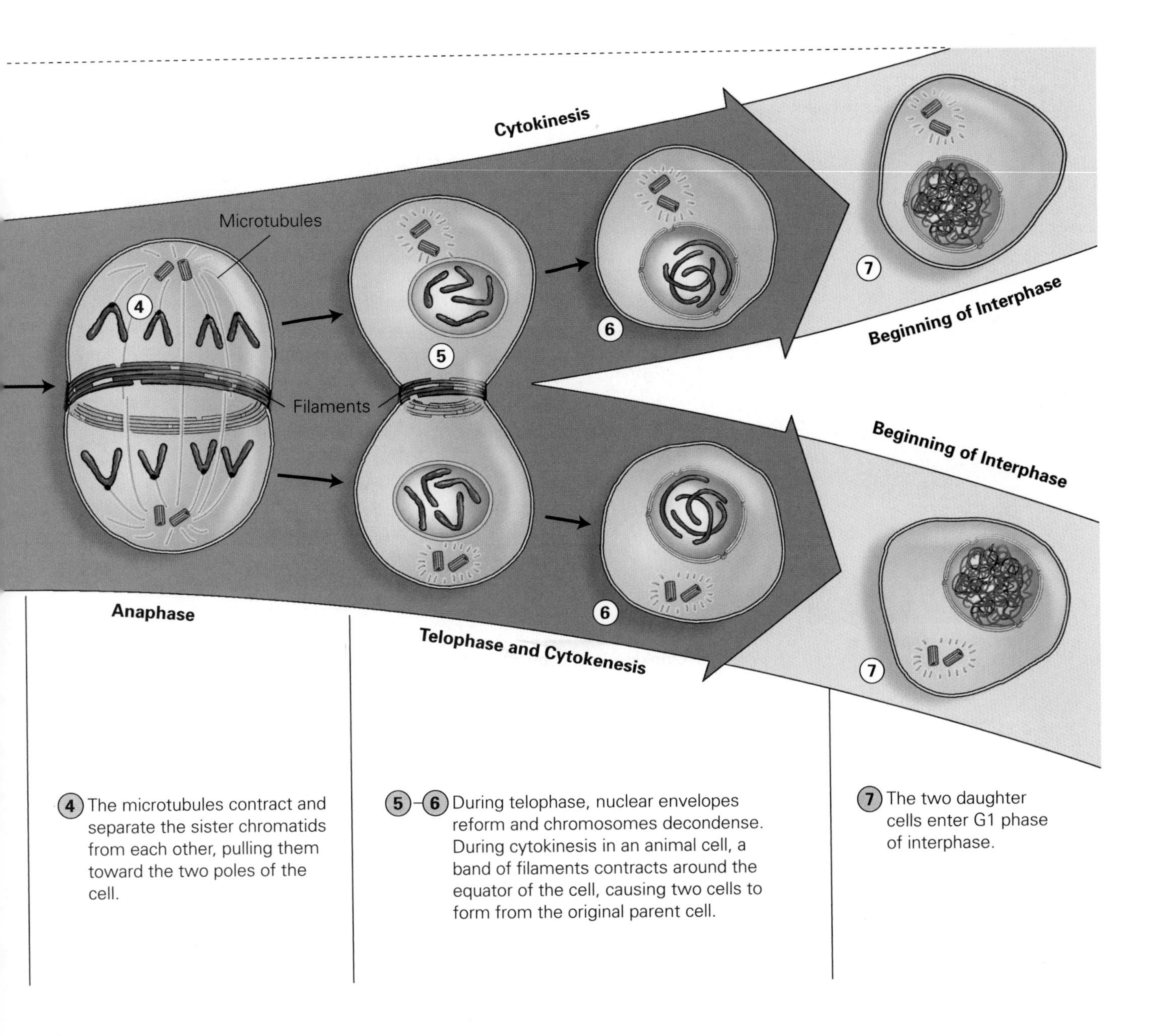

During cytokinesis in animal cells, a band of proteins encircles the cell at the equator and divides the cytoplasm. This band of proteins contracts to pinch apart the two cells that have formed from the original parent cell *(Figure 5.9b)*.

After cytokinesis, the cell reenters interphase, and if the conditions are favorable, the cell cycle may repeat itself. Cells that go through the cell cycle even if conditions are unfavorable can give rise to tumors.

5.4 Cell Cycle Control and Mutation

When cell division is working properly, it is a tightly controlled process. Cells are given signals for when and when not to divide. The normal cells in Nicole's ovary and the rest of her body were responding properly to the signals telling them when and how fast to divide. However, the cell that started her tumor was not responding properly to these signals.

(a) Cytokinesis in a plant cell

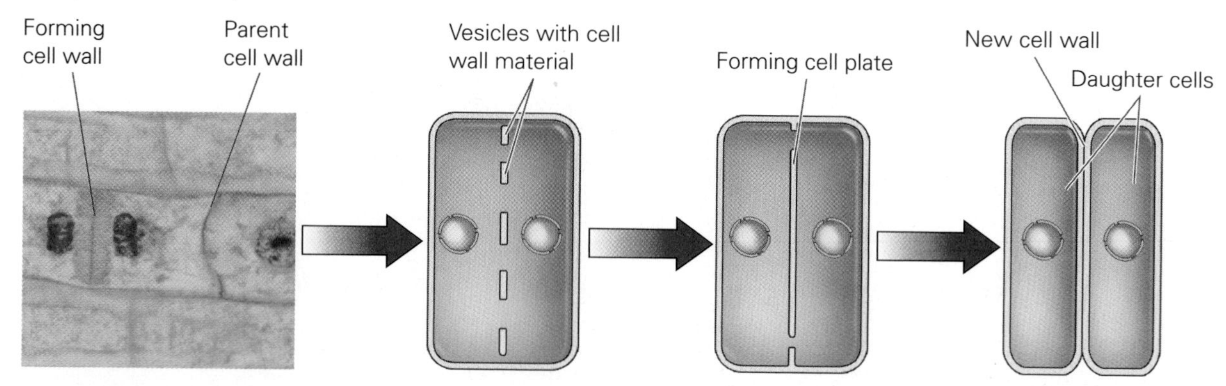

(b) Cytokinesis in an animal cell

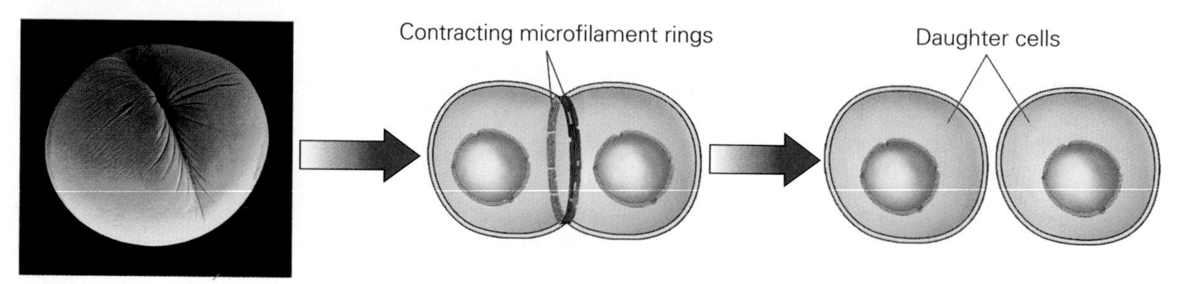

Figure 5.9 **Cytokinesis in plant and animal cells.** (a) Plant cells undergoing mitosis must do so within the confines of a rigid cell wall. Plant cells undergo prophase, metaphase, anaphase, and telophase in a manner similar to these cycles in an animal cell. During cytokinesis, plant cells form a cell plate that grows down the middle of the parent cell and eventually forms a new cell wall. (b) Animal cells, such as the frog cell shown here, produce a band of filaments that divide the cell in half.

An Overview of Controls in the Cell Cycle

Instead of proceeding in lockstep through the cell cycle, normal cells halt cell division at a series of **checkpoints.** During this stoppage, proteins survey the cell to ensure that conditions for a favorable cellular division have been met. Three checkpoints must be passed before cell division can occur; one takes place during G_1, one during G_2, and the last during metaphase *(Figure 5.10)*.

Proteins at the G_1 checkpoint determine whether it is necessary for a cell to divide. To do this, they survey the cell environment for the presence of other proteins called **growth factors** that stimulate cells to divide. When growth factors are limited in number, cell division does not occur. If enough growth factors are present to trigger cell division, then other proteins check to see if the cell is large enough to divide and if all the nutrients required for cell division are available. At the G_2 checkpoint, other proteins ensure that the DNA has replicated properly and double-check the cell size, again making sure that the cell is large enough to divide. The third and final checkpoint occurs during metaphase. Proteins present at metaphase verify that all the chromosomes have attached themselves to microtubules so that cell division can proceed properly.

If proteins surveying the cell at any of these three checkpoints determine that conditions are not favorable for cell division, the process is halted. When this happens, the cell may die.

Proteins that regulate the cell cycle, like all proteins, are coded for by genes. When these proteins are normal, cell division is properly regulated. When these cycle-regulating proteins are unable to perform their jobs, unregulated cell division leads to large masses of cells called *tumors*. Mistakes

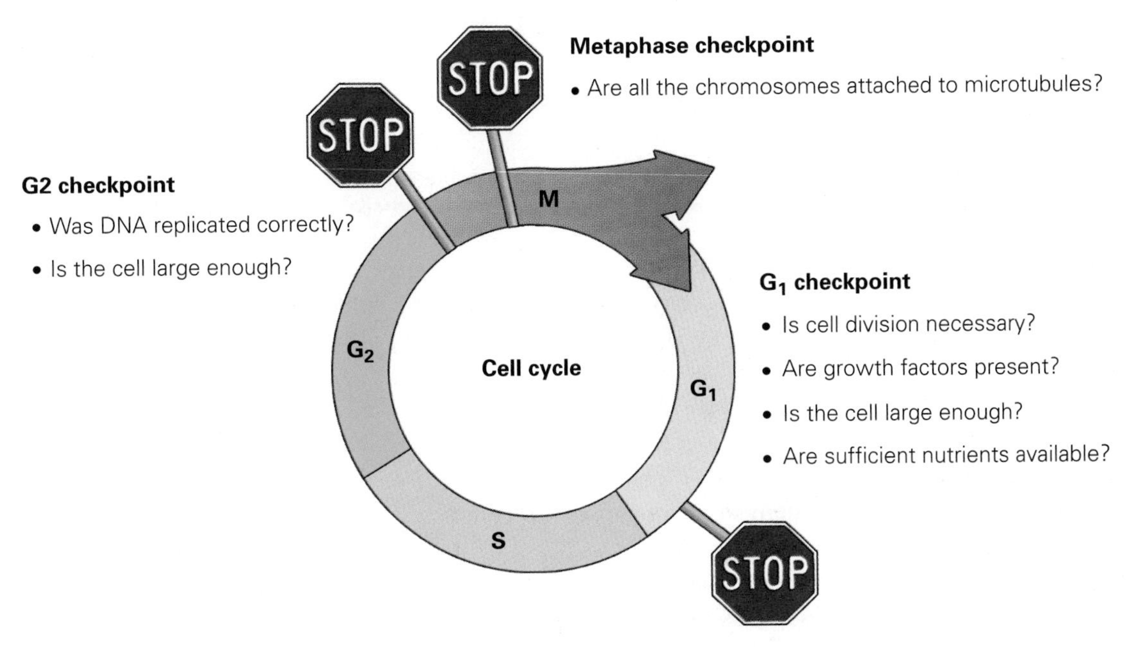

Figure 5.10 **Controls of the cell cycle.** Checkpoints at G_1, G_2, and metaphase determine whether a cell will continue to divide.

in cell cycle regulation arise when the genes controlling the cell cycle are altered, or mutated, versions of the normal genes. A **mutation** is a change in the sequence of DNA. Changes to DNA can change a gene and in turn can alter the protein that the gene encodes, or provides instructions for. Mutant proteins do not perform their required cell functions in the same way that normal proteins do. If mutations occur to genes that encode the proteins regulating the cell cycle, cells can no longer regulate cell division properly. One or more cells in Nicole's ovary must have accumulated mutations in the cell-cycle control genes, leading to the development of cancer.

How Nicole acquired cancer-causing mutations is unknown. Mutations may be inherited or induced by exposure to substances called **carcinogens** that damage DNA and chromosomes. For a substance to be considered carcinogenic, exposure to the substance must be correlated with an increased risk of cancer. Examples of carcinogens include cigarette smoke, radiation, ultraviolet light, asbestos, and some viruses.

A Closer Look: At Mutations to Cell-Cycle Control Genes

Those genes that encode the proteins regulating the cell cycle are called **proto-oncogenes** (*proto* means "before," and *onco* means "cancer"). Proto-oncogenes are normal genes located on many different chromosomes that enable organisms to regulate cell division. When they become mutated, these genes are called **oncogenes.** It is when the normal proto-oncogenes undergo mutations and become oncogenes that they become capable of causing cancer. A wide variety of organisms carry protooncogenes, which means that many different types of organisms can develop cancer (*Figure 5.11*).

Many proto-oncogenes provide the cell with instructions for building growth factors. A normal growth factor stimulates cell division only when the cellular environment is favorable and all conditions for division have been met. Oncogenes can overstimulate cell division (*Figure 5.12a*).

One gene involved in many cases of ovarian cancer is called *HER2*. (Names of genes are italicized, while names of the proteins they produce are not.) The *HER2* gene carries instructions for building a **receptor** protein, which "receives" a growth factor. When the shape of the receptor on the

(continued on the next page)

(A Closer Look continued)

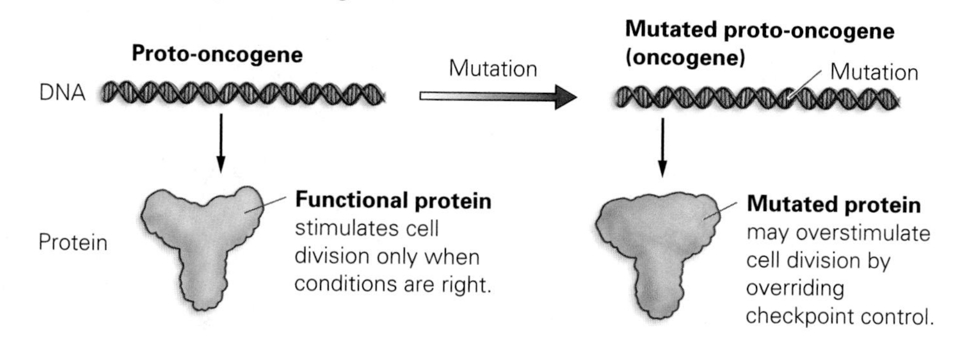

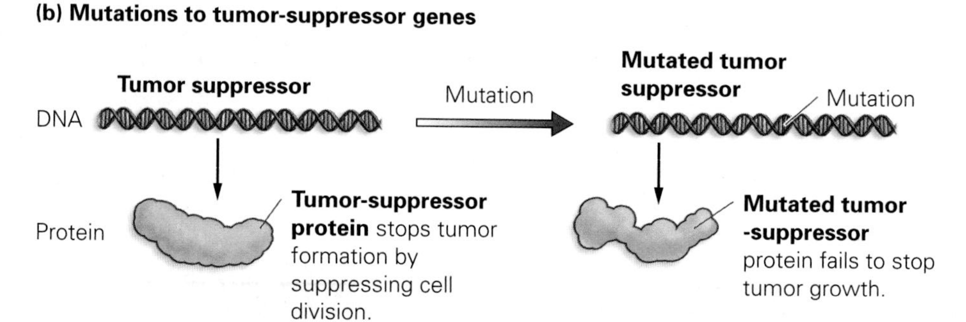

Figure 5.12 **Mutations to proto-oncogenes and tumor-suppressor genes.** (a) Mutations to proto-oncogenes and (b) tumor-suppressor genes can increase the likelihood of cancer developing.

Figure 5.11 **Cancer in nonhuman organisms.** Many organisms carry proto-oncogenes, which can mutate into oncogenes and cause the development of tumors or cancer.

cell's surface is normal, it signals the inside of the cell to allow division to occur. Mutations to the gene that encodes this receptor can result in a receptor protein with a different shape from that of the normal receptor protein. When mutated or misshapen, the receptor protein functions as if many growth factors are present, even when there are actually few-to-no growth factors.

Another class of genes involved in cancer are **tumor suppressors.** These genes, also present in all humans and many other organisms, carry the instructions for producing proteins that suppress or stop cell division if conditions are not favorable. These proteins can also detect and repair damage to the DNA. For this reason, normal tumor suppressors serve as backups in case the proto-oncogenes undergo mutation. If a growth factor overstimulates cell division, the normal tumor suppressor impedes tumor formation by preventing the mutant cell from moving through a checkpoint *(Figure 5.12b)*.

When a tumor-suppressor protein is not functioning properly, it does not force the cell to stop dividing even though conditions are not favorable. Mutated tumor suppressors also allow cells to override cell-cycle checkpoints. One well-studied tumor suppressor, named p53, helps to determine whether cells will repair damaged DNA or commit cellular suicide if the damage is too severe. Mutations to the gene that encodes p53 result in damaged DNA being allowed to proceed through mitosis, thereby passing on even more mutations. Over half of all human cancers involve mutations to the gene that encodes p53.

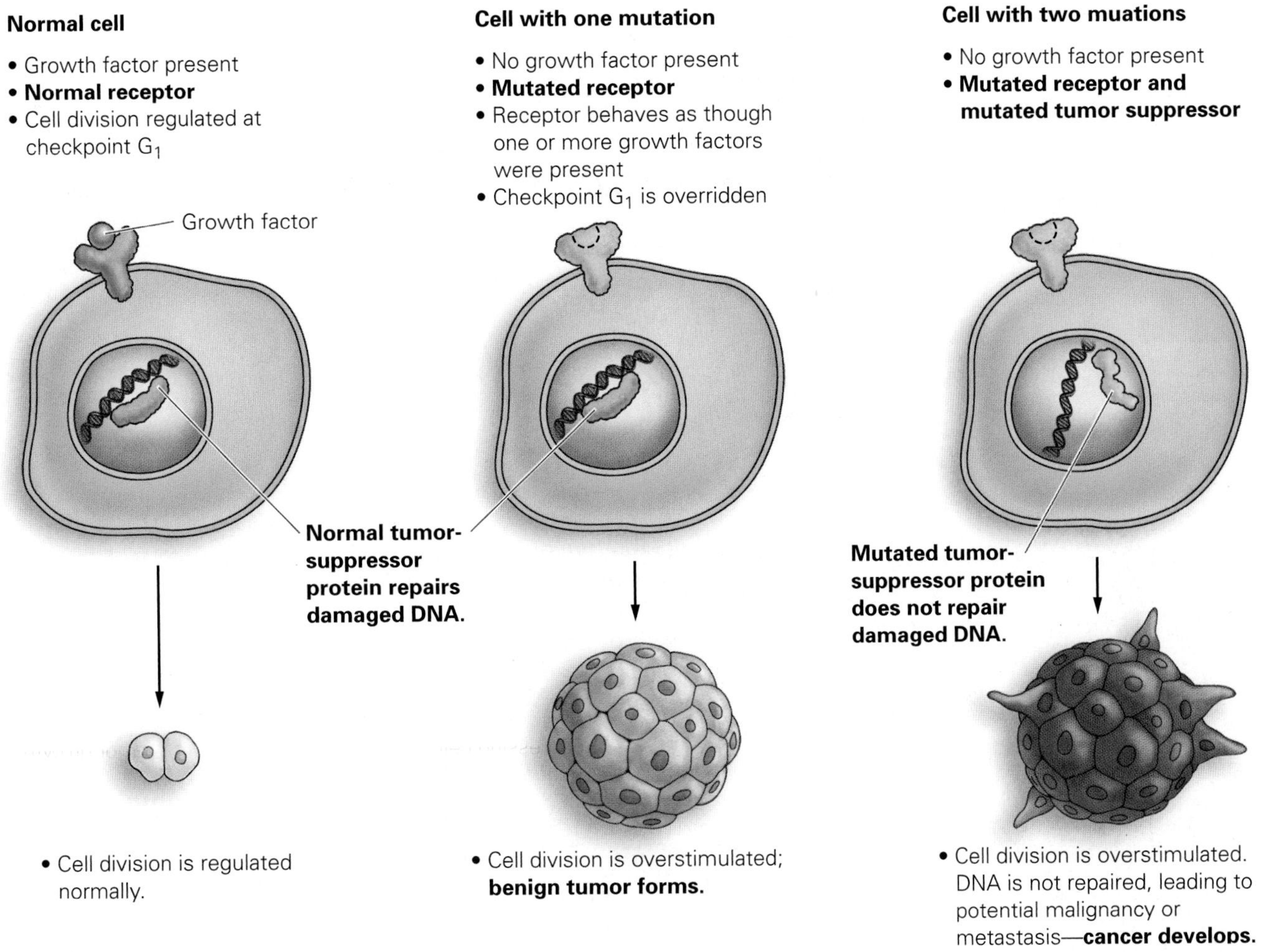

Normal cell

- Growth factor present
- **Normal receptor**
- Cell division regulated at checkpoint G_1

Growth factor

Cell with one mutation

- No growth factor present
- **Mutated receptor**
- Receptor behaves as though one or more growth factors were present
- Checkpoint G_1 is overridden

Cell with two muations

- No growth factor present
- **Mutated receptor and mutated tumor suppressor**

Normal tumor-suppressor protein repairs damaged DNA.

Mutated tumor-suppressor protein does not repair damaged DNA.

- Cell division is regulated normally.

- Cell division is overstimulated; **benign tumor forms.**

- Cell division is overstimulated. DNA is not repaired, leading to potential malignancy or metastasis—**cancer develops.**

Figure 5.13 **Growth factor receptors and tumor suppressors.** When a cell has normal growth factors and tumor suppressors, cell division is properly regulated. Mutations to growth factor receptors can cause cell division to be overstimulated, and a benign tumor can form. Additional mutations to tumor suppressor genes increase the likelihood of malignancy and metastasis.

Mutations to a tumor-suppressor gene are often found in ovarian cells that have become cancerous. Researchers believe that a normal *BRCA2* gene encodes a protein that is involved in helping to repair damaged DNA. The misshapen, mutant version of the protein cannot help to repair damaged DNA. This means that damaged DNA will be allowed to undergo mitosis, thus passing new mutations on to their daughter cells. As more and more mutations occur, the probability that a cell will become cancerous increases. *Figure 5.13* summarizes the roles of growth factors and tumor suppressors in the development of cancer.

Cancer Development Requires Many Mutations

Some mutations that occur as a result of damaged DNA being allowed to undergo mitosis are responsible for the progression of a tumor from a benign state, to a malignant state, to metastasis. For example, some cancer cells can stimulate the growth of surrounding blood vessels through a process called **angiogenesis.** These cancer cells secrete a substance that attracts and reroutes blood vessels so that they supply a developing tumor with oxygen (necessary for cellular respiration) and other nutrients. When a tumor has its own blood supply, it can grow at the expense of other, noncancerous cells.

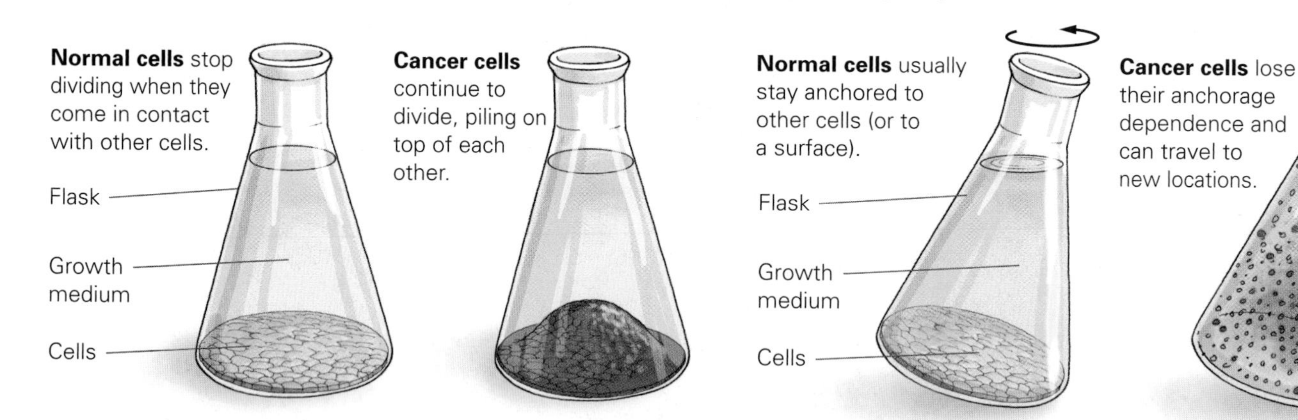

Figure 5.14 **Contact inhibition and anchorage dependence.** (a) When normal cells are grown on a solid support such as the bottom of a flask, they grow and divide until they cover the bottom of the flask.(b) Cancer cells lose the requirement that they adhere to other cells or a solid support.

Because the growth of rapidly dividing cancer cells occurs more quickly than the growth of normal cells in this process, entire organs can eventually become filled with cancerous cells. When this occurs, an organ can no longer work properly, leading to weakened functioning or organ failure. Damage to organs also explains some of the pain associated with cancer.

Normal cells also display a property called **contact inhibition,** which prevents them from dividing when doing so would require them to pile up on each other. Cancer cells, conversely, continue to divide and form a tumor *(Figure 5.14a)*. In addition, normal cells do need some contact with an underlayer of cells to stay in place. This phenomenon is the result of a process called **anchorage dependence** *(Figure 5.14b)*. Cancer cells override this requirement for some contact with other cells because cancer cells are dividing too quickly and do not expend enough energy to secrete adhesion molecules that glue the cells together. Once a cell loses its anchorage dependence, it may leave the original tumor and move to the blood, lymph, or surrounding tissues.

Most cells are programmed to divide a certain number of times—usually 60 to 70 times—and then they stop dividing. This limits most developing tumors to a small mole, cyst, or lump, all of which are benign. Cancer cells, however, do not obey these life-span limits. Instead, they are **immortal.** They achieve immortality by activating a gene that is usually turned off after early development. This gene produces an enzyme called **telomerase** that helps prevent the degradation of chromosomes. As chromosomes degrade with age, a cell loses its ability to divide. In cancer cells, telomerase is reactivated, allowing the cells to divide without limit.

Stop and Stretch 5.4: Telomerase is turned off early in development, causing chromosomes to age and eventually lose their ability to replicate. What does this fact imply about the life span of an individual?

In Nicole's case, the progression from normal ovarian cells to cancerous cells may have occurred as follows: (1) One single cell in her ovary may have acquired a mutation to its *HER2* growth factor receptor gene. (2) The descendants of this cell would have been able to divide faster than neighboring cells, forming a small, benign tumor. (3) Next, a cell within the tumor may have undergone a mutation to its *BRCA2* tumor suppressor gene, resulting in the inability of the BRCA2 protein to fix damaged DNA in the cancerous cells. (4) Cells produced by the mitosis of these doubly mutant cells would continue to divide even though their DNA is damaged, thereby enlarging the tumor and producing cells with more mutations. (5) Subsequent mutations could result in angiogenesis, lack of contact inhibition, reactivation of the telomerase enzyme, or overriding of anchorage dependence. If Nicole were very unlucky, the end result of these mutations could be that cells carrying many mutations would break away from the original ovarian tumor and set up a cancer at one or more new locations in her body.

Multiple-Hit Model. Because multiple mutations are required for the development and progression of cancer, scientists describe the process of cancer development using the phrase **multiple-hit model.** Nicole may have inherited some of these mutations, or they may have been induced by environmental exposures. Even though cancer is a disease caused by malfunctioning genes, most cancers are not caused only by the inheritance of mutant genes. In fact, scientists estimate that close to 70% of cancers are caused by mutations that occur during a person's lifetime.

Most of us will inherit few if any mutant cell-cycle control genes. Our level of exposure to risk factors will determine whether enough mutations will accumulate during our lifetime to cause cancer. Risk factors for ovarian cancer include smoking and uninterrupted ovulation. Ovarian cancer risk is thought to decrease for many years if ovulation is prevented for periods of time, as it is when a woman is pregnant, breastfeeding, or taking the birth control pill. When an egg cell is released from the ovary during ovulation, some tissue damage occurs. In the absence of ovulation, there is no damaged ovarian tissue and hence no need for extra cell divisions to repair damaged cells. Because mutations are most likely to occur when DNA is replicating before division, fewer divisions equal a lower likelihood of the appearance of cancer-causing mutations.

Regardless of its origin, once cancer is suspected, detection and treatment follow.

5.5 Cancer Detection and Treatment

Early detection and treatment of cancer increase the odds of survival dramatically. Being on the lookout for warning signs *(Figure 5.15)* can help alert individuals to developing cancers.

Detection Methods: Biopsy

Different cancers are detected using different methods. Some cancers are detected by the excess production of proteins that are normally produced by a particular cell type. Ovarian cancers often show high levels of CA125 protein in the blood. Nicole's level of CA125 led her gynecologist to think that a tumor might be forming on her remaining ovary. Once he suspected a tumor, Nicole's physician scheduled a biopsy.

A **biopsy** is the surgical removal of cells, tissue, or fluid that will be analyzed to determine whether they are cancerous.

When viewed under a microscope, benign tumors consist of orderly growths of cells that resemble the cells of the tissue from which they were taken. Malignant or cancerous cells do not resemble other cells found in the same tissue; they are dividing so rapidly that they do not have time to produce all the proteins necessary to build normal cells. This leads to the often-abnormal appearance of cancer cells as seen under a microscope *(Figure 5.16)*.

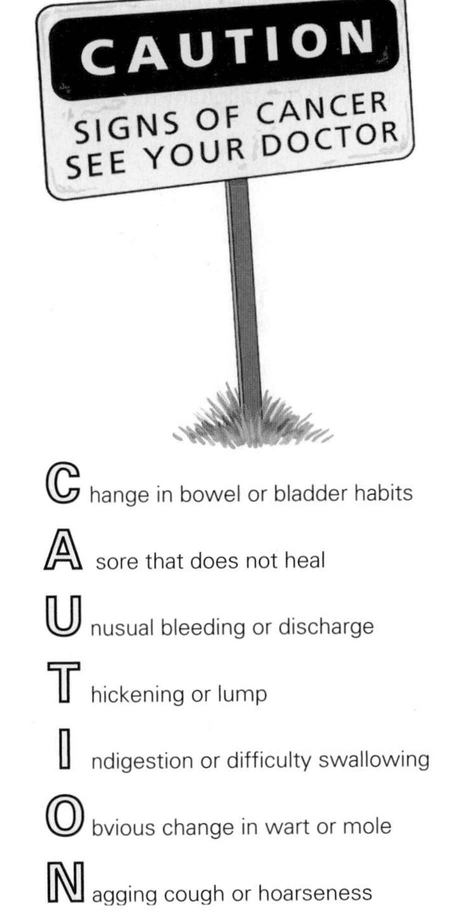

Figure 5.15 **Warning signs of cancer.** Self-screening for cancer can save your life. If you experience one or more of these warning signs, see your doctor.

(a) Normal ovarian tissue

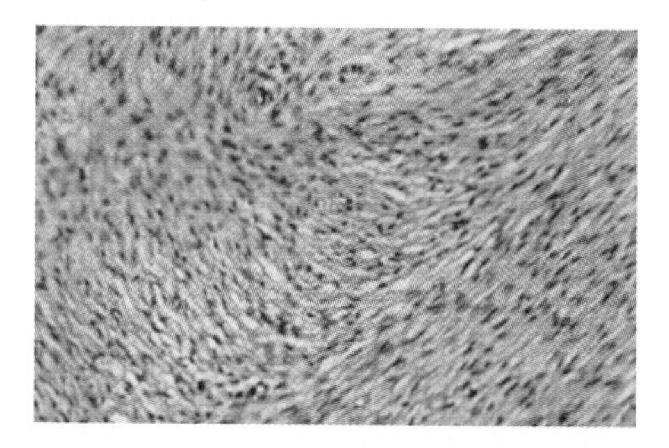

(b) Malignant ovarian tumor

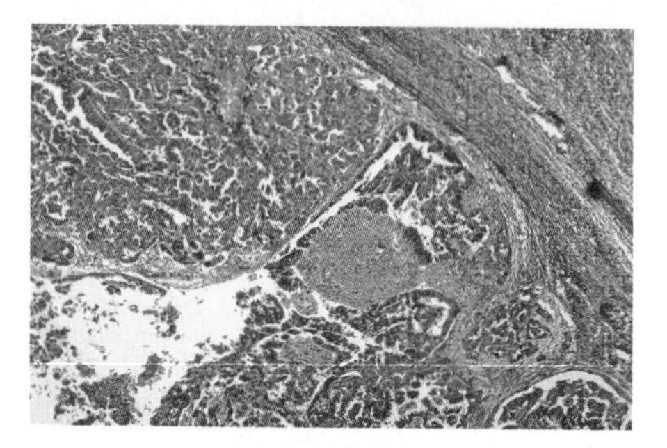

Figure 5.16 **Biopsy.** When stained and viewed under a microscope, (a) normal cells have a different appearance than benign and (b) malignant tumors.

A needle biopsy is usually performed if the cancer is located on or close to the surface of the patient's body. For example, breast lumps are often biopsied with a needle to determine whether the lump contains fluid and is a noncancerous cyst or whether it contains abnormal cells and is a tumor. When a cancer is diagnosed, surgery is often performed to remove as much of the cancerous growth as possible without damaging neighboring organs and tissues.

In Nicole's case, getting at the ovary to find tissue for a biopsy required the use of a surgical instrument called a **laparoscope.** For this operation, the surgeon inserted a small light and a scalpel-like instrument through a tiny incision above Nicole's navel.

Nicole's surgeon preferred to use the laparoscope since he knew Nicole would have a much easier recovery from laparoscopic surgery than she had from the surgery to remove her other, cystic ovary. Laparoscopy had not been possible when removing Nicole's other ovary—the cystic ovary had grown so large that her surgeon had to make a large abdominal incision to remove it.

A laparoscope has a small camera that projects images from the ovary onto a monitor that the surgeon views during surgery. These images showed that Nicole's tumor was a different shape, color, and texture from the rest of her ovary. They also showed that the tumor was not confined to the surface of the ovary; in fact, it appeared to have spread deeply into her ovary. Nicole's surgeon decided to shave off only the affected portion of the ovary and leave as much intact as possible, with the hope that the remaining ovarian tissue might still be able to produce egg cells. He then sent the tissue to a laboratory so that the pathologist could examine it. Unfortunately, when the pathologist looked through the microscope this time, she saw the disorderly appearance characteristic of cancer cells. Nicole's ovary was cancerous, and further treatment would be necessary.

Treatment Methods: Chemotherapy and Radiation

A treatment that works for one woman with ovarian cancer might not work for another ovarian cancer patient because a different suite of mutations may have led to the cancer in each woman's ovary. Luckily for Nicole, her ovarian cancer was diagnosed very early. Regrettably, this is not the case for most women with ovarian cancer, many of whom are diagnosed when symptoms become severe, leading them to see their physician. The symptoms of ovarian cancer tend to be vague and slow to develop. These symptoms include abdominal swelling, pain, bloating, gas, constipation, indigestion, menstrual disorders, and fatigue. Unfortunately, many women simply overlook these discomforts. The difficulty of diagnosis is compounded because no routine screening tests are available. For instance, CA125 levels are checked only when ovarian cancer is suspected because (1) ovaries are not the only tissues that secrete this protein; (2) CA125 levels vary from individual to individual; and (3) these levels depend on the phase of the woman's menstrual cycle. Elevated CA125 levels usually mean that the cancer has been developing for a long time. Consequently, by the time the diagnosis is made, the cancer may have grown quite large and metastasized, making it much more difficult to treat.

Nicole's cancer was caught early. However, her physician was concerned that some of her cancerous ovarian cells may have spread through blood vessels or lymph ducts on or near the ovaries or spread into her abdominal cavity, so he started Nicole on chemotherapy after her surgery.

Chemotherapy. During **chemotherapy,** chemicals are injected into the bloodstream. These chemicals selectively kill dividing cells. A variety of chemotherapeutic agents act in different ways to interrupt cell division.

Chemotherapy involves many drugs since most chemotherapeutic agents affect only one type of cellular activity. Cancer cells are rapidly dividing and do not take the time to repair mistakes in replication that lead to mutations. These cells are allowed to proceed through the G_2 checkpoint with many mutations. Therefore, cancer cells can randomly undergo mutations, a few of which might allow them to evade the actions of a particular chemotherapeutic agent. Cells that are resistant to one drug proliferate when the chemotherapeutic agent clears away the other cells that compete for space and nutrients. Cells with a preexisting resistance to the drugs are selected for and produce more daughter cells with the same resistant characteristics, requiring the use of more than one chemotherapeutic agent.

Scientists estimate that cancer cells become resistant at a rate of approximately one cell per million. Because tumors contain about one billion cells, the average tumor will have close to 1000 resistant cells. Therefore, treating a cancer patient with a combination of chemotherapeutic agents aimed at different mechanisms increases the chances of destroying all the cancerous cells in a tumor.

Unfortunately, normal cells that divide rapidly are also affected by chemotherapy treatments. Hair follicles, cells that produce red blood cells, white blood cells, and cells that line the intestines and stomach are often damaged or destroyed. The effects of chemotherapy therefore include temporary hair loss, anemia (dizziness and fatigue due to decreased numbers of red blood cells), and lowered protection from infection due to decreases in the number of white blood cells. In addition, damage to the cells of the stomach and intestines can lead to nausea, vomiting, and diarrhea.

Several hours after each chemotherapy treatment, Nicole became nauseous; she often had diarrhea and vomited for a day or so after her treatments. Midway through her chemotherapy treatments, Nicole lost most of her hair.

Radiation Therapy. Cancer patients often undergo radiation treatments as well as chemotherapy. **Radiation therapy** uses high-energy particles to injure or destroy cells by damaging their DNA, making it impossible for these cells to continue to grow and divide. Radiation is applied directly to the tumor when possible. A typical course of radiation involves a series of 10 to 20 treatments performed after the surgical removal of the tumor, although sometimes radiation is used before surgery to decrease the size of the tumor. Radiation therapy is typically used only when cancers are located close to the surface of the body because it is difficult to focus a beam of radiation on internal organs such as an ovary. Therefore, Nicole's physician recommended chemotherapy only.

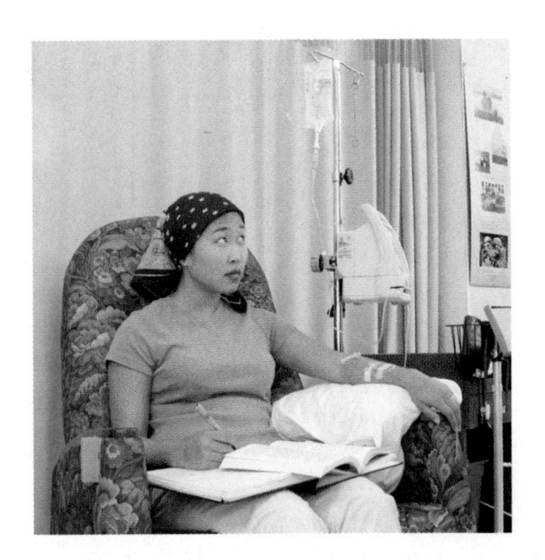

Figure 5.17 **Chemotherapy.** Many chemotherapeutic agents, such as Nicole's Taxol, are administered through an intravenous (IV) needle.

Stop and Stretch 5.5: One risk of radiation therapy is an increased likelihood of secondary cancer emerging 5 to 15 years later. Why might this treatment increase cancer risk?

Nicole's treatments consisted of many different chemotherapeutic agents, spread over many months. The treatments took place at the local hospital on Wednesdays and Fridays. She usually had a friend drive her to the hospital very early in the morning and return later in the day to pick her up. The drugs were administered through an intravenous (IV) needle into a vein in her arm *(Figure 5.17)*. During the hour or so that she was undergoing chemotherapy, Nicole usually studied for her classes. She did not mind the actual chemotherapy treatments that much. The hospital personnel were kind to her, and she got some studying done. It was the aftermath of these treatments that she hated. Most days during her chemotherapy regimen, Nicole was so exhausted that she did not get out of bed until late morning, and on the day after her treatments, she often slept until late afternoon. Then she would get up and try to get some work done or make some phone calls before going back to bed early in the evening. After 6 weeks of chemotherapy, Nicole's CA125 levels started to drop. After another two months of chemotherapy, her CA125 levels were back down to their normal, precancerous level. If Nicole has normal CA125 levels for five years, she will be considered to be in **remission,** or no longer suffering negative impacts from cancer. After ten years of normal CA125 levels, she will be considered cured of her cancer. Because Nicole's cancer responded to chemotherapy, she was spared from having to undergo any other, more experimental treatments.

Even though her treatments seemed to be going well, Nicole had other worries. She worried that her remaining ovary would not recover from the surgery and chemotherapy, which meant that she would never be able to have children. Nicole had always assumed that she would have children someday, and although she did not currently have a strong desire to have a child, she wondered if her feelings would change. Even though she was not planning to marry anytime soon, she also wondered how her future husband would feel if she were not able to become pregnant.

In addition to her concerns about being able to become pregnant, Nicole also became worried that she might pass on mutated, cancer-causing genes to her children. For Nicole, or anyone, to pass on genes to his or her children, reproductive cells must be produced by another type of nuclear division called meiosis.

web animation 5.4 *Meiosis*

BioFlix™ Meiosis

5.6 Meiosis

Meiosis is a form of cell division that occurs only in specialized cells within the **gonads** or sex organs. In humans, and in most animals, the male gonads are the testes, and the female gonads are the ovaries. During meiosis, specialized sex cells called **gametes** are produced. In animals, the male gametes are the sperm cells, while the gametes produced by the female are the egg cells.

Gametes differ from other cells of the body, called **somatic cells,** in that they contain half the number of chromosomes. Animal somatic cells include cells of the skin, muscle, liver, stomach tissues, and so on. Because human somatic cells have 46 chromosomes and meiosis reduces the chromosome number by one-half, the gametes produced during meiosis contain 23 chromosomes each. When an egg cell and a sperm cell combine their 23 chromosomes at fertilization, the developing embryo will then have the required 46 chromosomes.

The placement of chromosomes into gametes is not random; that is, meiosis does not simply place any 23 of the 46 human chromosomes into a gamete. Instead, meiosis apportions chromosomes in a very specific manner. The 46 chromosomes in human body cells are actually 23 different pairs of chromosomes, and meiosis produces cells that contain one chromosome of every pair.

It is possible to visualize chromosome pairs by preparing a **karyotype** *(Figure 5.18)*, a highly magnified photograph of the chromosomes arranged in pairs. A karyotype is usually prepared from chromosomes that have been removed from the nuclei of white blood cells, which have been treated with chemicals to stop mitosis at metaphase. Because these chromosomes are at metaphase of mitosis, they are composed of replicated sister chromatids and are shaped like the letter X. It is possible to photograph chromosomes and then digitally arrange them in pairs. The 46 human chromosomes can be arranged into 22 pairs of nonsex chromosomes, or **autosomes,** and one pair of **sex chromosomes** (the X and Y chromosomes) to make a total of 23 pairs. Human males have an X and a Y chromosome, while females have two X chromosomes. Each chromosome is paired with a mate that is the same size and shape and has its centromere in the same position.

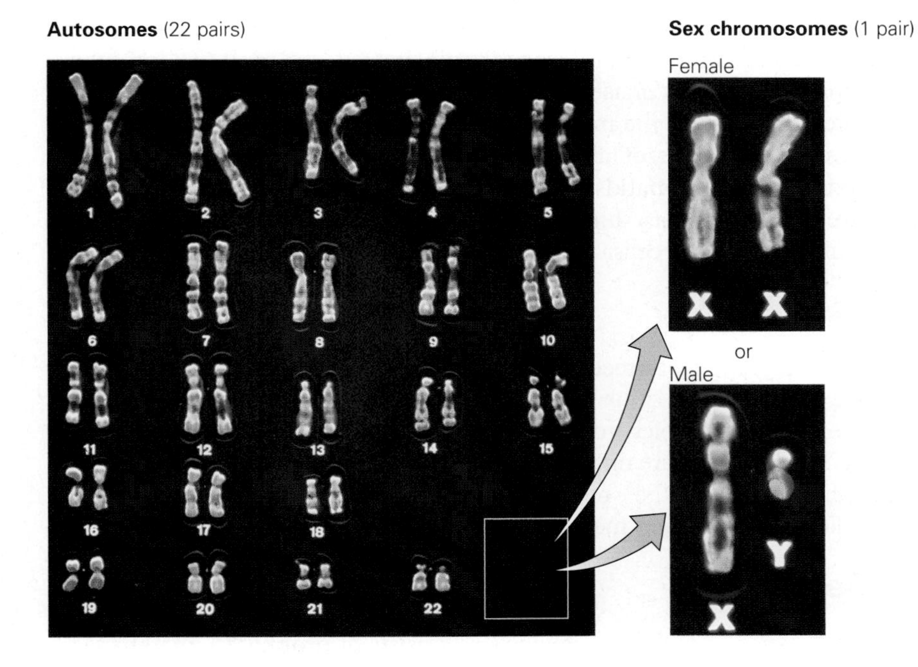

Figure 5.18 **Karyotype.** The pairs of chromosomes in this karyotype are arranged in order of decreasing size and numbered from 1 to 22. The X and Y sex chromosomes are the 23rd pair. The sex chromosomes from a female and a male are shown in the insets. **Visualize This:** How do the X and Y chromosomes differ in terms of structure?

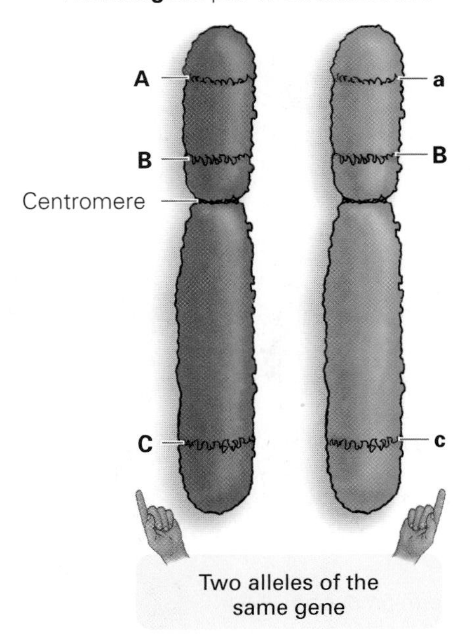

Figure 5.19 **Homologous and non-homologous pairs of chromosomes.** (a) Homologous pairs of chromosomes have the same genes (shown here as A, B, and C) but may have different alleles. The dominant allele is represented by an uppercase letter, while the recessive allele is shown with the same letter in lowercase. Note that the chromosomes of this pair each have the same size, shape, and positioning of the centromere. One member of each pair is inherited from one's mother (and colored pink), while the other is inherited from one's father (and colored blue).

The pairs of nonsex chromosomes are called **homologous pairs.** Each member of a homologous pair of chromosomes carries the same genes along its length, although not necessarily the same versions of those genes *(Figure 5.19)*. In your cells, one member of each pair was inherited from your mother and the other from your father.

Different versions of the same gene are called **alleles** of a gene. For example, there are normal and mutant alleles of the *BRCA2* gene. Note that there is a difference between the same type of information in the sense that both alleles of this gene code for a cell-cycle control protein, but they happen to code for different versions of the same protein. Alleles are alternate forms of a gene in the same way that chocolate and vanilla are alternate forms of ice cream.

When a chromosome is replicated, during the S phase of the cell cycle, the DNA is duplicated. Replication results in two copies, called *sister chromatids*, that are genetically identical. For this reason, we would find exactly the same information on the sister chromatids that comprise a replicated chromosome *(Figure 5.20)*.

Meiosis separates the members of a homologous pair from each other. Once meiosis is completed, there is one copy of each chromosome (1–23) in every gamete. When only one member of each homologous pair is present in a cell, we say that the cell is **haploid (*n*)**—both egg cells and sperm cells are haploid. All somatic cells in humans contain homologous pairs of chromosomes and are therefore diploid. For a diploid cell in a person's testes or ovary to become a haploid gamete, it must go through meiosis. After the sperm and egg fuse, the fertilized cell, or **zygote,** will contain two sets of chromosomes and is said to be **diploid (2*n*)** *(Figure 5.21)*. It is not just humans that undergo meiosis to produce gametes; all cells with nuclei and membrane-bound

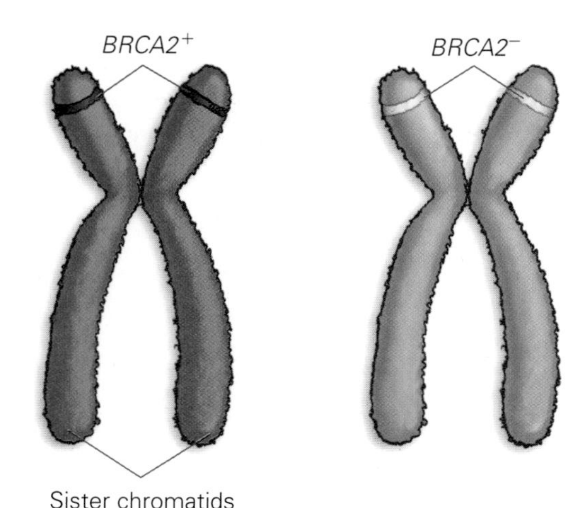

Figure 5.20 **Replicated chromosomes.** This homologous pair of chromosomes has been replicated. Note that the normal version of the BRCA2 gene (symbolized *BRCA2*$^+$) is present on both sister chromatids of the chromosome on the left. Its homologue on the right carries the mutant version of this allele (symbolized *BRCA2*$^-$).

organelles—eukaryotic cells—undergo meiosis. Like mitosis, meiosis is preceded by an interphase stage that includes G_1, S, and G_2. Interphase is followed by two phases of meiosis, called meiosis I and meiosis II, in which divisions of the nucleus take place *(Figure 5.22)*. Meiosis I separates the members of a homologous pair from each other. Meiosis II separates the chromatids from each other. Both meiotic divisions are followed by cytokinesis, during which the cytoplasm is divided between the resulting daughter cells.

Interphase

The interphase that precedes meiosis consists of G_1, S, and G_2. This interphase of meiosis is similar in most respects to the interphase that precedes mitosis. The centrioles from which the microtubules will originate are present. The G phases are times of cell growth and preparation for division. The S phase is when DNA replication occurs. Once the cell's DNA has been replicated, it can enter meiosis I.

Meiosis I

The first meiotic division, meiosis I, consists of prophase I, metaphase I, anaphase I, and telophase I *(Figure 5.23)*.

During prophase I of meiosis, the nuclear envelope starts to break down, and the microtubules begin to assemble. The previously replicated chromosomes condense so that they can be moved around the cell without becoming entangled. The condensed chromosomes can be seen under a microscope. At this time, the homologous pairs of chromosomes

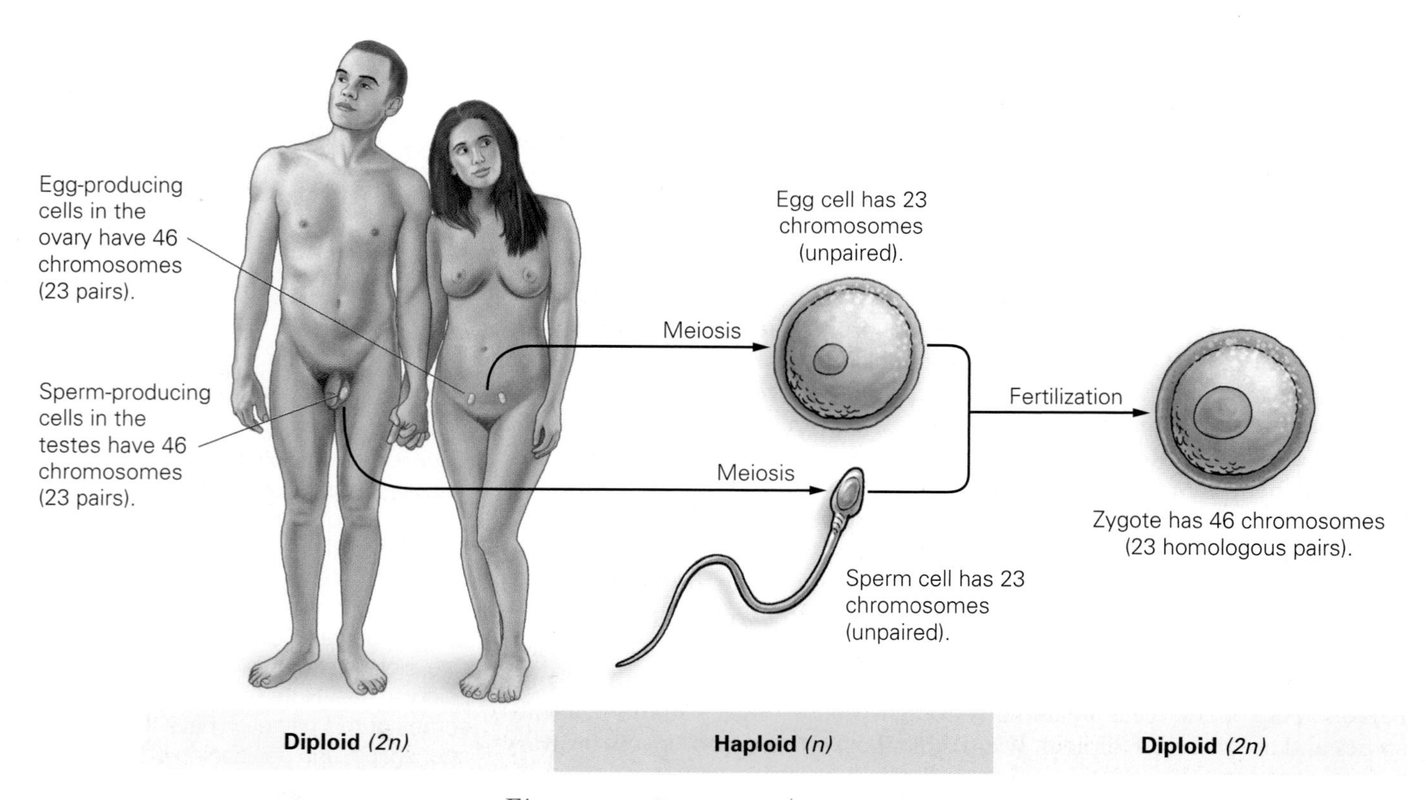

Figure 5.21 **Gamete production.** The diploid cells of the ovaries and testes undergo meiosis and produce haploid gametes. At fertilization, the diploid condition is restored.

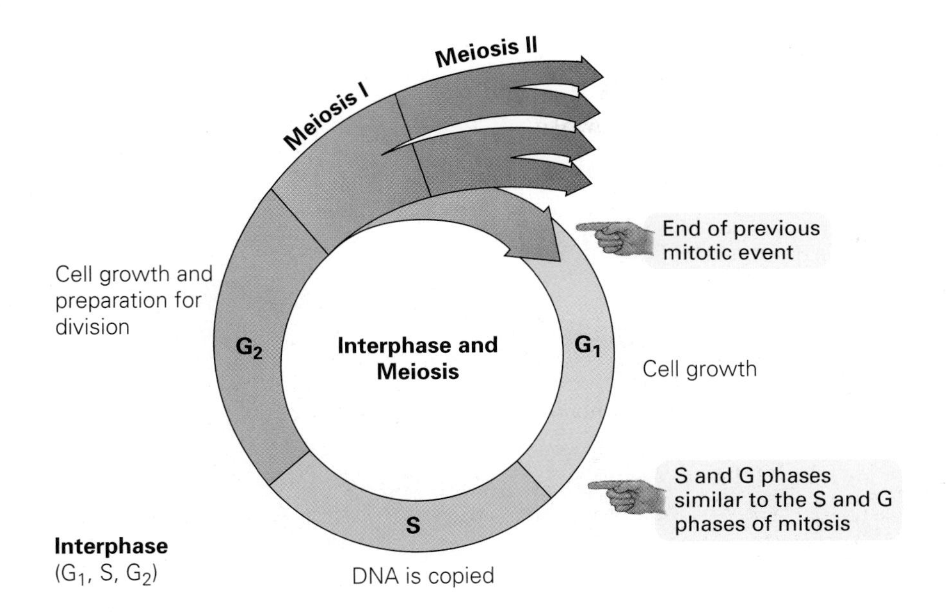

Figure 5.22 **Interphase and meiosis.** Interphase consists of G_1 S, and G_2 and is followed by two rounds of nuclear division, meiosis I and meiosis II.

exchange genetic information in a process called *crossing over*, which will be explained in a moment.

At metaphase I, the homologous pairs line up at the cell's equator, or middle of the cell. Microtubules bind to the metaphase chromosomes near the centromere. Homologous pairs are arranged arbitrarily regarding which member faces which pole. This process is called *random alignment*. At the end of this section, you will find detailed descriptions of crossing over and random alignment along with their impact on genetic diversity.

At anaphase I, the homologous pairs are separated from each other by the shortening of the microtubules, and at telophase I, nuclear envelopes re-form around the chromosomes. DNA is then partitioned into each of the two daughter cells by cytokinesis. Because each daughter cell contains only one copy of each member of a homologous pair, at this point the cells are haploid. Now both of these daughter cells are ready to undergo meiosis II.

Stop and Stretch 5.6: How is meiosis I similar to mitosis? How is it different?

Meiosis II

Meiosis II consists of prophase II, metaphase II, anaphase II, and telophase II. This second meiotic division is virtually identical to mitosis and serves to separate the sister chromatids of the replicated chromosome from each other.

At prophase II of meiosis, the cell is readying for another round of division, and the microtubules are lengthening again. At metaphase II, the chromosomes align across the equator in much the same way as they do during mitosis—not as pairs, as was the case with metaphase I. At anaphase II, the sister chromatids separate from each other and move to opposite poles of the cell. At telophase II, the separated chromosomes each become enclosed in their own nucleus. In this fashion, half of a person's genes are physically placed into each gamete; thus, children carry one-half of each parent's genes.

Each parent can produce millions of different types of gametes due to two events that occur during meiosis I—crossing over and random alignment. Both of these processes greatly increase the number of different kinds of gametes that

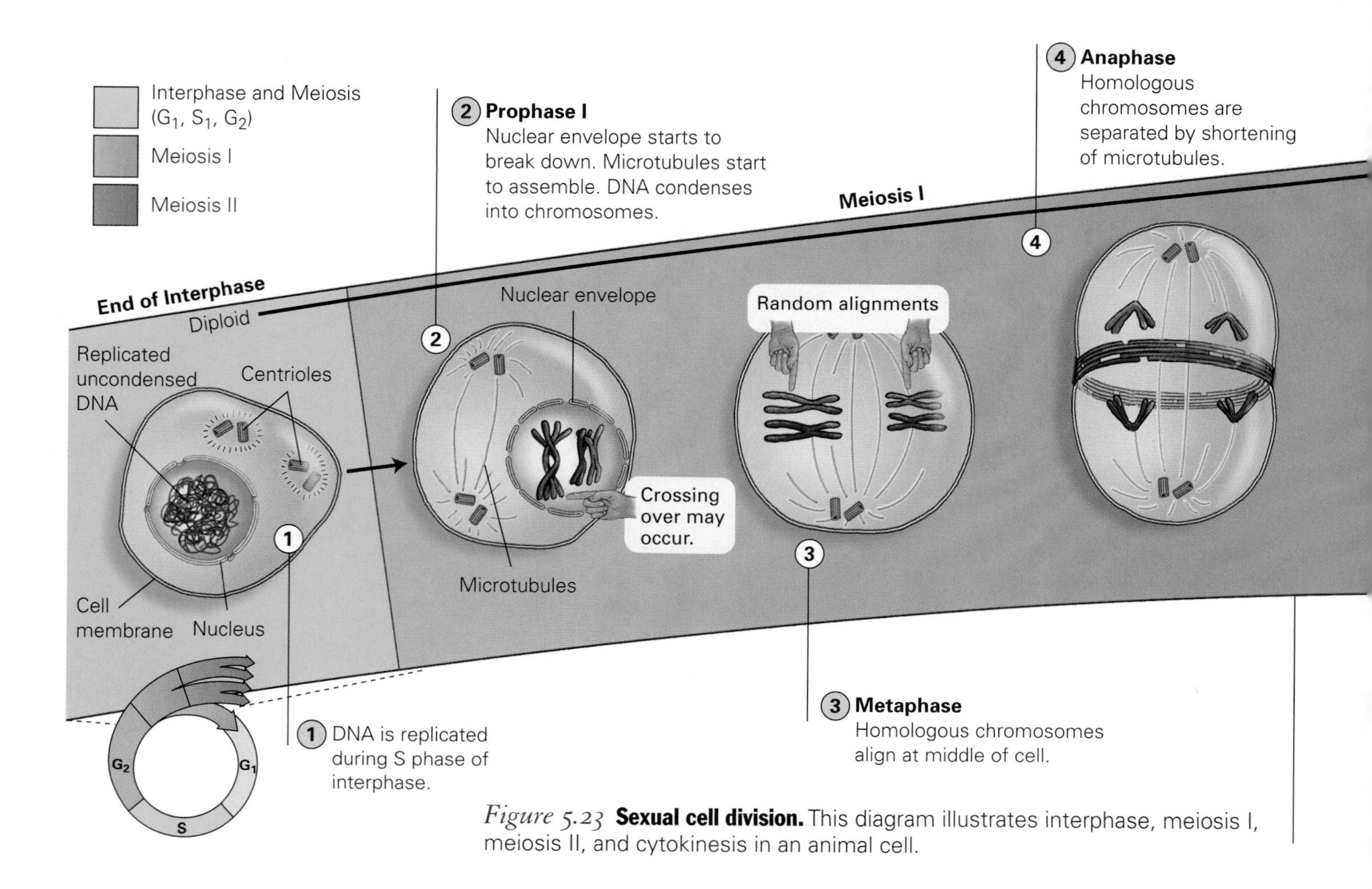

Figure 5.23 **Sexual cell division.** This diagram illustrates interphase, meiosis I, meiosis II, and cytokinesis in an animal cell.

an individual can produce and therefore increase the variation in individuals that can be produced when gametes combine.

Crossing Over and Random Alignment

Crossing over occurs during prophase I of meiosis I. It involves the exchange of portions of chromosomes from one member of a homologous pair to the other member. Crossing over is believed to occur several times on each homologous pair during each occurrence of meiosis.

To illustrate crossing over, consider an example using genes involved in the production of flower color and pollen shape in sweet pea plants. These two genes are on the same chromosome and are called **linked genes.** Linked genes move together on the same chromosome to a gamete, and they may or may not undergo crossing over.

If a pea plant has red flowers and long pollen grains, the chromosomes may appear as shown in *Figure 5.24*. It is possible for this plant to produce four different types of gametes with respect to these two genes. Two types of gametes would result if no crossing over occurred between these genes—the gamete containing the red flower and long pollen chromosome and the gamete containing the white flower and short pollen chromosome. Two additional types of gametes

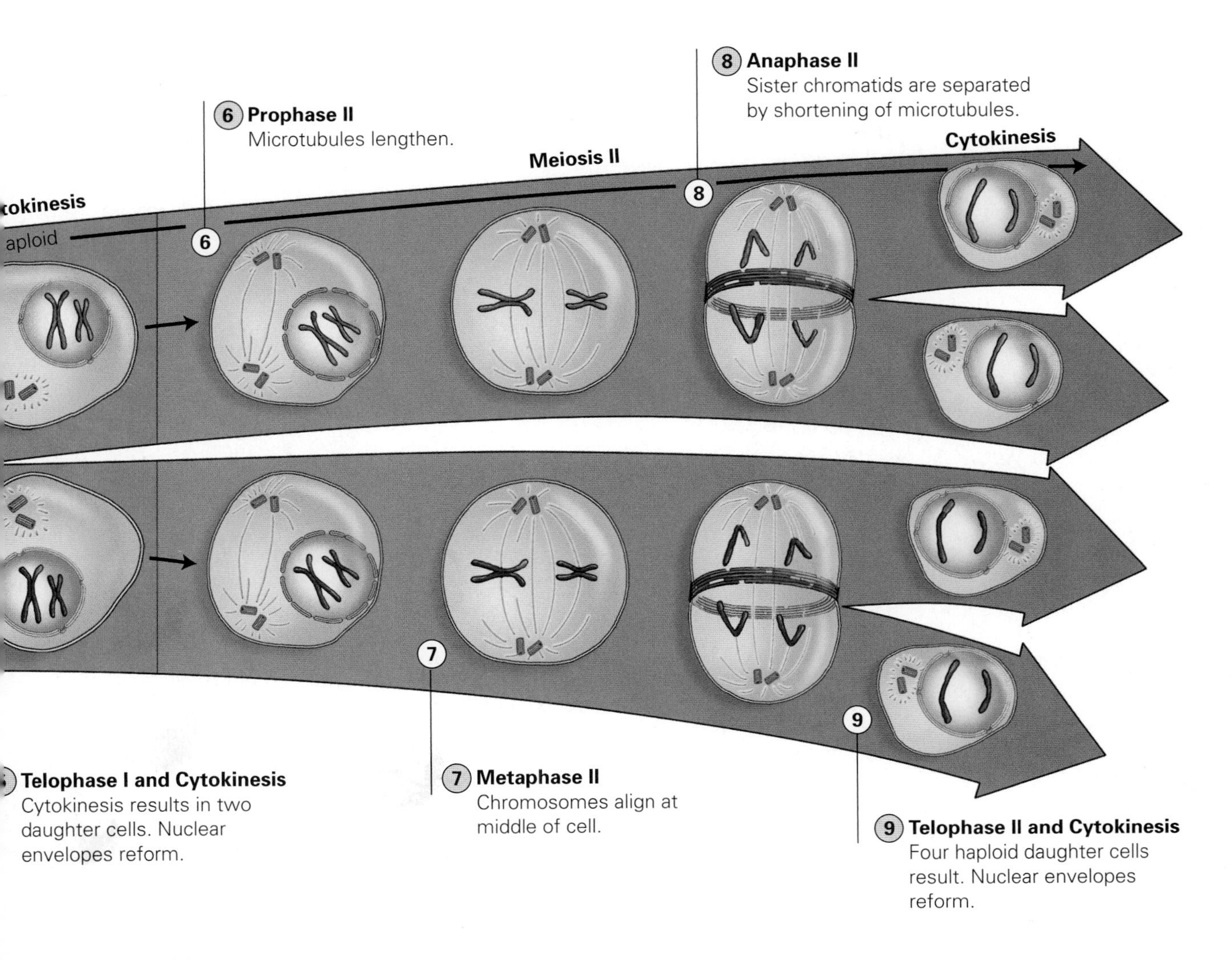

could be produced if crossing over did occur—one type containing the red flower and short pollen grain chromosome and the other type containing the reciprocal white flower and long pollen grain chromosome. Therefore, crossing over increases genetic diversity by increasing the number of distinct combinations of genes that may be present in a gamete.

Random alignment of homologous pairs also increases the number of genetically distinct types of gametes that can be produced. Using Nicole's chromosomes as an example *(Figure 5.25)*, let us assume that she did in fact inherit mutant versions of both the *BRCA2* and *HER2* genes and that these genes are located on different chromosomes. The arrangement of homologous pairs of chromosomes at metaphase I determines which chromosomes will end up together in a gamete. If we consider only these two homologous pairs of chromosomes, then two different alignments are possible, and four different gametes can be produced. For example, when Nicole produces egg cells, the two chromosomes that she inherits from her dad could move together to the gamete, leaving the two chromosomes she inherited from her mom to move to the other gamete. It is equally probable that Nicole could undergo meiosis in which one chromosome from each parent will align randomly together, resulting in two more types of gametes being produced. In Chapters 6 and 7, you will see how random alignment and crossing over affect the inheritance of genes in greater detail.

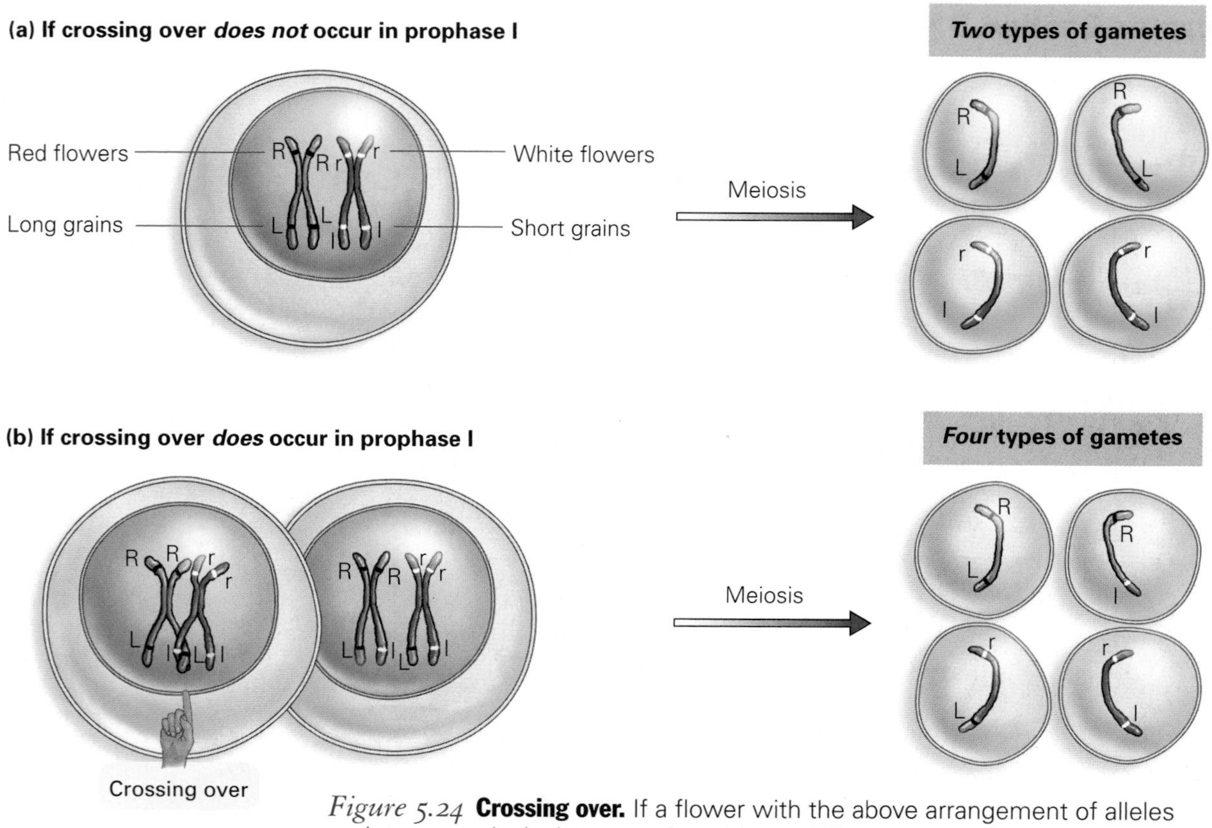

(a) If crossing over *does not* occur in prophase I

***Two* types of gametes**

Red flowers

White flowers

Long grains

Short grains

Meiosis

(b) If crossing over *does* occur in prophase I

***Four* types of gametes**

Meiosis

Crossing over

Figure 5.24 **Crossing over.** If a flower with the above arrangement of alleles undergoes meiosis, it can produce (a) two different types of gametes for these two genes if crossing over does not occur or (b) four different types of gametes for these two genes if crossing over occurs at L.

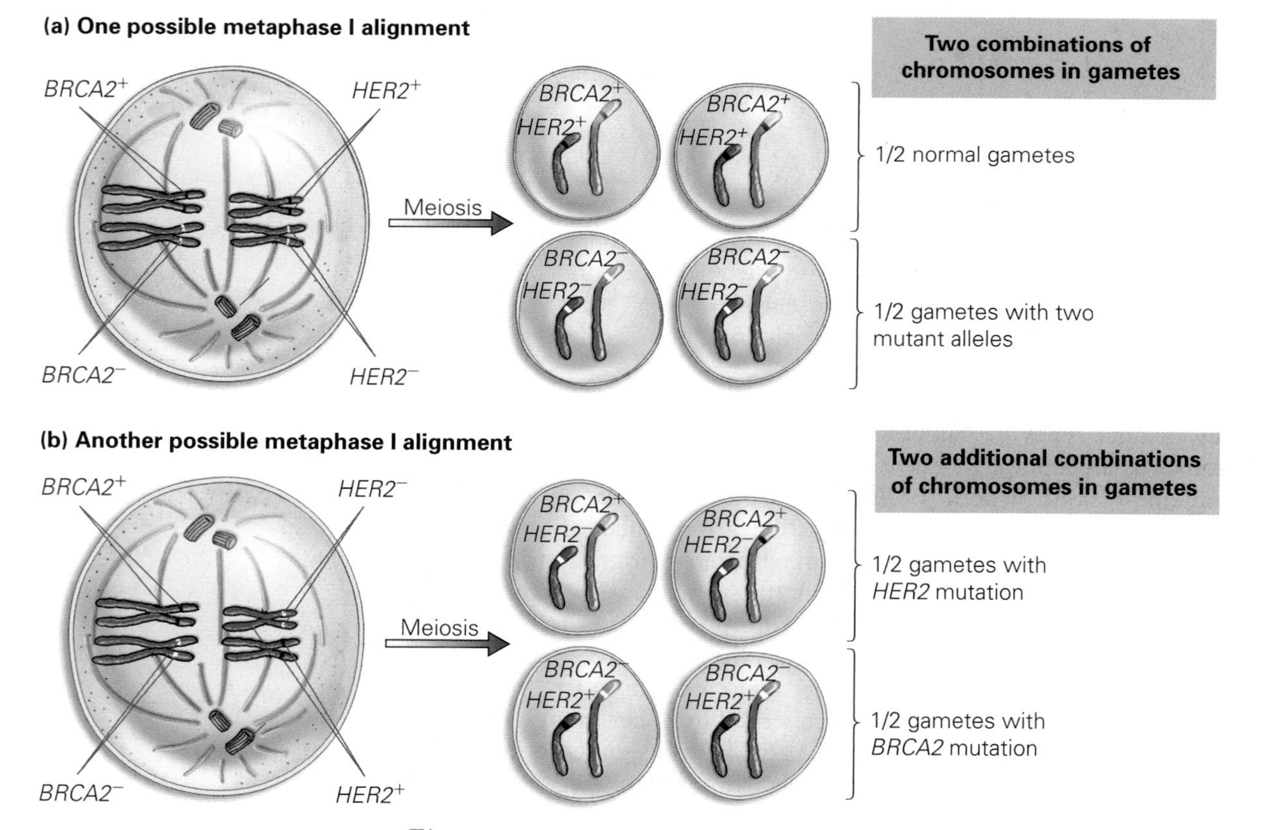

Figure 5.25 **Random alignment.** Two possible alignments, (a) and (b), can occur when there are two homologous pairs of chromosomes. These different alignments can lead to novel combinations of genes in the gametes. **Visualize This:** How many different alignments are possible with three homologous pairs of chromosomes?

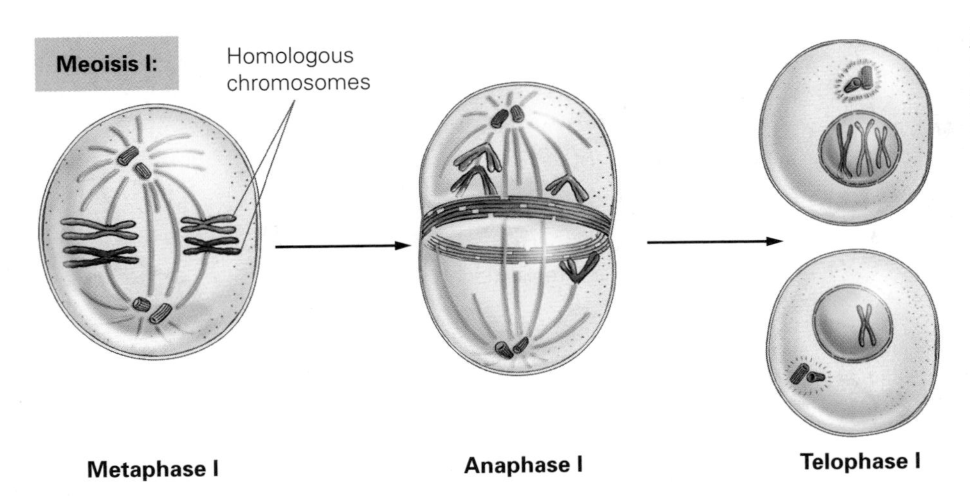

Figure 5.26 **Nondisjunction.** Nondisjunction during meiosis I produces gametes with too many and too few chromosomes.

Mistakes in Meiosis

Sometimes mistakes occur during meiosis that result in the production of offspring with too many or too few chromosomes. Too many or too few chromosomes can result when there is a failure of the homologues (or sister chromatids) to separate during meiosis. This failure of chromosomes to separate is termed **nondisjunction** *(Figure 5.26)*. The presence of an extra chromosome is termed **trisomy.** The absence of one chromosome of a homologous pair is termed **monosomy.**

Since the X and Y chromosomes do not carry the same genes and are not the same size and shape, they are not considered to be a homologous pair. There is, however, a region at the tip of each chromosome that is similar enough so that they can pair up during meiosis. Nondisjunction can occur on autosomes or sex chromosomes.

Typically, early embryos with too many or too few chromosomes will die because they have too much or too little genetic information. However, in some situations, such as when the extra or missing chromosome is very small (such as in chromosomes 21, 13, and 18) or contains very little genetic information (such as the Y chromosome or an X chromosome that will later be inactivated), the embryo can survive. *Table 5.2* lists some chromosomal anomalies in humans that are compatible with life.

From the previous discussions, you have learned that cells undergo mitosis for growth and repair and meiosis to produce gametes. *Figure 5.27* compares the significant features of mitosis and meiosis.

It is now possible to revisit the question of whether Nicole will pass on cancer-causing genes to any children she may have. Because Nicole developed cancer at such a young age, it seems likely that she may have inherited at least one mutant cell-cycle control gene; thus, she may or may not pass that gene on. If Nicole has both a normal and a mutant version of a cell-cycle control gene, then she will be able to make gametes with and without the mutant allele. Therefore, she could pass on the mutant allele if a gamete containing that allele is involved in fertilization. We have also seen that it takes many "hits" or mutations for a cancer to develop. Therefore, even if Nicole does pass on one or a few mutant versions of cell-cycle control genes to a child, environmental conditions will dictate whether enough other mutations will accumulate to allow a cancer to develop.

Mutations caused by environmental exposures are not passed from parents to children unless the mutation happens to occur in a cell of the gonads that will be used to produce a gamete. Nicole's cancer occurred in the ovary,

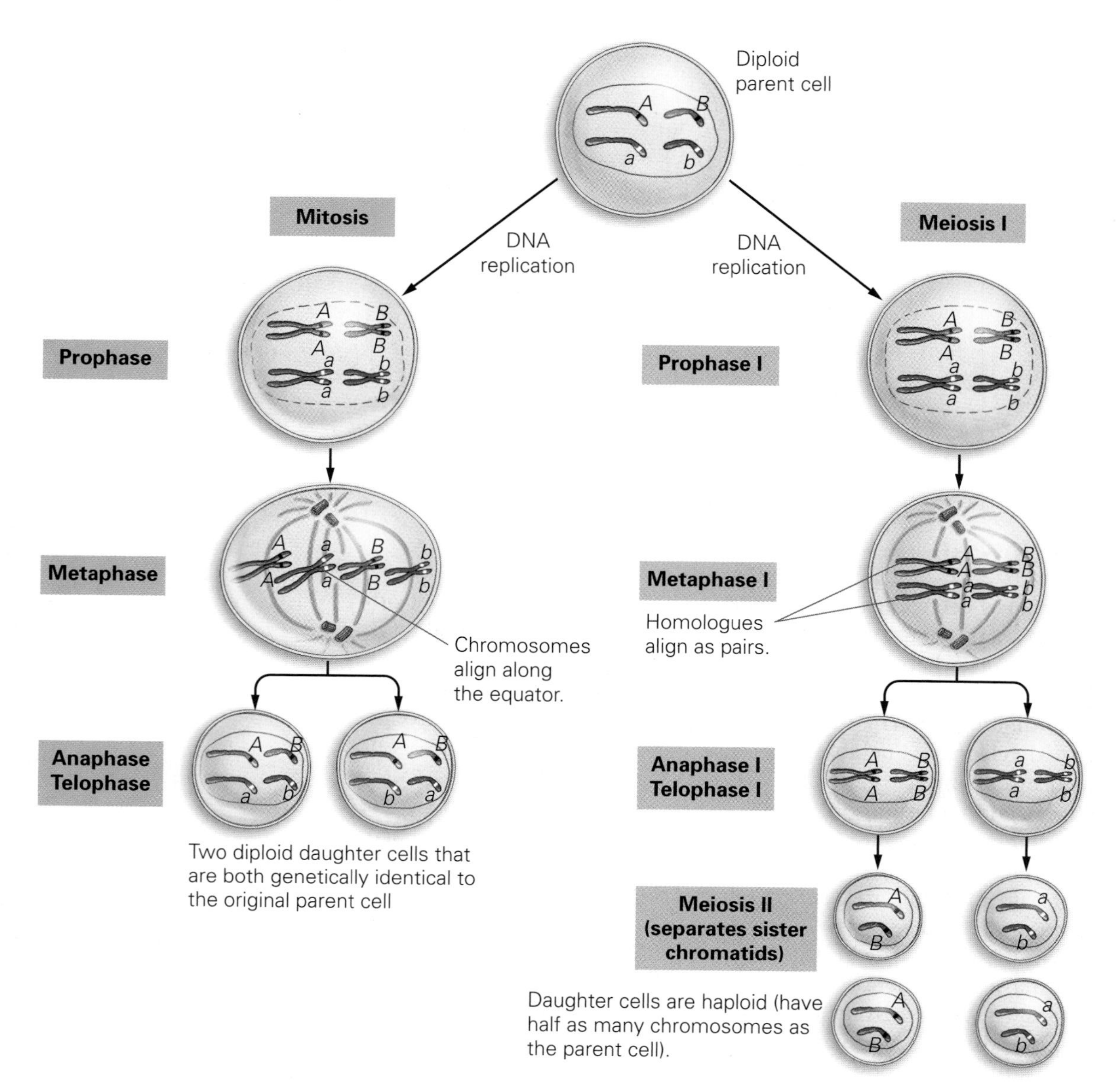

Figure 5.27 **Comparing mitosis and meiosis.** Mitosis is a type of cell division that occurs in somatic cells and gives rise to daughter cells that are exact genetic copies of the parent cell. Meiosis occurs in cells that will give rise to gametes and decreases the chromosome number by one-half. To do this and still ensure that each gamete receives one member of each homologous pair, the two members of each homologous pair align across the equator at the metaphase I of meiosis and are separated from each other during anaphase I.

the site of meiosis, but not all cells in the ovary undergo meiosis. Nicole's cancer originated in the outer covering of the ovary, a tissue that does not undergo meiosis. The cells involved in ovulation are located inside the ovary. A skin cancer that develops from exposure to ultraviolet light will not be passed on, and the same is true for most of the mutations that Nicole obtained from environmental exposures. Therefore, for any children that Nicole (or any of us) might have, it is the combined effects of inherited mutant alleles and any mutations induced by environmental exposures that will determine whether cancers will develop.

TABLE 5.2 **Autosomal and sex-linked chromosomal anomalies**

Conditions Caused by Nondisjunction of Autosomes	Comments
Trisomy 21—Down syndrome 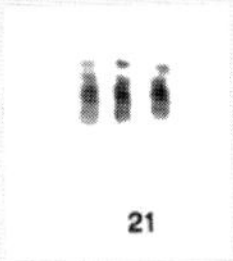	Affected individuals are mentally retarded and have abnormal skeletal development and heart defects. The frequency of Down syndrome children increases with maternal age. 1 in 1000 children of mothers under 35 are affected, while 4 in 1000 children of mothers over 45 are affected.
Trisomy 13—Patau syndrome 	Affected individuals are mentally retarded, deaf, and have a cleft lip and palate. Approximately 1 in 5000 newborns are affected.
Trisomy 18—Edwards syndrome 	Affected individuals have malformed organs, ears, mouth, and nose, leading to an elfin appearance. These babies usually die within 6 months of birth. Approximately 1 in 6000 newborns are affected.
Conditions Caused by Nondisjunction of Sex Chromosomes	**Comments**
XO—Turner Syndrome 	Females with one X chromosome can be sterile if their ovaries fail to develop. Webbing of the neck, shorter stature, and hearing impairment are also common. Approximately 1 in 5000 female newborns are born with only one X chromosome.
Trisomy X: Meta female 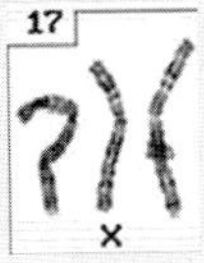	Females with three X chromosomes tend to develop normally. Approximately 1 in 1000 females are born with an extra X chromosome.
XXY—Kleinfelter syndrome 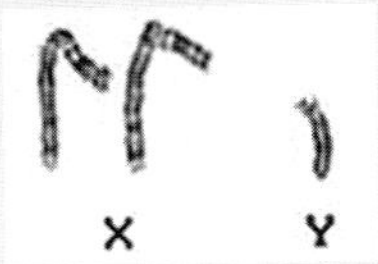	Males with the XXY genotype are less fertile than XY males; have small testes, sparse body hair, some breast enlargement; and may have mental retardation. Testosterone injections can reverse some of the anatomical abnormalities in the approximately 1 in 1000 males with this condition.
XYY condition 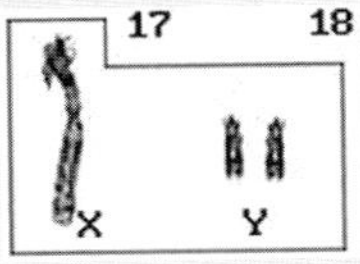	Males with two Y chromosomes tend to be taller than average but have an otherwise normal male phenotype. Approximately 1 in 1000 newborn males has an extra Y chromosome.

Savvy Reader An Alternative Cancer Treatment

SHARK CARTILAGE THERAPY EVALUATED

Dynamic Chiropractic

VOLUME 12, ISSUE 5,

William Lane, MS, PhD

One-hundred percent pure shark cartilage was used as the only therapy with 29 terminal cancer patients in a 16-week clinical trial in Cuba.

The patients were all hospitalized and doctors and nurses were in constant attendance during the study. These patients included six breast, five prostate, five central nervous system (brain), two stomach, two liver, two ovarian, two uterus, two esophageal, two tonsil, and one urinary bladder stage II or IV cases. Eight died during the 16-week study, and six have died since the study was completed. Fifteen of the 29 patients, diagnosed as terminal before the study began, are still alive a year later—a remarkable result by any measure.

1. This excerpt was taken from a website written by Dr. Lane. Is there any guarantee that something written on a website is true? Does the fact that the journal in which Dr. Lane's findings were published is not peer reviewed add or subtract from the credibility of this article?
2. Would the fact that Dr. Lane owns a company that sells shark cartilage (a product called Benefin) make you more or less skeptical of his findings? What about the fact that sharks actually do get cancer?
3. While most medical doctors have dedicated their lives to helping people in need, a few will promote products they might not believe in. Should you always believe the word of someone with an advanced degree?
4. Do scientists use phrases like "A remarkable result by any measure," or do they prefer the use of statistics?
5. Several years after Lane began his marketing campaign, the Federal Trade Commission forced him to stop making claims about the anticancer nature of Benefin and fined him over half a million dollars. However, his product is still on the market. Why might this be the case?

CHAPTER Review

For study help, animations, and more quiz questions go to www.mybiology.com.

Summary

5.1 What Is Cancer?

- Unregulated cell division can lead to the formation of a tumor. Benign or noncancerous tumors stay in one place and do not prevent surrounding organs from functioning. Malignant tumors are those that are invasive or those that metastasize to surrounding tissues, starting new cancers (pp. 111–115).

5.2 Passing Genes and Chromosomes to Daughter Cells

- Cell division is a process required for growth and development (pp. 116).
- For cell division to occur, the DNA must be copied and passed on to daughter cells (pp. 116–117).

- Once copied, the duplicated chromosome is composed of two sister chromatids that are attached to each other at the centromere (pp. 117–118).
- During DNA replication or synthesis, one strand of the double-stranded DNA molecule is used as a template for the synthesis of a new daughter strand of DNA. The newly synthesized DNA strand is complementary to the parent strand. The enzyme DNA polymerase ties together the nucleotides on the daughter strand (pp. 117–118).

web animation 5.1
The Structure of DNA

5.3 The Cell Cycle and Mitosis

- The cell cycle includes all of the events that occur as one cell gives rise to daughter cells (p. 118).
- Interphase consists of two gap phases of the cell cycle (G_1 and G_2), during which the cell grows and prepares to enter mitosis or meiosis. During the S (synthesis) phase, the DNA replicates. The S phase of interphase occurs between G_1 and G_2 (p. 119).
- During mitosis, the sister chromatids are separated from each other into daughter cells. During prophase, the replicated DNA condenses into linear chromosomes. At metaphase, these replicated chromosomes align across the middle of the cell. At anaphase, the sister chromatids separate from each other and align at opposite poles of the cells. At telophase, a nuclear envelope re-forms around the chromosomes lying at each pole (pp. 119–121).
- Cytokinesis is the last phase of the cell cycle. During cytokinesis, the cytoplasm is divided into two portions, one for each daughter cell (pp. 120–121).

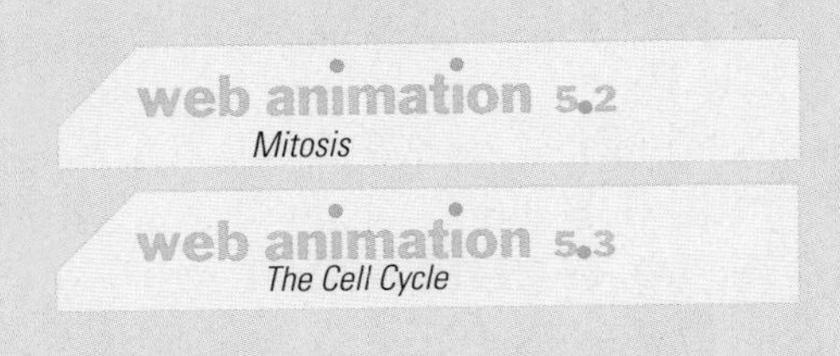

BioFlix™ Mitosis

5.4 Cell Cycle Control and Mutation

- When cell division is working properly, it is a tightly controlled process. Normal cells divide only when conditions are favorable (pp. 121).
- Proteins survey the cell and its environment at checkpoints as the cell moves through G_1, G_2, and metaphase and can halt cell division if conditions are not favorable (pp. 122–123).
- Mistakes in regulating the cell cycle arise when genes that control the cell cycle are mutated. Mutated genes can be inherited, or mutations can be caused by exposure to carcinogens (pp. 122–123).
- Proto-oncogenes regulate the cell cycle. Oncogenes are mutated versions of these genes. Many proto-oncogenes encode growth factors (pp. 123–125).
- Tumor suppressors are normal genes that can encode proteins to stop cell division if conditions are not favorable and can repair damage to the DNA. They serve as backups in case the proto-oncogenes undergo mutation (pp. 124–125).
- As a tumor progresses from benign to malignant, it often undergoes angiogenesis, loses contact inhibition and anchorage dependence, and becomes immortal. Thus, many changes or hits to the cancer cell are required for malignancy (pp.125–127).

5.5 Cancer Detection and Treatment

- Early cancer detection and treatment increase survival odds (pp. 127).
- A biopsy is a common method for detecting cancer. It involves removing some cells or tissues suspected of being cancerous and analyzing them (pp. 127–128).
- Typical cancer treatments include chemotherapy, which involves injecting chemicals that kill rapidly dividing cells, and radiation, which involves killing tumor cells by exposing them to high-energy particles (pp. 128–130).

Summary (continued)

5.6 Meiosis

- Meiosis is a type of sexual cell division, occurring in cells, that gives rise to gametes. Gametes contain half as many chromosomes as somatic cells do. The reduction division of meiosis begins with diploid cells and ends with haploid cells (pp. 130–131).
- Meiosis is preceded by an interphase stage in which the DNA is replicated (pp. 132).
- During meiosis I, the members of a homologous pair of chromosomes are separated from each other (pp. 132–134).
- During meiosis II, the sister chromatids are separated from each other (pp. 133–134).
- Homologous pairs of chromosomes exchange genetic information during crossing over at prophase I of meiosis, thereby increasing the number of genetically distinct gametes that an individual can produce (pp. 134–135).
- The alignment of members of a homologous pair at metaphase I is random with regard to which member of a pair faces which pole. This random alignment of homologous chromosomes increases the number of different kinds of gametes an individual can produce (pp. 135–137).

web animation 5.4 *Meiosis*

BioFlix™ Meiosis

Roots to Remember

The following roots of words come mainly from Latin and Greek and will help you decipher terms:

cyto- and **-cyte** relate to cells
-kinesis means motion
meio- means to make smaller.
mito- means a thread.
onco- means cancer.
proto- means before.
soma-, **somato-**, and **-some** mean body.
telo- means end or completion.

Learning the Basics

1. List the ways in which mitosis and meiosis differ.

2. What property of cancer cells do chemotherapeutic agents attempt to exploit?

3. A cell that begins mitosis with 46 chromosomes produces daughter cells with _______________.
A. 13 chromosomes; **B.** 23 chromosomes; **C.** 26 chromosomes; **D.** 46 chromosomes

4. The centromere is a region at which _______________.
A. sister chromatids are attached to each other; **B.** metaphase chromosomes align; **C.** the tips of chromosomes are found; **D.** the nucleus is located

5. Mitosis _______________.
A. occurs in cells that give rise to gametes; **B.** produces haploid cells from diploid cells; **C.** produces daughter cells that are exact genetic copies of the parent cell; **D.** consists of two separate divisions, mitosis I and mitosis II

6. At metaphase of mitosis, _______________.
A. the chromosomes are condensed and found at the poles; **B.** the chromosomes are composed of one sister chromatid; **C.** cytokinesis begins; **D.** the chromosomes are composed of two sister chromatids and are lined up along the equator of the cell

7. Sister chromatids _______________.
A. are two different chromosomes attached to each other; **B.** are exact copies of one chromosome that are attached to each other; **C.** arise from the centrioles; **D.** are broken down by mitosis; **E.** are chromosomes that carry different genes

8. DNA polymerase _______________.
A. attaches sister chromatids at the centromere; **B.** synthesizes daughter DNA molecules from fats and phospholipids; **C.** is the enzyme that facilitates DNA synthesis; **D.** causes cancer cells to stop dividing

9. After telophase I of meiosis, each daughter cell is ______________.

A. diploid, and the chromosomes are composed of one double-stranded DNA molecule; **B.** diploid, and the chromosomes are composed of two sister chromatids; **C.** haploid, and the chromosomes are composed of one double-stranded DNA molecule; **D.** haploid, and the chromosomes are composed of two sister chromatids

10. State whether the chromosomes depicted in parts (a)–(d) of *Figure 5.28* are haploid or diploid.

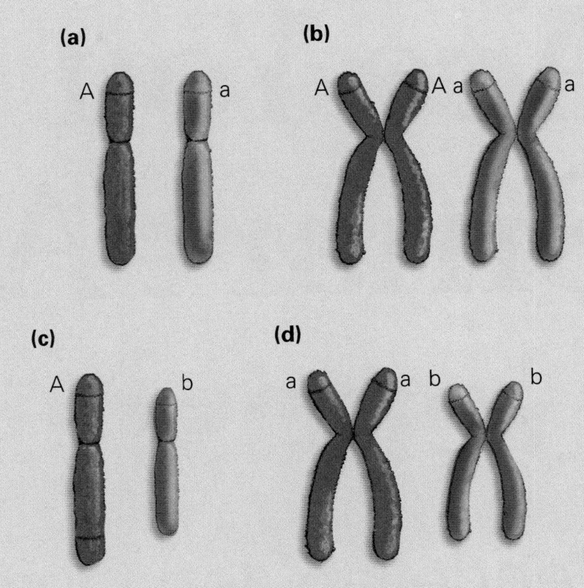

Figure 5.28 **Haploid or diploid chromosomes?**

Analyzing and Applying the Basics

1. Would a skin cell mutation that your father obtained from using tanning beds make you more likely to get cancer? Why or why not?

2. Why do cancer rates increase with age?

3. Why are some cancers treated with radiation therapy while others are treated with chemotherapy?

Connecting the Science

1. Should members of society be forced to pay the medical bills of smokers when the cancer risk from smoking is so evident and publicized? Explain your reasoning.

2. Would you want to be tested for the presence of cell-cycle control mutations? How would knowing whether you had some mutated proto-oncogenes be of benefit or harm?

CHAPTER

6

Are You Only As Smart As Your Genes?

Mendelian and Quantitative Genetics

Can a woman create the perfect child if she chooses the right sperm?

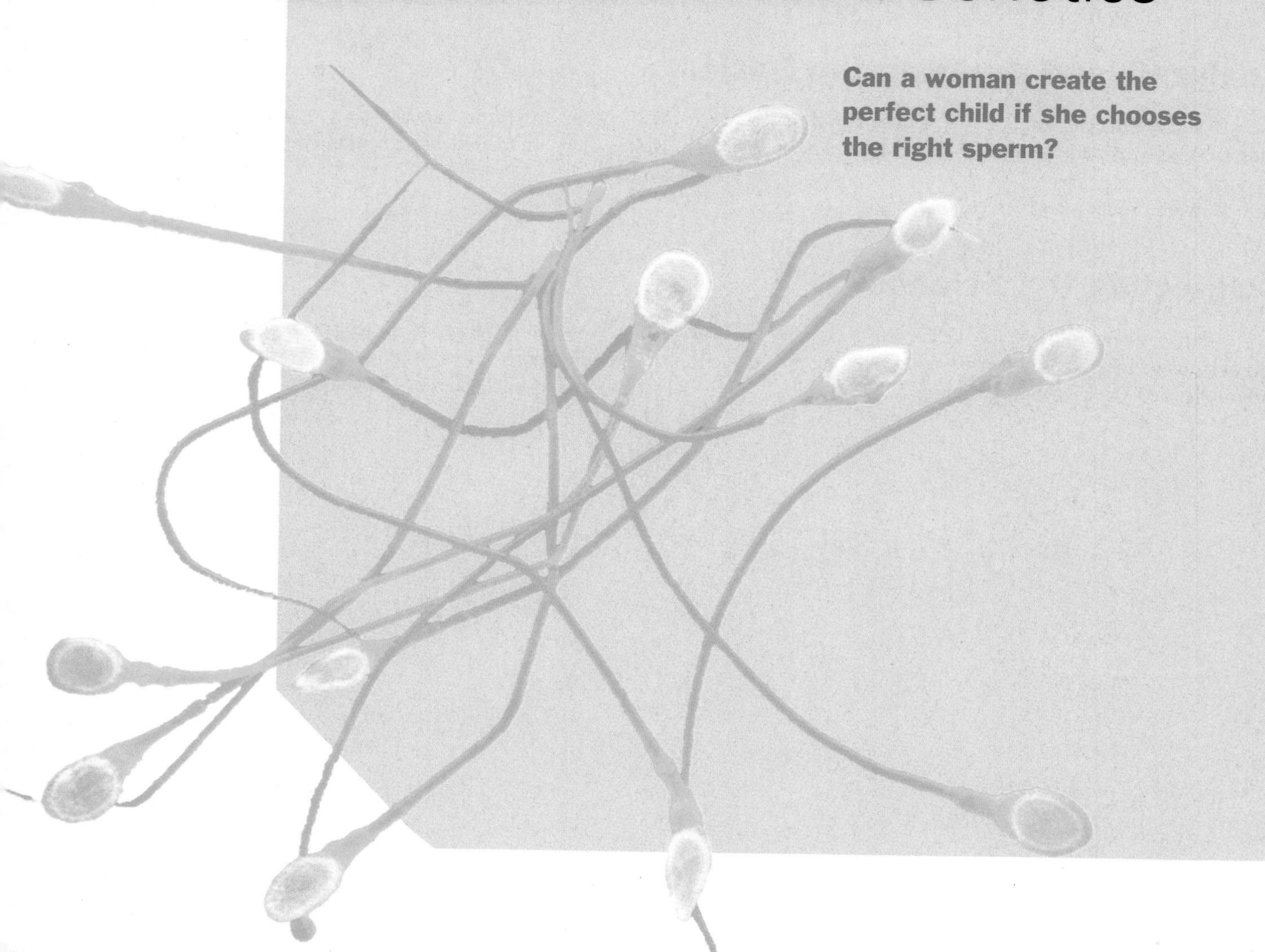

If a woman chooses a donor with the right genes, her child may look like her partner.

If she chooses a donor with the right genes, will her child be a genius?

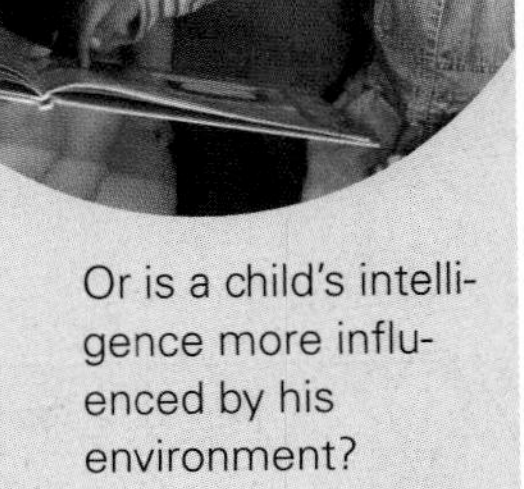

Or is a child's intelligence more influenced by his environment?

The Fairfax Cryobank is a nondescript brick building located in a quiet, tree-lined suburb of Washington, DC. Stored inside this unremarkable edifice are the hopes and dreams of thousands of women and their partners. The Fairfax Cryobank is a sperm bank; inside its many freezers are vials containing sperm collected from hundreds of men. Women can order these sperm for a procedure called artificial insemination, which may allow them to conceive a child despite the lack of a fertile male partner.

Women who purchase sperm from the Fairfax Cryobank can choose from hundreds of potential donors. The donors are categorized into 3 classes, and their sperm is priced accordingly. Most women who choose artificial insemination want detailed information about the donor before they purchase a sample; while all Fairfax Cryobank donors submit to comprehensive physical exams and disease testing and provide a detailed family health history, not all provide childhood pictures, audio CDs of their voices, or personal essays. Sperm samples from men who did not provide this additional information are sold at a discount because most women seek a donor who seems compatible in interests and aptitudes.

However, in addition to the information-rich donors, there is also a set of premium sperm donors referred to by Fairfax Cryobank as its "Doctorate" category. These men either are in the process of earning or have completed a doctoral degree in medicine, law, or academia. A sperm sample from this donor category is 30% more expensive than sperm from the standard donor, and because several samples are typically needed, ensuring pregnancy from one of these donors is likely to cost hundreds of additional dollars.

Why would some women be willing to pay significantly more for sperm from a donor who has an advanced degree? Because academic achievement is associated with intelligence. These women want intelligent children, and they are willing to pay more to provide their offspring with "extra smart" genes.

But, are these women putting their money in the right place? Is intelligence about genes, or is it a function of the environmental conditions in which a baby is raised? In other words, is who we are a result of our "nature" or our "nurture"? As you read this chapter, you will see that the answer to this question is not a simple one—our characteristics come both from our biological inheritance and the environment in which we developed.

6.1 The Inheritance of Traits

Most of us recognize similarities between our birth parents and ourselves. Family members also display resemblances—for instance, all the children of a single set of parents may have dimples. However, it is usually quite easy to tell siblings apart. Each child of a set of parents is unique, and none of us is simply the "average" of our parents' traits. We are each more of a combination—one child may be similar to her mother in eye color and face shape, another similar to mom in height and hair color.

To understand how your parents' traits were passed to you and your siblings, you need to revisit the human life cycle, described in Chapter 5. A **life cycle** is a description of the growth and reproduction of an individual *(Figure 6.1)*.

A human baby is typically produced from the fusion of a single **sperm** cell produced by the male parent and a single egg cell produced by the female parent. These gametes are produced by the process of *meiosis*. Egg and sperm fuse at **fertilization,** and the resulting *zygote* duplicates all of the genetic information it contains and undergoes mitosis to produce 2 identical daughter cells. Each of these daughter cells divides dozens of times in the same way. The cells in this resulting mass then differentiate into specialized cell types, which continue to divide and organize to produce the various structures of a developing human, called an *embryo*. Continued division of this same cell and its progeny leads to the production of a full-term infant and eventually an adult.

Figure 6.1 **Instructions inside.** Both parents contribute genetic information to their offspring via their sperm and egg. The single cell that results from the fusion of 1 sperm with 1 egg contains all of the instructions necessary to produce an adult human.

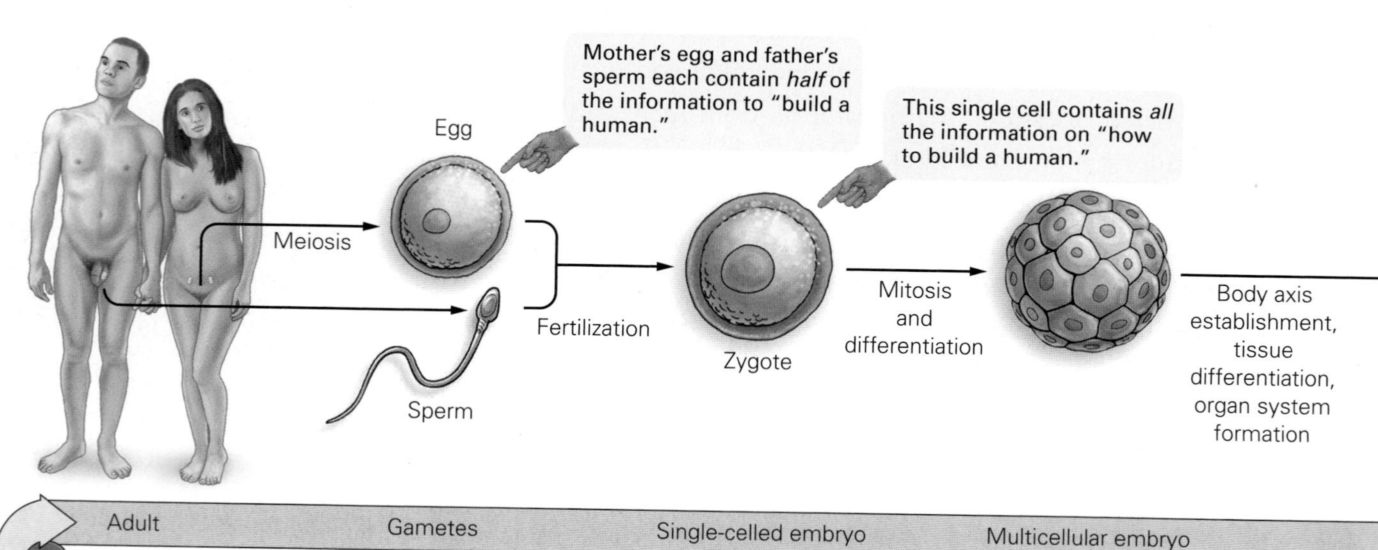

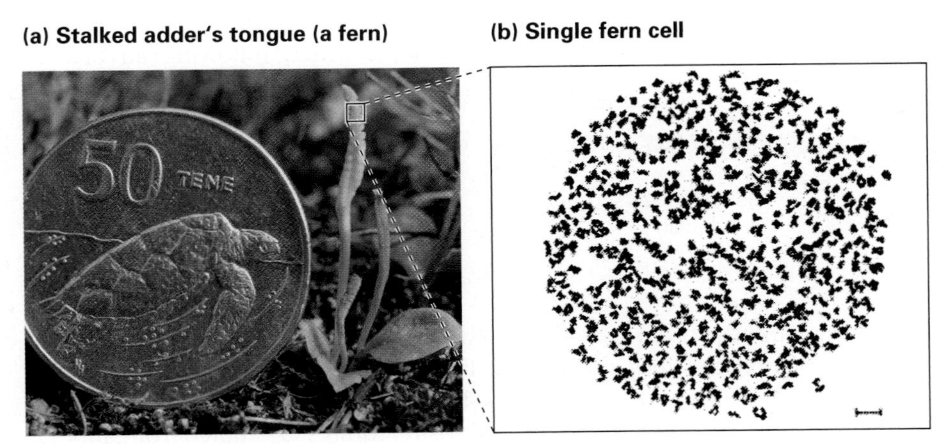

Figure 6.2 **Variation in chromosome number.** The amount of genetic information in an organism does not correlate to its complexity, as can be seen by examination of the chromosomes contained in a single cell of the stalked adder's tongue fern.

We are made up of trillions of individual cells, all of them the descendants of that first product of fertilization and nearly all containing exactly the same information originally found in the zygote. All of our traits are influenced by the information contained in that tiny cell.

Genes and Chromosomes

Each normal sperm and egg contains information about "how to build an organism." A large portion of that information is in the form of genes. As you learned in Chapter 5, genes are segments of DNA that code for proteins.

Imagine genes as roughly equivalent to the words used in an instruction manual. These words are contained on chromosomes, which are roughly analogous to pages in the manual.

Prokaryotes such as bacteria typically contain a single, circular chromosome that floats freely inside the cell and is passed in its entirety to each offspring. In contrast, eukaryotes carry their genes on more than 1 linear chromosome. The number of chromosomes in eukaryotes can vary greatly, from 2 in the jumper ant (*Myrmecia pilosula*) to an incredible 1260 in the stalked adder's tongue, a species of fern (*Ophioglossum reticulatum, Figure 6.2*). Human cells contain 46 chromosomes, most of which carry thousands of genes. Thus, each cell has 46 pages of instructions, with each page containing thousands of words.

The instruction manual inside a cell is different from the instruction manual that comes with, for instance, a kit for building a model car. You would

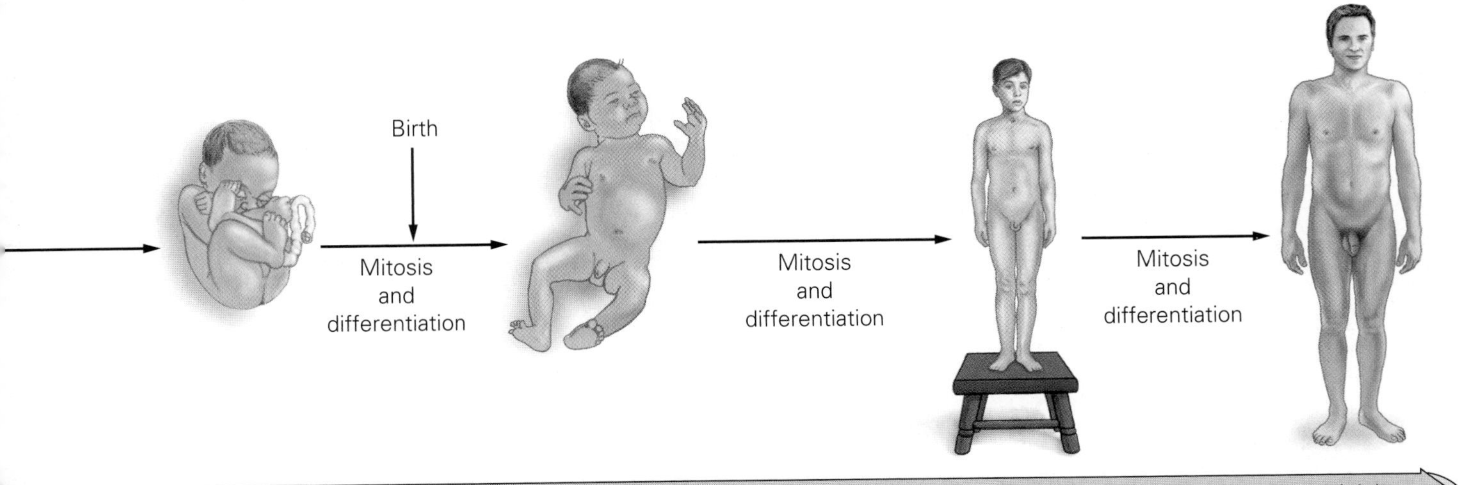

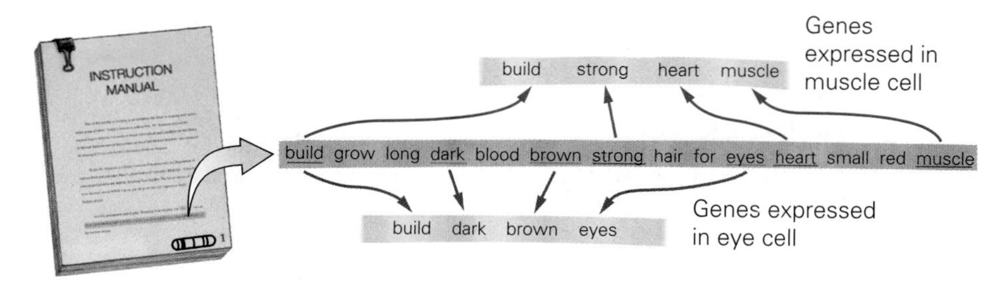

Figure 6.3 **Genes as words in an instruction manual.** Different words from the manual are used in different parts of the body, and identical words may be used in distinctive combinations in different cells. **Visualize This:** Write several more instructions that can be extracted from this group of 14 words.

read the manual for building the car in page order, following an orderly set of steps to produce the final product. The human instruction manual is much more complicated—the pages and words you read are different for different types of cell and may even change from time to time, according to the situation. The "final product" of any given cell depends on the words used and the order in which the words are read.

For instance, eye cells and heart cells in mammals both carry instructions for the protein rhodopsin, which helps detect light, but rhodopsin is produced only in eye cells, not in heart cells. Rhodopsin requires assistance from another protein to translate the light that strikes it into the actions of the eye cell. This second protein may also be produced in heart cells, but there it is combined with a third protein to help coordinate contractions of the heart muscle.

Thus, a protein may serve 2 different functions depending on its context. Because genes, like words, can be used in many combinations, the instruction manual for building a living organism is very flexible *(Figure 6.3)*.

Producing Diversity in Offspring

During reproduction, each parent contributes instructions to a child. The information must be copied before it is transmitted to the next generation, and it is this process that introduces variation among genes. It is this **genetic variation** that most interests the women seeking sperm donors; the genes in the sperm they select will provide information about the traits of their offspring.

Gene Mutation Creates Genetic Diversity. You can imagine how a page of instructions might change over many generations if each parent had to type a new copy of the instructions for each offspring. Typographical errors made by a father would be passed on in the manual he gave to his child. Likewise, the child's instructions may be changed during copying, and those changes would be passed on to the next generation.

Recall the process of DNA replication, covered in Chapter 5. Like a retyped instruction manual page, copies of chromosomes are rewritten rather than "photocopied." As a result, there is a chance of a typographical error, or *mutation*, every time a cell divides. Mutations in genes lead to different versions, or *alleles*, of the gene. Because mutations are random and not expected to occur in the same genes in various individuals, different families should have slightly different alleles for various genes. The various types of mutation are described in *Figure 6.4*.

Repeated mutation creates genetic variation in a population and contributes to differences between families. As we will discuss in Chapter 10, when a novel characteristic increases an individual's chance of survival and reproduction, the mutation contributes to a population's adaptation to its environment. Genetic misspellings are thus the engine that drives evolution itself.

Stop and Stretch 6.1: Not all of the DNA in a cell codes for proteins—some DNA functions as places for gene promoters (proteins that increase expression of a gene), other segments may provide structural support for chromosomes, and other parts may be meaningless. What do you think the effect of "misspellings" are on these nongene segments of DNA?

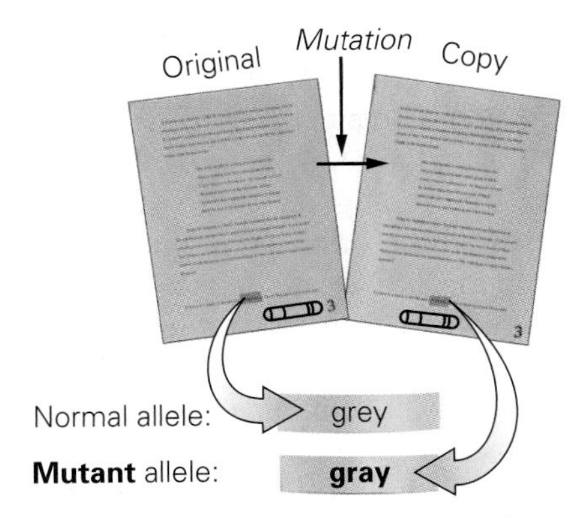

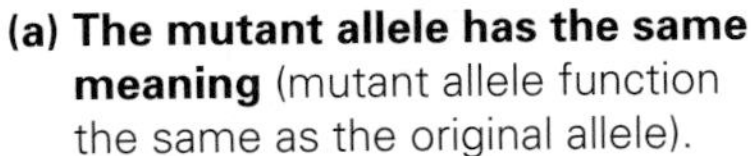

(a) The mutant allele has the same meaning (mutant allele function the same as the original allele).

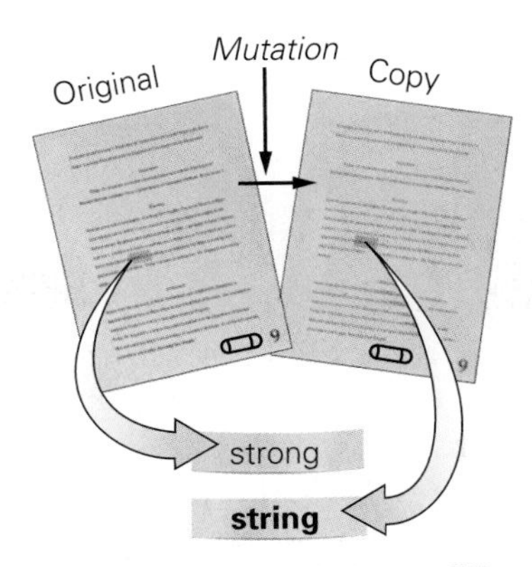

(b) The mutant allele has a different meaning (mutant allele functions differently than the original allele).

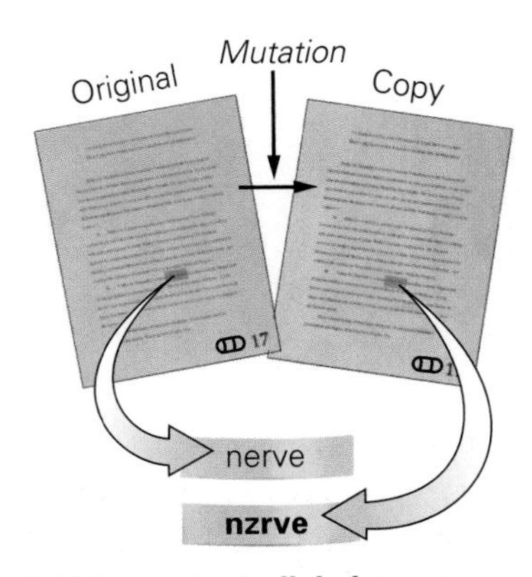

(c) The mutant allele has no meaning (mutant allele is no longer functional).

Figure 6.4 **The formation of different alleles.** Different alleles for a gene may form as a result of copying errors. Some of these misspellings do not change the meaning of the word, but some may result in altered meanings or have no meaning at all.

Segregation and Independent Assortment Create Gamete Diversity. Both parents contribute genetic instructions to each child, but they do not contribute their entire manual. If they did, the genetic instructions carried in human cells would double every generation, making for a pretty crowded cell. Instead, meiosis reduces the number of chromosomes carried in gametes by one-half.

Although they are only transmitting half of their genetic information in a gamete, each parent actually gives a complete copy of the instruction manual to each child *(Figure 6.5)*. This can occur because, in effect, our body cells each contain 2 copies of the manual—that is, each has 2 versions of each page, with each version containing essentially the same words. In other words, the 46 chromosomes each cell contains are actually 23 pairs of chromosomes, with each member of a pair containing essentially the same genes. Each set of 2 equivalent chromosomes is referred to as a **homologous pair.** The members of a homologous pair are equivalent, but not identical, because even though both have the same genes, each contains a unique set of mutations inherited from one or the other parent.

The process of meiosis separates homologous pairs of chromosomes and places chromosomes independently into each gamete. These 2 processes explain why siblings are not identical (with the exception of identical twins). For the most part, it is because parents do not give all of their offspring exactly the same set of alleles.

When homologous pairs are separated, the alleles they carry are separated as well. The separation of pairs of alleles during the production of gametes is called **segregation.** Thus, a parent with 2 different alleles of a gene will produce gametes with a 50% probability of containing 1 version of the allele and a 50% probability of containing the other version.

The segregation of chromosomes during meiosis leads to *independent assortment.* Independent assortment

Figure 6.5 **Equivalent information from parents.** Each parent provides a complete set of instructions to each offspring.

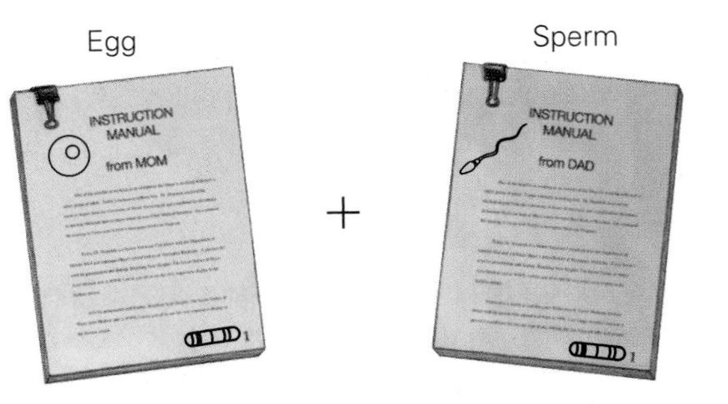

The 23 pages of each instruction manual are roughly equivalent to the 23 chromosomes in each egg and sperm.

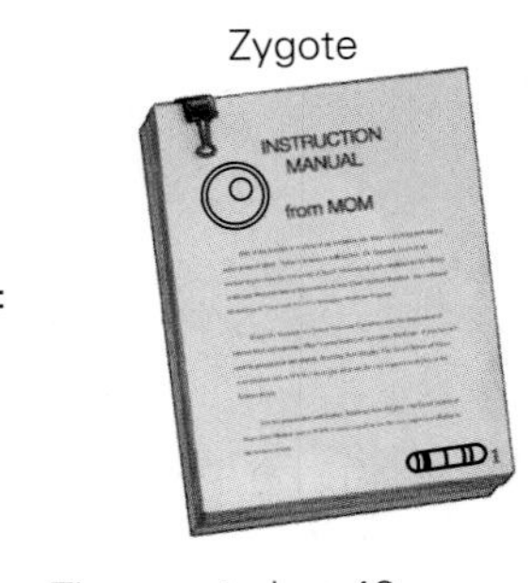

The zygote has 46 pages, equivalent to 46 chromosomes.

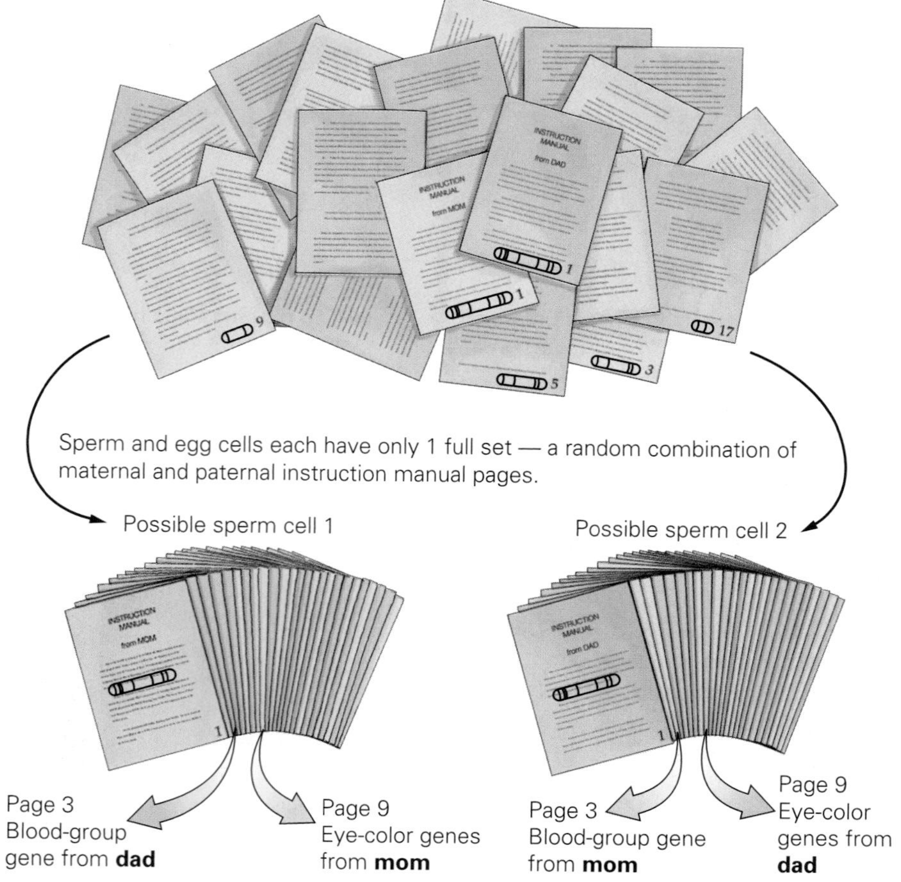

Figure 6.6 **Each egg and sperm is unique.** Since each sperm is produced independently, the set of pages in each sperm will be a unique combination of the pages that the man inherited from his mom and dad. **Visualize This:** This man could produce 4 distinctly different kinds of sperm cell when considering these 2 genes and 4 alleles. List the other 2 possibilities not pictured here.

arises from the *random alignment* of chromosomes during meiosis, as described in Chapter 5. Because each homologous chromosome pair is segregated into daughter cells independently of all the other pairs during the production of gametes, genes that are on different chromosomes are inherited independently of each other.

As a result of independent assortment, the instruction manual contained in a single sperm cell is made up of a unique combination of pages from the manuals a man received from each of his parents. In fact, almost every sperm he makes will contain a unique subset of chromosomes—and thus a unique subset of his alleles. *Figure 6.6* illustrates this. In the figure, you can see that independent assortment causes an allele for an eye-color gene to end up in a sperm cell independently from an allele for the blood-group gene.

Because the independent assortment of segregated chromosomes into daughter cells is repeated every time a sperm is produced, the set of alleles that each child receives from a father is different for all of his offspring. The sperm that contributed half of your genetic information might have carried an eye-color allele from your father's mom and a blood-group allele from his dad, while the sperm that produced your sister might have contained both the allele for eye color and the allele for blood group from your paternal grandmother. As a result of independent assortment, only about 50% of an individual's alleles are identical to those found in another offspring of the same parents—that is, for each gene, you have a 50% chance of being like your sister or brother.

Stop and Stretch 6.2: Figure 6.6, you can see that a man can produce 4 different combinations of alleles in his sperm when considering 2 genes on 2 different chromosomes. How many different allele combinations within gametes can be produced when considering 3 genes on 3 different chromosomes? What is the relationship between chromosome number and the number of possible combinations that can be produced via independent assortment?

Random Fertilization Results in a Large Variety of Potential Offspring. As a result of the independent assortment of 23 pairs of chromosomes, each individual human can make at least 8 million different types of egg and sperm. Consider now that each of your parents was able to produce such an enormous diversity of gametes. Further, any sperm produced by your father had an equal chance (in theory) of fertilizing any egg produced by your mother.

In other words, gametes combine without regard to the alleles they carry, a process known as **random fertilization**. Hence, the odds of your receiving your particular combination of chromosomes are 1 in 8 million times 1 in 8 million—or 1 in 64 trillion. Remarkably, your parents together could have made more than 64 trillion genetically different children, and you are only one of the possibilities.

Mutation creates new alleles, and independent assortment and random fertilization result in unique combinations of alleles in every generation. These processes help to produce the diversity of human beings.

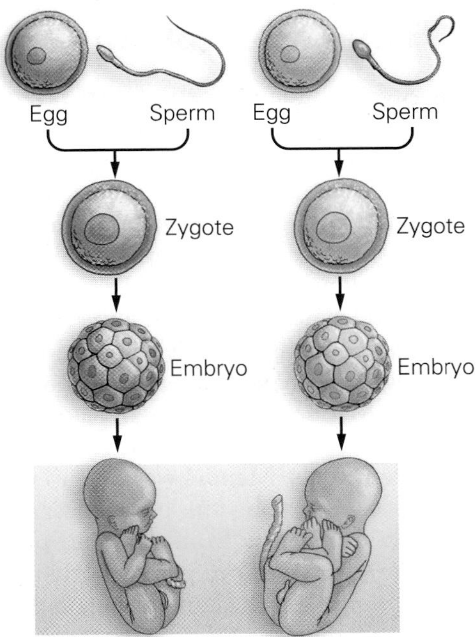

50% identical (no more similar than siblings born at different times)

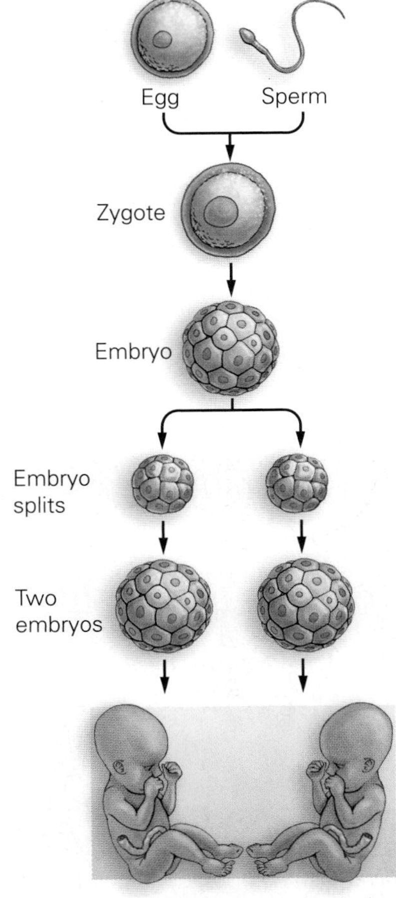

100% genetically identical

A Special Case—Identical Twins. Although the process of sexual reproduction can produce 2 siblings who are very different from each other, an event that may occur after fertilization can result in the birth of 2 children who share 100% of their genes. These children are identical twins and are the result of a single fertilization event—the fusion of 1 egg with 1 sperm.

Nonidentical twins, or fraternal twins, are called **dizygotic** (*di* means "two"), and they occur when 2 separate eggs fuse with different sperm. The resulting embryos, which develop together, are genetically no more similar than siblings born at different times *(Figure 6.7a)*.

Identical twins are referred to as **monozygotic twins** because they develop from a single egg and sperm. Recall that after fertilization the zygote grows and divides, producing an embryo made up of many daughter cells containing the same genetic information. Monozygotic twinning occurs when cells in an embryo separate from each other. If this happens early in development, each cell or clump of cells can develop into a complete individual, yielding twins who carry identical genetic information *(Figure 6.7b)*. In humans, about 1 in every 80 pregnancies produces dizygotic twins, while only approximately 1 of every 285 pregnancies results in identical twins.

Identical twins provide a unique opportunity to study the relative effects of our genes and environment in determining who we are. Because identical twins carry the same genetic information, researchers are able to study how important genes are in determining these twins' health, tastes, intelligence, and personality.

We will examine the results of some twin studies later in the chapter, as we continue to explore the question of predicting the heredity of the complex genetic traits possessed by a particular sperm donor. Our review of the relationship among parents, offspring, and genetic material has now prepared us to examine the inheritance of traits controlled by a single gene.

Figure 6.7 **The formation of twins.** (a) Dizygotic twins form from 2 independent fertilizations, resulting in 2 embryos who are as genetically similar as any other siblings. (b) Monozygotic twins form from 1 fertilization event and thus are genetically identical.

web animation 6.1
Mendel's Experiments

6.2 Mendelian Genetics: When the Role of Genes Is Clear

A few human genetic traits have easily identifiable patterns of inheritance. These traits are said to be "Mendelian" because Johann Gregor Mendel *(Figure 6.8)* was the first person to accurately describe their inheritance.

Mendel was born in Austria in 1822. Because his family was poor and could not afford private schooling, he entered a monastery to obtain an education. After completing his monastic studies, Mendel attended the University of Vienna. There he studied math and botany in addition to other sciences. Mendel attempted to become an accredited teacher but was unable to pass the examinations. After leaving the university, he returned to the monastery and began his experimental studies of inheritance in garden peas.

Mendel studied close to 30,000 pea plants over a 10-year period. His careful experiments consisted of controlled matings between plants with different traits. Mendel was able to control the types of mating that occurred by hand-pollinating the peas' flowers—for example, by applying pollen, which produce sperm, from a tall pea plant to the carpel (the egg-containing structure) of a short pea plant and then growing the seeds resulting from that cross. By doing this, he could evaluate the role of each parent in producing the traits of the offspring *(Figure 6.9)*.

Figure 6.8 **Gregor Mendel.** The father of the science of genetics.

Although Mendel himself did not understand the chemical nature of genes, he was able to determine how traits were inherited by carefully analyzing the appearance of parent pea plants and their offspring. His patient, scientifically sound experiments demonstrated that both parents contribute equal amounts of genetic information to their offspring.

Mendel published the results of his studies in 1865, but his contemporaries did not fully appreciate the significance of his work. Mendel eventually gave up his genetic studies and focused his attention on running the monastery until his death in 1884. His work was independently rediscovered by 3 scientists in 1900; only then did its significance to the new science of genetics become apparent.

The pattern of inheritance Mendel described occurs primarily in traits that are the result of a single gene with a few distinct alleles. *Table 6.1* lists some of the traits Mendel examined in peas; we will examine the principles he discovered, such as dominance and recessiveness, by looking at human disease genes that are of interest to prospective parents.

Genotype and Phenotype

We call the genetic composition of an individual his **genotype** and his physical traits his **phenotype**. The genotype is a description of the alleles for a particular gene carried on each member of a homologous pair. An individual who carries 2 different alleles for a gene has a **heterozygous** genotype. An individual who carries 2 copies of the same allele has a **homozygous** genotype.

The effect of an individual's genotype on her phenotype depends on the nature of the alleles she carries. Some alleles are **recessive,** meaning that their effects can be seen only if a copy of a dominant allele (described below) is not also present. For example, in pea plants, the allele that codes for wrinkled seeds is recessive to the allele for round seeds. Wrinkled seeds will only appear when seeds carry only the wrinkled allele and no copies of the round allele.

A typical recessive allele is one that codes for a nonfunctional protein. Homozygotes having 2 copies of such an allele produce no functional protein. In contrast, heterozygotes carrying 1 copy of the functional allele have normal phenotypes because the normal protein is still produced. Wrinkled seeds result

TABLE 6.1 **Pea traits studied by Mendel.**

Character Studied	Dominant Trait	Recessive Trait
Seed shape	Round	Wrinkled
Seed color	Yellow	Green
Flower color	Purple	White
Stem length	Tall	Dwarf

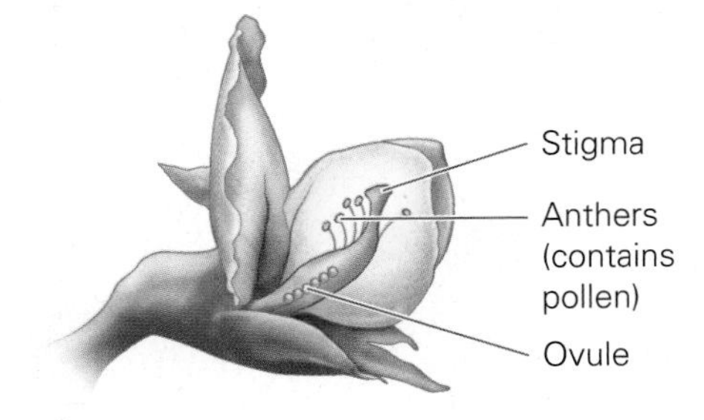

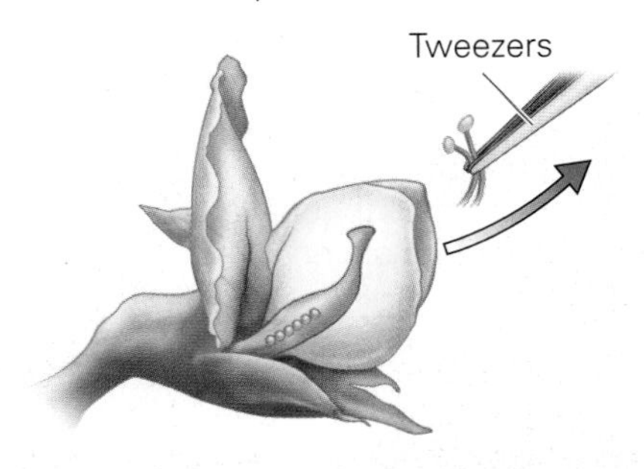

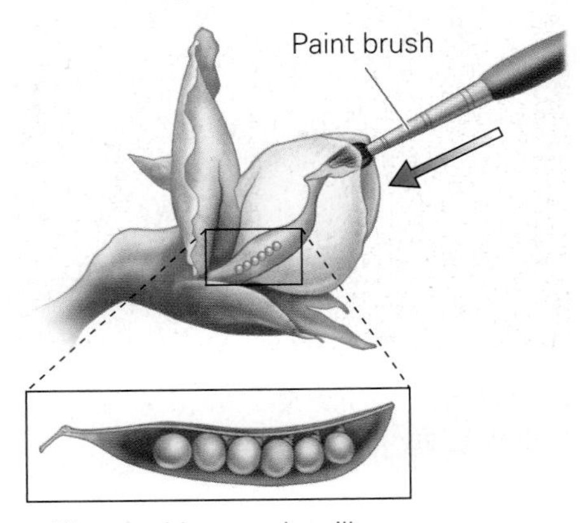

Figure 6.9 **Peas and genes.** Pea plants were an ideal study organism for Mendel because their reproduction is easy to control, they complete their life cycle in a matter of weeks, and a single plant can produce thousands of offspring.

from a recessive, nonfunctional allele. The functional allele in this case prevents water from accumulating in the seed. When seeds containing 1 or 2 copies of this allele dry, they look much the same as when they first matured. However, when the functional allele is missing from a seed, water flows into it, inflating it and causing its coat to increase in size. When this seed dries, it deflates and wrinkles, much like the surface of a balloon becomes wrinkly after it is blown up once and then deflated.

Dominant alleles are so named because their effects are seen even when a recessive allele is present. In the wrinkled seed example, the dominant allele was the one that produced a functional protein. However, sometimes mutations can create abnormal dominant alleles that essentially mask the effects of the recessive, normal allele. "American Albino" horses—known for their snow-white coats, tails, and manes; pink skin; and dark eyes—result from an allele that stops a horse's hair-color genes from being expressed during the horse's development. Because the allele prevents normal coat color development, it has its effect even if the animal carries only 1 copy—in other words, albinism in horses is dominant to the normal coat color *(Figure 6.10)*.

Mendel worked with traits that expressed only simple dominance and recessive relationships. However, for some genes, more than 1 dominant allele

Figure 6.10 **A dominant allele.** American Albino horses occur when an individual carries an allele that prevents normal coat color development. Because the product of this allele actively interferes with a biochemical pathway, horses that have only 1 copy of the allele are albino. (Photo by Linda Gordon.)

may be produced, and for others, a dominant allele may have different effects in a heterozygote than a homozygote. These situations are referred to as *codominance* and *incomplete dominance*, respectively, and are explored in detail in Chapter 7. For now, like Mendel, we will focus on simple dominant and recessive traits only.

Genetic Diseases in Humans

Most alleles in humans do not cause disease or dysfunction; they are simply alternative versions of genes. The diversity of alleles in the human population primarily contributes to diversity among us in our appearance, physiology, and behaviors.

While women who use sperm banks are likely interested in a wide variety of traits in donors, sperm banks are primarily concerned with providing sperm that do not carry a risk of common genetic diseases. Some genetic diseases are produced by recessive alleles, while others are the result of dominant alleles.

Cystic Fibrosis Is a Recessive Condition. Individuals with cystic fibrosis (CF) cannot transport chloride ions into and out of cells lining the lungs, intestines, and other organs. As a result of this dysfunction, the balance between sodium and chloride in the cell is disrupted, and the cell produces a thick, sticky mucus layer instead of the thin, slick mucus produced by cells with the normal allele. Affected individuals suffer from progressive deterioration of their lungs and have difficulties absorbing nutrients across the lining of their intestines. Most children born with CF suffer from recurrent lung infections and have dramatically shortened life spans *(Figure 6.11)*.

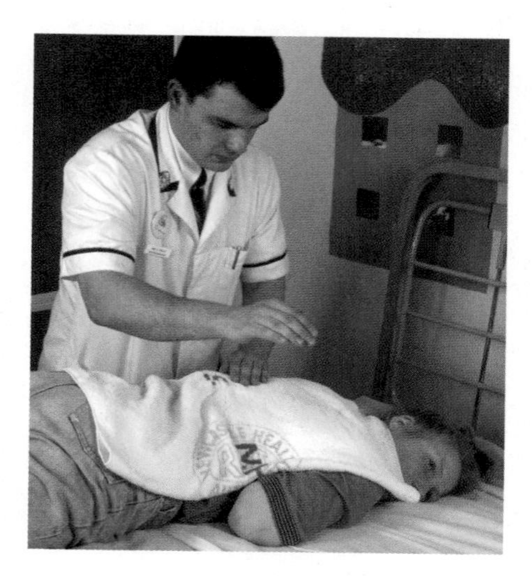

Figure 6.11 **Treatment for cystic fibrosis.** Percussive therapy consists of pounding a CF patient's back to loosen mucus inside the lungs. The mucus is then coughed up, reducing opportunities for bacterial infections to take hold.

Cystic fibrosis results from the production of a mutant chloride ion transporter protein that cannot embed in a cell membrane. The CF allele is recessive because individuals who carry only 1 copy of the normal allele can still produce the functional chloride transporter protein. The disease affects only homozygous individuals with 2 mutant alleles and thus no functional proteins.

Cystic fibrosis is among the most common genetic diseases in European populations; nearly 1 in 2500 individuals in these populations is affected with the disease, and 1 in 25 is heterozygous for the allele. Heterozygotes for a recessive disease are called **carriers** because even though they are unaffected, these individuals can pass the trait to the next generation. Sperm banks can test donor sperm for several recessive disorders, including CF; any men who are carriers of CF are excluded from most donor programs, including Fairfax Cryobank.

Huntington's Disease Is Caused by a Dominant Allele. Huntington's disease is an example of a fatal genetic condition caused by a dominant allele. Early symptoms of Huntington's disease include restlessness, irritability, and difficulty in walking, thinking, and remembering. These symptoms typically begin to manifest in middle age. Huntington's disease is progressive and incurable—the nervous, mental, and muscular symptoms gradually become worse and eventually result in the death of the affected individual.

The Huntington's allele causes production of a protein that forms clumps inside the nuclei of cells. Nerve cells in areas of the brain that control movement are especially likely to contain these protein clumps, and these cells gradually die off over the course of the disease *(Figure 6.12)*. Because this allele produces a protein that damages cells, the presence of the normal allele cannot compensate or correct for this mutant version. An individual needs only 1 copy of the allele to be affected by the disease; that is, even heterozygotes exhibit the symptoms of Huntington's.

Stop and Stretch 6.3: Typically, dominant mutations that result in death are not passed on from one generation to the next. What characteristic of Huntington's disease allows this allele to persist in the human population?

Only since the mid-1980s has genetic testing allowed people with a family history of Huntington's disease to learn whether they are affected before they show signs of the disease. Although most sperm banks do not test for the presence of the Huntington's allele because it is a rare condition, the detailed family medical histories required of sperm bank donors enable Fairfax Cryobank to exclude men with a family history of Huntington's disease from their donor list.

(a) Diseased brain

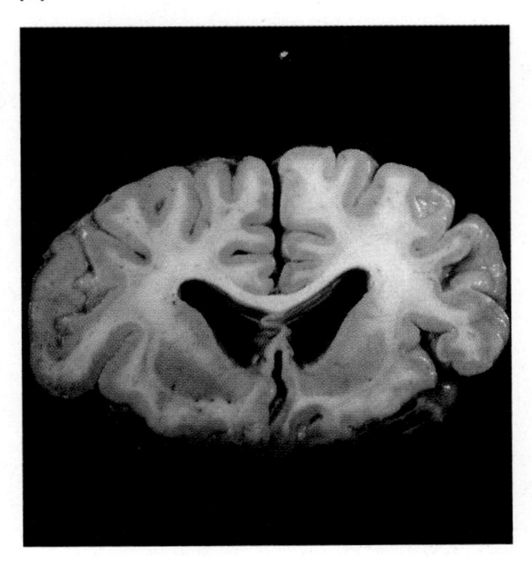

(b) Normal brain

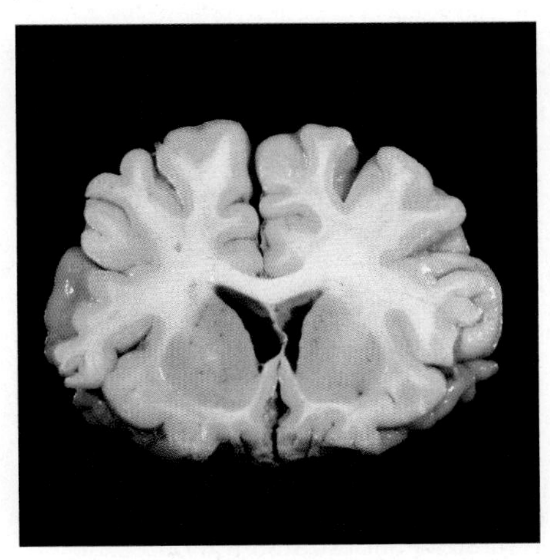

Figure 6.12 **Huntington's and the brain.** Cells containing mutant proteins die, shrinking and deforming the brain. This image compares the brain of (a) a Huntington's sufferer with (b) an unaffected individual.

Using Punnett Squares to Predict Offspring Genotypes

Traits such as cystic fibrosis and Huntington's disease are the result of a mutation in a single gene, and the inheritance of these conditions and of other single-gene traits is relatively easy to understand. We can predict the likelihood of inheritance of small numbers of these single-gene traits by using a tool developed by Reginald Punnett, a British geneticist. A **Punnett square** is a table that lists the different kinds of sperm or eggs parents can produce relative to the gene or genes in question and then predicts the possible outcomes of a **cross**, or mating, between these parents *(Figure 6.13)*.

Using a Punnett Square with a Single Gene.

Imagine a woman and a sperm donor who are both carriers of the cystic fibrosis allele. Different alleles for a gene are symbolized with letters or number codes that refer to a trait that the gene affects. For instance the cystic fibrosis gene is symbolized *CFTR* for *cystic fibrosis transmembrane regulator*. The dysfunctional *CFTR* allele is called *CFTR-ΔF508*, so both would have the genotype *CFTR/CFTR-ΔF508*. However, to make this easier to follow, we will use a

Possible types of eggs

Sperm sample *Ff*

Female carrier *Ff*

Possible types of sperm

	F	*f*
F	*FF*	*Ff*
f	*Ff*	*ff*

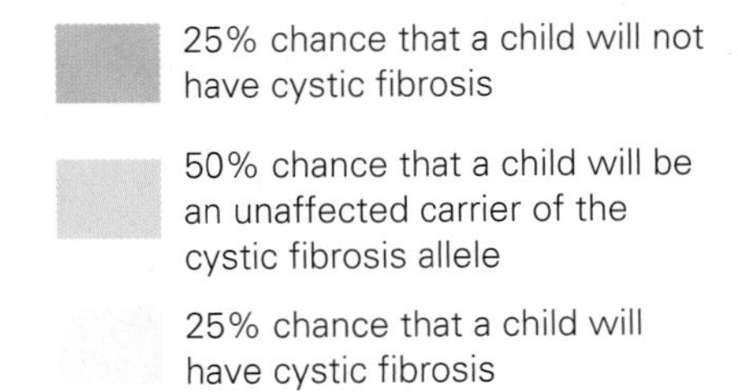

Figure 6.13 **What are the risks of accepting sperm from a carrier of cystic fibrosis?** This Punnett square illustrates the likelihood that a woman who carries the cystic fibrosis allele would have a child with cystic fibrosis if the sperm donor were also a carrier.

simpler key: the letters *F* and *f*, representing the dominant functional allele and recessive nonfunctional allele, respectively.

A carrier for cystic fibrosis would have the genotype *Ff*. A genetic cross between 2 carriers could then be symbolized as follows:

$$Ff \times Ff$$

We know that the female in this cross can produce eggs that carry either the *F* or *f* allele since the process of meiosis will segregate the 2 alleles from each other. We place these 2 egg types across the horizontal axis of what will become a Punnett square. The male can also make 2 types of sperm, containing either the *F* or the *f* allele. We place these along the vertical axis. Thus, the letters on the horizontal and vertical axes represent all the possible types of eggs and sperm that the mother and father can produce by meiosis if we consider only the gene that codes for the chloride transport protein.

Inside the Punnett square are all the genotypes that can be produced from a cross between these 2 heterozygous individuals. The content of each box is determined by combining the alleles from the egg column and the sperm row.

Note that for a cross involving a single gene with 2 different alleles, there are 3 possible offspring types. The chance of this cross producing a child affected with cystic fibrosis is one in four, or 25%, because the *ff* combination of alleles occurs once out of the four possible outcomes. You can see why sperm banks would exclude carriers of cystic fibrosis from their donor rolls: The risk of an affected child born to a woman who might not know she is a carrier is too high. The *FF* genotype is also represented once out of four times, meaning that the probability of a homozygous unaffected child is also 25%. The probability of producing a child who is a **carrier** of cystic fibrosis is one in two, or 50%, since 2 of the possible outcomes inside the Punnett square are unaffected heterozygotes—one produced by an *F* sperm and an *f* egg and the other produced by an *f* sperm and an *F* egg.

When parents know which alleles they carry for a single-gene trait, they can easily determine the probability that a child they produce will have the disease phenotype *(Figure 6.14)*. You should note that this probability is generated independently for each child. In other words, each offspring of 2 carriers has a 25% chance of being affected.

Punnett squares can also be employed to predict the likelihood of a particular genotype when considering multiple genes, as described in Chapter 7. In fact, as long as each gene of interest is carried on a separate chromosome and the number of alleles for each gene is known, we can predict the likelihood of a particular genotype.

Figure 6.14 **Calculating the likelihood of genetic traits in children.** This Punnett square illustrates the outcome of a cross between a man who carries a single copy of the dominant Huntington's disease allele and an unaffected woman. **Visualize This:** If one child produced by these parents carries the Huntington's allele, what is the likelihood that a second child produced by this couple will have Huntington's disease?

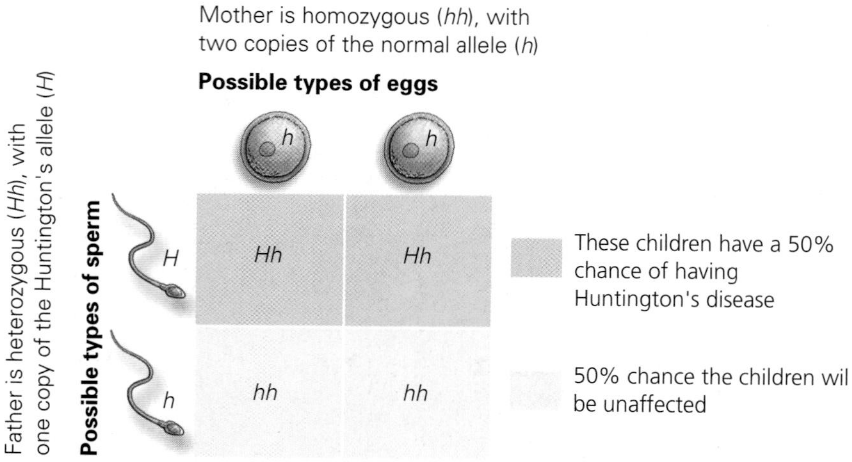

Mother is homozygous (*hh*), with two copies of the normal allele (*h*)

Possible types of eggs

Father is heterozygous (*Hh*), with one copy of the Huntington's allele (*H*)

Possible types of sperm

	h	*h*
H	*Hh*	*Hh*
h	*hh*	*hh*

These children have a 50% chance of having Huntington's disease

50% chance the children wil be unaffected

As you might imagine, as the number of genes in a Punnett square analysis increases, the number of boxes in the square increases, as does the number of possible genotypes. With 2 genes, each with 2 alleles, the number of unique gametes produced by a heterozygote is 4, the number of boxes in the Punnett square is 16, and the number of unique genotypes that can be produced is 9. With 3 genes, each with 2 alleles, the Punnett square has 64 boxes and 22 different possible genotypes. With 4 genes, the square has 256 boxes, and with 5 genes, there are over 1000 boxes! Predicting the outcome of a cross becomes significantly more difficult as the number of genes we are following increases.

As scientists identify more genes and alleles, the amount of information about the genes of sperm donors or any potential parent will also increase. Identifying and testing for particular genes in potential parents will allow us to predict the *likelihood* of numerous genotypes in their offspring. Unfortunately, this increase in genetic testing is not necessarily equaled by an increase in our understanding of how more complex traits develop, as we shall see in the next section.

6.3 Quantitative Genetics: When Genes and Environment Interact

The single-gene traits discussed in the previous section have a distinct "off-or-on" character; individuals have either one phenotype (for example, the pea seed is round) or the other (the pea seed is wrinkled). Traits like this are known as qualitative traits. However, many of the traits that interest women who are choosing a sperm donor do not have this off-or-on character.

An Overview of Quantitative Trait Inheritance

Traits such as height, weight, eye color, musical ability, susceptibility to cancer, and intelligence are called **quantitative traits**. Quantitative traits show **continuous variation**; that is, we can see a large range of phenotypes in a population—for instance, from very short people to very tall people. Wide variation in quantitative traits leads to the great diversity we see in the human population.

Variation among individuals in most quantitative traits comes from differences among them in both their genetic makeup and their environment. For instance, it is clear that the ability to complete a doctoral degree depends in some part on innate traits related to brain structure and function as well as environmental factors such as early learning experiences, exposure to environmental toxins, and access to higher education opportunities. To understand the inheritance of quantitative traits, scientists must determine what proportion of the variation within a population can be explained by genetic differences among individuals—a value called the **heritability** of the trait.

Heritability in humans is typically measured by examining correlations between groups. These studies calculate how similar or different parents are to their children, or siblings are to each other, in the value of a particular trait. When examined across an entire population, the strength of a correlation provides a measure of heritability *(Figure 6.15)*. The uses and limitations of heritability are addressed more specifically in section 6.4.

Figure 6.15 **What is the genetic component of a quantitative trait?** Comparisons between parents and children can provide information about the heritability of a particular trait, such as height.

For most traits, such as body size, people come in a wide variety of types.

How much of our differences is due to the environment, and how much is a result of different genes?

Correlations between relatives can provide information about the importance of genes in determining variation among individuals.

A Closer Look: Quantitative Traits and Heritability

The distribution of phenotypes of a quantitative trait in a population can be displayed on a graph. These data often take the form of a bell-shaped curve called a *normal distribution. Figure 6.16a* illustrates the normal distribution of heights in a college class. Each individual is standing behind the label (at the bottom) that indicates their height. The curved line drawn across the photo is the idealized bell-shaped curve that summarizes these data.

A bell-shaped curve contains two important pieces of information. The first is the highest point on the curve: the average, or **mean,** value for data. The mean is calculated by adding all of the values for a trait in a population and dividing by the number of individuals in that population. The second is in the width of the bell itself, which illustrates the variability of a population. The variability is described with a mathematical measure called **variance,** which is essentially the average distance any 1 individual in the population is from the mean. If a low variance for a trait indicates a small amount of variability in the population, a high variance indicates a large amount of variability *(Figure 6.16b)*.

Why Traits Are Quantitative. A range of phenotypes may be generated because numerous genotypes exist among the individuals in the population. This happens when a trait is influenced by more than 1 gene; traits influenced by many genes are called **polygenic traits**.

As we saw above, when a single gene with 2 alleles determines a trait, only 3 possible genotypes are present: *FF, Ff,* and *ff,* for example. But when more than 1 gene, each with more than 1 allele, influence a trait, many genotypes are possible. For example, eye color in humans is a polygenic trait influenced by at least 3 genes, each with more than 1 allele. These genes help produce and distribute the pigment melanin, a brown color, to the iris. When different alleles for the genes for eye pigment production and distribution interact, a range of eye colors, from dark brown (lots of melanin produced) to pale blue (very little melanin produced), is found in humans. The continuous variation in eye color among people is a result of several genes, each with several alleles, influencing the phenotype.

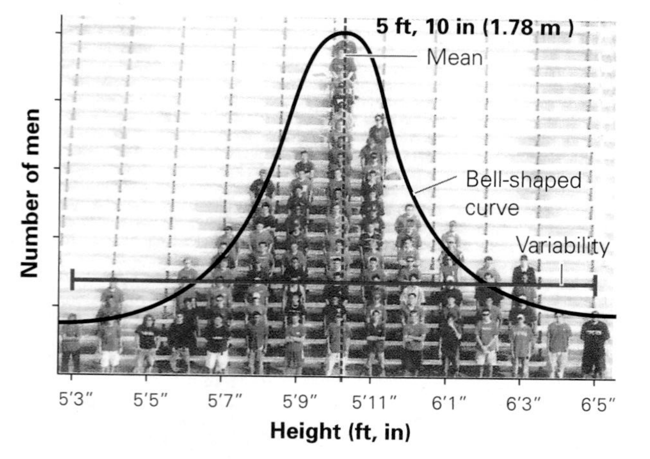

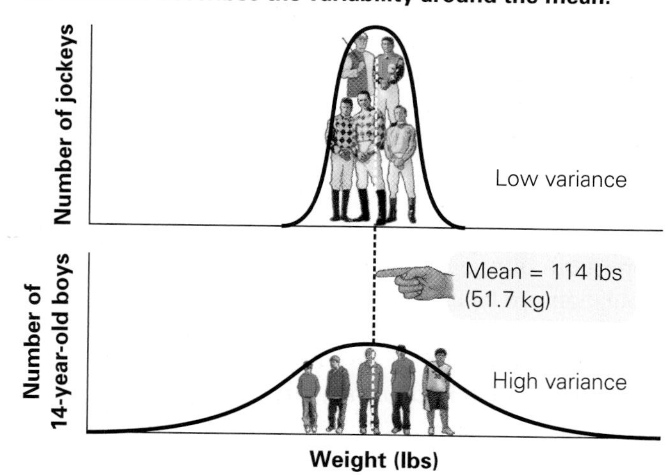

Figure 6.16 **A quantitative trait.** (a) This photo of men arranged by height illustrates a normal distribution. The highest point of the bell curve is also the mean height of 5 feet, 10 inches. (b) Fourteen-year-old boys and professional jockeys have the same average weight—approximately 114 pounds. However, to be a jockey, you must be within about 4 pounds of this average. Thus the variance among jockeys in weight is much smaller than the variance among 14-year-olds. **Visualize This:** Examine Figure 6.16a closely: Does an average height of 5 feet, 10 inches in this particular population imply that most men were this height? Were most men in this population close to the mean, or was there a wide range of heights?

(A Closer Look continued)

Continuous variation also may occur in a quantitative trait due to the influence of environmental factors. In this case, each genotype is capable of producing a range of phenotypes depending on outside influences. For a clear example of the effect of the environment on phenotype, see *Figure 6.17*. These identical twins share 100% of their genes but are quite different in appearance. Their difference is entirely due to variations in their environment—the twin on the right smoked cigarettes and had much greater sun exposure than did the twin on left.

Most traits that show continuous variation are influenced by both genes and the environment. Skin color in humans is an example of this type of trait. The shade of an individual's skin is dependent on the amount of melanin present near the skin's surface. As with eye color, a number of genes have an effect on skin-color phenotype—those that influence melanin production and those that affect melanin distribution. However, the environment, particularly the amount of exposure to the sun during a season or lifetime, also influences the skin color of individuals *(Figure 6.18)*. Melanin production increases, and any melanin that is present darkens in sun-exposed skin. In fact, after many years of intensive sun exposure, skin may become permanently darker.

Figure 6.17 **The effect of the environment on phenotype.** These identical twins have exactly the same genotype, but they are quite different in appearance due to environmental factors.

Stop and Stretch 6.4: The average height of American men has stayed the same over the past 50 years, while average heights in other countries (for example, Denmark) have increased. Height is a classic quantitative trait. Provide both a genetic and an environmental hypothesis for why the height of the American population is not increasing along with other countries.

Women choosing Doctorate category sperm donors from Fairfax Cryobank are presumably interested in having smart, successful children, but intelligence has both a genetic and an environmental component. With an important role for both influences, how can we predict if the child of a father with a doctorate will also be capable of earning a doctorate?

Measuring Heritability in Animals. Researchers trying to improve the production of domestic animals and crop plants were the first to develop the scientific model used to measure heritability. Farmers need to know if it is more effective to boost production by changing the environment (for example, how dairy cows are fed and housed) or to change the genetics of the crop itself (for example, by choosing only the offspring of the best milk producers for the next-generation herd). The technique of controlling the reproduction of individual organisms to influence the phenotype of the next generation is known as **artificial selection.** Artificial selection is similar to natural selection, which we describe more thoroughly in Chapter 10 and which is a primary cause of evolution in natural populations.

If the quantitative trait of milk production in cows has a high heritability, then artificial selection is an effective way to boost milk output. In fact, heritability of milk production can easily be measured by how well a herd responds to artificial selection *(Figure 6.19)*. If milk production increases as a result of artificial selection, it is because alleles that increase milk production in an individual (for instance, alleles for genes that control the size of the udder or those that influence the activity of milk-producing cells) have become more common in the herd. In other words, the trait must have a high heritability. Conversely, if milk production does *not* increase as a result of artificial selection, then the alleles of high-production cows must not be very important in determining milk output. In this case, heritability is low.

(continued on the next page)

(a) Genes

(b) Environment

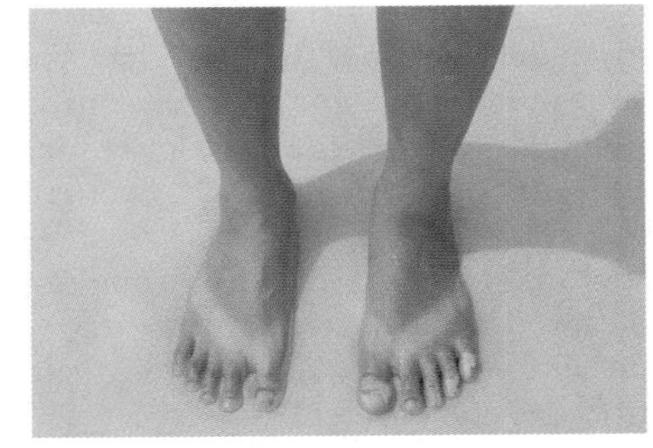

Figure 6.18 **Skin color is influenced by genes and environment.** (a) The difference in skin color between these 2 women is due primarily to variations in several alleles that control skin pigment production. (b) The difference in color between the sun-protected and sun-exposed portions of the individual in this picture is entirely due to environmental effects.

(A Closer Look continued)

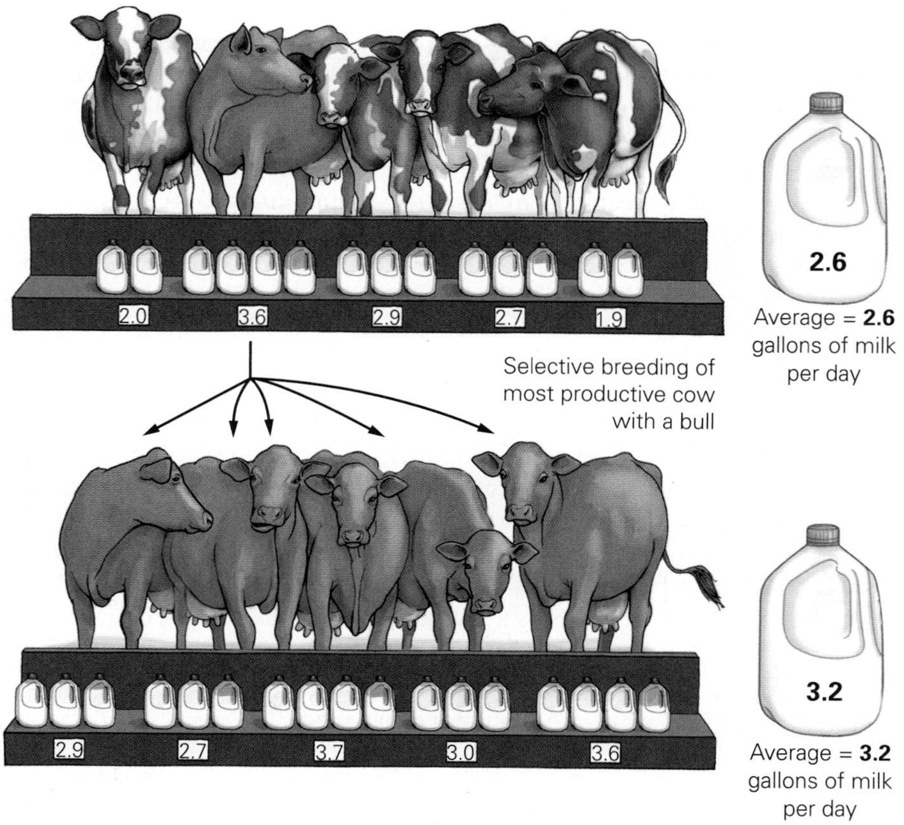

Figure 6.19 **Artificial selection increases milk production in cows.** Cows that produce exceptional amounts of milk are bred to produce the next generation of dairy cattle to increase production of the whole herd. Because milk production is only partially due to genes, the offspring of the best-producing cow do not necessarily produce as much milk as she did.

Scientists have calculated an average heritability of milk production in dairy cattle at a relatively low 0.30. This means that, in a typical dairy herd, about 30% of the variation among cows in milk production is due to differences in their genes. The remainder of their production variation is due to differences among the cows in their environmental conditions. Heritability has been estimated in this manner for many traits in multiple different crops and livestock animals.

To estimate heritability in populations in which artificial selection is impossible, researchers use correlations between individuals with varying degrees of genetic similarity. A correlation determines how accurately one can predict the measure of a trait in an individual when its measure in a related individual is known. For example, *Figure 6.20* shows a correlation between parent birds and their offspring in the strength of their response to tetanus vaccine. An individual that responded strongly produced a large number of antitetanus proteins, called antibodies, while ones that responded weakly produced a lower number of antibodies.

As you can see from the graph, parents with weak responses tended to have offspring with weak responses, and parents with strong responses had offspring with strong responses. This strong correlation indicates that the ability to respond to tetanus is highly heritable—most of the difference between birds in their immune system response results from genetic differences.

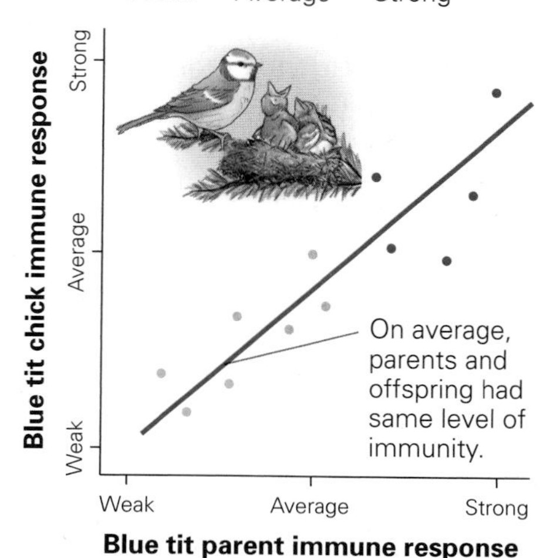

Figure 6.20 **Using correlation to calculate heritability.** The close correlation of immune system response between blue tit parents and offspring indicates that immune response is highly heritable. **Visualize This:** How would a graph that showed a low correlation between parents and offspring differ from this graph?

Calculating Heritability in Human Populations. Obviously, scientists cannot use artificial selection to choose the men and women who will produce the next generation of people, so correlation is the only technique for measuring heritability in human populations. However, interpreting the results of correlative studies in humans is tricky. We can see this as we examine attempts to calculate the heritability of intelligence.

Intelligence is often measured by performance on an IQ test. The French psychologist Alfred Binet developed the intelligence quotient (or IQ) in the early 1900s to identify schoolchildren who were in need of remedial help. Binet's IQ test was not based on any theory of intelli-

(A Closer Look continued)

gence and was not meant to comprehensively measure mental ability, but the tests remain a commonly used way to measure innate or "natural" intelligence. There is significant controversy over the use of IQ tests in this way.

Even if IQ tests do not really measure general intelligence, IQ scores have been correlated with academic success—meaning that individuals at higher academic levels usually have higher IQs. So, even without knowing their IQ scores, we can reasonably expect that donors in the Doctorate category have higher IQs than do other available sperm donors. However, the question of whether the high IQ of a prospective sperm donor has a genetic basis still remains.

The average correlation between IQs of parents and their children—in other words, the estimated heritability of IQ by this method—is 0.42. However, parents and the children who live with them are typically raised in a similar social and economic environment. As a result, correlations of IQ between the 2 groups cannot distinguish the relative importance of genes from the importance of the environment. This is the problem found in most arguments about "nature versus nurture"—do children resemble their parents because they are "born that way" or because they are "raised that way"?

To avoid the problem of overlap in environment and genes between parents and children, researchers seek situations that remove one or the other overlap. These situations are called **natural experiments** because one factor is "naturally" controlled, even without researcher intervention. Human twins are one source of a natural experiment to test hypotheses about the heritability of quantitative traits in humans.

By comparing monozygotic twins, who share all of their alleles, to dizygotic twins, who share 50% of their alleles, researchers can begin to separate the effects of shared genes from the effects of shared environments. Because twins raised in the same family have similar childhood experiences, one would expect that the only real difference between monozygotic and dizygotic twins is their genetic similarity. The average heritability of IQ calculated from a number of twin studies of this type is about 0.52. According to these studies, 52% of the variability in IQ among humans is due to differences in genotypes. Surprisingly, this value is even higher than the 42% calculated from the correlation between parents and children.

However, monozygotic twins and dizygotic twins likely *do* differ in more than just genotype. In particular, identical twins are treated more similarly than nonidentical twins. This occurs both because they look very similar and because other people may presume that they are identical in all other respects. If monozygotic twins are *expected* to be more alike than dizygotic twins, their IQ scores may be similar because they are encouraged to have the same experiences and to achieve at the same level.

There is one natural experiment that can address this problem, however. By comparing identical and nonidentical twins raised apart, the problem of differential treatment of the 2 *types* of twins is minimized because no one would know that the individual members of a pair have a twin (*Figure 6.21*). If variation in genes does not explain much of the variation among peoples' IQ scores, then identical twins raised apart should be no more similar than any 2 unrelated people of the same age and sex.

Unfortunately for scientists (but perhaps fortunately for children), the frequency of early twin separation is extremely low. Researchers have estimated the heritability of IQ at a remarkable 0.72 in a small sample of twins raised apart. This study and the other correlations appear to support the hypothesis that differences in our genes explain the majority of the variation in IQ among people. *Table 6.2* summarizes the estimates of IQ heritability and previews the cautions discussed in the next section of this chapter.

(continued on the next page)

Figure 6.21 **Twins separated at birth.** Tamara Rabi (left) and Adriana Scott were reunited at age 20 when a mutual friend noticed their remarkable resemblance. Tamara was raised on the Upper West Side of Manhattan and Adriana in suburban Long Island, but they both have similar outgoing personalities, love to dance, and even preferred the same brand of shampoo.

(A Closer Look continued)

TABLE 6.2 **To what extent is IQ heritable?** A summary of various estimates of IQ heritability, their shortcomings, and the problems with using them to understand the role of genes in determining an individual's potential intelligence.

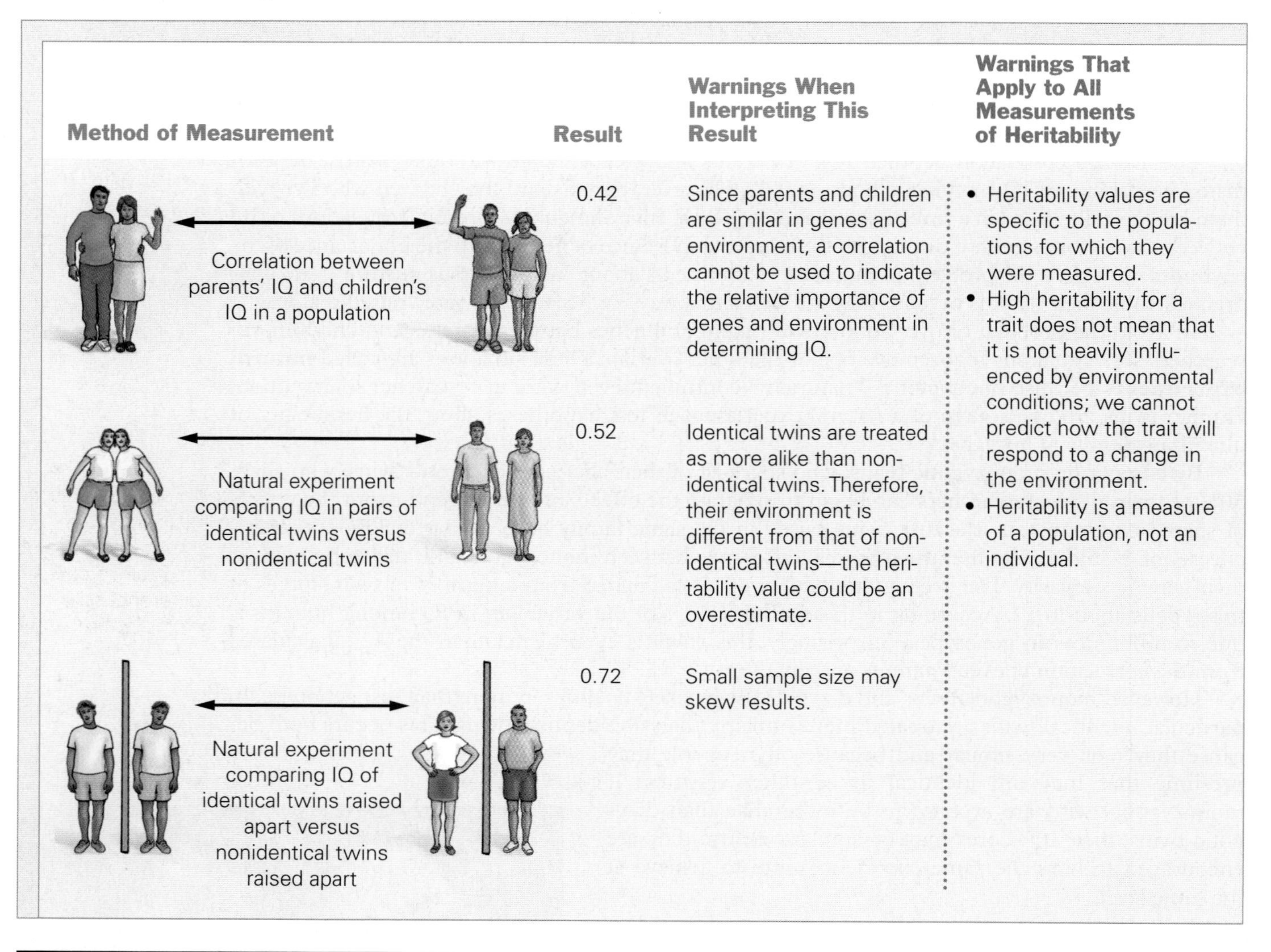

Method of Measurement	Result	Warnings When Interpreting This Result	Warnings That Apply to All Measurements of Heritability
Correlation between parents' IQ and children's IQ in a population	0.42	Since parents and children are similar in genes and environment, a correlation cannot be used to indicate the relative importance of genes and environment in determining IQ.	• Heritability values are specific to the populations for which they were measured. • High heritability for a trait does not mean that it is not heavily influenced by environmental conditions; we cannot predict how the trait will respond to a change in the environment. • Heritability is a measure of a population, not an individual.
Natural experiment comparing IQ in pairs of identical twins versus nonidentical twins	0.52	Identical twins are treated as more alike than nonidentical twins. Therefore, their environment is different from that of nonidentical twins—the heritability value could be an overestimate.	
Natural experiment comparing IQ of identical twins raised apart versus nonidentical twins raised apart	0.72	Small sample size may skew results.	

6.4 Genes, Environment, and the Individual

Perhaps we can now determine the importance of a sperm donor father who has earned a doctorate to his child's intellectual development. We know that a sperm donor will definitely influence some of his child's traits—eye and skin color and perhaps even susceptibility to certain diseases. In addition, according to the studies discussed above, the donor will probably pass on some intellectual traits to the child. In fact, with a heritability of IQ at above 50%, it appears that genes are primary in determining an individual's intelligence. Perhaps it is a good idea to pay a premium price for Doctorate category sperm after all.

However, we need to be very careful when applying the results of twin studies to questions about the individual sperm donors. To understand why, we will take a closer look at the practical significance of heritability.

The Use and Misuse of Heritability

Remember that heritability is a measure of the relative importance of genes in determining variation in quantitative traits among individuals. For example, with a heritability of 0.30, we can say that only 30% of the variation in dairy cows' milk production is due to variation in genes among these cows.

However, the calculated heritability value is unique to the population in which it was measured and to the environment of that population. We should be very cautious when using heritability to measure the *general* importance of genes to the development of a trait. The headings below illustrate why.

Differences Between Groups May Be Environmental.

A "thought experiment" can help illustrate this point. Body weight in laboratory mice has a strong genetic component, with a calculated heritability of about 0.90. In a population of mice in which weight is variable, bigger mice have bigger offspring, and smaller mice have smaller offspring.

Imagine that we randomly divide a population of variable mice into 2 groups—one group is fed a rich diet, and the other group is fed a poor diet. Otherwise, the mice are treated identically. As you might predict, the well-fed mice become fat, while the poorly fed mice become thin regardless of their genetic predispositions. Consider the outcome if we were to keep the mice in the same conditions, and allowed the population to reproduce. Not surprisingly, the second generation of well-fed mice is likely to be much heavier than the second generation of poorly fed mice.

Now imagine that another researcher came along and examined these 2 populations of mice without knowing their diets. Knowing that body weight is highly heritable, the researcher might logically conclude that the groups are genetically different. However, we know this is not the case—both are grandchildren of the same original source population. It is the environment of the 2 populations that differs *(Figure 6.22)*.

Figure 6.22 **The environment can have powerful effects on highly heritable traits.** If genetically similar populations of mice are raised in radically diverse environments, then differences between the populations are entirely due to environment.

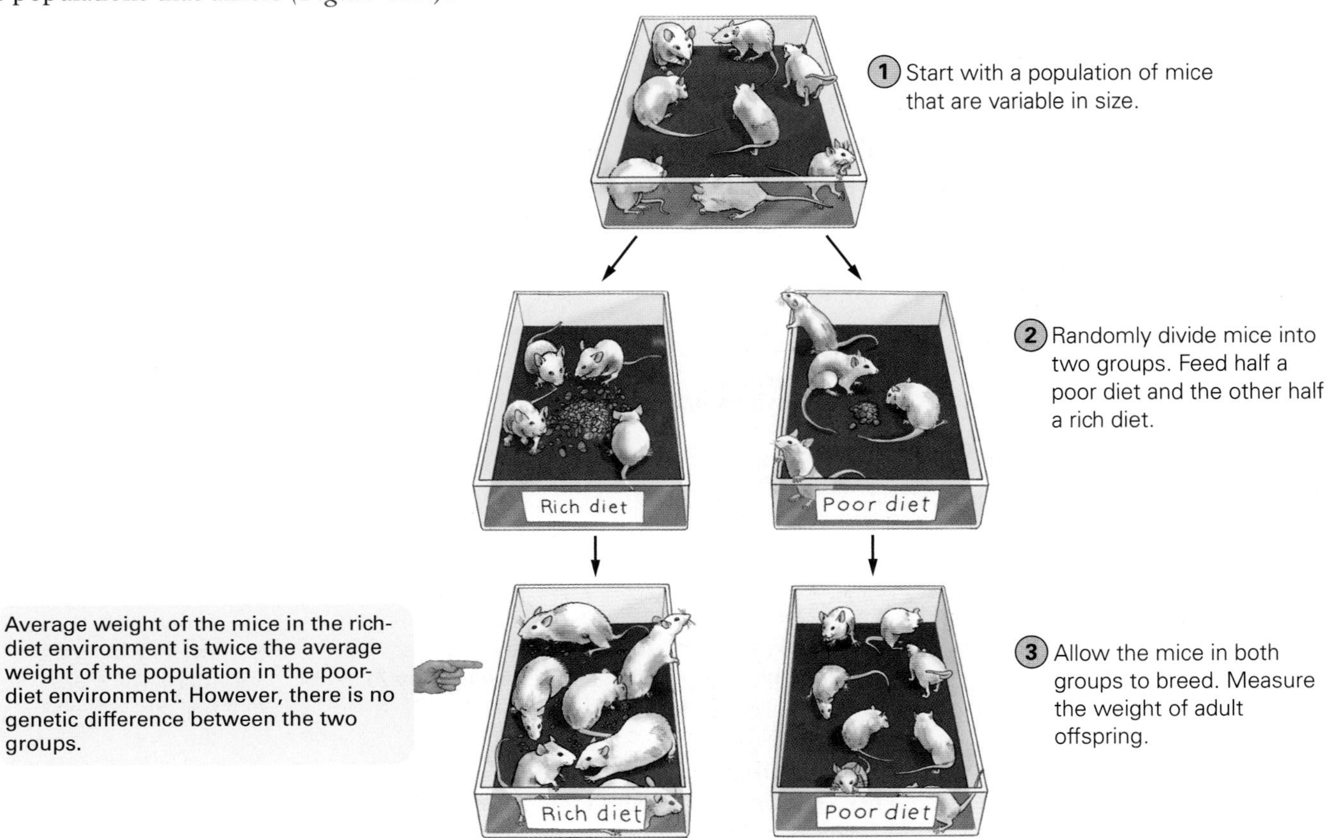

Now extend the same thought experiment to human groups. Imagine that we have 2 groups of humans, and we have determined that IQ had high heritability. In this case, people in one group were affluent, and their average IQ was higher. The other group was impoverished, and their average IQ was lower. What conclusions could you draw about the genetic differences between these 2 populations?

The answer to the question above is none—as with the laboratory mice, these differences could be entirely due to environment. The high heritability of IQ cannot tell us if 2 human groups in differing social environments vary in IQ because of variations in genes or because of differences in environment.

A Highly Heritable Trait Can Still Respond to Environmental Change. A high heritability for IQ might seem to imply that IQ is not strongly influenced by environmental conditions. However, intelligence in other animals can be demonstrated to be both highly heritable and strongly influenced by the environment.

Rats can be bred for maze-running ability, and researchers have produced rats that are "maze bright" and rats that are "maze dull." Maze-running ability is highly heritable in the laboratory environment; that is, bright rats have bright offspring, and dull rats have dull offspring. The results of an experiment that measured the number of mistakes made by maze-bright and maze-dull rats raised in different environments are presented in *Table 6.3*.

In the typical lab environment, bright rats were much better at maze running than dull rats. But in both a very boring or restricted environment and a very enriched environment, the 2 groups of rats did about the same. In fact, no rats excelled in a restricted environment, and all rats did better at maze running in enriched environments, with the duller rats improving most dramatically.

What this example demonstrates is that we cannot predict the response of a trait to a change in the environment, even when that trait is highly heritable. Thus, even if IQ has a strong genetic component, environmental factors affecting IQ can have big effects on an individual's intelligence.

TABLE 6.3 **A highly heritable trait is not identical in all environments.** This table describes the average number of mistakes made by rats of 2 different genotypes in 3 different environments.

		Number of Mistakes in ...		
	Phenotype	**Normal Environment**	**Restricted Environment**	**Enriched Environment**
	Maze-bright rats	115	170	112
	Maze-dull rats	165	170	122
Explanation of Results		Maze-dull rats made more mistakes than maze-bright rats when running a maze.	Both groups made the same number of mistakes when running a maze.	Both groups made fewer mistakes when running the maze. The maze-dull rats improved the most.

Stop and Stretch 6.5: Some commentators have argued that given IQ's high heritability, policies that increase financial resources to failing schools will ultimately fail to increase achievement since such a predominantly genetic trait will not respond well to environmental change. Use your understanding of the proper application of heritability to refute this argument.

Heritability Does Not Tell Us Why Two Individuals Differ. High heritability of a trait is often presumed to mean that the difference between 2 individuals is mostly due to differences in their genes. However, even if genes explain 90% of the population variability in a particular environment, the reason one individual differs from another may be entirely a function of environment (as an example of this, look back at the twins in Figure 6.17).

Currently, there is no way to determine if a particular child is a poor student because of genes, a poor environment, or a combination of both factors. There is also no way to predict whether a child produced from the sperm of a man with a doctorate will be an accomplished scholar. All we can say is that given our current understanding of the heritability of IQ and the current social environment, the alleles in Doctorate category sperm *may* increase the probability of having a child with a high IQ.

How Do Genes Matter?

We know that genes can have a strong influence on eye color, risk of genetic diseases such as cystic fibrosis, susceptibility to a heart attack, and even the structure of the brain. But what really determines who we are—nature or nurture?

Even with single-gene traits, the outcome of a cross between a woman and a sperm donor is not a certainty; it is only a probability. Couple this with traits being influenced by more than 1 gene, and independent assortment greatly increases the offspring types possible from a single mating. Knowing the phenotype of potential parents gives you relatively little information about the phenotype of their children. So, even if genes have a strong effect on traits, we cannot "program" the traits of children by selecting the traits of their parents *(Figure 6.23)*.

Figure 6.23 **Genes are not destiny.** Even when the traits of both parents are known, the child they produce may be very different in interests and aptitudes.

In truth, we are really asking ourselves the wrong question when we wonder if nature or nurture has a more powerful influence on who we are. Both nature and nurture play an important role. Our cells carry instructions for all the essential characteristics of humanity, but the process of developing from embryo to adult takes place in a physical and social environment that influences how these genes are expressed. Scientists are still a long way from answering questions about how all of these complex, interacting circumstances result in who we are.

What is the message for women and couples who are searching for a sperm donor from Fairfax Cryobank? Donors in the Doctorate category may indeed have higher IQs than donors in the cryobank's other categories, but there is no real way to predict if a particular child of 1 of these donors will be smarter than average. According to the current data on the heritability of IQ, sperm from high-IQ donors may increase the odds of having an offspring with a high IQ but only if parents provide them with a stimulating, healthy, and challenging environment in which to mature. This, of course, would be good for children with any alleles.

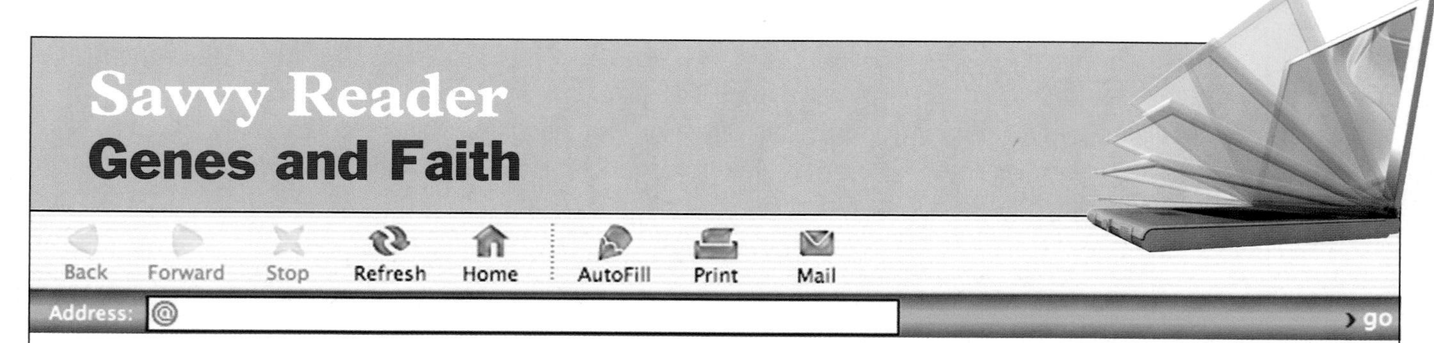

NATURE HELPS CREATE RELIGIOUS ADULTS

APRIL 5, 2005

Science Daily.com

A study published in the current issue of *Journal of Personality* studied adult male monozygotic (MZ) and dizygotic (DZ) twins to find that differences in religiousness are influenced by both genes and environment. But during the transition from adolescence to adulthood, genetic factors increase in importance while shared environmental factors decrease. . . .

Participants for this study were 169 MZ and 104 DZ male twin pairs from Minnesota. . . . The twins graded the frequency in which they partook in religious activities such as reading scripture or other religious material and the importance of religious faith in daily life. They also reported on their mother's, their father's, and their own religiousness when they were growing up. They were also asked to report on the current and past religiousness of their [twin] brother.

The factors were divided into subscales—external aspects of religion, like observing religious holidays, that might be the most susceptible to environmental influence and internal aspects, like seeking help through prayer, that might be the most susceptible to heritable influence. The external items were found to be more environmentally and less genetically influenced during childhood, but more genetically influenced in adulthood. The internal scale showed a similar pattern, but the genetic influences seemed to be slightly larger in childhood compared to the external scale and so more consistent across the 2 ages. "Like other personality traits, adult religiousness is heritable, and though changes in religiousness occur during development, it is fairly stable," the authors conclude.

1. Decode this article. What do the researchers' results indicate about the role of genetic factors versus environmental factors in determining our religious practices and beliefs as we age?
2. What evidence do the researchers give that "external" aspects of religion are more likely to be affected by the environment than "internal" aspects? Does your answer to this question influence your analysis of the study?
3. Does the headline accurately reflect the results of the study? What does the headline imply, and how might it be misleading?
4. Put this study in context with what you learned about heritability in this chapter. Is whether you are religious or not affected by your genes, your environment, or both?

CHAPTER Review

For study help, animations, and more quiz questions go to www.mybiology.com.

Summary

6.1 The Inheritance of Traits

- Children resemble their parents in part because they inherit their parents' genes, segments of DNA that contain information about how to make proteins (pp. 146–147).
- Mutations in gene copies can cause slightly different proteins to be produced; different gene versions are called alleles (p. 148).
- Due to independent assortment, parents contribute a unique subset of alleles to each of their nonidentical twin offspring (p. 149).
- On average, 2 offspring of the same set of parents share 50% of their alleles, although identical twins share 100% of their alleles (pp. 149–151).

6.2 Mendelian Genetics: When the Role of Genes Is Clear

- The phenotype of a given individual for a particular gene depends on which alleles it carries (its genotype) and whether the alleles are dominant or recessive (pp. 152–153).
- Diseases such as cystic fibrosis and Huntington's disease all have a genetic basis and illustrate the effects of recessive and dominant alleles, respectively (p. 154).
- A Punnett square helps us determine the probability that 2 parents of known genotype will produce a child with a particular genotype (pp. 155–156).

web animation 6.1 *Mendel's Experiments*

6.3 Quantitative Genetics: When Genes and Environment Interact

- Many traits—such as height, IQ, and musical ability—show quantitative variation, which results in a range of values for the trait within a given population (p. 157).
- The role of genes in determining the phenotype for a quantitative trait is estimated by calculating the heritability of the trait (p. 157).
- Quantitative variation in a trait may be generated because the trait is influenced by several genes, because the trait can be influenced by environmental factors, or due to a combination of both factors (pp. 158–159).
- Heritability is calculated by examining the correlation between parents and offspring or by comparing pairs of monozygotic twins to pairs of dizygotic twins (pp. 159–162).

6.4 Genes, Environment, and the Individual

- Calculated heritability values are unique to a particular population in a particular environment. The environment may cause large differences among individuals, even if a trait has high heritability (pp. 163–164).
- Knowing the heritability of a trait does not tell us why 2 individuals differ for that trait (p. 165).
- Our current understanding of the relationship between genes and complex traits does not allow us to predict the phenotype of a particular offspring from the phenotype of its parents (p. 165).

Roots to Remember

The following roots of words come mainly from Latin and Greek and will help you decipher terms:

di- means two.
hetero- means the other, another, or different.
homo- means the same.
mono- means one.
pheno- comes from a verb meaning "to show."
poly- means many.
-zygous derives from zygote, the "yoked" cell resulting from the union of an egg and sperm.

Learning the Basics

1. What is the relationship between genotype and phenotype?

2. What factors cause quantitative variation in a trait within a population?

3. Which of the following statements correctly describes the relationship between genes and chromosomes?
A. Genes are chromosomes; **B.** Chromosomes contain many genes; **C.** Genes are made up of hundreds or thousands of chromosomes; **D.** Genes are assorted independently during meiosis, but chromosomes are not; **E.** More than 1 of the above is correct.

4. An allele is a ______________.
A. version of a gene; **B.** dysfunctional gene; **C.** protein; **D.** spare copy of a gene; **E.** phenotype

5. Sperm or eggs in humans always ______________.
A. each have 2 copies of every gene; **B.** each have 1 copy of every gene; **C.** each contain either all recessive alleles or all dominant alleles; **D.** are genetically identical to all other sperm or eggs produced by that person; **E.** each contain all of the genetic information from their producer

6. A mistake or "misspelling" that occurs during the copying of a gene and results in a change in a gene is called a(n) ______________.
A. dominant allele; **B.** mutation; **C.** mistakes never occur in gene copying; **D.** dysfunction; **E.** improvement

7. What is the physical basis for the independent assortment of alleles into offspring?
A. There are chromosome divisions during gamete production; **B.** Homologous chromosomes are separated during gamete production; **C.** Sperm and eggs are produced by different sexes; **D.** Each gene codes for more than 1 protein; **E.** The instruction manual for producing a human is incomplete.

8. Which genotype can be present in an individual without causing the disease phenotype in that individual?
A. heterozygous for a dominant disease; **B.** homozygous for a dominant disease; **C.** heterozygous for recessive disease; **D.** homozygous for a recessive disease; **E.** all of the above

9. A quantitative trait ______________.
A. may be one that is strongly influenced by the environment; **B.** varies continuously in a population; **C.** may be infl-nced by many genes; **D.** has more than a few values in a population; **E.** all of the above are correct

10. When a trait is highly heritable, ______________.
A. it is influenced by genes; **B.** it is not influenced by the environment; **C.** the variance of the trait in a population can be explained primarily by variance in genotypes; **D.** A and C are correct; **E.** A, B, and C are correct

Genetics Problems

1. A single gene in pea plants has a strong influence on plant height. The gene has 2 alleles: tall (*T*), which is dominant, and short (*t*), which is recessive. What are the genotypes and phenotypes of the offspring of a cross between a *TT* and a *tt* plant?

2. What are the genotypes and phenotypes of the offspring of $TT \times Tt$?

3. The "*D*" gene controls height in pea plants. A plant with either the *DD* or *Dd* genotype is tall, while a plant with the *dd* genotype is short. What is the relationship between *D* and *d*?

4. Albinism occurs when individuals carry 2 recessive alleles (*aa*) that interfere with the production of melanin, the pigment that colors hair, skin, and eyes. If an albino child is born to 2 individuals with normal pigment, what is the genotype of each parent?

5. Pfeiffer syndrome is a dominant genetic disease that occurs when certain bones in the skull fuse too early in the development of a child, leading to distorted head and face shape. If a man with 1 copy of the allele that causes Pfeiffer syndrome marries a woman who is homozygous for the nonmutant allele, what is the chance that their first child will have this syndrome?

6. A cross between a pea plant that produces yellow peas and a pea plant that produces green peas results in 100% yellow pea offspring.
A. Which allele is dominant in this situation? **B.** What are the genotypes of the yellow pea and green pea plants in the initial cross?

7. A cross between a pea plant that produces yellow peas and a pea plant that produces green peas results in 50% yellow pea offspring and 50% green pea offspring. What are the genotypes of the plants in the initial cross?

8. A woman who is a carrier for the cystic fibrosis allele marries a man who is also a carrier.
A. What percentage of the woman's eggs will carry the cystic fibrosis allele? **B.** What percentage of the man's sperm will carry the cystic fibrosis allele? **C.** The probability that they will have a child who carries 2 copies of the cystic fibrosis allele is equal to the percentage of eggs that carry the allele times the percentage of sperm that carry the allele. What is this probability? **D.** Is this the same result you would generate when doing a Punnett square of this cross?

9. The allele *BRCA2* was identified in families with unusually high rates of breast and ovarian cancer. Up to 80% of women with 1 copy of the *BRCA2* allele develop 1 of these cancers in their lifetime.
A. Is *BRCA2* a dominant or a recessive allele? **B.** How is *BRCA2* different from the typical pattern of Mendelian inheritance?

Analyzing and Applying the Basics

1. Two parents both have brown eyes, but they have 2 children with brown eyes and 2 with blue eyes. How is it possible that 2 people with the same eye color can have children with different eye color? If eye color in this family is determined by differences in genotype for a single gene with 2 alleles, what percent of the children are expected to have blue eyes? If the ratio of brown to blue eyes in this family does not conform to expectations, why does this result not refute Mendelian genetics?

2. Does a high value of heritability for a trait indicate that the average value of the trait in a population will not change if the environment changes? Explain your answer.

3. The heritability of IQ has been estimated at about 72%. If John's IQ is 120 and Jerry's IQ is 90, does John have stronger "intelligence" genes than Jerry does? Explain your answer.

Connecting the Science

1. If scientists find a gene that is associated with a particular "undesirable" personality trait (for instance, a tendency toward aggressive outbursts), will it mean greater or lesser tolerance toward people with that trait? Will it lead to proposals that those affected by the "disorder" should undergo treatment to be "cured," and that measures should be taken to prevent the birth of other individuals who are also afflicted?

2. The higher price for Doctorate sperm at the Fairfax Cryobank seems to imply that these donors are rare and highly desirable. If you were a woman who was looking for a sperm donor, would you focus your selection process on the Doctorate donors? What might you miss by focusing only on these donors?

3. Down syndrome is caused by a mistake during meiosis and results in physical characteristics such as a short stature and distinct facial features as well as cognitive impairment (also known as mental retardation). Does the fact that Down syndrome is a genetic condition that results in low IQ mean that we should put fewer resources into education for people with Down syndrome? How does your answer to this question relate to questions about how we should treat individuals with other genetic conditions?

CHAPTER 7

DNA Detective

Complex Patterns of Inheritance and DNA Fingerprinting

The Romanov family ruled Russia until their overthrow, exile, and 1918 execution.

The fall of communist Soviet Union prompted the desire for a proper burial of the Romanov family.

Photo by Dr. Sergey Nikitin

People believed that bones found in a grave in Ekaterinburg were those of the slain Romanovs.

Forensic evidence, like that used on popular crime shows, helped confirm that the bones buried in this shallow grave belonged to the Romanovs.

On the night of July 16, 1918, the tsar of Russia, Nicholas II, his wife Alexandra Romanov, their five children, and four family servants were executed in a small room in the basement of the house to which they had been exiled. These murders ended three centuries of rule by the Romanov family over the Russian Empire.

In February 1917, in the wake of protests throughout Russia, Nicholas II had relinquished his power by abdicating for both himself and on behalf of his only son Alexis, then 13 years old. The tsar hoped that these abdications would protect his son, the heir to the throne, as well as the rest of the family from harm.

The political climate in Russia at that time was explosive. During the summer of 1914, Russia and other European countries became embroiled in World War I. This war proved to be a disaster for the imperial government. Russia faced severe food shortages, and the poverty of the common people contrasted starkly with the luxurious lives of their leaders. The Russian people felt deep resentment toward the tsar's family. This sentiment sparked the first Russian Revolution in February 1917. Following Nicholas's abdication, the imperial family was kept under guard at one of their palaces outside St. Petersburg.

In November 1917, the Bolshevik Revolution brought the communist regime, led by Vladimir Lenin, to power. Ridding the country of the last vestige of Romanov rule became a priority for Lenin and his political party. Lenin believed that doing so would solidify his regime as well as garner support among people who felt that the exiled Romanovs and their opulent lifestyle had come to represent all that was wrong with Imperial Russia. Fearing any attempt by pro-Romanov forces to save the family, Lenin ordered them to the town of Ekaterinburg in Siberia.

Shortly after midnight on July 16, the family was awakened and asked to dress. Nicholas, Alexandra, and their children—Olga, Tatiana, Maria, Anastasia, and Alexis—along with the family physician, cook, maid, and valet, were escorted to a room in the basement of the house in which they had been kept. Believing they were to be moved, the family waited. A soldier entered the room and read a short statement indicating they were to be killed. Armed men stormed into the room, and after a hail of bullets, the royal family and their entourage lay dead.

After the murders, the men loaded the bodies of the Romanovs and their servants into a truck and drove to a remote, wooded area in Ekaterinburg. Historical accounts differ regarding whether the bodies were dumped down a mine shaft, later to be removed, or were immediately buried. There is also some disagreement regarding the burial of two of the people who were executed. Some reports indicate that all eleven people were buried together, and two of them either were badly decomposed by acid placed on the ground of the burial site or were burned to ash. Other reports indicate that two members of the family were buried separately. Some people even believe that two victims escaped the execution. In any case, the bodies of at least nine people were buried in a shallow grave, where they lay undisturbed until 1991.

The bodies were not all that remained buried. For decades, details of the family's murder were hidden in the Communist Party archives in Moscow. However, after the dissolution of the Soviet Union, postcommunist leaders allowed the bones to be exhumed so that they could be given a proper burial. This exhumation took on intense political meaning because the people of Russia hoped to do more than just give the family a proper burial. The event took on the symbolic significance of laying to rest the brutality of the communist regime that took power after the murders of the Romanov family.

Because all that remained of these bodies when they were exhumed was a pile of bones, it was difficult to know if these were the remains of the royal family. A great deal of circumstantial evidence provided by forensic science pointed to that conclusion.

7.1 Forensic Science

Forensic science is a branch of science that helps answer questions of interest to the legal system. Many of the techniques typically used by forensic scientists were not possible to use in the case of the Romanovs due to the decay of the bodies and the crime scene. For instance, no fingerprint, toxicology, footprint, or ballistic (firearm) evidence remained at the scene. Evidence that did remain at the scene included dental and bone structure. Dental evidence showed that five of the bodies had gold, porcelain, and platinum dental work, which had been available only to aristocrats.

The bones seemed to indicate that they belonged to six adults and three children. Investigators electronically superimposed the photographs of the skulls on archived photographs of the family. They compared the skeletons' measurements with clothing known to have belonged to the family. These and other data were consistent with the hypothesis that the bodies could be those of the tsar, the tsarina, three of their five children, and the four servants.

Karyotype analysis (Chapter 5) requires dividing cells, which also did not exist at the crime scene. Therefore scientists tried to determine the sex of the buried individuals based on pelvic bone structure. Females have evolved to have wider pelvic openings to accommodate the passage of a child through the birth canal. Russian scientists thought that all three of the children's skeletons and two of the adult's skeletons were probably female (and four of the adult skeletons were male). However, the pelvises had decayed, so it was impossible to be certain.

By using the forensic evidence available to them, Russian scientists had shown only that these skeletons *might* be the Romanovs. They had not yet shown with any degree of certainty that these bodies *did* belong to the slain royals. The new Russian leaders did not want to make a mistake when symbolically burying a former regime. Unassailable proof was necessary because so much was at stake politically. As you will learn below, a more modern forensic technique, DNA fingerprinting, was finally used to help solve the mystery of the Romanovs.

Before this technique became available to scientists, they had to rely on more conventional techniques, such as showing relatedness between parents and children, for certain obvious genetic traits like hair texture and eye color.

For example, by looking at photographs we can see that the tsar had straight hair and dark eyes, and the tsarina had wavy hair and dark eyes. How could they produce offspring with such a wide variety of hair texture and eye color?

7.2 Dihybrid Crosses

Dihybrid crosses are genetic crosses involving two traits. The reason that so many different kinds of offspring could be produced, relative to two genes, is due to random alignment of homologues, leading to independent assortment of genes, which you learned about in Chapters 5 and 6.

Let's use Mendel's peas as an example. Seed color and seed shape are each determined by a single gene, and each are carried on different chromosomes. The two seed-color gene alleles are *Y*, which is dominant and codes for yellow color, and *y*, the recessive allele, which results in green seeds when homozygous. The two seed-shape alleles are *R*, the dominant allele, which codes for a smooth, round shape, and *r*, which is recessive and codes for a wrinkled shape.

Because the genes for seed color and seed shape are on different chromosomes, they are placed in eggs and sperm independently of each other. In other words, a pea plant that is heterozygous for both genes (genotype *Yy Rr*) can make 4 different types of eggs: one carrying dominant alleles for both genes (*Y R*), one carrying recessive alleles for both genes (*y r*), one carrying the dominant allele for seed color and the recessive allele for seed shape (*Y r*), and one carrying the recessive allele for color and the dominant allele for shape (*y R*).

A Punnett square for a cross between two individuals who are heterozygous for both seed-color and seed-shape genes would contain 16 boxes with

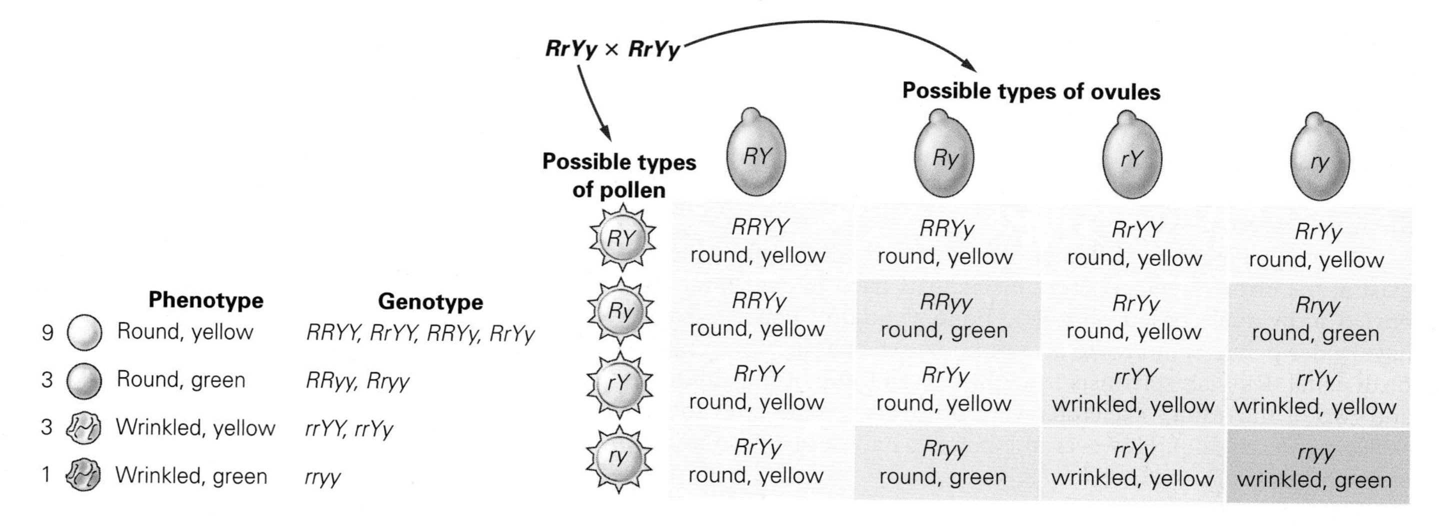

Figure 7.1 **A dihybrid cross.** Punnett squares can be used to predict the outcome of a cross involving two different genes. This cross involves two pea plants that are both heterozygous for the seed-color and seed-shape genes.

four different phenotypes *(Figure 7.1)*. The phenotypes produced result in a 9:3:3:1 **phenotypic ratio** where 9/16 include both dominant alleles (Y_R_); 3/16 include dominant alleles of one gene only (Y_rr); 3/16 include dominant alleles of the other gene (yyR_); and 1/16 carry recessive alleles only.

Recall from these earlier chapters that members of a homologous pair of chromosomes align at the cell's equator during meiosis, and that their alignment is random with respect to which member of a homologous pair faces which pole. *Figure 7.2* is a review of random alignment illustrating the process for two chromosomes that carry some of the genes for hair texture and eye color. Curly hair (*CC*) is dominant over wavy (*Cc*) or straight hair (*cc*), and darkly pigmented eyes (*DD* or *Dd*) are dominant over blue eyes (*dd*). Eye color is actually determined by several different genes, but for simplicity, we will follow the inheritance of only one of these genes.

Because there are color photos of the Romanovs that allow us to determine their eye color and hair texture, we will use these traits to illustrate how random alignment leads to independent assortment of these genes. The tsar had straight hair and dark eyes (*ccDd*), while the tsarina had wavy hair and dark eyes (*CcDd*). We know that the tsar and tsarina were heterozygous for the eye color gene since they had children with blue eyes. Likewise, each member of the royal couple must also have had at least one recessive hair-texture allele since they had children with straight hair. Because these genes are located on different chromosomes, together the royal couple could produce children with wavy hair and brown eyes (Tatiana and Maria), wavy hair and blue eyes (Anastasia and Olga), straight hair and blue eyes (Alexis), or straight hair and brown eyes.

Stop and Stretch 7.1: Punnett squares can be produced for any number of genes, each of which has two alleles. Imagine a cross between two heterozygous tall, yellow, and round seeded pea plants (TtYyRr). How many different types of gametes can each plant make? How many different cells would the resulting Punnett square contain?

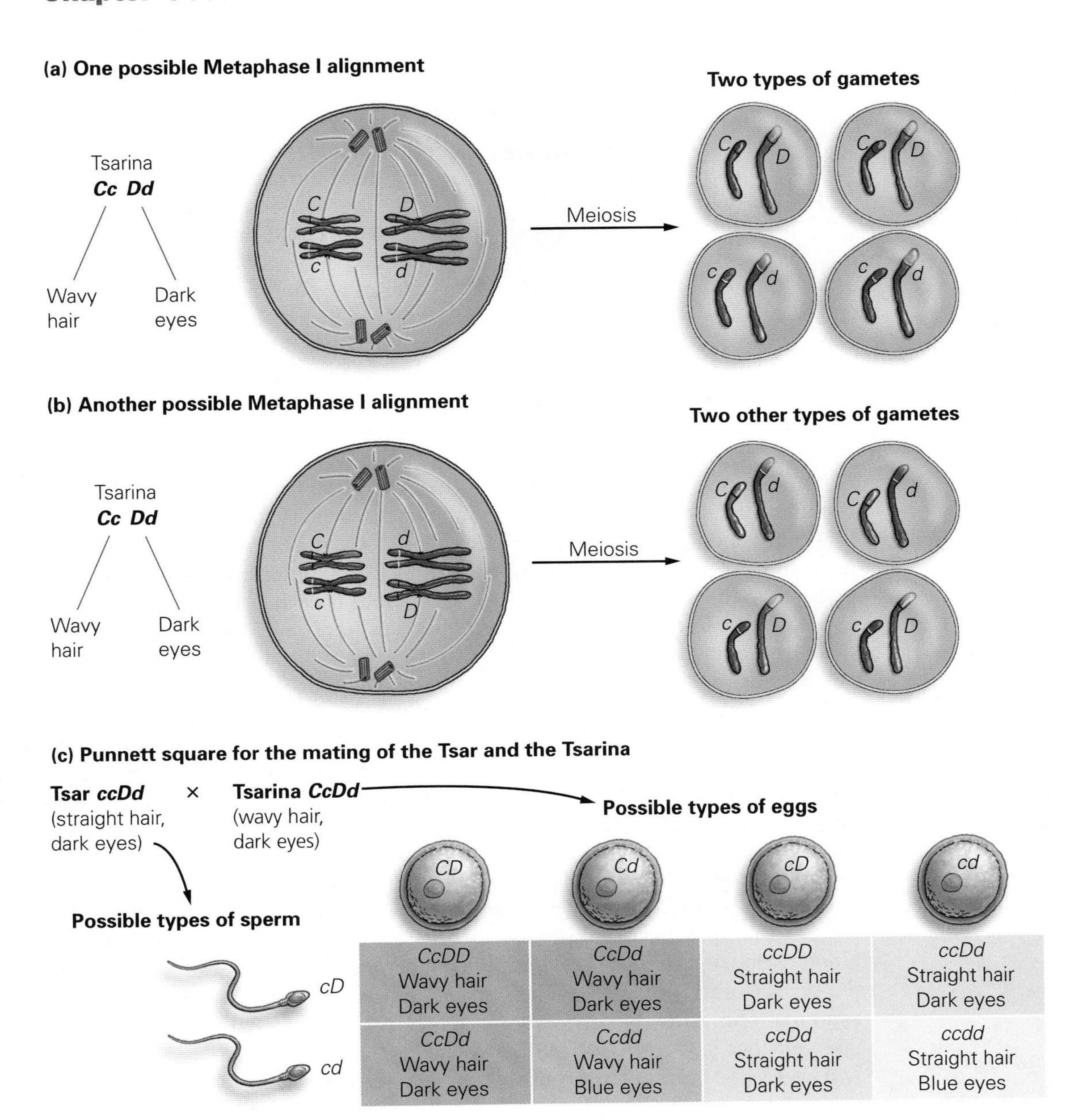

Figure 7.2 **Random alignment.** Two possible alignments, (a) and (b), can occur when there are two homologous pairs of chromosomes. (c) Due to random alignment, the tsarina could produce four different types of gametes relative to these two genes. Since the tsar had straight hair and dark eyes (genotype ccDd), he could produce cD and cd gametes. Possible offspring from the mating of the tsar and tsarina are shown inside the Punnett square.

7.3 Extensions of Mendelian Genetics

Solving the puzzle of the Ekaterinburg bones would ultimately require that scientists be able to show relatedness between the tsar and tsarina's skeletons and their children's skeletons. Patterns of inheritance are fairly simple to predict when genes are inherited in a straightforward manner—as you learned above, predictions about traits that are controlled by one or two genes with dominant and recessive alleles can be made by using Punnett squares. Patterns of inheritance that are a little more complex are called *extensions* of Mendelian genetics.

Flower color in snapdragons

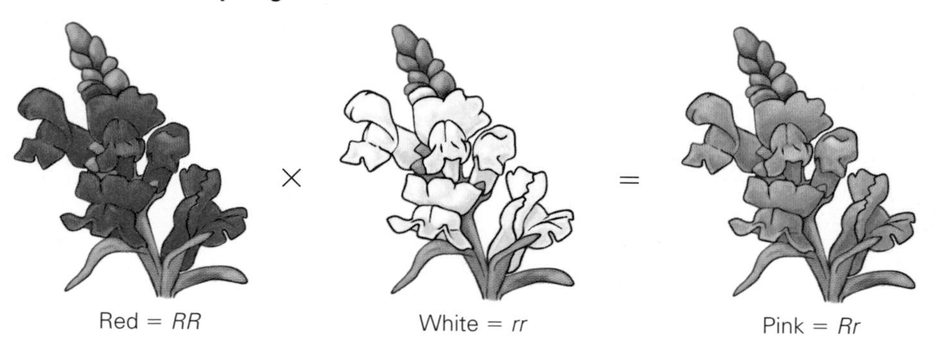

Figure 7.3 **Incomplete dominance.** Snapdragons show incomplete dominance in the inheritance of flower color. The heterozygous flower has a phenotype that is in between that of the two homozygotes.

For some genes, two identical copies of a dominant allele are required for expression of the full effect of a phenotype. In this case, the phenotype of the heterozygote is intermediate between both homozygotes—a situation called **incomplete dominance.** For example, the alleles that determine flower color in snapdragons are incompletely dominant: One homozygote produces red flowers, the other, presumably carrying two nonfunctional copies of a color gene, produces white flowers, and the heterozygote, carrying one "red" and one "white" allele, produces pink flowers *(Figure 7.3)*.

Alternatively, the phenotype of a heterozygote in which neither allele is dominant to the other may be a combination of both fully expressed traits. This situation, by which two different alleles of a gene are both expressed in an individual, is known as **codominance.** In cattle, for example, the allele that codes for red hair color and the allele that codes for white hair color are both expressed in a heterozygote. These individuals have patchy coats that consist of an approximately equal mixture of white hairs and red hairs *(Figure 7.4)*.

Because all that was left of the Romanovs was a pile of bones, the scientists could study only a few genetic traits to show the relatedness of the adult skeletons to two of the four children's skeletons. Genetic traits that were obvious, such as bone size and structure, are controlled by many genes and affected by environmental components like nutrition and physical activity level. Traits that are affected by the interactions between genes and environment are called **polygenic traits.** This made using bone size and structure to predict which of the adult's skeletons were related to the children's skeletons a matter of guesswork. The scientists had to use more sophisticated analyses.

A technique that scientists use to help determine relatedness of people is blood typing, which involves determining if certain carbohydrates are located on the surface of red blood cells. These surface markers are part of the **ABO blood system.** The ABO blood system displays two extensions of Mendelism—codominance and **multiple allelism,** which occurs when there are more than two alleles of a gene in the population. In fact, three distinct alleles of one blood-group gene code for the enzymes that synthesize the sugars found on the surface of red blood cells. Two of the three alleles display codominance to each other, and one allele is recessive to the other two.

Figure 7.5 summarizes the possible genotypes and phenotypes for the ABO blood system. The three alleles of this blood-type gene are I^A, I^B, and i. A

Figure 7.4 **Codominance.** Roan coat color in cattle is an example of codominance. Both alleles are equally expressed in the heterozygote, so the conventional uppercase and lowercase nomenclature for alleles no longer applies.

Coat color in cattle

Red = R^1R^1

×

White = R^2R^2

=

Roan = R^1R^2
(patchy red and white coat)

given individual will carry only two alleles, one on each of his or her homologous pairs of chromosomes, even though three alleles are being passed on in the entire population. In other words, one person may carry the I^A and I^B alleles, and another might carry the I^A and i alleles. There are three different alleles, but each individual can carry only two alleles.

The symbols used to represent these alternate forms of the blood-type gene tell us something about their effects. The lowercase i allele is recessive to both the I^A and I^B alleles. Therefore, a person with the genotype I^Ai has type A blood, and a person with the genotype I^Bi has type B blood. A person with both recessive alleles, genotype ii, has type O blood. The uppercase I^A and I^B alleles display codominance in that neither masks the expression of the other. Both of these alleles are expressed. Thus, a person with the genotype I^AI^B has type AB blood.

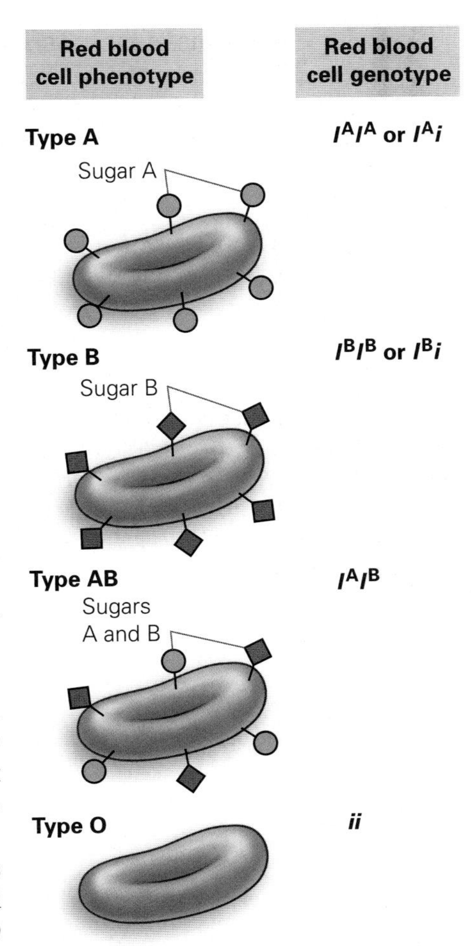

Figure 7.5 **ABO blood system.** Red blood cell phenotypes and corresponding genotypes.

Stop and Stretch 7.2: Imagine a new allele I^C that is codominant with I^A and I^B and dominant to i. List all of the possible blood group genotypes and phenotypes in a population containing all four alleles.

Another molecule on the surface of red blood cells is called the **Rh factor.** Someone who is positive (+) for this trait has the Rh factor on their red blood cells, while someone who is negative (−) does not. This trait, unlike the ABO blood system, is inherited in a straightforward two-allele, completely dominant manner with Rh^+ dominant to Rh^-. Persons who are Rh^+ can have the genotype Rh^+Rh^+ or Rh^+Rh^-. An Rh^- individual has the genotype Rh^-Rh^-.

Blood typing is often used to help establish whether a given set of parents could have produced a particular child. For example, a child with type AB blood and parents who are type A and type B could be related, but a child with type O blood could not have a parent with type AB blood. Likewise, if a child has blood type B and the known mother has blood type AB, then the father of that child could have type AB, A, B, or O blood, which does not help to establish parentage.

If a child has a blood type consistent with alleles that he or she may have inherited from a man who might be his or her father, this finding does not mean that the man is the father. Instead, it is only an indication that the man could be. In fact, many other men would also have that blood type *(Table 7.1)*. Therefore, blood-type analysis can be used only to eliminate people from consideration. Blood typing cannot be used to positively identify someone as the parent of a particular child.

Clinicians must take ABO blood groups into account when performing blood transfusions. Persons receiving transfusions from incompatible blood groups will mount an immune response against those sugars that they do not carry on their own red blood cells. The presence of these foreign red blood cell sugars causes a severe reaction in which the donated, incompatible red

TABLE 7.1 **Frequency of blood types in U.S. population.** Percentages of the population with a given blood type are listed from least to most common. The negative and positive superscripts refer to the absence or presence of the Rh factor.

Blood Type	AB^-	B^-	AB^+	A^-	O^-	B^+	A^+	O^+
Frequency in U.S. Population (%)	0.5	1.5	3	5	7	11	32	40

TABLE 7.2 **Blood transfusion compatibilities.**

Recipient	Recipient Can Receive	Recipient Cannot Receive
Type O	Type O	Type A Type B Type AB
Type A	Type O Type A	Type B Type AB
Type B	Type O Type B	Type A Type AB
Type AB	Type O Type A Type B Type AB	None

blood cells form clumps. This clumping can block blood vessels and kill the recipient. *Table 7.2* shows the types of blood transfusions individuals of various blood types can receive.

Blood-typing analysis can sometimes provide scientists with some information about potential relatedness of victims. However, it was not an option in the case of the Romanovs. The very old remains contained no blood, so blood typing was not possible.

The Romanov family was known to carry a trait that demonstrates another extension of Mendelism: pleiotropy. **Pleiotropy** is the ability of a single gene to cause multiple effects on an individual's phenotype. Alexis was the last of several Romanovs to have **hemophilia,** a pleiotropic blood-clotting disorder. A person with the most common form of hemophilia cannot produce a protein called clotting factor VIII. When this protein is absent, blood does not form clots to stop bleeding from a cut or internal blood vessel damage. Affected individuals bleed excessively, even from small cuts.

Due to the direct effects of excessive bleeding, hemophilia can lead to excessive bruising, pain and swelling in the joints, vision loss from bleeding into the eye, and anemia, resulting in fatigue. In addition, neurological problems may occur if bleeding or blood loss occurs in the brain.

Historical records indicate that Alexis, heir to the throne, was so ill with hemophilia that his father actually had to carry him to the basement room where he was executed.

web animation 7.1
X-Linked Recessive Traits

7.4 Sex Determination and Sex Linkage

It appears that Alexis inherited the hemophilia allele from his mother. We can deduce this pattern of inheritance because we now know that the clotting factor gene is inherited in a sex-specific manner. The clotting factor VIII gene (the gene that, when mutated, causes hemophilia) is located in the X chromosome. Of the 23 pairs of chromosomes present in the cells of human males, 22 pairs are **autosomes,** or nonsex chromosomes, and one pair, X and Y, are the **sex chromosomes.** Males have 22 pairs of autosomes and one X and one Y sex chromosome. Females also have 22 pairs of autosomes, but their sex chromosomes are comprised of two X chromosomes.

Chromosomes and Sex Determination

The X and Y chromosomes are involved in producing the sex of an individual through a process called **sex determination.** When men produce sperm and the chromosome number is divided in half through meiosis, their sperm cells contain one member of each autosome and either an X or a Y chromosome. Females produce gametes with 22 unpaired autosomes and one of their two X chromosomes. Therefore, human egg cells normally contain one copy of an X chromosome, but sperm cells can contain either an X or a Y chromosome.

The sperm cell determines the sex of the offspring resulting from a particular fertilization. If an X-bearing sperm unites with an egg cell, the resulting offspring will be female (XX). If a sperm bearing a Y chromosome unites with an egg cell, the resulting offspring will be male (XY). *Figure 7.6.* summarizes the process of sex determination in humans, and *Table 7.3* outlines some mechanisms of sex determination in nonhumans.

Sex Linkage

Genes located on the X or Y chromosome are called **sex-linked genes** because biological sex is inherited along with, or "linked to," the X or Y chromosome. Sex-linked genes found on the X chromosomes are said to be X linked, while those on the Y chromosome are Y linked. The X chromosome is much larger than the Y chromosome, which carries very little genetic information *(Figure 7.7)*.

X-Linked Genes.

X-linked genes are located on the X chromosome. The fact that males have only one X chromosome leads to some peculiarities in inheritance of sex-linked genes. Males always inherit their X chromosome from their mother because they must inherit the Y chromosome from their father to be male. Thus, males will inherit X-linked genes only from their mothers. Males are more likely to suffer from diseases caused by recessive alleles on the X chromosome because they have only one copy of any X-linked gene. Females are less likely to suffer from these diseases, because they carry two copies of the X chromosome and thus have a greater likelihood of carrying at least one functional version of each X-linked gene.

A **carrier** of a recessively inherited trait has one copy of the recessive allele and one copy of the normal allele and will not exhibit symptoms of the disease. Only females can be carriers of X-linked recessive traits because males with a copy of the recessive allele will have the trait. Both males and females can be carriers of non-sex-linked, autosomal traits.

Even though female carriers of an X-linked recessive trait will not display the recessive trait, they can pass the trait on to their offspring. For this reason, most women carrying the hemophilia allele will not even realize that they are a carrier until their son becomes ill. *Figure 7.8a* illustrates that a cross between a male who does not have hemophilia and a female carrier can produce unaffected females, carrier females, unaffected males, and affected males. *Figure 7.8b* illustrates that no male children produced by a cross between an affected male and an unaffected female would have hemophilia. All daughters produced by this cross would be carriers of the trait. In the United States, there are over 20,000 hemophiliacs, nearly all of whom are male.

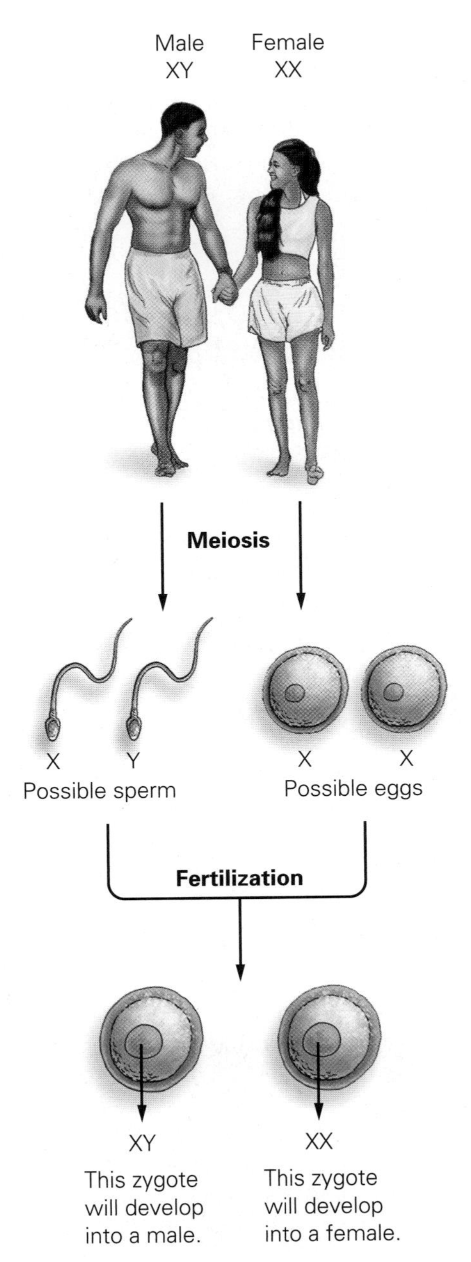

Figure 7.6 **Sex determination in humans.** In humans, sex is determined by the male since males produce sperm that carry either an X or a Y chromosome (in addition to 22 autosomes). The egg cell always carries an X chromosome along with the 22 autosomes. When an X-bearing sperm fertilizes the egg cell, a female (XX) results. When a Y-bearing sperm fertilizes an egg cell, an XY male results. **Visualize This:** What is the probability that a couple will have a boy? What is the probability that a couple with four boys will have a girl for their fifth child?

TABLE 7.3 Sex determination in nonhuman organisms.

Type of Organism	Mechanism of Sex Determination
Vertebrates (fish, amphibians, reptiles, birds, and mammals) 	The male may have two of the same chromosomes and the female two different chromosomes. In these cases, the female determines the sex of the offspring.
Egg-laying reptiles 	In many egg-laying species, two organisms with the same suite of sex chromosomes could become different sexes. Sex depends on which genes are activated during embryonic development. For example, the sex of some reptiles is determined by the incubation temperature of the egg.
Wasps, ants, bees 	Sex is determined by the presence or absence of fertilization. In bees, males (drones) develop from unfertilized eggs. Females (workers or queens) develop from fertilized eggs.
Bony fishes 	Some species of bony fishes change their sex after maturation. All individuals will become females unless they are deflected from that pathway by social signals such as dominance interactions.
Caenorhabditis elegans 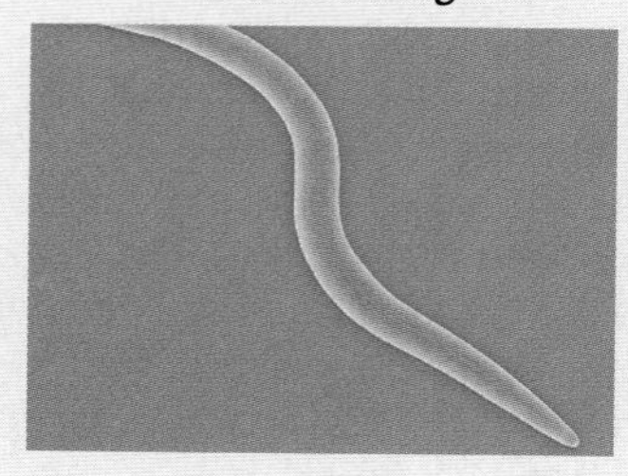	The nematode *C. elegans* can either be male or have both male and female reproductive organs. Such individuals are called hermaphrodites.

Figure 7.7 **The X and Y chromosomes.** The Y chromosome (left) is smaller than the X chromosome and carries fewer genes.

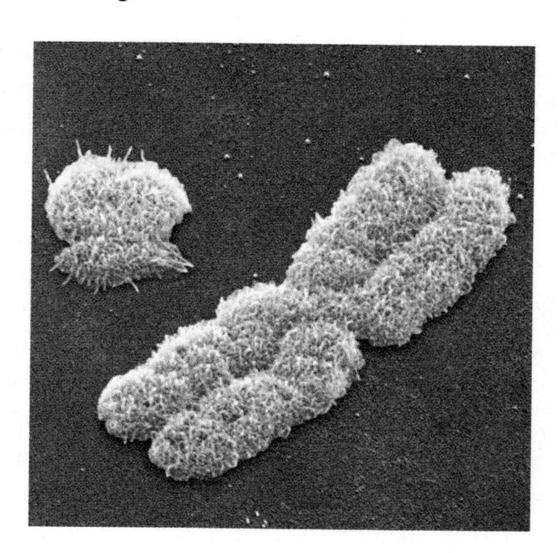

(a) Unaffected male × Carrier female

X^HY × X^HX^h

Possible types of eggs: X^H, X^h

Possible types of sperm: X^H, Y

	X^H	X^h
X^H	X^HX^H Unaffected female	X^HX^h Carrier female
Y	X^HY Unaffected male	X^hY Hemophiliac male

$\frac{1}{4}$ Unaffected females
$\frac{1}{4}$ Carrier females
$\frac{1}{4}$ Hemophiliac males
$\frac{1}{4}$ Unaffected males

(b) Hemophiliac male × Unaffected female

X^hY × X^HX^H

Possible types of eggs: X^H

Possible types of sperm: X^h, Y

	X^H
X^h	X^HX^h Carrier female
Y	X^HY Unaffected male

$\frac{1}{2}$ Carrier females
$\frac{1}{2}$ Unaffected males

Figure 7.8 **Genetic crosses involving the X-linked hemophilia trait.** Cross (a) shows the possible outcomes and associated probabilities of a mating between a non-hemophilic male and a female carrier of hemophilia. Cross (b) shows the possible outcomes and associated probabilities of a cross between a hemophilic male and an unaffected female. **Visualize This:** What genetic cross would result in the highest frequency of affected males?

Stop and Stretch 7.3: What must be true of the parents of a female who has an X-linked recessive disease?

Red-green colorblindness is another X-linked trait. This trait affects approximately 4% of males. Red blindness is the inability to see red as a distinct color. Green blindness is the inability to see green as a distinct color. When normal (in this case the dominant alleles are normal), these genes code for the production of proteins called opsins that help absorb different wavelengths of light. A lack of opsins causes insensitivity to light of red and green wavelengths.

Duchenne muscular dystrophy is a progressive, fatal X-linked disease of muscle wasting that affects approximately 1 in 3500 males. The onset of muscle wasting occurs between 1 and 12 years of age, and by age 12, affected boys are often confined to a wheelchair. The affected gene is one that normally codes for the dystrophin protein. When at least one allele is normal, dystrophin stabilizes cell membranes during muscle contraction. It is thought that the absence of normal dystrophin protein causes muscle cells to break down and muscle tissue to die.

Most of the protein products of over 100 genes on the X chromosome have nothing at all to do with the production of biological sex differences. Accordingly, females and males should require equal doses of the products of X-linked genes. How can we account for the fact that females, with their two X chromosomes, could receive two doses of X-linked genes, while males receive only one? The answer comes from a phenomenon called **X inactivation** that occurs in all of the cells of a developing female embryo. This inactivation guarantees that all females actually receive only one dose of the proteins produced by genes on the X chromosomes. Inactivation of the genes on one of the two X chromosomes takes place in the embryo at about the time that the embryo implants in the uterus. One chromosome is inactivated when a string of RNA is wrapped around it (*Figure 7.9*).

Active X chromosome

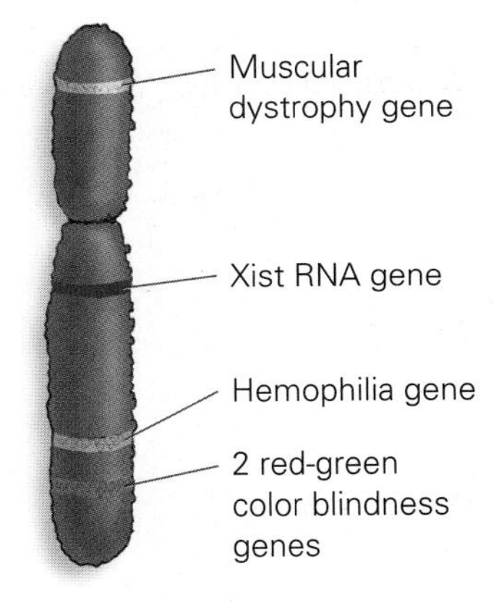

Inactive X chromosome

Figure 7.9 **X inactivation.** One of a female's two X chromosomes is inactivated when a strand of RNA wraps itself around the chromosome.

This inactivation is random with respect to the parental source—either of the two X chromosomes can be inactivated in a given cell. Inactivation is also irreversible and, as such, is inherited during cell replication. In other words, once a particular X chromosome is inactivated in a cell, all descendants of that cell continue inactivating the same chromosome.

Cats with tortoiseshell coat coloring illustrate the effects of X inactivation. The coats of tortoiseshell cats are a mixture of black and orange patches.

The genes for fur color are located on the X chromosomes. If a cat with orange fur mates with a cat with black fur, a female kitten could have one X chromosome with the gene for orange fur, and one X chromosome with the gene for black fur *(Figure 7.10a)*. Early in development, when the embryo consists of about 16 cells, one of the two X chromosomes is randomly inactivated in each cell. Thus, some cells will be expressing the orange fur-color gene, and others will be expressing the black fur-color gene. The pattern of inactivation (the X chromosome that the kitten inherited from its mother or the X chromosome that the kitten inherited from its father) is passed on to the daughter cells of the 16-celled embryo, resulting in the patches of orange and black fur color seen in tortoiseshell cats *(Figure 7.10b)*. Because this pattern

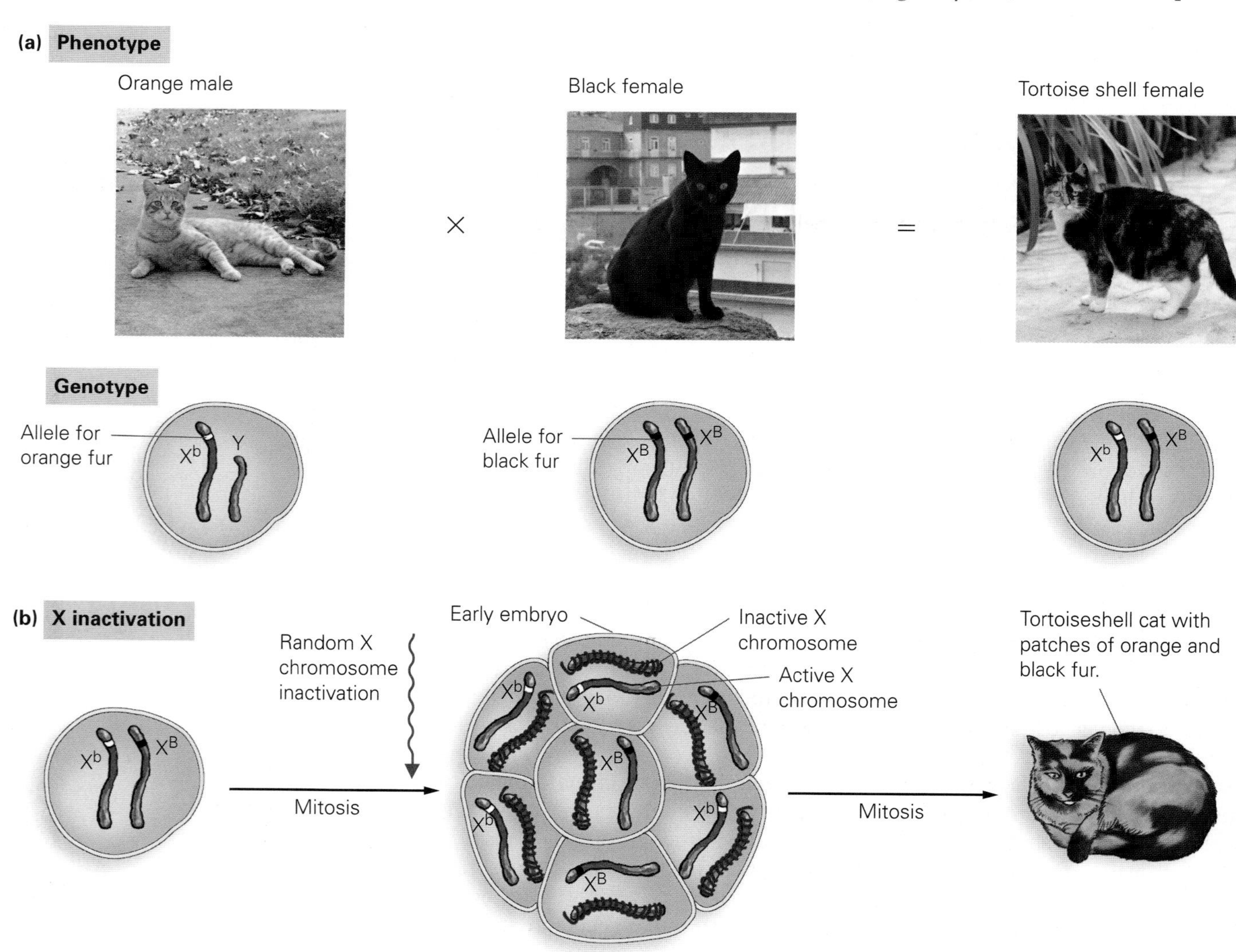

Figure 7.10 **X inactivation and patchy gene expression.** (a) An orange male cat and a black female cat can produce female offspring with tortoiseshell coats. (b) Random inactivation of one X chromosome early in development leads to patches of coat color.

of coat coloration requires the expression of both alleles of the color gene in different patches of cells, a cat must have two different alleles of this X-linked gene. Therefore, these cats are almost always female. On rare occasions, male cats can have this pattern of coloring. This can happen only if a male cat has two X chromosomes and two Y chromosome—a situation that does occur, though infrequently, via nondisjunction (Chapter 5)—and results in sterility.

Y-Linked Genes. **Y-linked genes** are located on the Y chromosome and are passed from fathers to sons. Although this distinctive pattern of inheritance should make Y-linked genes easy to identify, very few genes have been localized to the Y chromosomes. One gene known to be located exclusively on the Y chromosome is called the *SRY* gene (for sex-determining region of the Y chromosome). The expression of this gene triggers a series of events leading to development of the testes and some of the specialized cells required for male sexual characteristics. Genes other than *SRY*, on chromosomes other than the Y, code for proteins that are unique to males but are not expressed unless testes develop.

Stop and Stretch 7.4: Sometimes chromosome sex does not match biological sex. For instance, some males are known to have two X chromosomes and no Y chromosome. Most of these males carry a copy of the SRY gene on one X chromosome. What event during meiosis can explain this phenomenon? Some genes on the Y chromosome code for sperm production. What does this imply about XX males?

In the case of the bones thought to belong to the Romanovs, since karyotype and pelvic bone analyses were difficult to perform on the decayed bones, scientists analyzed DNA from the bones for sequences known to be present only on the Y chromosome. When DNA that was isolated from the children's remains was analyzed, it became clear that the children's bones all belonged to girls. If these bones did belong to the Romanovs, one of the two missing children was Alexis, the Romanovs' only son.

Another line of evidence was provided by the extensive family trees of the Romanovs and their relatives.

7.5 Pedigrees

Because the hemophilia gene is X linked, Alexis Romanov inherited the disease from his mother, who must have been a carrier of the disease. We can trace the lineage of this disease through the Romanov family by using a chart called a pedigree. A **pedigree** is a family tree that follows the inheritance of a genetic trait for many generations of relatives. Pedigrees are often used in studying human genetics since it is impossible and unethical to set up controlled matings between humans the way one can with fruit flies or plants. Pedigrees allow scientists to study inheritance by analyzing matings that have already occurred. *Figure 7.11* identifies some of the symbols used in pedigrees, and *Figure 7.12* shows how scientists can use pedigrees to determine whether a trait is inherited as autosomal dominant or recessive or as sex-linked recessive.

Information is available about the Romanovs' ancestors because they were royalty and because scientists interested in hemophilia had kept very good records of the inheritance of that trait. Hemophilia was common among European royal families but rare among the rest of the population. This was because members of the royal families intermarried to preserve the royal bloodlines. The tsarina must have been a carrier of the hemophilia allele because her son

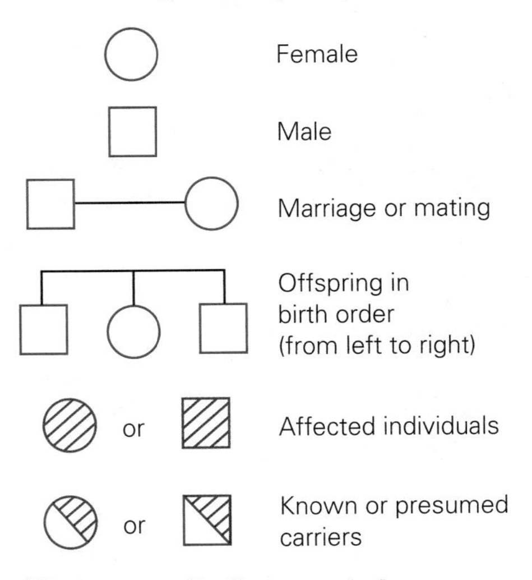

Figure 7.11 **Pedigree analysis.** Symbols used in pedigrees. **Visualize This:** Draw a pedigree showing a female who has a boy with her first husband and a girl with her second husband.

had the trait. Her mother, Alice, must also have been a carrier because the tsarina's brother Fred had the disease, as did two of her sister Irene's sons, Waldemar and Henry. The tsarina's grandmother, Queen Victoria, seems to be the first carrier of this allele in this family because there is no evidence of this disease before her eighth child, Leopold, is affected. Queen Victoria's mother, Princess Victoria, most likely incurred a mutation to the clotting factor VIII gene while the cells of her ovary were undergoing DNA synthesis to produce egg cells. The egg cell that carried the mutant clotting factor VIII gene was passed to her daughter Victoria. The fertilized egg cell that produced Queen Victoria carried this mutation.

When the cell divided by mitosis to produce her body cells, the mutation was passed on to each of her cells. When the cells in her ovary underwent meiosis to produce gametes, the Queen passed the mutant version on to three of her nine children *(Figure 7.13)*. The extensive pedigree available to the scientists working on the Romanov case, in concert with a powerful technique called DNA fingerprinting, would provide the key data in solving the mystery of the buried bones.

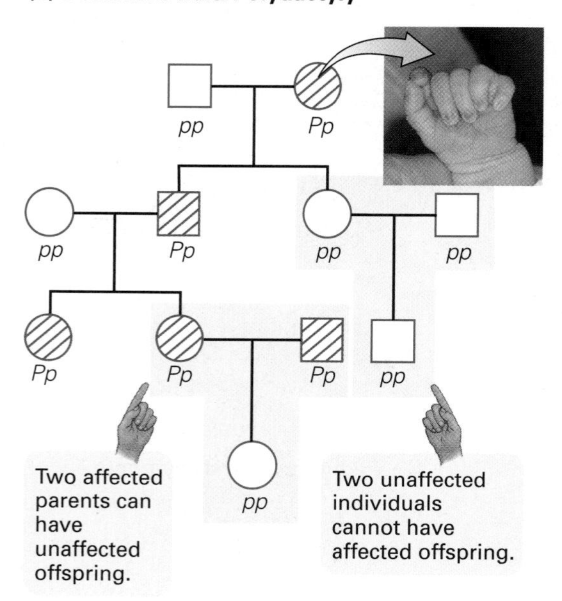

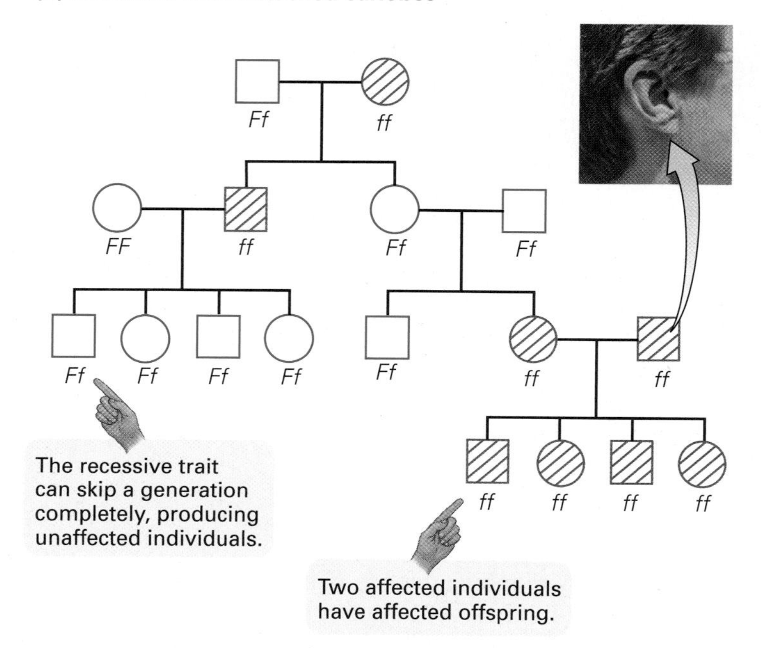

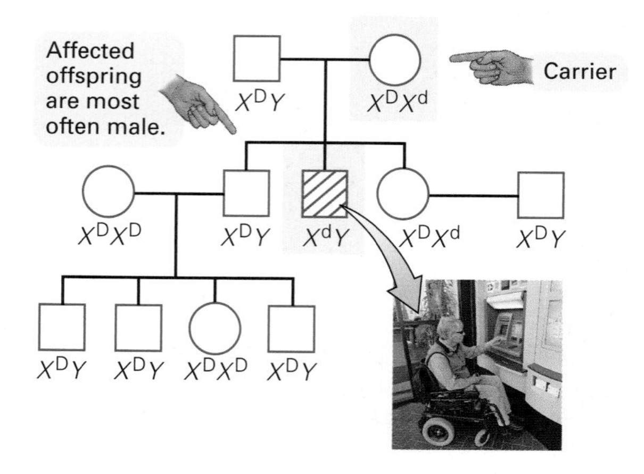

Figure 7.12 **Pedigrees showing different modes of inheritance.** (a) Polydactyly is a dominantly inherited trait. People with this condition have extra fingers or toes. (b) Having attached earlobes is a recessively inherited trait. (c) Muscular dystrophy is inherited as an X-linked recessive trait.

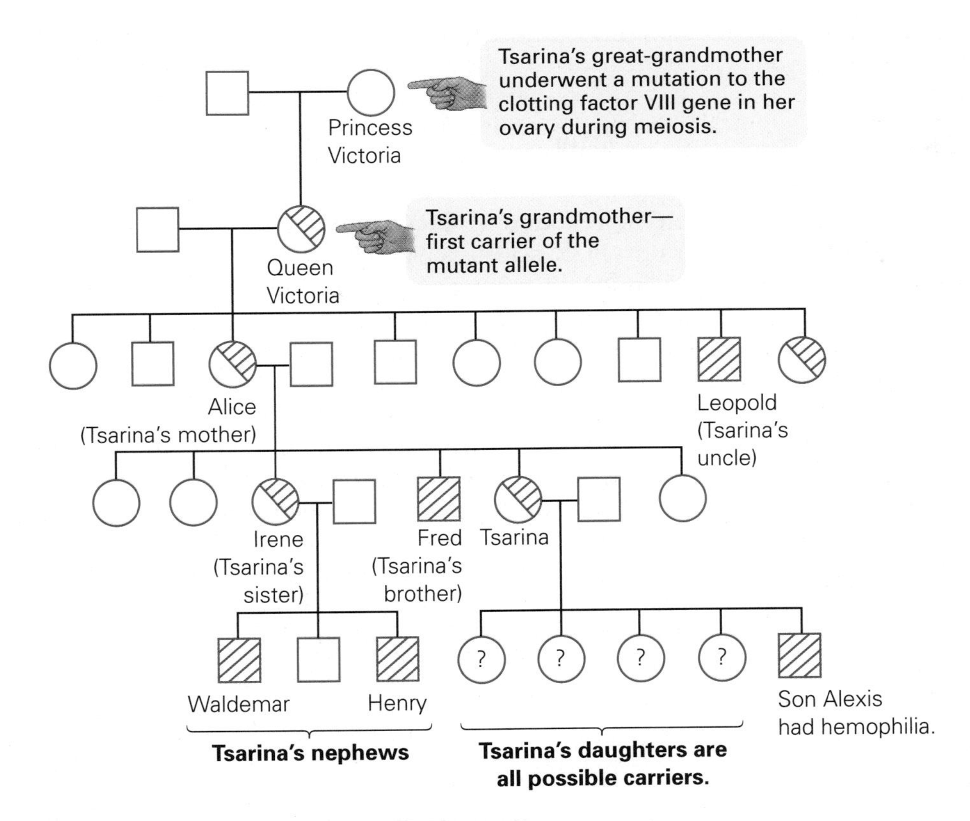

Figure 7.13 **Origin and inheritance of the hemophilia allele.** This abbreviated pedigree shows the origin of the hemophilia allele and its inheritance among the tsarina's family. It appears that the tsarina's great-grandmother underwent a mutation that she passed on to her daughter, who then passed it on to three of her nine children—one of whom was the tsarina's mother, Alice.

7.6 DNA Fingerprinting

web animation 7.2
Polymerase Chain Reaction (PCR)

Limits on the power of conventional forensic techniques such as blood typing and karyotyping to identify the bones found in the Ekaterinburg grave necessitated the use of more sophisticated techniques. To do so, scientists took advantage of the fact that any two individuals who are not identical twins have small differences in the sequences of nucleotides that comprise their DNA. To test the hypothesis that the bones buried in the Ekaterinburg grave belonged to the Romanov family, the scientists had to answer the following questions:

1. Which of the bones from the pile are actually different bones from the same individuals?
2. Which of the adult bones could have been from the Romanovs, and which bones could have belonged to their servants?
3. Are these bones actually from the Romanovs, not some other related set of individuals?

An Overview of DNA Fingerprinting

All of these questions were answered using **DNA fingerprinting.** This technique allows unambiguous identification of people in the same manner that traditional fingerprinting has been used in the past.

To begin this process, it is necessary to isolate the DNA to be fingerprinted. Scientists can isolate DNA from blood, semen, vaginal fluids, a hair root, skin, and even (as was the case in Ekaterinburg) degraded skeletal remains. This DNA is then cut apart with special molecular scissors that cleave each individual's DNA at specific locations. Because each person has a unique sequence of DNA, this cutting of the DNA generates DNA fragments of different sizes. When the cut DNA fragments are separated and stained, a unique pattern is produced for every individual except identical twins.

A Closer Look: DNA Fingerprinting

When very small amounts of DNA are available, as is typically the case, scientists can make many copies of the DNA by first performing a DNA-synthesizing reaction.

Polymerase Chain Reaction (PCR). The **polymerase chain reaction (PCR)** is used to amplify or produce large quantities of DNA *(Figure 7.14)*. To perform PCR, scientists place the double-stranded DNA to be amplified, along with the individual building-block subunits of DNA—the nucleotides adenine (A), cytosine (C), guanine (G), and thymine (T)—in a small test tube. Next an enzyme called *Taq* **polymerase** is added to the tube containing the DNA. This enzyme was given the first part of its name (*Taq*) because it was first isolated from the single-celled organism *Thermus aquaticus*, which lives in hydrothermal vents and can withstand very high temperatures. The second part of the enzyme's name (polymerase) describes its synthesizing activity—it acts as a DNA polymerase. DNA polymerases use one strand of DNA as a template for the synthesis of a daughter strand that carries complementary nucleotides (A:T base pairs are complementary as are G:C base pairs). The main difference between human DNA polymerase and *Taq* polymerase is that the *Taq* polymerase is resistant to extremely high temperatures, temperatures at which human DNA polymerase would be inactivated. The heat-resistant abilities of *Taq* polymerase thus allow PCR reactions to be run at very high temperatures. High temperatures are necessary since the DNA molecule being amplified must first be **denatured,** or split up the middle of the double helix, to produce single strands. After heating, the DNA solution is allowed to cool, and the *Taq* polymerase adds complementary nucleotides to the

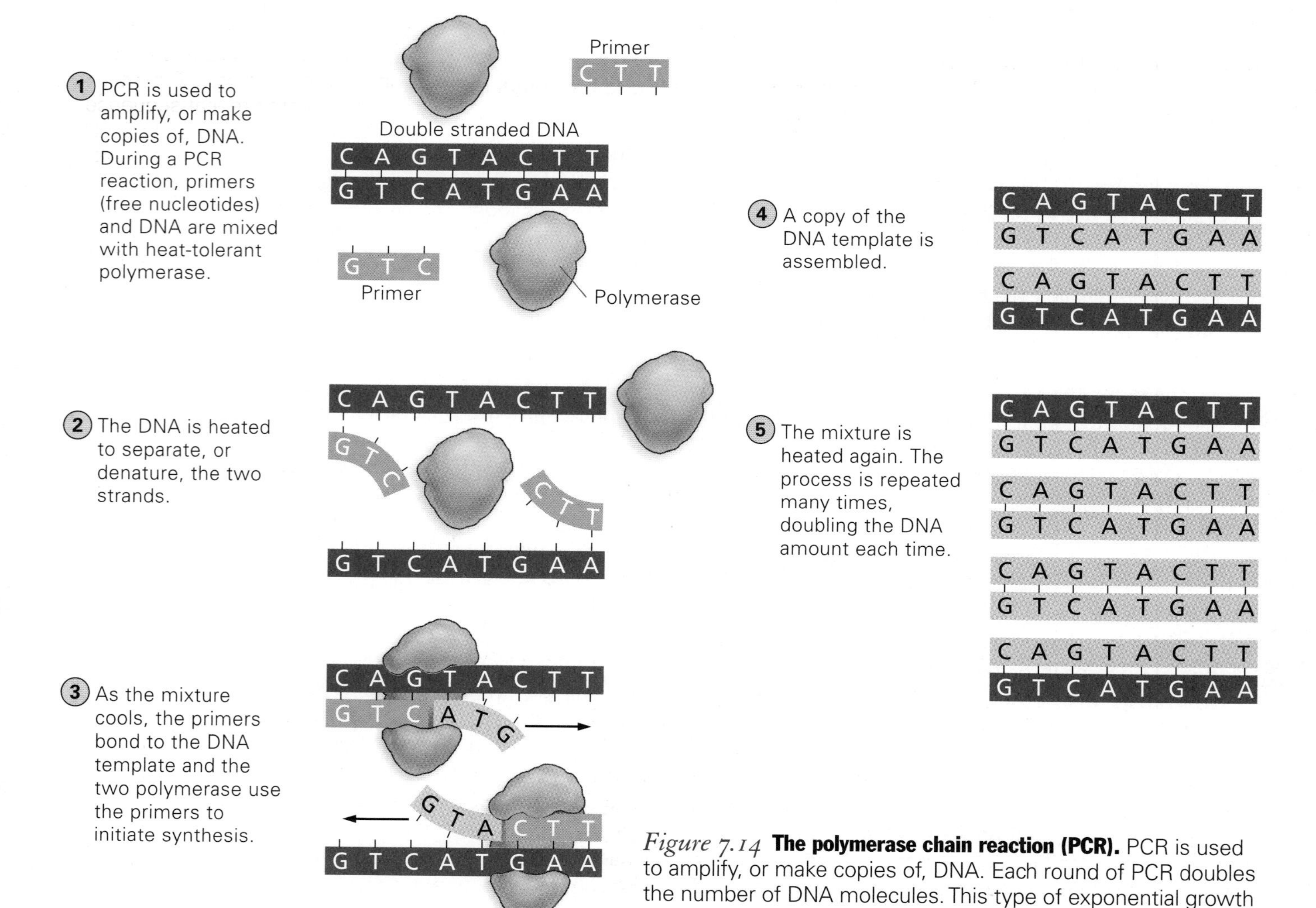

Figure 7.14 **The polymerase chain reaction (PCR).** PCR is used to amplify, or make copies of, DNA. Each round of PCR doubles the number of DNA molecules. This type of exponential growth can yield millions of copies of DNA.

(A Closer Look continued)

single strands of the DNA molecule, producing double-stranded DNA molecules. This cycle of heating and cooling is repeated many times, with each round of PCR doubling the amount of double-stranded DNA present in the test tube.

Once scientists have produced enough DNA by PCR, they can treat the DNA with enzymes that cleave, or cut, the DNA at specific nucleotide sequences. These enzymes are called **restriction enzymes,** and they act like highly specific molecular scissors. As you will see more fully in Chapter 8, individual restriction enzymes cut DNA only at specific nucleotide sequences. Because each individual has distinct nucleotide sequences, cutting different people's DNA with the same enzymes produces fragments of different sizes.

Restriction Fragment Length Polymorphism (RFLP) Analysis. The differently sized fragments produced by restriction digestion of different individuals' DNA are called **restriction fragment length polymorphisms (RFLPs).** During DNA fingerprinting, restriction fragments are produced in a variety of lengths due to the presence of DNA sequences that vary in number, called **variable number tandem repeats (VNTRs).** These are nucleotide sequences that all of us carry but in different numbers. For example, one person may have four copies of the following sequence (CGATCGA) on one chromosome and five copies of the same sequence (CGATCGA) on the other, homologous, chromosome. Another person may have six copies of that same repeat sequence in the same location on the same chromosome and three copies on the other member of the homologous pair *(Figure 7.15)*. Thus, within a population, there are variable numbers of these known tandem repeat sequences. When restriction enzymes cut DNA around these VNTRs, those segments of DNA that carry more repeats will be longer (and heavier) than those that carry fewer repeats. Since it is impossible to look at a test tube with digested DNA and determine the size of fragments, techniques that allow for the separation and visualization of the DNA are required.

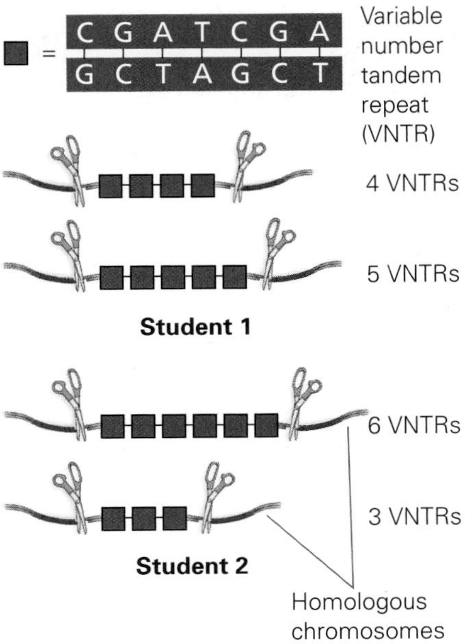

Figure 7.15 **Variable number tandem repeats (VNTRs).** Student 1 has four repeat sequences comprising the VNTR on one of his chromosomes and five on the other. Student 2 has six copies of the same repeat sequence on one of her chromosomes and three on the other member of the homologous pair. The repeat sequence is represented as a box. Restriction enzymes (represented here as scissors) cut around the VNTRs, generating fragments of different sizes.

Stop and Stretch 7.5: VNTRs do not code for proteins. If two individuals have different numbers of VNTRs around a particular gene, does that mean that they carry different alleles for the gene? Explain your answer.

Gel Electrophoresis. The fragments of DNA generated by restriction enzyme cleavage can be separated from each other by allowing the fragments to migrate through a solid support called an **agarose gel,** which resembles a thin slab of gelatin. When an electric current is applied, the gel impedes the progress of the larger DNA fragments more than it does the smaller ones, facilitating the size-based separation. The size-based separation of molecules when an electric current is applied to a gel is a technique called **gel electrophoresis.** Segments of DNA with more repeats would be heavier than those with fewer repeats. Heavier DNA segments will not migrate as far in the gel as lighter DNA fragments. Thus, agarose gel electrophoresis separates the DNA fragments on the basis of their size *(Figure 7.16a)*.

For further analysis, the DNA fragments must be physically removed from the gel. Removing the DNA from the gel is accomplished by drawing the DNA up through the gel and onto a piece of filter paper. When the filter paper is placed on top of the gel, the DNA is wicked up through the gel and adheres to the filter paper. Thus the DNA has been removed from the gel and is now attached to the filter paper. The filter paper is then treated with chemicals that break the hydrogen bonds between the two strands of the DNA helix. The resulting single-stranded DNA now resembles a ladder that has been sliced up the middle, each rung having been cut in half.

The single-stranded DNA on the filter paper is then mixed with specially prepared single-stranded DNA that has been radioactively labeled, called a **probe.** The probe is synthesized so that

(continued on the next page)

(A Closer Look continued)

the phosphate molecules in its sugar-phosphate backbone are the radioactive form of phosphorus, ^{32}P. The probe DNA can be designed so that it will base-pair with the VNTR sequences present. Such a probe will bind to each repeat sequence on the DNA. For example, DNA sequences with three repeats will bind to three separate probes. When the single-stranded probe molecules come into contact with the single-stranded repeat sequences, complementary base-pairing ensues, and a double-stranded, radioactive DNA molecule is formed. Note that the only radioactive DNA portion of a DNA molecule would be the segment containing VNTRs. Since not all of the DNA present in a particular cell will have the repeat sequence, most fragments will be unlabeled *(Figure 7.16b)*.

Student 1 has DNA sequences that carry 4 and 5 repeat sequences. Student 2 has 3 and 6 repeats. The remaining DNA is DNA that does not carry repeat sequences. Even though the DNA is visible in this figure, DNA is not visible to the unaided eye.

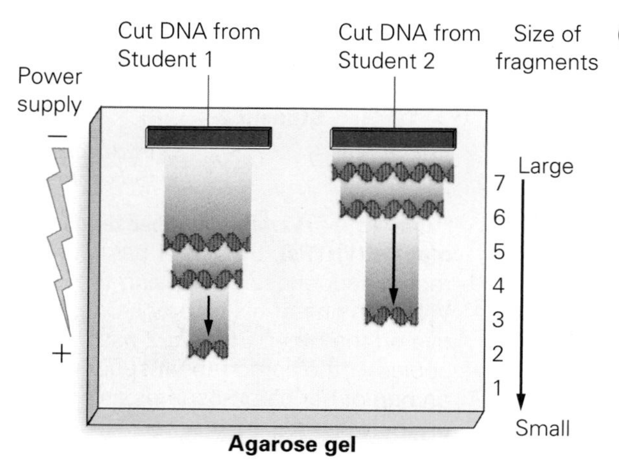

(a) DNA from two different individuals is cut with restriction enzymes and loaded on an agarose gel. When these fragments are subjected to an electric current, shorter fragments migrate through the gel faster than do larger fragments.

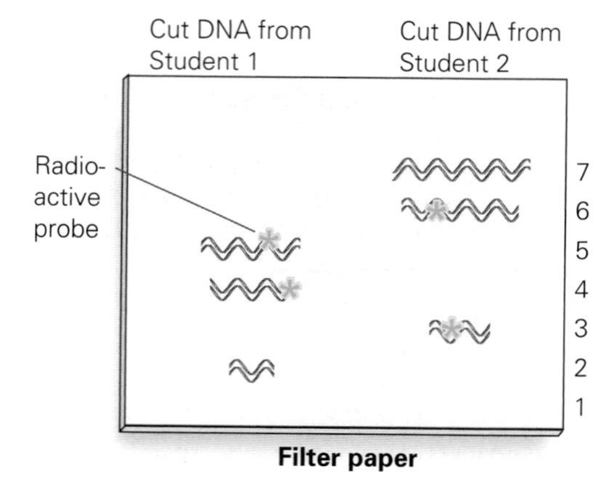

(b) The DNA is transferred from the gel onto filter paper and chemically treated to make it single-stranded. The DNA is not visible on the filter paper and must be probed with single-stranded, radioactively labeled DNA that is complementary to the repeat sequence. DNA that does not contain the repeat sequence will not bind to the probe.

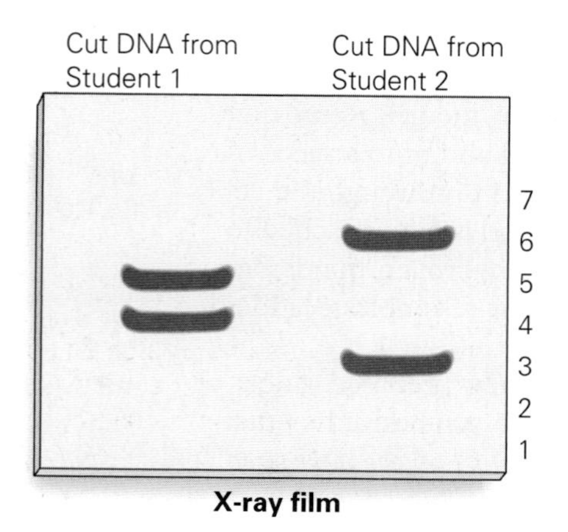

(c) When the filter is exposed to X-ray film, radioactive DNA sequences (where the probe is bound to the filter paper) produce a characteristic banding pattern, or DNA fingerprint. Student 1's DNA has 4 and 5 VNTRs, and Student 2's DNA has 3 and 6 VNTRs. The two bands represent the number of VNTR repeats.

Figure 7.16 **DNA fingerprinting.** Cut DNA can be electrophoresed and probed to produce a DNA fingerprint.

(A Closer Look continued)

Once it is bound to the labeled probe, the presence of radioactive DNA can be detected on photographic film, which records the location of DNA when the radiation emitted from radioactive DNA molecules bombards the film. A piece of X-ray film placed over the filter paper shows the locations of the radioactive DNA as a series of bands. Different individuals have different bands since the probe binds to each repeat in a VNTR location. The specific banding pattern that is produced makes up the fingerprint *(Figure 7.16c)*.

DNA Fingerprinting Evidence Helped Solve the Romanov Mystery

In 1992, a team of Russian and English scientists used the DNA fingerprinting technique to determine which of the bones discovered at Ekaterinburg belonged to the same skeleton. They used a slightly different DNA fingerprinting technique in that they isolated DNA from mitochondria located in the bone cells to confirm that the pile of decomposed bones in the Ekaterinburg grave belonged to nine different individuals.

Once scientists had established that the bones from nine different people were buried in the grave, they tried to determine which bones might belong to the adult Romanovs and which belonged to the servants. For the answer to these questions, scientists took advantage of the fact that Romanov family members would have more DNA sequences in common with each other than they would with the servants.

In fact, each DNA region that a child carries is inherited from one of his or her parents. Therefore, each band produced in a DNA fingerprint of a child must be present in the DNA fingerprint of one of that child's parents *(Figure 7.17)*. By comparing DNA fingerprints made from the smaller skeletons, scientists were able to determine which of the six adult skeletons could have been the tsar and tsarina. *Figure 7.18* shows a hypothetical DNA fingerprint that illustrates how the banding patterns produced can be used to determine which of the bones belonged to the parents of the smaller skeletons and which bones may have belonged to the unrelated servants.

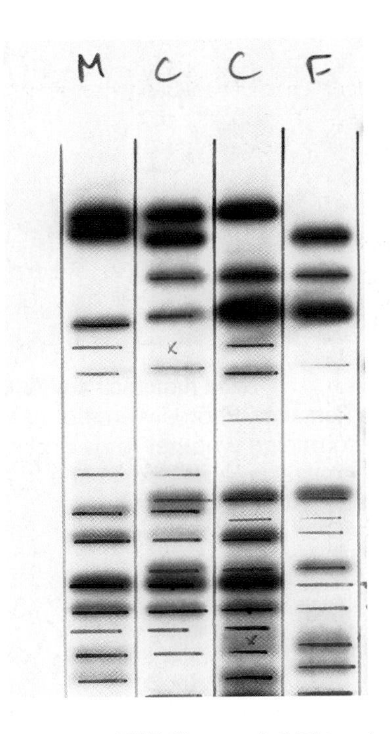

Figure 7.17 **DNA fingerprint.** This photograph of a DNA fingerprint shows a mother (M), a father (F), and their two children (C). Note that every band present in a child must also be present in one of the parents.

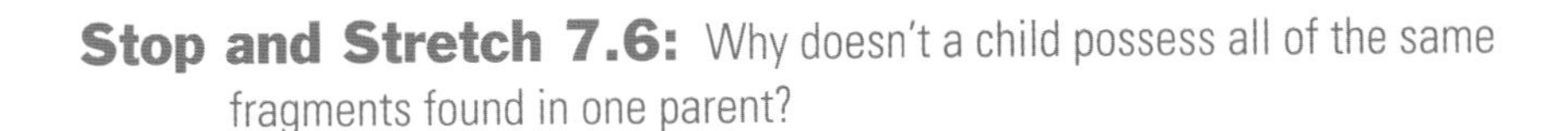

Stop and Stretch 7.6: Why doesn't a child possess all of the same fragments found in one parent?

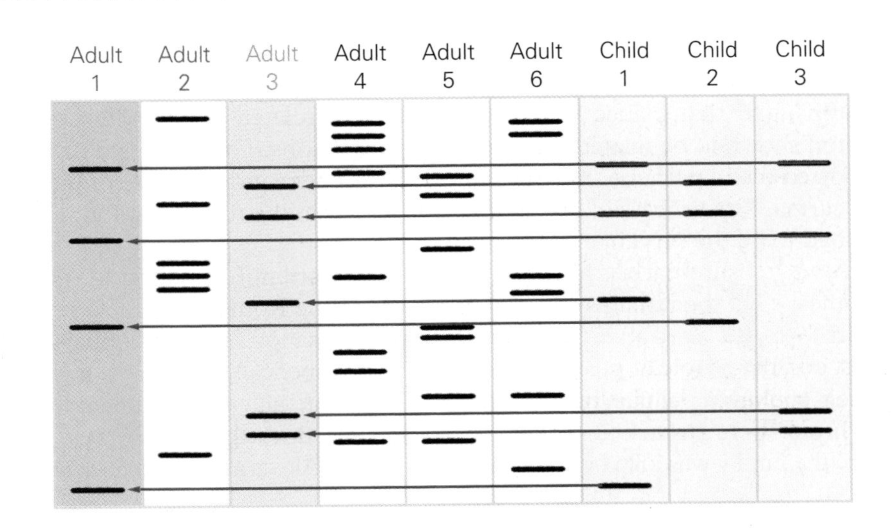

Figure 7.18 **Hypothetical fingerprint of adult- and child-sized skeletons.** Shown is a hypothetical DNA fingerprint made by using many probes to the bone cells of individuals found in the Ekaterinburg grave. From the results of this fingerprint, it is evident that children 1, 2, and 3 are the offspring of adults 1 and 3. Note that each band from each child has a corresponding band in either adult 1 or adult 3. The remaining DNA from adults does not match any of the children, so these adults are not the parents of any of these children.

(a) Anna Anderson

(b) Anastasia Romanov

Figure 7.19 **Anna Anderson and Anastasia Romanov.** Photos were not useful in determining whether Anna Anderson was Anastasia Romanov, as she claimed. DNA evidence was.

DNA evidence was further used to help put to rest claims made by many pretenders to the throne. People from all over the world had alleged that they were either a Romanov who had escaped execution or a descendant of an escapee.

The most compelling of these claims was made by a young woman who was rescued from a canal in Berlin, Germany, two years after the murders. This young woman suffered from amnesia and was cared for in a mental hospital, where the staff named her Anna Anderson (*Figure 7.19*). She later came to believe that she was Anastasia Romanov, a claim she made until her death in 1984. The 1956 Hollywood film *Anastasia*, starring Ingrid Bergman, made Anna Anderson's claim seem plausible. A more recent animated version of the story of an escaped princess, also titled *Anastasia*, convinced many young viewers that Anna Anderson was indeed the Romanov heiress.

Because the sex-typing analysis showed only that one daughter was missing from the grave, but not which daughter, scientists again looked to the fingerprinting data. DNA fingerprinting had been done in the early 1990s on intestinal tissue removed during a surgery performed before Anna Anderson's death. The analysis showed that Anna was not related to anyone buried in the Ekaterinburg grave. She could not be Anastasia.

Thus far, scientists had answered two of the questions posed at the beginning of this section. They had determined (1) that nine different individuals were buried in the Ekaterinburg grave; (2) that two of the adult skeletons were the parents of the three children. The last question was still unanswered. How would the scientists show that these were bones from the Romanov family, not just some other set of related individuals?

To answer this question, the scientists turned to living relatives of the Romanovs. DNA testing was performed on England's Prince Philip, who is a grandnephew of Tsarina Alexandra. In addition, Nicholas II's dead brother George was exhumed, and his DNA was tested. *Figure 7.20* shows the Romanov family pedigree, so that you can see how these individuals are related to each other. The DNA testing performed on these individuals showed that George was genetically related to the adult male skeleton related to the children's skeletons, and that Prince Philip was genetically related to the adult female skeleton shown to be related to the children's skeletons. This evidence strongly supported the hypothesis that the adult skeletons were indeed those of the tsar and tsarina. The process of elimination suggests that the remaining four skeletons had to be the servants.

Table 7.4 summarizes how scientists used the scientific method to test the hypothesis that the remains were indeed those of the Romanovs.

In 1998, 80 years after their execution, the Romanov family was finally laid to rest and the people of postcommunist Russia symbolically laid to rest this part of their country's political history. 20 years after that, all cases of pretenders to the throne were dismissed when bone fragments unearthed in a forest near where the family was killed were found to belong to Alexis and his sister Maria.

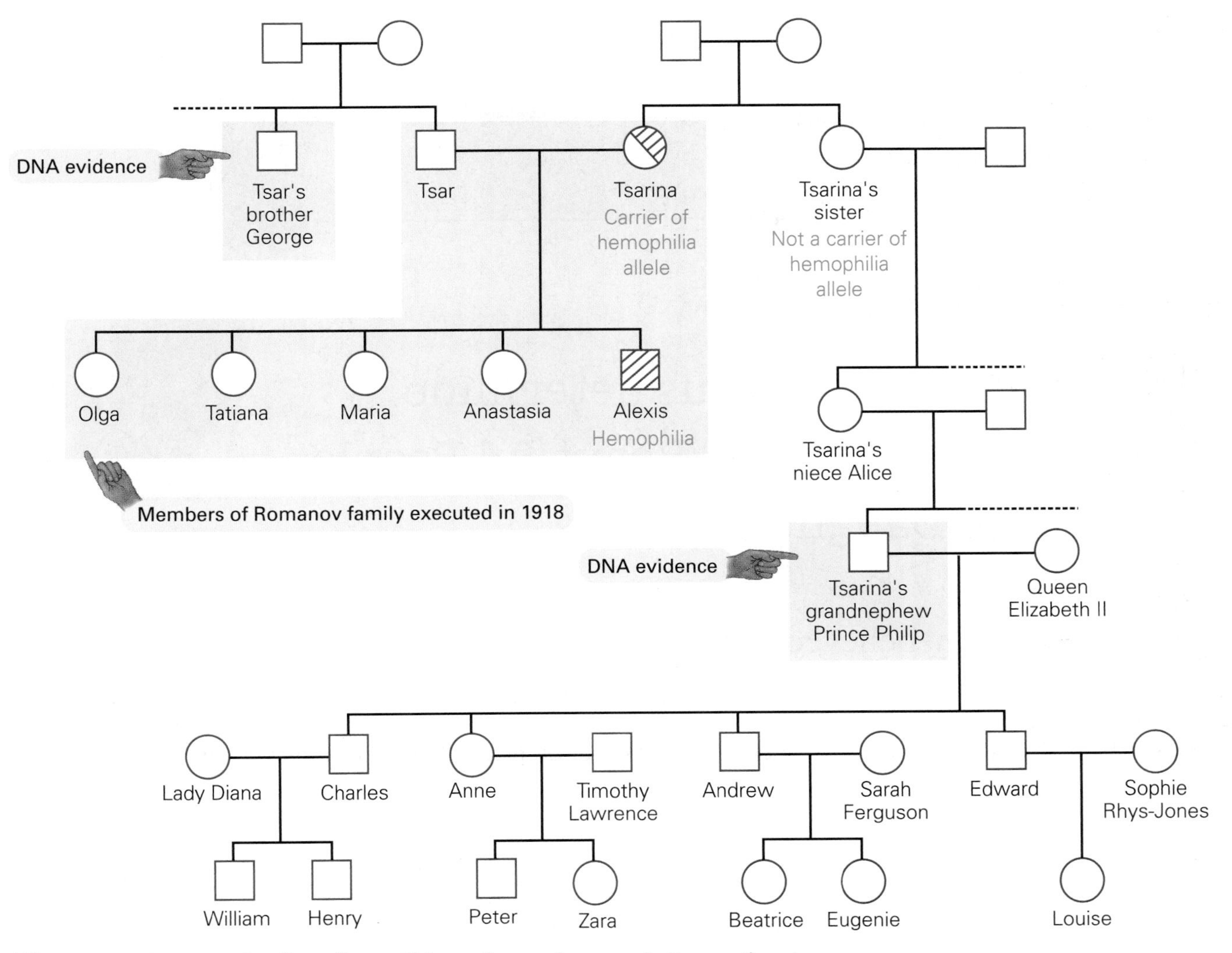

Figure 7.20 **Romanov family pedigree.** This pedigree shows only the pertinent family members. DNA from the tsar's brother George showed that he was related to the tsar. Note that Prince Philip is the tsarina's grandnephew. Prince Philip married Queen Elizabeth II. Together they had four children, Charles, Anne, Andrew, and Edward, the current British royal family. The tsarina's sister does not appear to have been a carrier of the hemophilia allele because none of her descendants have been affected by the disease.

TABLE 7.4 **The scientific method.** A summary of tests and the conclusions that were drawn from them.

Hypothesis: The bones found in the Ekaterinburg grave belonged to the Romanov family and their servants.

Test	Description of results
Analyze teeth	Expensive dental work was typically seen only in royalty.
Measure skeletons	The skeletons are those of six adults and three children.
Sex typing	The two children missing from the grave are a male and a female.
DNA fingerprinting	Children in the grave are related to two adults in the grave.
DNA fingerprinting	Claims to be one of the missing Romanov children or their descendants are disproved.
DNA fingerprinting	The two adults related to the children are related to known Romanov relatives.

Conclusion: When you look at each result individually, the evidence is less compelling than when you look at all the evidence together. As a whole, the evidence strongly supports the hypothesis that it was indeed the Romanovs who were buried in the Ekaterinburg grave.

Savvy Reader Social Status and Sex Ratios

Psychology Today *August 2007*

Does your Social Status determine the sex of your child? Beautiful People have more Daughters

It is commonly believed that whether parents conceive a boy or a girl is up to random chance. The normal sex ratio at birth is 105 boys for every 100 girls. But the sex ratio varies slightly in different circumstances and for different families. There are factors that subtly influence the sex of an offspring. If parents have any traits that they can pass on to their children and that are better for sons than for daughters, then they will have more boys. Conversely, if parents have any traits that they can pass on to their children that are better for daughters, then they will have more girls.

Physical attractiveness, while a universally positive quality, contributes even more to women's reproductive success than to men's. This generalized hypothesis would then predict that physically attractive parents should have more daughters than sons. Americans who are rated very attractive have a 56% chance of having a daughter for their first child compared with 48% for everyone else.

1. Do any data presented in this article support the idea that physical attractiveness is more important to female than male reproductive success? Can we make any decisions about the validity of this assertion without data and information about the significance of any reported differences?
2. No biological mechanism has been proposed for this purported difference. Can you think of any way your gametes could be directed to produce a male or female offspring?
3. This article states that "If parents have any traits that they can pass on to their children and that are better for sons than for daughters, then they will have more boys." Based on your understanding of natural selection and gametogenesis, does this seem possible?

CHAPTER Review

For study help, animations, and more quiz questions go to www.mybiology.com.

Summary

7.1 Forensic Science

- Forensic science is a branch of science that helps answer questions when a crime has been committed (pp. 172–173).

7.2 Dihybrid Crosses

- Dihybrids are heterozygous for two traits. It is possible to predict the outcome of a dihybrid cross by using a Punnett square (pp. 173–174).

7.3 Extensions of Mendelian Genetics

- Incomplete dominance is an extension of Mendelian genetics by which the phenotype of the progeny is intermediate to that of both parents (p. 175).
- Polygenic inheritance occurs when many genes control one trait (p. 176).
- Codominance occurs when both alleles of a given gene are expressed (p. 176).
- Genes that have more than two alleles segregating in a population are said to have multiple alleles (p. 176).
- The ABO blood system displays both multiple allelism (alleles I^A, I^B, and i) and codominance since both I^A and I^B are expressed in the heterozygote (pp. 176–178).
- The genetic blood-clotting disorder hemophilia can trigger a phenomenon known as pleiotropy, which occurs when a single gene leads to multiple effects (p. 178).

7.4 Sex Determination and Sex Linkage

- One mechanism of sex determination involves the suite of sex chromosomes present. In humans, males have an X and a Y chromosome and can produce gametes containing either sex chromosome. Females have two X chromosomes and always produce gametes containing an X chromosome. When an X-bearing sperm fertilizes an egg cell, a female baby will result. When a Y-bearing sperm fertilizes an egg cell, a male baby will result (pp. 178–179)
- Genes linked to the X and Y chromosomes show characteristic patterns of inheritance. Males need only one recessive X-linked allele to display the associated phenotype. Females can be carriers of an X-linked recessive allele and may pass an X-linked disease on to their sons (pp. 179–181).
- One of the two X chromosomes in females is inactivated, and the genes residing on it are not expressed. This inactivation is faithfully propagated to all daughter cells (pp. 181–183).
- Y-linked genes are passed from fathers to sons (p. 183).

web animation 7.1
X-Linked Recessive Traits

Summary (continued)

7.5 Pedigrees

- Pedigrees are charts that scientists use to study the transmission of genetic traits among related individuals (pp. 183–185).

7.6 DNA Fingerprinting

- DNA fingerprinting is used to show the relatedness of individuals based on similarities in their DNA sequences (p. 185).
- The polymerase chain reaction (PCR) utilizes a special temperature-resistant polymerase called *Taq* polymerase to make millions of copies of a DNA sequence (p. 186).
- Length differences in DNA fragments, generated by restriction digestion of DNA, are characteristic of a given individual (p. 187).
- DNA samples in an agarose gel subjected to an electric current will separate according to their size (p. 187).
- Radioactive probes can be used to identify sequences of DNA that have been lifted from a gel (pp. 187–188).

web animation 7.2
Polymerase Chain Reaction (PCR)

Roots to Remember

The following roots of words come mainly from Latin and Greek and will help you decipher terms:

forensic comes from a Latin word, *forum*, which was the place where legal disputes and lawsuits were heard.

hemo means blood.

pleio- and **poly-** both come from Greek words for many.

Learning the Basics

1. What does it mean for a trait to have multiple alleles? Can an individual carry more than two alleles of one gene?

2. How is sex determined in humans?

3. Describe the technique of DNA fingerprinting.

4. If a man with blood type A and a woman with blood type B have a child with type O blood, what are the genotypes of each parent?

5. A man with type A^+ blood whose father had type O^- blood and a woman with type AB^- blood could produce children with what phenotypes relative to these blood-type genes?

6. Which of the following is *not* part of the procedure used to make a DNA fingerprint?

A. DNA is treated with restriction enzymes; **B.** Cut DNA is placed in a gel to separate the various fragments by size; **C.** The genes that encode fingerprint patterns are cloned into bacteria; **D.** DNA from blood, semen, vaginal fluids, or hair-root cells can be used for analysis; **E.** An electrical current is used to separate DNA fragments.

7. Which of the following statements is consistent with the DNA fingerprint shown in *Figure 7.21*?
A. B is the child of A and C; **B.** C is the child of A and B; **C.** D is the child of B and C; **D.** A is the child of B and C; **E.** A is the child of C and D.

8. What is the probability that a family with 2 children will have 1 boy and 1 girl?
A. 100%; **B.** 75%; **C.** 50%; **D.** 25%

9. The pedigree in *Figure 7.22* illustrates the inheritance of hemophilia (sex-linked recessive trait) in the royal family. What is the genotype of individual II-5 (Alexis)?
A. $X^H X^H$; **B.** $X^H X^h$; **C.** $X^h X^h$; **D.** $X^H Y$; **E.** $X^h Y$

10. A woman is a carrier of the X-linked recessive colorblindness gene. She mates with a man with normal color vision. Which of the following is true of their offspring?
A. All the males will be colorblind; **B.** All the females will be carriers; **C.** Half the females will be colorblind; **D.** Half the males will be colorblind.

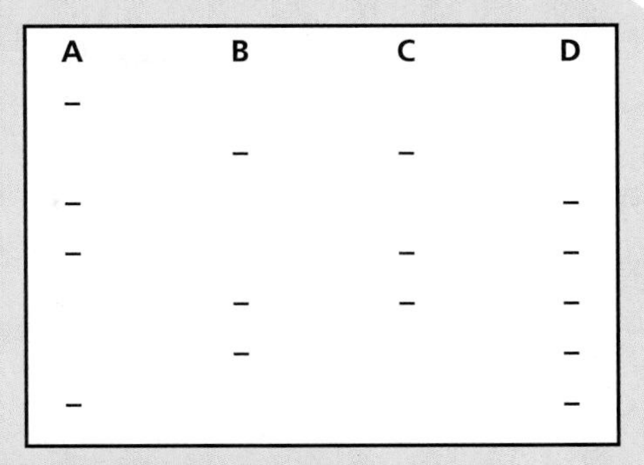

Figure 7.21 **DNA fingerprint.**

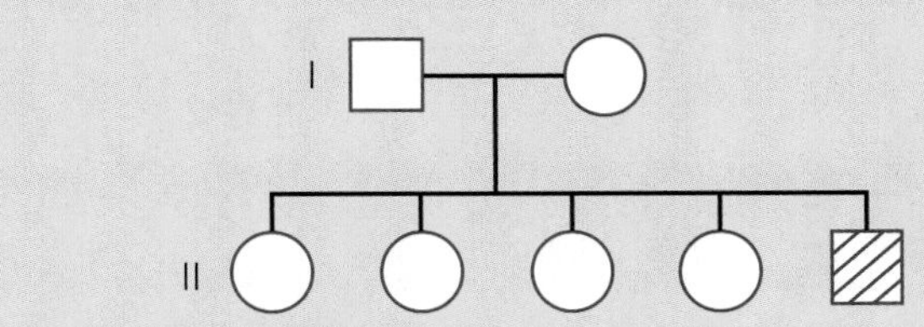

Figure 7.22 **Pedigree.**

Analyzing and Applying the Basics

1. Compare and contrast codominance and incomplete dominance.

2. Draw a DNA fingerprint that might be generated by two sisters and their parents.

3. Draw a pedigree of a mating between first cousins and use the pedigree to explain why matings between relatives can lead to an increased likelihood of offspring with rare recessive diseases.

Connecting the Science

1. Science helped solve the riddle of who was buried in the Ekaterinburg grave, leading to a church burial for some of the Romanov family. In this manner, science played a role in helping the people of Russia come to terms with the brutal communist regime that followed the deaths of the royal family. Can you think of other examples for which science has been used to help answer a question with great social implications?

2. The Innocence Project is a nonprofit legal clinic at the Benjamin Cardozo School of Law in New York City that attempts to help prisoners whose claims of innocence can be verified by DNA testing because biological evidence from their cases still exists. Close to 200 inmates have been exonerated since the project started in 1992. Thousands of inmates await the opportunity for testing. Does the fact that the Innocence Project has shown that the criminal justice system convicts innocent people change your opinion about the death penalty? Why or why not.

CHAPTER 8

Genetically Modified Organisms

Gene Expression, Mutation, and Cloning

Many people are concerned about genetic engineering.

Can you tell which of these foods have been genetically modified?

Is it acceptable to modify animals such as these pigs that can produce an essential nutrient for human consumption?

Or animals such as these luminescent mice?

Genetic engineering is the use of one or more techniques to alter hereditary traits in an organism. Genetically modified foods, also called GM foods, have had one or more of their genes modified or a gene from another organism inserted alongside their normal complement of genes. Emotions run high in the debate over the altering of our food supply. People on both sides of the issue are making their cases dramatically.

Demonstrators dressed in biohazard suits tossed genetically modified crackers, cereal, and pasta into a garbage can in front of a supermarket while shareholders inside voted on whether to remove such foods from the store shelves. Late one summer night, a group calling itself Seeds of Resistance hacked down a plot of genetically modified corn at the University of Maine. Other detractors destroyed the world's first outdoor trial of genetically engineered coffee.

In response to concerns over genetically modified foods, some parents refuse to send their kids to day-care centers that use milk from cows treated with growth hormone, and most natural food coops won't sell the so-called frankenfoods.

On the other side of the issue are proponents of genetically modifying foods who believe the technology holds tremendous promise. Advocates claim that genetically modified foods have the potential to help wipe out hunger. They see a future in which all children will be vaccinated through the use of edible vaccines, and common human diseases like heart disease will be prevented when people begin to eat genetically modified animals such as pigs that produce omega-3 fatty acids.

Proponents even see environmental benefits because some genetically modified crops allow farmers to use far smaller amounts of chemicals that damage the environment.

Even more controversy surrounds the idea of genetically modifying humans. It might be possible to fix humans with genetic diseases by replacing defective genes with normal ones or giving them new healthy cells to replace malfunctioning cells. Scientists have already cloned many different animals; can humans be far behind?

How can the average citizen determine whether it is okay to purchase and consume modified foods or whether the benefits of genetic engineering outweigh the drawbacks? To answer these questions for yourself, you must learn about how genes typically produce their products and then how they can be modified to suit human desires.

web animation 8.1 *Transcription*

web animation 8.2 *Translation*

8.1 Protein Synthesis and Gene Expression

To genetically modify foods, scientists can move a gene known to produce a certain protein from one organism to another. Alternatively, the scientists can change the amount of protein a gene produces. Regulating the amount of protein produced by a cell is also referred to as regulating gene expression.

One of the first examples of scientists controlling gene expression occurred in the early 1980s when genetic engineers at Monsanto® Company began to produce **recombinant bovine growth hormone (rBGH)** in their laboratories. Recombinant (r) bovine growth hormone is a protein that has been made by genetically engineered bacteria. These bacterial cells have had their DNA manipulated so that it carries the instructions for, or encodes, a cow growth hormone that can be produced in the laboratory. Hormones are substances that are secreted from specialized glands and travel through the bloodstream to affect their target organs. Growth hormones act on many different organs to increase the overall size of the body and, in cows, to increase milk production.

Genetic engineers at Monsanto realized that genetic technology would allow them to produce large quantities of bovine growth hormone in the laboratory, inject it into dairy cows, and increase their milk production. These scientists understood that if they were successful, Monsanto would stand to make a healthy profit from the dairy farmers who would buy the engineered growth hormone to increase the milk yield of their herds.

To increase milk yield, scientists needed to turn up the synthesis of the growth hormone protein. To produce more growth hormone protein, or any protein, cells use the genetic information coded in their DNA.

From Gene to Protein

Protein synthesis involves using the instructions carried by a gene to build a particular protein. Genes do not build proteins directly; instead, they carry the instructions that dictate how a protein should be built. Understanding protein synthesis requires that we review a few basics about DNA, genes, and RNA. First, DNA is a polymer of nucleotides that make chemical bonds with each other based on their complementarity (A to T, and C to G). Second, a gene is a sequence of DNA that encodes a protein. Proteins are large molecules composed of amino acids. Each protein has a unique function that is dictated by its particular structure. The structure of a protein is the result of the order of

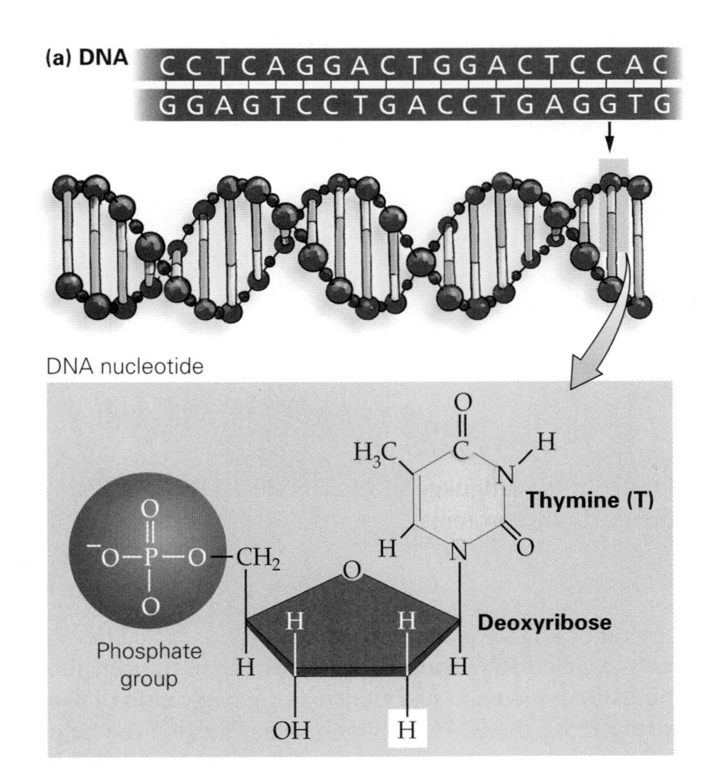

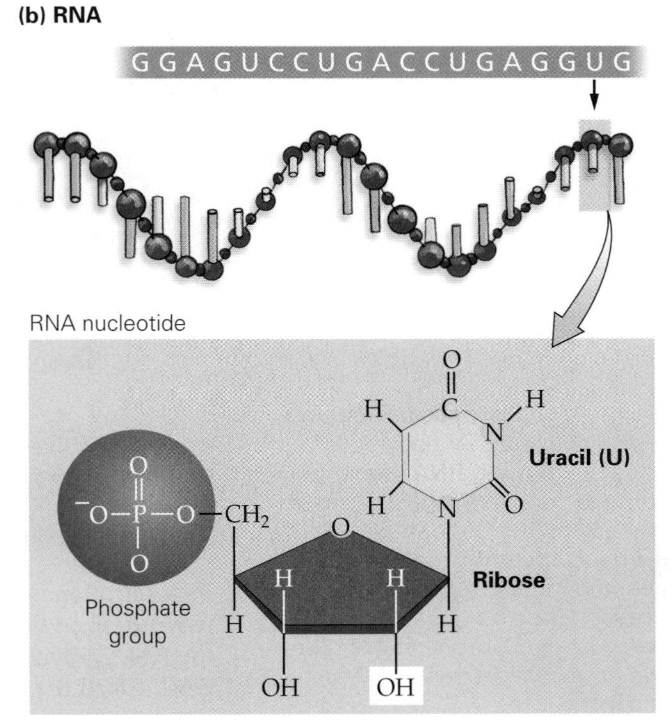

Figure 8.1 **DNA and RNA.** (a) DNA is double stranded. Each DNA nucleotide is composed of the sugar deoxyribose, a phosphate group, and a nitrogen-containing base (A, G, C, or T). (b) RNA is single stranded. RNA nucleotides are composed of the sugar ribose, a phosphate group, and a nitrogen-containing base (A, G, C, or U). **Visualize This:** Point out the chemical difference between the sugar in DNA and the one in RNA and the difference between the nitrogenous bases thymine and uracil.

amino acids that constitute it because the chemical properties of amino acids cause a protein to fold in a particular manner.

Before a protein can be built, the instructions carried by a gene are first copied. When the gene is copied, the copy is made up not of DNA (deoxyribonucleic acid) but of **RNA (ribonucleic acid).**

RNA, like DNA, is a polymer of nucleotides. A nucleotide is composed of a sugar, a phosphate group, and a nitrogen-containing base. Whereas the sugar in DNA is deoxyribose, the sugar in RNA is ribose. RNA has the nitrogenous base uracil (U) in place of thymine. RNA is usually single stranded, not double stranded like DNA *(Figure 8.1)*.

When a cell requires a particular protein, a strand of RNA is produced using DNA as a guide or template. RNA nucleotides are able to make base pairs with DNA nucleotides. C pairs with G, and A pairs with U.

The RNA copy then serves as a blueprint that tells the cell which amino acids to join together to produce a protein. Thus, the flow of genetic information in a eukaryotic cell is from DNA to RNA to protein *(Figure 8.2)*.

How does this flow of information actually take place in a cell? Going from gene to protein involves two steps. The first step, called **transcription,** involves producing the copy of the required gene. In the same way that a transcript of a speech is a written version of the oral presentation, transcription inside a cell produces a transcript of the original gene, with the RNA nucleotides substituted for DNA nucleotides. The second step, called **translation,** involves decoding the copied RNA sequence and producing the protein for which it codes. In the same way that a translator deciphers one language into another, translation

DNA CGTAGACAAGTG / GCATCTGTTCAC — Polymer of nucleotides (two complementary strands)
Transcription
RNA GCAUCUGUUCAC — Polymer of nucleotides (single strand)
Translation
Protein ala ser val his — Polymer of amino acids

Figure 8.2 **The flow of genetic information.** Genetic information flows from a DNA to an RNA copy of the DNA gene, to the amino acids that are joined together to produce the protein coded for by the gene.

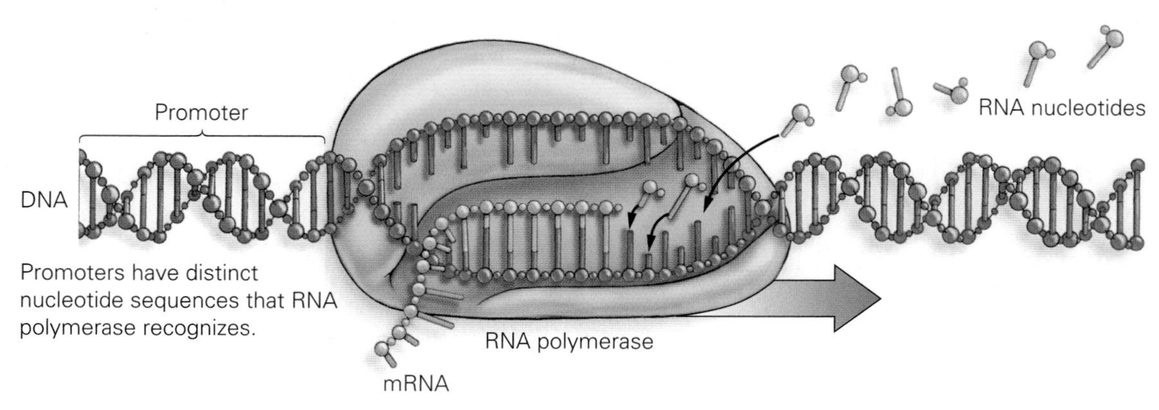

Figure 8.3 **Transcription.** RNA polymerase ties together nucleotides within the growing RNA strand as they hydrogen bond to their complementary base on the DNA. When the RNA polymerase reaches the end of the gene, the mRNA transcript is released. **Visualize This:** Propose a sequence for the DNA strand that is being used to produce the mRNA. Keep in mind that purines (A and G) are composed of two rings and thus represented by longer pegs in this illustration. Once you have proposed a DNA sequence, determine the mRNA sequence that would be produced by transcpription.

in a cell involves moving from the language of nucleotides (DNA and RNA) to the language of amino acids and proteins.

Transcription

Transcription is the copying of a DNA gene into RNA *(Figure 8.3)*. The copy is synthesized by an enzyme called **RNA polymerase.** To begin transcription, the RNA polymerase binds to a nucleotide sequence at the beginning of every gene, called the **promoter.** Once the RNA polymerase has located the beginning of the gene by binding to the promoter, it then rides along the strand of the DNA helix that comprises the gene. As it is traveling along the gene, the RNA polymerase unzips the DNA double helix and ties together RNA nucleotides that are complementary to the DNA strand it is using as a template. This results in the production of a single-stranded RNA molecule that is complementary to the DNA sequence of the gene. This complementary RNA copy of the DNA gene is called **messenger RNA (mRNA)** since it carries the message of the gene that is to be expressed.

Translation

The second step from gene to protein requires that the mRNA be used to produce the actual protein for which the gene encodes through translation. For translation to occur, a cell needs mRNA, a supply of amino acids to join in the proper order, and some energy in the form of ATP. Translation also requires structures called ribosomes and transfer RNA molecules.

Ribosomes. **Ribosomes** are subcellular, globular structures *(Figure 8.4)* that are composed of another kind of RNA called **ribosomal RNA (rRNA),** which is wrapped around many different proteins. Each ribosome is composed of two subunits—one large and one small. When the large and small subunits of the ribosome come together, the mRNA can be threaded between them. In addition, the ribosome can bind to structures called **transfer RNA (tRNA)** that carry amino acids.

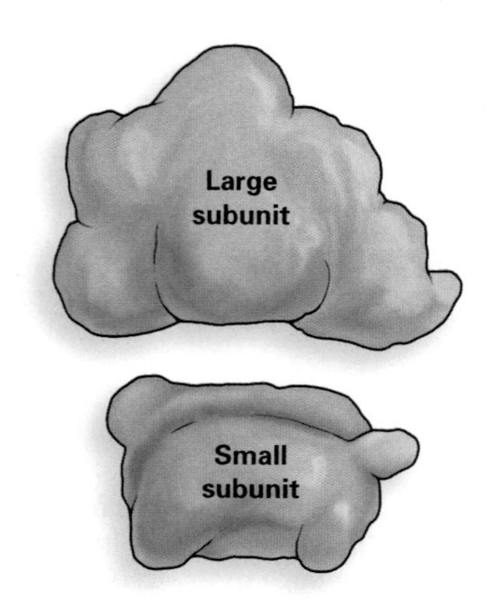

Figure 8.4 **Ribosome.** Ribosomes are composed of two subunits. Each subunit in turn is composed of rRNA and protein.

Transfer RNA (tRNA). Transfer RNA *(Figure 8.5)* is yet another type of RNA found in cells. An individual transfer RNA molecule carries one specific amino acid and interacts with mRNA to place the amino acid in the correct location of the growing polypeptide.

As mRNA moves through the ribosome, small sequences of nucleotides are sequentially exposed. These sequences of mRNA, called **codons,** are three nucleotides long and encode a particular amino acid. Transfer RNAs also have a set of three nucleotides, which will bind to the codon if the right sequence is present. These three nucleotides at the base of the tRNA are therefore called the **anticodon** since they complement a codon on mRNA. The anticodon on a particular tRNA binds to the complementary mRNA codon. In this way, the codon calls for the incorporation of a specific amino acid. The ribosome moves along the mRNA sequentially exposing codons for tRNA binding.

When a tRNA anticodon binds to the mRNA codon, a peptide bond is formed. The ribosome adds the amino acid that the tRNA is carrying to the growing chain of amino acids that will eventually constitute the finished protein. The transfer RNA functions as a sort of cellular translator, fluent in both the language of nucleotides (its own language) and the language of amino acids (the target language).

To help you understand protein synthesis, let us consider its similarity to an everyday activity such as baking a cake (*Figure 8.6*). To bake a cake, you would consult a recipe book (genome) for the specific recipe (gene) to make your cake (protein). You may copy the recipe (mRNA) out of the book so that the original recipe (gene) does not become stained or damaged. The original recipe (gene) is left in the book (genome) on a shelf (nucleus) so that you can make another copy when you need it. The original recipe (gene) can be copied again and again. The copy of the recipe (mRNA) is placed on the kitchen counter (ribosome) while you assemble the ingredients (amino acids). The ingredients (amino acids) for your cake (protein) include flour, sugar, butter, milk, and eggs. The ingredients are measured in measuring spoons and cups (tRNAs). (While in baking you might use the same cups and spoons for several ingredients, in protein synthesis we are using tRNAs that are dedicated to one specific ingredient.) The measuring spoons and cups bring the ingredients to the kitchen counter. Like the amino acids that are combined in different orders to produce a specific protein, the ingredients in a cake can be used in many ways to produce a variety of foods. The ingredients (amino acids) are always added according to the instructions specified by the original recipe (gene).

Inside of cells, the sequence of bases in the DNA dictates the sequence of bases in the RNA, which in turn dictates the order of amino acids that will be

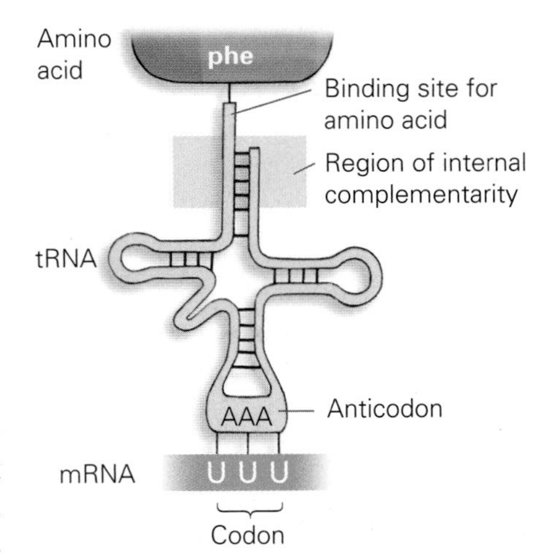

Figure 8.5 **Transfer RNA (tRNA).** Transfer RNAs translate the language of nucleotides into the language of amino acids. **Visualize This:** What nitrogenous bases might be involved in bonding in the highlighted region of internal complementarity?

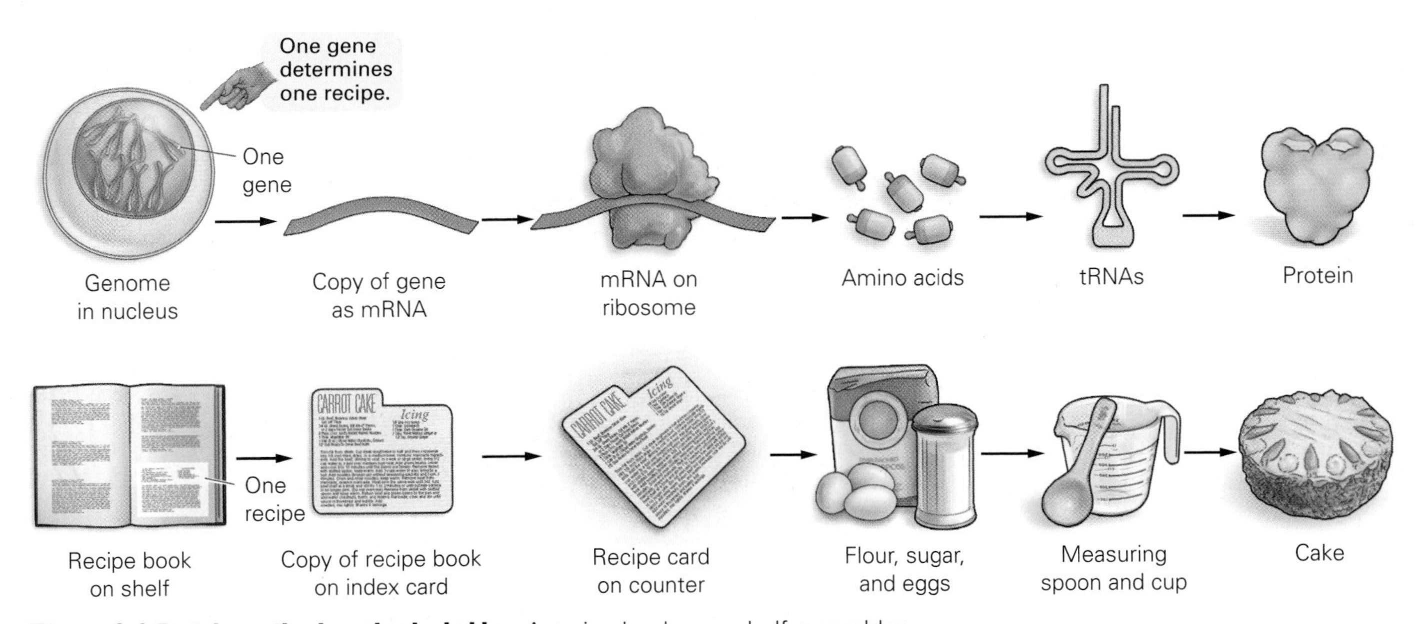

Figure 8.6 **Protein synthesis and cake baking.** A recipe book on a shelf resembles the genome in the nucleus of a cell. Copying one recipe onto an index card is similar to copying the gene to produce mRNA. The recipe card is placed on the kitchen counter, and the mRNA is placed on the ribosome. The ingredients for a cake are flour, sugar, butter, milk, and eggs. The ingredients to make a protein are various amino acids. The ingredients are measured in measuring spoons and cups (tRNAs) that are dedicated to one specific ingredient. The measuring spoons and cups bring the ingredients to the kitchen counter. The ingredients (amino acids) are always added according to the instructions specified by the original recipe (gene).

joined together to produce a protein. Protein synthesis ends when a codon that does not code for an amino acid, called a **stop codon,** moves through the ribosome. When a stop codon is present in the ribosome, no new amino acid can be added, and the growing protein is released. Once released, the protein folds up on itself and moves to where it is required in the cell. A summary of the process of translation is shown in *Figure 8.7*.

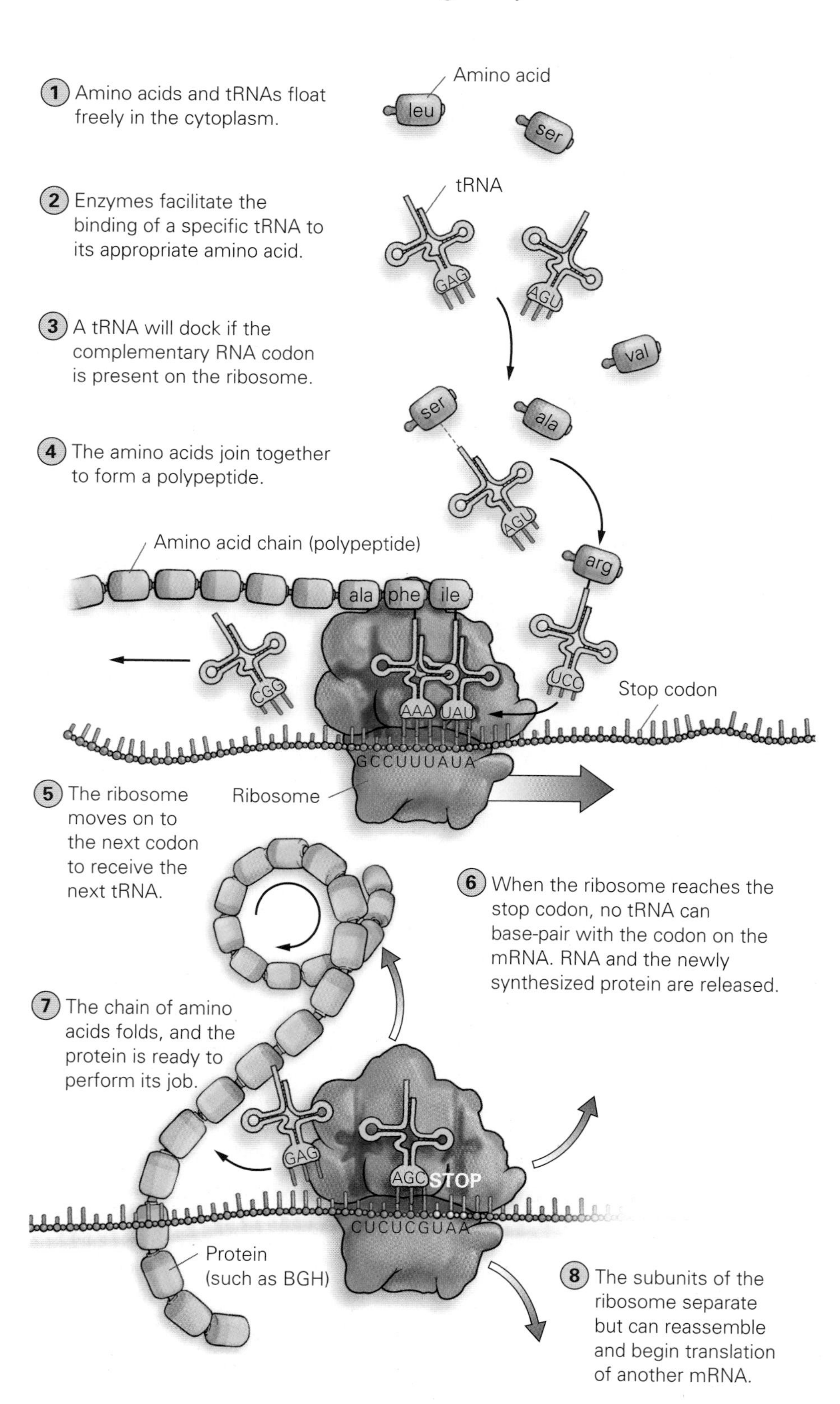

Figure 8.7 **Translation.** During translation, mRNA directs the synthesis of a protein.

The process of translation allows cells to join amino acids in the sequence coded by the gene. Scientists can determine the sequence of amino acids that a gene calls for by looking at the **genetic code.**

Genetic Code. The genetic code determines which mRNA codons code for which amino acids *(Table 8.1)*. As Table 8.1 shows, there are 64 codons, 61 of which code for amino acids. Three of the codons are stop codons that occur near the end of an mRNA. Since stop codons do not code for an amino acid, protein synthesis ends when a stop codon enters the ribosome. In the table, you can see that the codon AUG functions both as a start codon (and thus is found near the beginning of each mRNA) and as a codon dictating that the amino acid methionine (met) be incorporated into the protein being synthesized. This initial methionine is often removed later. Notice also that the same amino acid can be coded for by more than one codon. For example, the amino acid threonine (thr) is incorporated into a protein in response to the codons ACU, ACC, ACA, and ACG. The fact that more than one codon can code for the same amino acid is referred to as *redundancy* in the genetic code. There is, however, no situation where a given codon can call for more than one amino acid. For example, AGU codes for serine (ser) and nothing else. Therefore, there is no *ambiguity* in the genetic code regarding which amino acid any codon will call for. The genetic code is also *universal* in the sense that organisms typically decode the same gene to produce the same protein. This

TABLE 8.1 **The genetic code.** It is possible to determine which amino acid is coded for by each mRNA codon using a chart called the genetic code. Look at the left-hand side of the chart for the first-base nucleotide in the codon; there are 4 rows, 1 for each possible RNA nucleotide—A, C, G, or U. By then looking at the intersection of the second-base columns at the top of the chart and the first-base rows, you can narrow your search for the codon to 4 different codons. Finally, the third-base nucleotide in the codon on the right-hand side of the chart determines the amino acid that a given mRNA codon codes for. Note the 3 codons, UAA, UAG, and UGA, that do not code for an amino acid; these are stop codons. The codon AUG is a start codon, found at the beginning of most protein-coding sequences.

First base	Second base: U		C		A		G		Third base
U	UUU	Phenylalanine (phe)	UCU	Serine (ser)	UAU	Tyrosine (tyr)	UGU	Cysteine (cys)	U
	UUC		UCC		UAC		UGC		C
	UUA	Leucine (leu)	UCA		UAA	**Stop codon**	UGA	**Stop codon**	A
	UUG		UCG		UAG	**Stop codon**	UGG	Tryptophan (trp)	G
C	CUU	Leucine (leu)	CCU	Proline (pro)	CAU	Histidine (his)	CGU	Arginine (arg)	U
	CUC		CCC		CAC		CGC		C
	CUA		CCA		CAA	Glutamine (gln)	CGA		A
	CUG		CCG		CAG		CGG		G
A	AUU	Isoleucine (ile)	ACU	Threonine (thr)	AAU	Asparagine (asn)	AGU	Serine (ser)	U
	AUC		ACC		AAC		AGC		C
	AUA		ACA		AAA	Lysine (lys)	AGA	Arginine (arg)	A
	AUG	Methionine (met) **Start codon**	ACG		AAG		AGG		G
G	GUU	Valine (val)	GCU	Alanine (ala)	GAU	Aspartic acid (asp)	GGU	Glycine (gly)	U
	GUC		GCC		GAC		GGC		C
	GUA		GCA		GAA	Glutamic acid (glu)	GGA		A
	GUG		GCG		GAG		GGG		G

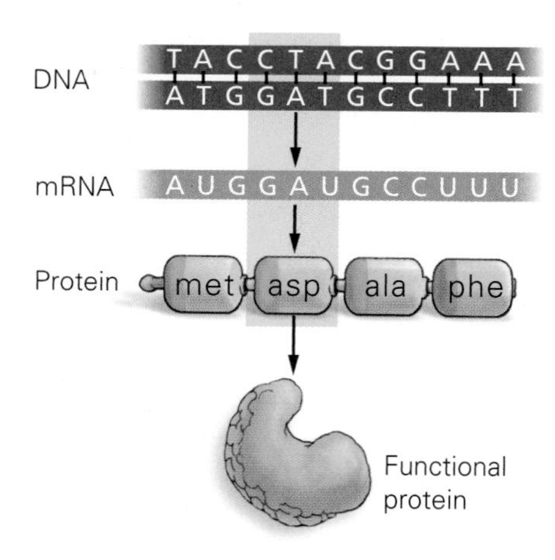

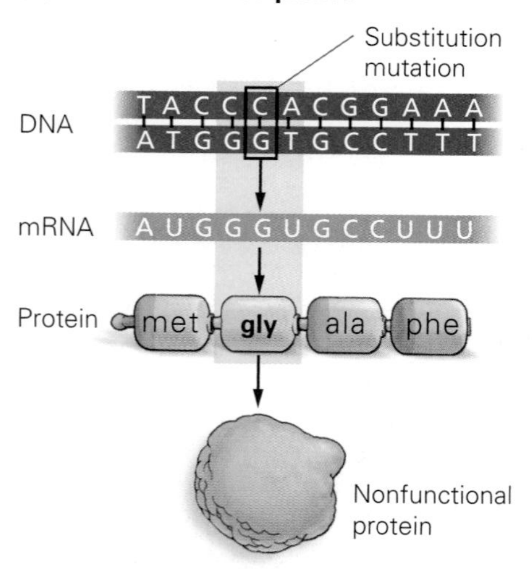

Figure 8.8 **Substitution mutation.** A single nucleotide change from the normal DNA sequence (a) to the mutated sequence (b) can result in the incorporation of a different amino acid. If the substituted amino acid has chemical properties different from those of the original amino acid, then the protein may assume a different shape and thus lose its ability to perform its job.

is why genes can be moved from one organism to another, as was case with the mice in the chapter opener photo, which can express a gene normally expressed in jellyfish.

Stop and Stretch 8.1: Humans require 20 amino acids. Could the 4 DNA bases produce an unambiguous genetic code if the codons were only 2 bases in length?

Mutations

Changes to the DNA sequence, called **mutations,** can affect the order of amino acids incorporated into a protein during translation. Mutations to a gene can result in the production of different forms, or alleles, of a gene. Different alleles result from changes in the DNA that alter the amino acid order of the encoded protein. Mutations can result in the production of either a nonfunctional protein or a protein different from the one previously required. If this protein does not have the same amino acid composition, it may not be able to perform the same job *(Figure 8.8)*. For instance, a substitution of a single nucleotide results in the incorporation of a new amino acid in the hemoglobin protein and compromises the ability of cells to carry oxygen, producing sickle-cell disease.

There are also cases when a mutation has no effect on a protein. They may occur when changes to the DNA result in the production of a mRNA codon that codes for the same amino acid as was originally required. Due to the redundancy of the genetic code, a mutation that changes the mRNA codon from ACU to ACC will have no impact because both of these codons code for the amino acid threonine. This is called a **neutral mutation** *(Figure 8.9a)*. In addition, mutations can result in the substitution of one amino acid for another with similar chemical properties, which may have little or no effect on the protein.

Inserting or deleting a single nucleotide can have a severe impact since the addition (or deletion) of a nucleotide can change the groupings of nucleotides in every codon that follows *(Figure 8.9b)*. Changing the triplet groupings is called altering the **reading frame.** All nucleotides located after an insertion or deletion will be regrouped into different codons, producing a **frameshift mutation.** For example, inserting an extra letter "H" after the fourth letter of the sentence, "The dog ate the cat," could change the reading frame to the nonsensical statement, "The dHo gat eth eca t." Inside cells, this often results in the incorporation of a stop codon and the production of a shortened, nonfunctional protein.

Stop and Stretch 8.2: What effect does the formation of new alleles have on natural selection?

All cells in all organisms undergo this process of protein synthesis, with different cell types selecting different genes from which to produce proteins. *Figure 8.10a* shows the coordination of these two processes as they occur in cells with nuclei, that is, eukaryotic cells. In eukaryotic cells, transcription and translation are spatially separate, with transcription occurring in the nucleus

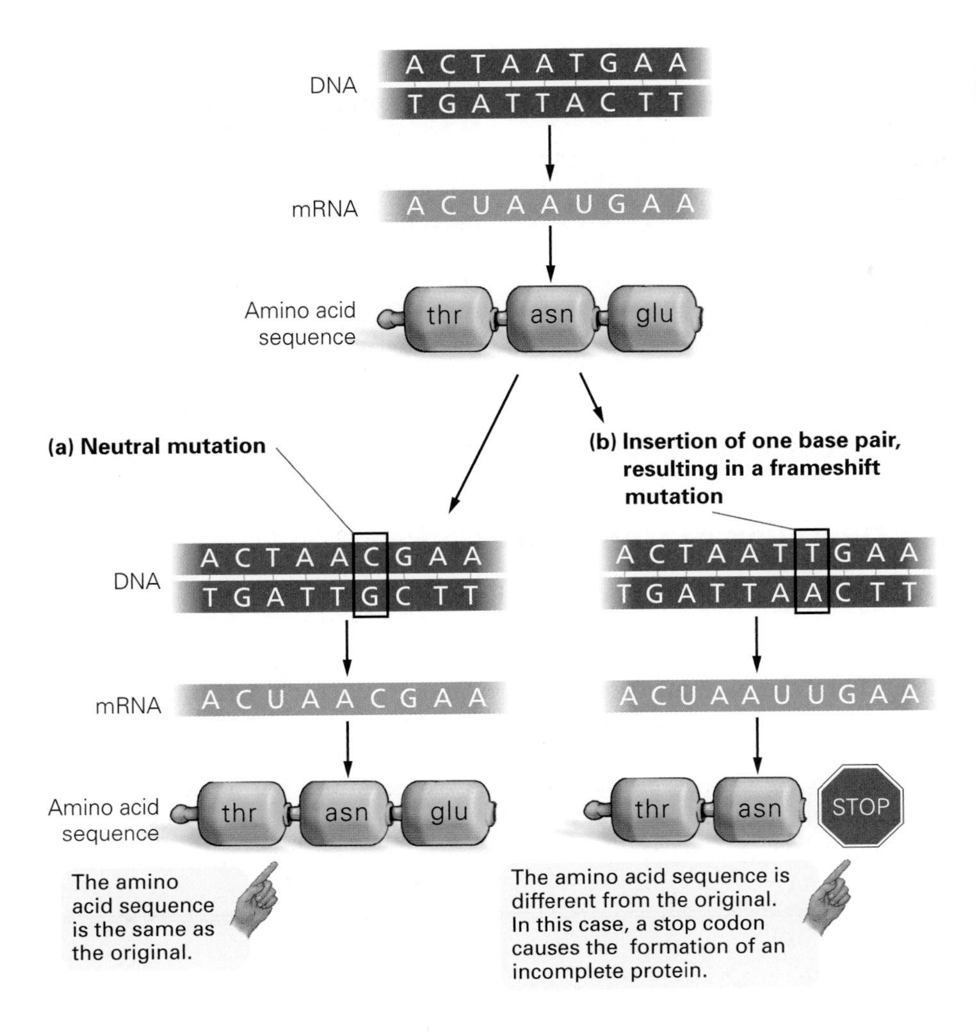

Figure 8.9 **Neutral and frameshift mutations.** (a) Neutral mutations result in the incorporation of the same amino acid as was originally called for. (b) The insertion (or deletion) of a nucleotide can result in a frameshift mutation. **Visualize This:** Is the top or bottom strand of the DNA serving as the template for RNA synthesis in this figure?

and translation occurring in the cytoplasm. Cells lacking a membrane-bound nucleus and organelles are called prokaryotic cells. Prokaryotic cells (such as bacterial cells) also undergo protein synthesis, but transcription and translation occur simultaneously in the same location instead of occurring in separate places. As a mRNA is being transcribed, ribosomes attach and begin translating *(Figure 8.10b)*.

Figure 8.10 **Protein synthesis in eukaryotic and prokaryotic cells.** (a) In eukaryotes, transcription occurs in the nucleus and translation in the cytoplasm. (b) In prokaryotic cells, which lack nuclei, transcription and translation occur simultaneously.

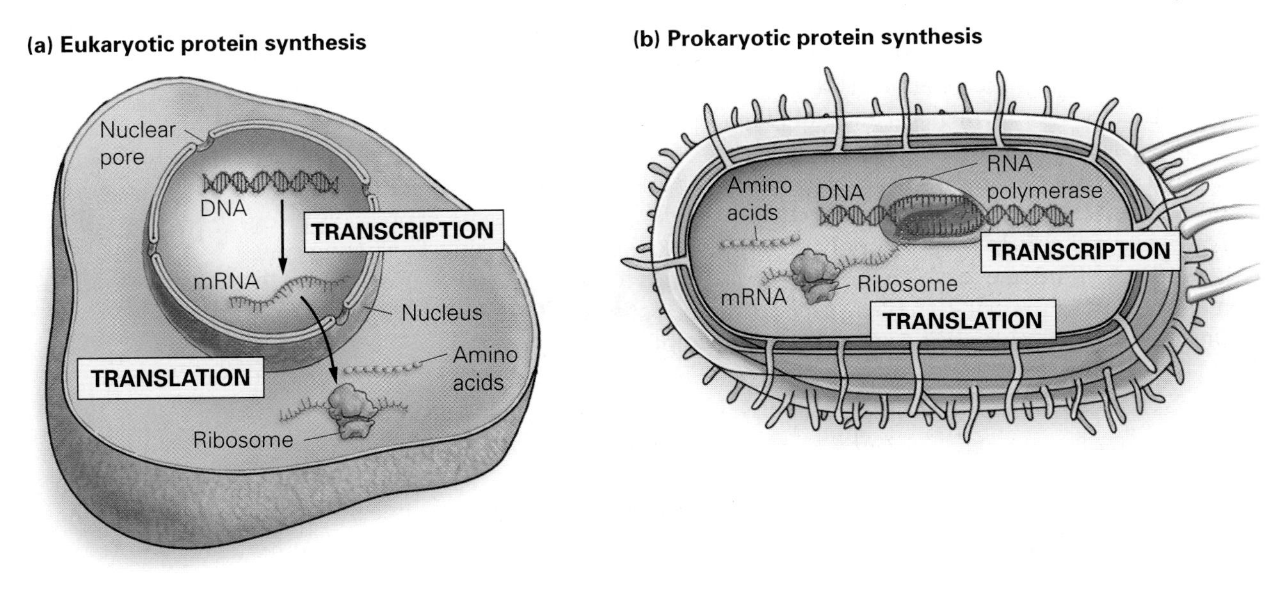

Figure 8.11 **Gene expression differs from cell to cell.** (a) A muscle cell performs different functions and expresses different genes than a (b) nerve cell. Both of these cells have the same suite of genes in their nucleus but express different subsets of genes.

(a) Muscle cells

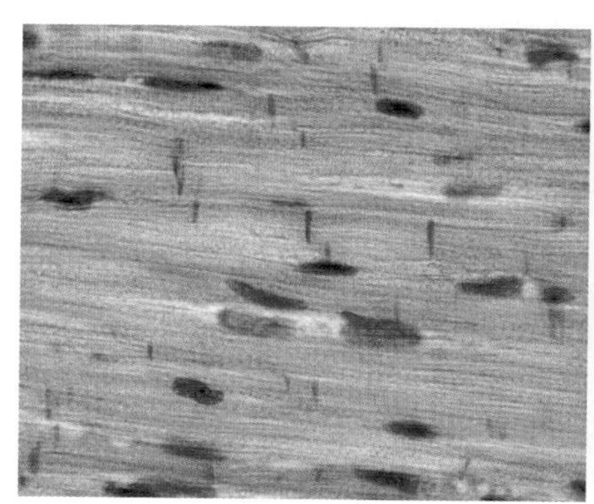

(b) Nerve cells

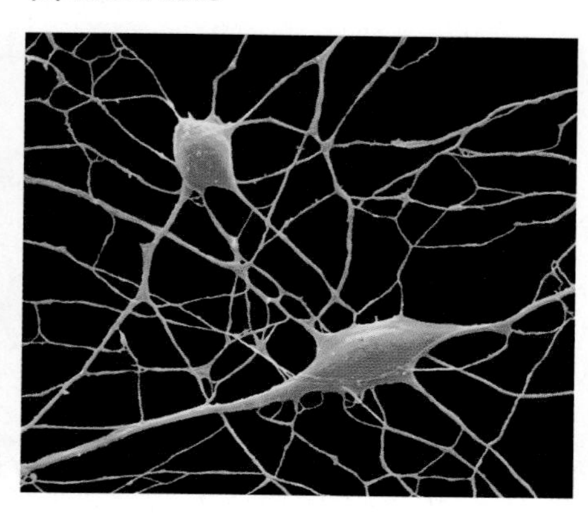

An Overview of Gene Expression

Different cell types transcribe and translate different genes. Each cell in your body, except sperm or egg cells, has the same complement of genes you inherited from your parents but expresses only a small percentage of those genes. For example, since your muscles and nerves each perform a specialized suite of jobs, muscle cells turn on or express one suite of genes and nerve cells another *(Figure 8.11)*. Turning a gene on or off, or modulating it more subtly, is called **regulating gene expression.** The expression of a given gene is regulated so that it is turned on and turned off in response to the cell's needs.

A Closer Look: Gene Expression

Gene expression can be regulated by regulating the rate of transcription or translation. It is also possible to regulate how long an mRNA or protein remains functional in a cell.

Regulation of Transcription. Gene expression is most commonly regulated by controlling the rate of transcription. Regulation of transcription can occur at the promoter, the sequence of nucleotides adjacent to a gene to which the RNA polymerase binds to initiate transcription. When a cell requires a particular protein, the RNA polymerase enzyme binds to the promoter for that particular gene and transcribes the gene.

Prokaryotic and eukaryotic cells both regulate gene expression by regulating transcription but have different strategies for doing so. Prokaryotic cells typically regulate gene expression by blocking transcription via proteins called **repressors** that bind to the promoter and prevent the RNA polymerase from binding. When the gene needs to be expressed, the repressor will be released from the promoter so that the RNA polymerase can bind *(Figure 8.12a)*. This is the main mechanism by which simple single-celled prokaryotes regulate gene expression.

The more complex eukaryotic cells have evolved more complex mechanisms to control gene expression. To control transcription, eukaryotic cells more commonly enhance gene expression using proteins called **activators** that help the RNA polymerase bind to the promoter, thus facilitating gene expression *(Figure 8.12b)*. The rate at which the polymerase binds to the promoter is also affected by substances that are present in the cell. For example, the presence of alcohol in a liver cell might result in increased transcription of a gene involved in the breakdown of alcohol.

Stop and Stretch 8.3: How would a mutation in a repressor that makes it nonfunctional likely affect the individual with this mutation?

Regulation by Chromosome Condensation. It is also possible to regulate gene expression by condensing all or part of a chromosome. This prevents RNA polymerase from being able to access

(A Closer Look continued)

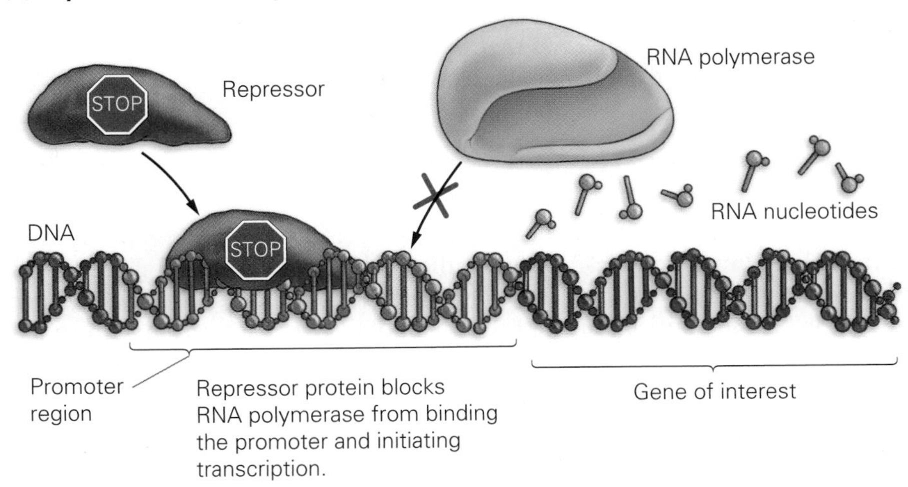

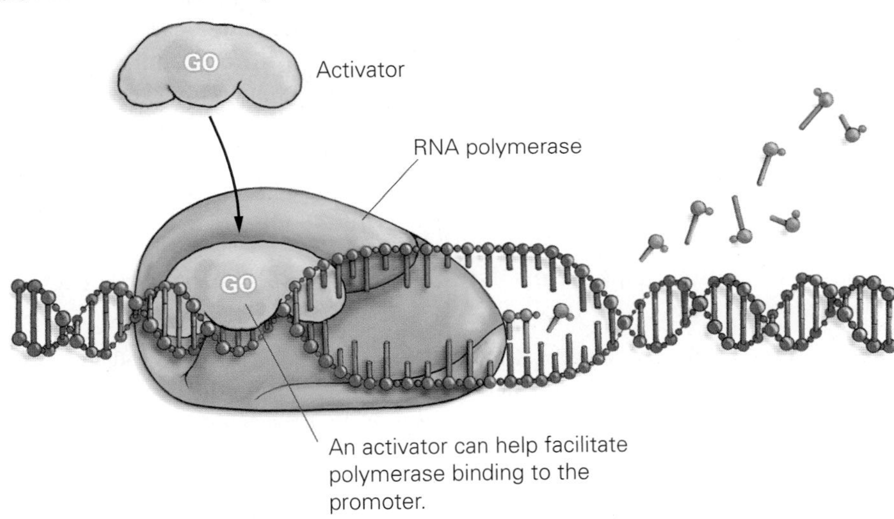

Figure 8.12 **Regulation of gene expression.** Gene expression can be regulated by (a) repression or (b) activation. Prokaryotes more typically use repression and eukaryotes activation of transcription to regulate gene expression.

genes. In Chapter 7 you learned that the inactivation of an X chromosome turns off the expression of X-linked genes in organisms that have two X chromosomes. Entire chromosomes are also inactivated when they condense during mitosis.

Regulation by mRNA Degradation. Eukaryotic cells can also regulate the expression of a gene by regulating how long a messenger RNA is present in the cytoplasm. Enzymes called **nucleases** roam the cytoplasm, cutting RNA molecules by binding to one end and breaking the bonds between nucleotides. If a particular mRNA has a long adenosine nucleotide "tail," it will survive longer in the cytoplasm and be translated more times. All mRNAs are eventually degraded in this manner; otherwise, once a gene had been transcribed one time, it would be expressed forever.

Regulation of Translation. It is also possible to regulate many of the steps of translation. For example, the binding of the mRNA to the ribosome can be slowed or hastened, as can the movement of the mRNA through the ribosome.

(continued on the next page)

(A Closer Look continued)

Regulation of Protein Degradation. Once a protein is synthesized, it will persist in the cell for a characteristic amount of time. Like the mRNA that provided the instructions for its synthesis, the life of a protein can be affected by cellular enzymes called *proteases* that degrade the protein. Speeding up or slowing down the activities of these enzymes can change the amount of time that a protein is able to be active inside a cell.

Genetic engineering permits precise control of gene expression in many different circumstances. For example, the problem of regulating gene expression is easily solved in the case of rBGH. Farmers can simply decide how much protein to inject into the bloodstream of a cow. However, they must first synthesize the protein.

web animation 8.3 *Producing Bovine Growth Hormone*

8.2 Producing Recombinant Proteins

The first step in the production of the rBGH protein is to transfer the BGH gene from the nucleus of a cow cell into a bacterial cell. Bacteria are single-celled prokaryotes that copy themselves very rapidly. They can thrive in the laboratory if they are allowed to grow in a liquid broth containing the nutrients necessary for survival. Bacteria with the BGH gene can serve as factories to produce millions of copies of this gene and its protein product. Making many copies of a gene is called **cloning** the gene.

Cloning a Gene Using Bacteria

The following three steps are involved in moving a BGH gene into a bacterial cell *(Figure 8.13)*.

Step 1. Remove the Gene from the Cow Chromosome. The gene is sliced out of the cow chromosome on which it resides by exposing the cow DNA to enzymes that cut DNA. These enzymes, called **restriction enzymes,** act like highly specific molecular scissors. Most restriction enzymes cut DNA only at specific sequences, called *palindromes*, such as:

TAT CGTACG AAC
ATA GCATGC TTG

Note that the bottom middle sequence is the reverse of the top middle sequence. Many restriction enzymes cut the DNA in a staggered pattern, leaving "sticky ends" such as.

The unpaired bases form bonds with any complementary bases with which they come in contact. The enzyme selected by the scientist cuts on both ends of the *BGH* gene but not inside the gene.

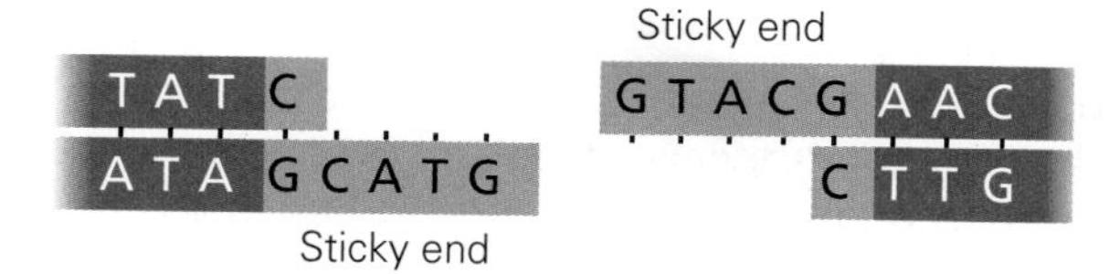

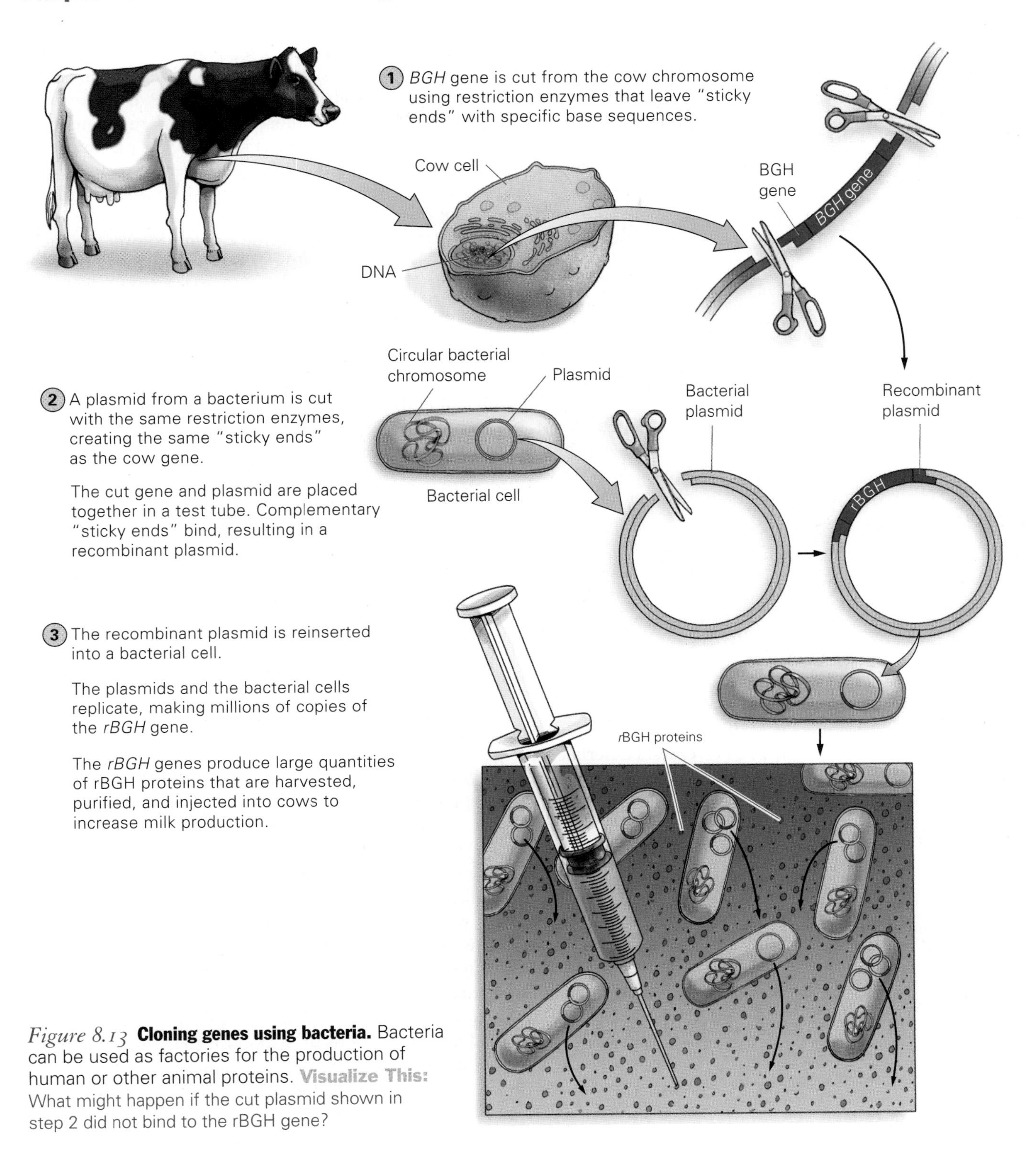

Figure 8.13 **Cloning genes using bacteria.** Bacteria can be used as factories for the production of human or other animal proteins. **Visualize This:** What might happen if the cut plasmid shown in step 2 did not bind to the rBGH gene?

Since different individual restriction enzymes cut DNA only at specific points, scientists need some information about the entire suite of genes present in a particular organism, called the **genome,** to determine which restriction enzyme cutting sites surround the gene of interest. Cutting the DNA generates many different fragments, only one of which will carry the gene of interest.

Step 2. Insert the BGH Gene into the Bacterial Plasmid. Once the gene is removed from the cow genome, it is inserted into a bacterial structure called a plasmid. A plasmid is a circular piece of DNA that normally exists separate from the bacterial chromosome and can replicate independently of the bacterial chromosome. Think of the plasmid as a ferry that carries the gene into the bacterial cell where it can be replicated. To incorporate the BGH gene into the plasmid, the plasmid is also cut with the same restriction enzyme used to cut the gene. Cutting both the plasmid and gene with the same enzyme allows the sticky ends that are generated to base-pair with each other (A to T and G to C). When the cut plasmid and the cut gene are placed together in a test tube, they re-form into a circular plasmid with the extra gene incorporated.

The bacterial plasmid has now been genetically engineered to carry a cow gene. At this juncture, the BGH gene is referred to as the rBGH gene, with the r indicating that this product is genetically engineered, or recombinant, because it has been removed from its original location in the cow genome and recombined with the plasmid DNA.

Step 3. Insert the Recombinant Plasmid into a Bacterial Cell. The recombinant plasmid is now inserted into a bacterial cell. Bacteria can be treated so that their cell membranes become porous. When they are placed into a suspension of plasmids, the bacterial cells allow the plasmids back into the cytoplasm of the cell. Once inside the cell, the plasmids replicate themselves, as does the bacterial cell, making thousands of copies of the *rBGH* gene. Using this procedure, scientists can grow large amounts of bacteria capable of producing BGH.

Once scientists successfully clone the BGH gene into bacterial cells, the bacteria produce the protein encoded by the gene. Then the scientists are able to break open the bacterial cells, isolate the BGH protein, and inject it into cows. Bacteria can be genetically engineered to produce many proteins of importance to humans. For example, bacteria are now used to produce the clotting protein missing from people with hemophilia as well as human insulin for people with diabetes.

Close to one-third of all dairy cows in the United States now undergo daily injections with recombinant bovine growth hormone. These injections increase the volume of milk that each cow produces by around 20%.

Prior to marketing the recombinant protein to dairy farmers, the Monsanto Company had to demonstrate that its product would not be harmful to cows or to humans who consume the cows' milk. This involved obtaining approval from the U.S. Food and Drug Administration (FDA).

The FDA is the governmental organization charged with ensuring the safety of all domestic and imported foods and food ingredients (except for meat and poultry, which are regulated by the U.S. Department of Agriculture). The manufacturer of any new food that is not **generally recognized as safe (GRAS)** must obtain FDA approval before marketing its product. Adding substances to foods also requires FDA approval unless the additive is GRAS.

According to both the FDA and Monsanto, there is no detectable difference between milk from treated and untreated cows and no way to distinguish between the two. Even if there were increased levels of rBGH in the milk of treated cows, there should be no effect on the humans consuming the milk because we drink the milk and do not inject it. Drinking the milk ensures that any protein in it will be digested by the body, just like any other protein that is present in food. Therefore, in 1993, the FDA deemed the milk from rBGH-treated cows as safe for human consumption.

Stop and Stretch 8.4: Not all anti-GM organism activists are convinced that milk produced by rBGH-treated cows is identical to that from nontreated cows. Given the effects of rBGH, how might the milk produced by these cows be different, even if it is safe?

In addition, since the milk from treated and untreated cows is indistinguishable, the FDA does not require that milk obtained from rBGH-treated cows be labeled in any manner. Vermont is the only state that requires labeling of rBGH-treated milk. However, many distributors of milk from untreated cows label their milk as "hormone free," even though there is no evidence of the hormone in milk from treated cows *(Figure 8.14)*.

But what about the welfare of the cows? There is some evidence that cows treated with rBGH are more susceptible to certain infections than untreated cows. Due in part to concerns about the health of these animals, Europe and Canada have banned rBGH use.

The rBGH story is a little different from that of genetically modified crop foods because rBGH protein is produced by bacteria and then administered to cows. When foods are genetically modified, the genome of the food itself is altered.

Figure 8.14 **Hormone-free milk.** Some manufacturers label their products as hormone free.

8.3 Genetically Modified Foods

Whether you realize it or not, you have probably been eating genetically modified foods for your entire life. Some of these modifications have occurred over the last several thousand years due to farmers' use of selective breeding techniques—breeding those cattle that produce the most milk or crossing crop plants that are easiest to harvest *(Figure 8.15)*. While this artificial selection does not involve moving a gene from one organism to another, it does change the overall frequency of certain alleles for a gene in the population.

The genetic engineering techniques described earlier have allowed scientists to move genes between organisms. Unless you eat only certified organic foods, you have been eating food that has been modified in this way for some time. This may lead you to wonder why and how plants are genetically modified, what impact eating them has on your health, and whether growing them affects the environment.

Why Genetically Modify Crop Plants?

Crop plants are genetically modified to increase their shelf life, yield, and nutritive value. For example, tomatoes have been engineered to soften and ripen more slowly. The longer ripening time meant that tomatoes would stay on the vine longer, thus making them taste better. The slower ripening also increased the amount of time that tomatoes could be left on grocery store shelves without becoming overripe and mushy.

Genetic engineering techniques increase crop yield when plants are engineered to be resistant to pesticides, herbicides, drought, and freezing. The nutritive value of crops can also be increased through genetic engineering. A gene that regulates the synthesis of β-carotene has been inserted into rice, a staple food for many of the world's people. Scientists hope that the engineered rice will help decrease the number of people who become blind in underdeveloped nations due to a deficiency of beta-carotene, a chemical necessary for the synthesis of vitamin A. Eating this genetically modified rice, called *golden rice*, increases a

Figure 8.15 **Artificial selection in corn.** Selective-breeding techniques resulted in the production of modern corn (right) from ancient teosinite corn.

Figure 8.16 **Golden rice.** Golden rice has been genetically engineered to produce beta-carotene, which causes the rice look more gold in color than the unmodified rice. This picture shows a mixture of modified and unmodified rice.

person's ability to synthesize vitamin A *(Figure 8.16)*. However, golden rice is not yet approved for human consumption, and there is debate about how effective the rice will actually be in preventing blindness.

Modifying Crop Plants with the Ti Plasmid and Gene Gun

To modify crop plants, the gene must be able to gain access to the plant cell, which means it must be able to move through the plant's rigid, outer cell wall. One "ferry" for moving genes into flowering plants is a naturally occurring plasmid of the bacterium *Agrobacterium tumefaciens*. In nature, this bacterium infects plants and causes tumors called **galls** *(Figure 8.17a)*. The tumors are induced by a plasmid, called **Ti plasmid** (Ti for tumor inducing).

Genes from different organisms can be inserted into the Ti plasmid by using the same restriction enzyme to cut the Ti plasmid and the gene, resulting in identical sticky ends and then connecting the gene and plasmid. The recombinant plasmid can then be inserted into a bacterium. The bacterium, *A. tumefaciens*, with the recombinant Ti plasmid, is then used to infect plant cells. During infection, the recombinant plasmid is transferred into the host-plant cell *(Figure 8.17b)*. For genetic engineering purposes, scientists use only the portion of a plasmid that does not cause tumor formation.

Moving genes into other agricultural crops such as corn, barley, and rice can also be accomplished by using a device called a **gene gun.** A gene gun shoots metal-coated pellets covered with foreign DNA into plant cells *(Figure 8.18)*. A small percentage of these genes may be incorporated into the plant's genome. When a gene from one organism is incorporated into the genome of another organism, a **transgenic organism** is produced. A transgenic organism is more commonly referred to as a **genetically modified organism (GMO).**

Effect of GMOs on Health

Concerns about the potential negative health effects of consuming GM crops have led some citizens to fight for legislation requiring that genetically modified foods be labeled, enabling consumers to make informed decisions about the foods they eat. Manufacturers of GM crops argue that labeling foods is expensive, and the labels will be viewed by consumers as a warning, even in the absence of any proven risk. Those manufacturers believe that GM food labeling will decrease sales and curtail further innovation.

Figure 8.17 **Genetically modifying plants using the Ti plasmid.** (a) Plants infected by *Agrobacterium tumefaciens* in nature show evidence of the infection by producing tumorous galls. (b) The Ti plasmid from *A. tumefaciens* serves as a shuttle for incorporating genes into plant cells. The recombinant plasmid is then used to infect developing plant cells, producing a genetically modified plant. When the plant cell reproduces, it may pass on the engineered gene to its offspring.

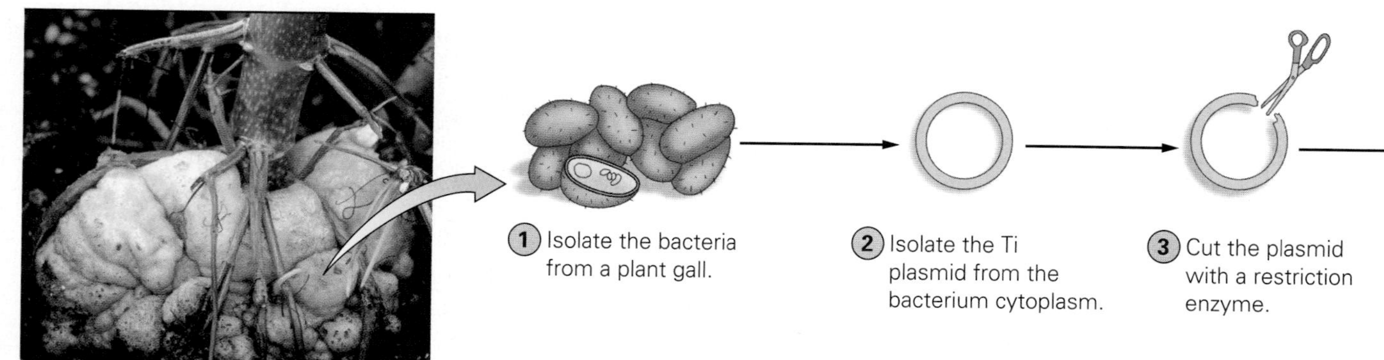

As the labeling controversy continues, most of us are already eating GM foods. Scientists estimate that over half of all foods in U.S. markets, including virtually all processed foods, contain at least small amounts of GM foods. Products that do not contain GMOs are often labeled to promote that fact. Concern about GM foods is not limited to their consumption. Many people are also concerned about the effects of GM crop plants on the environment.

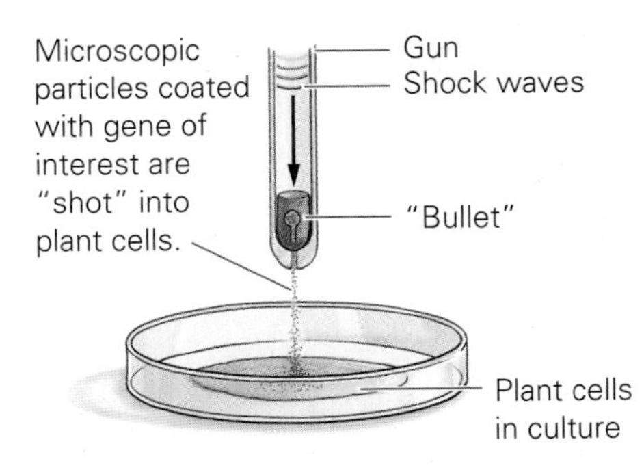

Figure 8.18 **Genetically modifying plants using a gene gun.** A gene gun shoots a plastic bullet loaded with tiny DNA-coated pellets into a plant cell. The bullet shells are prevented from leaving the gun, but the DNA-covered pellets penetrate the cell wall, cell membrane, and nuclear membrane of some cells.

GM Crops and the Environment

Many genetically modified crops have been engineered to increase their yield. For centuries, farmers have tried to increase yields by killing the pests that damage crops and by controlling the growth of weeds that compete for nutrients, rain, and sunlight. In the United States, farmers typically spray high volumes of chemical pesticides and herbicides directly onto their fields *(Figure 8.19)*. This practice concerns people worried about the health effects of eating foods that have been treated by these often toxic or cancer-causing chemicals. In addition, both pesticides and herbicides may leach through the soil and contaminate drinking water.

To help decrease farmers' reliance on pesticides, agribusiness companies have engineered plants that are genetically resistant to pests. Unfortunately, this is a short-term solution because the pest populations will eventually evolve resistance. Application of a pesticide does not always kill all of the targeted organisms. The few pests that have preexisting resistance genes and are not susceptible survive and produce resistant offspring. Eventually, widespread resistance develops, and a new pesticide must be developed and applied.

Figure 8.19 **Application of chemicals.** The application of herbicides and pesticides may decrease with the use of genetic technologies.

The continued need for the development of new pesticides in farming is paralleled by farmers' reliance on herbicides. Herbicide-resistant crop plants, such as Roundup Ready® soybeans, have been engineered to be resistant to Roundup®, or glyophosphate, a herbicide used to control weeds in soybean fields. Glyophosphate inhibits the synthesis of certain amino acids required by growing plants. Roundup Ready plants are engineered to synthesize more of these amino acids than unengineered plants and thus can grow in the presence of the herbicide while nearby plants and weeds will die. Therefore, farmers can spray their fields of genetically engineered soybeans with herbicides that will kill everything but the crop plant. Some people worry that this resistance gene will allow farmers to spray more herbicide on their crops since there is no chance of killing the GM plant, thereby exposing consumers and the environment to even more herbicide.

There is also concern that GM crop plants may transfer engineered genes from modified crop plants to their wild or weedy relatives. Wind, rain, birds, and bees carry genetically modified pollen to related plants near fields containing GM crops (or even to farms where no GM crops are being grown). Many cultivated crops have retained the ability to interbreed with their wild relatives; in these cases, genes from farm crops can mix with genes from the wild crops.

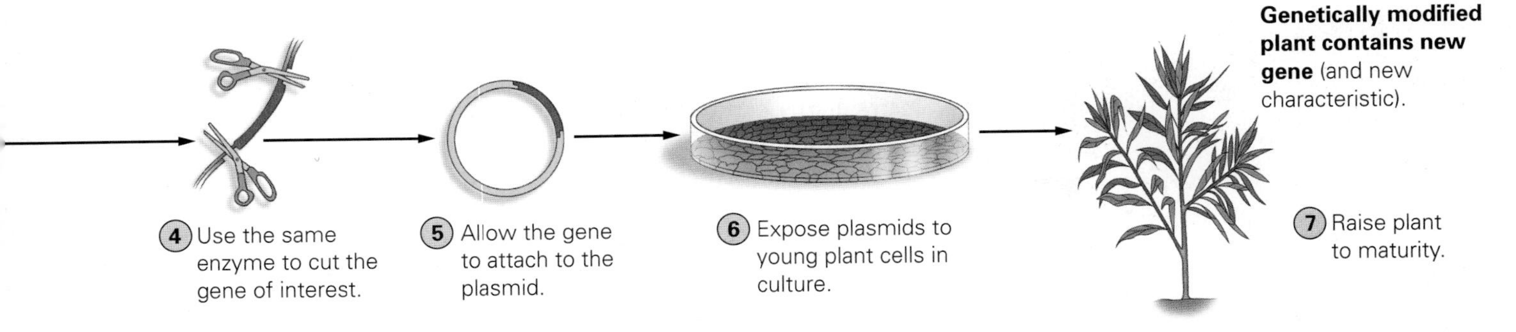

While there is hope that genetic engineers will be able to help solve hunger problems by making farming more productive, there are also concerns about any negative health and environmental effects of GM foods. It remains to be seen whether genetic engineering will constitute a lasting improvement to agriculture.

8.4 Genetically Modified Humans

Humans can also be genetically modified. Scientists may someday be able to cure or alleviate some of the symptoms of many different diseases. It may also be possible for physicians to diagnose genetic defects in early embryos and fix them, allowing the embryo to develop into an adult without any genetic diseases.

Stem Cells

Stem cells, unlike most of the cells in your body, do not perform a specific function; instead, they are able to produce many different kinds of cells and tissues. Because stem cells do not have a particular function, they are said to be **undifferentiated.** Although they are undifferentiated, they can be pressed into service as many different cell types.

Imagine that you are remodeling an old home, and you have a type of material that you can mold into anything you might need for the remodeling job—brick, tile, pipe, plaster, and so forth. Having a supply of this material would help you fix many different kinds of damage. Scientists believe that stem cells may serve as this type of all-purpose repair material in the body. If cells are nudged in a particular developmental direction in the laboratory, they can be directed to become a particular tissue or organ. Using stem cells from early embryos to produce healthy tissues as replacements for damaged tissues is called **therapeutic cloning.** Tissues and organs grown from stem cells in the laboratory may someday be used to replace organs damaged in accidents or organs that are gradually failing due to **degenerative diseases.** Degenerative diseases start with the slow breakdown of an organ and progress to organ failure. In addition, when one organ is not working properly, other organs are affected. Degenerative diseases include diabetes, liver and lung diseases, heart disease, multiple sclerosis, and Alzheimer's and Parkinson's diseases.

Stem cells could provide healthy tissue to replace those tissues damaged by spinal cord injury or burns. New heart muscle could be produced to replace muscle damaged during a heart attack. A diabetic could have a new pancreas, and people suffering from some types of arthritis could have replacement cartilage to cushion their joints. Thousands of people waiting for organ transplants might be saved if new organs were grown in the lab.

Stem cells are usually isolated from early embryos that are left over after fertility treatments. **In vitro** (Latin, meaning *in glass*) fertilization procedures often result in the production of excess embryos because many egg cells are harvested from a woman who wishes to become pregnant. These egg cells are then mixed with her partner's sperm in a petri dish, resulting in the production of many fertilized eggs that grow into embryos. A few of the embryos are then implanted into the woman's uterus. The remaining embryos are stored so that more attempts can be made if pregnancy does not result or if the couple desires more children. When the couple achieves the desired number of pregnancies, the remaining embryos are discarded or, with the couple's consent, used for stem cell research.

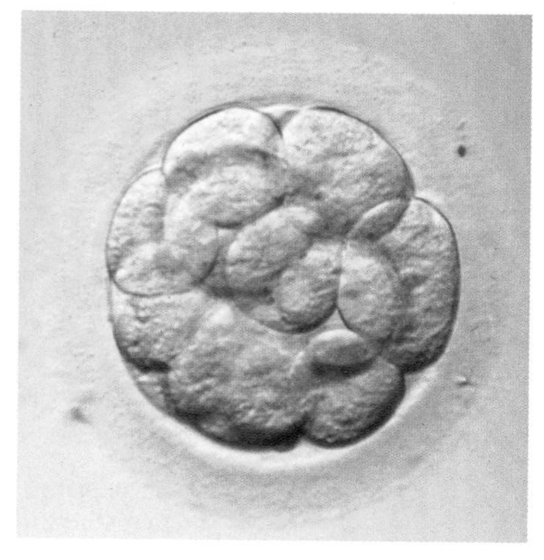

Figure 8.20 **Human embryo.** Stem cells can be obtained from an early embryo such as the one in this petri dish.

Early embryonic cells are harvested because stem cells are **totipotent** directly after fertilization; in other words, these stem cells can become any other cell type *(Figure 8.20)*. As the embryo develops, its cells become less and less able to produce other cell types. As a human embryo grows, the early cells start dividing and forming different, specialized cells such as heart cells, bone cells, and muscle cells. Once formed, specialized non-stem cells can divide only to produce replicas of themselves. They cannot backtrack and become a different type of cell.

Stop and Stretch 8.5: For parents known to carry a harmful genetic mutation, fertility clinics will take a cell from an 8-cell preembryo to test it for the presence of the defective gene. Taking a single cell causes no apparent harm to the developing structures. What does this tell you about the capabilities of preembryonic cells?

Laws in the United States restrict government funding for embryonic stem cell research. Embryos are destroyed when stem cells are harvested from them—a result that many find objectionable. In 2001, President Bush signed an executive order to ban scientists from using government money for studies involving human embryonic stem cells unless those cells were created before the 2001 ban.

There is, however, a small supply of stem cells present in adult tissues, probably so that these tissues can repair themselves. Scientists have found adult stem cells in many different tissues, and work is under way to determine whether these cells can be used for transplants and treatment of degenerative diseases.

Genetic modifications may one day include replacing defective or nonfunctional alleles of a gene with a functional copy of the gene. This requires detailed information about the human genome, much of which has been obtained.

The Human Genome Project

The **Human Genome Project** involves determining the nucleotide-base sequence (A, C, G, or T) of the entire human genome and the location of each of the 20,000 to 25,000 human genes.

The scientists involved in this multinational effort also sequenced the genomes of the mouse, the fruit fly, a roundworm, baker's yeast, and a common intestinal bacterium named *E. coli*. Scientists thought it was important to sequence the genomes of organisms other than humans because these **model organisms** are easy to manipulate in genetic studies and because important genes are often found in many different organisms. In fact, 90% of human genes are also present in mice; 50% are in fruit flies, and 31% are in baker's yeast. Therefore, understanding how a certain gene functions in a model organism helps us understand how the same gene functions in humans.

To sequence the human genome, scientists first isolated DNA from white blood cells. They then cleaved the chromosomes into more manageable sizes using restriction enzymes, cloned them into plasmids, and determined the base sequence using automated DNA sequencers. These sequencing machines distinguish between nucleotides based on structural differences in the nitrogenous bases. Sequence information was then uploaded to the Internet. Scientists working on this, or any other project, could search for regions of sequence information that overlapped with known sequences, a process called chromosome walking *(Figure 8.21)*. Using overlapping regions, scientists in laboratories all over the world worked together to patch together DNA sequence information. DNA sequence information obtained by means of the Human Genome Project may someday enable medical doctors to take blood samples from patients and determine which genetic diseases are likely to affect them.

Many people worry about having these types of tests performed because this personal information may get back to their insurance companies or employers, but there is a positive side to having all of this information available. Once the genetic basis of a disease has been worked out—that is, how the gene of a healthy person differs from the gene of a person with a genetic disease—the information can be used to develop treatments or cures.

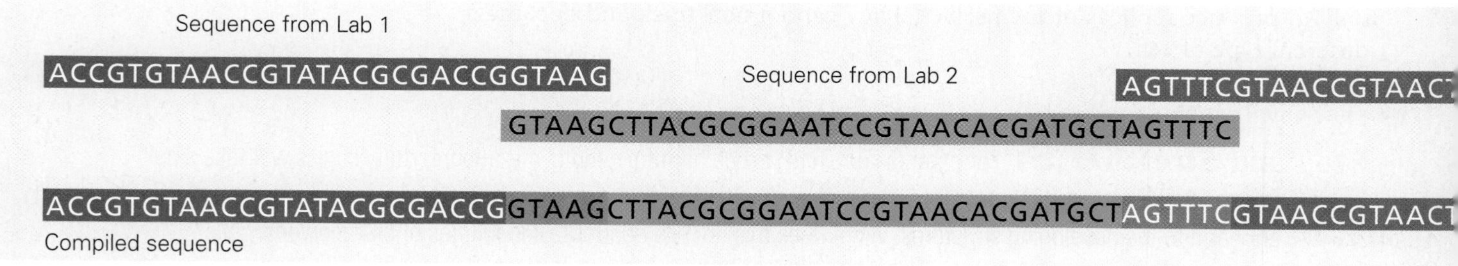

Sequence from Lab 1

ACCGTGTAACCGTATACGCGACCGGTAAG

Sequence from Lab 2

AGTTTCGTAACCGTAAC

GTAAGCTTACGCGGAATCCGTAACACGATGCTAGTTTC

ACCGTGTAACCGTATACGCGACCGGTAAGCTTACGCGGAATCCGTAACACGATGCTAGTTTCGTAACCGTAACT

Compiled sequence

Figure 8.21 **Chromosome walking.** Scientists from many different labs worked together to sequence the human genome. Sequence information is uploaded to a common database, and scientists search for overlapping regions to fill in gaps in the sequence, much like assembling a jigsaw puzzle.

Gene Therapy

Scientists who try to replace defective human genes (or their protein products) with functional genes are performing **gene therapy.** Gene therapy may someday enable scientists to fix genetic diseases in an embryo. To do so, the scientists would supply the embryo with a normal version of a defective gene; this so-called **germ-line gene therapy** would ensure that the embryo and any cells produced by cell division would replicate the new, functional version of the gene. Thus, most of the cells would have the corrected version of the gene, and when these genetically modified individuals have children, they will pass on the corrected version of the gene. If scientists can fix genetic defects in early embryos, some genetic diseases can be prevented.

Another type of gene therapy, called **somatic cell gene therapy,** can be performed on body cells to fix or replace the defective protein in only the affected cells. Using this method, scientists introduce a functional version of a defective gene into an affected individual cell in the laboratory, allow the cell to reproduce, and then place the copies of the cell bearing the corrected gene into the diseased person.

This treatment may seem like science fiction, but it is likely that this method of treating genetic diseases will be considered a normal procedure in the not-too-distant future. In fact, genetic engineers already have successfully treated a genetic disorder called **severe combined immunodeficiency (SCID),** a disease caused by a genetic mutation that results in the absence of an important enzyme and severely weakens the individual's immune system. Because their immune systems are compromised, people with SCID are incapable of fighting off any infection, and they often suffer severe brain damage from the high temperatures associated with unabated infection. Any exposure to infection can kill or disable someone with SCID, so most patients must stay inside their homes and often live inside protective bubbles that separate them from everyone, even family members.

To devise a successful treatment for SCID, or any disease treated with gene therapy, scientists had to overcome a major obstacle—getting the therapeutic gene to the right place.

Proteins break down easily and are difficult to deliver to the proper cells, so it is more effective to replace a defective gene than to continually replace a defective protein. Delivering a normal copy of a defective gene only to the cell type that requires it is a difficult task. SCID, a disorder that has been treated successfully, was chosen by early gene therapists in part because defective immune system cells could be removed from the body, treated, and returned to the body.

Immune system cells that require the enzyme missing in SCID patients circulate in the bloodstream. Blood removed from a child with SCID is infected with nonpathogenic (non-disease-causing) versions of a virus. This virus is first engineered to carry a normal copy of the defective gene in SCID patients. After the immune system cells are infected with the virus, these recombinant cells, which now bear copies of the functional gene, are returned to the SCID patient *(Figure 8.22a)*.

Sequence from Lab 3

CCGTTACGGATATGCTTACTGTAC

Sequence from Lab 4

TATGCTTACTGTACCTTCAAAACTGACTTTGGCTAACCGTACTCTGTACCTTCAAAACTGACTTTT

CCGTTACGGATATGCTTACTGTACCTTCAAAACTGACTTTGGCTAACCGTACTCTGTACCTTCAAAACTGACTTTT

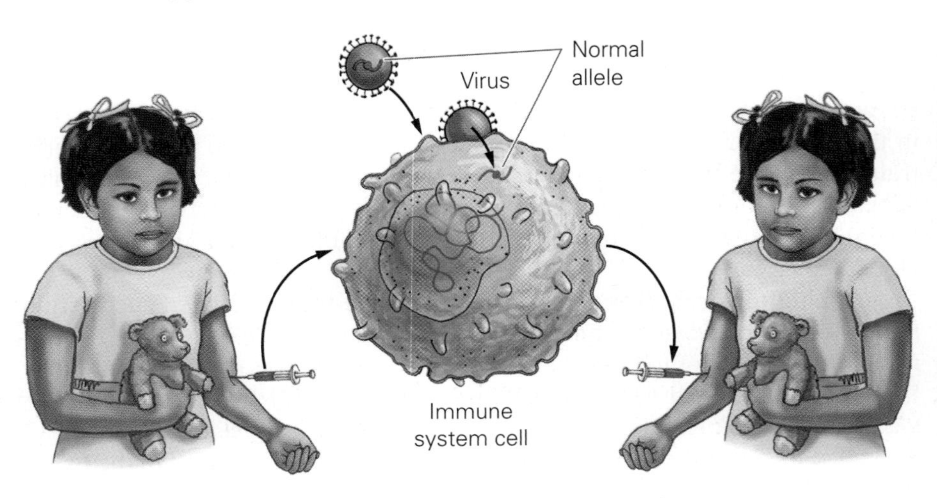

Figure 8.22 **Gene therapy in a SCID patient.** (a) A virus carrying the normal gene is allowed to infect immune system cells that have been removed from a person with SCID. The virus inserts the normal copy of the gene into some of the cells, and these cells are then injected into the SCID patient. (b) Ashi DiSilva, the first gene therapy patient.

In 1990, a 4-year-old girl named Ashi DiSilva *(Figure 8.22b)* was the first patient to receive gene therapy for SCID. Ashi's parents were willing to face the unknown risks to their daughter because they were already far too familiar with the risks of SCID—the couple's two other children also had SCID and were severely disabled. Ashi is now a healthy adult with an immune system that is able to fight off most infections.

However, Ashi must continue to receive treatments because blood cells, whether genetically engineered or not, have limited life spans. When most of Ashi's engineered blood cells have broken down, she must be treated again; thus, she undergoes this gene therapy a few times each year. Since Ashi's gene therapy turned out well, many other SCID patients have been successfully treated and can live normal lives. Unfortunately, Ashi's gene therapy does not prevent her from passing on the defective allele to her biological children because this therapy is not "fixing" the allele in her ovaries.

Although things worked out well for Ashi, successful gene therapy is far from routine. Two of 11 French boys treated with gene therapy for SCID developed leukemia that is thought to be related to their treatment, and an American teenager died from complications of experimental gene therapy meant to cure his relatively mild genetic disorder.

In addition to the risks involved in conducting any experimental therapy, not many genetic diseases can be treated with gene therapy. Gene therapy to date has focused on diseases caused by single genes for which defective cells can be removed from the body, treated, and reintroduced to the body. Most genetic diseases are caused by many genes, affect cells that cannot be removed and replaced, and are influenced by the environment.

Most people support the research of genetic engineers in their attempts to find better methods for delivering gene sequences to the required locations and for regulating the genes once they are in place. A far more controversial type of genetic engineering involves making an exact copy of an entire organism by a process called reproductive **cloning.**

Cloning Humans

Human cloning occurs commonly in nature via the spontaneous production of identical twins. These clones arise when an embryo subdivides itself into 2 separate embryos early in development. This is not the type of cloning that many people find objectionable; people are more likely to be upset by cloning that involves selecting which traits an individual will possess. Natural cloning of an early embryo to make identical twins does not allow any more selection for specific traits than does fertilization. However, in the future it may be possible to select adult humans who possess desired traits and clone them. Since cloning does not actually alter an individual's genes, it is more of a reproductive technology than a genetic engineering technology. However, it may someday be possible to alter the genes of a cloned embryo.

Cloning offspring from adults with desirable traits has been successfully performed on cattle, goats, mice, cats, pigs, rabbits, and sheep. In fact, the animal that brought cloning to the attention of the public was a ewe named Dolly.

Dolly was cloned when Scottish scientists took cells from the mammary gland of an adult female sheep and fused it with an egg cell that had previously had its nucleus removed. Treated egg cells were then placed in the uterus of an adult ewe that had been hormonally treated to support a pregnancy. Scientists had to try many times before this **nuclear transfer** technique worked. In all, 277 embryos were constructed before one was able to develop into a live lamb *(Figure 8.23)*. Dolly was born in 1997.

The research that led to Dolly's birth was designed to provide a method of ensuring that cloned livestock would have the genetic traits that made them most beneficial to farmers. Sheep that produced the most high-quality wool and cattle that produced the best beef would be cloned. This technique is more efficient than allowing two prize animals to breed because each animal gives only half of its genes to the offspring. There is no guarantee that the offspring of 2 prize animals will have the desired traits. Even when a genetic clone is produced, there is no guarantee that the clone produced will be identical in the appearance and behavior to the original because of environmental differences.

No one knows if nuclear transfer will work in humans—or if cloning is safe. If Dolly is a representative example, cloning animals may not be safe. Dolly was euthanized at age six to relieve her from the discomfort of arthritis and a progressive lung disease, conditions usually found only in older sheep. The fact that Dolly developed these conditions has led scientists to believe that she had aged prematurely.

Stop and Stretch 8.6: Devise a hypothesis about why Dolly may have aged more rapidly than a sheep that was produced naturally. How could you test this hypothesis?

The debate about human cloning mimics the larger debate about genetic engineering *(Table 8.2)*. As a society, we need to determine whether the potential for good outweighs the potential harm for each application of these technologies.

Figure 8.23 **Dolly.** This sheep is the result of a successful attempt to clone a mammal. She was genetically identical to her cell-donor mother.

TABLE 8.2 **Some pros and cons of genetic engineering.**

Why the work of genetic engineers is important
• GM crops may make farms more productive. • GM crops may be made to taste better. • GM crops can be made to have longer shelf lives. • GM crops can me made to contain more nutrients. • Genetic engineers hope to cure diseases and save lives.
Why the work of genetic engineers is controversial
• GM crops encourage agribusiness, which may close down some small farms. • GM animals and crops may cause health problems in consumers. • GM crops might have unexpected adverse effects on the environment. • Lack of genetic diversity of GM crops could lead to destruction of food supply worldwide by pest or environmental change. • Present research might lead to the unethical genetic modification of humans.

Savvy Reader Political Cartoons

Both of the above political cartoons are about stem cells.

1 What clues point to the position each of the cartoonists takes on the issue of stem cell research ?
2 Can we trust that cartoons are scientifically accurate?
3 Is the issue of whether or not an embryo is a human one that science can answer?

CHAPTER Review

For study help, animations, and more quiz questions go to www.mybiology.com.

Summary

8.1 Protein Synthesis and Gene Expression

- Genes carry instructions for synthesizing proteins (p. 198).
- Protein synthesis involves the processes of transcription and translation (pp. 198–200).
- Transcription occurs in the nucleus of eukaryotic cells when an RNA polymerase enzyme binds to the promoter, located at the start site of a gene, and makes an mRNA that is complementary to the DNA gene. RNA differs from DNA in that the sugar in RNA is ribose (not deoxyribose) and the nitrogenous bases are adenine, guanine, cytosine, and uracil (no thymines in RNA) (p. 200).
- Translation occurs in the cytoplasm of eukaryotic cells and involves mRNA, ribosomes, and tRNA. Messenger RNA carries the code from the DNA, and ribosomes are the site where amino acids are assembled to synthesize proteins. Transfer RNA (tRNA) carries amino acids, which bind to triplet nucleotide sequences on the mRNA called codons (p. 200).
- A particular tRNA carries a specific amino acid. Each tRNA has its unique anticodon that binds to the codon and carries instructions for its particular amino acid (p. 200).
- The amino acid coded for by a particular codon can be determined using the genetic code (p. 203).
- The flow of genetic information is from the DNA sequence to the mRNA transcript to the encoded protein (p. 203).
- Mutations are changes to DNA sequences that can affect protein structure and function. Neutral

mutations are changes to the DNA that do not result in a different amino acid being incorporated. Insertions or deletions of nucleotides can result in frameshift mutations that change the protein drastically (pp. 204–205).

- A given cell type expresses only a small percentage of the genes that an organism possesses (p. 206).
- Turning the expression of a gene up or down is accomplished in different ways in prokaryotes and eukaryotes. Prokaryotes typically block the promoter with a repressor protein to keep gene expression turned off. Eukaryotes regulate gene expression in any of 5 ways: (1) increasing transcription through the use of proteins that stimulate RNA polymerase binding; (2) varying the time that DNA spends in the uncondensed, active form; (3) altering the mRNA life span; (4) slowing down or speeding up translation; and (5) affecting the protein life span (pp. 206–208).

web animation 8.1 *Transcription*

web animation 8.2 *Translation*

8.2 Producing Recombinant Proteins

- Bovine growth hormone is a protein produced by the pituitary glands of cows. To increase the quantity of milk that a cow produces, additional growth hormone is injected into the cow (p. 208).
- Modern genetic engineering techniques enable scientists to produce recombinant BGH in the lab by placing the gene for growth hormone into plasmids, which then clone the gene by making millions of copies of it as they replicate themselves inside their bacterial hosts. Bacteria can then express the gene by transcribing an mRNA copy and translating the mRNA into a protein. The recombinant bovine growth hormone is then isolated and injected into cows (pp. 208–210).
- FDA approval is required for any new food or additive that is not generally recognized as safe (p. 210).

web animation 8.3 *Producing Bovine Growth Hormone*

BioFlix™ **Protein Synthesis**

8.3 Genetically Modified Foods

- Genetic engineering techniques allow for the modification of foods much more quickly than do selective breeding techniques (p. 211).
- Crop plants are genetically modified to increase their shelf life, yield, and nutritive value (p. 211).
- A Ti plasmid or gene gun can be used to insert a particular gene into plant cells (pp. 212–213).
- Although there have been no documented incidents of negative health effects from GM food consumption, there is concern that GM foods may be unhealthy (pp. 212–213).
- Concerns about GM crops include their impacts on surrounding organisms, the evolution of resistances, and transfer of modified genes to wild and weedy relatives (pp. 213–214).

Summary (continued)

8.4 Genetically Modified Humans

- Stem cells are undifferentiated cells that can be reprogrammed to act as a variety of cell types. Stem cells may someday allow scientist to treat degenerative diseases. (pp. 214–215)
- Information about genes obtained from the Human Genome Project can be used to help scientists replace genes that are defective or missing in people with genetic diseases (pp. 215–216).
- Gene therapy involves replacing defective genes or their products in an embryo or in the affected adult tissue (p. 216).
- Gene therapy is considered experimental but may hold promise once scientists determine how to target genes to the right locations and express them in the proper amounts (pp. 216–217).
- Cloning animals with desirable agricultural traits has occurred. It may someday be possible to clone humans, but it is unclear if these humans would be healthy (pp. 218–219).

Roots to Remember

The following roots of words come mainly from Latin and Greek and will help you decipher terms:

chromo- means color.
-ic is a common ending of acids.
nucleo- or **nucl-** refer to a nucleus.
-ose is a common ending for sugars, such as ribose, sucrose, and fructose.
re-: a reminder that "re-" means again.
-some or **-somal** relate to a body, whether the whole body (somatic) or a small body (ribosome).

Learning the Basics

1. List the order of nucleotides on the mRNA that would be transcribed from the following DNA sequence: CGATTACTTA

2. Using the genetic code (Table 8.1 on p. 203), list the order of amino acids encoded by the following mRNA nucleotides: CAACGCAUUUUG.

3. List the subcellular structures that participate in translation.

4. Transcription ______________.
A. synthesizes new daughter DNA molecules from an existing DNA molecule; **B.** results in the synthesis of an RNA copy of a gene; **C.** pairs thymines (T) with adenines (A); **D.** occurs on ribosomes

5. Transfer RNA (tRNA) ______________.
A. carries monosaccharides to the ribosome for synthesis; **B.** is made of messenger RNA; **C.** has an anticodon region, which is complementary to the mRNA codon; **D.** is the site of protein synthesis

6. During the process of transcription, ______________.
A. DNA serves as a template for the synthesis of more DNA; **B.** DNA serves as a template for the synthesis of RNA; **C.** DNA serves as a template for the synthesis of proteins; **D.** RNA serves as a template for the synthesis of proteins

7. Translation results in the production of ______________.
A. RNA; **B.** DNA; **C.** protein; **D.** individual amino acids; **E.** transfer RNA molecules

8. The RNA polymerase enzyme binds to ______________, initiating transcription.
A. amino acids; **B.** tRNA; **C.** the promoter sequence; **D.** the ribosome

9. A particular triplet of bases in the coding sequence of DNA is TGA. The anticodon on the tRNA that binds to the mRNA codon is ______________.
A. TGA; **B.** UGA; **C.** UCU; **D.** ACU

10. RNA and DNA are similar because ______________.
A. they are both double-stranded helices; **B.** uracil is found in both of them; **C.** both contain the sugar deoxyribose; **D.** both are made up of nucleotides consisting of a sugar, a phosphate, and a base

Analyzing and Applying the Basics

1. Take another look at the genetic code (Table 8.1, p. 203). Do you see any similarities between codons that code for the same amino acid? Based on this difference, why might a mutation that affects the nucleotide in the third position of the codon be less likely to affect the structure of the protein than a mutation that affects the codon in the first position?

2. Genes encode RNA polymerase molecules. What would happen to a cell that has undergone a mutation to its RNA polymerase gene?

3. Draw a box around the 6-base-pair site at which a restriction enzyme would most likely be cut:
ATGAATTCCGTCCG
TACTTAAGGCAGGC

Connecting the Science

1. The first "test-tube baby," Louise Brown, was born over 30 years ago. Sperm from her father was combined with an egg cell from her mother. The fertilized egg cell was then placed into the mother's uterus for the period of gestation. At the time of Louise's conception, many people were very concerned about the ethics of scientists performing these in vitro fertilizations. Do you think human cloning will eventually be as commonplace as in vitro fertilizations are now? Why or why not?

2. Do you think it is acceptable to grow genetically modified foods if health risks turn out to be low but environmental effects are high?

CHAPTER 9

Where Did We Come From?

The Evidence for Evolution

Should public school students be taught about alternative hypotheses to evolution?

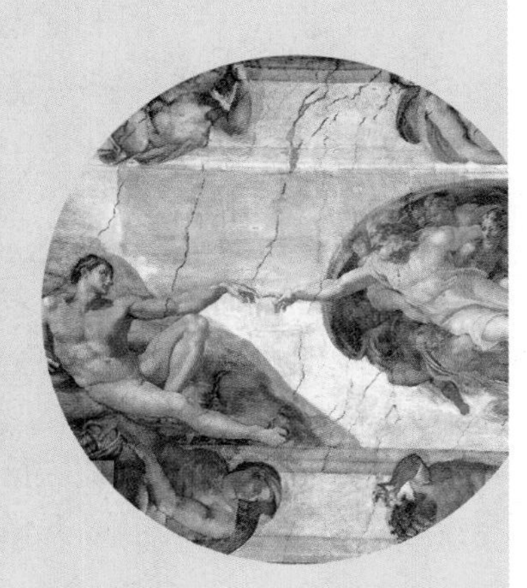

One idea about the origin of humans: special creation.

Another idea: evolution.

Why do biologists insist that only evolution be taught?

In Dover, Pennsylvania, sometime in August 2003, at least 2 people presided over a strange, quiet ceremony. A small bonfire was lit and was fed by a mural that had been hanging in Dover High School's science classroom for 5 years, ever since a former student had painted it. The mural's subject: human evolution. The participants in the ceremony: the head of maintenance for the school district and at least one member of board of education. While this event was upsetting to many members of the Dover community, it passed from the headlines quickly. Today it stands out as the initial incident in a series of events that threatened to tear this tiny, peaceful community apart.

By October 2004, the views of the majority of Dover's school board members were made clear to everyone, as they agreed to add the following statement to their biology curriculum:

> Students will be made aware of the gaps/problems in Darwin's theory and of other theories of evolution including, but not limited to, intelligent design. Note: Origins of life is not taught.

Later that fall, the school board directed the science faculty at Dover High School to read a statement to their students encouraging them to "keep an open mind." The statement also emphasized that although the state requires students to take a standardized test that includes evolution, Dover students should know that intelligent design is a viable alternative theory.

By December, 11 parents had filed suit against the Dover district, claiming that the statement was harming their children's right to a public education free from religious indoctrination. The parents made this claim because the theory of intelligent design claims that certain features of the natural world are too complex to have evolved and must have been designed—while the designer is typically unspecified, intelligent design advocates are generally referring to God. In apparent support of the parents' position, the school board received free legal representation in defense of this suit from a conservative Christian law center with the stated purpose of being the "sword and shield for people of faith."

The case became national news, debated by talking heads on cable channels around the country. In Dover, friends and families were bitterly divided by the issue; children of the plaintiffs were teased as "monkey girls" and school board members derided as clueless rubes. By November of the following year, both sides were facing off in the courtroom of U.S. Federal Judge John E. Jones III.

Did the school board actions in Dover represent a revolution in science education—the recognition that Darwin's ideas are as much a matter of faith as religious doctrine? Or does it represent a dangerous trend in U.S. education—one that sets religious belief on equal footing with scientific understanding? In this chapter, we examine these questions by exploring the theory of evolution and the origin of humans as a matter of science.

web animation 9.1
Principles of Evolution

9.1 What Is Evolution?

Evolution really has two different meanings to biologists. The term can refer to either a process or an organizing principle, that is, a theory.

The Process of Evolution

Generally, the word evolution means "change," and the process of biological evolution reflects this definition. **Biological evolution** is a change in the characteristics of a population of organisms that occurs over the course of generations. The changes in populations that are considered evolutionary are those that are inherited via genes. Other changes that may take place in populations due to environmental change are not evolutionary. For instance, the average dress size for women in the United States has increased from 8 to 14 over the past 50 years because of an increase in average calorie intake and average age, not because of a change in genes.

As an example of the process of biological evolution, consider the species of organism commonly known as head lice. A **species** consists of a group of individuals that can regularly breed together, producing fertile offspring, and that is generally distinct from other species in appearance or behavior. Most species are subdivided by geography into smaller groups, or **biological populations**, that are somewhat independent of other populations.

Some populations of head lice in the United States have become resistant to the pesticide permethrin, found in over-the-counter delousing shampoos. Initially, lice infections were readily controlled through treatment with these products; however, over time, populations of lice changed to become less susceptible to the effects of these chemicals. The evolution of resistance can occur rapidly—a study in Israel demonstrated that populations of head lice were four times less susceptible to permethrin only 30 months (or 40 lice generations) after the pesticide was introduced in that country.

Note that in this example, individual head lice did not "evolve" or change; instead, the population as a whole changed from one in which most lice were susceptible to one in which most lice were resistant to the pesticide. Individual lice were either naturally susceptible or naturally resistant to permethrin.

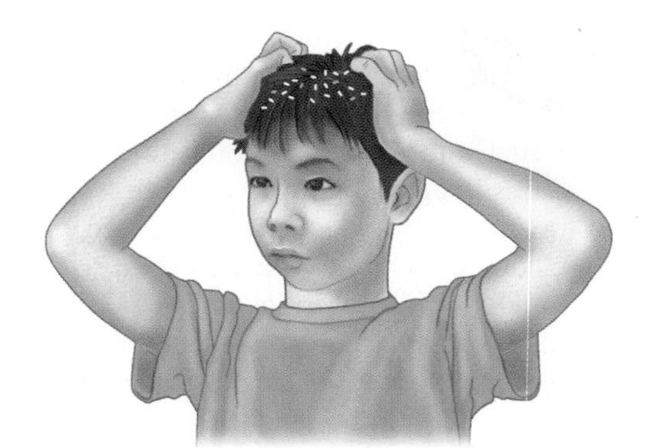

Initial lice infestation consists of both susceptible and resistant lice.

After permethrin treatment, most lice are dead, but a few that are resistant to the pesticide survive.

Reinfestation with the offspring of the resistant lice. The population of lice is now more resistant to permethrin.

Figure 9.1 **The process of evolution.** The evolution of pesticide resistance in lice occurs as a result of natural selection, one of the mechanisms by which traits in a population can change over time. **Visualize This:** How would this figure be different if no lice were resistant to pesticide? Could evolution occur?

The differences in resistance to permethrin among individuals in the population resulted from genetic variation. In particular, some lice carried gene variants (that is, alleles) that conferred resistance, and others carried nonresistant alleles. Because individuals with the resistant alleles survived permethrin treatment, they passed these resistant alleles on to their offspring. As a result, a population made up primarily of individual lice that carried the susceptible alleles changed into one in which most of the lice carried the resistant alleles. This change in the characteristics of the population took many generations *(Figure 9.1)*.

According to the definition of evolution presented earlier, the population of lice has evolved. In this case, the process of **natural selection**, the differential survival and reproduction of individuals in a population, brought about the evolutionary change. Natural selection is the process by which populations adapt to their changing environment. We will discuss natural selection in more detail in Chapter 10. Other forces, including chance, can cause evolutionary changes in the genetic makeup of populations as well. We discuss those forces in Chapter 11.

Most people accept that traits in populations can evolve. Evolutionary change in biological populations, such as the development of pesticide resistance in insects and antibiotic resistance in bacteria (discussed in Chapter 10), has been observed multiple times. Changes that occur within a species in the characteristics of populations are referred to as **microevolution**. The controversy about teaching evolution instead comes from discussions of macroevolution, the changes that occur as a result of microevolution over long periods of time and result in the origin of new species.

Stop and Stretch 9.1: Not every change in the characteristics of a population over time is a result of microevolution. For instance, in the past 50 years, the average woman's dress size in the United States has increased from 8 to 14. How is this change different from a microevolutionary change?

The Theory of Evolution

Although some members or supporters of the Dover school board disagree that the process of microevolution occurs, theirs is a minority view. Many of the board members, as well as many people in and out of Dover, dispute whether the theory of evolution, which includes both the process of microevolution and its macroevolutionary results, is "scientific truth."

Some of the controversy is generated by the use of the word *theory*. When people use this word in everyday conversation, they often are referring to a tentative explanation with little systematic support. A sports fan might have

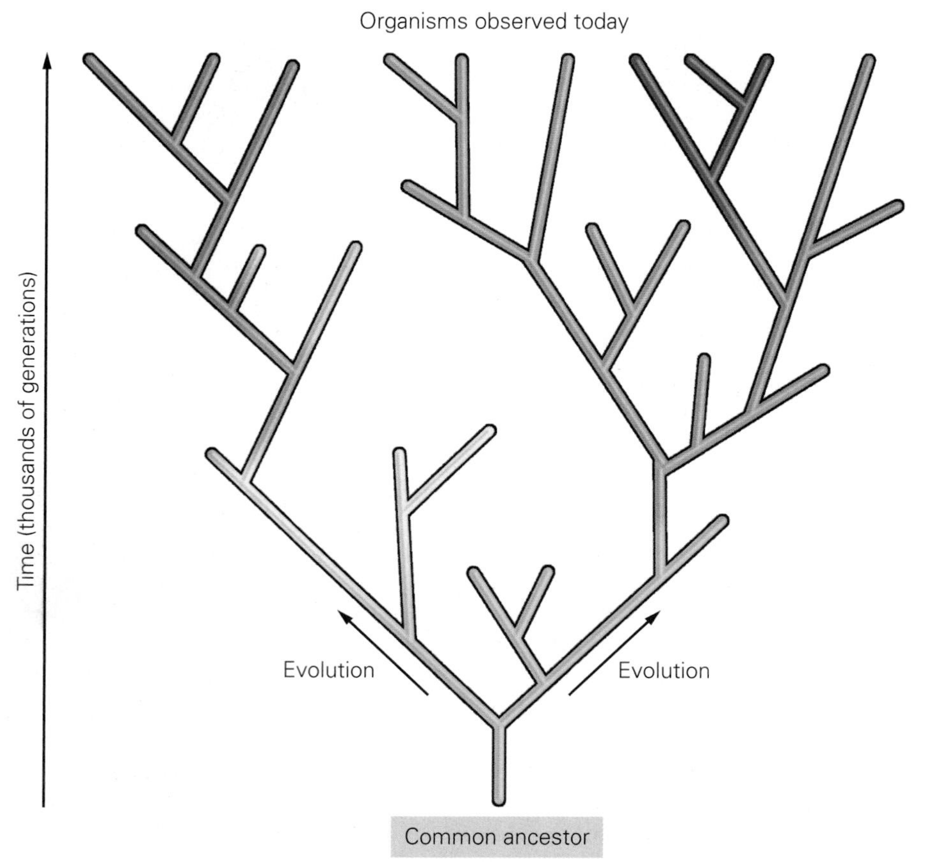

Figure 9.2 **The theory of common descent.** This theory states that all modern organisms descended from a single common ancestor. Each branching point on the tree represents the origin of new species from an ancestral form.

a theory about why her team is losing, or a gardener might have a theory about why his roses fail to bloom. Usually, these ideas amount to a "best guess" regarding the cause of some phenomenon. As first described in Chapter 1, a **scientific theory** is much more substantial—it is a body of scientifically acceptable general principles that help explain how the universe works. Scientific theories are supported by numerous lines of evidence and have withstood repeated experimental tests. For instance, the theory of gravity explains the motion of the planets, and germ theory explains the relationship between microorganisms and human disease.

The theory of evolution is a principle for understanding how species originate and why they have the characteristics that they exhibit. The **theory of evolution** can thus be stated:

> All species present on Earth today are descendants of a single common ancestor, and all species represent the product of millions of years of accumulated microevolutionary changes.

In other words, modern animals, plants, fungi, bacteria, and other living things are related to each other and have been diverging from their common ancestor, by various processes, since the origin of life on this planet.

Most nonscientists think only of the first phrase in this statement of theory—that modern species originate from a single common ancestor—when they hear the word *evolution.* This part of the theory of evolution, called the theory of **common descent**, is illustrated in *Figure 9.2* and underlies the opposition to what is often termed "Darwinism."

9.2 Charles Darwin and the Theory of Evolution

The theory of evolution is sometimes called "Darwinism" because Charles Darwin is credited with bringing it into the mainstream of modern science *(Figure 9.3)*.

The youngest son of a wealthy physician, Darwin spent much of his early life as a lackluster student. After dropping out of medical school, Darwin entered Cambridge University to study for the ministry at the urging of his father. Darwin hated most of his classes but did strike up friendships with several scientists at the college. One of his closest companions was Professor John Henslow, an influential botanist who nurtured Darwin's deep curiosity about the natural world. It was Henslow who secured Darwin his first "job" after graduation. In 1831, at age 22, Darwin set out on what would become his life-defining journey—the voyage of the *HMS Beagle.*

The *Beagle*'s mission was to chart the coasts and harbors of South America. The ship's personnel was to include a naturalist for "collecting, observing, and noting anything worthy to be noted in natural history." Henslow recommended Darwin to be the unpaid assistant naturalist (and socially appropriate dinner companion) to the *Beagle*'s aristocratic captain after two other candidates had turned it down.

Early Views of Evolution

The hypothesis that organisms, and even the very rock of Earth, had changed over time was not new when Darwin embarked on his voyage. The Greek poet Anaximander (611–546 B.C.) seems to have been the first Western philosopher to postulate that humans evolved from fish that had moved onto land. Darwin was also familiar with the work of the eighteenth-century geologist James Hutton and his own contemporary, Charles Lyell, who argued that geological features such as gorges and mountains were the result of slow, gradual change over the course of eons.

The first modern evolutionist, Jean Baptiste Lamarck, had published his ideas about evolution in 1809, the year of Darwin's birth. Lamarck was the first scientist to clearly state that organisms adapted to their environments. He proposed that all individuals of every species had an innate, inner drive for perfection, and that the traits they developed over their lifetimes could be passed on to their offspring. Lamarck used this principle in an attempt to explain the long legs of wading birds, which he argued arose when those animals' ancestors attempted to catch more distant fish. As they waded in deeper water, they would stretch their legs to their full extent, making them slightly longer in the process. These longer legs would be passed on to the next generation, who would in turn stretch their legs and pass on even longer legs to the next generation.

Figure 9.3 **Charles Darwin.** As a young naturalist, Darwin conceived of and developed a theory of evolution that remains 1 of the most well-supported ideas in biology.

Lamarck's contemporaries were unconvinced by his proposed mechanism for species change—for instance, it was easily seen that the children of muscular blacksmiths had similar-sized biceps to those of bankers' children and had not inherited their father's highly developed muscles. Lamarck's critics were also unwilling to question the more socially acceptable alternative hypothesis that Earth and its organisms were created in their current forms by God and did not change over time. It was this more acceptable hypothesis that Darwin found himself questioning on his around-the-world voyage.

The Voyage of the *Beagle*

In the 5 years that he spent on the expedition of the *HMS Beagle,* Darwin spent most of his time on land, luckily for him as it turns out since he never became accustomed to the ocean's swells and was nearly constantly seasick on board the ship *(Figure 9.4)*. The trip was a dramatic awakening for the young man, who was awed at the sight of the Brazilian rain forest, amazed by the scantily clothed natives in the chilly climate of Tierra del Fuego, and intrigued by the diversity of animals and plants he collected.

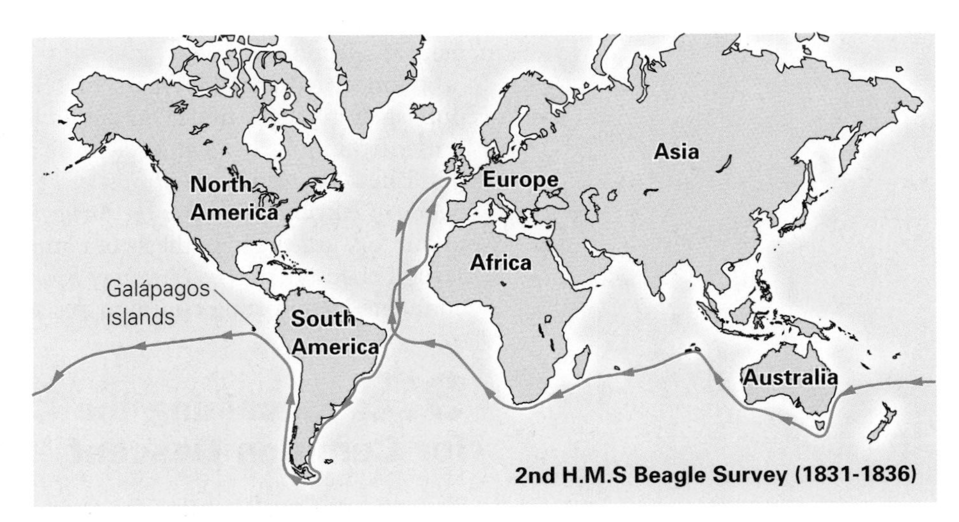

Figure 9.4 **The voyage of *HMS Beagle.*** Darwin's expedition took him to tropical locales from South America to Tahiti.

(a) Dome-shelled tortoise from Santa Cruz Island, an island with abundant vegetation

(b) Flat-shelled tortoise from Española Island, an island with sparse vegetation

Figure 9.5 **Giant tortoises of the Galápagos.** The subspecies of giant tortoises on the Galápagos Islands from different environments look distinct from each other.

On the ship, Darwin had ample time to read, including Charles Lyell's book *Principles of Geology,* which put forth the hypothesis that Earth's physical structure was constantly changing. In supporting this idea, Lyell argued that deep canyons resulted from the gradual erosion of rock by rivers and streams over enormous timescales. Lyell's hypothesis called into question the belief that Earth was less than 10,000 years old.

Darwin was also strongly influenced by a stop in the Galápagos, a small archipelago of volcanic islands off the coast of Ecuador. While at first look the islands seemed nearly lifeless, during the month that the *Beagle* spent sailing them, Darwin collected an astonishing variety of organisms. Many of the birds and reptiles he observed appeared to be unique to each island. For instance, while all islands had populations of tortoises, the type of tortoise found on one island was different from the types found on other islands *(Figure 9.5)*.

Darwin wondered why God would place different, unique subtypes of tortoise on islands in the same small archipelago. On his return to England, Darwin reflected on his observations and concluded that the populations of tortoises on the different islands must have descended from a single ancestral tortoise population. He noted a similar pattern of divergence between other groups of species on the islands and the closest mainland—for instance, prickly pear cacti in mainland Ecuador have the ground-hugging shape familiar to many of us, while on the Galápagos, these plants are tree sized *(Figure 9.6)*.

Developing the Hypothesis of Common Descent

Darwin's observations and portions of his fossil collection, sent periodically back to Henslow via other ships, made Darwin a scientific celebrity even before the *Beagle* returned to England. A journal of his travels was a best seller, and he settled into a comfortable life with his wife, Emma, heiress to a large fortune. However, Darwin saw that the observations he had made on his travels supported the view of his own grandfather, Erasmus, who had put forth the hypothesis of common descent. Erasmus Darwin had been ridiculed and derided as radical. Darwin's knowledge of his grandfather's treatment and his own fear of rejection prevented him from sharing his hypothesis with all but a few close friends.

Instead of publishing his ideas about evolution, Darwin spent the next 2 decades carefully collecting evidence and further developing his theory. He was finally spurred into publishing his ideas in 1858, after receiving a letter from Alfred Russel Wallace. With Wallace's letter was a manuscript detailing a mechanism for evolutionary change—nearly identical to Darwin's theory of natural selection.

Concerned that his years of scholarship would be forgotten if Wallace published his ideas first, Darwin had excerpts of both his and Wallace's work presented in July 1858 at a scientific meeting in London, and the next year he published *On the Origin of Species by Means of Natural Selection, or the Preservation of Favoured Races in the Struggle for Life*. The main point of this text was to put forward the hypothesis of natural selection. But Darwin devoted the last several chapters of *The Origin of Species* to describing the evidence he had accumulated supporting common descent.

9.3 Examining the Evidence for Common Descent

The evidence put forth in *The Origin of Species* was so complete, and from so many different areas of biology, that the hypothesis of common descent no longer appeared to be a tentative explanation. In response, scientists began to refer to this idea and its supporting evidence as the *theory* of common descent. Indeed, most biologists would agree that common descent is a scientific fact. Darwin's careful catalogue of evidence had revolutionized the science of biology.

Let us explore the statement, "common descent (or evolution) is a fact," more closely. When *The Origin of Species* was published, most Europeans believed that **special creation** explained how organisms came into being. According to this belief, God created organisms during the 6 days of creation described in the first book of the Bible, Genesis. This belief also states that organisms, including humans, have not changed significantly since creation. According to some biblical scholars, the Genesis story indicates that creation also occurred fairly recently, within the last 10,000 years.

Consider special creation as an alternative to the theory of common descent for explaining how modern organisms came to be. Is it an alternative that should be presented in public high school science classes? Since the idea of special creation requires the action of a supernatural entity, an all-powerful creator, it is not itself a scientific hypothesis.

As discussed in Chapter 1, for a hypothesis to be testable by science, it must be able to be evaluated through observations of the material universe. Since a supernatural creator is not observable or measurable, there is no way to determine the existence or predict the actions of such an entity via the scientific method. This is the essential problem with any statement that supposes a supernatural cause—including intelligent design. Part of Judge Jones's ruling on the Dover case rested on this point, that intelligent design is not science.

Stop and Stretch 9.2: Faith can be defined as the acceptance of ideals or beliefs that are not necessarily demonstrable through experimentation or direct observation of nature. How does faith differ from science? Can statements of faith be tested scientifically? Explain.

The belief of special creation does provide some scientifically testable hypotheses. For instance, the assertion that organisms came into being within the last 10,000 years and that they have not changed substantially since their creation is testable through observations of the natural world. We can call this hypothesis about the origin and relationships among living organisms the *static model hypothesis,* indicating that organisms are recently derived and unchanging *(Figure 9.7a)*.

There are also several intermediate hypotheses between the static model and common descent. One intermediate hypothesis is that all living organisms were created, perhaps even millions of years ago, and that changes have occurred in these species but brand-new species have not arisen; we will call this the *transformation hypothesis (Figure 9.7b)*. Another intermediate hypothesis is that different *types* of organisms (for example, plants and animals with backbones) arose separately and since their origin have diversified into numerous species; we will call this the *separate types hypothesis (Figure 9.7c)*.

Polls of the American public indicate that many people view these 3 alternatives and the theory of common descent *(Figure 9.7d)* as equally likely and reasonable explanations for the origin of biological diversity. We can see why the Dover school board thought it best to counsel their students to keep open minds.

But what about all those scientists who insist that the theory of common descent is fact? Why would so many scientists maintain this position? As you will soon see, the three alternative hypotheses are not equivalent to the theory of common descent. To understand why, we must examine the observations that help us test these hypotheses.

(a) Prickly pear cactus in mainland South America

(b) Prickly pear cactus in Galápagos

Figure 9.6 **Divergence from a common ancestor.** Prickly pear cactuses have very different forms in South America compared to the Galápagos, but they clearly share ancestry as evidenced by similar pad and flower structures. **Visualize This:** What environmental factors might have caused tree forms to evolve on the islands? Hint: Consider the possible environmental differences between the islands and the mainland.

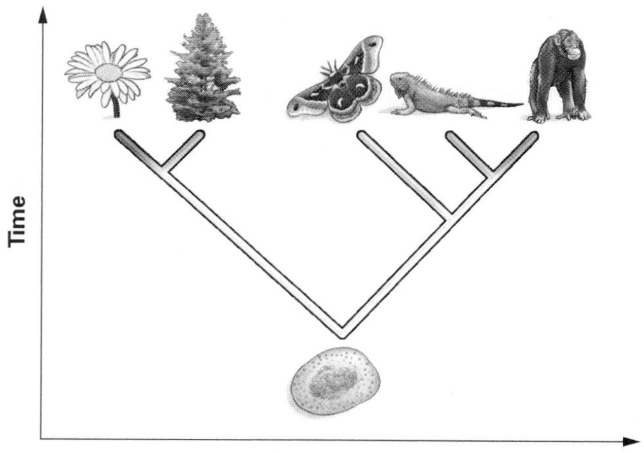

Figure 9.7 **Four hypotheses about the origin of modern organisms.** A graphical representation of the 4 hypotheses.

An Overview of the Evidence for Evolution

The evidence that all organisms share a common ancestor comes from several areas of biology and geology and are summarized below.

- **The biological classification system in Darwin's era.** The most logical and useful system for organizing diversity, even before evolutionary theory emerged, grouped organisms in a hierarchy. From broadest to narrowest groupings, these classification levels are

  DOMAIN
  KINGDOM
  PHYLUM (OR DIVISION)
  CLASS
  ORDER
  FAMILY
  GENUS
  SPECIES

Darwin noted that these levels of classification can be interpreted as different degrees of relationship. In other words, all species in the same family share a relatively recent common ancestor, while all families in the same class share a more distant common ancestor *(Figure 9.8)*.

- **Anatomical similarities among species.** Organisms that look quite different have surprisingly similar structures. Each of the mammal forelimbs illustrated in *Figure 9.9* shares a common set of bones that are in the same relationship to each other, even though each limb has a very different function. The simplest explanation for this observation is that each species inherited the basic structure from the same common ancestor, and evolution led to their modification in each group.
- **Useless traits in modern species.** For example, flightless birds such as

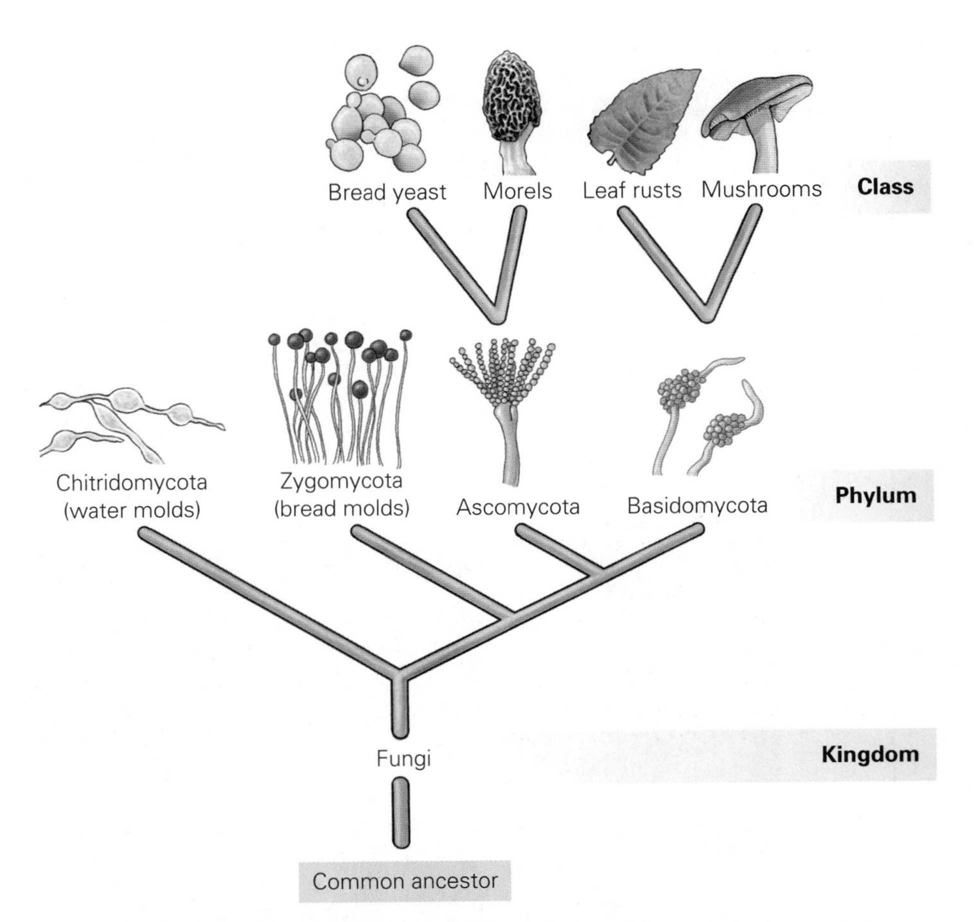

Figure 9.8 **Classification implies common ancestry.** The shared characteristics among fungal groups that help establish their classification in the Kingdom Fungi can be arranged in the form of a "tree of life."

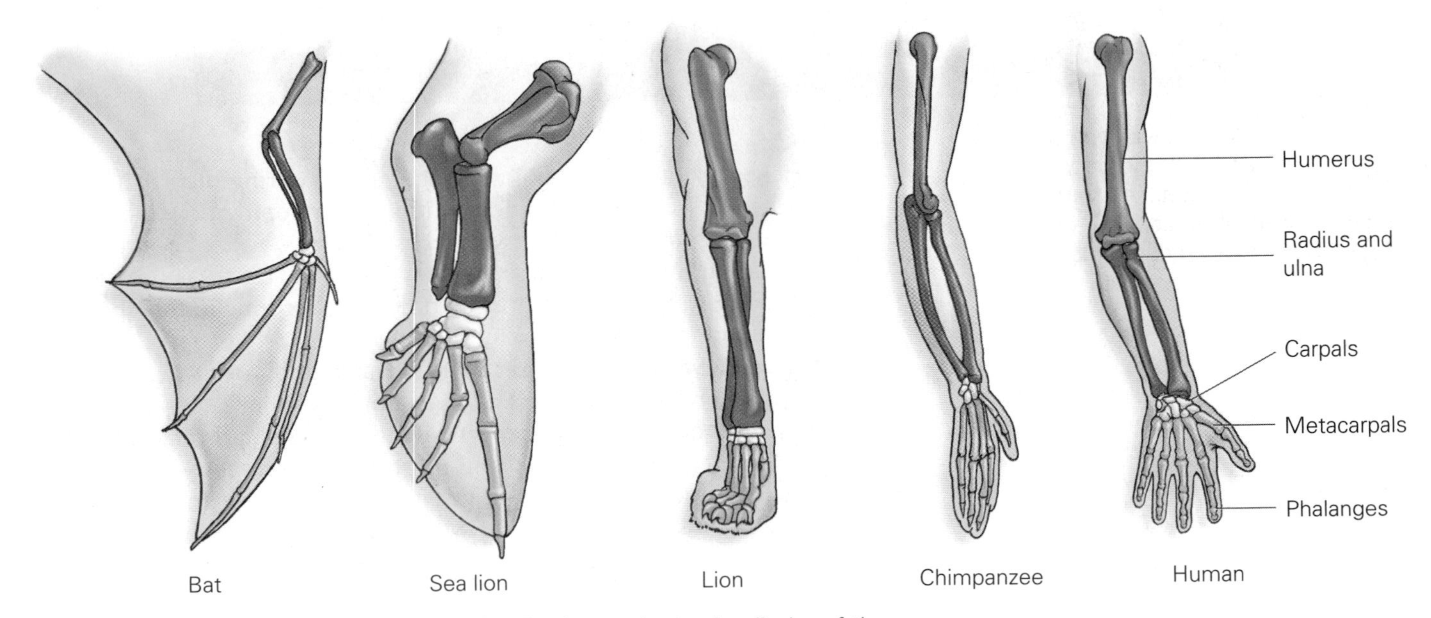

Figure 9.9 **Homology of mammal forelimbs.** The bones in the forelimbs of these mammals are very similar; equivalent bones in each organism are shaded the same color. The similarity in underlying structure despite great differences in function is evidence of shared ancestry.

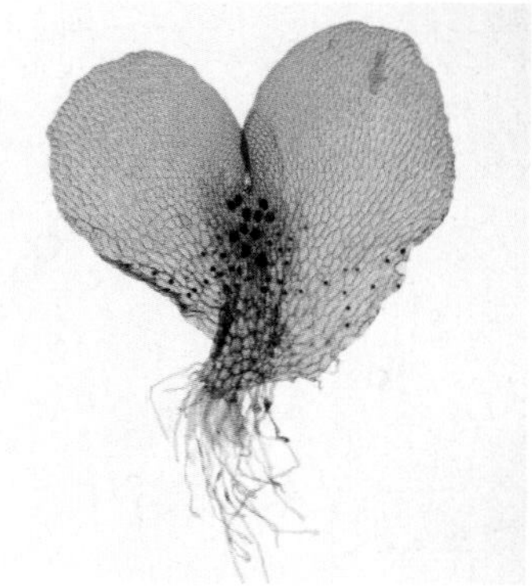

Figure 9.10 **Vestigial structures in plants.** (a) The ancestors of flowering plants were similar to modern ferns, which have two independent stages in their life cycle: the sporophyte stage with which we are familiar and a tiny, independent, gametophyte stage. (b) Flowering plants no longer have two independent stages but do produce a tiny gametophyte within the flower itself.

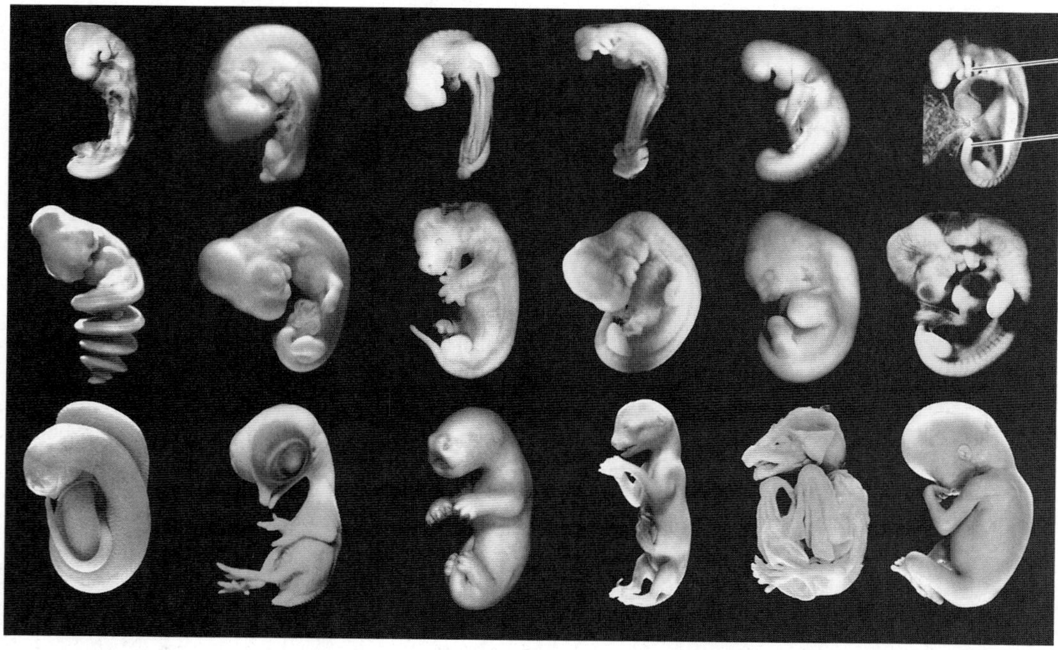

(a) Fern gametophyte

(b) Flower ovary

Snake Chicken Possum Cat Bat Human

Early embryo

Intermediate embryo

Late embryo

Pharyngeal slits

Tail

Figure 9.11 **Similarity among chordate embryos.** These diverse organisms appear very similar in the first stages of development (shown in the top row), evidence that they share a common ancestor that developed along the same pathway.

ostriches produce functionless wings, and flowering plants still produce a tiny "second generation" within a developing flower ovule, a vestige of their relationship with ferns *(Figure 9.10)*. The simplest explanation for these useless features is that the trait was functional in an ancestral species but lost its function over time in this species or group of species.

- **Shared developmental pathways.** Early embryos of very different species often look very similar. For example, all chordates—animals that have a backbone or closely related structure—produce structures called pharyngeal slits, and most have tails as early embryos *(Figure 9.11)*. These similarities suggest that these organisms derived from a single common ancestor with a particular developmental pathway that they all inherited.
- **DNA similarities.** The nucleotide sequence for the entire genome and for individual genes displays a hierarchical pattern of similarity. DNA sequence comparisons in a wide variety of organisms support evolutionary trees implied by their classification on physical similarities *(Figure 9.12)*.

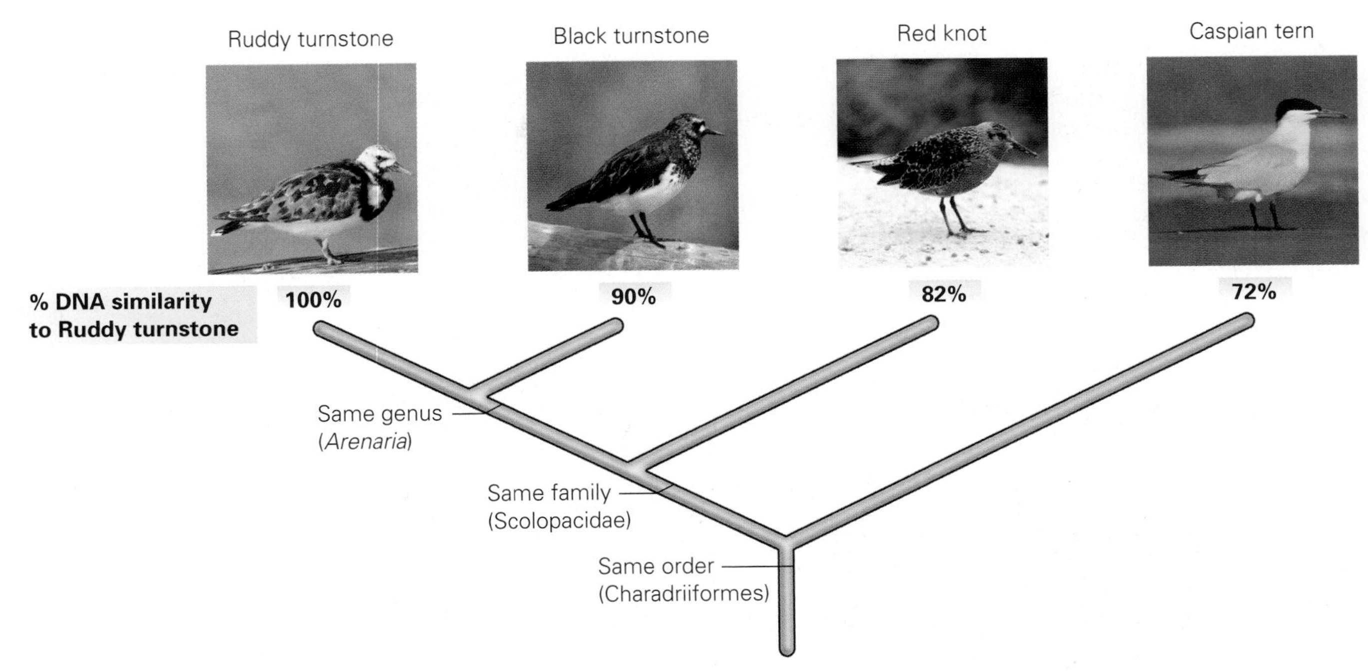

Figure 9.12 **DNA evidence of relationship.** Scientists used the physical characteristics of these 4 bird species to put them into different taxonomic categories. The tree here translates the taxonomic grouping into a hypothesis of evolutionary relationship. DNA studies later supported this hypothesis since species that share a more recent common ancestor (that is, those in the same genus) would be expected to share more similar DNA sequences than species that share a more distant ancestor (that is, in the same family but different genera).

(a) Charles mockingbird, Champion Island

(b) Hood mockingbird, Hood Island

(c) Longtailed mockingbird, mainland Ecuador

Figure 9.13 **Biogeographic patterns imply common descent.** Different mockingbird species are found on different islands in the Galápagos. The Galápagos species all resemble another species of mockingbird found on mainland Ecuador. This observation supports the hypothesis that the Galápagos mockingbirds share a common ancestor with the mainland species.

- **The distribution of organisms on Earth.** Darwin noticed that each island in the Galápagos had a unique species of mockingbird, all similar in appearance to a different species found on mainland Ecuador *(Figure 9.13)*. In contrast, the species on these islands were very different from species found on other, similar tropical islands that he visited. These observations suggest species in a geographic location are generally descended from ancestors in that geographic location. If the alternative hypotheses that species appeared independently were true, we would predict either that all tropical islands are very similar or that all are very different.
- **Fossil evidence.** The remains of extinct organisms form a record of ancient life and provide direct evidence of change in organisms

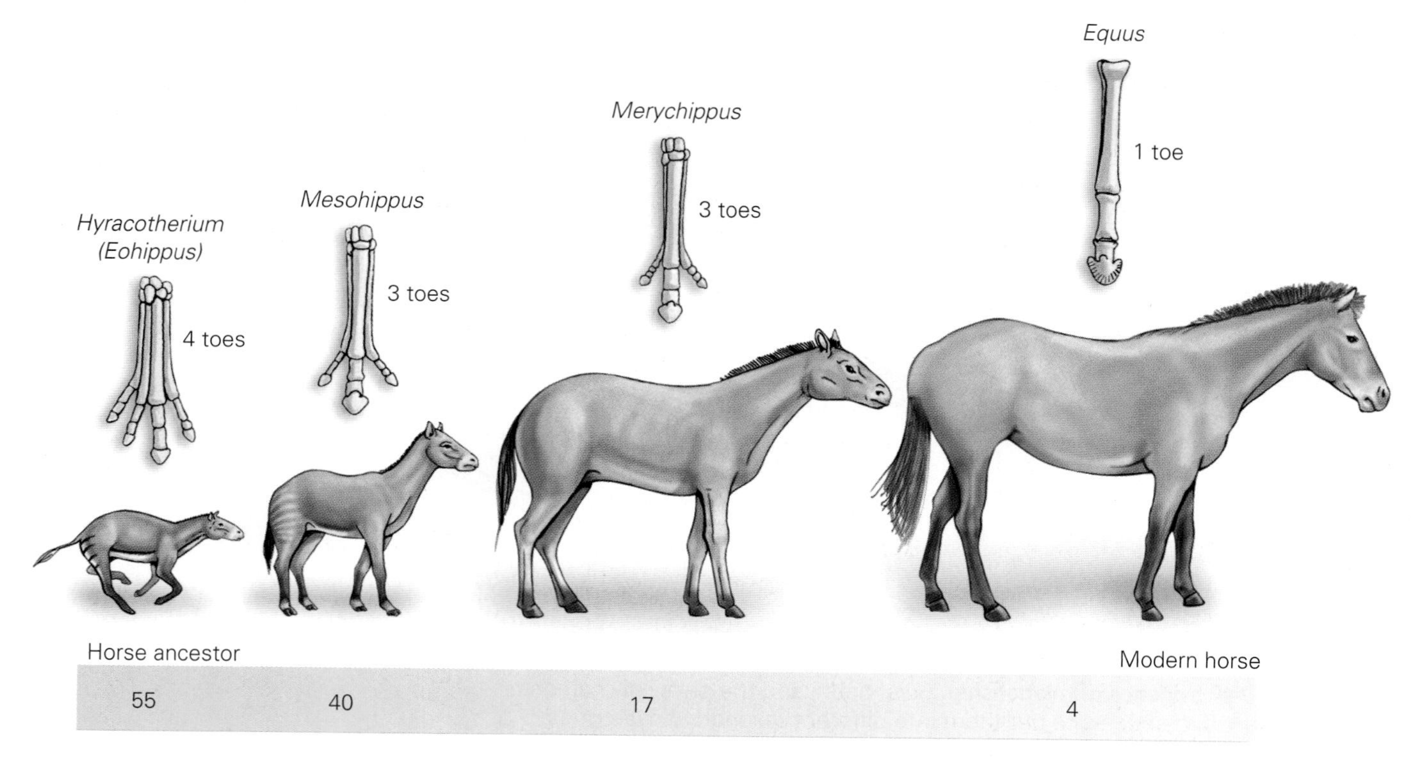

Figure 9.14 **The fossil record of horses.** Horse fossils provide a fairly complete sequence of evolutionary change from small, catlike animals with 4 toes to the modern horse with 1 massive toe.

over time. There are many examples of fossil series that show a progression from more ancient forms to more modern forms, as in the transition between ancient and modern horses *(Figure 9.14)*.

For a closer look at the evidence, we will address one of the most controversial questions underlying this debate: Are humans really related to apes?

A Closer Look: Humans and Apes Share a Common Ancestor

Any zookeeper will tell you that the primate house is their most popular exhibit. People love apes and monkeys. It is easy to see why—primates are curious, playful, and agile. In short, they are fun to watch. But something else drives our fascination with these wonderful animals: We see ourselves reflected in them. The forward placement of their eyes and their reduced noses appear humanlike. They have hands with fingernails instead of paws with claws. Some can stand and walk on 2 legs for short periods. They can finely manipulate objects with their fingers and opposable thumbs. They exhibit extensive parental care, and even their social relations are similar to ours—they tickle, caress, kiss, and pout *(Figure 9.15)*.

Why are primates, particularly the great apes (gorillas, orangutans, chimpanzees, and bonobos) so similar to humans? Scientists contend that it is because humans and apes are recent descendants of a common biological ancestor. This statement is supported by the biological classification of humans.

Figure 9.15 **Are humans related to apes?** Biologists contend that apes and humans are similar in appearance and behavior because we share a relatively recent common ancestor.

Linnaean Classification. As the modern scientific community was developing in the sixteenth and seventeenth centuries, various methods for organizing biological diversity were proposed. Many of these classification systems grouped organisms by similarities in habitat, diet, or behavior; some of these classifications placed humans with the great apes, and others did not.

Into the classification debate stepped Carl von Linné, a Swedish physician and botanist. Von Linné gave all species of organisms a 2-part, or binomial, name in Latin, which was the common language of science at the time. In fact,

(A Closer Look continued)

he adopted a Latinized name for himself by which he is more universally known—Carolus Linnaeus.

The Latin names that Linnaeus assigned to other organisms typically contained information about the species' traits—for instance, *Acer saccarhum* is Latin for "maple tree that produces sugar," the tree commonly known as the sugar maple, while *Acer rubrum* is Latin for "red maple." The scientific name of a species contains information about its classification as well. For example, species with the generic name *Acer* are all maples, and species with the generic name *Ursus* are bears.

Linneaus's hierarchical classification system was described in the previous section and is summarized in *Figure 9.16*. Although Linneaus himself believed in special creation and preceded the theory of evolution, Darwin noted that his classification system implied evolutionary relationships among organisms. Linneaus's classification system was so effective at capturing these relationships that, even after the widespread acceptance of evolutionary theory, it has only been modified slightly, with the addition of "sub" and "super" levels between his categories—such as superfamily between family and order—to better represent the complex relationships among groups of organisms. Modern classification science is discussed more thoroughly in Chapter 12.

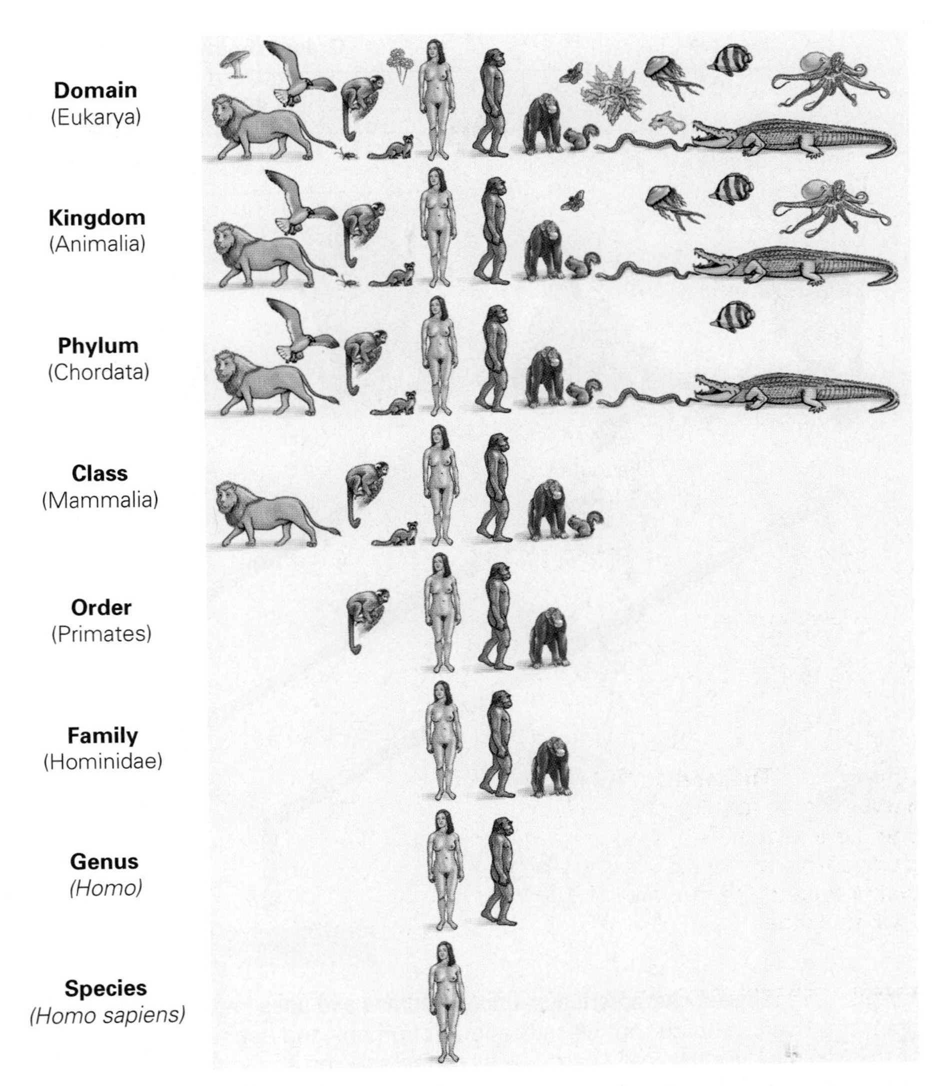

Figure 9.16 **The Linnaean classification of humans.** All organisms within a category share basic characteristics, and as the groups become smaller subdivisions toward the bottom of the figure, the organisms have more similarities. **Visualize This:** List one or more traits that all members of the Order Primates share.

Using his classification system, Linnaeus placed humans, monkeys, and apes in the same order, which he called Primates. The modern classification of humans reflects only refinements of Linnaeus's ideas. Among the primates, humans are most similar to apes. Humans and apes share a number of characteristics, including relatively large brains, erect posture, lack of a tail, and increased flexibility of the thumb. Scientists now place humans and apes in the same family, Hominidae.

Humans and the African great apes (gorillas, chimpanzees, and bonobos) share even more characteristics, including elongated skulls, short canine teeth, and reduced hairiness; they are placed together in the same *sub*family, Homininae. When the classification of humans, great apes, and other primates is shown in the form of a tree diagram *(Figure 9.17)*, it is easy to see why Darwin concluded that humans and modern apes probably evolved from the same ancestor.

The tree of relationship implied by Linnaeus's classification forms a hypothesis that can be tested. If modern species represent the descendants of ancestors that also gave rise to other species, we should be able to observe other, less-obvious similarities between humans and apes in anatomy, behavior, and genes. These similarities are referred to as **homology.**

(continued on the next page)

(A Closer Look continued)

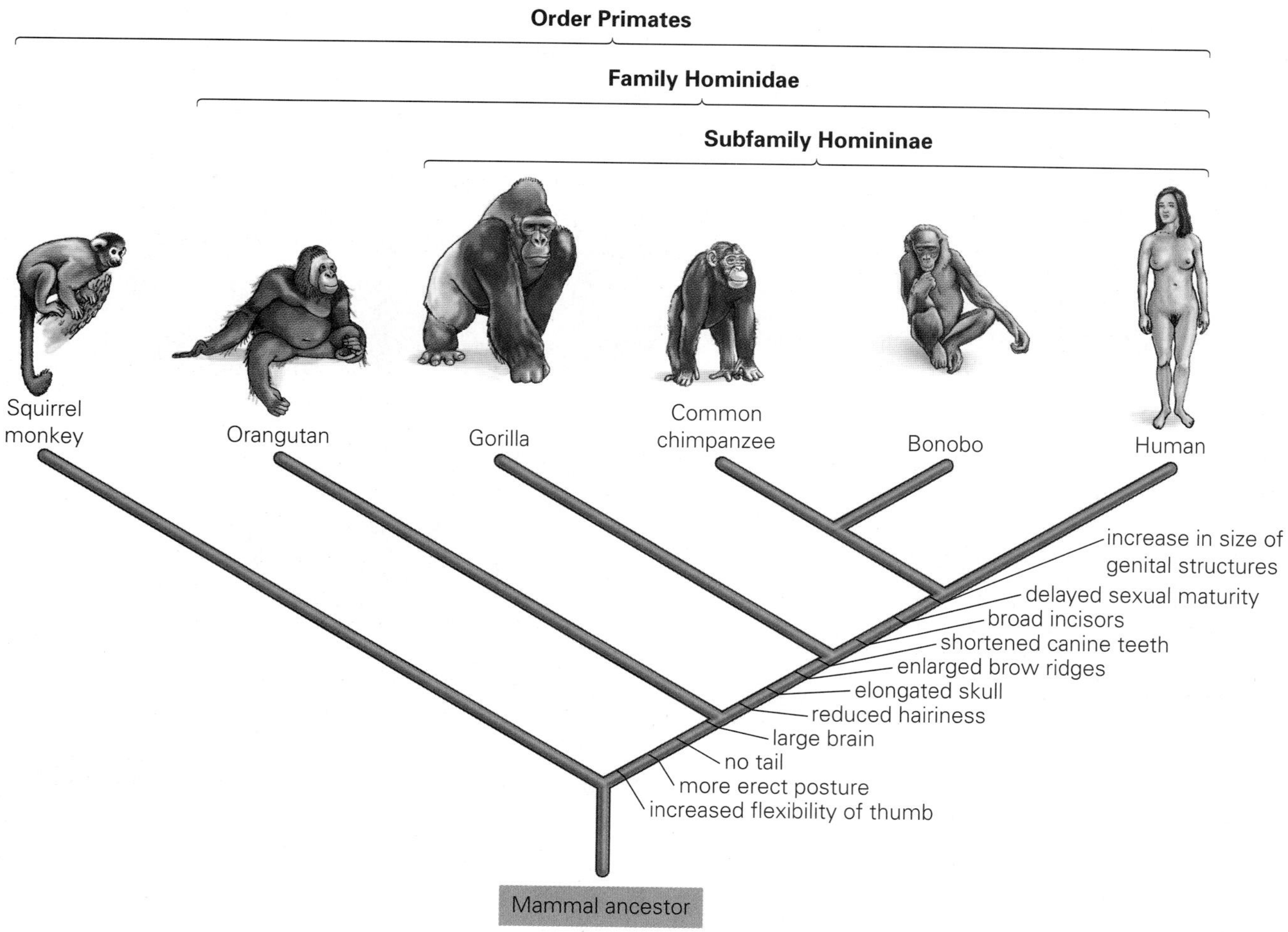

Order Primates
Family Hominidae
Subfamily Homininae

Squirrel monkey
Orangutan
Gorilla
Common chimpanzee
Bonobo
Human

increase in size of genital structures
delayed sexual maturity
broad incisors
shortened canine teeth
enlarged brow ridges
elongated skull
reduced hairiness
large brain
no tail
more erect posture
increased flexibility of thumb

Mammal ancestor

Figure 9.17 **Shared characteristics among humans and apes imply shared ancestry.** This tree diagram represents the current classification of humans and apes. Characteristics noted on the side of the evolutionary tree are shared by all of the species above that point. **Visualize This:** Add "squirrel" to this tree. Where is the shared ancestor of squirrels and other members of this group?

Evidence of Anatomical and Developmental Homology. Figure 9.9 illustrates in part the homology in skeletal anatomy between a human arm and a chimpanzee forelimb. As you can see in this figure, the similarities are striking, especially the presence of a distinct, opposable thumb.

Even more compelling evidence of shared ancestry comes from similarities between functional traits in one organism and nonfunctional features, or **vestigial traits,** in another. In other words, these vestigial traits represent a vestige, or remainder, of our biological heritage. *Figure 9.18* provides 2 examples of vestigial traits in humans that link us to other primates and great apes. Great apes and humans have a tailbone like other primates, yet neither great apes nor humans have a tail.

In addition, all mammals possess tiny muscles called arrector pili at the base of each hair. When the arrector pili contract under conditions of emotional stress or cold temperatures, the hair is elevated. In furry mammals, the arrector pili help to increase the perceived size of the animal, and they increase the insulating value of the hair coat. In humans, the same emotional or physical conditions produce only goose bumps, which provide neither benefit.

Darwin maintained that the hypothesis of evolution provided a better explanation for vestigial structures than did the hypothesis of special creation represented by the static model. A useless trait such as goose bumps is better explained as the result of inheritance from our biological ancestors than as a feature that appeared independently in our species.

(A Closer Look continued)

Stop and Stretch 9.3: Some vestigial structures persist because they may still have a function or did in the very recent past. Wisdom teeth, which erupt in early adulthood and often have to be removed because they do not "fit" in a modern person's jaw, are often considered vestigial. Why might wisdom teeth have persisted over the course of human evolution?

As illustrated in (Figure 9.11), homology even extends to embryonic structures during development. The presence in early human embryos of a tail that is later lost is another piece of evidence that the tailbones of human adults represent a vestige of our evolutionary history that links us to other primates.

Evidence of Molecular Homology. Scientists now understand that differences among individuals arise largely from differences in their genes. It stands to reason that differences among species must also derive from differences in their genes. If the hypothesis of common descent is correct, then species that appear to be closely related must have more similar genes than do species that are more distantly related. The most direct way to measure the overall similarity of two species' genes is to evaluate similarities in the DNA sequences of genes found in both organisms.

Many genes are found in nearly all living organisms. For instance, genes that code for histones, the proteins that help store DNA neatly inside cells, are found in algae, fungi, fruit flies, humans, and all other organisms that contain linear chromosomes. Among organisms that share many aspects of structure and function, such as humans and chimpanzees, many genes are shared. However, the *sequences* of these genes are not identical.

A comparison of the sequences of dozens of genes that are found in humans and other primates demonstrates the relationship between classification and gene sequence similarity (*Figure 9.19*). The DNA sequences of these genes in humans and chimpanzees are 99.01% similar, whereas the DNA sequences of humans and gorillas are identical over 98.9% of their length. More distantly related primates are less similar to humans in DNA sequence. This pattern of similarity in DNA sequence exactly matches the biological relationships implied by physical similarity. This result supports the hypothesis of common descent among the primates.

At first, a finding of similarities in DNA sequence may not seem especially surprising. If genes are instructions, then you would expect the instructions for building a human and a chimpanzee to be more similar than the instructions for building a human and a monkey. After all, humans and chimpanzees have many more physical similarities than humans and monkeys do, including reduced hairiness and lack of a tail.

(continued on the next page)

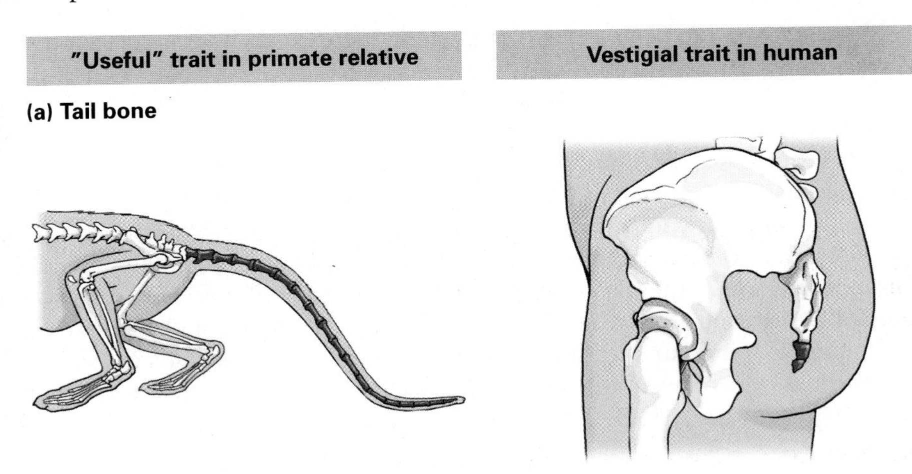

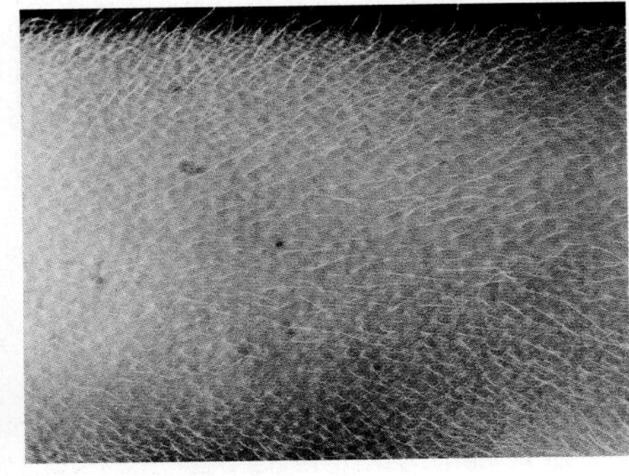

Figure 9.18 **Vestigial traits reflect our evolutionary heritage.** (a) Humans and other great apes do not have tails, but they do have a vestigial tailbone. (b) Goose bumps are reminders of our relatives' hairier bodies.

(A Closer Look continued)

However, remember that the genes being compared perform the same function in all of these species. For example, one of the genes in this DNA analysis is *BRCA1*, which in humans is associated with risk for breast cancer, giving the gene its name. However, *BRCA1* has the general function of helping repair damage to DNA in all organisms. Given this identical function in all organisms, there is no reason to expect that differences in *BRCA1* sequences among different species should display a pattern—unless the organisms are related by descent. But there is a pattern; the *BRCA1* gene of humans is more similar to the *BRCA1* gene of chimpanzees than to the same gene in monkeys. The best explanation for this observation is that humans and chimpanzees share a more recent common ancestor than either species shares with monkeys.

Differences in DNA sequence between humans and chimpanzees can also allow us to estimate when these 2 species diverged from their common ancestor. The estimate is based on a **molecular clock.** The principle behind a molecular clock is that the rate of change in certain DNA sequences, due to the accumulation of mutations within a species, seems to be relatively constant. According to one application of a molecular clock, the amount of time it takes for a 1% difference in DNA sequence (about the difference between humans and chimpanzees) to accumulate in diverging species is 5 to 6 million years. The direct evidence of human ancestors found in fossils can help confirm this date.

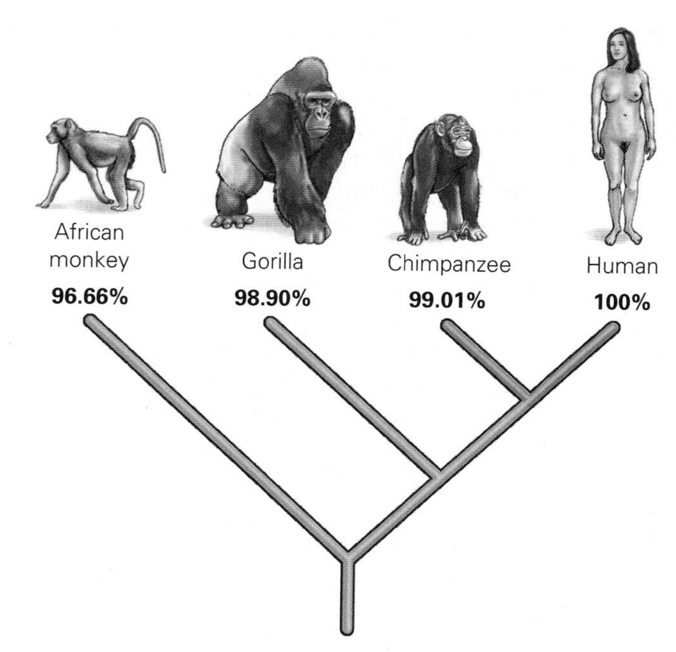

Figure 9.19 **Similar organisms have similar DNA sequences.** Species that appear to be physically similar to humans have more similar DNA sequences for the same genes when compared to species that are less physically similar.

Evidence from Biogeography. The distribution of species on Earth is known as the **biogeography** of life. Darwin's observations of the similarities among tortoises, cactuses, and mockingbirds in the Galápagos (see Figure 9.13 on p. 241) implied that related species should be found close to each other.

By the time Darwin made his observations, humans were distributed over the entire Earth. But, he reasoned, if humans and apes share a common ancestor, then highly mobile humans must have first appeared where their less-mobile relatives can still be found. Thus, Darwin predicted that evidence of human ancestors would be found in Africa, the home of chimpanzees and other great apes.

Evidence of human ancestors comes from **fossils,** the remains of living organisms left in soil or rock. Fossils form when the organic material in bone decomposes and minerals fill the space left behind *(Figure 9.20)*. This is more likely to occur when dead organisms are quickly buried by

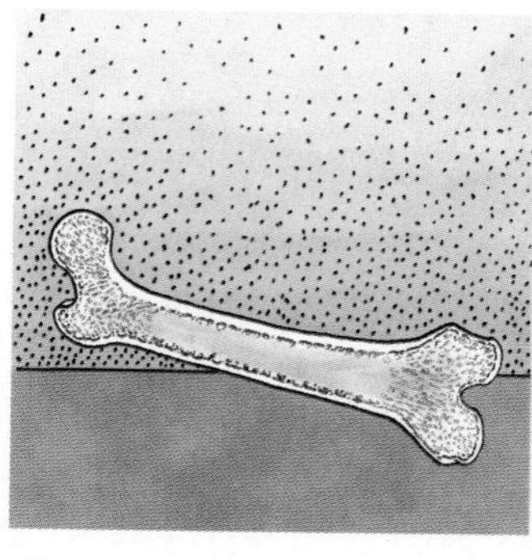

(1) An organism is rapidly buried in water, mud, sand, or volcanic ash. The tissues begin to decompose very slowly.

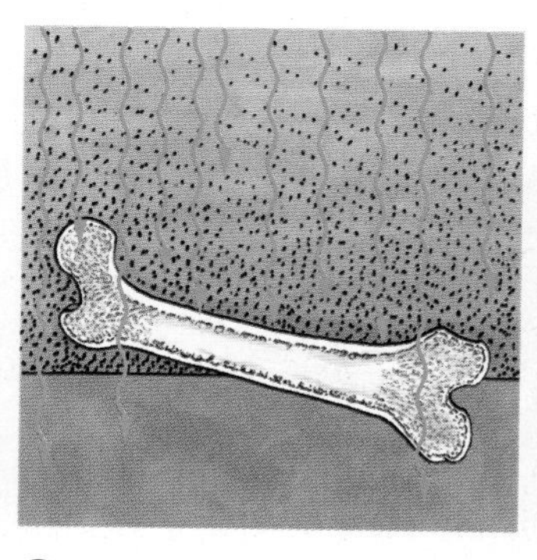

(2) Water seeping through the sediment picks up minerals from the soil and deposits them in the spaces left by the decaying tissue.

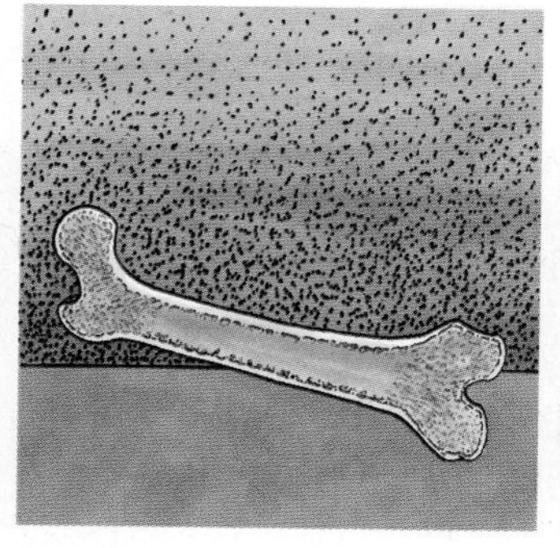

(3) After thousands of years, most or all of the original tissue is replaced by very hard minerals, resulting in a rock model of the original bone.

(4) When erosion or human disturbance removes the overlying sediment, the fossil is exposed.

Figure 9.20 **Fossilization.** A fossil is typically a rock "model" of an organic structure.

(A Closer Look continued)

sediment. Fortunately for scientists looking for fossils of **hominins**—humans and human ancestors—these organisms were likely to live near water and be rapidly buried, so their fossil record provides a compelling record of evolutionary history.

Hominin fossils can be distinguished from other primate fossils by some key characteristics. One essential difference between humans and other apes is our mode of locomotion. While chimpanzees and gorillas use all 4 limbs to move, humans are bipedal; that is, they walk upright on only 2 limbs. This difference results in several anatomical differences between humans and other apes *(Figure 9.21)*. In hominins, the face is on the same plane as the back instead of at a right angle to it; thus the foramen magnum, the hole in the skull through which the spinal cord passes, is found on the back of the skull in other apes but at the base of the skull in humans. In addition, the structures of the pelvis and knee are modified for an upright stance; the foot is changed from being grasping to weight bearing, and the lower limbs are elongated relative to the front limbs.

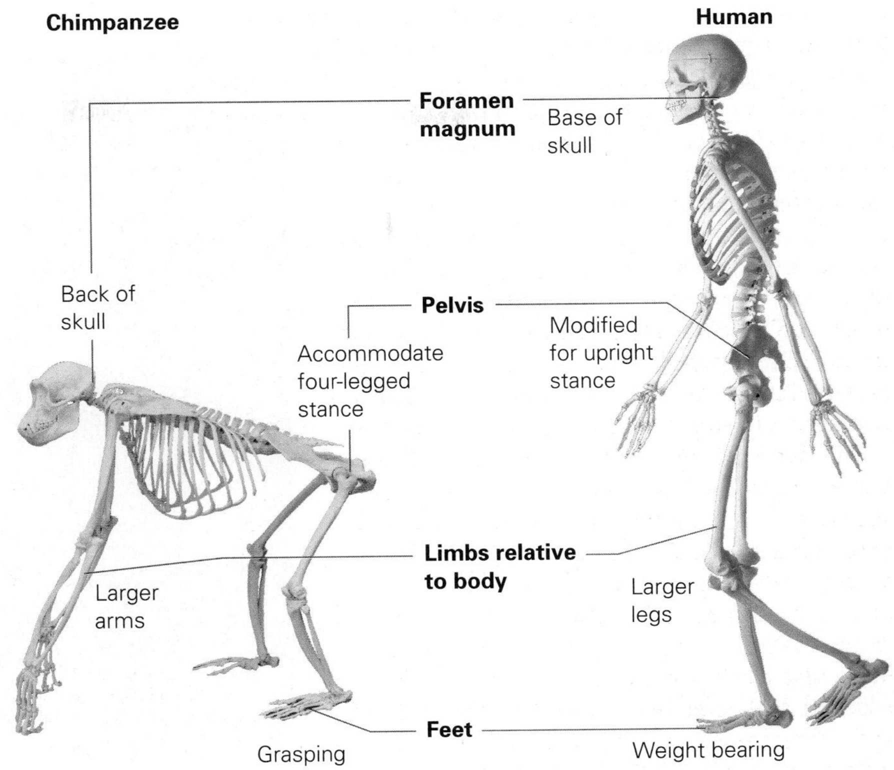

Figure 9.21 **Anatomical differences between humans and chimpanzees.** Humans are bipedal animals, while chimpanzees typically travel on all fours. If any bipedal features are present in a fossil primate, the fossil is classified as a hominin.

The first hominin fossils were found not in Africa, but in Europe. Remains of *Homo neanderthalensis* (Neanderthal man) were discovered in 1856 in a small cave within the Neander Valley of Germany. In 1891, fossils of older, human-like creatures now called *Homo erectus* (standing man) were discovered in Java in 1891. It was not until 1924 that the first *African* hominin fossil, the Taung child, was discovered in South Africa. This fossil was later placed in the species *Australopithecus africanus*. Paleontologists continue to discover new hominin fossils in southern and eastern Africa, including the famous "Lucy," a remarkably complete skeleton of the species *Australopithecus afarensis*, discovered in 1974 in Ethiopia *(Figure 9.22)*. Lucy's fossil skeleton included a large section of her pelvis, which clearly indicated that she walked upright.

By determining the age of these fossils and many other hominin species, scientists have confirmed Darwin's predictions—the earliest human ancestors arose in Africa.

Evidence from Fossil Pedigrees. Scientists can determine the date when an ancient fossil organism lived by estimating the age of the rock that surrounds the fossil. **Radiometric dating** relies on radioactive decay, which occurs as radioactive elements in rock spontaneously break down into different, unique elements known as *daughter products*. Each radioactive element decays at its own unique rate. The rate of decay is measured by the element's *half-life*—the amount of time required for one-half of the amount of the element originally present to decay into the daughter product.

When rock is newly formed from the liquid underlying Earth's crust, it contains a fixed amount of any radioactive element. When the rock hardens, some of these radioactive elements become trapped. As a trapped element decays over time, the amount of radioactive material in the rock declines, and correspondingly, the amount of daughter product increases. By determining the ratio of radioactive element to daughter product in a rock sample and knowing the half-life of the

(continued on the next page)

(A Closer Look continued)

radioactive element, scientists can estimate the number of years that have passed since the rock formed *(Figure 9.23)*.

Scientists have used radiometric dating to estimate the age of Earth and the time of origin of various groups of organisms. Using this technique, scientists have also determined that the most ancient hominin fossil, the species *Ardepithecus ramidus*, is 5.2 to 5.8 million years old. (Two even older fossil species, *Orrorin tugenensis* and *Sahelanthropus tchadensis*, were recently described as 6- and 7-million-year-old human ancestors, respectively. However most scientists are reserving judgment about these fossil specimens until more examples are found.) These very early fossils probably represent hominins that are quite similar to the common ancestor of humans and chimpanzees.

(a) Lucy's skeleton

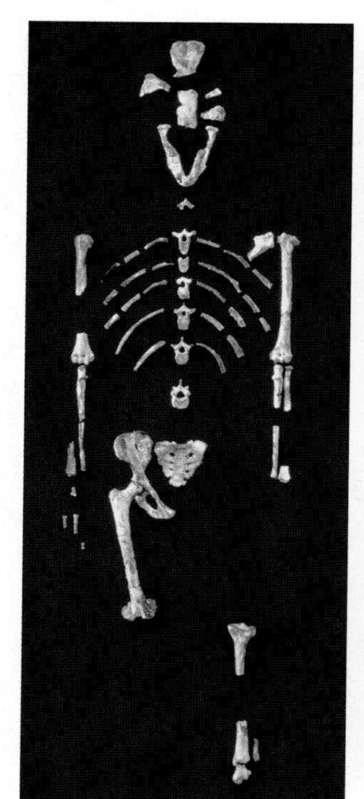

(b) Artist's reconstruction

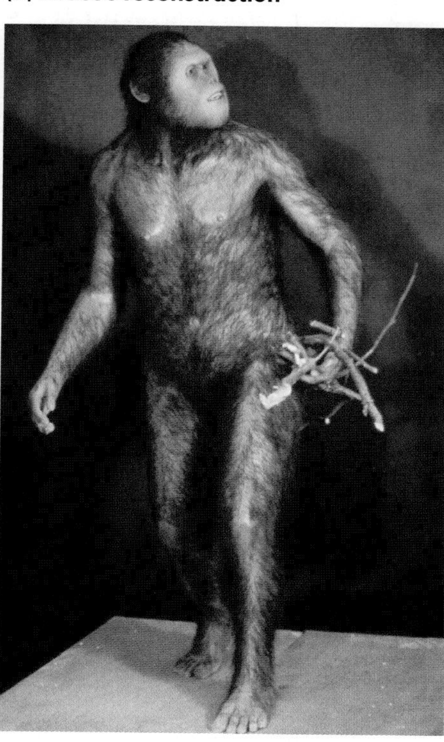

Figure 9.22 **A fossil ancestor.** (a) Lucy is still the most complete fossil of *Australopithecus afarensis* ever found. Her pelvis and knee joint provide evidence that she walked on 2 legs. (b) This artist's conception of what Lucy looked like in life is based on her fossil remains as well as other fossils of the same species.

Figure 9.23 **Radiometric dating.** (a) The age of rocks can be estimated by measuring the amount of radioactive material (designated by dark purple circles) with a known half-life and the amount of daughter material (designated by light blue circles) in a sample of rock. (b) The age of a fossil can be estimated when it is found between 2 layers of magma-formed rock. **Visualize This:** How old is a rock that contains 12.5% of the original parent element?

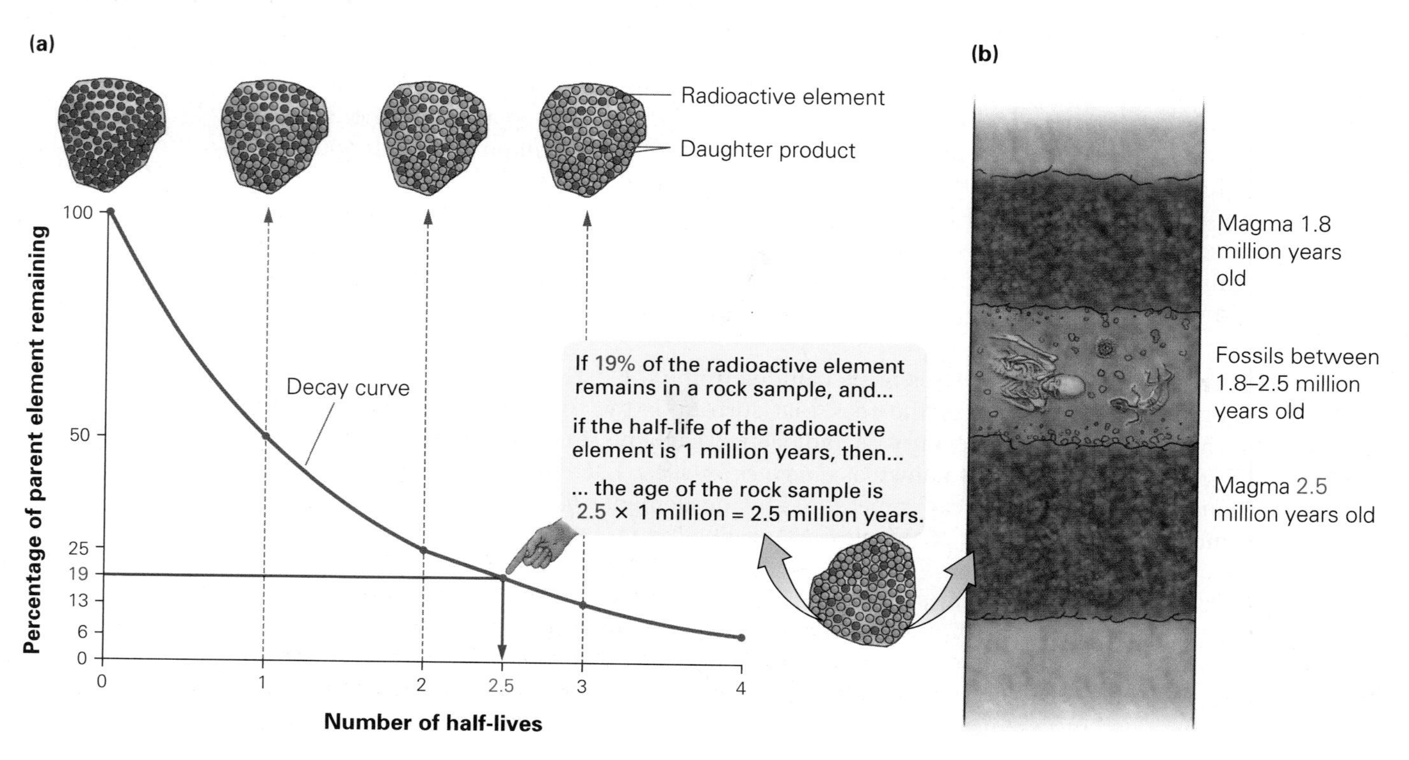

(A Closer Look continued)

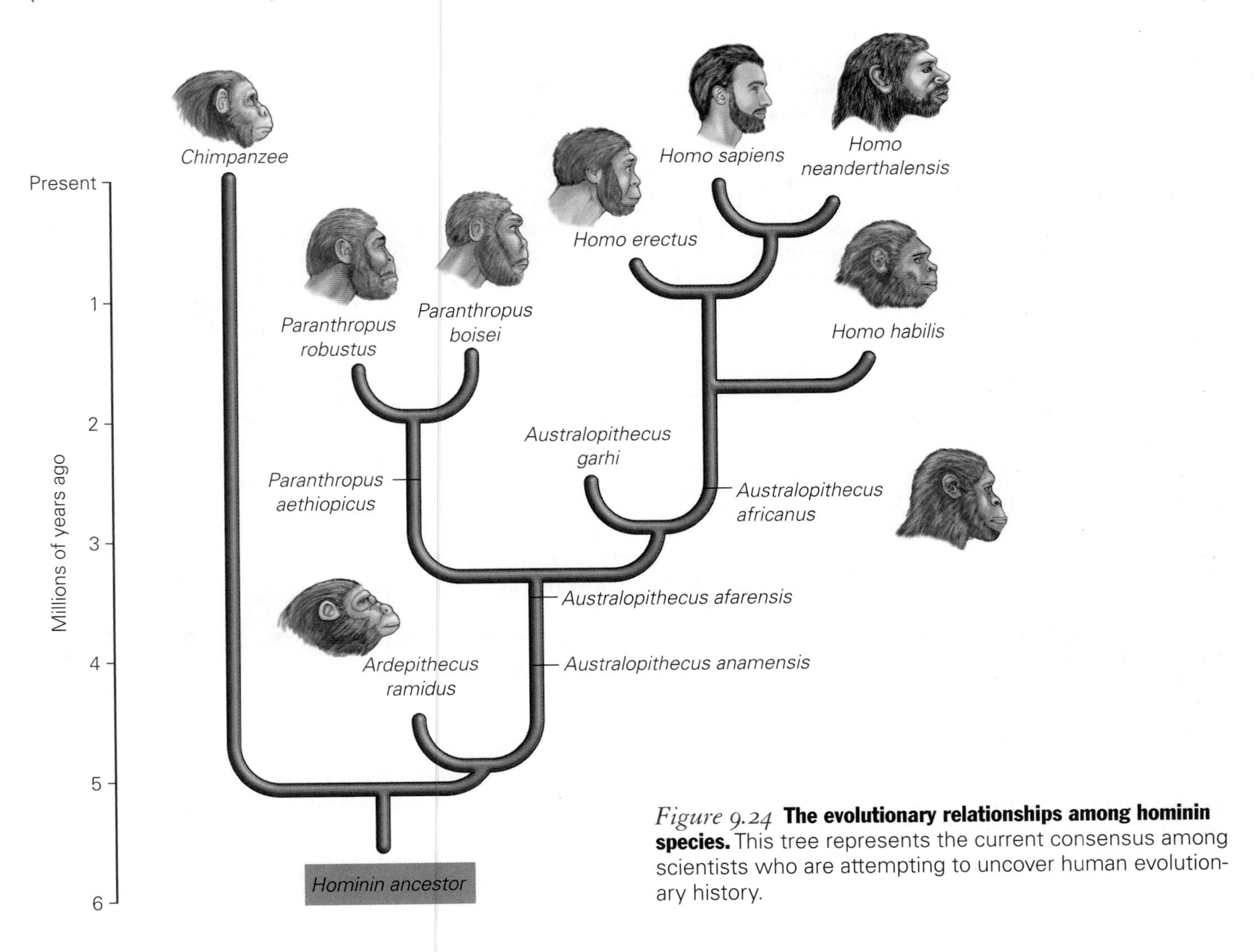

Figure 9.24 **The evolutionary relationships among hominin species.** This tree represents the current consensus among scientists who are attempting to uncover human evolutionary history.

Stop and Stretch 9.4: How does the age estimate for the oldest human ancestors compare to the age of divergence between chimpanzees and humans calculated by the molecular clock?

As the number of described hominin fossils has increased, a tentative genealogy of humans has emerged. The fossil species can be arranged in a pedigree from most ancient to most modern species by determining the age of a fossil and the anatomical similarities among organisms *(Figure 9.24)*. What this pedigree indicates is that modern humans are the last remaining branch of a once-diverse group of hominins.

The pedigree illustrates a common theme seen in the fossil record of other groups: When a new lifestyle (in this case, bipedal ape) evolves, many different types appear, but only a small number of lineages survive over the long term. Biologists often refer to the lineages that die out as "evolutionary experiments." But does the pedigree provide convincing evidence that modern humans evolved from a common ancestor with other apes?

The common ancestor of humans and chimpanzees is often called the "missing link" because it has not been identified. However, finding the fossilized common ancestor between chimpanzees and humans, or between any two species for that matter, is extremely difficult, if not impossible.

To identify a common ancestor, the evolutionary history of both species since their divergence must be clear. Like humans, modern chimpanzees have been evolving over the 5 million years since they diverged from humans. In other words, a missing link would not look like a modern chimpanzee

(continued on the next page)

(A Closer Look continued)

with some human features or like a cross between the two species—as suggested by nineteenth-century cartoons depicting Charles Darwin as an "ape man" *(Figure 9.25)*.

The hominin fossil record shows a clear progression from more "ape-like" to more "human-like" ancestors over time. Besides being bipedal, humans differ from other apes in having a relatively large brain, a flatter face, and a more extensive culture. The oldest hominins are bipedal but are otherwise similar to other apes in skull shape, brain size, and probable lifestyle. More modern hominins show greater similarity to modern humans, with flattened faces and increased brain size *(Figure 9.26)*. Evidence collected with fossils of the most recent human ancestors indicates the existence of symbolic culture and extensive tool use, trademarks of modern humans.

Figure 9.25 **The common ancestor of humans and chimpanzees?** A "missing link" between humans and chimpanzees would not look half human and half ape, as this cartoon suggests.

Observations of anatomical and genetic similarities among modern organisms and biogeographical patterns provide good evidence to support the theory of evolution. As with nearly all evidence in science, these observations allow us to infer the accuracy of the hypothesis but do not prove the hypothesis correct. This type of evidence is similar to the "circumstantial evidence" presented in a murder trial, such as finding the murder weapon in a car belonging to a suspect or the presence of the suspect's fingerprints on the victim's door. The European scientific community in the nineteenth century was convinced by the circumstantial evidence Darwin had collected and embraced the theory of common descent as the best explanation for the origin of species. Since Darwin's time, scientists have accumulated additional indirect evidence, such as the DNA sequence similarity discussed earlier, to support this theory.

But as in a murder trial, direct evidence is always preferred to establish the truth—for instance, the testimony of an eyewitness or a recording of the crime by a security camera. Of course, there are no human eyewitnesses to the evolution of humans, but we have a type of "recording" in the form of the fossil record. Because the fossilization requires specific conditions, the fossil record is not a complete recording of the of the history of life—it is more like a security video that captures only a small portion of the action, with many blank segments. Just as the blank segments in a security video do not make the video an incorrect record of events, "gaps" in the fossil record do not diminish its value. The evidence present in the fossil record provides convincing support for the theory of common descent. All of the lines of evidence for common descent are summarized in *Table 9.1*.

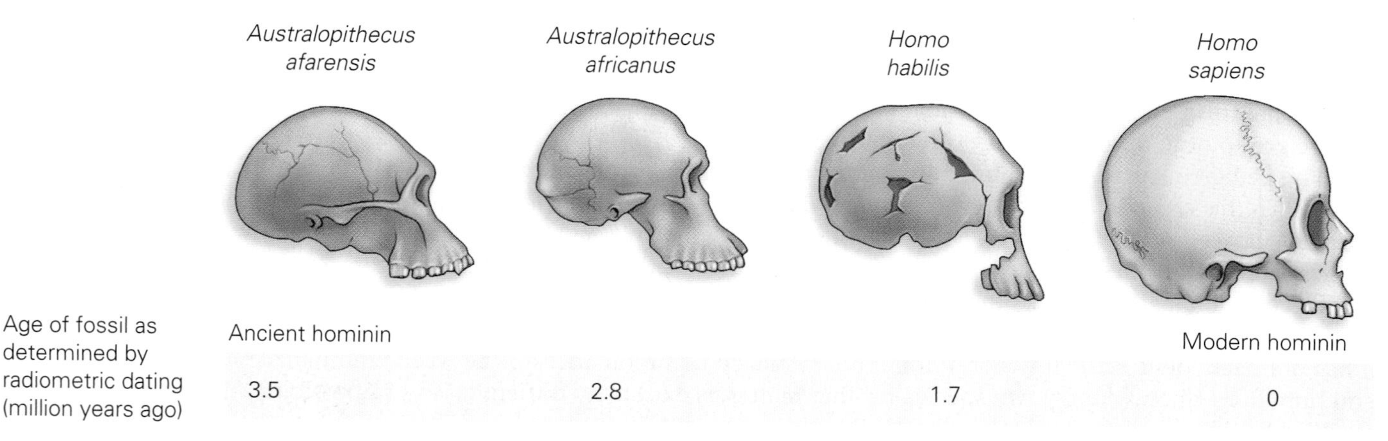

Figure 9.26 **The ape-to-human transition.** Ancient hominins display numerous ape-like characteristics, including a large jaw, small braincase, and receding forehead. More recent hominins have a reduced jaw, larger braincase, and smaller brow ridge, much like modern humans.

TABLE 9.1 **The evidence for evolution.**

Observation	Example	Why It Suggests Common Descent
Biological classification. The most logical and useful system groups organisms hierarchically.	The levels of classification can identify degrees of relationship. 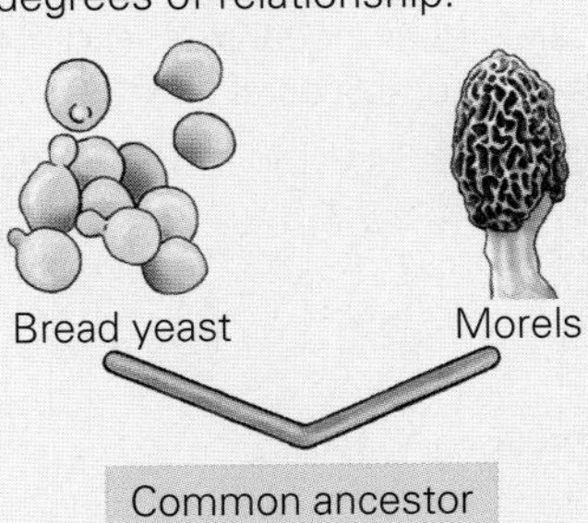	All species in the same family share a relatively recent common ancestor, while all families in the same class share a more distant common ancestor.
Anatomical homology. Organisms that look quite different have surprisingly similar structures.	Mammalian forelimbs share a common set of bones organized in the same way, despite their very different functions. 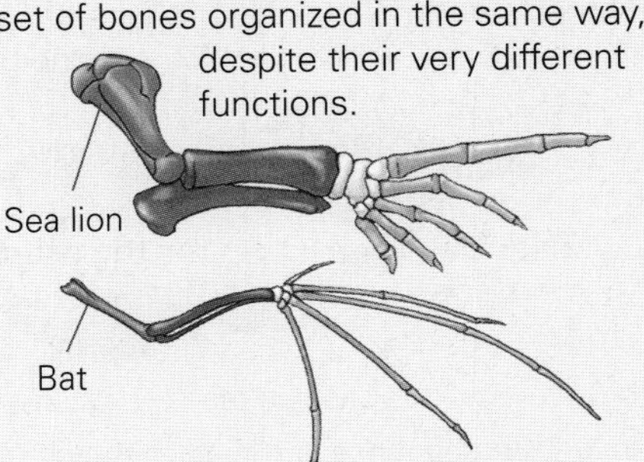	The simplest explanation is that each species inherited the basic structure from the same common ancestor, and evolution led to their modification in each group.
Vestigial traits. Some species display traits that are nonfunctional but have a functional equivalent in other species.	Flightless birds such as ostriches produce functionless wings. 	The simplest explanation is that the trait was functional in an ancestral species but lost its function over time in one branch of the evolutionary line.
Biogeography. The distribution of organisms on Earth corresponds, in part, to the relationship implied by biological classification.	The different species of tortoises on the Galápagos Islands are clearly related. 	Similarities among species in a geographic location imply divergence from ancestors in that geographic location.

(continued on the next page)

TABLE 9.1 **The evidence for evolution.** *(Continued)*

Observation	Example	Why It Suggests Common Descent
Homology in development. Early embryos of different species often look similar.	All chordates—animals that have a backbone or closely related structure—produce structures called pharyngeal slits, and most have tails as early embryos. 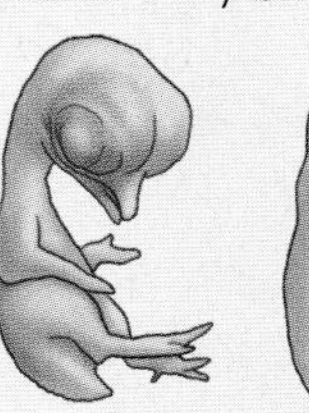 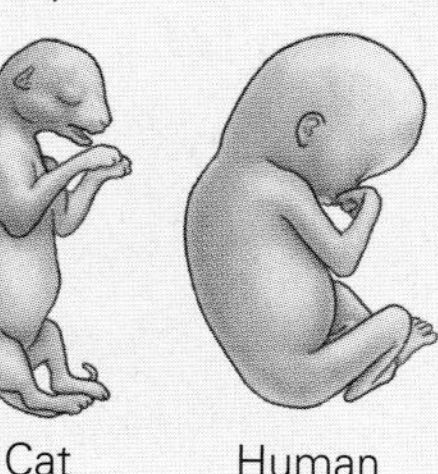 	Similarities in early development suggest that these organisms derived from a single common ancestor that developed along a similar pathway.
Homology of DNA. The DNA sequences of species that are closely related in a taxonomic grouping are more similar than those from more distantly related groups.	The closer 2 organisms are in classification, the more similar is the DNA sequence for the same genes. 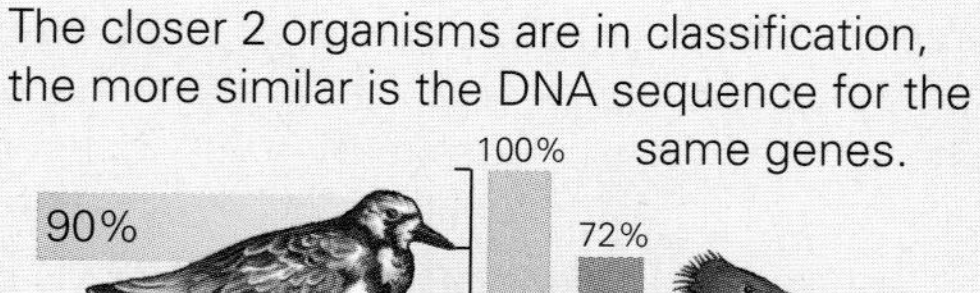	Similar DNA sequences in different species imply that the species evolved from a common ancestor with a particular sequence.
The fossil record. The remains of extinct organisms show progression from more ancient forms to more modern forms.	The transition between ancient and modern horses 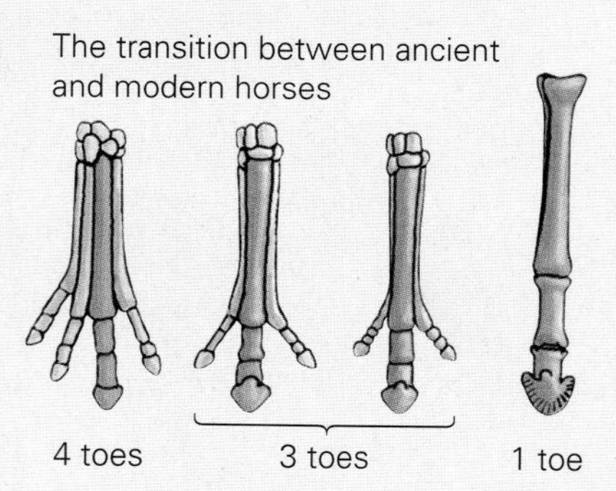	Fossils provide direct evidence of change in organisms over time and suggest relationships among modern species.

9.4 Are Alternatives to the Theory of Evolution Equally Valid?

Now we return to the 4 competing hypotheses: static model, transformation, separate types, and common descent. Do the observations described in the previous section allow us to reject any of these hypotheses? *Figure 9.27* summarizes our findings.

Weighing the Alternatives

The physical evidence we have discussed thus far allows us to clearly reject only 1 of the hypotheses—the static model. Radiometric dating indicates that Earth is

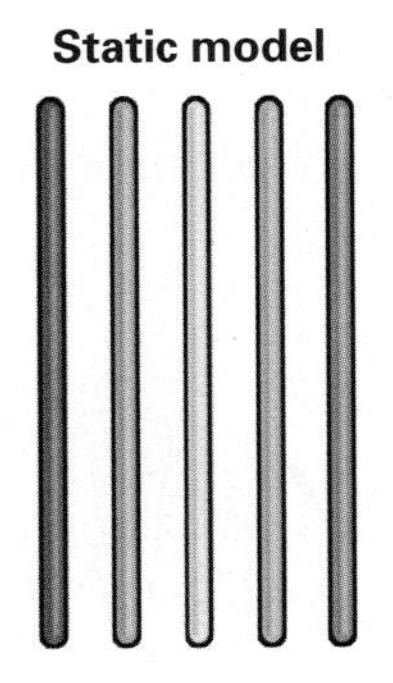

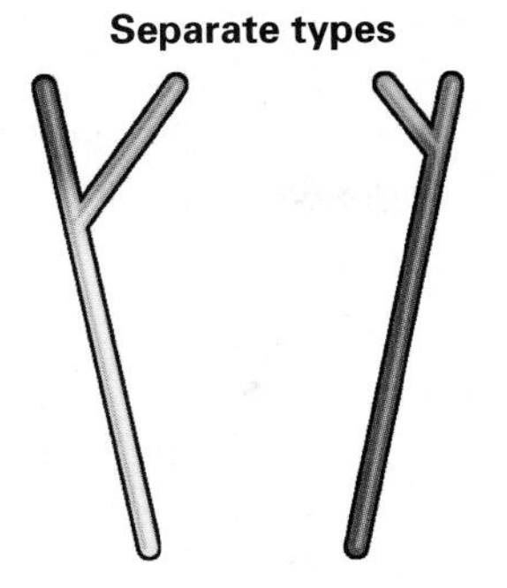

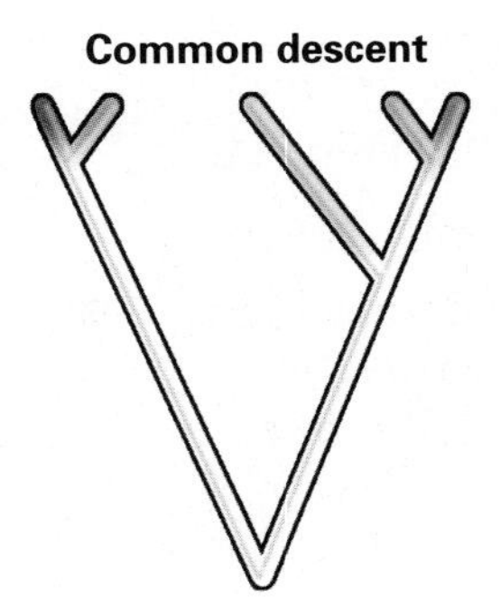

Figure 9.27 **Four hypotheses about the origin of modern organisms.** A scientific evaluation of these hypotheses leads to the rejection of all of them except for the theory of common descent.

far older than 10,000 years, and the fossil record provides unambiguous evidence that the species that have inhabited this planet have changed over time.

Of the remaining three hypotheses, transformation is the poorest explanation of the observations. If organisms arose separately, and each changed on its own path, there is no reason to expect that different species would share structures—especially if these structures are vestigial in some of the organisms. There is also no reason to expect similarities among species in DNA sequence. The hypothesis of transformation predicts that we will find little evidence of biological relationships among living organisms. As our observations have indicated, evidence of relationships abounds.

Both the hypothesis of common descent and the hypothesis of separate types contain a process by which we can explain observations of relationships. That is, both hypothesize that modern species are descendants of common ancestors. The difference between the two theories is that common descent hypothesizes a single common ancestor for all living things, while separate types hypothesizes that ancestors of different groups arose separately and then gave rise to different types of organisms. Separate types seems more reasonable than common descent to many people. It seems impossible that organisms as different as pine trees, mildew, ladybugs, and humans share a common ancestor. However, several observations indicate that these disparate organisms are all related.

The most compelling evidence for the single origin of all life is the universality of both DNA and of the relationship between DNA and proteins. As noted in Chapter 8, genes from bacteria can be transferred to plants, and the plants will make a functional bacterial protein. This is possible only because both bacteria and plants translate genetic material into functional proteins in a similar manner. If bacteria and plants arose separately, we could not expect them to translate genetic information similarly.

Units One and Two of this text describe biochemical processes and cell structures that are found in nearly all living things. The fact that organisms as different as pine trees, mildew, ladybugs, and humans contain cells with nearly all of the same components and biochemistry is also evidence of shared ancestry *(Figure 9.28)*. A mitochondrion could have many different possible structural forms and still perform the same function; the fact that the mitochondria in a plant cell and an animal cell are essentially identical implies that both groups of organisms inherited these mitochondria from a common ancestor.

Pine trees, mildew, ladybugs, and humans *are* very different. Proponents of the hypothesis of separate types argue that the differences among these organisms could not have evolved in the time since they shared a common

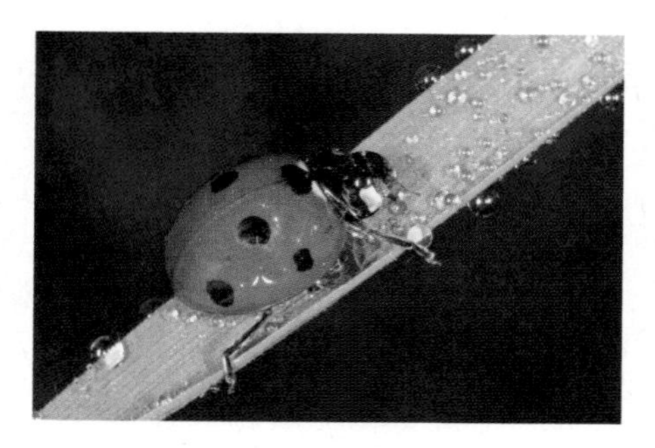

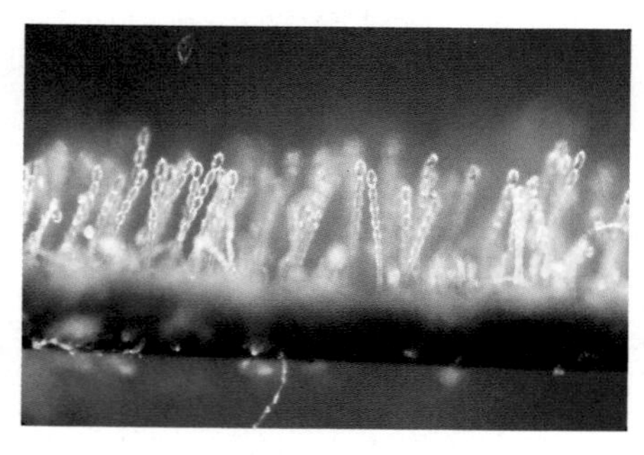

Figure 9.28 **The unity and diversity of life.** The theory of evolution, including the theory of common descent, provides the best explanation for how organisms as distinct as pine trees, humans, mildew, and ladybugs can look very different while sharing a genetic code and many aspects of cell structure and cell division.

ancestor. But the length of time during which pine trees, mildew, ladybugs, and humans have been diverging is immense—at least 1 billion years. The remaining basic similarities among all living organisms serve as evidence of their ancient relationship.

The Origin of Life

While the theory of evolution starts from the premise that a single common ancestor once existed, it does not address the origin of this single ancestor. The origin of life is a subject that is generally studied by biochemists and physicists rather than biologists. Recall the statement by the Dover school board in the introduction. They explicitly stated that "origins of life is not taught." The board could support this position because the subject is not even covered by Pennsylvania's standardized science exams. But without invoking a supernatural creator, the working hypothesis must be that the first living organisms must have derived from nonliving precursors.

According to this hypothesis, the process can be broken down into three basic steps:

1. Nonbiological processes assembled the simple molecules that were present early in the history of the solar system into more complex molecules.
2. These molecules then assembled themselves into chains that could store information and/or drive chemical reactions.
3. Collections of these complex molecules were assembled into a self-replicating "cell" with a membrane and energy source. This cell fed on other complex molecules.

There is some experimental evidence to support all three steps of this hypothesis. First, Stanley Miller, a graduate student working in the lab of his mentor Harold Urey in 1953, attempted to re-create conditions on early Earth within a laboratory apparatus *(Figure 9.29)*. After allowing the apparatus, which contained very simple molecules and an energy source, to "run" for one week, Miller found that complex organic molecules had formed spontaneously. These molecules included the building blocks of proteins and sugars. His results and other similar experiments support step 1 of this hypothesis.

More recent experiments have demonstrated that these building-block chemicals can be induced to form long chains when put into contact with hot sand, clay, or rock. Long chains of DNA and RNA nucleotides and of amino acids have been created using these methods, providing some experimental support for step 2.

Finally, in the early 1980s, 2 teams of scientists demonstrated that an information-carrying molecule, RNA, could also potentially copy itself, and so at least part of step 3, the capacity for self-replication, has experimental support. Although life as we would identify it has not been created in the lab from scratch, these results support the hypothesis that life could have formed spontaneously on Earth.

Stop and Stretch 9.5: Some scientists hypothesize that life arose on another planet and was seeded here by asteroids thrown off by that planet. Does this hypothesis cause a change in how scientists study the origin of life?

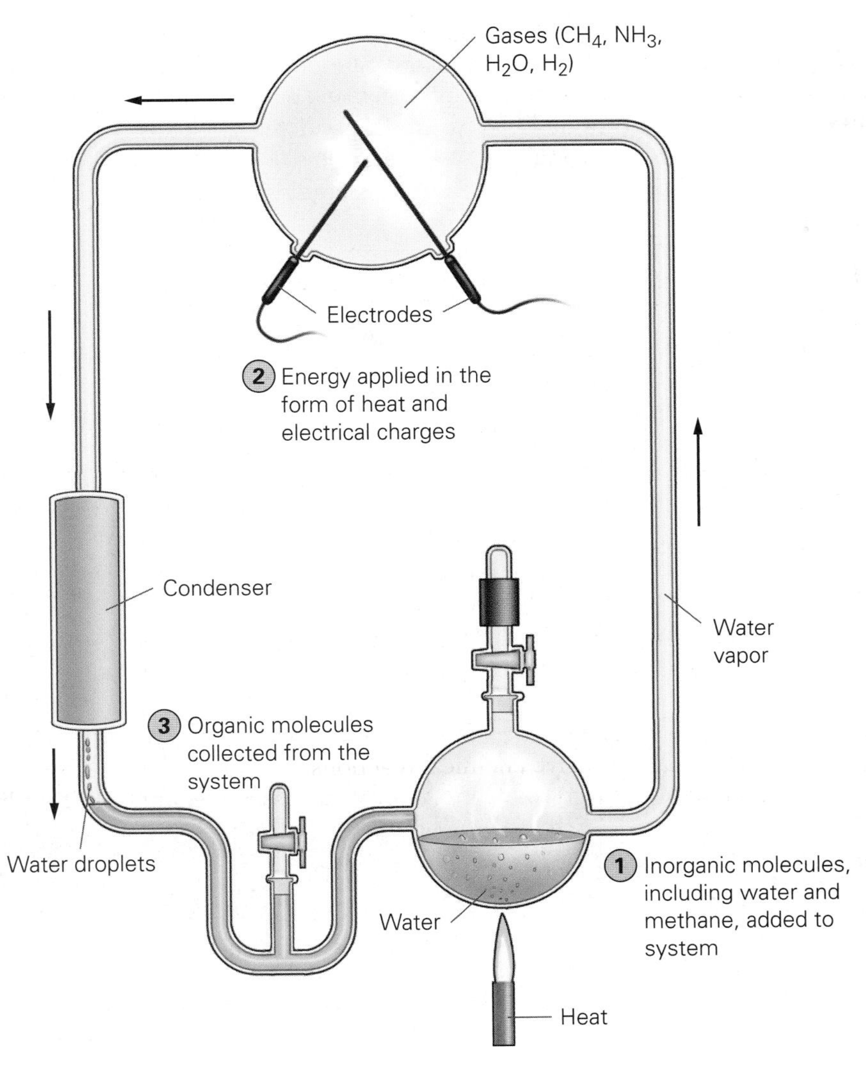

Figure 9.29 **The spontaneous formation of organic molecules.** This apparatus simulated conditions on early Earth and demonstrated an essential step in the formation of life from inorganic precursors.

Figure 9.30 **Why evolution matters.** The study of evolution informs all areas of biological science, from learning about human genes by experimenting on related genes in other animals (a), to using our understanding of how life responded to global changes in the past to help us prepare for the future (b).

The Best Scientific Explanation for the Diversity of Life

Scientists favor the theory of common descent because it is the best explanation for how modern organisms came about. The theory of evolution—including the theory of common descent—is robust, meaning that it is a good explanation for a variety of observations and is well supported by a wide variety of evidence from anatomy, geology, molecular biology, and genetics. Evidence for the theory of common descent demonstrates **consilience**, meaning that there is concurrence among observations derived from different sources. Consilience is a feature of all strongly supported scientific theories.

Similarities between humans and modern apes are seen in anatomy, behavior, development, and DNA sequence. The age, location, and appearance of hominin fossils also demonstrate a relationship between these species. The theory of common descent is no more tentative than is atomic theory; few scientists disagree with the models that describe the basic structures of atoms, and few disagree that the evidence for the theory of common descent is overwhelming. Most scientists would say that both of these theories are so well supported that we can call them fact.

Evolutionary theory helps us understand the functions of human genes, comprehend the interactions among species, and predict the consequences of a changing global environment for modern species *(Figure 9.30)*. Describing evolution as "*just* a theory" vastly understates the importance of evolutionary theory as a foundation of modern biology. Students who do not have a grasp of this fundamental biological principle may lack an appreciation of the basic unity and diversity of life and fail to understand the effects of evolutionary history and change on the natural world and ourselves. For example, we will explore why evolution is important for understanding and treating human disease in Chapter 10.

School boards and legislatures do not serve their students well by arguing that "alternative" scientific hypotheses that have been convincingly falsified through systematic observation should be included in their education. And unfalsifiable alternative theories that posit a supernatural creator do not belong in publicly funded classrooms. Judge Jones overturned the Dover school board's actions in his ruling, stating, "The overwhelming evidence at trial established that [intelligent design] is a religious view, a mere re-labeling of creationism, and not a scientific theory." Because they encouraged students and the publicly funded schools they administered to "explore" intelligent design, the Dover school board illegally inserted religion into a governmental function. For now, intelligent design cannot be presented as a viable alternative to evolution in public schools. But there is no doubt that this is not the last we will hear of a debate that is as old as the theory of evolution itself.

Savvy Reader Do Scientists Doubt Evolution?

Discovery Institute February 8, 2007

Another 100 scientists have joined the ranks of scientists from around the world publicly stating their doubts about the adequacy of Darwin's theory of evolution.

"Darwinism is a trivial idea that has been elevated to the status of the scientific theory that governs modern biology," says dissent list signer Dr. Michael Egnor. Egnor is a professor of neurosurgery and pediatrics at State University of New York, Stony Brook and an award-winning brain surgeon named one of New York's best doctors by New York Magazine.

Discovery Institute's Center for Science and Culture today announced that over 700 scientists from around the world have now signed a statement expressing their skepticism about the contemporary theory of Darwinian evolution. The statement, located online at www.dissentfromdarwin.org, reads: "We are skeptical of claims for the ability of random mutation and natural selection to account for the complexity of life. Careful examination of the evidence for Darwinian theory should be encouraged."

(continued on the next page)

Savvy Reader (continued)

1. The "dissent from Darwin" project can be thought of as an argument against the theory of evolution. Describe the essence of this argument.
2. Critique the argument. What counterpoints can be made?
3. The Internet contains volumes of information on topics that promote controversy, such as evolution. How would you determine if the Web site this article was drawn from was credible? If a group that sponsors a Web site has a clear agenda on an issue, how might that impact the material posted there?

CHAPTER Review

For study help, animations, and more quiz questions go to www.mybiology.com.

Summary

9.1 What Is Evolution?

- The process of evolution is the change that occurs in the characteristics of organisms in a population over time (p. 226).
- The theory of evolution, as described by Charles Darwin, is that all modern organisms are related to each other and arose from a single common ancestor (p. 228).

web animation 9.1
Principles of Evolution

9.2 Charles Darwin and the Theory of Evolution

- Scientists before Charles Darwin had hypothesized that species could change over time. Darwin's voyage on the *HMS Beagle* led him to suspect that this hypothesis was correct. Over the course of 20 years, he was able to gather enough evidence to support this hypothesis and the hypothesis of common descent so that most scientists accepted these theories as the best explanation for the diversity of life on Earth (pp. 228–230).

9.3 Examining the Evidence for Common Descent

- Linnaeus classified humans in the same order with apes and monkeys based on his observations of physical similarities between these organisms. Darwin argued that the pattern of biological relationships illustrated by Linnaeus's classification provided strong support for the theory of common descent (p. 231).
- Similarities in the underlying structures of a variety of organisms and the existence of vestigial structures are difficult to explain except through the theory of common descent (p. 231).
- Similarities in embryonic development among diverse organisms are best explained as a result of their common ancestry (p. 232).
- Modern data on similarities of DNA sequences among organisms match the hypothesized evolutionary relationships suggested by anatomical similarities and provide an independent line of evidence supporting the hypothesis of common descent (p. 233).
- Biogeographical patterns support the hypothesis of common descent because species that appear

Summary (continued)

related physically are also often close to each other geographically (p. 240).

- Radiometric dating of fossil remains indicates that human-like, bipedal primates appeared about 5 million years ago (p. 241).
- As predicted by the theory of common descent, ancient hominins have more ape-like characteristics than do more modern hominins (p. 241).

9.4 Are Alternatives to the Theory of Evolution Equally Valid?

- Evidence strongly supports the hypothesis that organisms have changed over time and are related to each other (p. 246).
- Shared characteristics of all life, especially the universality of DNA and the relationship between DNA and proteins, provide evidence that all organisms on Earth descended from a single common ancestor rather than from multiple ancestors (p. 248).
- While life has not been created from nonliving precursors in the laboratory, there is some experimental evidence that the spontaneous generation of life may have led to the common ancestor of all living things (p. 248)

Roots to Remember

These roots come from Greek or Latin and will help you decode the meaning of words:

evol- means to unroll.
homini- means human-like.
homolog- indicates similar or shared origin; from a word meaning "in agreement."

Learning the Basics

1. Describe the theory of common descent.

2. What observations did Charles Darwin make on the Galápagos Islands that helped convince him that evolution occurs?
A. the existence of animals that did not fit into Linneaus's classification system; **B.** the similarities and differences among mockingbirds and tortoises on the different islands; **C.** the presence of species he had seen on other tropical islands far from the Galápagos; **D.** the radioactive age of the rocks of the islands; **E.** fossils of human ancestors

3. The process of biological evolution _______________.
A. is not supported by scientific evidence; **B.** results in a change in the features of individuals in a population; **C.** takes place over the course of generations; **D.** B and C are correct; **E.** A, B, and C are correct

4. In science, a theory is a(n) _______________.
A. educated guess; **B.** inference based on a lack of scientific evidence; **C.** idea with little experimental support; **D.** a body of scientifically acceptable general principles; **E.** statement of fact

5. The theory of common descent states that all modern organisms _______________.
A. can change in response to environmental change; **B.** descended from a single common ancestor; **C.** descended from one of many ancestors that originally arose on Earth; **D.** have not evolved; **E.** can be arranged in a hierarchy from "least evolved" to "most evolved"

6. The DNA sequence for the same gene found in several species of mammals _______________.
A. is identical among all species; **B.** is equally different between all pairs of mammal species; **C.** is more similar between closely related species than between distantly related species; **D.** provides evidence for the hypothesis of common descent; **E.** more than 1 of the above is correct

7. Match the observation to the type of evidence it provides for the theory of evolution

___ fossils of the oldest hominins are found in Africa near modern great apes
___ both humans and chimpanzees have a vestigial appendix
___ the brain size of ancient hominins was less than that of modern humans

A. fossil record
B. homology
C. biogeography

8. A species of crayfish that lives in caves produces eyestalks like its above-ground relatives, but no eyes. Eyestalks in cave-dwelling crayfish are thus ______________.
A. an evolutionary error; **B.** a dominant mutation; **C.** biogeographical evidence of evolution; **D.** a vestigial trait; **E.** evidence that evolutionary theory may be incorrect.

9. What characteristics of a fossil can paleontologists use to determine whether the fossil is a part of the human evolutionary lineage?
A. the position of the foramen magnum; **B.** the structure of the pelvis; **C.** the structure of the foot; **D.** A and C are correct; **E.** A, B, and C are correct

10. The fossil record of hominins ______________.
A. does not indicate a relationship between humans and apes because a missing link has not been found; **B.** dates back at least 5 million years; **C.** indicates that bipedal apes first evolved in Africa; **D.** B and C are correct; **E.** A, B, and C are correct

Analyzing and Applying the Basics

1. The classification system devised by Linnaeus can be "rewritten" in the form of an evolutionary tree. Draw a tree that illustrates the relationship among these flowering species, given their classification (note that "subclass" is a grouping between class and order):

Pasture rose (*Rosa carolina*, Family Rosaceae, Order Rosales, Subclass Rosidae)
Live forever (*Sedum purpureum*, Family Crassulaceae, Order Rosales, Subclass Rosidae)
Spring avens (*Geum vernum*, Family Rosaceae, Order Rosales, Subclass Rosidae)
Spring vetch (*Vicia lathyroides*, Family Fabaceae, Order Fabales, Subclass Rosidae)
Multiflora rose (*Rosa multiflora*, Family Rosaceae, Order Rosales, Subclass Rosidae)

2. DNA is not the only molecule that is used to test for evolutionary relationships among organisms. Proteins can also be used, and the sequences of their building blocks (called amino acids) can be compared in much the same way that DNA sequences are compared. *Cytochrome c* is a protein found in nearly all living organisms; it functions in the transformation of energy within cells. The percent difference in amino acid sequence between humans and other organisms can be summarized as follows:

Cytochrome c sequence of	*Percent difference from human sequence*
Chimpanzee	0.0
Mouse	8.7
Donkey	10.6
Carp	21.4
Yeast	32.7
Corn	33.3
Green algae	43.4

Draw the evolutionary tree implied by these data that illustrates the relationship between humans and the other organisms listed.

3. Look at the tree you generated above. It does not imply that yeast is more closely related to corn than it is to green algae. Why not?

Connecting the Science

1. Humans and chimpanzees are more similar to each other genetically than many very similar-looking species of fruit fly are to each other. What does this similarity imply regarding the usefulness of chimpanzees as stand-ins for humans during scientific research? What do you think it implies regarding our moral obligations to these animals?

2. Creationists have argued that if students learn that humans descended from animals and are, in fact, a type of animal, these impressionable youngsters will take this fact as permission to act on their "animal instincts." What do you think of this claim?

CHAPTER 10

An Evolving Enemy

AN EVOLVING ENEMY AN EVOLVING ENEMY AN EV

Natural Selection

What should we do with someone who has a dangerous, deadly disease?

Andrew Speaker found out one answer to that question.

He may have knowingly exposed thousands of fellow airline passengers to his disease.

In the early summer of 2007, the U.S. Centers for Disease Control and Prevention (CDC), the federal agency responsible for monitoring and protecting public health, did something that hasn't occurred for over 40 years. They ordered a man to be placed under involuntary quarantine in a hospital.

The man was Andrew Speaker, a 31-year-old personal injury lawyer from Atlanta. In spring of that year, Speaker had been diagnosed with a disease that is relatively harmless in healthy Americans—tuberculosis. He had no apparent symptoms, and the diagnosis and subsequent drug treatment didn't interfere with his plans for later that spring, which were to fly from Atlanta to a Greek island to get married and then travel to Rome, Italy, for his honeymoon. But while he was in Rome, the CDC contacted him to inform him that on further study, they had determined that he had a very dangerous form of the disease.

Depending on which side tells the tale, the CDC either ordered or asked Speaker and his wife to remain in Rome until they could arrange travel back to the United States. Speaker instead booked a flight to Prague, Czechoslovakia, and from there to Montreal, Canada, before renting a car and driving it back to Atlanta. In his travels, everyone he had contact with incurred a risk of infection by the bacteria that causes this dangerous variety of tuberculosis.

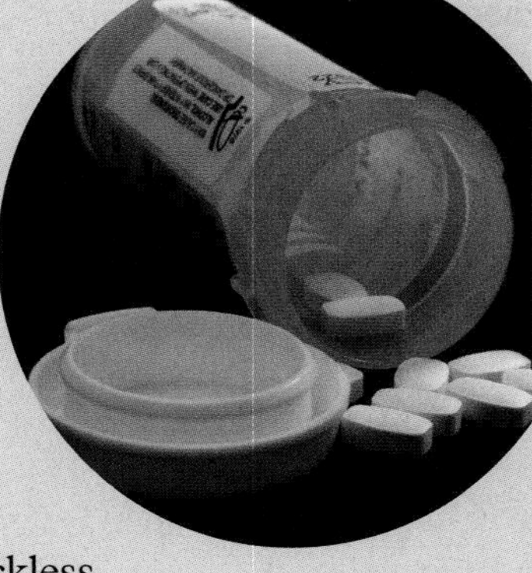

But it is the unconsciously reckless behavior of thousands of us that is the ultimate cause of the problem.

People around the world were outraged at Andrew Speaker's seemingly reckless behavior. How could he put so many people at risk of a devastating and often deadly disease? But others were outraged by the treatment he received from the CDC. Why would they pluck a seemingly healthy man from his honeymoon and place him in involuntary confinement? As it turns out, the CDC has some cause to be very concerned about this form of tuberculosis—but few of us are threatened by the actions of Speaker. Instead, it is the recklessness of millions of ordinary people around the world that put us at risk of this, and other, deadly microbes.

10.1 Return of an Ancient Killer

Tuberculosis (TB) has been plaguing humans for thousands of years. Tubercular decay has been found in the spines of Egyptian mummies from 3000 to 2400 B.C. Around 460 B.C., the Greek physician Hippocrates identified a condition that appears to be tuberculosis as the most widespread disease of the times. In 1906, 2 of every 1000 deaths in the United States were due to TB infection.

Thanks to advances in science and medicine, TB now accounts for only 1.5 of every 100,000 deaths in the United States. In the 1980s, the dramatic drop in infection and death had many public health professionals convinced that this ancient scourge could be completely eliminated. Today, those hopes have largely faded.

What Is Tuberculosis?

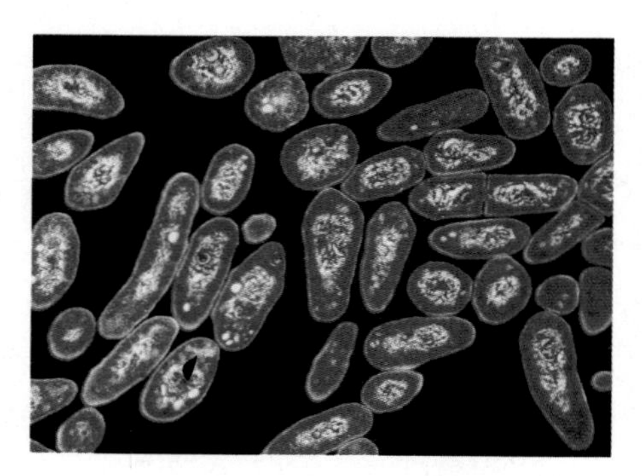

Figure 10.1 **The organism that causes tuberculosis.** *Mycobacterium tuberculosis.*

Tuberculosis is caused by the single-celled bacterium *Mycobacterium tuberculosis (Figure 10.1)*. Tuberculosis is a threat not because it is an especially deadly infection, but instead because it affects so many people. Two billion people, over one-third of the world's population, carry *M. tuberculosis,* and new infections are estimated to occur at a rate of 1 per second. Ninety percent of *M. tuberculosis* infections are symptomless, and most of these resolve as the bacteria are destroyed by the infected individual's immune system. However, the remaining individuals develop active disease, and more than one-half of these individuals will die without treatment. Tuberculosis now causes approximately 2 million deaths worldwide every year.

The symptoms of tuberculosis include a cough that produces blood, fever, fatigue, and a long relentless wasting in which the patient gradually becomes weaker and thinner. These symptoms explain the antiquated name for this disease, consumption, because the infection seemed to consume people from within.

We now understand that the consumptive symptoms arise from destruction of lung tissue caused by the body's reaction to active *M. tuberculosis* infection. Colonies of the bacteria are walled off by immune system cells in structures called tubercles *(Figure 10.2)*; while this may slow the spread of the disease in the body, it permanently damages the lung tissue. Physical wasting is a result of a reduction in the lung's capacity to provide oxygen to the body. *Mycobacterium tuberculosis* can lie dormant for months inside these tubercles, which may degrade and then release bacteria back into the lung tissue. Once released from the tubercule, the disease is considered active again, and these bacteria can cause additional infections within the lungs and be spread to other individuals.

Tuberculosis infection is a challenge to diagnose, primarily because *M. tuberculosis* is difficult to grow in the lab from saliva or lung tissue. Typically, infection is confirmed by an X-ray examination for tubercles in the lungs and a skin test that looks for the presence of an immune response to the bacteria

(a) (b)

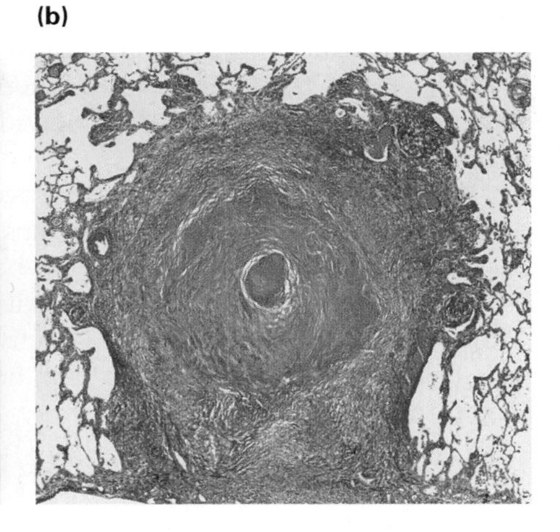

Figure 10.2 **Lung tubercles.** (a) The white spots on this lung X-ray are nodules, or tubercles, within the lung tissue. (b) The tubercles contain *M. tuberculosis* encased by immune system cells and other tissues.

(Figure 10.3). A significant reaction to the skin test indicates that the individual was infected with *M. tuberculosis* but cannot determine whether the infection is current or if the individual has cleared the bacterium from his body.

Although you can be infected without symptoms, transmission of tuberculosis is almost entirely from people with active disease. When these individuals cough, sneeze, speak, or spit, they expel infectious droplets. A single sneeze can release about 40,000 of these droplets *(Figure 10.4)*. People with prolonged, frequent, or intense contact with infected individuals are at highest risk of becoming infected themselves.

Most individuals can fight off the infection, but many cannot. Those at highest risk of active TB are young children; the elderly; individuals in poor overall health due to poor nutrition, other illnesses, or drug abuse; and people with AIDS. Until about 60 years ago, these individuals had little chance of surviving tuberculosis. Today, their prognosis—especially in Western countries where high-quality health care is readily available—is much better. But that may be changing.

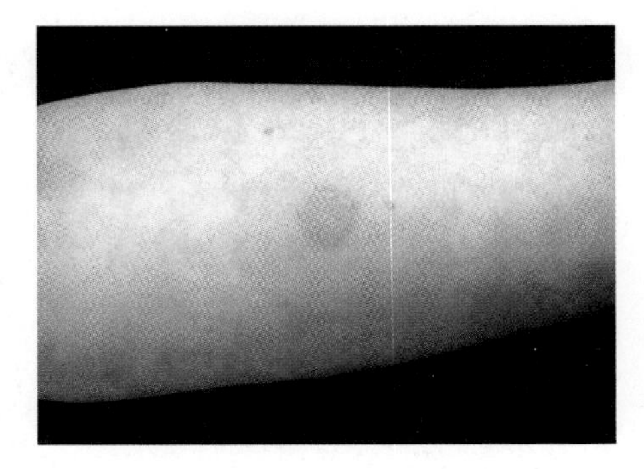

Figure 10.3 **Tuberculosis test.** The red spot on this individual's forearm marks the site where a small amount of disassembled *M. tuberculosis* was injected just under the surface of the skin. The cell fragments cannot cause disease, but if the individual has been exposed to the bacteria in the past, the spot will become inflamed, as seen here.

Treatment—and Treatment Failure

The treatments for tuberculosis in the nineteenth and early twentieth century, at least among the wealthy, consisted primarily of long stays in environments where the air was fresh and unpolluted. These tuberculosis sanatoriums *(Figure 10.5)* were useful for 2 reasons. By moving patients from areas where the air was thick with lung-damaging particles and chemicals, they preserved lung function for longer periods. And by isolating these patients, they reduced the spread of TB to the rest of the community. In poorer communities, individuals with active TB were often forcibly isolated in much more grim conditions.

The discovery of **antibiotics,** drugs that kill certain nonhuman (primarily bacterial) cells, revolutionized tuberculosis treatment in the 1940s. Since then, infected patients with active TB are typically kept in isolation for only 2 weeks until antibiotics kill off most of the *M. tuberculosis* in the lungs. At this point, the patient is no longer contagious and can return to the community. However, because the *M. tuberculosis* can hide inside immune system cells for long periods, antibiotic treatment must be maintained for 6 to 12 months to completely eliminate the organism.

Figure 10.4 **Transmission through the air.** Like cold and flu viruses, the tuberculosis bacteria can be transmitted in droplets that are emitted in the sneeze or cough of an infected individual.

Since the 1980s, however, scientists have chronicled a disturbing rise in the number of **antibiotic-resistant** tuberculosis infections—ones that cannot be cured by the standard drug treatment. According to the CDC, from 2000 to 2004, 20% of reported TB cases did not respond to standard treatments

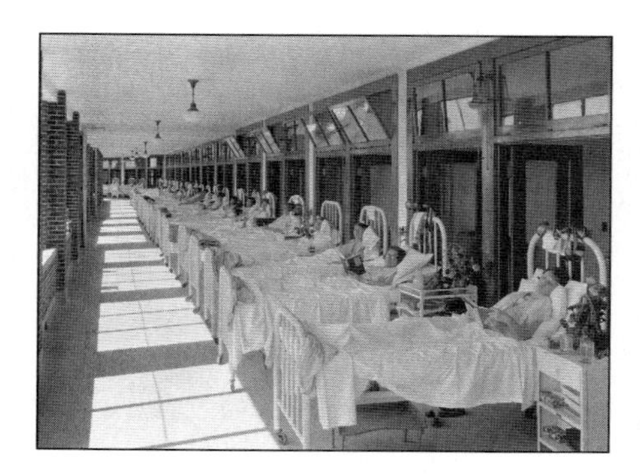

Figure 10.5 **A tuberculosis sanatorium.** Typically, patients at sanatoriums spent many hours every day exposed to fresh air.

(and thus are called multidrug-resistant TB, or MDR-TB), and 2% were resistant to treatment with second-line drugs (called extensively drug-resistant TB, or XDR-TB). Even in the United States, with abundant access to resources and drugs, one-third of individuals diagnosed with active XDR-TB have died of the disease.

As control has become less effective, the number of cases of TB in both developed and developing countries has begun to rise. In countries with fewer resources, the toll of XDR-TB could be much greater. In a recent XDR-TB outbreak in South Africa, 52 of 53 individuals diagnosed with the strain died within a month of showing signs of active disease. The resurgence of tuberculosis has now been declared a global health emergency by the World Health Organization.

Stop and Stretch 10.1: Why is it a problem for patients with active TB if the first antibiotic treatment they try is ineffective? How do you think this affects their overall health and chances of survival?

Why are we losing the battle against tuberculosis? And what can be done to stop it? Answering these questions requires an understanding of an important force for evolutionary change: natural selection.

web animation 10.1 *Natural Selection*

10.2 Natural Selection Causes Evolution

In *The Origin of Species*, Charles Darwin put forth two major ideas: the theory of common descent and the theory of **natural selection.** We discussed the theory of common descent in detail in Chapter 9 and learned that all species living today appear to have descended from a single ancestor that arose in the distant past. Darwin's presentation of this theory was thorough and convincing. Within 20 years of his book's publication, Darwin's principle that all living organisms are related to each other through common descent had been accepted by most scientists. However, it was another 60 years before the scientific community accepted Darwin's ideas about *how* the great variety of living organisms had come about—the process he called natural selection.

Today, natural selection is considered one of the most important causes of evolution, although others, such as the processes of genetic drift and sexual selection as described in Chapter 11, also cause populations to change over time.

Four Observations and an Inference

The theory of natural selection is elegantly simple. It is an inference based on 4 general observations:

1. **Individuals within Populations Vary.** Observations of groups of humans support this statement—people do come in an enormous variety of shapes, sizes, colors, and facial features. It may be less obvious that there is variation in nonhuman populations as well. For example, within a litter of gray wolves born to a single female, some individuals may be black, tawny, or reddish in color *(Figure 10.6)*.

(a) Variation in coat color

(b) Variation in blooming time

Figure 10.6 **Observation 1: Individuals within populations vary.** (a) Gray wolves vary in coat color, even within a single litter of animals. (b) Flowers may vary in blooming time, with some individual plants blooming much earlier than others of the same species.

We can add all kinds of less-obvious differences to this visible variation; for example, the amount of caffeine produced in the seeds of a coffee plant varies among individuals in a wild population. Each different type of individual in a population is called a **variant.**

2. **Some of the Variation Among Individuals Can Be Passed on to Their Offspring.** Although Darwin did not understand how it occurred, he observed many examples of the general resemblance between parents and offspring. He also noticed that people can take advantage of the inheritance of variation in other animals.

   For example, when pigeon breeders wanted a particular fancy variety, they encouraged breeding among the birds with traits they desired. This allowed the breeders to change their flock because, for instance, pigeons with fan-shaped tails were more likely to produce offspring with fan-shaped tails than were pigeons with straight tails (*Figure 10.7*). Darwin hypothesized that offspring tend to have the same characteristics as their parents in natural populations as well.

   For several decades after *The Origin of Species* was published, the observation that some variations were inherited was the most controversial part of the theory of natural selection. Since scientists could not adequately explain the origin and inheritance of variation, many were unwilling to accept that natural selection could be a mechanism for evolutionary change. When Gregor Mendel's work on inheritance in pea plants (discussed in Chapter 6) was rediscovered in the 1800s, the mechanism for this observation became clear—natural selection operates on genetic variation that is passed from one generation to the next.

Figure 10.7 **Observation 2: Some of the variation among individuals can be passed on to their offspring.** Darwin noted that breeders could create flocks of pigeons with fantastic traits by using as parents of the next generation only those individuals that displayed these traits.

3. **Populations of Organisms Produce More Offspring Than Will Survive.** This observation is clear to most of us—the trees in the local park make literally millions of seeds every summer, but only a few of the much smaller number that sprout live for more than a year or two.

   In *The Origin of Species*, Darwin gave a graphic illustration of the difference between offspring production and survival. In his example, he used elephants, animals that live long lives and are very slow breeders. A female elephant does not begin breeding until age 30, and she produces about 1 calf every 10 years until around age 90. Darwin calculated that even at this very low rate of reproduction, if all the descendants of a single pair of African elephants survived and lived full, fertile lives, after

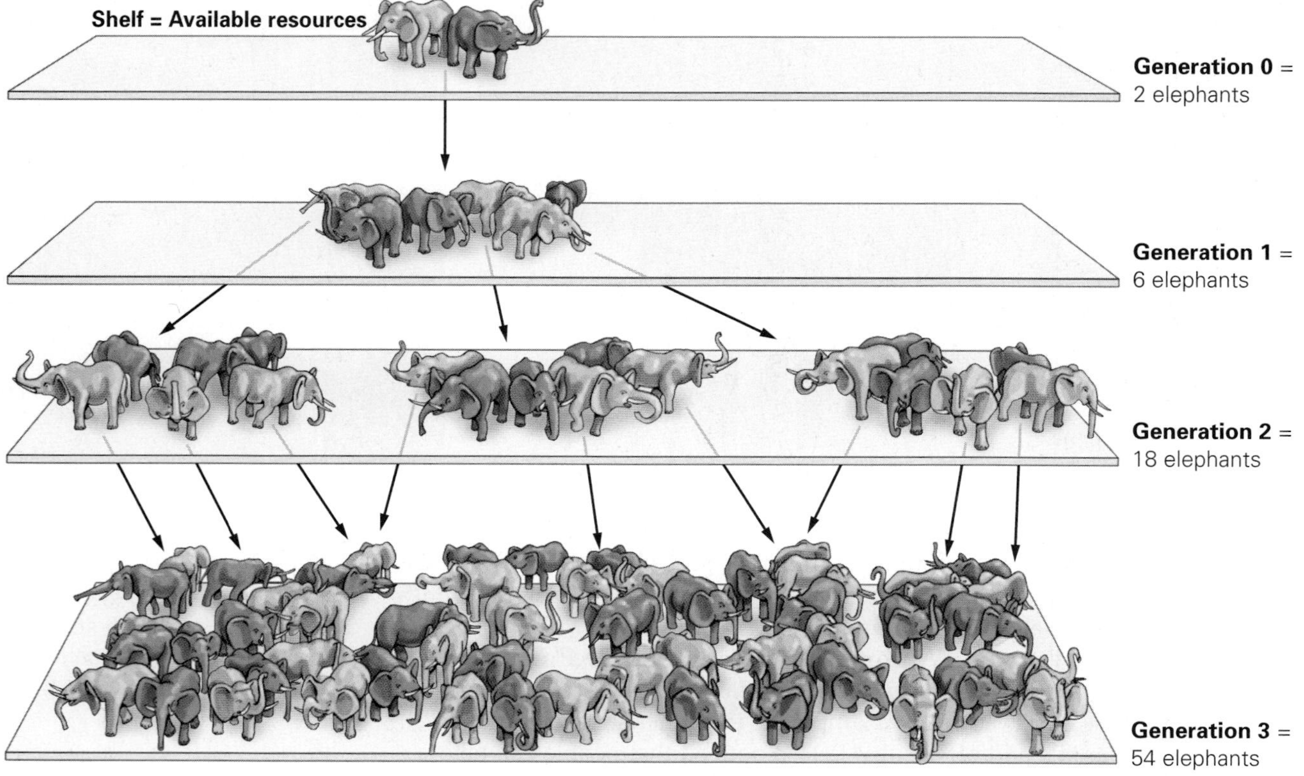

Figure 10.8 **Observation 3: Populations of organisms produce more offspring than will survive.** Even slow-breeding animals like elephants are capable of producing huge populations relatively quickly. **Visualize This:** What happens to the elephants when generation 4 is produced?

about 500 years their family would have more than 15 million members (*Figure 10.8*)—many more than can be supported by all the available food resources on the African continent!

Clearly, only a fraction of the elephants born in every generation survives long enough to reproduce. The same is true for nearly every species; the capacity for reproduction far outstrips the resources of the environment, so many individuals do not survive to maturity.

4. **Survival and Reproduction Are Not Random.** In other words, the subset of individuals that survive long enough to reproduce is not an arbitrary group. Some variants in a population have a higher likelihood of survival and reproduction than other variants do. The relative survival and reproduction of 1 variant compared to others in the same population is referred to as its **fitness**. Traits that increase an individual's fitness in a particular environment are called **adaptations**. Individuals with adaptations to a particular environment are more likely to survive and reproduce than are individuals lacking such adaptations.

Darwin referred to the results of differential survival and reproduction as natural selection. Although Darwin used the word *selection*, which implies some active choice, natural selection is a passive process. Adaptations are "selected for" in the sense that individuals possessing them survive and contribute offspring to the next generation.

For example, among the birds called medium ground finches living on an island in the Galápagos archipelago, scientists have observed that when rainfall

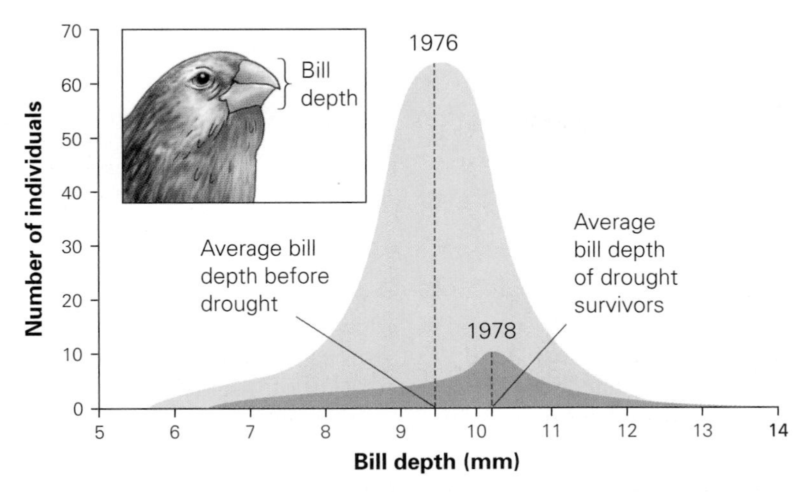

Figure 10.9 **Observation 4: Survival and reproduction are not random.** The pale purple curve summarizes bill depth in ground finches on Daphne Island in the Galápagos in 1976. The dark purple curve below it represents the population in 1978, after the drought of 1977. These data indicate that survivors of the drought had an average bill depth 0.6 mm greater than that of the predrought population. The change in the population's average bill size occurred because finches with larger-than-average bills had higher fitness than did small-billed birds during the drought. **Visualize This:** How did total population size change between 1976 and 1978?

is scarce, a large bill is an adaptation. This is because birds with larger bills are able to crack open large, tough seeds—the only food available during severe droughts. As shown in *Figure 10.9*, the 90 survivors of a 1977 drought had an average bill depth that was 6% greater than the average bill depth of the original population of 751 birds.

Adaptations are not only traits that increase survival. Any trait that increases the number of offspring produced relative to others in a population is also an adaptation. For example, flowers in a crowded mountain meadow may have a relatively limited number of potential insect pollinators. More pollinator visits generally result in more seeds being produced by a single flower, so any trait that increases a flower's attractiveness to pollinators, such as a brighter color or greater nectar production, should be favored by natural selection *(Figure 10.10)*.

Figure 10.10 **Adaptations are not about survival only.** Variations that increase a flower's attractiveness to a pollinator can increase its reproductive success by increasing the number of seeds it produces.

Darwin's Inference: Natural Selection Causes Evolution.

The result of natural selection is that favorable inherited variations tend to increase in frequency in a population over time, while unfavorable variations tend to be lost. In other words, adaptations become more common in a population as the individuals who possess them contribute larger numbers of their offspring to the succeeding generation. Natural selection results in a change in the traits of individuals in a population over the course of generations—voilà, evolution.

It is a testament to the power of the idea of natural selection that today it seems self-evident to us—we might even wonder why Darwin is considered a brilliant scientist for explaining something that seems so obvious. But it is only obvious to us because the idea of natural selection influences our understanding of so much of our world, from the success of particular brands of soft drinks to the relationships among nations. The idea of natural selection only became so powerful after it was tested and shown to work—in nature—in the manner Darwin described.

Stop and Stretch 10.2: Some biologists argue that human evolution as a result of natural selection on our physical traits has nearly stopped. What characteristics of humans support this view? In what ways might humans still be subject to natural selection?

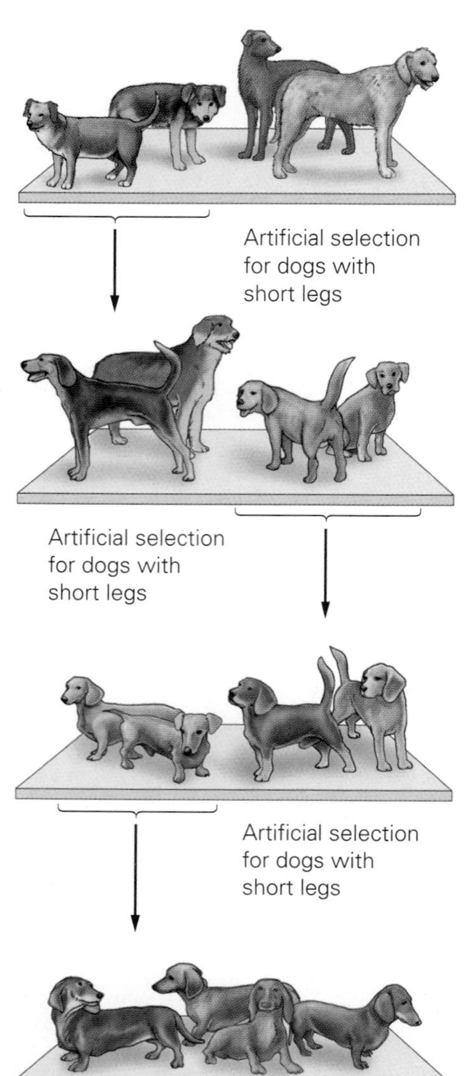

Figure 10.11 **Artificial selection can cause evolution.** When breeders select dogs with certain traits to produce the next generation of animals, they increase the frequency of that trait in the population. Over generations, the trait can become quite exaggerated. Dachshunds are descendants of dogs that were selected for the production of very short legs. **Visualize This:** What would this sequence of changes look like if the selection was for dogs with long legs and coats, as in Afghan hounds?

Testing Natural Selection

Darwin proposed a scientific explanation of how evolution occurs, and like all good hypotheses, it needed to be tested. All of the tests described below illustrate that natural selection is an effective mechanism for evolutionary change.

Artificial Selection. Selection imposed by human choice is called artificial selection. It is artificial in the sense that humans deliberately control the survival and reproduction of individual plants and animals with favorable characteristics to change the characteristics of the population.

The fancy pigeons that Darwin studied arose by artificial selection, just as dog varieties resulted from this process. In each case, different types evolved through selection by breeders for various traits *(Figure 10.11)*. These examples demonstrate that differential survival and reproduction change the characteristics of populations. However, due to the direct intervention of humans on the survival and reproduction of these organisms, artificial selection is not exactly equivalent to natural selection. Can change in populations occur without direct human intervention?

Natural Selection in the Lab. Another test of the effectiveness of natural selection is to look for the adaptation of populations to environments manipulated in laboratory settings. An example of this kind of experiment is one performed on fruit flies placed in environments containing different concentrations of alcohol.

High concentrations of alcohol cause cell death. Many organisms, including fruit flies and humans, produce enzymes that metabolize alcohol—that is, they break it down, extract energy from it, and modify it into less-toxic chemicals. There is variation among fruit flies in the rate at which they metabolize alcohol. In a typical laboratory environment, most flies process alcohol relatively slowly, but about 10% of the population possesses an enzyme variant that allows those flies to metabolize alcohol twice as rapidly as the more common variant.

In the experiment described in *Figure 10.12*, scientists divided a population of fruit flies into 2 random groups. Initially, these 2 groups had the same percentage of fast and slow alcohol metabolizers. One group of flies was placed in an environment containing typical food sources; the other group was placed in an environment containing the same food spiked with alcohol. After 57 generations, the percentage of fast-metabolizing flies in the normal environment was the same as at the beginning of the experiment—10%. But after the same number of generations, 100% of the flies in the alcohol-spiked environment were the fast-metabolizing variety. Because all of the flies in this environment were now the fast-metabolizing variety, the *average* rate of alcohol metabolism in the population was much higher in generation 57 than in generation 1. The population had evolved.

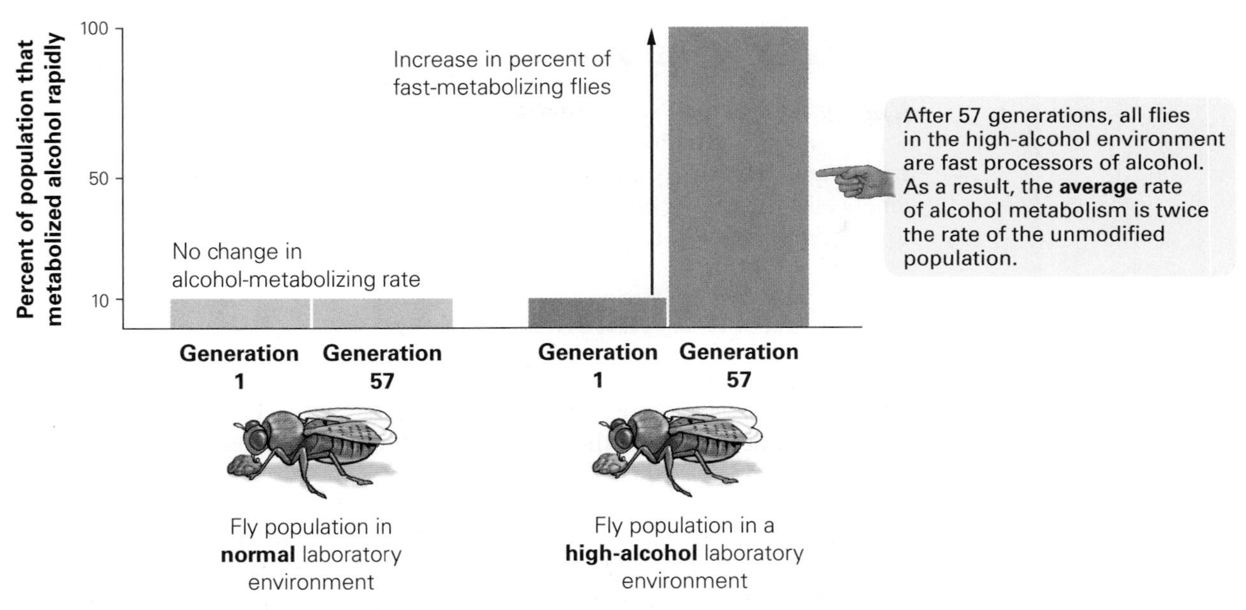

Figure 10.12 **An observation of natural selection.** When fruit flies are placed in an alcohol-spiked environment, the percentage of flies that can rapidly metabolize alcohol increases over many generations because of natural selection. In the normal laboratory environment, there is no selection for faster alcohol processing, so the average rate of alcohol metabolism does not change.

The evolution of the fruit flies in this experiment was a result of natural selection. In an environment where alcohol concentrations were high, individuals that were able to metabolize alcohol relatively rapidly had higher fitness. Since they lived longer and were less affected by alcohol, they left more offspring than the slow metabolizers did. Thus, each generation had a higher frequency of fast-metabolizing individuals than the previous generation did. After many generations, flies that could metabolize alcohol rapidly predominated in the population.

Selection can change populations in highly regulated laboratory environments. But does it have an effect in natural wild populations?

Natural Selection in Wild Populations. The evolution of *M.tuberculosis* from susceptible to antibiotics to resistant is 1 example of natural selection in a wild population; clearly, a change in the environment (that is, the introduction of antibiotics) caused a change in the bacteria population. Dozens of other pathogens have become resistant to drugs and pesticides in the last 50 years as well. But even these changes may seem suspect since the adaptation is to a human-imposed environmental change. Although studying adaptation to natural environmental changes in the field is a significant challenge, the effects of natural selection have been observed in dozens of wild populations.

A classic example of natural selection in a natural setting is the evolution of bill size in Galápagos finches in response to drought, described in Figure 10.9. The survivors of the drought tended to be those with the largest bills, which could more easily handle the tough seeds that were available in the dry environment.

The survival of this nonrandom subset of birds resulted in a change in the next generation. The population of birds that hatched from eggs in 1978—the descendants of the drought survivors—had an average bill depth 4% to 5% larger than that of the predrought population.

10.3 Natural Selection Since Darwin

While tests of natural selection seemed to support the theory of natural selection, it took 60 years for the rest of biological science to catch up to Darwin's insight and adequately explain it. Scientists working in the then-new field of genetics in the 1920s began to recognize that genetic concepts could explain how natural selection causes population change.

The Modern Synthesis

The union between genetics and evolution was termed "the modern synthesis" during its development in the 1930s and 1940s. The principles of the modern synthesis outline the model of evolutionary change accepted by the vast majority of biologists.

As we discussed in Chapter 8, genes are segments of genetic material (typically DNA, but RNA in some viruses) that contain information about the structure of molecules called proteins. The actions of proteins within an organism help determine its physical traits. Different versions of the same gene are called alleles, and variation in traits among individuals in a population is often due to variation in the alleles they carry.

We can apply these genetic principles to the fruit flies exposed to a high-alcohol environment. In this population there are 2 variants, or alleles, of the gene that controls alcohol processing. One allele produces an enzyme we will refer to as "fast," and the other produces an enzyme we will call "slow." Flies that make the fast enzyme can metabolize alcohol rapidly. To make this enzyme, they must carry 2 copies of the fast allele.

As described in detail in Chapter 6, half of the alleles carried by a parent are passed to their offspring via their eggs or sperm. In the high-alcohol environment, flies with the fast enzyme had more offspring than did flies with the slow enzyme. Since they carry 2 copies of the fast allele, each of the offspring of a fast metabolizer received a copy of this allele. Therefore, in the next generation, a higher percentage of individuals carried the fast allele. We can now describe the evolution of a population as an increase or decrease in the frequency of an allele for a particular gene.

Understanding the nature of genes also explains the origin of their variations. Different alleles for the same gene arise through mutation—changes in the DNA sequence. As described in Unit Two, mutations that can be passed on to offspring occur by chance when DNA is copied during the production of eggs and sperm. If a mutation results in an allele that has a function different from that of the original allele, the resulting variation could become subject to the process of natural selection.

The existence of 2 different alleles for alcohol metabolism in fruit flies suggests that 1 of these alleles is a mutated version of the other. In the normal laboratory environment, neither of these alleles appears to have a strong effect on fitness. Since the slow metabolizers are more numerous than the fast metabolizers, it appears that there might be a slight disadvantage to carrying the fast enzyme in a low-alcohol environment. However, in the high-alcohol environment, the mutation resulting in the fast allele gives a strong advantage, and its presence in the population allows for the population's evolution *(Figure 10.13)*.

Scientists now understand that the random process of gene mutation generates the raw material—variations—for evolution, and that natural selection acts as a filter that selects for or against new alleles produced by mutation.

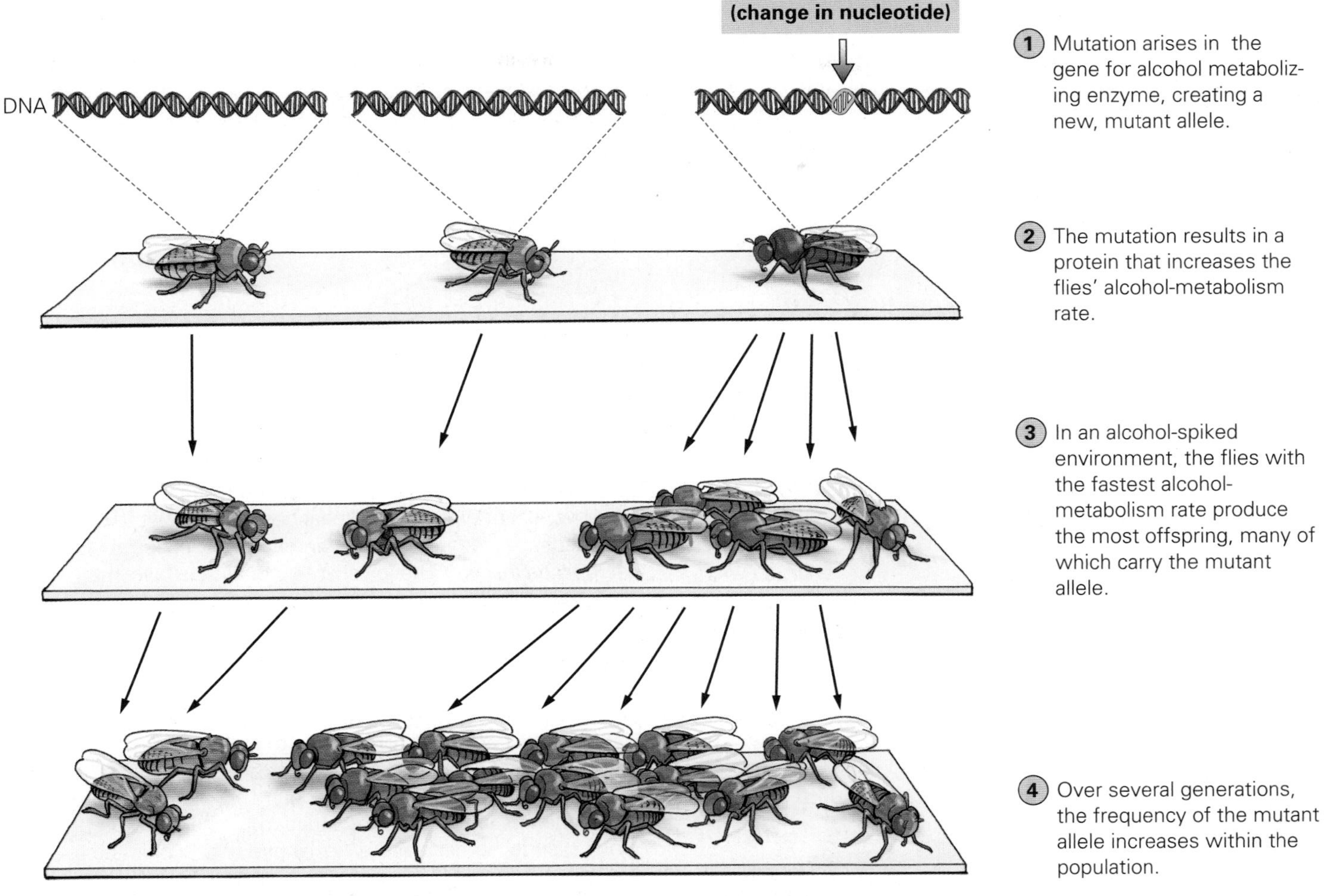

Figure 10.13 **Mutation and natural selection.** When a gene has mutated, its product may have a slightly different activity. If this new activity leads to increased fitness in individuals carrying the mutated gene, it will become more common in the population through the process of natural selection. **Visualize This:** How would this image differ if the flies were not in a high alcohol environment?

Stop and Stretch 10.3: The allele responsible for cystic fibrosis (discussed in Chapter 6) appears to protect carriers (individuals with 1 copy of the allele) from *M. tuberculosis* infection. In addition, it appears that the cystic fibrosis allele is more common in humans whose recent ancestors lived in crowded, urban environments. Use the modern understanding of evolution via natural selection to develop a hypothesis about why the cystic fibrosis is allele is more common in these populations.

Overview: The Subtleties of Natural Selection

Because the idea of natural selection has been applied to the realm of human society such as to the success or failure of a particular company or technology—misunderstandings of how it works in nature are common. Common misunderstandings of natural selection fall into 3 categories: the relationship between the individual and the population, the limitations on the traits that can be selected, and the ultimate result of selection. *Table 10.1*

TABLE 10.1 **Misconceptions about natural selection.**

A Misunderstanding of Natural Selection		How Natural Selection Really Works
"The Dodo was too stupid to adapt to human hunters, so it had to go extinct."		Only traits that are present in the population can be selected for. The Dodo was not "stupid" or unworthy of survival; the population simply did not contain variants with hunter-avoiding traits.
"Some animals are not adapted to their environments very well. For instance, if swallows were well adapted to North America, they wouldn't have to migrate to the tropics every winter."		Natural selection does not lead to some sort of idealized perfect state. There are always trade-offs between traits. A swallow is very well adapted to catching flying insects. Any swallow variant with physical changes that allow it to eat the seeds that are available in winter would be less able to catch flying insects. These variants would thus lose in the competition for food during the summer.
"If natural selection improves populations, why aren't chimpanzees evolving into humans?"		Natural selection adapts organisms to their current environment; it is not a process that moves organisms from simpler to more complex. Evolution in chimpanzees fits these animals to their current situation. In their environment, bipedal humans already exist and fully exploit the resources in which our ancestors were able to first specialize. Traits that make chimpanzees more "human-like" would not be successful because the resources humans use are already taken.

provides examples of statements that you might hear in reference to natural selection and an explanation of how each demonstrates a misunderstanding of this process.

In short, natural selection can only act on traits present in the population; the presence of any adaptation sometimes means that a slightly less-valuable trait must be forgone; and natural selection is only capable of fitting an organism to the environment it is currently experiencing, not some future condition.

A Closer Look: Subtleties of Natural Selection

Natural selection does not result in the evolution of "perfect" organisms. To make sense of that statement, you need to have a good understanding of not only the power, but also the limitations of this important evolutionary process.

Natural Selection Cannot Cause New Traits to Arise. One mistaken idea about natural selection is that certain species are doomed to extinction because they "will not" adapt to a new environment. Polar bears are threatened with extinction because they rely on sea ice for hunting,

(A Closer Look continued)

and sea ice in the Arctic is rapidly diminishing as a result of global warming. Why don't they just switch to another food? After all, grizzly bears eat a wide variety of foods.

This idea that species "should" change to survive results from our understanding of the selection process in business. A car company will fail if it does not change its designs to meet the changing needs of its customers. But change in the company occurs because businesses can *invent* new "adaptations" that respond to rapid changes in tastes and gas prices. In contrast, in living populations, evolution can occur only when traits that influence survival are already present in a population and have a genetic basis.

The example of the alcohol-metabolizing fruit flies illustrates this point. Selection did not cause change in individual flies; either a fly could rapidly metabolize alcohol or it could not. It was the differential survival and reproduction of these types of flies in an alcohol-laced environment that caused the population to change.

Natural Selection Does Not Result in Perfection. Natural selection does cause populations to become better fit to their environment, but the result of that process is not necessarily "better" organisms that can survive over a wider range of conditions—simply organisms that are better adapted to the current situation. Changes in traits that increase survival and reproduction in one environment may be liabilities in another environment. In other words, most adaptations are trade-offs between success in one situation versus another.

For example, Richard Lenski and his coworkers at Michigan State University found that certain bacteria would adapt to an environment where food levels were low by evolving chemicals that were deadly to other bacteria. Individual bacteria that produced this chemical had an advantage over those that did not; by killing off their nearby competitors, the poisonous bacteria had more of the very limited food available to them and thus became prevalent in the population. However, when the poisonous bacteria were grown in a food-rich environment, they did not grow as well as nonpoisonous bacteria did and were outcompeted for the available food. This occurs because the poisonous bacteria were using energy to produce their toxin—a trade-off that reduced the amount of energy that could have been used for growth and reproduction. In a food-rich environment, individual bacteria that expended the maximum amount of energy on growth had more offspring and thus were better adapted.

Adaptations of organisms are also constrained by their underlying biology. This is apparent throughout nature in what evolutionary biologists call *jury-rigged design*, meaning adaptations made using whatever is available.

Figure 10.14 **The panda's thumb.** In addition to the 5 digits on its paw, the giant panda has an opposable "thumb" made up of elongated wrist bones.

Stephen Jay Gould described one of the most famous examples of jury-rigged design—the thumb found on a giant panda's front paws (*Figure 10.14*). These animals apparently have 6 digits: 5 fingers composed of the same bones as our fingers and a thumb constructed from an enlarged bone equivalent to 1 found in our wrist. The muscles that operate this opposable thumb are rerouted hand muscles. This structure in pandas appears to be an adaptation that increases their ability to strip leaves from bamboo shoots, their primary food source. A more effective design for an opposable thumb is our own, adapted from 1 of the basic 5 digits. However, in the panda population, this variation did not exist. Individuals with enlarged wrist bones did exist, so what evolved in giant pandas was a jury-rigged thumb that does its job but is not as flexible as our own thumb.

Another, more familiar, example of jury-rigged design is our own upright posture—judging by the number of lower back and knee injuries reported every year, humans are not perfectly designed for bipedal walking. This is because our skeleton evolved from a 4-footed ancestor, and the changes that make us upright are adaptations of that ancestral design.

Natural Selection Does Not Cause Progression Toward a Goal. You may have heard natural selection described as "survival of the fittest"; however, it is important to recognize

(continued on the next page)

(A Closer Look continued)

Figure 10.15 **Natural selection does not imply progression.** Grasses are flowering plants that evolved from ancestors with much showier flowers.

that natural selection favors those variants with the most appropriate adaptations to the current environment. The survivors are not fittest in an absolute sense, only relative to others in the same population. And natural selection is a passive process; in no sense are individual organisms "choosing" to change or adapt to environment. Instead, the fittest individuals are simply born with particular traits that increase their chances of survival or reproduction in a particular situation. Natural selection thus does not result in the progress of a population toward a particular predetermined goal; instead, it depends on the particular situation of the population.

The example of the alcohol-metabolizing flies again helps to illustrate this point. Only the population of flies in the high-alcohol environment evolved a faster rate of alcohol metabolism. Without a change in the environment, the alcohol-metabolizing rate of the population of flies in the normal environment did not evolve.

The situational nature of natural selection can lead to evolutionary patterns that defy our sense that species are evolving toward a "more perfect" condition. For example, flowering plants evolved from nonflowering plants, and the flower is, in part, an adaptation to attract bees and other pollinators to help increase seed production. However, some species of flowering plants, especially grasses, have adapted to environments where wind pollination is particularly effective and the need to attract insect pollinators is much reduced. In these plants, natural selection has favored individuals that have very reduced flower parts and generate primarily pollen-producing and egg-producing structures—much like the reproductive structures of their distant nonflowering ancestors (*Figure 10.15*). Grasses have not regressed but instead have simply adapted to the environment they experience.

Patterns of Selection

As Darwin noted, natural selection is a force that causes the traits in a population to change over time. A more modern understanding of the process has helped scientists recognize that different environmental conditions may lead to no change in the population or even cause it to split into 2 species.

The type of natural selection experienced by the flies in the alcohol-laden environment is called **directional selection** because it causes the population traits to move in a particular direction (*Figure 10.16a*). Directional selection is typically the type of selection that leads to change in a population over time.

In certain environments, however, the average variant in the population may have the highest fitness. This results in **stabilizing selection**, in which the extreme variants in a population are selected against and the traits of the population stay the same (*Figure 10.16b*). For example, in humans, the survival of newborns is correlated to birth weight—both extremely small and extremely large babies are selected against, causing the average birth weight of babies to be relatively stable over time. Stabilizing selection causes populations to tend to resist change in unchanging environments.

Finally, in some situations, the most common variant may have the lowest fitness, resulting in **diversifying selection**. Diversifying selection causes

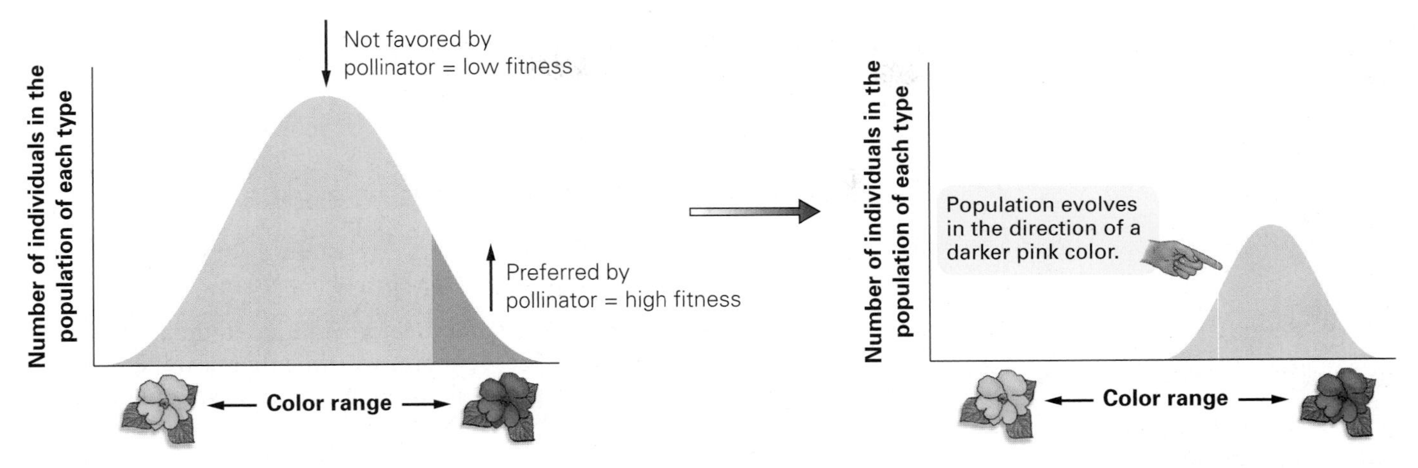

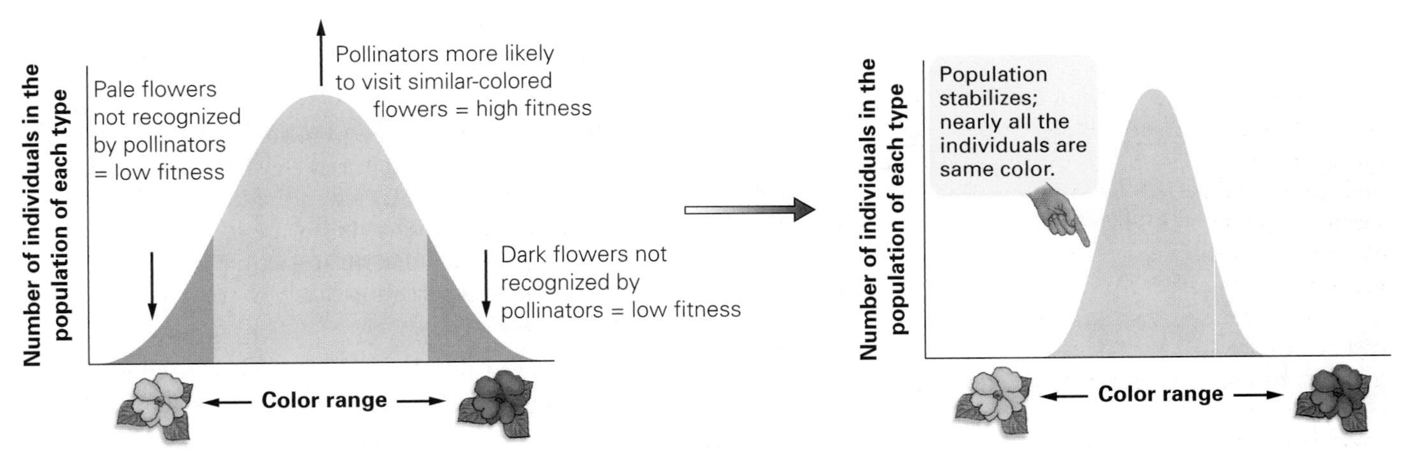

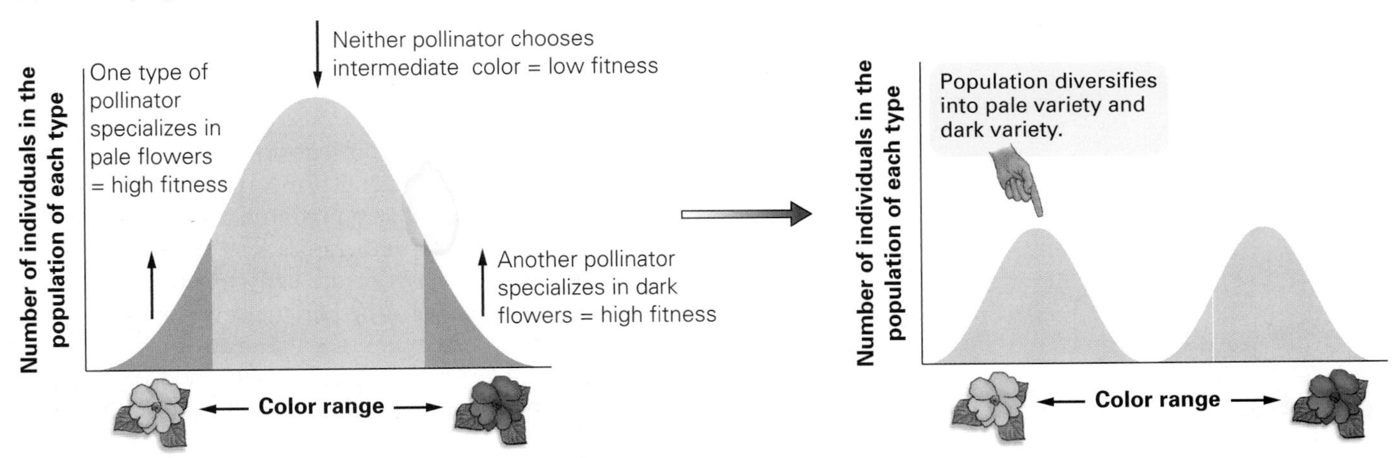

Figure 10.16 **Directional, stabilizing, and diversifying selection.** In a variable population, different environmental conditions cause different types of selection.

the evolution of a population consisting of 2 or more variants *(Figure 10.16c)*. Diversifying selection is especially likely if different subpopulations are experiencing different environmental conditions. Chapter 11 examines the role of diversifying selection, among other mechanisms, in the origin of new species.

web animation 10.2
Drug Resistance and Natural Selection

10.4 Natural Selection and Human Health

Knowing how natural selection works allows us to understand how populations of *M. tuberculosis* have become resistant to all of our most effective antibiotics. It also provides insight into how best to combat these evolving enemies.

Tuberculosis Fits Darwin's Observations

Mycobacterium tuberculosis evolved to become resistant to our antibiotics via natural selection because it fulfills all of Darwin's observations.

1. **Bacteria in the population vary.** Bacteria hidden in the lungs reproduce as they feed on the tissue there; any time there is reproduction, mutation can occur. As a result, even during drug treatment, new variants of *M. tuberculosis* continually arise. Some of these variants have proteins that disable or counteract certain antibiotics, making them more resistant to the drugs.
2. **The variation among bacteria can be passed on to offspring.** The traits that cause drug resistance are coded for in a bacterium's DNA. When a cell divides to reproduce, it copies this DNA and passes it—and the trait—on to its daughter cells.
3. **More bacteria are produced than survive.** The antibiotic treatment eliminates most of the bacteria in the infected individual's body.
4. **Bacterial survival is not random.** Bacterial cells with traits that make them more resistant to the antibiotic are more likely to survive than those that are less resistant.

Because of the increased fitness of particular variants, subsequent bacterial generations consist of a greater percentage of individuals that are resistant to the antibiotic. In other words, the population evolves to become resistant to the drug treatment.

Immediately after scientists began using antibiotics against tuberculosis, they noticed that some individuals would become ill again after seemingly successful treatment. Even more puzzling was that these recurrent infections were much more difficult to treat than the initial one. It was only until scientists incorporated their understanding of natural selection into tuberculosis treatments that effective, long-term therapies were developed.

Two characteristics of the early treatment strategies for tuberculosis actually sped up the development of drug resistance. The first was using the drugs for too short a time period. The second was using only one type of antibiotic at a time in patients with active tuberculosis disease.

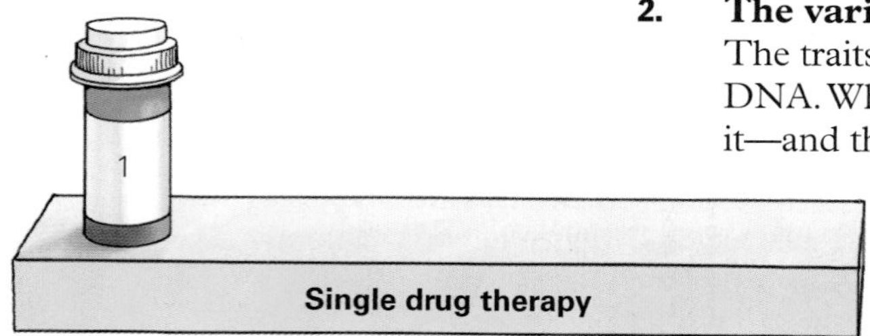

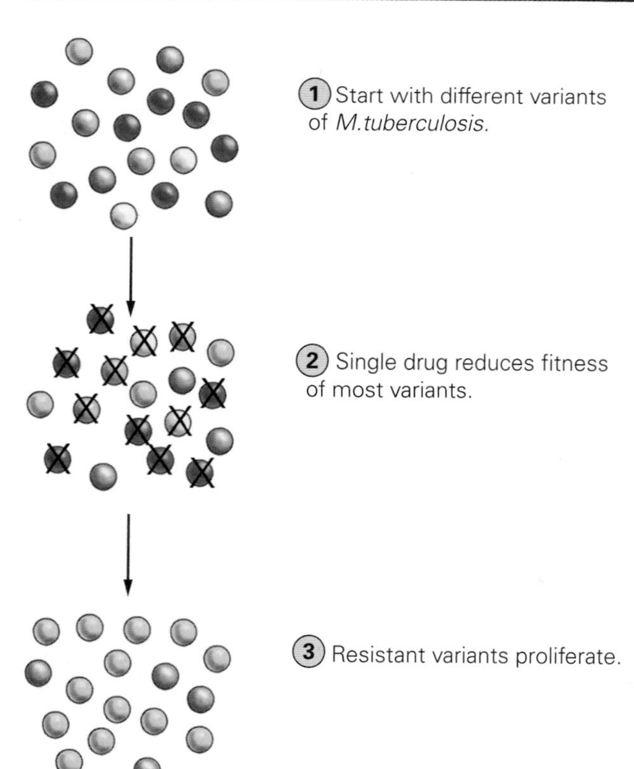

Figure 10.17 **Directional selection in tuberculosis.** In a variable population of *M. tuberculosis* bacteria, some individuals may be more resistant to a particular antibiotic. When this antibiotic is used, these resistant variants survive and continue to reproduce. The result is a population that is resistant to the antibiotic.

Selecting for Drug Resistance

Because most of the *M. tuberculosis* cells in an infected individual are susceptible to the antibiotic, a few days of treatment will eliminate the majority of these organisms. Once most of the bacteria are dead, the patient feels much better as the most debilitating aspects of the disease (e.g., fever, severe cough) are reduced or eliminated.

However, a small number of bacteria that are more resistant to the antibiotic take longer to kill. If drug treatment stops as soon as the patient feels better—the typical pattern in early treatment protocols—any of the more resistant bacteria that remain can multiply and restart the infection. A drug-resistant strain is born. And because this new population is more resistant, the resurgent infection is much more difficult to control *(Figure 10.17)*.

The number of bacteria in the lungs of an individual with an active infection is in the billions. In some patients, among these billions are some individual cells that possess a mutation that make them strongly resistant to a given antibiotic. When that antibiotic is administered, these resistant cells can take advantage of the resources left by the dying bacteria and form a ready-made, highly drug-resistant population. If the patient returns to the community with this active, resistant, infection, a disease that is highly resistant to the most commonly effective drugs could begin to spread.

Stop and Stretch 10.4: Why would some individual bacteria carry a mutation that makes them more resistant to an antibiotic even though they have never been exposed to this antibiotic?

Stopping Drug Resistance

Combating the development of drug resistance in *M. tuberculosis* required removing the factors that promoted that development in the first place. One strategy is to maintain drug therapy for months, until all signs of the bacterial infection are cleared from the body. The other is to use multiple drugs on active infections to avoid selecting for strongly drug-resistant variants that already exist in the population.

Combination drug therapy, also called drug cocktail therapy, is commonly used on diseases for which resistance to a single drug can develop rapidly. HIV, the virus that causes AIDS, is one example. The effectiveness of combination therapy is based on the following fact: The greater the number of drugs used, the greater the number of changes that are required in the bacterial genome for resistance to develop.

The likelihood of a bacterial variant arising that is resistant to a single drug is relatively small but still very possible in a patient with 1 billion different bacterial variants. However, the likelihood of a bacterial variant arising with resistance to 2 or 3 drugs in a cocktail is extremely small. Put another way, the chance that any bacterium exists that is resistant to a single drug is analogous to the likelihood that in 1 billion lottery ticket holders, 1 person will hold the winning combination—in other words, relatively likely. The likelihood of a variant being resistant to several different drugs is analogous to that same ticket holder winning the lottery several times in a row—incredibly unlikely. Just as it is exceedingly rare to win the lottery twice in a row, it is very difficult for *M. tuberculosis* to adapt to an environment where it faces 2 "killer drugs" at once *(Figure 10.18)*.

If scientists learned these lessons about drug resistance in the 1940s, why is drug-resistant *M. tuberculosis* making a comeback only now? Primarily, public health researchers let down their guard and didn't follow TB patients to ensure that they continued taking their drugs for the 6 to 12 months required to clear out all of the invading cells. It wasn't just Andrew Speaker's recklessness that exposed his fellow airline passengers to XDR-TB. It was thousands of ordinary TB patients who failed to complete their treatment regimen that led to the development of this dangerous variant. With this understanding, public health strategies that focus on maintaining the long-term treatment of individuals, and bringing that treatment to underserved communities, can go a long

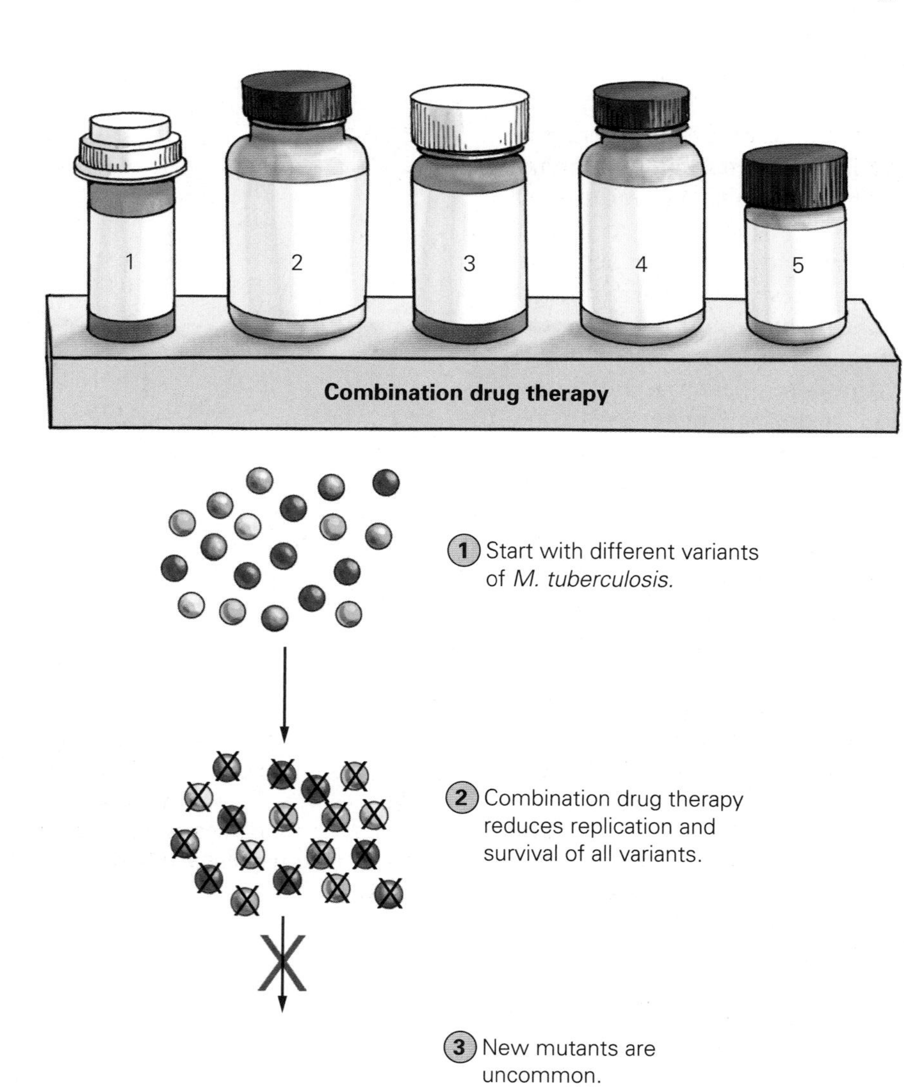

Figure 10.18 **Combination drug therapy prevents antibiotic resistance.** Using multiple antibiotics makes the environment much harsher for *M. tuberculosis* and decreases the likelihood that a variant with multiple resistances will evolve.

way in stemming the development of more cases of XDR-TB.

Not surprisingly, tuberculosis is not the only bacterial disease in which resistance has developed, largely as a result of inadequate treatment with antibiotics. MRSA, methicillin-resistant *Staphylococcus aureus*, is another formerly easily treated bacteria that has evolved into a more dangerous and deadly type *(Figure 10.19)*.

The rise of antibiotic resistance stems not only from failure to follow drug treatment but also from the overuse of these drugs. Antibiotics are often erroneously prescribed to individuals with viral infections, such as the common cold, on which they have no effect and only serve to increase the likelihood that formerly easily controlled bacteria may develop resistance. Antibiotics are also commonly and liberally used in animal agriculture, where they are given to poultry, cows, and pigs to prevent bacterial disease outbreaks in crowded feedlots. Drug resistance is also not confined to bacteria; the same evolutionary process has occurred in viruses, like HIV, and other pathogens, such as the protozoan that causes malaria.

To stop the development of these antibiotic-resistant "superbugs," patients and physicians must use antibiotics judiciously and wisely. But perhaps evolution has another trick up its sleeve that gives us some hope as well.

Can Natural Selection Save Us from Superbugs?

Bacteria can rapidly evolve resistance to our antibiotics. Why can't we evolve resistance to the bacteria? Can't natural selection save us from these pathogens?

There is clearly a difference between healthy individuals in their susceptibility to long-term tuberculosis infection and even to active disease. Given the existence of this heritable variation and differences in survival among those exposed to this killer virus, can we expect natural selection to cause the human population to evolve resistance to *M. tuberculosis*?

Eventually, perhaps—but remember that natural selection occurs because of differential survival and reproduction of individuals over time. In short, for the human population as a whole to become resistant to this bacterium, nonresistant variants must die out of the population. This process has not occurred in the more than 6000 years since tuberculosis has been in human populations. Because a majority of nonresistant people are not even exposed to *M. tuberculosis* and thus will continue to survive and reproduce, nonresistant variants will probably never be lost from the population. Clearly, future *human* evolution is not a likely solution to the problem of antibiotic-resistant superbugs.

However, recall that natural selection does not result in perfect organisms, only those with traits that are effective in the current environmental conditions. In *M. tuberculosis,* variants that are resistant to multiple antibiotics are also much less likely to spread to other individuals. In other words, antibiotic resistance is a trade-off that reduces a bacterial cell's ability to survive and reproduce under normal conditions. This raises hopes that quick response to TB outbreaks in vulnerable populations can contain the spread of this dangerous pathogen.

Natural selection has assisted in our fight against tuberculosis in another way as well. By shaping human brains so that we are able to make predictions and test hypotheses, it has provided us with a powerful tool for fighting this disease—our own ingenuity. Understanding how the disease is spread has helped prevent thousands of new infections, and the development of antibiotics has fought off millions more. A vaccine that can stimulate the immune system to prevent initial infection is in development that may eventually eliminate the disease entirely. Before that happens, we need to apply our ingenuity fully to address the alarming spread of this resistant disease.

The good news for Andrew Speaker is that later tests of the *M. tuberculosis* variant in his lungs confirmed that it was the more treatable MDR-TB and not XDR-TB. He has since received antibiotic treatment and lung surgery and been released from the hospital. This is good news for the people on Andrew Speaker's airplane flights, as well as the fact that in the absence of active disease, his ability to infect other individuals was very low. Tuberculosis does not need to return to the unbeatable enemies list. And with attention to the process of natural selection guiding our use and application of antibiotics, it shouldn't become one.

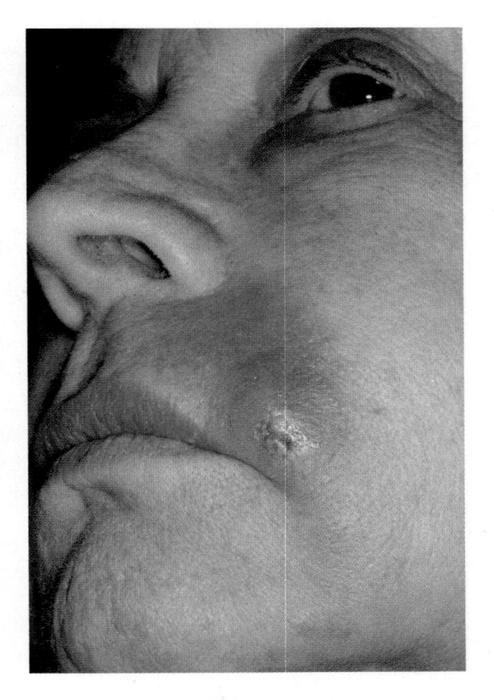

Figure 10.19 **MRSA.** Staph infections of the skin are relatively common, causing boils like these. However, if untreatable by antibiotics, staph can become a systemic disease that may be deadly.

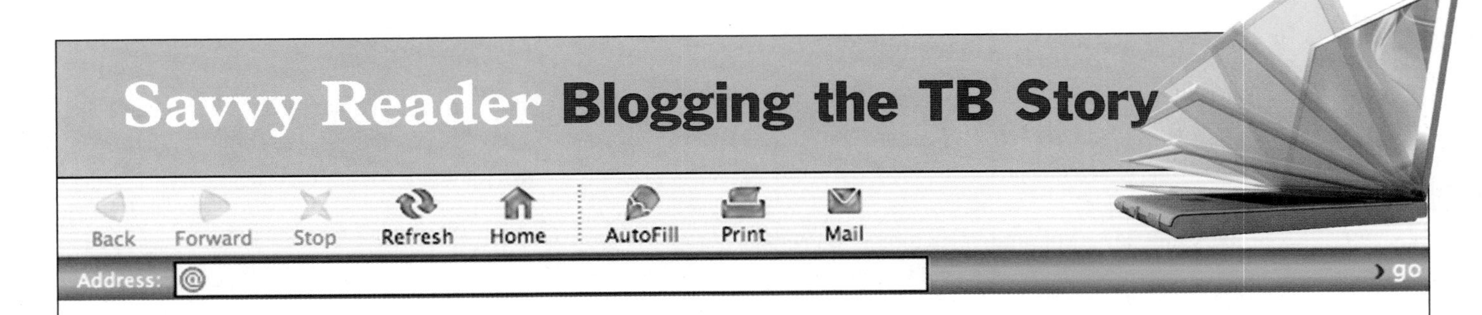

What gall!

From the Wordpress blog site of "fwtandy" (no further identification) posted June 8, 2007

Who does Andrew Speaker think he is? Everyone knows that tuberculosis is a pretty bad disease and is spread through the air. Everyone knows that the air on an airplane is pretty poluted [sic] on the best days, and that one is more likely than not to come off an airplane with the virus du jour. So he knows he has TB, gets on a plane anyway, or make that several planes, long international flights. Runs from the CDC when they track him down—on another plane, no less. And his comment now is that he really didn't think he was putting anyone at risk?

(continued on the next page)

Savvy Reader (continued)

Right. Maybe he should have stopped at "I didn't think." or maybe changed the sentence to, "I didn't care." What incredible disregard for other people. What arrogance!

I'm afraid it rather sums up the way too many people have become these days. The thought is ME, ME, ME. Speaker is perhaps the quintessential ME guy. He is even willing to put his bride at risk, for goodness sakes! Not to mention all the hundreds of people he came in contact with. Shame on him.

1. List the facts presented within this opinion piece. Does the author provide any supporting evidence for them?
2. No opinion piece presents evidence for all the statements contained within it. What resources do you have to evaluate the credibility of this author?
3. This blog posting is a reasonable summary of the reactions many people had to initial reports about Andrew Speaker. How do these initial reactions appear now that more of the facts of the case have been revealed?

CHAPTER Review

For study help, animations, and more quiz questions go to www.mybiology.com.

Summary

10.1 Return of an Ancient Killer

- Tuberculosis is an ancient and widespread disease caused by bacterial infection and can result in lung disease and death in about 10% of cases (p. 256).
- The advent of antibiotics turned tuberculosis from a possible death sentence to a readily curable condition. However, beginning about 25 years ago, variants of the tuberculosis bacteria appeared that are resistant to most antibiotics (p. 257).

10.2 Natural Selection Causes Evolution

- Individuals in a population vary, and some of this variation can be passed on to offspring (p. 258).
- Not all individuals born in a population survive to adulthood, and not all adults produce the maximum number of offspring possible (p. 259).
- Advantageous traits, called adaptations, increase an individual's fitness, his or her chance of survival or reproduction (p. 260).
- The increased fitness of individuals with particular adaptations causes the adaptation to become more prevalent in a population over generations (p. 260).
- Natural selection is a mechanism for evolutionary change in populations (p. 261).
- Artificial selection, when humans deliberately control an organism's fitness, causes the evolution of different breeds of animals and varieties of plants (p. 262).

- Populations exposed to environmental changes, both in the lab and in nature, have been shown to evolve traits that make them better fitted to the environment (p. 262).

web animation 10.1 *Natural Selection*

10.3 Natural Selection Since Darwin

- The modern definition of evolution is a genetic change in a population of organisms (p. 264).
- Alleles that code for adaptations become more common in a population over generations as a result of natural selection (pp. 264–266).
- Natural selection can act only on the variants currently available in the population. Natural selection results in a population that is better adapted to its environment but usually not perfectly adapted as a result of trade-offs. Natural selection does not push a population in the direction of a predetermined "goal" (p. 267).
- Selection can cause the traits in a population to change in a particular direction. However, in some environments it may cause certain traits to resist change and in other environments cause multiple variants to evolve (pp. 268–269).

10.4 Natural Selection and Human Health

- The tuberculosis bacterium can evolve resistance to antibiotics because it consists of multiple variants that have differential survival when exposed to various drugs; thus, a population can evolve drug resistance via natural selection (p. 270).
- Mutant bacteria that are resistant to several different antibiotics are relatively unlikely, so combination drug therapy can reduce the risk of antibiotic resistance evolving (p. 271).
- Resistance to tuberculosis is unlikely to become commonplace soon in the human population as a result of natural selection (p. 272).
- Varieties of the tuberculosis bacteria that are multiply drug resistant are less transmissible than nonresistant varieties (p. 272).
- Patients and doctors can help prevent the evolution of antibiotic-resistant bacterial diseases by using these essential drugs judiciously and according to direction (p. 272).

web animation 10.2 *Drug Resistance and Natural Selection*

Roots to Remember

These roots come mainly from Latin and Greek and will help you understand terms:

adapt- means to fit in.
anti- means in opposition to.
-biotic indicates pertaining to life.
-osis indicates a condition.
tubercul- means a small swelling.

Learning the Basics

1. Define *fitness* as used in the context of evolution and natural selection.

2. Define *artificial selection* and compare and contrast it with natural selection.

3. Describe how *Mycobacterium tuberculosis* evolves when it is exposed to an antibiotic.

4. Which of the following observations is not part of the theory of natural selection?
A. Populations of organisms have more offspring than will survive;
B. There is variation among individuals in a population;
C. Modern organisms descended from a single common ancestor;
D. Traits can be passed on from parent to offspring;
E. Some variants in a population have a higher probability of survival and reproduction than other variants do.

5. The best definition of *evolutionary fitness* is ___________.
A. physical health;
B. the ability to attract members of the opposite sex;
C. the ability to adapt to the environment;
D. survival and reproduction relative to other members of the population;
E. overall strength

6. An adaptation is a trait of an organism that increases ___________.
A. its fitness;
B. its ability to survive and replicate;
C. in frequency in a population over many generations;
D. A and B are correct;
E. A, B, and C are correct

7. The heritable differences among organisms are a result of ___________.
A. differences in their DNA;
B. mutation;
C. differences in alleles;
D. A and B are correct;
E. A, B, and C are correct

8. Since the modern synthesis, the technical definition of evolution is a change in ______ in a _______ over the course of generations.
A. traits, species;
B. allele frequency, population;
C. natural selection, natural environment;
D. adaptations, single organism;
E. fitness, population

9. Ivory from elephant tusks is a valuable commodity on the world market. As a result, male African elephants with large tusks have been heavily hunted for the past few centuries. Today, male elephants have significantly shorter tusks at full adulthood than male elephants in the early 1900s. This is an example of ___________.
A. disruptive selection;
B. stabilizing selection;
C. directional selection;
D. evolutionary regression;
E. more than 1 of the above is correct

10. Antibiotic resistance is becoming common among organisms that cause a variety of human diseases. All of the following strategies help reduce the risk of antibiotic resistance evolving in a susceptible bacterial population except _____.
A. using antibiotics only when appropriate, for bacterial infections that are not clearing up;
B. using the drugs as directed, taking all the antibiotic over the course of days prescribed;
C. using more than 1 antibiotic at a time for difficult-to-treat organisms;
D. preventing natural selection by reducing the amount of evolution the organisms can perform;
E. reducing the use of antibiotics in non-health care settings, such as agriculture

Analyzing and Applying the Basics

1. Most domestic fruits and vegetables are a result of artificial selection from wild ancestors. Use your understanding of artificial selection to describe how domesticated strawberries must have evolved from their smaller wild relatives. What tradeoffs do domesticated strawberries exhibit relative to their wild ancestors?

2. The striped pattern on zebras' coats is considered to be an adaptation that helps reduce the likelihood of a lion or other predator identifying and preying on an individual animal. The ancestors of zebras were probably not striped. Using your understanding of the processes of mutation and natural selection, describe how a population of striped zebras might have evolved from a population of zebras without stripes.

3. Are all features of living organisms adaptations? How could you determine if a trait in an organism is a product of evolution by natural selection?

Connecting the Science

1. The theory of natural selection has been applied to human culture in many different realms. For instance, there is a general belief in the United States that "survival of the fittest" determines which people are rich and which are poor. How are the forces that produce differences in wealth among individuals like natural selection? How are they different?

2. Ninety-five percent of worldwide tuberculosis cases occur in developing countries, where most of the population cannot afford effective antibiotic therapy. Does the United States have an obligation to provide people in the developing world with low-cost, effective anti-TB therapy? In countries where the needs of daily survival often overshadow the requirement to take the drugs for an adequate amount of time, drug-resistant strains of TB are more likely to develop. How might governments and funding agencies combat this problem?

CHAPTER 11

Who Am I?

Species and Races

What race is Indigo?

Do the races on Indigo's census form represent different "basic types" of humans?

Are we more similar to people of the same race than to people of different races?

Do racial categories even matter?

Indigo pondered the choices in front of her: White, Black, American Indian or Alaskan Native, Vietnamese, Other Asian. As the daughter of a man with African American and Choctaw ancestry and a woman with a white American father and a Laotian mother, Indigo was not sure what race she should report on the U.S. census form.

The 2000 census was different from all previous censuses in that it allowed people to check multiple boxes under the category "race." While this change satisfied many multiracial individuals who felt that it was impossible to classify themselves as 1 particular race, the opportunity to do so was disquieting to others. In Indigo's case, despite having a half-Asian mother, she knew that most people saw her as black, and her connection to her mother's white father was weak at best; she didn't feel a kinship to other whites at all. And while her dad was proud of his Choctaw heritage, Indigo did not know a single Native American. Indigo did not see herself as belonging to any of the races on the census form. Maybe "human" was what she was looking for. She wondered, "Why do I need to specify my racial category? And what does it mean? If I'm part white and part black, am I somehow different from each group?"

Indigo's questions reflect those posed by many people over the years. Why do human groups differ from each other in skin color, eye shape, and stature? Do these physical differences reveal underlying basic biological differences among these groups?

web animation 11.1
The Process of Speciation

11.1 What Is a Species?

All humans belong to the same species. Before we can understand the concept of race, we first need to understand both what is meant by this statement and what is known about how species originate.

In the mid-1700s, the Swedish scientist Carolus Linnaeus began the task of cataloging all of nature. As described in Chapter 9, Linnaeus developed a classification scheme that grouped organisms according to shared traits. The primary category in his classification system was the **species,** a group whose members have the greatest resemblance. Linnaeus assigned a 2-part name to each species—the first part of the name indicates the **genus,** or broader group; the second part is specific to a particular species within that genus. For example, lions, the species *Panthera leo*, are classified in the same genus with other species of roaring cats such as the leopard, *Panthera pardus (Figure 11.1)*.

Linnaeus coined the binomial name *Homo sapiens* (*Homo* meaning "man" and *sapiens* meaning "knowing or wise") to describe the human species. Although Linnaeus recognized the variability among humans, by placing all of us in the same species he acknowledged our basic unity.

Modern biologists have kept the basic Linnaean classification, although they have added a **subspecies** name, *Homo sapiens sapiens*, to distinguish modern humans from earlier humans who appeared approximately 250,000 years ago.

(a) Lion

(b) Leopard

Figure 11.1 **Same genus, different species.** (a) *Panthera leo*, the lion and (b) *Panthera pardus*, the leopard.

The Biological Species Concept

While most people intuitively grasp the differences between most species—lions and leopards are both definitely cats but not the same species—biologists have had difficulty finding a single definition that can be consistently applied. Several useful concepts have been proposed and used. The definition most commonly used is the biological species concept.

Biological Species Are Reproductively Isolated. According to the **biological species concept,** a species is defined as a group of individuals that, in nature, can interbreed and produce fertile offspring but cannot reproduce with members of other species. In practice, this definition can be difficult to apply. For example, species that reproduce asexually (such as some bacteria) and species known only via fossils do not easily fit into this species concept. However, the biological species concept does help us understand why species are distinct from each other.

Recall that differences in traits among individuals arise partly from differences in their genes. By the process of evolution, a particular allele can become more common in a species. If interbreeding does not occur, then this allele cannot spread from 1 species to the other. In this way, 2 species can evolve differences from each other. For example, evidence suggests that the common ancestor of lions and leopards had a spotted coat. The allele that eliminated the spots from lions arose and spread within this species, but the allele is not found in leopards because lions and leopards cannot interbreed.

Stop and Stretch 11.1: Why might the allele for spotlessness have become common in the lion population?

Scientists refer to the sum total of the alleles found in all the individuals of a species as the species' **gene pool.** A single species makes up an impermeable container for its gene pool—a change in the frequency of an allele in a gene pool can take place only within a biological species.

The Nature of Reproductive Isolation. The spread of an allele throughout a species' gene pool is called gene flow. Gene flow cannot occur between different biological species because pairing between them fails to produce fertile offspring. This reproductive isolation can take 2 general forms: prefertilization barriers or postfertilization barriers, summarized in *Table 11.1.*

Prefertilization barriers to reproduction include failure to mate and failure to produce an embryo. The most obvious impediment to mating is that individuals from different biological species simply never contact each other; that is, they are separated by distance, a reproductive barrier known as **spatial isolation.**

Among species that do contact each other, differences in mating behavior can impede reproduction, a mechanism known as **behavioral isolation.** For example, many of the songs and displays produced by birds serve as prefertilization barriers. Male blue-footed boobies, seabirds that look almost as goofy as their name implies, perform an elaborate dance for the female before mating *(Figure 11.2a)*. A female blue-footed booby will not respond until she has witnessed the dance, at which time she will join the male in a pointing display *(Figure 11.2b)*. The male's dance and the pair's pointing display provide a way for both birds to recognize that they belong to the same species. Neither will mate unless the ritual is performed correctly.

TABLE 11.1 **Mechanisms of reproductive isolation.**

Type	Effect	Example
Prefertilization barriers: prevent fertilization from occurring		
Spatial isolation	Individuals from different species do not come in contact with each other.	Polar bear (Arctic) and spectacled bear (South America)
Behavioral	Ritual behaviors that prepare partners for mating are different in different species.	Many birds with premating songs and/or "dances"
Mechanical	Sex organs are incompatible between different species, so sperm cannot reach egg.	Many insects with "lock-and-key" type genitals
Temporal	Timing of readiness to reproduce is different in different species.	Plants with different flowering periods
Gamete incompatibility	Proteins on egg that allow sperm binding do not bind with sperm from another species.	Animals with external reproduction, such as sponges
Postfertilization barriers: Fertilization occurs, but hybrid cannot reproduce.		
Hybrid inviability	Zygote cannot complete development because genetic instructions incomplete.	A sheep crossed with a goat can produce an embryo, but it dies in the early developmental stages.
Hybrid sterility	Hybrid organism cannot produce offspring because chromosome number is odd.	Mules

(a) Courting dance

(b) Pointing display

Figure 11.2 **A behavioral prefertilization barrier to reproduction.** (a) Female blue-footed boobies will not mate with males who fail to perform this dance. (b) Male blue-footed boobies will not mate with females who do not engage in the pointing display with them.

Mating can also be prevented if the sexual organs of 2 individuals are incompatible, a mechanism known as **mechanical isolation.** The genitals of male and female insects of the same species often fit together like a lock and a key, making matings with other species with different equipment impossible.

Differences in the timing of reproduction, called **temporal isolation,** can also form a prefertilization barrier between species. Temporal isolation is common in flowering plants, different species of which have distinct flowering periods *(Figure 11.3)*.

The most common prefertilization barrier between species that will mate with each other is an incompatibility between eggs and sperm. In this case, sperm from 1 species cannot fuse with eggs from another species. Among species that release their gametes into the environment for fertilization—such as fish, amphibians, sponges, and many plants—this method of reproductive isolation, called **gamete incompatibility,** is widespread *(Figure 11.4)*.

Postfertilization barriers between species occur when fertilization takes place but the resulting offspring does not survive or is sterile. Most **interspecies hybrids**—that is, the offspring of parents from 2 different species—die before birth. Placing genes from different species together provides a hybrid offspring with incomprehensible information about how to build a body. Thus, normal development cannot take place.

A few related biological species can produce living hybrid offspring, but these organisms are typically sterile. A well-known example of an interspecies

Figure 11.3 **A temporal prefertilization barrier to reproduction.** These photos show the completely different suite of species present in (a) spring and (b) summer in the same meadow. Flowers blooming in spring cannot fertilize flowers blooming in summer and vice versa.

(a)

(b)

hybrid is the mule, resulting from a cross between a horse and a donkey. Mules have a well-earned reputation as tough and sturdy farm animals, but they are also sterile and cannot produce their own offspring.

In mules, hybrid sterility occurs because hybrid individuals cannot produce proper sperm or egg cells. Recall from Chapter 5 that during the production of eggs and sperm, homologous chromosomes pair up and separate during the first cell division of meiosis. Since a hybrid forms from the chromosome sets of 2 different species, the chromosomes cannot pair up correctly during this process.

In the case of mules, the horse parent has 64 chromosomes and therefore produces eggs or sperm with 32; the donkey parent has 62 chromosomes, producing eggs or sperm with 31. The mule will therefore have 63 chromosomes and no way to effectively sort these into pairs during the first division of meiosis (*Figure 11.5*). While a tiny number of female mules have, surprisingly, produced offspring, this event is so rare that the gene pools of donkeys and horses have remained separate.

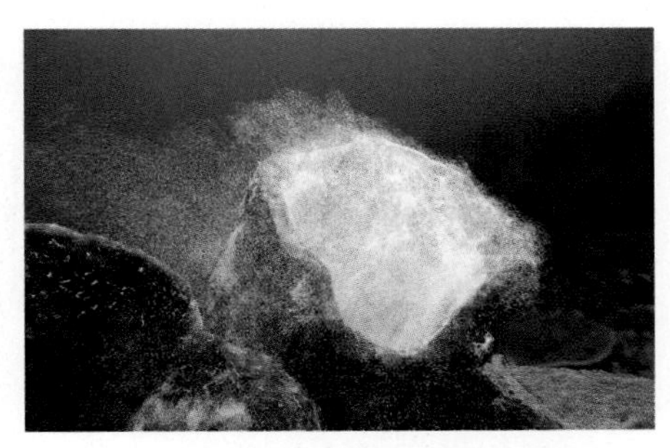

Figure 11.4 **Gamete incompatibility.** Sponges release their eggs and sperm into the water. Receptors on the egg surface permit only sperm from the same species to fertilize.

Stop and Stretch 11.2: Hybrid corn is produced when 2 different varieties of the species *Zea mays* (for example, a short variety and a sweet variety) are crossed. Do you expect that hybrid corn is sterile? Why or why not?

Given the definition of a biological species, it is clear that all humans belong to the same biological species. Indigo herself demonstrates that no real barriers to reproduction occur between human groups. To understand the concept of races within a species, however, we must first examine how species form.

(a) A mule results from the mating of a horse and a donkey.

Horse cells: 64 chromosomes

Donkey cells: 62 chromosomes

Meiosis

Horse egg has 32 chromosomes.

Donkey sperm has 31 chromosomes.

Mule: 63 chromosomes

(b) Why mules are sterile

Mule cell

Horse chromosome

Donkey chromosome

Metaphase I of meiosis

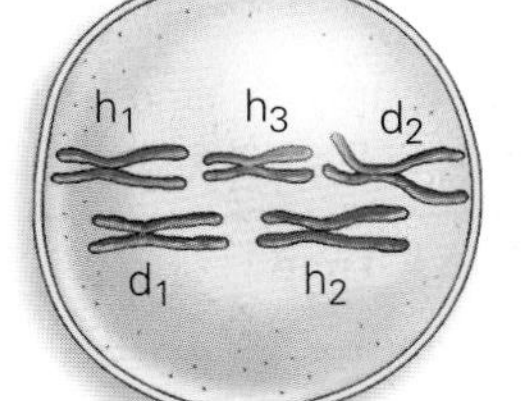

The chromosomes are from different species, so they are unable to pair during the first part of meiosis.

Figure 11.5 **Reproductive isolation between horses and donkeys.** (a) A cross between a female horse and a male donkey produces a mule with 63 chromosomes. (b) Mules produce only very few eggs or sperm because their chromosomes cannot pair properly during meiosis. Only a small number of chromosomes are illustrated to simplify the drawing. **Visualize This:** If donkeys had 60 chromosomes so that their sperm contained 30, a mule would have 62 chromosomes. Would having an even number of chromosomes permit mules to be fertile?

Speciation: An Overview

According to the theory of common descent, all modern organisms descended from a common ancestral species. This evolution of 1 or more species from an ancestral form is called **speciation.**

For 1 species to give rise to a new species, most biologists agree that 3 steps are necessary:

1. Isolation of the gene pools of subgroups, or **populations,** of the species;
2. Evolutionary changes in the gene pools of 1 or both of the isolated populations; and
3. The evolution of reproductive isolation between these populations, preventing any future gene flow.

Recall that gene flow occurs when reproduction is occurring within a species. Now imagine what would happen if 2 populations of a species became physically isolated from each other, so that the movement of individuals between these 2 populations was impossible. Even without genetic or behavioral barriers to mating between these 2 populations, gene flow between them would cease.

What is the consequence of eliminating gene flow between 2 populations? It is identical to what occurs in separate biological species. New alleles that arise in 1 population may not arise in the other. Thus, a new allele may become common in 1 population, but it may not exist in the other. Even among existing alleles, 1 may increase in frequency in 1 population but not in the other. In this way, each population would be evolving independently.

Over time, the traits found in 1 population begin to differ from the traits found in the other population. In other words, the populations begin to **diverge** *(Figure 11.6)*. When the divergence is great enough, reproductive isolation can occur. The 3 steps of speciation are discussed in detail in the Closer Look section.

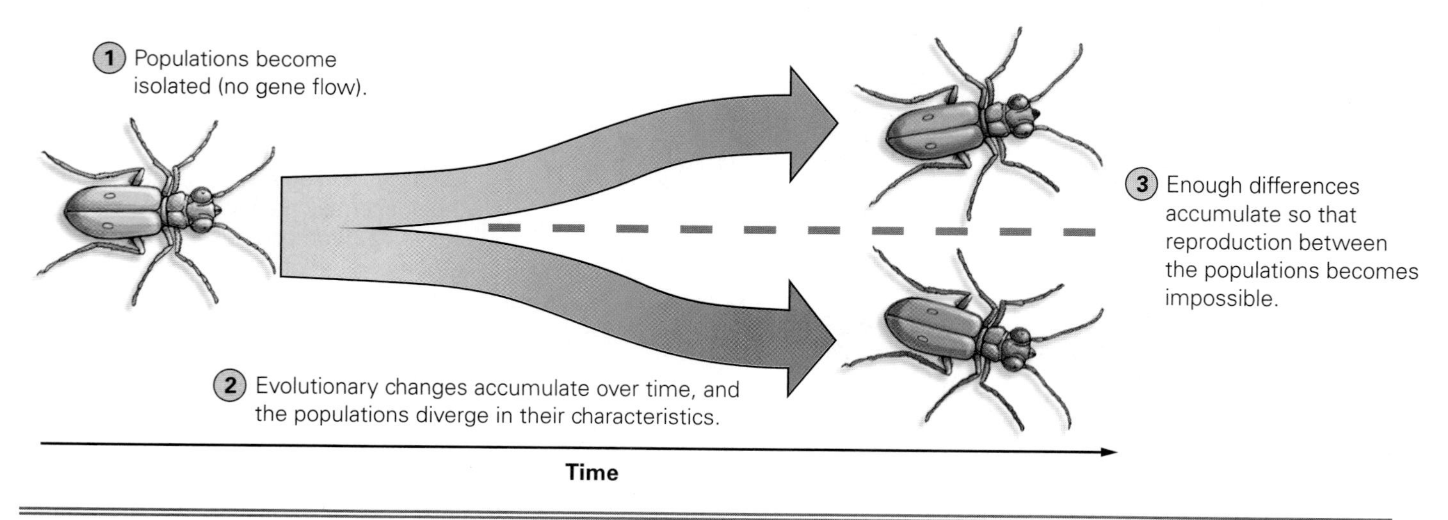

Figure 11.6 **Speciation.** Isolated populations diverge in traits. Divergence can lead to reproductive isolation and thus the formation of new species.

A Closer Look: Speciation

Speciation is a historical process and thus likely to be different and somewhat unique in every group of organisms. However, there are patterns in nature that give us some insight into how and why speciation may happen.

Isolation and Divergence of Gene Pools. The gene pools of populations may become isolated from each other for several reasons. Often, a small population becomes isolated when it migrates to a location far from the main population. This is the case on many oceanic islands, including the Galápagos and Hawaiian

(A Closer Look continued)

islands. Bird, reptile, plant, and insect species on these islands appear to be the descendants of species from the nearest mainland. The original ancestral migrants arrived on the islands by chance.

Because it is rare for organisms to find their way across hundreds of miles of open ocean to these islands, populations at each site are nearly completely isolated from each other *(Figure 11.7)*. In addition, because migrant populations are small, their gene pools can change rapidly via the process of genetic drift, as described in section 11.4.

The establishment of a new population far from related species may lead to the evolution of several new species—an idea known as the **founder hypothesis.** According to this hypothesis, the diversity of unique species on oceanic islands, as well as in isolated bogs, caves, and lakes, resulted from colonization of these once "empty" environments by 1 species that rapidly speciated, taking advantage of many different resources. For example, more than 50 species of silversword are now found on the Hawaiian islands, all descendants of the original founding population.

Populations may also become isolated from each other by the intrusion of a geologic barrier. This could be an event as slow as the rise of a mountain range or as rapid as a sudden change in the course of a river.

The emergence of the Isthmus of Panama between 3 and 6 million years ago represents 1 such intrusion event. This land bridge finally connected South and North America but divided the ocean gulf between them. Scientists have described at least 6 pairs of snapping shrimp on both sides of the isthmus that apparently diverged during and after this event. These shrimp species appear to be related to each other because of similarities in appearance, DNA sequence, and lifestyle. In each case, 1 member of each of the 6 pairs is found on the Caribbean side of the land bridge, while the other member of each pair is found on the Pacific side *(Figure 11.8)*. This geographic pattern indicates that each pair descended from ancestral species that each had been separated into 2 populations by the land bridge.

Populations that are isolated from each other by distance or a barrier are known as **allopatric.** However, separation between 2 populations' gene pools may occur even if the populations are living near each other, that is, if they are **sympatric.** This appears to be the case in populations of the apple maggot fly.

Apple maggot flies are notorious pests of apples grown in northeastern North America. However, apple trees are not native to North America; they were first introduced to this continent less than 300

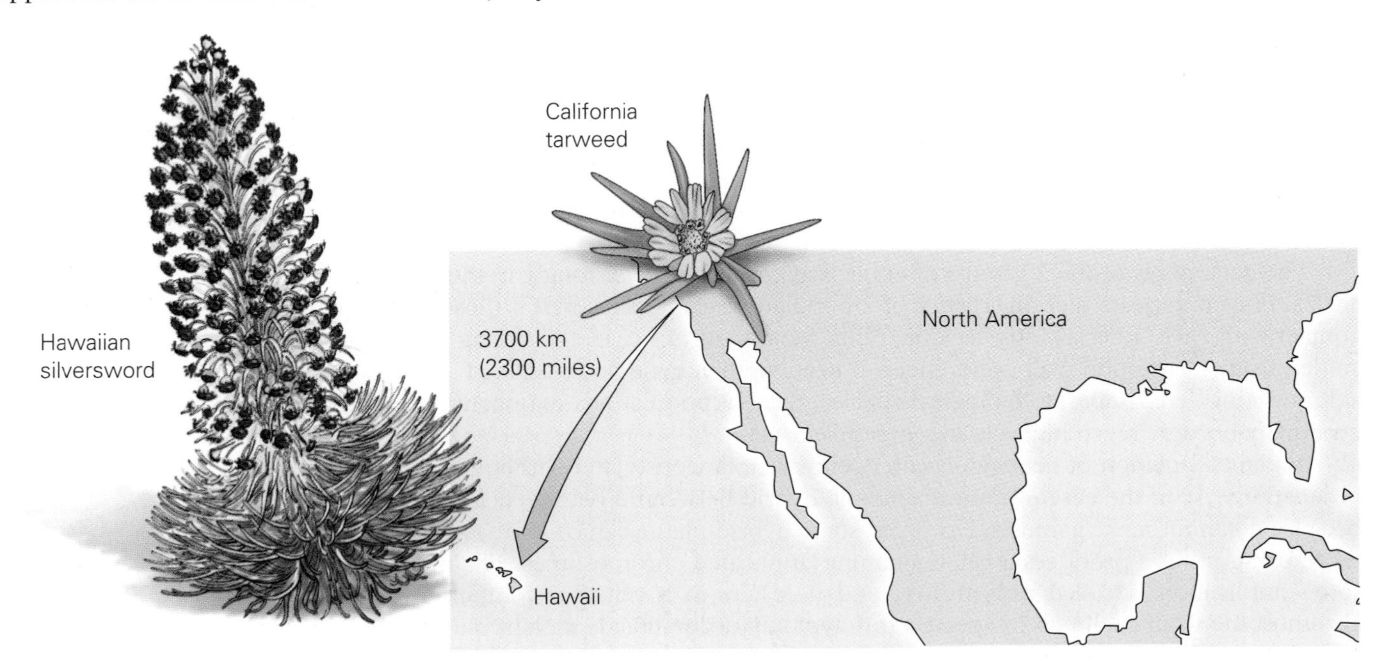

Figure 11.7 **Migration leads to speciation.** California tarweed seeds were blown or carried by birds to the Hawaiian Islands, creating an isolated population. With no gene flow between the 2 populations, a very different group of species, Hawaiian silverswords, evolved.

(continued on the next page)

(A Closer Look continued)

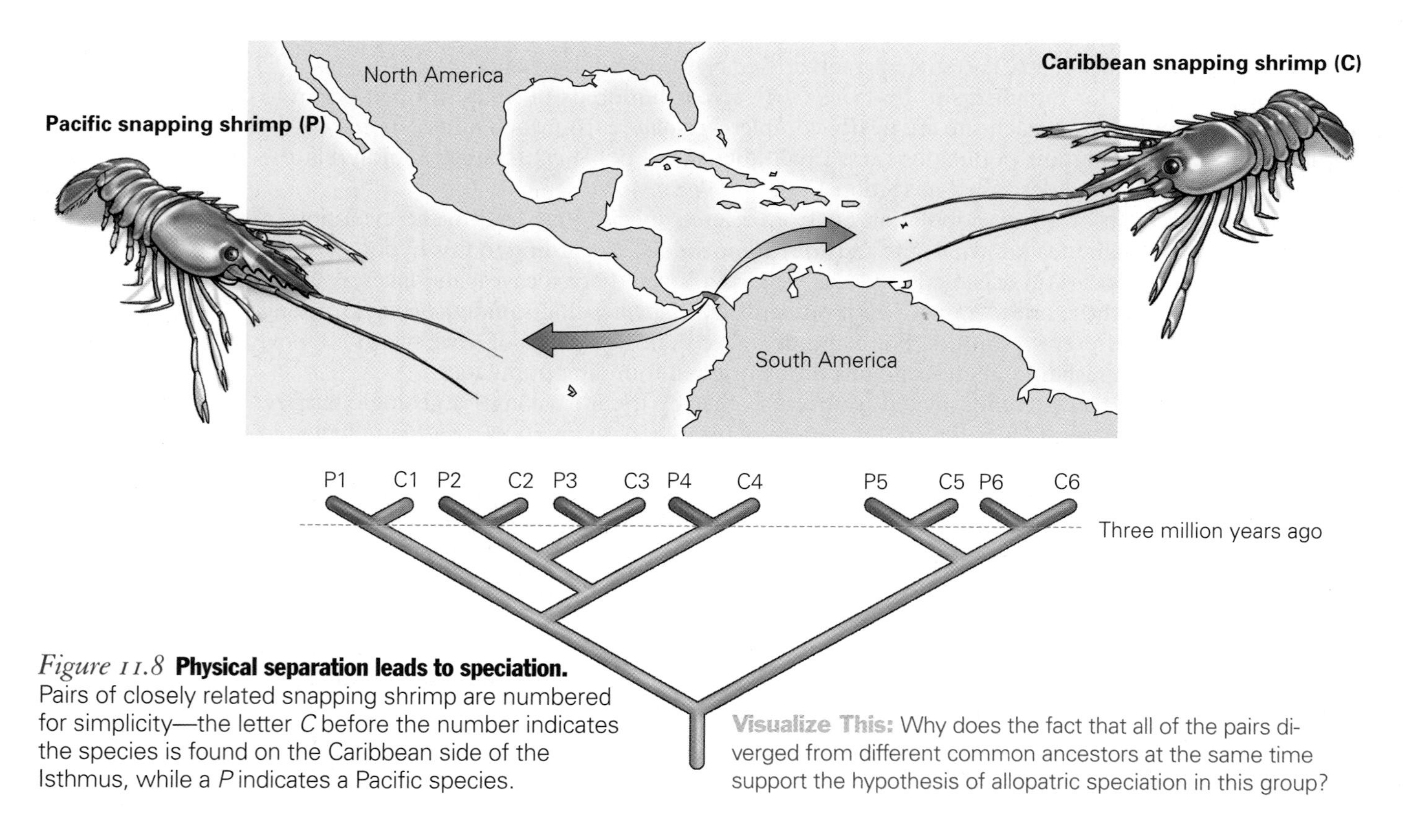

Figure 11.8 **Physical separation leads to speciation.** Pairs of closely related snapping shrimp are numbered for simplicity—the letter *C* before the number indicates the species is found on the Caribbean side of the Isthmus, while a *P* indicates a Pacific species.

Visualize This: Why does the fact that all of the pairs diverged from different common ancestors at the same time support the hypothesis of allopatric speciation in this group?

years ago. Apple maggot flies also infest the fruit of hawthorn shrubs, a group of species that are native to North America. Apples and hawthorns live in close proximity, and apple maggot flies clearly have the ability to fly between apple orchards and hawthorn shrubs. At first glance, it does not appear that the apple maggot flies that eat apples and those that eat hawthorn fruit could be isolated from each other.

However, on closer inspection, it is clear that flies on apples and those on hawthorns actually have little opportunity to mate. Flies mate on the fruit where they will lay their eggs, and hawthorns produce fruit approximately 1 month after apples do. Each population of fly has a strong preference regarding the fruit on which it will mate, and flies that lay eggs on hawthorns develop much faster than flies that lay eggs on apples. There appears to be little gene flow between the apple-preferring and hawthorn-preferring populations.

The gene pools of the 2 groups of apple maggot flies differ strongly in the frequency of some alleles. Thus it appears that divergence of 2 populations can occur even if those populations are in contact with each other as long as some other factor—in this case, the timing of reproduction resulting from variation in fruit preference—is keeping their gene pools isolated *(Figure 11.9)*. While still not considered separate biological species, these 2 populations may continue to diverge and eventually become reproductively incompatible.

In plants, isolation of gene pools can occur instantaneously and without any barriers between populations. As in the case of mules, a simple hybrid between 2 plant species is typically infertile because it cannot make gametes. However, some hybrid plants can become fertile again—if a mistake during mitosis produces a cell containing duplicated chromosomes. The process of chromosome duplication is called **polyploidy,** and it results in a cell that contains 2 copies of each chromosome from each parent species. If polyploidy occurs inside a plant bud, all the cells of the branch that arises from that bud will be polyploid.

Because polyploid cells now contain pairs of chromosomes, meiosis can proceed, and thus flowers produced on the branch can produce eggs and sperm. Since most plants produce both types of gametes in a flower, a polyploid flower can self-fertilize and give rise to hundreds of offspring, representing a brand new species that is isolated from its parent plants. One example of species that formed via polyploidy is canola, an important agricultural crop grown for the oil produced in its seeds

(A Closer Look continued)

(Figure 11.10). Canola developed as a result of chromosome duplication in a hybrid of kale (*Brassica oleracea*) and turnip (*Brassica campestris*).

Polyploidy can result in speciation even in nonhybrid plants. For example, in his research garden the geneticist Hugo de Vries discovered individual evening primrose plants that had 28 chromosomes—twice as many as other plants in the same population. On investigation, he found that these plants were reproductively isolated and unable to produce viable offspring with evening primrose plants having 14 chromosomes. Recent research suggests that this process of "instantaneous speciation" may have been a key factor in the evolution of diversity in plants. As many as 50% of flowering plant species may have resulted from this process. Polyploidy appears to occur in some animal groups, such as insects and frogs, as well.

Figure 11.9 **Differences in the timing of reproduction can lead to speciation.** This graph illustrates the life cycle of 2 populations of the apple maggot fly: one that lives on apple trees and one that lives on hawthorn shrubs. The mating period for these 2 populations differs by a month, resulting in little gene flow between them.

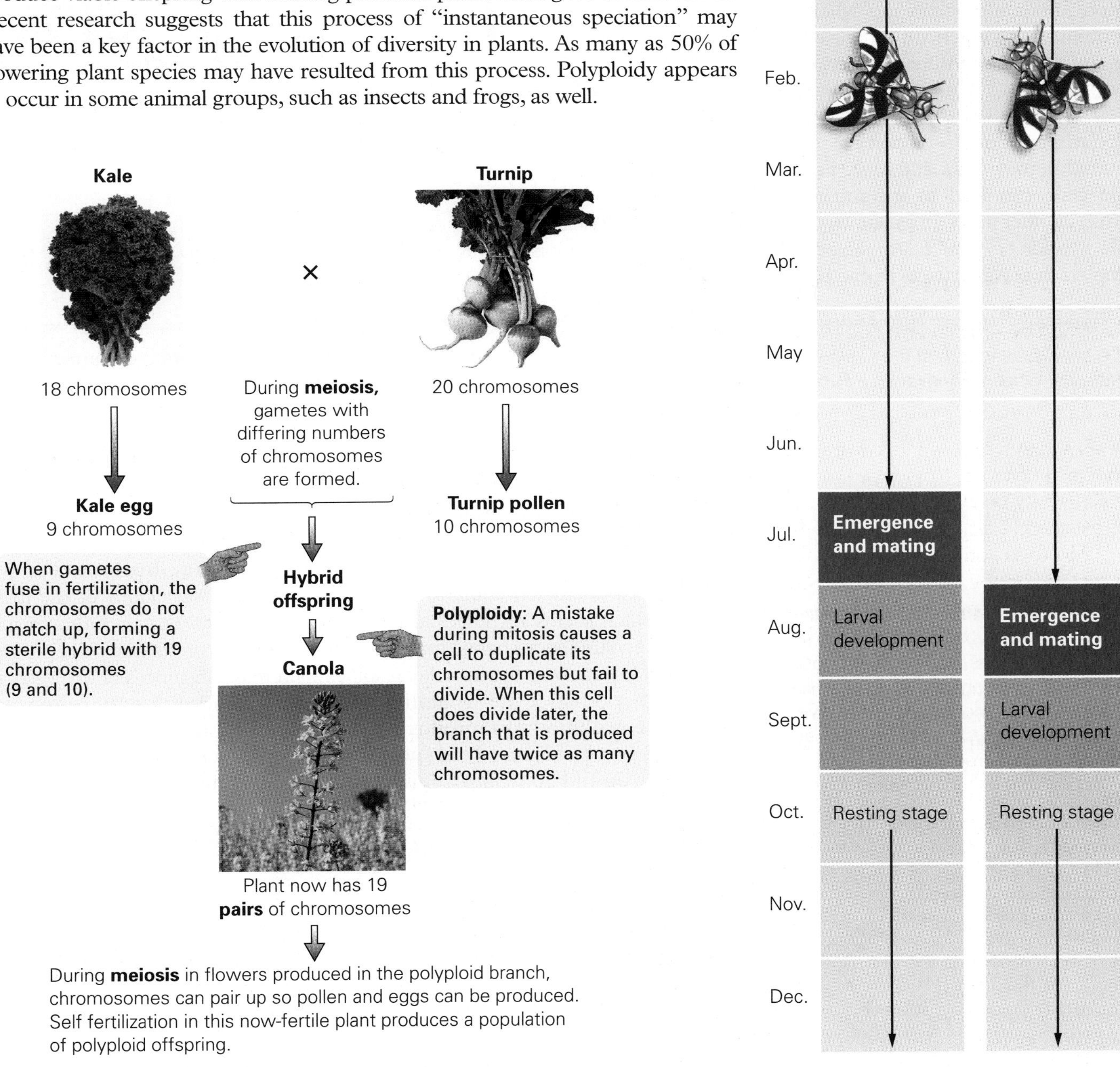

Figure 11.10 **Instantaneous speciation.** Canola evolved from a hybrid of kale and turnip via chromosome duplication. Because it has a different number of chromosomes from either of its parents, canola pollen cannot fertilize kale or turnip plant eggs and vice versa.

(continued on the next page)

(A Closer Look continued)

The Evolution of Reproductive Isolation. To become truly distinct biological species, diverging populations must become reproductively isolated either by their behavior or by genetic incompatibility. In the case of canola, genetic incompatibility occurs immediately—a cross between canola and kale does not result in fertilization. In most animals, the process may be more gradual, occurring when the amount of divergence has caused numerous genetic differences between 2 populations.

There is no hard-and-fast rule about how much divergence is required; sometimes a difference in a single gene can lead to incompatibility, while at other times, populations demonstrating great physical differences can produce healthy and fertile hybrids *(Figure 11.11)*. Exactly how reproductive isolation evolves on a genetic level is still unknown and is an actively researched question in biology.

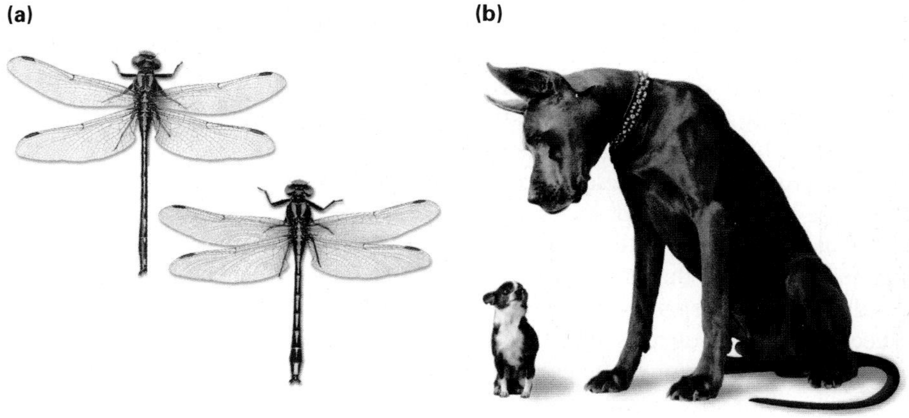

(a) (b)

Figure 11.11 **How different are 2 species?** (a) These 2 species of dragonfly look alike but cannot interbreed. (b) Dog breeds provide a dramatic example of how the evolution of large physical differences does not always result in reproductive incompatibility.

Once reproductively isolated, species that derived from a common ancestor can accumulate many differences, even completely new genes. How rapidly and smoothly different forms evolve is another intriguing question in biology.

Darwin assumed that speciation occurred over millions of years as tiny changes gradually accumulated. This hypothesis is known as **gradualism.** Other biologists, most notably Stephen Jay Gould, have argued that most speciation events are sudden, result in dramatic changes in form within the course of a few thousand years, and are followed by many thousands or millions of years of little change—a hypothesis known as **punctuated equilibrium.** The hypothesis of punctuated equilibrium is supported by observations of the fossil record, which seems to reflect just this pattern *(Figure 11.12)*. Although the tempo of evolutionary change may not match Darwin's predictions, the process of natural selection he described can still explain many instances of divergence.

The period after the separation of the gene pools of 2 populations but before the evolution of reproductive isolation could be thought of as a period during which races of a species may form. Determining if the racial groupings on Indigo's census form came about via this process is our focus in the next sections.

Figure 11.12 **Punctuated equilibrium.** The pattern of evolutionary change in groups of species may be (a) smooth, representing a constant level of small changes, or (b) more punctuated, with hundreds of thousands of years of stasis followed unpredictably by rapid, large changes occurring within a few thousand years.

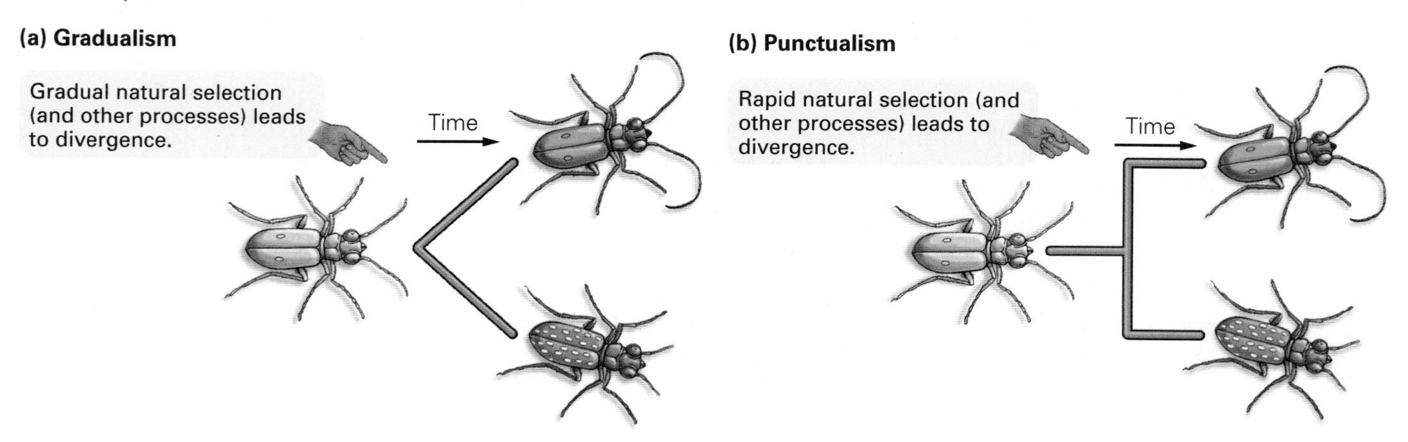

11.2 Races and Genealogical Species

Biologists do not agree on a standard definition of *biological race*. In fact, not all biologists feel that *race* is a useful term; many prefer to use the term *subspecies* to describe subgroups within a species, and others feel that race is not a useful biological concept at all. When the term is applied, it is often inconsistent. For example, populations of birds with slightly different colorations might be called different races by some bird biologists, while other biologists would argue that the same contrasts in color are meaningless.

However, Indigo's question about how race matters in her life leads us to a specific definition of race. What Indigo wants to know is: If she identifies herself as a member of a particular race, does that mean she is more closely related and thus biologically more similar to other members of the same race than she is to members of other races? The definition of **biological race** that addresses this question is the following: Races are populations of a single species that have diverged from each other. With little gene flow among them, evolutionary changes that occur in one race may not occur in a different race. This definition of biological race is actually very similar to another commonly used species definition.

According to the **genealogical species concept,** a species is defined as the smallest group of reproductively compatible organisms containing all of the known descendants of a single common ancestor. More so than the biological species concept, the genealogical species concept emphasizes unique evolutionary lineages; thus, it greatly increases the number of identifiable species.

For example, the spotted owl, *Strix occidentalis*, is described under the biological species concept as a single species with 3 distinct populations: northern, California, and Mexican *(Figure 11.13)*. The northern population is the well-known endangered owl of old-growth forests in the northwestern United States. Although northern owls could theoretically reproduce with Californian and Mexican birds, the northern population is physically isolated from the California and Mexican populations. As a result, its gene pool is separate, and any trait that evolved in the northern spotted owl population is found only within that population. According to the genealogical species concept, the northern spotted owl is a unique species because it is a unique lineage, representing all the descendants of the first spotted owls to colonize the Pacific Northwest.

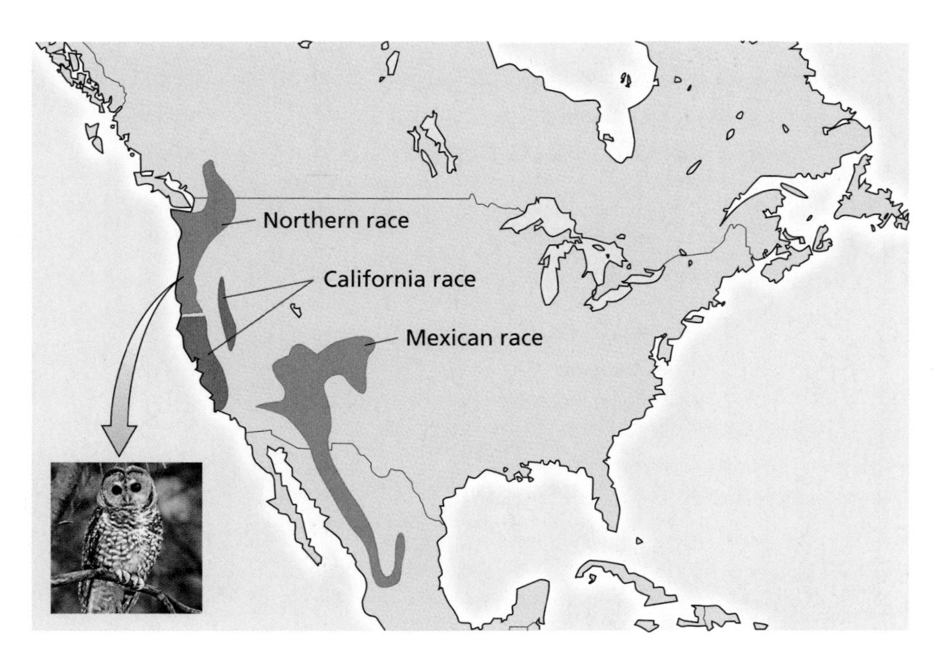

Figure 11.13 **Genealogical species.** The geographic isolation of the northern spotted owl has resulted in the development of a unique gene pool in this population. **Visualize This:** Should the northern, California, and Mexican owl races be classified as independent genealogical species?

The advantage of the genealogical species concept is that species are more easily delineated. If the gene pool of a population is consistently different from other related populations, then that population represents a different species. The genealogical concept can apply to all groups of living species whether they reproduce sexually or not. Its disadvantage is that it can be difficult to apply in practice—different populations of the same biological species have to be studied carefully to determine if their gene pools differ.

Indigo's question about whether race matters, at least biologically, can now be restated as: Do the racial groups on the census form represent populations of the human species whose gene pools have been isolated? In other words, should we consider human races to be different genealogical species? To answer this question, we must first understand how these racial categories came to be identified in American society.

11.3 Humans and the Race Concept

Until the height of the European colonial period in the seventeenth and eighteenth centuries, few cultures distinguished between groups of humans according to shared physical characteristics. People primarily identified themselves and others as belonging to particular cultural groups with different customs, diets, and languages.

As northern Europeans began to contact people from other parts of the world, being able to set groups of people "apart" made colonization and slavery less morally troublesome. Thus, when Linnaeus classified all humans as a species, he was careful to distinguish definitive varieties (what we would now call races) of humans. Linnaeus recognized 5 races of *Homo sapiens*. Not only did Linnaeus describe physical characteristics, he ascribed particular behaviors and aptitudes to each race; reflecting his Western bias, he set the European race as the superior form *(Figure 11.14)*.

The classification shown in Figure 11.14 is one of dozens of examples of how scientists' work has been used to legitimize cultural practices. In this case, hundreds of years of injustice were supported by a scientific classification that made the oppression experienced by nonwhite groups appear "natural."

Figure 11.14 **Linnaean classification of human variety.** Linnaeus published this classification of the varieties of humans in the tenth edition of *Systema Naturae* in 1758. The text reflects a widespread bias among his contemporaries that the European race was superior to other races. Interestingly, one variety of humans recognized by Linnaeus—the *Wild Man* in his classification—is the chimpanzee!

1. HOMO.

Sapiens. Diurnal; varying by education and situation.

2. Four-footed, mute, hairy. — Wild Man.

3. Copper-coloured, choleric, erect. — American.
Hair black, straight, thick; nostrils wide, face harsh; beard scanty; obstinate, content free. Paints himself with fine red lines. Regulated by customs.

4. Fair, sanguine, brawny. — European.
Hair yellow, brown, flowing; eyes blue; gentle, acute, inventive. Covered with close vestments. Governed by laws.

5. Sooty, melancholy, rigid. — Asiatic.
Hair black; eyes dark; severe, haughty, covetous. Covered with loose garments. Governed by opinions.

6. Black, phlegmatic, relaxed. — African.
Hair black, frizzled; skin silky; nose flat; lips tumid; crafty, indolent, negligent. Anoints himself with grease. Governed by caprice.

Scientists since Linnaeus have also proposed hypotheses about the number of races of the human species. The most common number is 6 (white, black, Pacific Islander, Asian, Australian Aborigine, and Native American), although some scientists have described as many as 26 different races of the human species. The physical characteristics used to identify these races are typically skin color, hair texture, and eye, skull, and nose shape.

To answer Indigo's question about the biological meaning of race, we must determine if the physical characteristics used to delineate hypothesized human races developed because these groups evolved independently of each other. We can test this hypothesis by looking at the fossil record for evidence of isolation during human evolution and by looking at the gene pools of modern populations of these proposed races for consistent differences among groups.

The Morphological Species Concept

The ancestors of humans are known only through the fossil record. We cannot delineate fossil species using either the biological or genealogical species concepts. Instead, paleontologists, scientists who study fossils, use a more practical definition: A species is defined as a group of individuals with some reliable physical characteristics distinguishing them from all other species. In other words, individuals in the same species have similar morphology—they look alike in some key feature. This is known as the **morphological species concept.** The morphological differences among species are assumed to correlate with isolation of gene pools. Table 11.2 compares and contrasts the 3 species concepts.

Stop and Stretch 11.3: Which species concept do you think is more likely to be used by biologists who are trying to count the number of bird species in a local park?

TABLE 11.2 **Comparison of 3 species concepts.**

Species Concept	Definition	Pros	Cons
Biological	Species consist of organisms that can interbreed and produce fertile offspring and are reproductively isolated from other species.	Useful in identifying boundaries between populations of similar organisms. Relatively easy to evaluate for sexually reproducing species.	Cannot be applied to organisms that reproduce asexually or to fossil organisms. May not be meaningful when 2 populations of the same species are separated by large geographical distances.
Genealogical	Species consist of organisms that can interbreed and are all descendants of a common ancestor and represent independent evolutionary lineages.	Most evolutionarily meaningful because each species has its own unique evolutionary history. Can be used with asexually reproducing species.	Difficult to apply in practice. Requires detailed knowledge of gene pools of populations within a biological species. Cannot be applied to fossil organisms.
Morphological	Species consist of organisms that share a set of unique physical characteristics that is not found in other groups of organisms.	Easy to use in practice on both living and fossil organisms. Only a few key features are needed for identification.	Does not necessarily reflect evolutionary independence from other groups.

Using the morphological species concept, scientists have identified the fossils of our direct human ancestors. This has allowed them to reconstruct the movement of humans since our species' first appearance.

Modern Humans: A History

The immediate predecessor of *Homo sapiens* was *Homo erectus*, a species that first appeared in east Africa about 1.8 million years ago and spread to Asia and Europe over the next 1.65 million years. Fossils identified as early *H. sapiens* appear in Africa in rocks that are approximately 250,000 years old. The fossil record shows that these early humans rapidly replaced *H. erectus* populations in the Eastern Hemisphere.

There is debate among paleontologists about whether *H. sapiens* evolved just once, in Africa (this is called the *out-of-Africa hypothesis*), or throughout the range of *H. erectus* (known as the *multiregional hypothesis*). Even if *H. sapiens* evolved in Africa and then migrated to Europe and Asia, it is unclear whether populations of early humans hybridized with *H. erectus* in different areas of the globe (the *hybridization-and-assimilation hypothesis*). Because this scientific question is still unresolved, it is difficult to know when the ancestral population of modern humans split into regional populations; it could have been anytime from 150,000 to 1.8 million years ago *(Figure 11.15)*.

Most data support the hypothesis that all modern human populations descended from African ancestors within the last few hundred thousand years. One line of evidence is that humans have much less genetic diversity (measured by the number of different alleles that have been identified for any gene) than any other great ape, indicating that we are a young species that has had little time to accumulate many different gene variants. Among human populations, those in Africa are more genetically diverse than others around the

Figure 11.15 **Three models of modern human origins.** All of these hypotheses may explain the origin of different human groups. The out-of-Africa hypothesis has the most support and indicates that human groups have not been separate for long.

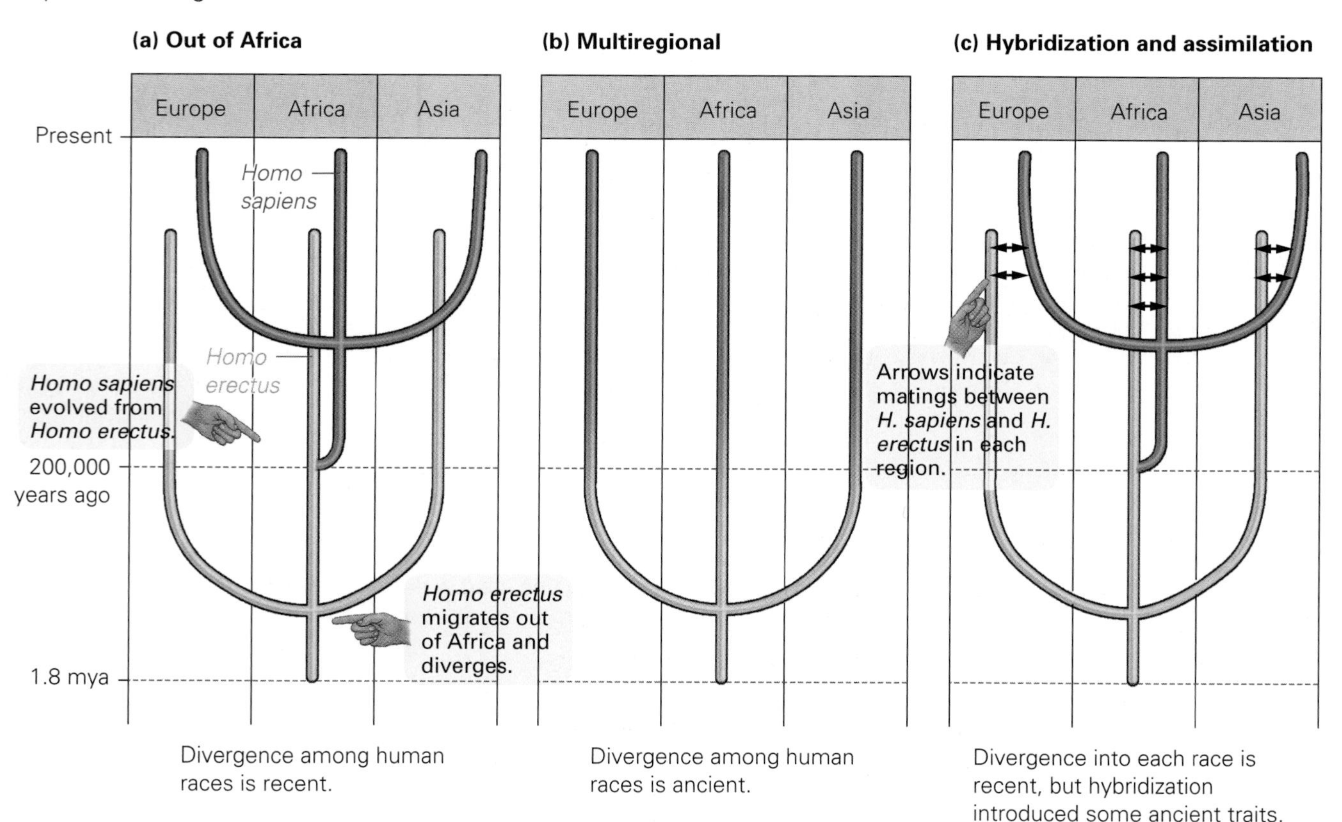

world. In other words, African populations must be the oldest human populations and thus are likely the source of all other human populations. On balance, the out-of-Africa hypothesis has the strongest support, although with the evidence accumulated to date, none of the other hypotheses can be completely rejected.

According to the out-of-Africa hypothesis, the physical differences we see among human populations must have arisen in the last 150,000 to 200,000 years or in about 10,000 human generations. In evolutionary terms, this is not much time. All humans shared a common ancestor very recently; thus the defined human races cannot be very different from each other.

Stop and Stretch 11.4: Which of the hypotheses of modern human origins implies that human groups may be very different from each other?

Genetic Evidence of Divergence

Even with little genetic difference among populations, Indigo's question about the meaning of race is still relevant. After all, even if 2 races differ from one another only slightly, if the difference is consistent, then people are biologically more similar to members of their own race than to people of a different race.

We can examine the gene pool of a population to determine if it is isolated from other groups. Remember that when populations are isolated from each other, little gene flow occurs between them. If an allele appears in one population, it cannot spread to another, and evolutionary changes that occur in one population do not necessarily occur in others.

Recall from Chapter 9 that when a trait becomes more common in a population due to evolution, it is because the allele for that trait has become more common. In other words, evolution results in a change in **allele frequency** in a population, that is, in the percentage of copies of any given gene that are a particular allele. For example, in a population of 50 people, imagine that 2 individuals carry 1 copy of the allele that codes for blood-type B, and the remainder carry 2 copies of the allele that codes for blood type O. Since every person carries 2 copies of each gene, there are actually 100 copies of the blood type gene in the population of 50 people. Because 2 of these copies are allele *B*, the frequency of the *B* allele is 2 out of 100, or 2%, in this population. An evolutionary change in blood type would be seen as an increase or decrease in the frequency of the *B* allele in the next generation.

We can now make 2 predictions to test a hypothesis of whether biological races exist within a species. If a race has been isolated from other populations of the species for many generations, it should have these 2 traits:

1. Some unique alleles
2. Differences in allele frequency for some genes relative to other races

The tree diagram in *Figure 11.16* illustrates these predictions of the hypothesis of biological races. In the figure, butterfly populations colored teal, navy, and sky blue are all part of the same race ("blues"), and populations colored rust, magenta, and pink are part of a separate race ("reds"). The grid at the bottom of the tree illustrates the frequency of alleles for 3 genes in the ancestral butterfly population. For instance, there are 2 alleles for gene 1—one that is very common (allele *a*) and thus high in frequency and one that is rare (allele *b*) and low in frequency. The 2 races described at the top of the tree originated when

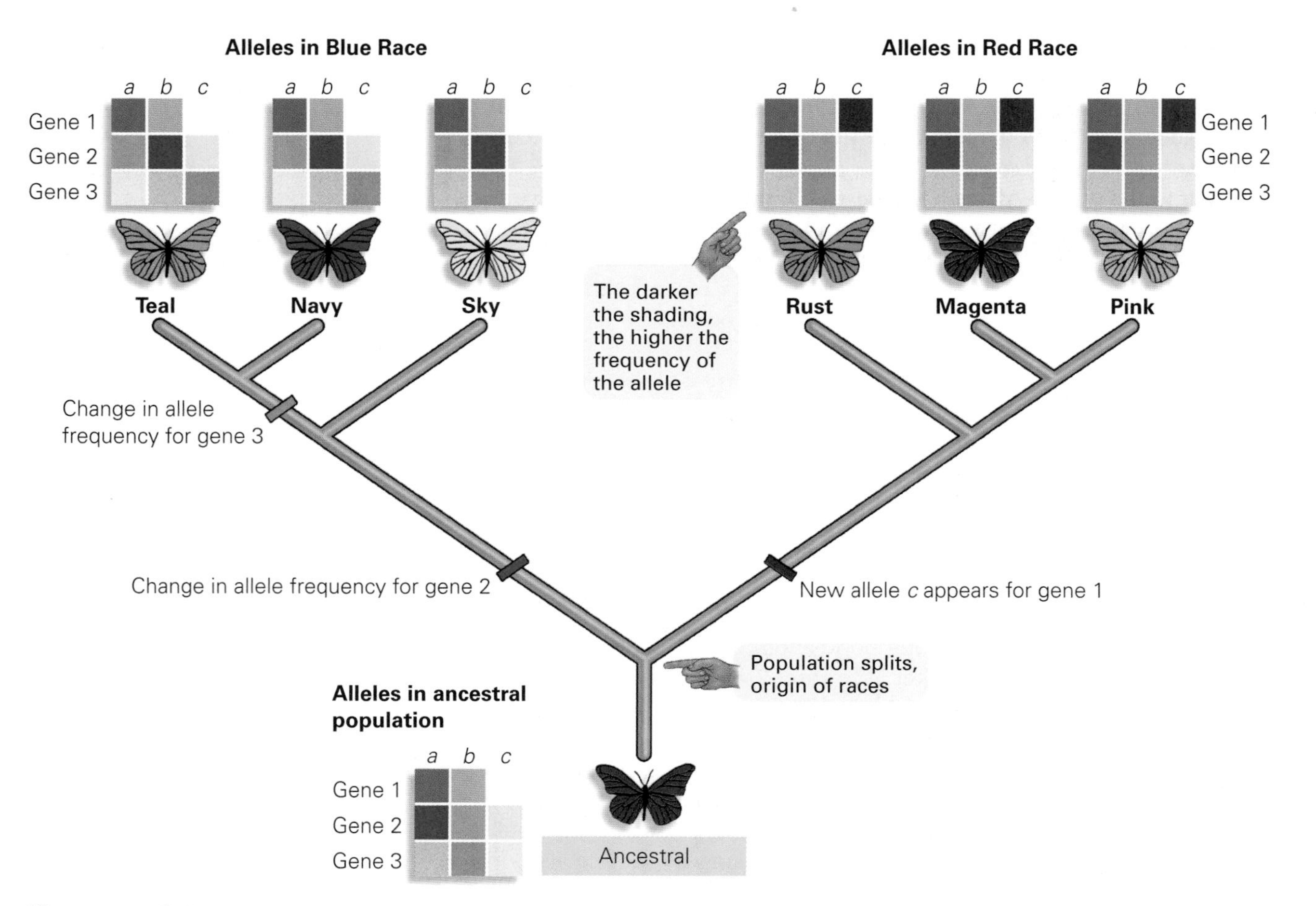

Figure 11.16 **Genetic evidence of isolation between races.** New alleles that appear in 1 population will not spread to the other populations because gene flow is restricted; thus, each race should have unique alleles. The frequency of alleles for various genes will also be more similar among populations within a race than between races. **Visualize This:** Explain the evolutionary change in gene 3 and whether this change supports the hypothesis that the blues and reds are different races.

the ancestral population split, and the 2 resulting populations became isolated from each other. Notice the following patterns:

- **A race-specific allele.** Not long after the divergence between the blue and red races, mutation causes a new allele for gene 1 (that is, allele *c*) to arise *but only in the red race.* Because reds and blues are isolated, this allele does not spread to the gene pool of the blue race—there is no allele *c* for gene 1 in any of the blue populations. In addition, because the populations colored rust, magenta, and pink diverge *after* this allele appears in reds, all of these populations contain individuals that carry allele *c* for gene 1.
- **Similar allele frequencies in populations within races.** Also not long after the divergence between reds and blues, natural selection causes a change in the allele frequency of gene 2. Perhaps the environment inhabited by blues is different from that inhabited by reds, and thus allele *b* is only favored in 1 branch. Thus allele *b* becomes more common in the blue race. This evolutionary change occurred before the divergence of the populations colored teal, navy, and sky blue. Therefore, all of these blue race populations have a similar pattern of allele frequency for this gene, but the pattern differs from that in the populations that make up the red race.

As a result of the evolutionary independence of the blue race and the red race, if you compare the allele frequency grids of all the populations, you will notice that populations colored teal, navy, and sky blue are similar (although not identical) to each other, and populations colored rust, magenta, and pink are similar to each other. However, the allele frequencies for the genes in the teal population are distinctly different from those in the rust, magenta, and pink populations.

Observing a pattern of unique allele frequencies in different populations of the same species supports the hypothesis that the populations have been isolated from each other. For example, scientists have observed that certain alleles are more common in apple-eating populations of apple maggot flies than in the hawthorn-eating populations. This has led researchers to conclude that these populations are genetically isolated from each other and should be considered different races.

Using the Hardy-Weinberg Theorem to Calculate Allele Frequencies

Allele frequencies in a population can be determined using a mathematical relationship independently discovered by Godfrey Hardy of Cambridge University and German physician Wilhelm Weinberg. The **Hardy-Weinberg theorem** states that allele frequencies will remain stable in populations that are large in size, randomly mating, and experiencing no migration or natural selection. This rule provides a baseline for predicting how allele frequencies will change if any of the theorem's assumptions are violated. In other words, the Hardy-Weinberg theorem enables scientists to quantify the effect of evolutionary change on allele frequencies. Today, the Hardy-Weinberg theorem forms the basis of the modern science of **population genetics.**

In the simplest case, the Hardy-Weinberg theorem (which we abbreviate to HW) describes the relationship between allele frequency and genotype frequency for a gene with 2 alleles in a stable population. The frequency of these 2 alleles are written as p and q.

Let us say that 70% of the alleles in the population are dominant (*A*), and 30% are recessive (*a*). Thus, $p = 0.7$ and $q = 0.3$ Each gamete produced by members of the population carries 1 copy of the gene. Therefore, 70% of the gametes produced by this entire population will carry the dominant allele, and 30% will carry the recessive allele. The frequency of gametes produced of each type is equal to the frequency of alleles of each type *(Figure 11.17a)*.

HW assumes that every member of the population has an equal chance of mating with any member of the opposite sex. The fertilizations that occur in this situation are analogous to the result of a lottery drawing. In this analogy, we can imagine individuals in a population each contributing an equal number of gametes to a "bucket." Fertilizations result when gamete drawn from the sperm bucket fuses with another drawn from the egg bucket. Since the frequency of gametes carrying the dominant allele in the bucket is equal to the frequency of the dominant allele in the population, the chance of drawing an egg that carries the dominant allele is 70%.

In *Figure 11.17b*, a modified Punnett square illustrates the relationship between allele frequency and genotype frequency in a stable population. On the horizontal axis of the square, we place the 2 types of gametes that can be produced by females in the population (*A* and *a*), while on the vertical axis we place the 2 types of gametes that can be produced by males. In addition, on each axis is an indication of the frequency of these types of egg and sperm in the population: 0.7 for *A* eggs and *A* sperm, 0.3 for *a* eggs and *a* sperm.

Used like the typical Punnett square in Chapter 6, the grid of the square also shows the frequency of each genotype in this population. The frequency of the *AA* genotype in the next generation will be equal to the frequency of *A* sperm being drawn (0.7) times the frequency that *A* eggs will be drawn (0.7), or 0.49. This calculation can be repeated for each genotype. The frequency of the *AA* genotype is $p \times p$ ($= p^2$), the *aa* genotype $q \times q$ ($= q^2$), and the *Aa* genotype $p \times q \times 2$ ($= 2pq$). Hardy and Weinberg mathematically proved that the frequency of genotypes in 1 generation of a population depends on the frequency of genotypes in the previous generation of the same population.

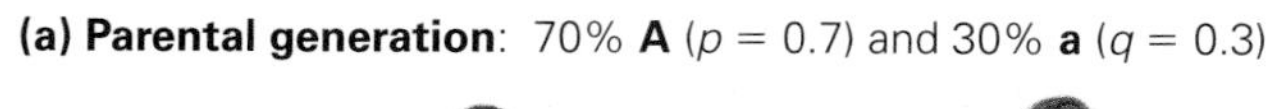

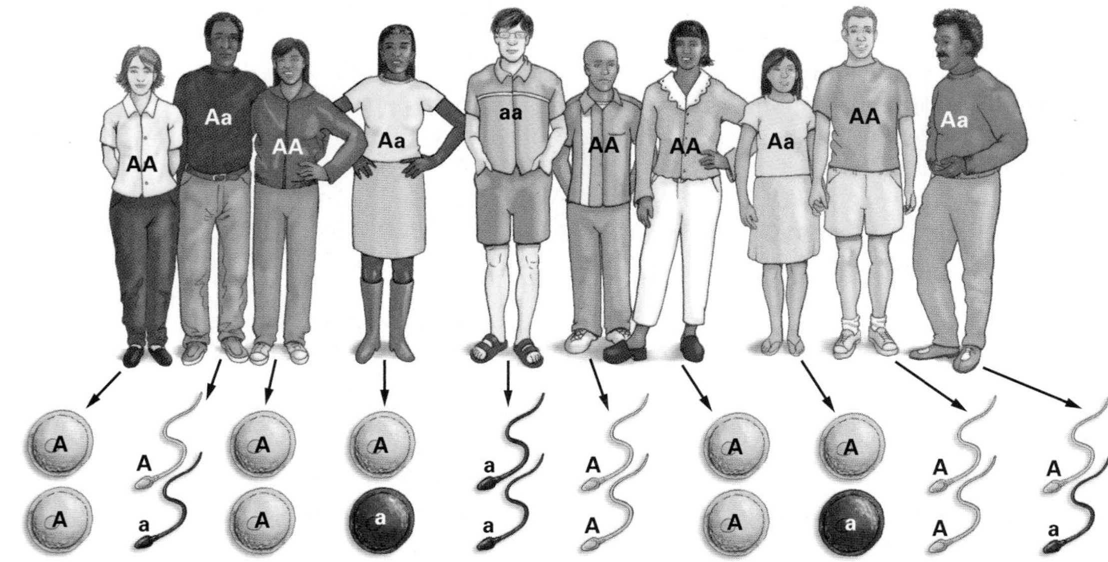

Gametes: 70% **A** ($p = 0.7$) and 30% **a** ($q = 0.3$)

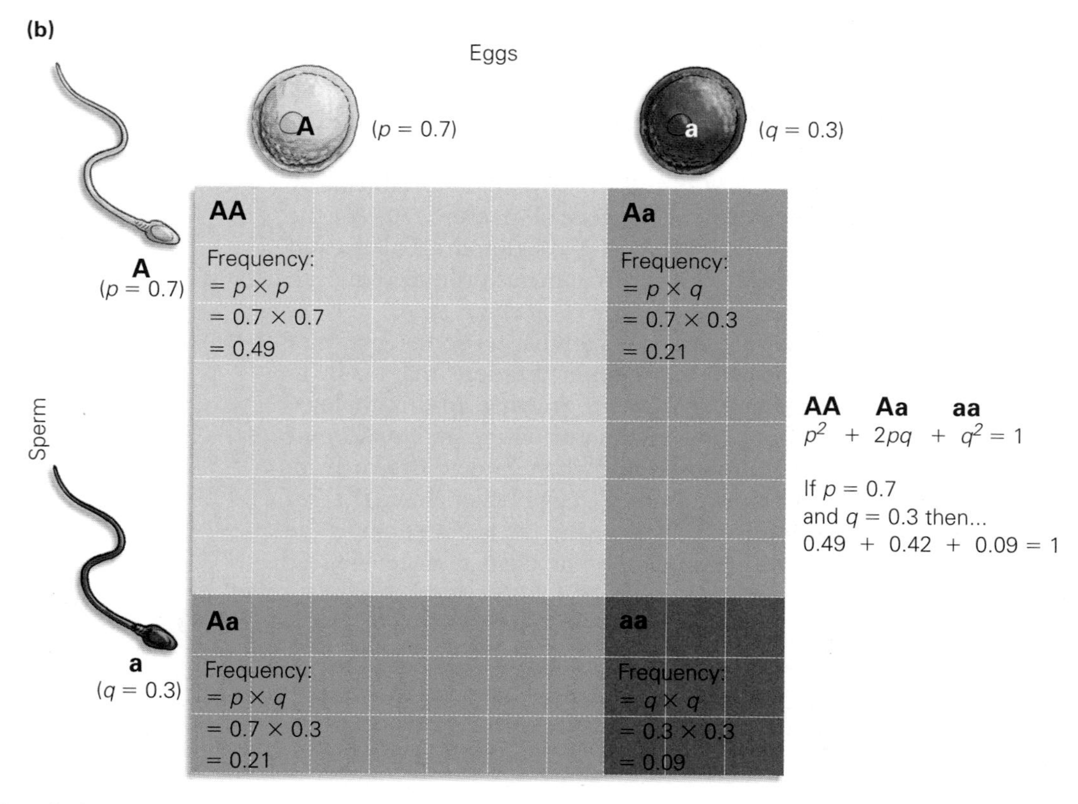

AA Aa aa

$p^2 + 2pq + q^2 = 1$

If $p = 0.7$
and $q = 0.3$ then...
$0.49 + 0.42 + 0.09 = 1$

Figure 11.17 **The relationship between allele frequency and gamete frequency.** (a) The frequency of any allele in a population of adults is equal to the frequency of that allele in gametes produced by that population. (b) Knowing the allele frequency of gametes allows us to predict the frequency of various genotypes in the next generation. **Visualize This:** Why is the frequency of heterozygotes 2pq instead of just pq?

Scientists rarely have information about allele frequency; however, they often have information about genotype frequency. When scientists know the frequency of a phenotype produced by a recessive allele, they know the frequency of that genotype. They can then use HW to calculate the allele frequency in a population. For instance, if the frequency of individuals with sickle-cell anemia is 1 in 100 births (0.01), we know that q^2 —the frequency of homozygous recessive individuals in the population—is equal to 0.01. Therefore, q is simply the square root of this number, or 0.1.

For traits with a phenotype that is not so obvious, scientists use the DNA fingerprinting technology described in Chapter 7 to determine allele frequencies in a subsample of the population.

Human Races Are Not Biological Groups

Recall the 6 major human races described by many authors: white, black, Pacific Islander, Asian, Australian Aborigine, and Native American. Do these groups show the predicted pattern of race-specific alleles and unique patterns of allele frequency? In a word, no.

No Race-Specific Alleles Have Been Identified. Sickle-cell anemia has long been thought of as a "black" disease. This illness occurs in individuals who carry 2 copies of the sickle-cell allele, resulting in red blood cells that deform into a sickle shape under certain conditions. The consequences of these sickling attacks include heart, kidney, lung, and brain damage. Many individuals with sickle-cell anemia do not live past childhood.

Nearly 10% of African Americans and 20% of Africans carry 1 copy of the sickle-cell allele. However, if we examine the distribution of the sickle-cell allele more closely, we see that the pattern is not quite so simple. Not all human populations classified as black have a high frequency of the sickle-cell allele. In fact, in populations from southern and north-central Africa, which are traditionally classified as black, this allele is very rare or absent. Among populations that are classified as white or Asian, there are some in which the sickle-cell allele is relatively common, such as white populations in the Middle East and Asian populations in northeast India *(Figure 11.18)*. Thus, the sickle-cell allele is not a characteristic of all black populations or unique to a supposed "black race."

Similarly, cystic fibrosis, a disease that results in respiratory and digestive problems and early death, was often thought of as a "white" disease. Cystic fibrosis occurs in individuals who carry 2 copies of the cystic fibrosis allele. As with sickle-cell anemia, it has become clear that the allele that causes cystic fibrosis is not found in all white populations and is found, in low frequency, in some black and Asian populations.

These examples of the sickle-cell allele and cystic fibrosis allele demonstrate the typical pattern of gene distribution. Scientists have not identified a single allele that is found in all (or even most) populations of a commonly described race but not found in other races. The hypothesis that human races represent independent evolutionary groups is not supported by these observations.

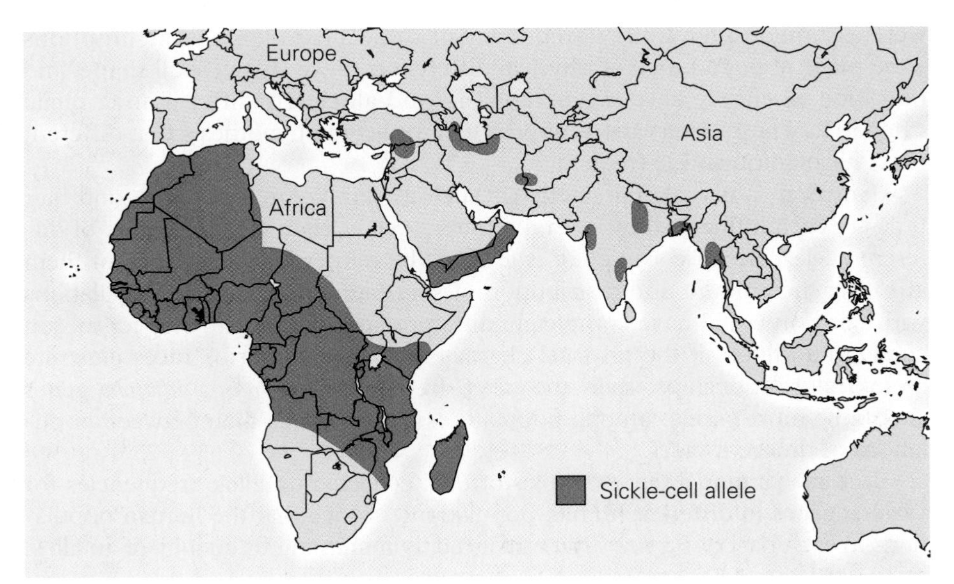

Figure 11.18 **The sickle-cell allele: Not a "black gene."** The map illustrates where the sickle-cell allele is found in human populations. Note that it is not found in all African populations but is found in some European and Asian populations.

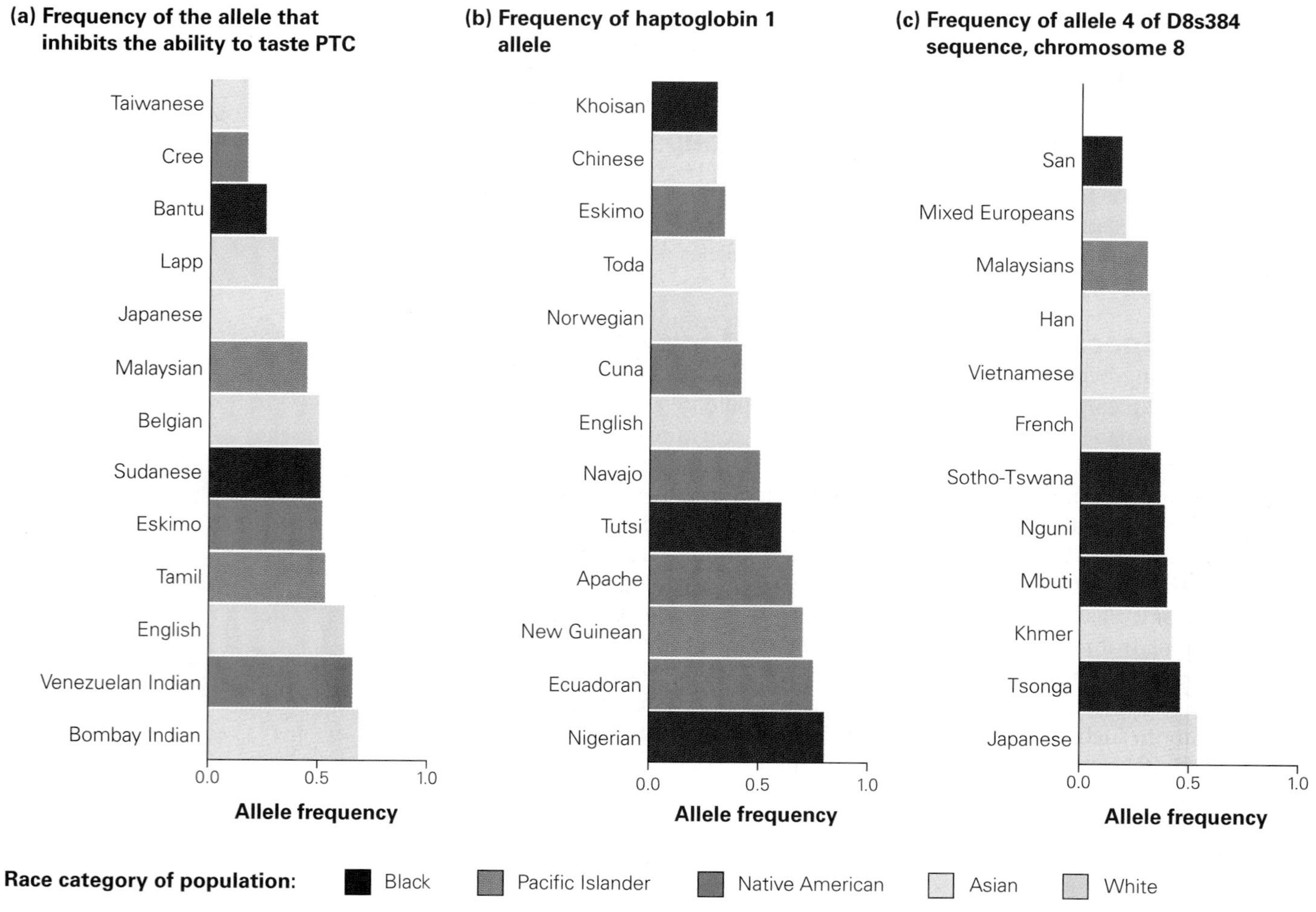

Figure 11.19 **Do human races show genetic evidence of isolation?** The bars on each of these histograms illustrate the frequency of the described allele in many different human populations. These histograms illustrate that populations within these "races" are not necessarily more similar to each other than they are to populations in different races.

Populations Classified in the Same Race Do Not Have Similar Allele Frequencies. Until the advent of modern techniques allowing scientists to isolate genes and the proteins they produce, there was no way of directly measuring the frequency of alleles for most of the genes in a population. However, scientists could evaluate the racial categories already in place and assume that their average physical differences reflected genetic differences among them. Thus populations with dark skin were assumed to have a high frequency of "dark skin" alleles, while populations with light skin were assumed to have a low frequency of these alleles. Similar assumptions were made about a range of physical differences—eye shape, skull shape, and hair type all clearly have a genetic basis, and all clearly differ among racial categories. These observations appear to support the hypothesis that different races have unique allele frequencies.

However, physical characteristics such as skin color, eye shape, and hair type are each influenced by several different genes, each with a number of different alleles. Because skin color is affected by numerous genes, each of them affecting the amount and distribution of skin pigment, 2 human populations with fair skin could have completely different gene pools with respect to skin color. In addition, if the physical characteristics that describe races illustrate biological relationships, then the allele frequency for *many different* genes should be more similar among populations within a race than between populations of different races.

Let us examine a few examples of data collected on allele frequencies for various genes in different human populations. Notice that the human populations in each part of *Figure 11.19* are listed by increasing frequency of an allele in the population. The color coding of each population group in each of the graphs corresponds to the racial category in which the population is typically placed. For example, at the top of part a, the Taiwanese population is categorized as Asian, while the Cree Indian population of eastern Canada is categorized as Native American. If the hypothesis that human racial groups have a biological basis is correct, then populations from the same racial group should be clustered together on each bar graph.

Figure 11.19a shows the frequency of the allele that interferes with an individual's ability to taste the chemical phenylthiocarbamide (PTC) in several populations. People who carry 2 copies of this recessive allele cannot detect PTC, which tastes bitter to people who carry 1 or no copies of the allele. Note that members of a single "race," such as Asian, vary widely in frequency of this allele.

Figure 11.19b lists the frequency of 1 allele for the gene *haptoglobin 1* in a number of different human populations. Haptoglobin 1 is a protein that helps scavenge the blood protein hemoglobin from old, dying red blood cells. Again, we see a wide distribution in allele frequency within the race categories.

Figure 11.19c illustrates variation among human populations in the frequency of a repeating DNA sequence on chromosome 8. Repeating sequences are common in the human genome, and differences among individuals in the number of repeats create the unique signatures of DNA fingerprints. The frequency of 1 pattern of repeating sequence in a segment of chromosome 8, called allele 4 of the D8s384 sequence, is illustrated for a number of populations.

What we see in the 3 graphs in Figure 11.19 is that allele frequencies for these genes are *not* more similar within racial groups than between racial groups. In fact, in 2 of the 3 graphs, Figure 11.19a and 11.19b, the populations with the highest and lowest allele frequencies belong to the same race—for these genes, there is more variability *within* a race than there are average differences *among* races. Scientists have observed this same pattern for every gene they have studied in humans.

Human races fail to meet any of the criteria for identifying populations as isolated from each other. Both the fossil evidence and genetic evidence indicate that the 6 commonly listed human racial groups do *not* represent biological races.

Stop and Stretch 11.5: We are accustomed to looking at differences in skin color and eye shape and presuming they signify underlying biological relationships. Can you think of another trait that varies greatly in human populations but that is not used as evidence for biological relationship? How would using this trait affect the classification of races?

Human Races Have Never Been Truly Isolated

The evidence that human populations have been "mixing" since modern humans first evolved is contained within the gene pool of human populations. For instance, the frequency of the B blood group decreases from east to west across Europe *(Figure 11.20)*. The I^B allele that codes for this blood type apparently evolved in Asia, and the pattern of blood group distribution seen in Figure 11.20 corresponds to the movement of Asians into Europe beginning about 2000 years ago. As the Asian immigrants mixed with the European residents, their alleles became a part of the European gene pool. Populations closest to Asia experienced a large change in their gene pools, while populations who were more distant encountered a more "diluted" immigrant gene pool made up of the offspring between the Asian immigrants and their European neighbors.

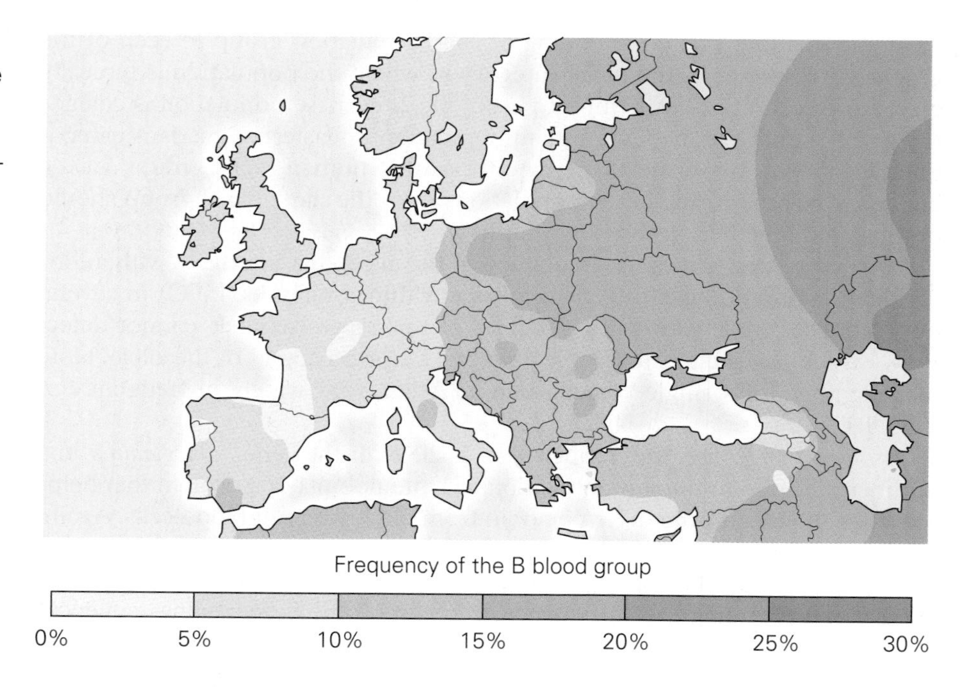

Figure 11.20 **Genetic mixing in humans.** The decline in frequency of type B blood from west to east in Europe reflects the movement of alleles from Asian populations into European populations over the past 2000 years.

Other genetic analyses have led to similar maps. For example, one indicates that populations that practiced agriculture arose in the Middle East and migrated throughout Europe and Asia about 10,000 years ago. These data indicate that there are no clear boundaries within the human gene pool. Interbreeding of human populations over hundreds of generations has prevented the isolation required for the formation of distinct biological races.

11.4 Why Human Groups Differ

Human races are not true biological races. However, as is clear to Indigo and to all of us, human populations do differ from each other in many traits. In this section, we explore what is known about why populations share certain superficial traits and differ in others.

Natural Selection

Recall the distribution of the sickle-cell allele in human populations shown in Figure 11.18 (p. 303). This allele is found in some populations of at least 3 of the typically described races. Why is it higher in certain populations? Because natural selection favors individuals who carry 1 copy of the sickle-cell allele in particular environments.

The sickle-cell anemia allele is an *adaptation,* a feature that increases fitness, within populations in malaria-prone areas. Malaria is a disease caused by a parasitic, single-celled organism that spends part of its life cycle feeding on red blood cells, eventually killing the cells. Because their red blood cells are depleted, people with severe malaria suffer from anemia, which may result in death. When individuals carry a single copy of the sickle-cell allele, their blood cells deform when infected by a malaria parasite. These deformed cells quickly die, reducing the parasite's ability to reproduce and infect more red blood cells and therefore reducing a carrier's risk of anemia.

The sickle-cell allele reduces the likelihood of severe malaria, so natural selection has caused it to increase in frequency in susceptible populations. The protection that the sickle-cell allele provides to heterozygote carriers is demonstrated

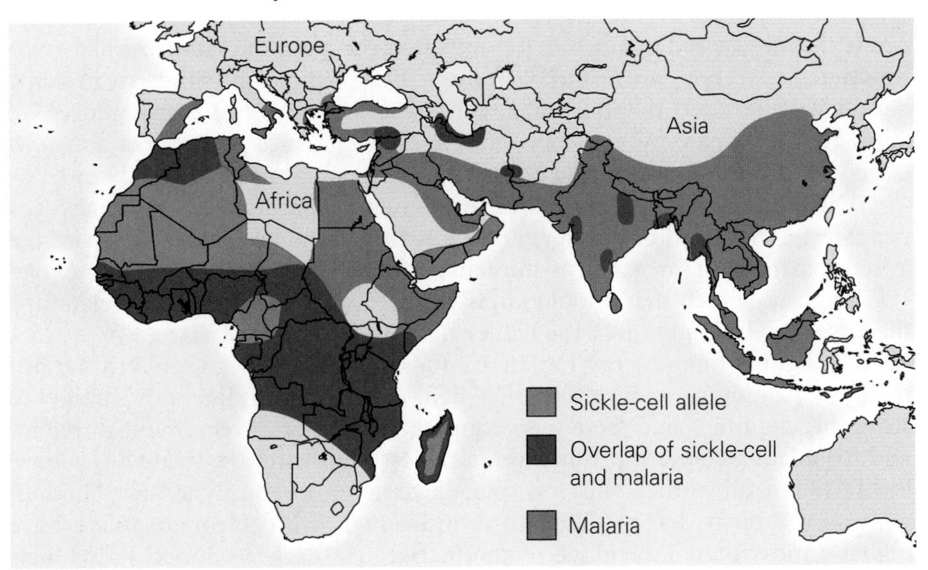

Figure 11.21 **The sickle-cell allele is common in malarial environments.** This map shows the distributions of the sickle-cell allele and malaria in human populations.

by the overlap between the distribution of malaria and the distribution of sickle-cell anemia *(Figure 11.21)*.

Another physical trait that has been affected by natural selection is nose form. The pattern of nose shape in populations generally correlates to climate factors—populations in dry climates tend to have narrower noses than do populations in moist climates. A long, narrow nose has a greater internal surface area, exposing inhaled air to more moisture, reducing lung damage and increasing the fitness of individuals in dry environments. Among tropical Africans, people living at drier high altitudes have much narrower noses than do those living in humid rain-forest areas *(Figure 11.22)*.

Interestingly, our preconception puts 2 populations of Africans in the same race and explains differences in their nose shape as a result of natural selection, but we place white and black populations into different races and explain their skin color differences as evidence of long isolation from each other. However, like nose shape, skin color is a trait that is strongly influenced by natural selection.

(a) Ethiopian with a narrow nose

(b) Bantu with a broad nose

Figure 11.22 **Nose shape is affected by natural selection.** (a) Long, narrow noses are more common among populations in cold, dry environments. (b) Broad, flattened noses are more common in warm, wet environments.

Figure 11.23 **Convergence.** The similarity in shape between dolphins and sharks results from similar adaptations to life as an oceanic predator of fish, not shared ancestry.

Convergent Evolution

Traits that are shared by unrelated populations because they share similar environmental conditions are termed *convergent*. For example, the similarity in shape between white-sided dolphins and reef sharks is a result of convergence. We know by their anatomy and reproductive characteristics that sharks are most closely related to other fish, and dolphins to other mammals *(Figure 11.23)*.

The pattern of skin color in human populations around the globe also appears to be the result of **convergent evolution.** When scientists compare the average skin color in a native human population to the level of ultraviolet (UV) light to which that population is exposed, they see a close correlation—the lower the UV light level, the lighter the skin *(Figure 11.24)*.

UV light is high-energy radiation in a range that is not visible to the human eye. Among its many effects, UV light interferes with the body's ability to store the vitamin folate. Folate is required for proper development in babies and for adequate sperm production in males. Men with low folate levels have low fertility, and women with low folate levels are more likely to have children with severe birth defects. Therefore, individuals with adequate folate have higher fitness than individuals without. Because darker-skinned individuals absorb less UV light, they have higher folate levels in high-UV environments than light-skinned individuals do. In other words, in environments where UV light levels are high, dark skin is favored by natural selection.

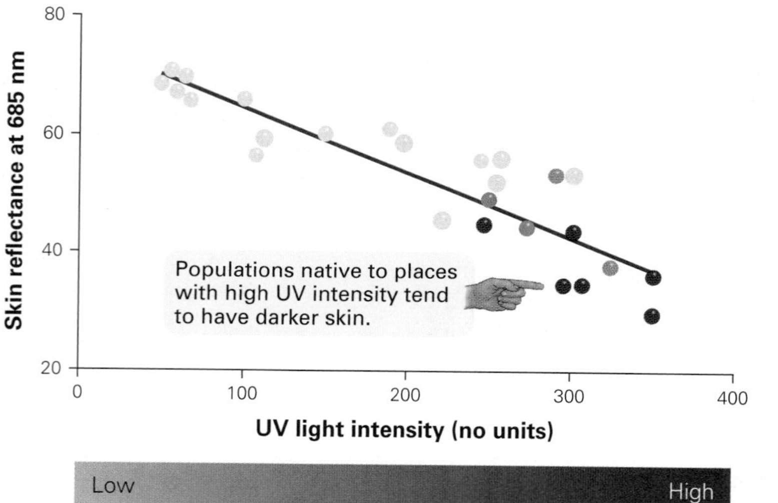

Figure 11.24 **There is a strong correlation between skin color and exposure to UV light.** Reflectance is an indication of color—higher reflectance indicates lighter skin. The color of the dots on the graph specifies the racial category of each population. **Visualize This:** Find 2 populations classified as the same race that have very different skin color. Find 2 populations classified in different races that have the same skin color.

Human populations in low-UV environments face a different challenge. Absorption of UV light is essential for the synthesis of vitamin D. Vitamin D is crucial for the proper development of bones. Women are especially harmed by low vitamin D levels—inadequate development of the pelvic bones can make giving birth deadly. There is no risk of not making enough vitamin D when UV light levels are high, regardless of skin color. However, in areas where levels of UV light are low, individuals with lighter skin absorb a larger fraction of UV light and thus have higher levels of vitamin D. In darker environments, light skin has been favored by natural selection *(Figure 11.25)*.

Because UV light has important effects on human physiology, it has driven the evolution of skin color in human populations. Where UV light levels are high, dark skin is an adaptation, and populations become dark skinned. Where UV light levels are low, light skin is usually an adaptation, and populations evolve to become light skinned. The pattern of skin color in human populations is a result of the convergence of different populations in similar environments, not necessarily evidence of separate races of humans.

Stop and Stretch 11.6: The word for *mother* in many different languages is a variant on "ma." This appears to be an example of convergent evolution in language. Think about the first word-like sounds babies make. Why might many different cultures share this characteristic name for mother?

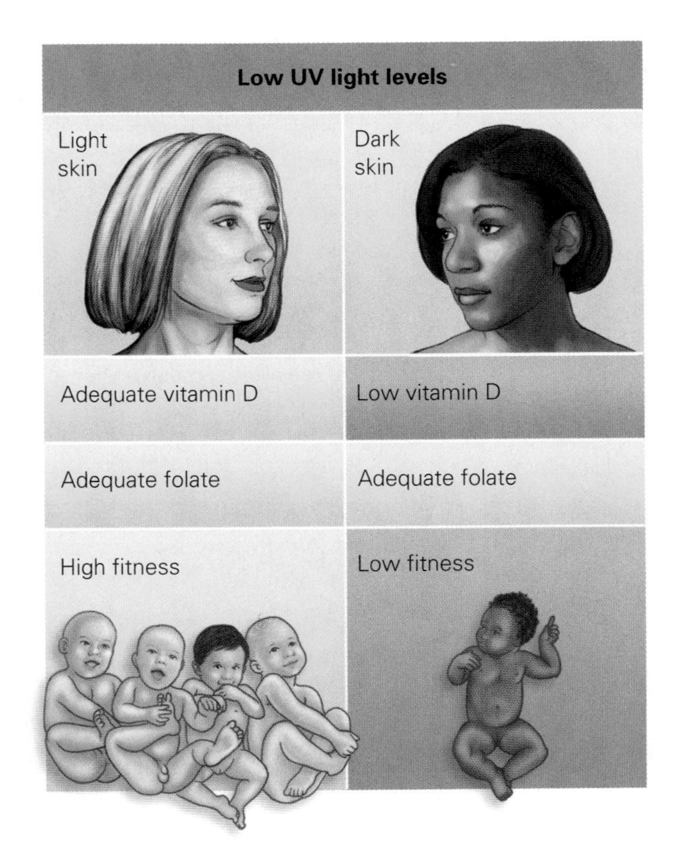

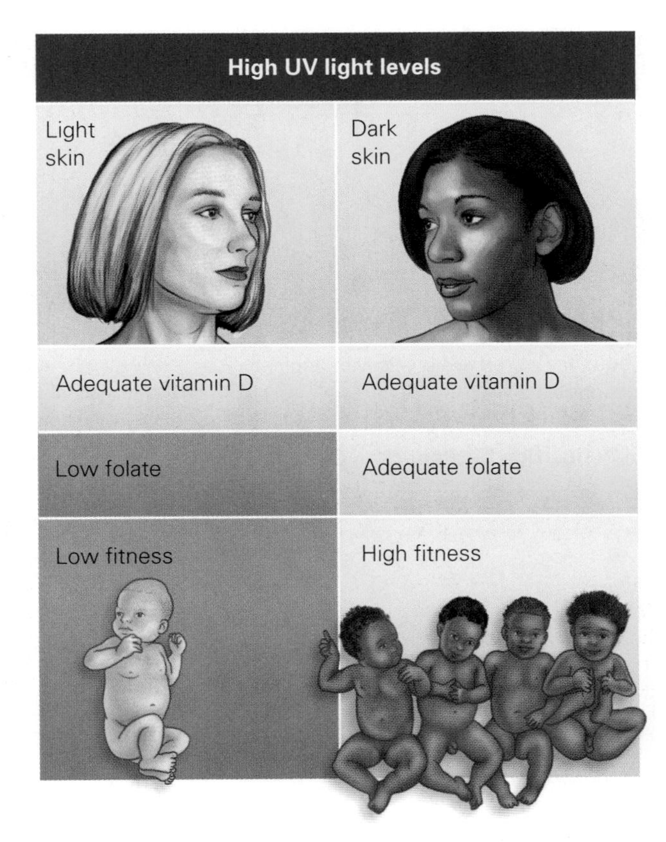

Figure 11.25 **The relationship between UV light levels, folate, vitamin D, and skin color.** Populations in regions where UV light levels are high experience selection for darker, UV-resistant skin. Populations in regions where UV light levels are low experience selection for lighter, UV-transparent skin.

Natural selection has caused differences among human populations, but it has also resulted in some populations superficially appearing more similar to some other human populations. In contrast, populations who appear on the surface to be similar may be quite different, simply by chance.

Genetic Drift

A change in allele frequency that occurs due to chance is called **genetic drift.** Human populations tend to travel and colonize new areas, and so we seem to be especially prone to evolution via genetic drift. Genetic drift occurs in 3 different types of situations.

Founder Effect. Genetic differences can occur when a small sample of a larger population establishes a new population. The gene pool of a sample is rarely an exact reflection of the source population's gene pool. The difference between the 2 gene pools is called *sampling error.* As discussed in Chapter 1, sampling error is more severe for smaller subsets of a population, such as those that typically found new settlements. This type of sampling error is often referred to as the **founder effect** *(Figure 11.26a)*.

Rare genetic diseases that are more common in certain populations often result from the founder effect. For example, the Amish of Pennsylvania are descended from a population of 200 German founders established 200 years ago. Ellis-van Creveld syndrome, a recessive disease that causes dwarfism (among other effects), is 5000 times more common in the Pennsylvania Amish population than in other German American populations. This difference is a result of a single founder in that original population who carried the allele. Since the Pennsylvania Amish usually marry others within their small religious community, the allele has stayed at a high level—1 in 8 Pennsylvania Amish are carriers of the Ellis-van Creveld allele, compared to fewer than 1 in 100 non-Amish German Americans.

(a) Founder effect: A small sample of a large population establishes a new population.

Frequency of red allele is low in original population.

Several of the travelers happen to carry the red allele.

Frequency of red allele is much higher in new population.

(b) Population bottleneck: A dramatic but short-lived reduction occurs in population size.

Frequency of red allele is low in original population.

Many survivors of tidal wave happen to carry red allele.

Frequency of red allele is much higher in new population.

(c) Chance events in small populations: The carrier of a rare allele does not reproduce.

Frequency of red allele is low in original population.

The only lizard with red allele happens to fall victim to an eagle and dies.

Red allele is lost.

Figure 11.26 **The effects of genetic drift.** A population may contain a different set of alleles because (a) its founders were not representative of the original population; (b) a short-lived drop in population size caused a change in allele frequency in 1 of the populations; (c) 1 population is so small that low-frequency alleles are lost by chance.

Plants with animal-dispersed seeds appear to be especially prone to the founder effect. For example, cocklebur, a widespread weed that produces hitchhiker fruit designed to grab onto the fur of a passing mammal *(Figure 11.27)*, consists of populations that are quite variable in size and shape. The variation among populations appears to have been caused by differences in the hitchhikers that happen to be carried to a new colony.

Population Bottleneck. Genetic drift may also occur as the result of a **population bottleneck,** a dramatic but short-lived reduction in population size followed by a rapid increase in population *(Figure 11.26b)*. Bottlenecks may occur as a result of natural disasters. As with the founder effect, the new population differs from the original because the gene pool of the survivors is not an exact model of the source population's gene pool.

A sixteenth-century bottleneck on the island of Puka Puka in the South Pacific resulted in a human population that is clearly different from other Pacific island populations: The 17 survivors of a tsunami on Puka Puka were all relatively petite, and modern Puka Pukans are significantly shorter in stature compared to people native to nearby islands.

Bottlenecks are experienced by nonhuman populations as well. For example, the genetic similarity among individuals in a large population of Galápagos tortoises on the island of Isabela seems to suggest genetic drift. Geological evidence indicates that most of the tortoise population was wiped out during a volcanic eruption about 88,000 years ago. The current population apparently descended from a tiny group of survivors and thus has little genetic diversity.

Genetic Drift in Small Populations. Even without a population bottleneck, allele frequencies may change in a population due to chance events. When an allele is in low frequency within a small population, only a few individuals carry a copy of it. If 1 of these individuals fails to reproduce, or if it passes on only the more common allele to surviving offspring, the frequency of the rare allele may drop in the next generation *(Figure 11.26c)*. If the population is small enough, even relatively high-frequency alleles may be lost after a few generations by this process.

A human population that illustrates the effects of genetic drift in small populations is the Hutterites, a religious sect with communities in South Dakota and Canada. Modern Hutterite populations trace their ancestry back to 442 people who migrated from Russia to North America between 1874 and 1877. Hutterites tend to marry other members of their sect, and so the gene pool of this population is small and isolated from other populations. Genetic drift in this population over the last century has resulted in a near absence of type B blood among the Hutterites, as compared to a frequency of 15% to 30% in other European migrants in North America.

Genetic drift in populations that remain small for many generations can lead to a rapid loss of many different alleles. While this problem is uncommon in humans, the effects of genetic drift on small populations of endangered species can lead to extinction and will be explored in Chapter 14.

Humans are a highly mobile species, and we have been founding new populations for millennia. Most early human populations were also probably quite small. These factors make human populations especially susceptible to the founder effect, genetic bottlenecks, and genetic drift and have contributed to the differences among modern human groups. However, in addition to natural selection and random genetic change, humans' highly social nature has contributed to superficial differences in the appearance of different populations.

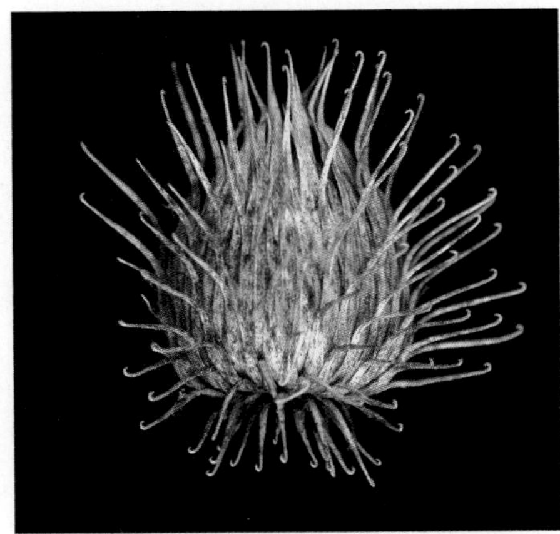

Figure 11.27 **An example of the founder effect in plants.** Cocklebur is a hitchhiker plant, dispersing by hooking the spikes on its fruits to the fur (or sock) of a passing animal. When these burrs are removed at a distant location, the plant can found a new population that is a genetic subset of the parent population and thus might be quite different in size and shape.

Sexual Selection

Men and women within a population may have preferences for particular physical features in their mates. These preferences can cause populations to differ in appearance. When a trait influences the likelihood of mating, that trait is under the influence of a form of natural selection called **sexual selection.**

Darwin hypothesized sexual selection in 1871 as an explanation for differences between males and females within a species. For instance, the enormous tail on a male peacock results from female peahens that choose mates with showier tails. Because large tails require so much energy to display and are more conspicuous to their predators, peacocks with the largest tails must be both physically strong and smart to survive. Peahens can use the size of the tail, therefore, as a measure of the "quality" of the male. Tail length does appear to be a good measure of overall fitness in peacocks; the offspring of well-endowed males are more likely to survive to adulthood than are the offspring of males with scanty tails. When a peahen chooses a male with a large tail, she is ensuring that her offspring will receive high-quality genes. Sexual selection explains the differences between males and females in many species *(Figure 11.28)*.

(a) Peacock

(b) Lion

(c) Blue morpho butterfly

Figure 11.28 **The effects of sexual selection.** Sexual selection is responsible for many unique and fantastic characteristics of organisms from (a) the peacock's tail to (b) the male lion's mane to (c) the bright colors of butterflies.

Stop and Stretch 11.7: In most cases of sexual selection in animals, it is female choice that drives the evolution of traits in the male. Why do you think there is more incentive for females to choose the right male than vice versa?

In humans, there is some evidence that the difference in overall body size between men and women is a result of sexual selection—namely, a widespread female preference for larger males—perhaps again because size may be an indication of overall fitness. However, some apparently sexually selected traits seem to have little or no relationship to fitness and reflect simply a "social preference." In our highly social species, this type of sexual selection may be common.

For example, some scientists have suggested that lack of thick facial and body hair in many Native American and Asian populations resulted from selection by both men and women for less-hairy mates. Some scientists even hypothesize that many physical features that are unique to particular human populations evolved as a result of these socially derived preferences. While intriguing, there is as yet little evidence to support this idea and no simple way to test it.

Assortative Mating

Differences between human populations may be reinforced by the ways in which people choose their mates. Individuals usually prefer to marry someone who is like themselves, by a process called positive **assortative mating.** For example, there is a tendency for people to mate assortatively by height—that is, tall women tend to marry tall men—and by skin color *(Figure 11.29)*.

When 2 populations differ in obvious physical characteristics, the number of matings between them may be small if the traits of 1 population are considered unattractive to members of the other population. Assortative mating has been observed in other organisms as well; for instance, sea horses choose mates that are similar in size to themselves, and in some species of fruit flies, females will mate only with males who have the same body color. Positive assortative mating tends to maintain and even exaggerate physical differences between populations. In highly social humans, assortative mating may be an important cause of differences between groups.

While human populations may show superficial differences due to natural selection in certain environments, genetic drift, sexual selection, and assortative mating, the genetic evidence indicates that many of these differences are literally no more than skin deep. Beneath a veneer of physical differences, humans are basically the same.

11.5 Race in Human Society

The discussion in this chapter may still leave Indigo unsatisfied. Scientific data indicate that the racial categories on her census form are biologically meaningless. Races that were thought of as unitary groups have been revealed to be diverse collections of populations. Two unrelated individuals of the "black race" are no more likely to be biologically similar than a black person and a white person. Yet everywhere she looks, Indigo sees evidence that the racial categories on her census form matter to people—from the existence of her college's Black Student Association to the heated discussions in her American Experience class about immigration policies in the United States.

Figure 11.29 **Assortative mating.** People tend to choose mates who look like them in socially important characteristics.

Part of the disparity between what recent science has revealed and what our common experience tells us about human races comes from the fact that racial categories are *socially* meaningful. In the United States, we all learn that skin color, eye shape, and hair type are the primary physical characteristics that signal differences among groups. These particular physical characteristics are significant due to the history of European colonization, slavery, immigration, and Native American oppression. In other words, race is a social construct—a product of history and learned attitudes. In every society, children learn which physical differences among people are significant in distinguishing "us" from "them." Even if a child is never explicitly taught racial categories, the fact that many communities are highly segregated provides a lesson about which physical characteristics mark someone as "different from me."

The construction of racial groups allowed some "races" to justify unethical and inhumane treatment of other "races." Thus, human races were described in the seventeenth century primarily to support **racism,** the idea that some groups of people are naturally superior to others *(Figure 11.30)*. The U.S. government collects information about race on the census form as part of its effort to measure and ameliorate the lingering effects of historical, state-supported racism, but on its form, the Census Bureau acknowledges that the races with which people identify "should not be interpreted as being primarily biological or genetic."

It may be easier to see that racial categories are socially constructed if you imagine what might have happened if Western history had followed a different path. If the origin of American slaves had been from around the Mediterranean Sea, we might now identify racial groups on the basis of some other physical difference besides skin color—perhaps height, weight, or the presence of thick facial hair.

Alternatively, compare the racial groupings in modern North America to those in modern Rwanda, where individuals are identified with different racial groups (Hutu and Tutsi) based on physical stature only. This classification reflects the differential social status attained by the typically taller Tutsi tribe and the typically shorter Hutu tribe under European colonization in the nineteenth century. In the United States, we would classify Hutu and Tutsi together in the same race—an assignment that many members of these 2 groups would vigorously reject.

Figure 11.30 **Institutionalized racism.** In some areas of the United States, so-called Jim Crow laws permitted the establishment of separate facilities for blacks and whites that were meant to discourage mixing between the races.

When socially constructed racial categories are considered biologically meaningful, they become traps that are extremely difficult for individuals to escape. We now understand that grouping human populations on the basis of skin color and eye shape is as arbitrary as grouping them on the basis of height and weight. However, arbitrary groupings are not necessarily bad. We all group ourselves into social categories: Christian or Muslim, baseball or football fan, cat or dog person.

Even if the racial categories on the census form were once part of a racist system, when people identify themselves as members of a particular race, they are acknowledging a shared history with others who also identify themselves as members of that race. The biological evidence tells Indigo that she is able to choose her racial category based on her own history and relationships—and that she should feel free to choose "human race" if she desires.

Savvy Reader

Race and Medicine

EXCERPTED FROM AN ARTICLE BY GREGORY M. LAMB, *THE CHRISTIAN SCIENCE MONITOR*, MARCH 3, 2005

In recent months, race-based medicine has gained momentum:

- The Food and Drug Administration is expected to approve the drug BiDil in June [2005], making it the first "ethnic drug" on the market. After failing in a broader study, BiDil was shown to be effective in treating heart failure in a clinical study that included only African-Americans.
- African-Americans need higher doses of one medication used to treat asthma than Caucasians, suggesting "an inherent predisposition" in blacks not to absorb the medicine as easily, says a study in the February [2005] issue of the journal *Chest*.

Much of what we're seeing in human illness can be explained by "issues as prosaic and mundane as access to healthy water and good nutrition," says Professor [Troy] Duster, who teaches at New York University and the University of California at Berkeley. For example: A 2002 report by the Institute of Medicine showed that racial and ethnic minorities tend to receive lower-quality healthcare than whites, even if factors such as insurance status, income, age, and severity of condition are similar.

Many in the field are calling for broader international studies to make sure the bigger picture emerges. Studies comparing white and black Americans, for example, have shown that blacks have higher rates of hypertension, suggesting a genetic difference. But earlier this year research by Dr. [Richard] Cooper and a team at Loyola showed that although African-Americans do show higher levels than North American whites, whites have higher levels than Nigerians and Jamaicans, who are "ethnically" black. Overall, the range of levels of hypertension among blacks in the study ranged from 14 to 44 percent while in whites it was higher, 27 to 55 percent.

"'Race' and 'ethnicity' are poorly defined terms that serve as flawed surrogates for multiple environmental and genetic factors in disease causation, including ancestral geographic origins, socioeconomic status, education, and access to health care," wrote Francis Collins, the head of the National Human Genome Research Institute (NHGRI), last fall. "Research must move beyond these weak and imperfect proxy relationships to define the more proximate factors that influence health."

But other researchers say they expect to keep finding genetic racial differences and that society must learn to become comfortable with the idea. "New forms of scientific knowledge will point out more and more ways in which we are diverse," wrote geneticist James Crow, a professor emeritus at the University of Wisconsin, in a 2002 article. "I hope that differences will be welcomed, rather than accepted grudgingly. Who wants a world of identical people, even if they are Mozarts or Jordans?"

(continued on the next page)

Savvy Reader (continued)

1. In this article, what is the evidence that racial categories may be important to medical practice?
2. What evidence is presented that argues against this idea?
3. What more would you need to know to understand whether race has a place in medicine?

CHAPTER Review

For study help, animations, and more quiz questions go to www.mybiology.com.

Summary

11.1 What Is a Species?

- All humans belong to the same biological species, *Homo sapiens sapiens*. A biological species is defined as a group of individuals that can interbreed and produce fertile offspring. Biological species are reproductively isolated from each other, thus separating the gene pools of species (p. 280).
- Reproductive isolation is maintained by prefertilization or postfertilization factors (p. 281).
- Speciation occurs when populations of a species become isolated from each other. These populations diverge from each other, and reproductive isolation between the populations evolves (pp. 281–288).

web animation 11.1
The Process of Speciation

11.2 Races and Genealogical Species

- Biological races are populations of a single species that have diverged from each other but have not become reproductively isolated. This definition of race corresponds to the genealogical species concept (pp. 289–291).

11.3 Humans and the Race Concept

- The fossil record provides evidence that the modern human species is approximately 200,000 years old, which is not much time for major differences between human groups to have evolved (p. 292).
- The genetic evidence for biological races includes alleles that are unique to a particular race, as well as similar allele frequencies for a number of genes among populations within races but differences in allele frequencies among populations in different races (pp. 293–294).
- The Hardy-Weinberg theorem gives a rule for calculating allele frequency from genotype frequency (pp. 295–296)
- Modern human groups do not show evidence that they have been isolated from each other and formed distinct races (pp. 297–298).
- Genetic evidence indicates that human groups have been mixing for thousands of years (p. 299).

Summary (continued)

11.4 Why Human Groups Differ

- Similarities among human populations may evolve as a result of natural selection. The sickle-cell allele is selected for in populations in which malaria incidence is high, and light skin is selected for in areas where the UV light level is low (pp. 300–302).
- Human populations may show differences due to genetic drift, which is defined as changes in allele frequency due to chance events such as founder effects or population bottlenecks (pp. 303–304).
- Sexual selection, by which individuals—typically females—choose mates that display some "attractive" quality, may also be responsible for creating differences among human populations (p. 305).
- Positive assortative mating, in which individuals choose mates who are like themselves, can reinforce differences between human populations (p. 306).

11.5 Race in Human Society

- While race in the human species has no biological meaning, it is an important social construct based on shared history and self-identity (p. 306)

Roots to Remember

The following roots come mainly from Latin and Greek and will help you decipher terms:

a- means without.
allo- means different.
morpho- means shape or form.
paleo- means ancient.
-patric means country or place of origin.
-ploid means the number of different types of chromosomes.
poly- means many.
sym- means same or united.

Learning the Basics

1. Describe the 3 steps of speciation.

2. Can speciation occur when populations are not physically isolated from each other? How?

3. Describe 3 ways that evolution can occur via genetic drift.

4. Which of the following is an example of a prefertilization barrier to reproduction?

A. A female mammal is unable to carry a hybrid offspring to term; **B.** Hybrid plants produce only sterile pollen; **C.** A hybrid between 2 bird species sings a song that is not recognized by either species; **D.** A male fly of 1 species performs a "wing-waving" display that does not convince a female of another species to mate with him; **E.** A hybrid embryo is not able to complete development.

5. According to the most accepted scientific hypothesis about the origin of 2 new species from a single common ancestor, most new species arise when _______________.

A. many mutations occur; **B.** populations of the ancestral species are isolated from each other; **C.** there is no natural selection; **D.** a Creator decides that 2 new species would be preferable to the old one; **E.** the ancestral species decides to evolve

6. For 2 populations of organisms to be considered separate biological species, they must be _______________.

A. reproductively isolated from each other; **B.** unable to produce living offspring; **C.** physically very different from each other; **D.** a and c are correct; **E.** a, b, and c are correct

7. Cystic fibrosis, a recessive disease, affects 1 out of every 2500 Caucasian babies born in the United States, a frequency of 0.0004. Use the Hardy-Weinberg theorem to calculate the frequency of the cystic fibrosis allele in this population.
A. 0.0004; **B.** 0.04; **C.** 0.02; **D.** 0.00000016; **E.** There is not enough information in the question to calculate the answer.

8. In the fossil record, species appear suddenly, persist without much change for millions of years, and then disappear. This observation supports the theory of
A. allopatric speciation; **B.** gradualism; **C.** genetic drift; **D.** punctuated equilibrium; **E.** the biological species concept.

9. Similarity in skin color among different human populations appears to be primarily the result of ______________.
A. natural selection; **B.** convergence; **C.** which biological race they belong to; **D.** a and b are correct; **E.** a, b, and c are correct

10. The tendency for individuals to choose mates who are like themselves is called ______________.
A. natural selection; **B.** sexual selection; **C.** assortative mating; **D.** the founder effect; **E.** random mating

Analyzing and Applying the Basics

1. Wolf populations in Alaska are separated by thousands of miles from wolf populations in the northern Great Lakes of the lower 48 states. Wolves in both populations look similar and have similar behaviors. However, the U.S. government has treated these 2 populations quite differently, listing the Great Lakes populations as endangered until recently but allowing hunting of wolves in Alaska. Some opponents of wolf protection have argued that the "wolf" should not be considered endangered at all in the United States because of the large population in Alaska, while supporters of wolf protection state that the Great Lakes population represents a unique population that deserves special status. Should these 2 populations be considered different races or species? What information would you need to test your answer?

2. Phenylketonuria (PKU) is the inability to metabolize the amino acid phenylalanine. The frequency of PKU in Irish populations is 1 in every 7000 births, while the frequency in urban British populations is 1 in 18,000 and only 1 in 36,000 in Scandinavian populations. PKU is found only in individuals who are homozygous recessive for the disease allele. Use the Hardy-Weinberg theorem to calculate the frequency of the disease allele in each population. Give 2 reasons that this allele, which results in severe mental retardation in homozygous individuals, may be found in different frequencies in these populations.

3. A species of aster normally blooms between mid-August and late September. The aster's range is located in a climate that usually produces freezing temperatures in late October. The aster's pollinators (bees, ants, and other insects) are active between late April and early October. Mutations can occur in the gene that controls flowering time for some individuals of this aster, causing them to flower earlier or longer than normal. The table shows possible flowering times.

Aster Type	Flowering Time
Normal	August 15–September 30
Early mutation	July 1–August 15
Expanded mutation	July 15– September 30

Describe how this population of asters might split into 2 species?

Connecting the Science

1. African Americans have higher rates of hypertension (high blood pressure), heart disease, and stroke than do whites in the United States. Is this difference likely to be biological? How could you test your hypothesis?

2. The only information that was collected from *every resident* of the United States in the 2000 census was name, place of residence, sex, age, and race. (Note that some residents received a "long form" with many additional questions, but everyone had to answer these 5 basic questions.) Do you think it is important to collect race data from all citizens? Why or why not? Is there some other piece of information that you think is more useful to the government?

CHAPTER 12

Prospecting for Biological Gold

Biodiversity and Classification

This strange Yellowstone hot spring . . .

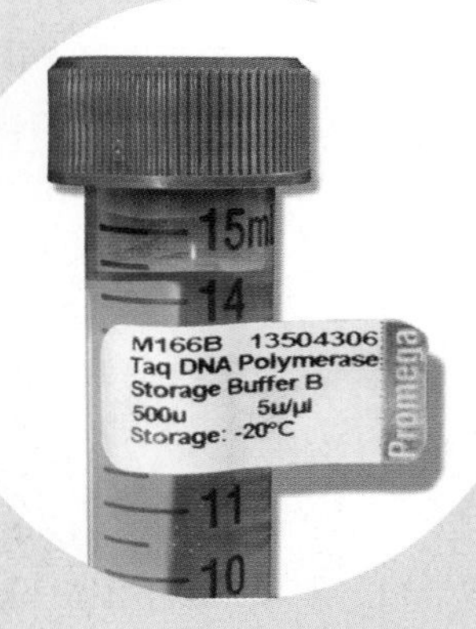

... is a source of this extremely valuable chemical, produced by microorganisms.

At the end of a narrow foot trail in Yellowstone National Park lies a natural curiosity—Octopus Spring. The boiling hot water of the spring is colored an otherworldly blue. A gooey white crust encircles its main pool, and along the banks of the drainage streams radiating from it are brightly colored mats and streamers of pink, yellow, green, and orange. Although Octopus Spring is certainly not the most beautiful or dramatic feature of Yellowstone, this relatively small spring looms large in the history of biological discovery.

The brilliant colors of Octopus Spring result from microscopic organisms living in the water and on nearby surfaces. In the 1960s, Dr. Thomas Brock of the University of Wisconsin was the first to describe this biological community. One of the species he discovered, which he named *Themus aquaticus* for its affinity for hot water, contained a protein new to science. This protein, an enzyme called *Taq* polymerase, allows *T. aquaticus* to replicate its DNA at much higher temperatures than other organisms can. *Taq* polymerase is now an essential part of the polymerase chain reaction (PCR), a high-temperature process used to prepare DNA samples for research or for the process of DNA fingerprinting. PCR has revolutionized DNA research and has made many of the advances in genetic technology possible.

Scientists are interested in finding other useful products in Yellowstone.

Until it was ruled invalid in 1999, the patent for *Taq* polymerase was held by the Swiss pharmaceutical firm Hoffman-LaRoche. Licensing agreements with users and producers of *Taq* polymerase netted the company over $100 million every year. Of this substantial sum, Yellowstone National Park, the National Park Service, and the U.S. Treasury received nothing in royalty payments. Even a small share of the royalties would have helped improve and manage this heavily used national park.

The managers of Yellowstone do not want to miss out on the financial rewards that may come from other valuable species protected by their borders. To capitalize on future discoveries in the park, Yellowstone entered into an agreement in 1997 with Diversa Corporation to identify and describe some of the microscopic species in the park. In return, Diversa agreed to make a one-time payment of $100,000 to Yellowstone National Park and

Are they likely to succeed? Is there much left to discover about life?

to provide several thousand dollars of research services. Diversa also agreed to share a percentage of royalties from any profitable products that result.

The announcement of this deal set off a flurry of criticisms—from environmentalists who fear the exploitation of the park, to government watchdogs concerned about a few private stockholders profiting from a park maintained for the entire public. Diversa, which has since merged with another company and changed its name to Verenium, has yet to begin exploration in Yellowstone's hot springs, pending the resolution of several legal challenges to their agreement.

Even without the legal challenges, the agreement between Yellowstone's managers and Diversa is a calculated risk by both parties. The company was to invest millions of dollars in this venture, and Yellowstone risks damage to the wildlife and scenery that the park was designed to protect. Why are the parties to this agreement willing to take these risks? What is the likelihood of success in the search for valuable species within Yellowstone National Park? Can there be many organisms that humankind has yet to discover? What do we know about the organisms we have identified? And how can we screen newly discovered species for their valuable traits? We can answer many of these questions by applying evolutionary theory to investigations of the amazing variety of life on Earth.

web animation 12.1 *The Tree of Life*

12.1 Biological Classification

Diversa Corporation's proposed hunt for new organisms and new uses of known organisms is called **bioprospecting.** Bioprospectors seek to strike biological "gold" by finding the next penicillin (originally discovered in a fungus), aspirin (produced by willow trees), or *Taq* polymerase in the living world.

Yellowstone isn't the only potentially rich source of biological gold—drug companies are also bioprospecting in vast Amazonian rain forests, strange hydrothermal vents of the deep ocean *(Figure 12.1)*, and bleak expanses of Antarctic ice. Other scientists are surveying more commonly encountered organisms such as airborne molds and the bacteria that cause tooth decay. To understand the challenges associated with this survey, we must know something about the diversity of life on Earth.

Figure 12.1 **A source of biological riches?** The organisms surrounding this deep-sea volcanic site have been known to science for fewer than 30 years.

How Many Species Exist?

A characteristic of life on Earth is that it is full of variety. Scientists refer to the variety within and among living species as **biodiversity.** Studies of biodiversity not only provide drugs and chemicals to solve human problems. More important to many biologists, these studies help us understand the evolutionary origins of species and discover their role in healthy biological systems.

To bioprospectors, the promise of biodiversity is its great variety, but great variety is also a source of challenge. The number of species described by science is between 1.4 and 1.8 million; even if prospectors spent only a single day screening each species for valuable products, examining them would take more than 5000 years. The variety of different species also greatly underestimates the biodiversity within a species. Just as the species of tomato can come in many shapes, sizes, and flavors, there are species in which individuals differ greatly in the amount and potency of a particular biological molecule *(Figure 12.2)*. Bioprospectors could miss a valuable molecule if they test only a handful of individuals from a single population.

Biologists disagree about the total number of species that have already been identified and described. This uncertainty stems from the method of storing and cataloguing known species. **Systematists** are scientists who specialize in describing and categorizing a particular group of organisms. For an organism to be considered a new species by the scientific community, a systematist must create a description of the species that clearly distinguishes it from similar species and must publish this description in a professional journal. The scientist also must collect individual specimens for storage in a specialized museum. Most animal collections are found in natural history museums, while plant repositories are called *herbaria* (*Figure 12.3*); microbes and fungi are kept in facilities called *type-collection centers.*

Figure 12.2 **Biological diversity.** These varieties of the garden tomato illustrate diversity within a species.

Because there are numerous large natural history museums and herbaria all over the world, along with many different journals, it can be unclear whether a species has already been described. Systematists evaluate collections at different centers to see if there is overlap, but this process is slow and further complicated by the continual discovery and description of new species. Because there is no central resource for species collections and descriptions, the total number of described species is only an estimate. The number of known species also represents only a fraction of the total number on Earth. Some estimates of the actual number are as large as 100 million unique species, but a typical estimate is around 10 million. Scientists also know that the diversity of life today is very different from the diversity in the past.

Paleontologists have been able to piece together the history of life on Earth by examining fossils and other ancient evidence. Early in this reconstruction, they recognized distinct "dynasties" of groups of organisms that appeared during different periods. The rise and fall of these dynasties allowed scientists to subdivide life's history into **geologic periods.** Each period is defined by a particular set of fossils. *Table 12.1* gives the names of major geologic periods, their age and length, and the major biological events that occurred during each period.

Despite the long history of diversity, it is very rare for scientists to describe either a modern or fossil organism that appears unrelated to all other species. In fact, species can be grouped into a few broad categories based on shared characteristics. The most general categories are kingdoms and domains.

Kingdoms and Domains: An Overview

Modern species are divided into 3 large groups, called **domains,** that are unified by similarities in cell structure. Within each of these domains are smaller groups, called **kingdoms.** Since most identified living organisms are in the domain Eukarya, we focus only on the kingdoms in this group. The other domains contain kingdoms as well, but these classifications are generally of interest only to specialists. *Table 12.2* summarizes the kingdoms and domains discussed in section 12.2.

(a) Natural history museum

(b) Herbarium

Figure 12.3 **Biological collections.** Most collections contain several examples of each species to show the range of variation within the species. (a) A collection of animals in a natural history museum. (b) Plant specimens are stored in herbaria.

Table 12.1 Geological periods. The history of life is divided into 4 major eras, with all but the first era divided into several periods. Periods are marked by major changes in the dominant organisms present on Earth.

Era	Period	Millions of Years Ago	Features of Life on Earth
Cenozoic	Quaternary	0	Most modern organisms present.
	Tertiary	1.8	After the extinction of the dinosaurs, mammals, birds, and flowering plants diversify.
Mesozoic	Cretaceous	65	Massive carnivorous and flying dinosaurs are abundant. Large cone-bearing plants dominate forests. Flowering plants appear.
	Jurassic	144	Huge plant-eating dinosaurs evolve. Forests are dominated by cycads and tree ferns.
	Triassic	206	Early dinosaurs, mammals, and cycads appear on land. Life "restarts" in the oceans.
Paleozoic	Permian	251	Early reptiles appear on land. Seedless plants abundant. Coral and trilobites abundant in oceans. Permian ends with extinction of 95% of living organisms.
	Carboniferous	290	Land is dominated by dense forests of seedless plants. Insects become abundant. Large amphibians appear.
	Devonian	354	Known as the age of fishes. Sharks and bony fish appear. Large trilobites are abundant in the oceans.
	Silurian	408	Life begins to invade land. The first colonists are small seedless plants, primitive insects, and soft-bodied animals.
	Ordovician	439	Life is diverse in the oceans. Cephalopods appear, and trilobites are common.
	Cambrian	495	All modern animal groups appear in the oceans. Algae are abundant.
Pre-Cambrian		543	Life is dominated by single-celled organisms in the ocean. Ediacaran fauna appear at the end of the era.
		4500	Formation of Earth. The first organisms, primitive bacteria, appear within 1000 million years.

TABLE 12.2 **The classification of life.** Until recently, most biologists used the 5-kingdom system to organize life's diversity. Now many use a 6-category system, which better reflects evolutionary relationships by acknowledging the existence of 3 major domains as well as 4 of the kingdoms.

Kingdom Name	Kingdom Characteristics	Examples	Approximate Number of Known Species	Domain Name and Characteristics
Plantae	Eukaryotic, multicellular, make own food, largely stationary	Pines, wheat, moss, ferns	300,000	**Eukarya** All organisms contain eukaryotic cells.
Animalia	Eukaryotic, multicellular, rely on other organisms for food, mobile for at least part of life cycle	Mammals, birds, fish, insects, spiders, sponges	1,000,000	
Fungi	Eukaryotic, multicellular, rely on other organisms for food, reproduce by spores, body made up of thin filaments called hyphae	Mildew, mushrooms, yeast, *Penicillium*, rusts	100,000	
Protista	Eukaryotic, mostly single-celled forms, wide diversity of lifestyles, including plant-like, fungus-like, and animal-like types	Green algae, *Amoeba, Paramecium,* diatoms, chytrids	15,000	
Bacteria	Prokaryotic, mostly single-celled forms, although some form permanent aggregates of cells	*Escherichia coli*, Salmonella, *Bacillus anthracis, Anabena*, sulfur bacteria	4,000	**Bacteria** Prokaryotes with cell wall containing peptidoglycan. Wide diversity of lifestyles, including many that can make their own food.
Archaea		*Thermus aquaticus, Halobacteria halobium,* methanogens	1,000	**Archaea** Approximate number of known species. Prokaryotes without peptidoglycan and with similarities to Eukarya in genome organization and control. Many known species live in extreme environments.

A Closer Look: Kingdoms and Domains

Systematists work in the field of **biological classification,** in which they attempt to organize biodiversity into discrete and logical categories. The task of classifying life is much like categorizing books in a library—books can be divided into "fiction" or "nonfiction," and within each of these divisions, more precise categories can be made (for example, nonfiction can be divided into biography, history, science, etc.). The book-cataloguing system used in most public libraries, the Dewey decimal system, is only one way of shelving books. For instance, academic and research libraries use a different system, developed by the U.S. Library of Congress. Librarians use the cataloguing system that is appropriate to the collection of books they work with and the needs of the library's users. Just as there are alternative methods of organizing books, there is more than one way to organize biodiversity to meet differing needs.

Biologists have traditionally subdivided living organisms into large groups that share some basic characteristic. Fifty years ago, most biologists divided life into 2 categories: plants, for organisms that were immobile and apparently made their own food; and animals, for organisms that could move about and relied on other organisms for food.

Many organisms did not fit easily into this neat division of life, so scientists began to use a system of 5 **kingdoms,** which categorized organisms according to cell type and method of obtaining energy: a group consisting of single-celled organisms without a nucleus; a group containing organisms with nuclei that spend some of their life cycle living as single, mobile cells, called the Protista; and Plantae, Animalia, and Fungi, groups of multicellular organisms that make their own food, rely on other organisms for food, or digest dead organisms, respectively. The 5-kingdom system is not perfect either; for instance, the Protista kingdom contains a wide diversity, from amoebas to seaweeds, with only superficial similarities.

More recently, many biologists have argued that the most appropriate way to classify life is according to evolutionary relationships among organisms. Recall that the theory of evolution states that all modern organisms represent the descendants of a single common ancestor. Evidence for this theory includes the universality of the genetic code and shared cell structures and biochemical pathways. Separate populations of this ancestor diverged as natural selection and genetic drift occurred in each group, resulting in evolutionary lineages. The process of divergence from early ancestors into the diversity of modern species has resulted in the modern "tree of life" *(Figure 12.4)*.

When life is classified according to the relationships among organisms, major groupings correspond to divergences that occurred very early in life's history, and minor groupings correspond to more recent divergences. Classifying life according to evolutionary relationships may be useful to bioprospectors if relationships provide clues to chemical similarity.

Determining the evolutionary relationship among *all* living organisms requires comparisons of their DNA. Since each species is unique, the sequence of nucleotides within DNA of each species is unique. However, because all species share a common ancestor, all organisms also have basic similarities in their DNA sequences. As evolutionary lineages diverged from each other, mutations in DNA sequences occurred independently in each lineage and are now a record of

Figure 12.4 **The tree of life.** This tree is a simplification of the current state of knowledge regarding evolutionary relationships among living organisms. Living organisms represent a small remnant of all the species that have appeared over Earth's history.

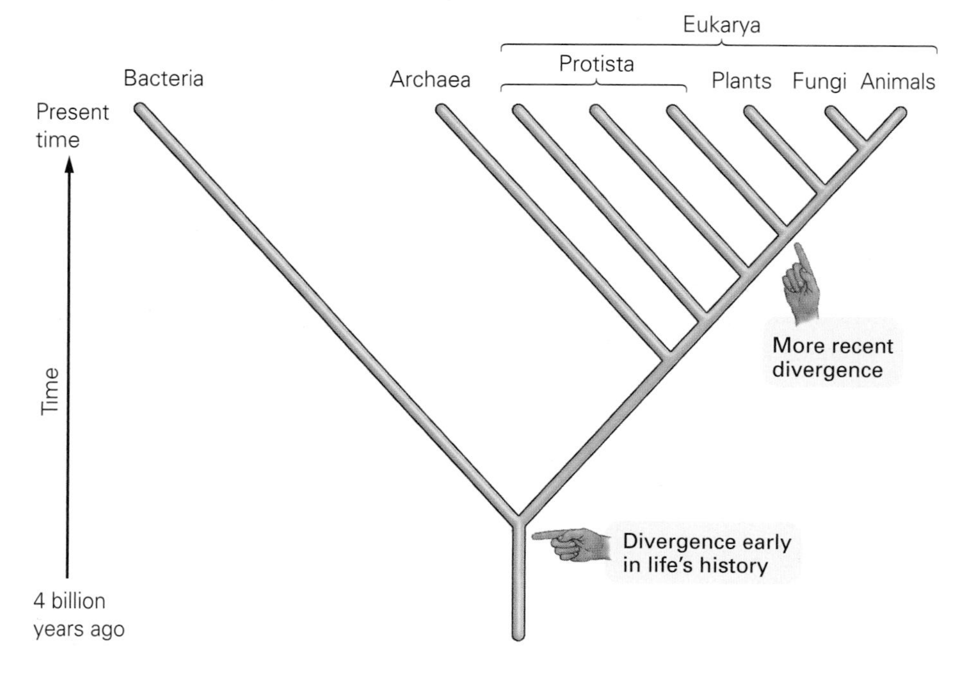

(A Closer Look continued)

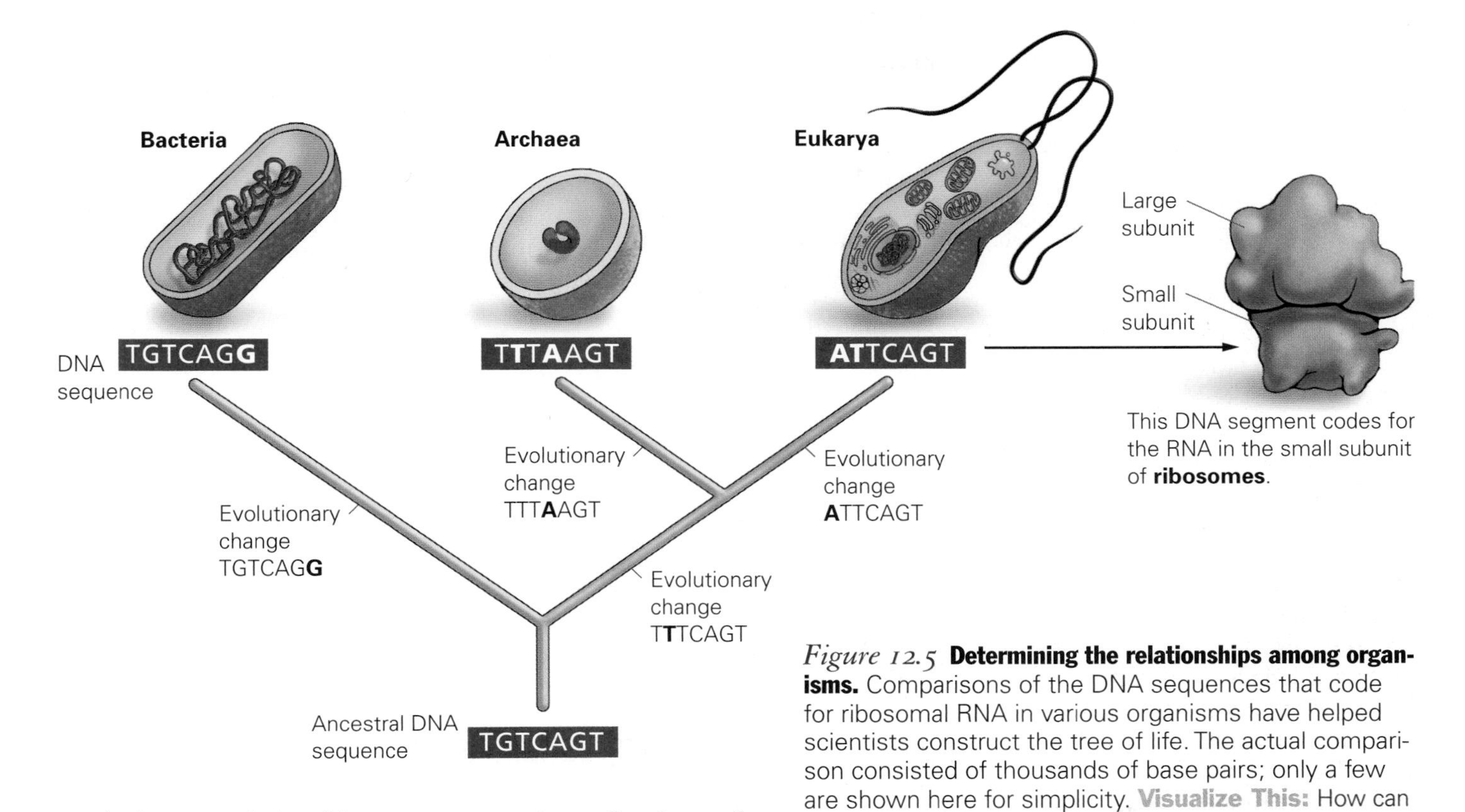

Figure 12.5 **Determining the relationships among organisms.** Comparisons of the DNA sequences that code for ribosomal RNA in various organisms have helped scientists construct the tree of life. The actual comparison consisted of thousands of base pairs; only a few are shown here for simplicity. **Visualize This:** How can scientists infer the ancestral sequence?

evolutionary relationship among organisms. In short, the DNA sequences of closely related organisms should be more similar than the DNA sequences of more distantly related organisms *(Figure 12.5)*.

When comparing all modern species, scientists must examine genes that perform a similar function among organisms as diverse as humans, willow trees, and *Thermus aquaticus*. The DNA sequence that best fits this requirement contains instructions for making ribosomal RNA (rRNA), a structural part of ribosomes. Recall that ribosomes are the factories found in all cells, translating genes into proteins. Each ribosome contains several rRNA molecules in both its large and small subunits. A comparison of the DNA coding for small-subunit rRNAs from myriad organisms yielded the tree diagram in Figure 12.4.

Note from Figure 12.4 that three of the kingdoms (Fungi, Animalia, and Plantae) represent relatively recently diverged groups of organisms. Single-celled organisms that do not contain a nucleus for storing DNA were once all placed in the same kingdom but are actually found in two different, quite distinct groups, the Archaea and Bacteria; and the kingdom Protista is a hodgepodge of many very different organisms. To better reflect such biological relationships, biologists categorize life into three **domains**—Bacteria, Archaea, and Eukarya. These domains represent the most ancient divergence of living organisms.

Systematists have begun delineating kingdoms in the Bacteria and Archaea domains using DNA sequence comparisons and revising the Eukarya kingdoms to more accurately reflect evolutionary history. However, just as the Library of Congress cataloguing system is not appropriate for a public library, an evolutionary classification is not the most effective organization for surveying diversity. In this chapter we use a common hybrid of the 5-kingdom and 3-domain system that specifies 6 categories: the domains Bacteria and Archaea and the 4 Eukarya kingdoms—Protista, Fungi, Plantae, and Animalia.

Stop and Stretch 12.1: Why is a classification based on superficial similarities a better tool for surveying diversity than an evolutionary classification?

web animation 12.2
Endosymbiotic Theory

12.2 The Diversity of Life

In reality, the number of kingdoms and domains into which life is appropriately classified is of minor importance to a bioprospector. (Understanding more recent evolutionary relationships may be very important, as we discuss in Section 12.3)

However, dividing life into 6 categories—those described in Table 12.2—simplifies *our* discussion of biodiversity and its bioprospecting potential. In this section we describe the 6 categories. In addition, we discuss entities that have many characteristics of living organisms but are not considered *alive*—the viruses.

Bacteria and Archaea

Life on Earth arose at least 3.6 billion years ago according to the fossil record. The most ancient fossilized cells *(Figure 12.6)* are remarkably similar in external appearance to modern bacteria and archaea. Both **bacteria** and **archaea** are **prokaryotes;** this means they do not contain a nucleus, which provides a membrane-bound, separate compartment for the DNA in other cells. Prokaryotes also lack other internal structures bounded by membranes, such as mitochondria and chloroplasts, which are found in more complex **eukaryotes.**

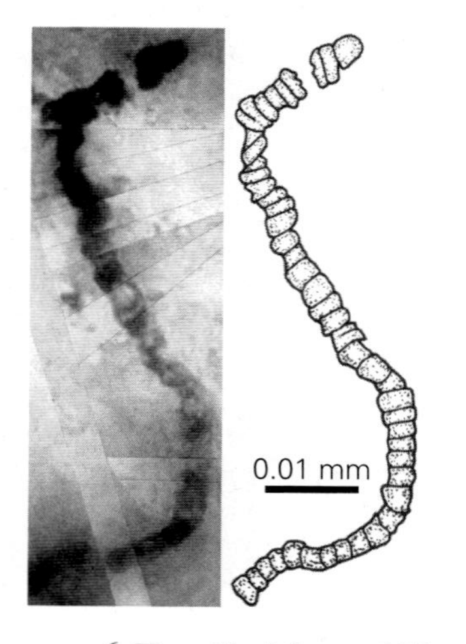

Figure 12.6 **The oldest form of life.** This photograph of a fossil is accompanied by an interpretive drawing showing the fossil's living form. It was found in rocks dated at 3.465 billion years old.

Although some species may be found in chains or small colonies as in *Figure 12.7*, most prokaryotes are **unicellular,** meaning that each cell is an individual organism. Individual prokaryotic cells are hundreds of times smaller than the cells that make up our bodies; for this reason, they are often called microorganisms or **microbes,** and biologists who study these organisms (as well as unicellular eukaryotes) are known as **microbiologists.** Their small size, easily accessible DNA, and simple structure make prokaryotes very attractive to bioprospectors because they are generally easier to grow, study, and manipulate than eukaryotes.

The relatively simple structure of prokaryotes belies their incredible chemical complexity and diversity. Some prokaryotes can live on petroleum, others on hydrogen sulfide emitted by volcanoes deep below the surface of the ocean, and some simply on water, sunlight, and air *(Figure 12.7)*. Prokaryotes are found in and on nearly every square centimeter of Earth's surface, including very hot and very salty places, and even thousands of feet below ground. Prokaryotes are also incredibly numerous; for instance, there are more prokaryotes living in your mouth right now than the total number of humans who have ever lived!

Domain Bacteria. Although most are likely harmless to humans, the majority of bacteria that have been identified are known because they cause disease in humans or our crops. While enormous efforts are expended to control these organisms, the fact that they can live in and on other living creatures is remarkable. To survive within a host organism, bacteria must escape eradication by the host's infection-fighting system. The molecules that allow bacteria to effectively colonize living humans could be useful in treating immune system diseases. Disease-causing organisms thus, surprisingly, represent a source of bacterial biological gold.

Many of the known bacteria obtain nutrients by decomposing dead organisms. Bacterial species that function as decomposers often have competitors—other species that also consume the same food source. When many individuals compete for the same resource, natural selection will favor traits that provide an edge over the competition. With competing bacterial decomposers, this edge is often in the form of **antibiotics** that kill or disable other bacteria. More than half of the antibiotic drugs we use to treat infections are derived from bacteria.

Bacterial competition has resulted in another class of valuable molecules called *restriction enzymes*, proteins that can chop up DNA at specific sequence sites and thus interfere with, or restrict, the growth of other organisms. Restriction enzymes are used in the production of DNA fingerprints. Bioprospectors are very interested in finding more of both of these extremely valuable compounds—antibiotics and restriction enzymes—within the domain Bacteria.

Domain Archaea. Although superficially similar to domain Bacteria because of its prokaryotic cells, the domain Archaea differs from Bacteria in many fundamental ways, including the structure of its cell membranes. The known Archaea are typically found in extreme environments, including high-salt, high-sulfur, and high-temperature habitats. *Thermus aquaticus*, the source of *Taq* polymerase and a hot-spring dweller, belongs to Archaea.

Taq polymerase is valuable because it operates at a high temperature—making it, along with compounds from other hot-spring archaeans, potentially useful in industrial settings. Natural selection in extreme environments has likely caused the evolution of other biological molecules that can operate at high temperatures, at high pressures, or in extremely salty environments in the Archaea.

Scientists and bioprospectors still have much to learn about bacteria and archaea, beginning with an understanding of exactly how diverse they are. For instance, while most archaeans are known from extreme environments, species in this domain are found everywhere they are sought. Some scientists estimate that the number of undescribed prokaryotic species could range up to 100 million. And prokaryotes are not the only microscopic organisms; most species in the diverse kingdom Protista also cannot be seen with the naked eye.

Protista

The kingdom **Protista** is made up of the simplest known eukaryotes. Most protists are single-celled creatures, although several have enormous multicellular (many-celled) forms. The most ancient fossils of eukaryotic cells are approximately 2 billion years old, nearly 1.5 billion years younger than the oldest prokaryotic fossils.

The earliest eukaryotic cells likely developed from prokaryotes that produced excess cell membrane that folded into the cell itself. In some cells, these internal membranes may have segregated the genetic material into a primitive nucleus and formed channels for translating, rearranging, and packaging proteins, much like the modern endoplasmic reticulum and Golgi bodies. According to the **endosymbiotic theory** for the origin of protists, modern eukaryotes are most likely the descendants of a confederation of these primitive eukaryotes with other prokaryotic cells. Mitochondria and chloroplasts appear to have descended from bacteria that took up residence inside larger primitive eukaryotes. When organisms live together, the relationship is known as a *symbiosis*. In this case the relationship was mutually beneficial, and over time, the cells became inextricably tied together *(Figure 12.8)*.

When biologist Lynn Margulis first popularized the endosymbiotic hypothesis in the United States in 1981, many of her colleagues were skeptical. But an examination of the membranes, reproduction, and ribosomes of mitochondria and chloroplasts shows clear similarities to the same features in certain bacteria. Even more convincingly, the sequence of DNA found in

(a) *Escherichia coli*

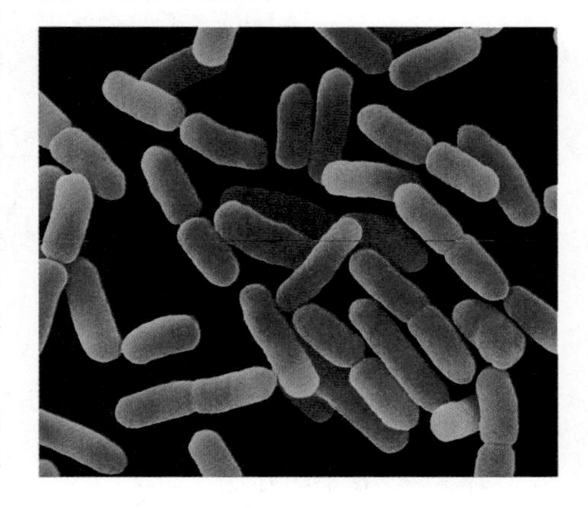

(b) *Streptomyces venezuelae* (whole culture and magnified)

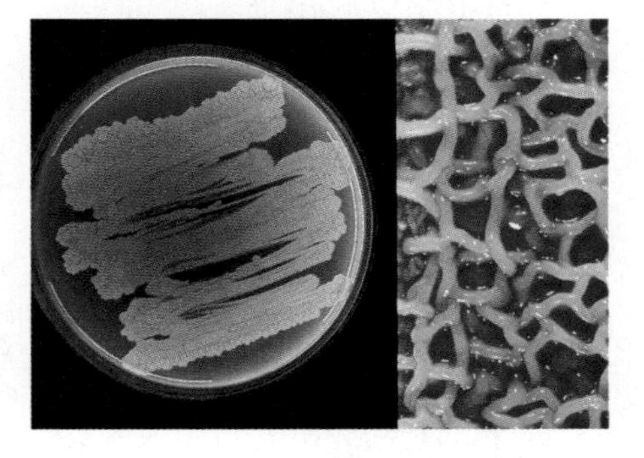

(c) *Halobacterium*

Figure 12.7 **A diversity of prokaryotes.** (a) *Escherichia coli*, the "lab rat" of basic genetic studies, lives on the partially digested food in our intestines. (b) A colony of billions of cells of *Streptomyces venezuelae*, source of the antibiotic chloramphenicol and a member of the group of bacteria that live in soil and decompose dead plant matter. (c) *Halobacterium*, a salt-loving archaean, is found in high populations in this salty pond. The red pigment in this bacterium's cells is used to convert energy from sunlight into cellular energy.

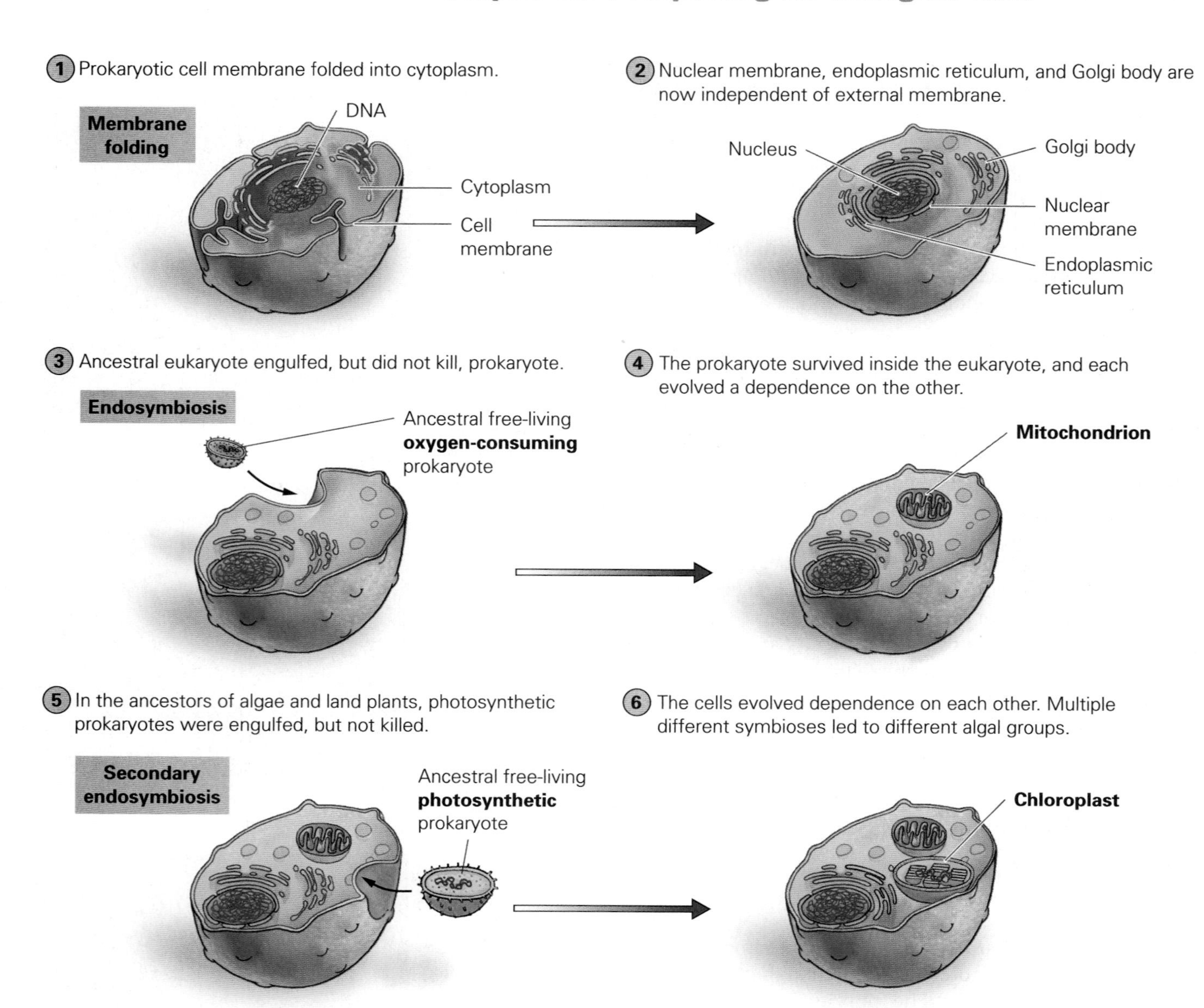

Figure 12.8 **The evolution of eukaryotes.** Mitochondria and chloroplasts appear to be descendants of once free-living bacteria that took up residence within an ancient nucleated cell.

mitochondria (mtDNA) is most similar to the DNA sequence found in species of the purple sulfur bacteria. After the evolution of mitochondria, independent endosymbioses between eukaryotic cells and various species of photosynthetic bacteria appeared. These relationships led to the evolution of chloroplasts—one line led to green algae and land plants, while other lines led to the unique chloroplasts in other modern groups, including red algae and brown algae.

Stop and Stretch 12.2: What do you think the evidence is that mitochondria evolved before chloroplasts?

The modern kingdom Protista contains organisms resembling animals, fungi, and plants. As with Bacteria and Archaea, most members of kingdom Protista remain unknown. There is currently no agreement among scientists regarding how many **phyla,** that is, groups just below the level of kingdom, are contained within Protista. Some argue as few as 8, and others propose as many as 80. *Table 12.3* lists a few of the more common phyla within the kingdom.

TABLE 12.3 **The diversity of Protista.** Protista contains animal-like, fungus-like, and plant-like organisms. A sampling of protistan phyla is described here.

Kingdom Protista: Common Names and Characteristics of Select Phyla		Example
Animal-like protists	**Ciliates** Free-living, single-celled organisms that use hair-like structures to move.	*Paramecium* 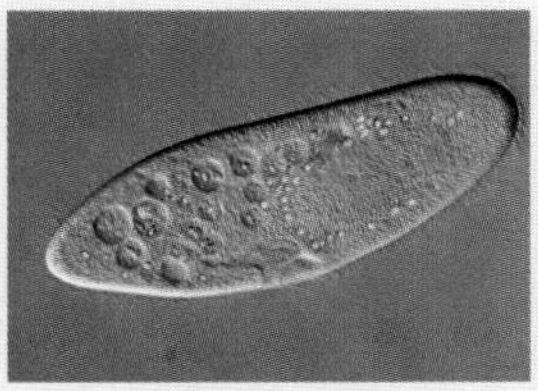
	Flagellates Use 1 or more long whip-like tail for locomotion. Most are free-living but some cause disease by infecting human organs.	*Giardia* 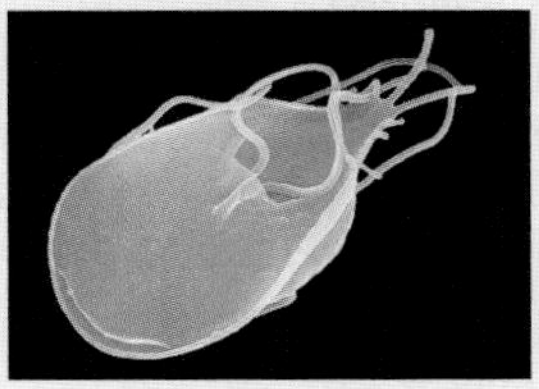
	Amoebas Flexible cells that can take any shape and move by extending pseudopodia ("false feet").	*Amoeba* 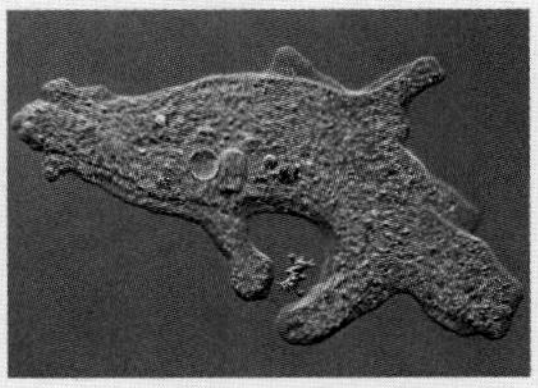
Fungus-like protists	**Slime molds** Feed on dead and decaying material by growing net-like bodies over a surface or by moving about as single amoeba-like cells.	*Physarum* 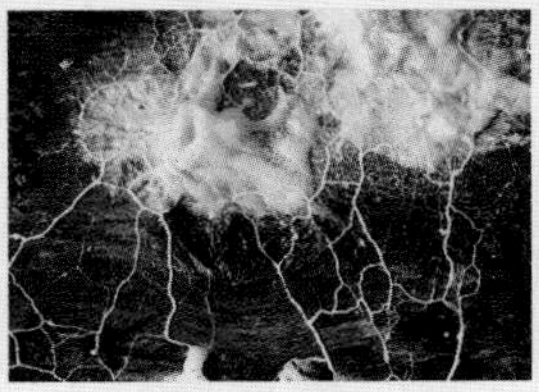
Plant-like protists	**Diatoms** Single cells encased in a silica (glass) shell.	Diatom 
	Brown algae Large multicellular seaweeds.	Kelp 
	Green algae Closest relatives to land plants. Single-celled to multicellular forms.	*Volvox* 

As a result of its diversity, kingdom Protista may contain a plethora of useful organisms and compounds *(Figure 12.9)*. The protists that bioprospectors have investigated in the most detail are **algae,** the only members of this kingdom with the ability to manufacture food. Like plants, algae make food via photosynthesis. And also like plants, algae represent a rich and tempting food source to nonphotosynthetic predators, organisms that will eat them.

Consequently, natural selection in most photosynthetic organisms has favored the evolution of defensive chemicals. These molecules make the algae distasteful or even poisonous to a potential predator. We humans can often use these chemicals to control *our* predators. For example, extracts of red algae stop the reproduction of several different viruses in human tissues grown in laboratory dishes. As yet, no drugs derived from algae are available to consumers, but with great interest in treatments for HIV (the virus that causes AIDS), influenza, and severe acute respiratory syndrome (SARS), these organisms remain a focus for bioprospectors.

The group of organisms we call algae is actually made up of several distinct, quite divergent, categories of organisms. Each of these algal phyla has its own methods of producing and storing food as each represents the descendants

Figure 12.9 **Protista.** Protista is the most diverse of life's kingdoms. Valuable organisms and compounds have been found in (a) animal-like forms, (b) fungus-like forms, and (c) and (d) plant-like forms.

(a) Animal-like protist

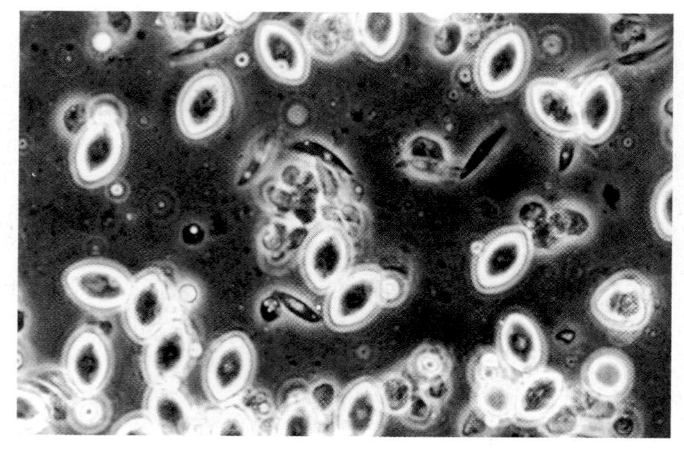

Mattesia oryzaephili is a parasite that infests and kills beetles in stored grain. *M. oryzaephili* may be useful in controlling these pests.

(b) Fungus-like protist

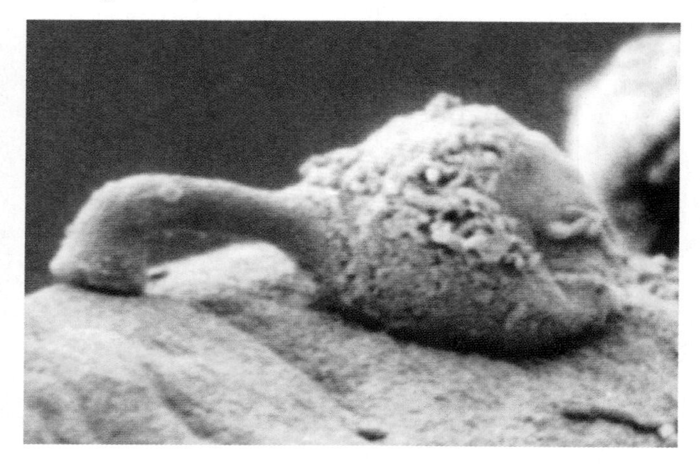

Lagenidium giganteum, sprayed on ponds to control mosquito populations

(c) Plant-like protist

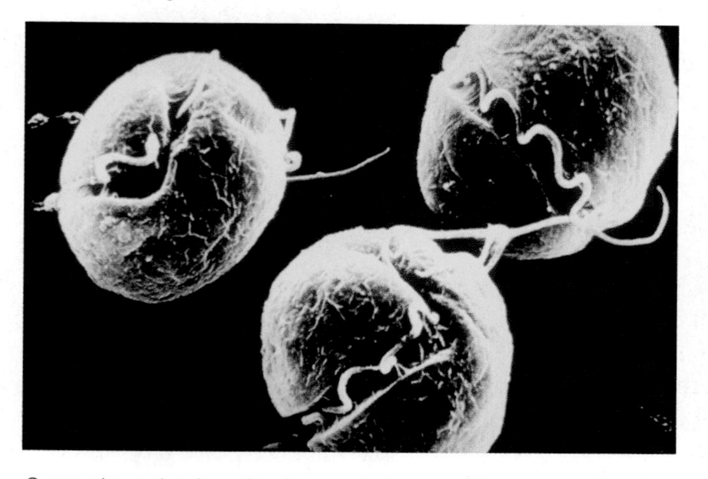

Gonyaulax polyedra, a luminescent alga used to measure levels of toxic materials in ocean sediments

(d) Plant-like protist

Chondrus crispus, a red alga that is the source of carageenan of a separate endosymbiosis. Some of the unique compounds produced by different algal phyla are useful to humans. For example, carageenan, a slimy carbohydrate harvested from red algae, is used as a stabilizer and thickener in foods, medicines, and cosmetics.

Animal-like protists include many notorious parasites of humans, including the organisms that cause malaria, African sleeping sickness, trichomonas, and giardiasis. Fungus-like protists include slime molds and some important crop pests, most notably potato blight, the organism responsible for the Irish famine of the late 1840s.

The animal-like and fungus-like protists are less interesting to most bioprospectors, presumably because animals and fungi have not been as rich a source of useful biological products as photosynthesizers have. However, as we shall see, both the kingdom Animalia and the kingdom Fungi have some intriguing characteristics and can be a source of useful biochemicals. It may be that within these categories of protists, bioprospectors also will find natural products that humans can use.

Animalia

From the origin of the first prokaryote until approximately 1.2 billion years ago, life on Earth consisted only of single-celled creatures. Then, multicellular organisms first began to appear in the fossil record. The ancient, many-celled creatures of 600 million years ago, called the Ediacaran fauna, were organisms unlike any modern species, including giant fronds and ornamented disks *(Figure 12.10)*.

Figure 12.10 **Ediacaran fauna.** This reconstruction of multicellular organisms that lived before the Cambrian explosion is based on 580-million-year-old fossil remains.

Biologists are unsure which of these ancient species is the common ancestor of modern **animals**—defined as multicellular organisms that make their living by ingesting other organisms and that are motile (have the ability to move) during at least 1 stage of their life cycle. What is clear from the fossil record is that by about 530 million years ago, *all* modern animal groups had dramatically emerged.

The remarkably sudden emergence of the modern forms of animals—a period comprising little more than 1% of the history of life on Earth—is referred to as the **Cambrian explosion,** named for the geologic period during which it occurred. Some scientists hypothesize that the evolution of the animal lifestyle itself—that is, as predators of other organisms—led to the Cambrian explosion.

It can be difficult to conceive of an animal as complex as a human evolving from a simple eukaryotic ancestor. However, humans are not very different from other eukaryotes. When the first cell containing a nucleus appeared, all of the complicated processes that take place in modern cells, such as cell division and cellular respiration, must have evolved. When the first multicellular animals appeared, many of the processes required to maintain these larger organisms, such as communication systems among cells and the formation of organs and organ systems, arose. Although a human and a starfish appear to be very different, the way we develop and the structures and functions of our cells and common organs are nearly identical.

In fact, there appears to be surprisingly little genetic difference between humans and starfish; most of that difference occurs in a group of genes that control **development,** the process of transforming from a fertilized egg into an adult creature. In addition, the amount of time since the divergence of the major evolutionary lineages of animals—530 million years—is a long time for differences among phyla to evolve. Put in more familiar terms, if the time since the Cambrian explosion was one 24-hour day, all of human history would fit in the last 2 seconds.

Zoologists, scientists who study the kingdom **Animalia,** have described more than 25 phyla. Some of the more well-known groups are illustrated in *Table 12.4*.

TABLE 12.4 **Phyla in the kingdom Animalia.** A sampling of the diversity of animals. The rows are arranged generally in order of appearance in evolutionary time—from the more ancient sponges to the more recent chordates.

Kingdom Animalia: Major Phyla	Description	Example
Porifera	The most ancient animal group. Fixed to underwater surface and filter bacteria from water that is drawn into their loosely organized body cavity.	Sponge 
Cnidaria	Radially symmetric (like a wheel) with tentacles. Some are fixed to a surface as adults (e.g., corals), while others are free floating in marine environments.	Anenome 
Platyhelminthes	Flatworms with a ribbon-like form. Live in a variety of environments on land and sea or as parasites of other animals.	Tapeworm 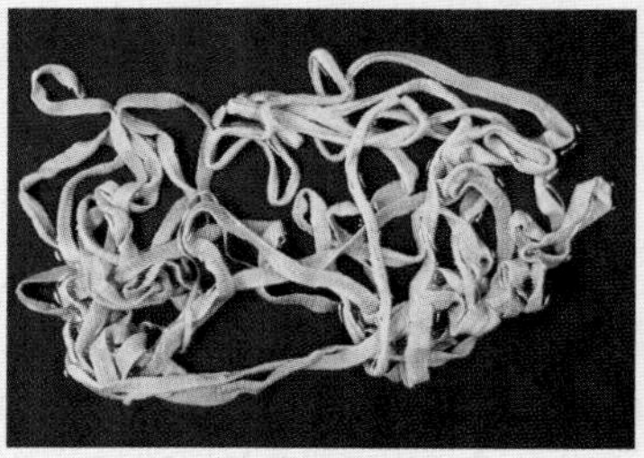
Mollusca	Soft-bodied animals often protected by a hard shell. Body plan consists of a single muscular foot and body cavity enclosed in a fleshy mantle. Phylum includes snails, clams, and squid.	Octopus 
Annelida	Segmented worms. Body divided into a set of repeated segments.	Earthworm 

(continued on the next page)

TABLE 12.4 **Phyla in the kingdom Animalia.** *(Continued)*

Kingdom Animalia: Major Phyla	Description	Example
Nematoda	Roundworms with a cylindrical body shape. Very diverse and widespread in many environments. Earliest animal to evolve a complete digestive tract including a mouth and anus.	Roundworm
Arthropoda	Segmented animals in which the segments have become specialized into different roles (such as legs, mouthparts, and antennae). Body completely enclosed in an external skeleton that molts as the animal grows. Phylum includes insects and spiders as well as crabs and lobsters.	Shrimp
Echinodermata	Slow-moving or immobile animals without segmentation and with radial symmetry. Internal skeleton with projections gives the animal a spiny or armored surface.	Sea urchin
Chordata	Animals with a spinal cord (or spinal cord-like structure). Includes all large land animals, as well as fish, aquatic mammals, and salamanders.	Duck-billed platypus

Most people typically picture mammals, birds, and reptiles when they think of animals, but species with backbones (including mammals, birds, reptiles, fish, and amphibians) represent only 4% of the total species in the kingdom. A small number of these **vertebrates** have traits interesting to a bioprospector. For instance, poison dart frogs *(Figure 12.11)* secrete high levels of toxins onto their skin. These toxins are nerve poisons that cause convulsions, paralysis, and even death to their potential predators. In fact, the

Figure 12.11 **Poison dart frog.** This brightly colored frog contains glands in its skin that release poison when the frog is handled. **Visualize this:** What is the likely purpose of the frog's bright color?

name of these frogs derives from the use of their excretions to coat the tips of hunting darts. Poison dart frog toxins are potentially valuable as sources of potent, nonaddictive painkillers—in low doses, of course.

Most of the bioprospecting work in kingdom Animalia focuses on the remaining 96% of known organisms—the **invertebrates** (animals without backbones, as shown in *(Figure 12.12)*. The vast majority of multicellular organisms on Earth are invertebrates, and most of these animals are insects. Many invertebrates contain chemical compounds not found elsewhere in nature. Species of beetles, ants, bees, wasps, and spiders produce venom to repel predators and competitors; these venoms are sources of potential drugs. For example, the tropical ant *Pseudomyrmex triplarinus* produces venom that may be useful for treating the joint swelling and pain associated with arthritis. In addition, ants, bees, wasps, and termites that live in crowded colonies have evolved protective molecules that reduce disease spread in these environments. These compounds may have the potential to reduce the spread of disease in human populations as well.

The animals that inhabit the oceans' incredibly diverse coral reefs are especially interesting to bioprospectors *(Figure 12.13)*. These biological communities are crowded with life, and the individuals within them continually interact with predators and competitors. As a result of these challenges, many successful coral-reef organisms contain defensive chemicals that might be useful as drugs. The number of unknown invertebrate species, especially in the oceans, is estimated to be anywhere from 6 to 30 million.

Stop and Stretch 12.3: One mantra of reef bioprospectors is, "If it is bright red, slow-moving, and alive, we want it." Why is such an animal probably a good source of defensive compounds?

Figure 12.12 **Examples of invertebrates.** Most animals are invertebrates, including these examples.

(a) Ant (*Pseudomyrmex triplarinus*)

Potential source of arthritis treatment

(b) Jellyfish (*Aequorea victoria*)

Source of fluorescent protein, a useful labeling tool in microbiology

(c) Horseshoe crab (*Limulus polyphemus*)

Source of blood proteins used to test for pathogens in humans

While our ignorance about the diversity of animals is great, another kingdom of multicellular eukaryotes is even less well known, although no less important to the functioning of ecosystems and as a source of molecules useful to humans—the fungi.

Figure 12.13 **A coral reef.** This extremely diverse biological community is a rich source of interesting biological chemicals.

Fungi

Like plants, fungi are immobile, and many produce fruit-like organs that disperse **spores,** cells that are analogous to plant seeds in that they germinate into new individuals. However, the mushroom you think of when you imagine a fungus is a misleading image of the kingdom. Most of the functional part of fungi is made up of very thin, stringy material called **hyphae,** which grows over and within a food source *(Figure 12.14)*.

Fungi feed by secreting chemicals from the hyphae; these chemicals break down the food into small molecules, which the fungi then absorb. The string-like form of fungal hyphae maximizes the surface over which feeding takes place, so the vast majority of the body of most fungi is microscopic and diffuse. Fungal food sources typically include dead organisms, and the actions of fungi are key to recycling nutrients from these organisms. Fungi are more like animals than plants in that they rely on other organisms for food. In fact, DNA sequence analysis by **mycologists,** biologists who study fungi, indicates that Fungi and Animalia are more closely related to each other than either kingdom is to the plants.

The phyla of fungi are distinguished by their method of spore formation *(Table 12.5)*. However, convergent evolution, in which unrelated species take similar forms because of similar environments, has led to the appearance of similar body shapes and lifestyles—which we call "fungal forms" among these different phyla. One of the most commercially important fungal forms is **yeast,** a single-celled type of fungi. Yeast forms are found in at least 2 different fungal phyla and live in liquids such as plant sap, soaked grains, or fruit juices. The activity of yeasts in oxygen-poor but sugar-rich environments results in the formation of alcohol. The metabolism of yeasts within flour batter leads to the production of carbon dioxide bubbles, which are trapped by wheat protein fibers and thus allow the dough to "rise" during bread making. Another fungal form known as **mold** is found in all phyla and is also commercially important. This quickly reproducing, fast-growing form can spoil fruits and other foods, although some produce certain types of flavorful cheese, including blue and Camembert, as they are "spoiling."

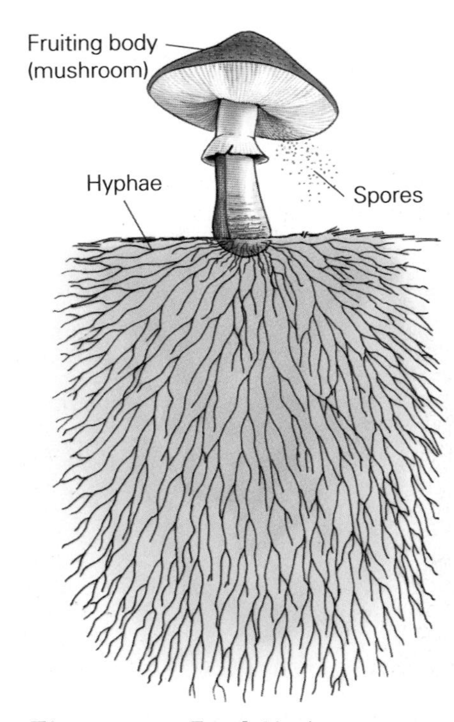

Figure 12.14 **Fungi.** Hyphae can extend over a large area. The familiar mushroom, as well as the fruiting structures of less-familiar fungi, primarily function only as methods of dispersing spores. **Visualize This:** Why is a mushroom produced outside the food resource for the hyphae?

About one-third of the bacteria-killing antibiotics in widespread use today are derived from fungi. Fungi produce antibiotics because bacteria compete with them for their dead food source. Penicillin, the first commercial antibiotic, is produced by a fungus *(Figure 12.15)*. Its discovery is one of the great examples of good fortune in science.

Before he went on vacation during the summer of 1928, British bacteriologist Alexander Fleming left a dish containing the bacteria *Staphylococcus aureus* on his lab bench. While he was away, this culture was contaminated by a spore from a *Penicillium* fungus, probably from a different laboratory in the same building. When Fleming returned to his laboratory, he noticed that the growth of *S. aureus* had been inhibited on the fungus-contaminated culture dish. Fleming inferred that some antibiotic chemical substance had diffused from the fungus; he named it penicillin, after the fungus itself. The first batches of this bacteria-slaying drug became available during World War II and greatly reduced the number of deaths of wounded soldiers from infection. Since the discovery of penicillin, hundreds of other antibiotics have been isolated from different fungus species.

TABLE 12.5 **Fungal diversity.** Fungi are classified into phyla based on their mode of spore production. The most common phyla are listed here.

Kingdom Fungi: Major Phyla	Description	Example
Zygomycota	Sexual reproduction occurs in a small resistant structure called a zygospore. Most reproduction is asexual—directly via mitosis.	*Rhizopus stolonifera*—bread mold
Ascomycota	Spores are produced in sacs on the tips of hyphae in fruiting structures.	Morel
Basidiomycota	Spores are produced in specialized club-shaped appendages on the tips of hyphae in fruiting structures.	*Amanita muscaria*—the poisonous fly agaric

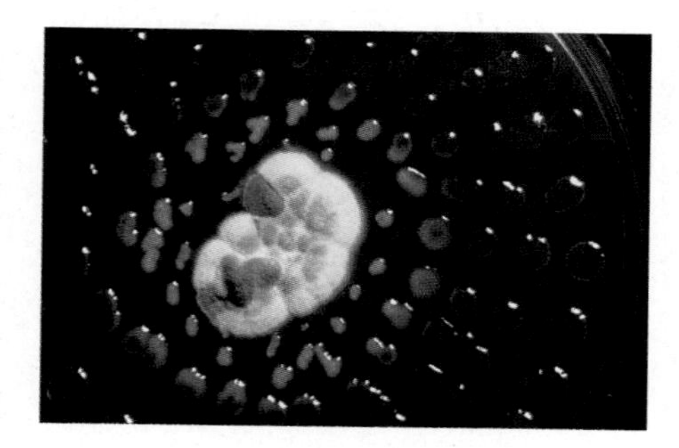

Figure 12.15 **Antibiotics from fungi.** The large white-and-gray mold in the center of this petri dish is a colony of *Penicillium*. The red dots on the edges of the dish are colonies of bacteria. The pink dots surrounding the *Penicillium* colony are bacterial colonies that are dying from contact with antibiotic secretions from the fungi.

Some fungi infect living animals and thus also have potential as sources of drugs. Cyclosporin is a molecule produced by a number of different parasitic fungi and has the effect of suppressing a host's immune response. Humans have used cyclosporin to prevent the immune systems of organ-transplant recipients from attacking foreign transplanted organs. Fungi are also the source of a powerful class of anticholesterol drugs, called statins, which help treat and prevent heart disease.

The fungus *Claviceps purpurea*, also known as ergot, has long been known to have powerful effects on the human body. Midwives throughout the nineteenth century used this pest of rye and wheat to stimulate uterine contractions and speed labor, and farmers throughout Europe knew that consuming grain infected with ergot could lead to neurological effects, including burning pain in the limbs, hallucinations, and convulsions. The symptoms of "demonic possession" that led to the Salem Witch Trials in 1691–92 may have been caused by consumption of ergot-contaminated grain. More recently, the drug LSD has been derived from this fungus, with similar effects on recreational users.

While fungi have been the third most important source of molecules useful to humans—right after bacteria in terms of numbers and impacts of derived compounds—the source of most naturally derived drugs has been the plant kingdom.

Plantae

The kingdom **Plantae** consists of multicellular eukaryotic organisms that make their own food via photosynthesis. Plants have been present on land for over 400 million years, and their evolution is marked by increasingly effective adaptations to the terrestrial environment *(Table 12.6)*.

The first plants to colonize land were small and close to the ground for they had no way to transport water from where it is available in the soil to where it is needed, in the leaves. The evolution of **vascular tissue**, made up

TABLE 12.6 **Plant diversity.** The 4 major phyla of plants are listed here in order of their appearance in evolutionary history.

Kingdom Plantae: Major Phyla	Description	Example
Bryophyta	Mosses. Lacking vascular tissue, these plants are very short and typically confined to moist areas. Reproduce via spores.	Moss
Pteridophyta	Ferns and similar plants. Contain vascular tissue and can reach tree size. Reproduce via spores.	Staghorn fern
Coniferophyta	Cone-bearing plants resembling the first seed producers	Cycad
Anthophyta	Flowering plants. Seeds produced within fruits, which develop from flowers. Advances in vascular tissues and chemical defenses contribute to their current dominance on Earth.	Orchid

(a) Foxglove (*Digitalis purpurea*)

Source of heart drug digitalis

(b) Aloe (*Aloe barbadensis*)

Source of aloe vera, used to treat burns and dry skin

(c) Willow (*Salix alba*)

Source of aspirin

Figure 12.16 **Diversity of flowering plants.** A few of the enormous variety of plants that provide important medicines.

of specialized cells that can transport water and other substances, allowed plants to reach tree-sized proportions and to colonize much drier areas. The evolution of **seeds**, structures that protect and provide a food source for young plants, represented another adaptation to dry conditions on land. However, most modern plants belong to a group that appeared only about 140 million years ago, the **flowering plants.** Like their ancestors, flowering plants possess vascular tissue and produce seeds; in addition, these plants evolved a specialized reproductive organ, the flower. Over 90% of the known plant species are flowering plants *(Figure 12.16)*.

From about 100 million to 80 million years ago, the number of distinct groups, or families, of flowering plants increased from around 20 to over 150. During this time, flowering plants became the most abundant plant type in nearly every habitat. The rapid expansion of flowering plants is called **adaptive radiation**—the diversification of one or a few species into a large and varied group of descendant species. Adaptive radiation typically occurs either after the appearance of an evolutionary breakthrough in a group of organisms or after the extinction of a competing group. For example, the radiation of animals during the Cambrian explosion may be a result of the evolution of the predatory lifestyle, and the radiation of mammals beginning about 65 million years ago occurred after the extinction of the dinosaurs. The radiation of flowering plants must be due to an evolutionary breakthrough—some advantage they had over other plants allowed them to assume roles that were already occupied by other species.

Plant biologists, or **botanists,** still debate which traits of flowering plants give them an advantage over nonflowering types. Some botanists believe that the unique reproductive characteristics of flowering plants led to their radiation. These unique characteristics include waiting to provision an embryo until successful fertilization of the egg (a process called *double fertilization*), as well the assistance of animals in transferring gametes *(Figure 12.17)*. Other botanists think that chemical defenses in these plants reduced their susceptibility to predators and provided their edge over other plant groups.

The diversity of chemical defenses in flowering plants makes them particularly interesting to bioprospectors. Because plants cannot physically escape from their predators, natural selection has favored the production of predator-deterring toxins via side, or secondary, reactions to primary biochemical pathways. These chemicals are known as **secondary compounds.** For instance, curare vines produce toxins that block the connection between nerves and muscles. Organisms that get this toxin, called curarine, in their bodies become

(1) **Flower petals** attract insects that move pollen from one flower to another, helping fertilization to occur.

(2) **Double fertilization** occurs. The pollen tube carries two sperm. One fertilizes the embryo, and the other fuses with two nuclei in another cell to produce the endosperm.

(3) **Fruit** consists of seeds packaged in a structure that aids their dispersal, such as tasty flesh or a parachute.

(4) **Seeds** contain an embryo and endosperm, and are highly resistant to drying. The endosperm is a tissue that nourishes the embryo.

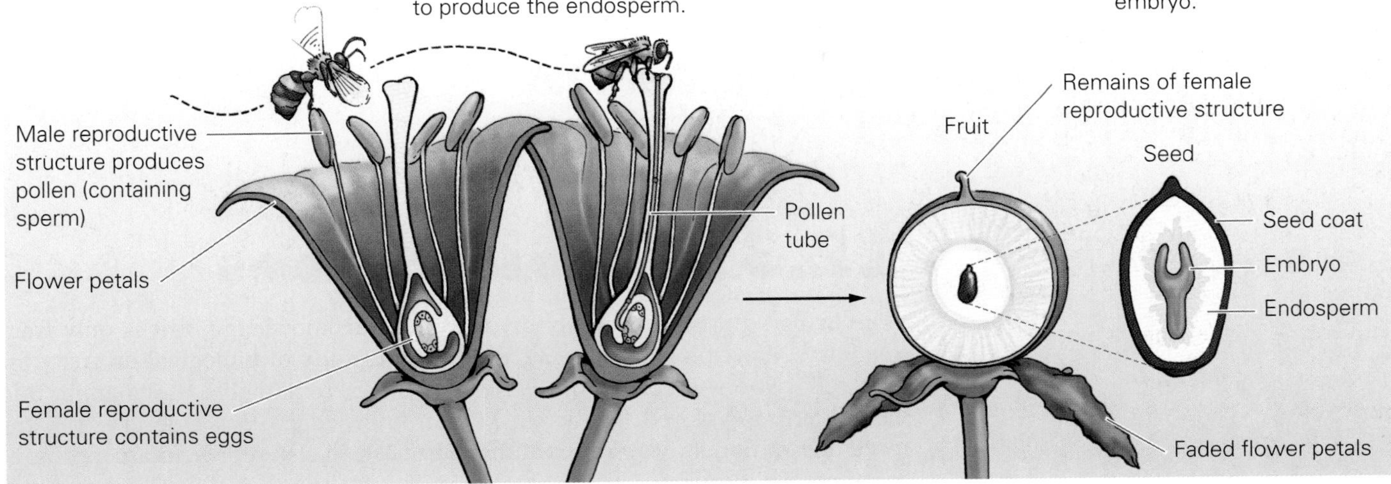

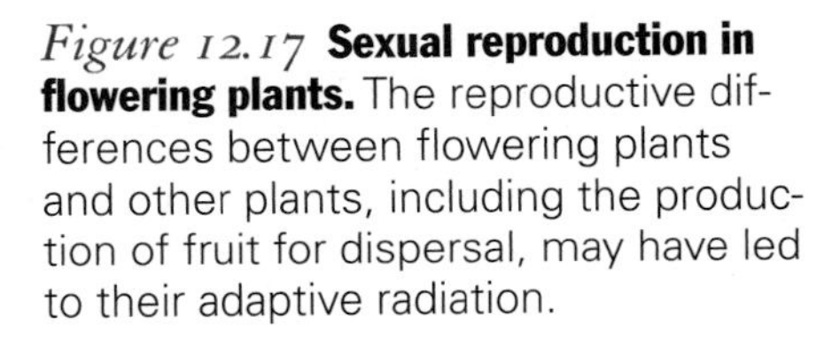

Figure 12.17 **Sexual reproduction in flowering plants.** The reproductive differences between flowering plants and other plants, including the production of fruit for dispersal, may have led to their adaptive radiation.

paralyzed and can do little damage to the vine. Curarine is produced via a secondary pathway of the process of amino acid synthesis. In this case, natural selection must have favored genetic variations leading to production of not only normal amino acids but also this toxic secondary compound. Today, doctors use curarine as a muscle relaxant during surgery.

The kingdom Plantae is the source of many other well-known, naturally derived drugs. Aspirin from willow, the heart drug digitalis from foxglove, the anticancer chemical vincristine from the rosy periwinkle, morphine from opium poppies, caffeine from coffee, and dozens of other pharmaceutical products are derived directly from plants. Pharmaceutical manufacturers now reproduce hundreds of other drugs first identified in plants. Many botanists believe that the number of unknown plant species is relatively small—probably a few thousand—but the potential drugs available in even the known species are mostly unknown.

Stop and Stretch 12.4: Why do you think most of the known nature-derived drugs used by humans come from plants?

Viruses

A virus typically consists of a strand of DNA or RNA surrounded by a protein shell, called a capsid. Some viruses are also enveloped in membranous coat. Viruses are basically rogue pieces of genetic material that hijack cells to reproduce *(Figure 12.18)*. When a virus enters a host cell, it uses the transcription machinery of the cell to make copies of its DNA or RNA. The replicated viral genomes then use the ribosome, transfer RNA, and amino acids of the host to make new protein capsules and other polypeptides. If the virus is enveloped, it even uses the host cell's own membrane to make envelopes for the daughter viruses.

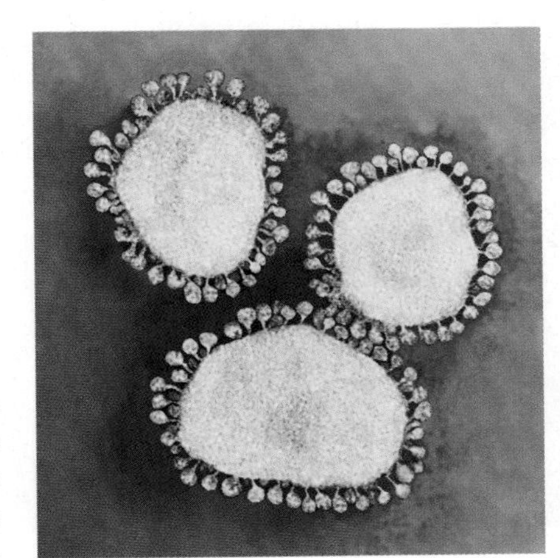

Figure 12.18 **Viruses.** The severe acute respiratory virus (SARS), viewed under very high magnification. Visible here is the protein capsid, studded with receptors that bind to host cells.

Once a cell has been hijacked by a virus, it cannot perform its own necessary functions and will die. But before it does, dozens of duplicated viruses can be released and move on to infect other cells. Some of the most devastating pathogens of humans are viruses, including polio, smallpox, influenza, and HIV. HIV is especially troublesome because it destroys cells of the immune system, which are essential for fighting off infection. An HIV-infected individual gradually loses the ability to combat other diseases, including the HIV infection itself.

Stop and Stretch 12.5: Why do you think viruses are not considered living organisms?

12.3 Learning About Species

The living world is amazingly diverse, and our knowledge of it is only fragmentary. Many biologists are attracted to the study of biological diversity for just this reason—because the variety of life is remarkable, fascinating, and largely an unexplored frontier in human knowledge. As we observed in the previous section, however, nonhuman species can also represent an enormous resource for humans.

Our survey of life's diversity illustrates the essential problem for a bioprospector—there are many more potentially valuable species than there are resources to find and evaluate them. This challenge has led many drug companies to abandon bioprospecting in favor of "rational drug design," which uses an understanding of the causes of a particular illness to create synthetic drugs used to treat it. However, some of the most effective plant-based drugs are so strange in structure that they would never have been discovered by rational design.

Bioprospectors continue to hope to find more of these strange natural compounds. What tools can biologists offer to bioprospectors who seek to mine biological gold?

Figure 12.19 **The Pacific yew.** The source of the anticancer drug paclitaxel was overharvested after the drug was discovered. Fortunately, the drug can be synthesized from other common yew species.

Fishing for Useful Products

The National Cancer Institute (NCI) has taken a brute-force approach to screening species for cancer-suppressing chemicals. NCI scientists receive frozen samples of organisms from around the world, chop them up, and separate them into a number of extracts, each probably containing hundreds of components. These extracts are tested against up to 60 different types of cancer cells for their efficacy in stopping or slowing growth of the cancer. Promising extracts are then further analyzed to determine their chemical nature, and chemicals in the extract are tested singly to find the effective compound. This approach is often referred to as the "grind 'em and find 'em" strategy.

To date, this strategy has been effective in identifying one major anticancer chemical—paclitaxel, also known by the trade name Taxol, from the Pacific yew *(Figure 12.19)*. Paclitaxel continues to be produced from the needles of yew trees and is effective against ovarian cancer, advanced breast cancers, malignant skin cancer, and some lung cancers. This strategy has identified dozens of other less well-known anticancer drugs as well.

Understanding Ecology

The grind 'em and find 'em approach works best when researchers are seeking treatments for a specific disease or set of diseases, such as cancer. But most bioprospectors are much more speculative—they are interested in determining whether an organism contains a chemical that is useful against any disease. Doing this effectively requires a more thoughtful approach, taking into account the biology of the species.

One aspect of an organism's biology that can be illuminating is its **ecology**—that is, its relationship to the environment and other living organisms. Our survey of diversity illustrated some ecological characteristics that increase the likelihood that a species contains valuable chemicals. These characteristics include survival in extreme environments, high levels of competition with bacteria and fungi, susceptibility to predation, ability to live in and on other living organisms, and high population density. In these cases, natural selection can lead to the evolution of biochemicals that function at high temperature or salinity; antibacterial, antifungal, or antiviral compounds; molecules that suppress or modify the effects of the immune system; and chemical defenses that may have physiological effects.

An understanding of ecology is useful even within a species. For instance, populations of plants experiencing high levels of insect attacks may produce more defensive compounds than populations that are not under attack *(Figure 12.20)*. Focusing on certain ecological situations can help bioprospectors increase their success.

Figure 12.20 **Induced defenses.** A study in Japan demonstrated that *Linden viburnum* damaged by deer were subsequently more resistant to insect attack.

Reconstructing Evolutionary History

Clues to an organism's chemical traits can come from knowing the traits found in its closest relatives. This is one reason some scientists argue that a classification system reflecting evolutionary relationships is more useful than one based on more superficial similarities.

The classification of certain birds helps illustrate this point. Vultures are birds that specialize in feeding on dead animals. These birds spend a large amount of time soaring on broad, flat wings, have sharp beaks for tearing meat, and regurgitate food to feed their offspring. A nonevolutionary classification places all vultures together. However, research published in the 1970s demonstrated that New World vultures in the Western Hemisphere *(Figure 12.21b)* are more closely related to storks *(Figure 12.21c)*—long-legged birds with long beaks that specialize in catching fish in shallow waters—than they are to Old World vultures from the Eastern Hemisphere *(Figure 12.21a)*. Even though species of New World vulture *look* like Old World vultures, they share a more

Figure 12.21 **The challenge of biological classification.** The evolutionary relationship between New World vultures and storks is not evident from their appearance.

(a) Old World vulture

Hooded vulture (*Necrosyrtes monachus*)

(b) New World vulture

Turkey vulture (*Cathartes aura*)

(c) Stork

Wood stork (*Mycteria americana*)

recent common ancestor with storks and are thus more similar to storks anatomically, physiologically, and genetically.

An **evolutionary classification** can be quite useful in the study of living organisms; for instance, if scientists wish to know more about the basic biology of New World vultures, then they might start by learning what is known about the biology of storks, their closest relatives. And if bioprospectors want to look for new valuable biological compounds, they could start by screening the close relatives of organisms producing known valuable chemicals.

Developing Evolutionary Classifications. Evolutionary classifications are based on the principle that the descendants of a common ancestor should share any biological trait that first appeared in that ancestor. For example, this principle has been used to uncover the evolutionary relationship among different species of sparrows.

Figure 12.22 illustrates a hypothesized **phylogeny**—the evolutionary relationship—of sparrows in the genus *Zonotrichia.* Scientists have used a technique called **cladistic analysis,** an examination of the variation in these sparrows' traits relative to a closely related species, to determine this phylogeny. For example, if we examine just the heads of the 4 sparrow species compared to their relative, the dark-eyed junco, we see that 3—all but the Harris's sparrow—have dark and light alternating stripes on the crown of their heads. This observation seems to indicate that crown striping evolved early in the radiation of these sparrows. Among the 3 species with crown stripes, 2 have 7 stripes, while the golden-crowned sparrow has only 3 stripes. An increase in the number of stripes appears to have evolved after the original striped crown pattern. Finally, of the 2 species with 7 crown stripes, only 1 has evolved a white throat—the aptly named white-throated sparrow. In the case of these 4 sparrows, it appears that every step in their radiation involved a visible change in their appearance.

Unfortunately, reconstructing evolutionary relationships is not as simple as the sparrow example suggests. Descendant species may lose a trait that evolved in their ancestor, or unrelated species may acquire identical traits via

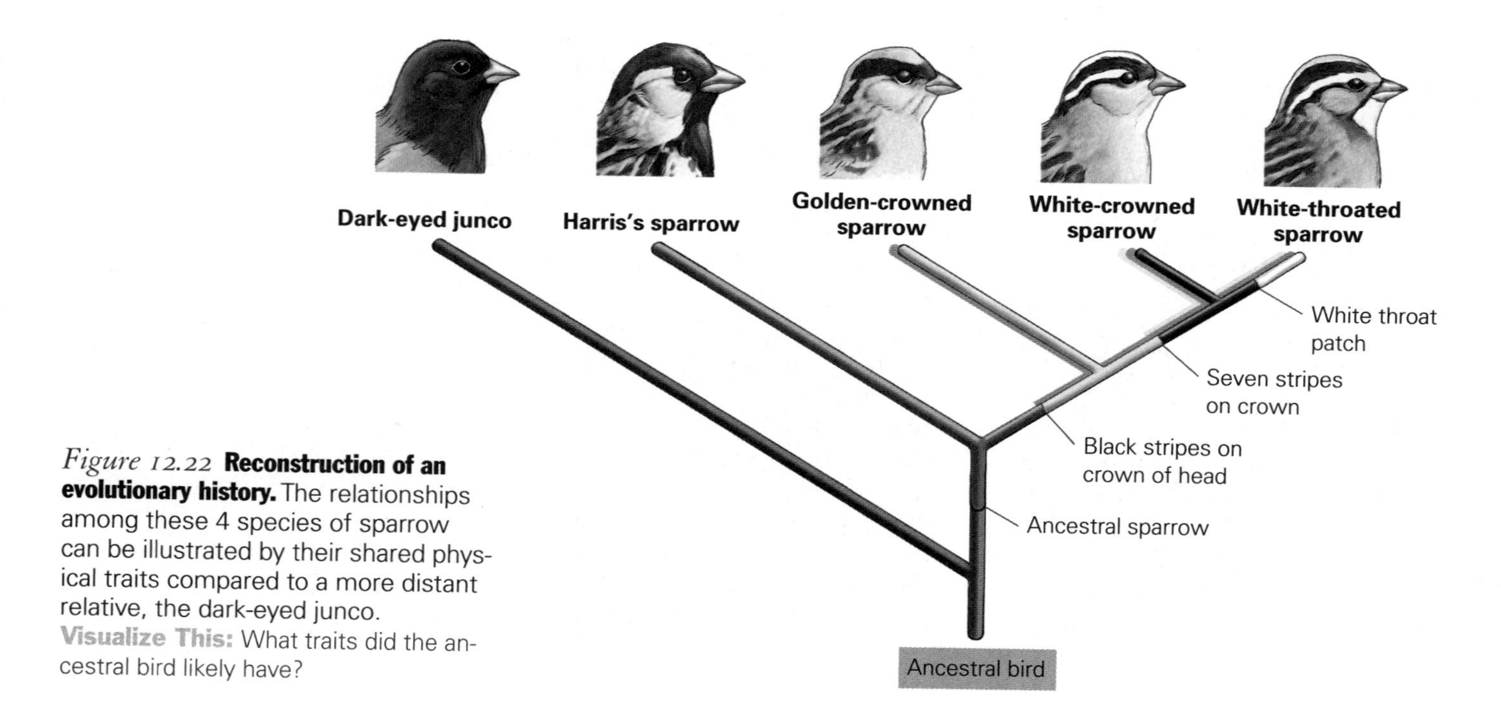

Figure 12.22 **Reconstruction of an evolutionary history.** The relationships among these 4 species of sparrow can be illustrated by their shared physical traits compared to a more distant relative, the dark-eyed junco.
Visualize This: What traits did the ancestral bird likely have?

convergent evolution. You can even see convergence in the phylogeny in Figure 12.22; golden-crowned sparrows, like the white-throated species, have a patch of golden feathers on their heads that appears to have evolved independently. The existence of convergent traits complicates the development of evolutionary classifications.

Stop and Stretch 12.6: Amateur bird watchers often consider vultures as birds of prey, along with hawks, falcons, eagles, and owls. What convergent traits do vultures share with these other birds that would cause them to classify the birds this way? Does this "folk classification" have a function for beginning birders?

Testing Evolutionary Classifications. An *evolutionary classification* is a hypothesis of the relationship among organisms. It is difficult to test this hypothesis directly—scientists have no way of observing the actual evolutionary events that gave rise to distinct organisms. However, scientists *can* test their hypotheses by using information from both fossils and living organisms.

By examining the fossils of extinct organisms, scientists can gather clues about the genealogy of various groups. For example, fossils of vulture-like birds clearly indicate that this lifestyle evolved independently in both the Old World and the New World.

Information from living organisms can provide an even finer level of detail about evolutionary relationships. As illustrated in Figure 12.5, closely related species should have similar DNA. If the pattern of DNA similarity matches a hypothesized evolutionary relationship among species, the phylogeny is strongly supported. This is the case with the hypothesized relationship between New World vultures and storks; DNA sequence comparisons indicate that the DNA of New World vultures is more similar to the DNA of storks than it is to the DNA of Old World vultures.

In contrast, DNA sequence comparisons do not support the sparrow phylogeny presented in Figure 12.22. Here, the data suggest that white-crowned birds and golden-crowned birds are closely related, and white-throated sparrows are a more distant relative. In this case, more observations are needed to discern the true evolutionary relationships among the *Zonotrichia*. In Chapter 9 we described multiple supportive tests of another phylogeny—in that case, the evolutionary relationship among humans and apes.

The examples of phylogeny reconstruction and testing as described here and in Chapter 9 provide nice illustrations of the process, but the species in them are not likely to contain valuable biological molecules. The evolutionary reconstructions that interest bioprospectors consist of groups of species that already contain valuable members. For instance, relatives of the curarine-producing curare vine are likely to contain similar secondary biochemical pathways that produce slightly different muscle relaxants *(Figure 12.23)*. Once a hypothesis of evolutionary relationship *is* reasonably well supported by additional data, bioprospectors can use the information gathered about one species to predict the characteristics of its relatives.

Although an evolutionary classification of groups that are likely to contain valuable chemicals would increase the speed of discovery, the slow pace of reconstructing and testing phylogenies means that bioprospectors do not always have this information. As an alternative, bioprospectors have turned to the humans who have the most intimate knowledge of useful organisms in their environments—healers in preindustrial cultures.

Figure 12.23 **Curare.** Its name translates from the native words "*bird*" and "*kill*." An evolutionary classification would help us identify its closest relatives, which may also have medicinal properties.

Figure 12.24 **Indigenous knowledge.** This shaman is of the Matses people of the Amazon rain forest. His intimate knowledge of the natural world is the product of the long history of his people in this diverse environment.

Learning from the Shamans

To this point, we have been describing the search for biologically active compounds in living organisms as an exploration of the unknown. This is true for organisms in Yellowstone's hot springs; chemicals derived from these organisms were probably truly unknown to humans. For many other species, however, people have known of their usefulness for thousands of years. This knowledge is maintained in the traditions of indigenous people in biologically diverse areas, people who use native organisms as medicines, poisons, and foods.

In many cultures, the repository of this traditional knowledge is the medicine man or woman. A shaman, as these healers are often called, can help direct bioprospectors to useful compounds by teaching about their culture's traditional methods of healing. Shamans employ many remedies that are highly effective against disease. Several bioprospectors have consulted with shamans to increase their chances of finding useful drugs (*Figure 12.24*).

Using the knowledge of native people in developing countries to discover compounds for use in wealthy, developed countries is highly controversial. This process is often referred to as **biopiracy** because organisms and active compounds thus identified can potentially provide enormous financial rewards to the bioprospector with no return to a shaman or the shaman's people.

The United Nations Convention on Biodiversity has sought to reduce biopiracy by asserting that each country owns the biodiversity within its borders. However, because the U.S. government has not signed this legally binding document, companies in the United States are not required to abide by its terms. In addition, even when the government of a country makes a bioprospecting agreement with a pharmaceutical firm, any rewards may not trickle back down to the indigenous community that was the source of the knowledge.

The bioprospecting agreement made between Diversa and Yellowstone National Park did not escape charges of biopiracy. While the company was not planning on relying on information from indigenous people to help locate valuable organisms, critics have charged that the managers of Yellowstone were essentially "selling off" organisms and chemical compounds that belong to the American public. In addition, they argue that the action of bioprospecting itself will damage the very resource that provides these remarkable discoveries. The federal courts dismissed lawsuits against Yellowstone National Park and Diversa that addressed these points, but the issue remains an ethical dilemma: What is the responsibility of individuals and corporations profiting from biological diversity to the source and survival of that diversity?

Biological diversity represents an enormous resource for humans, but it also comes with an awesome responsibility. Actions of the U.S. Congress protected Yellowstone National Park and perhaps ultimately enabled the discovery of *Thermus aquaticus* and *Taq* polymerase. But thousands of useful organisms are lost every year through the destruction of native habitat, and our ability to use these organisms is diminished by the loss of indigenous cultures and their shamans.

The dramatic rate of biodiversity loss not only denies humans still-undiscovered biological molecules. It also diminishes Earth's ability to sustain our population. At current rates of loss, our generation may be the last to truly enjoy the diverse wonder, beauty, and benefits of wild nature. Chapter 14 discusses the causes, consequences, and possible solutions for the current biodiversity crisis. Humans can help to reduce the rate at which biodiversity is being lost—but only if we begin to appreciate the value of the diversity that surrounds and sustains us.

Savvy Reader Is the Government Keeping the Truth from Us?

Self-help books dealing with health are very common. This promotional text comes from a sequel to one of the most popular in recent years.

"The United States Federal Trade Commission censored Kevin Trudeau's first book, *Natural Cures "They" Don't Want You to Know About.* That book still saved lives. Hundreds of thousands of people have reported better health and the curing of their diseases without drugs and surgery after reading it. Now, Kevin Trudeau takes on, and goes head-to-head against, governments worldwide, the international pharmaceutical medical cartels, and even the news media. He reveals for the first time never-before-released secret information about his covert involvement with Big Pharma, the food industry, governments in over sixty countries, and some of the richest and most powerful families, people, and private organizations in the world. He is being called the most daring whistleblower of corporate and government corruption of all time. Risking potential criminal prosecution, Trudeau now releases the material previously censored by the U.S. Government: the specific product brand names that Trudeau believes can be used to prevent, treat, and cure disease. This is the book the government wants to ban. This is the book "they" don't want you to read."

—Front book flap text from *More Natural Cures Revealed: Previously Censored Brand Name Products that Cure Disease*, by Kevin Trudeau, Alliance Publishing, May 2006

1. What is the apparent expertise of the author of this book?
2. What concerns and opinions of readers does this text appeal to get individuals to buy the book?
3. What strategies would you use to determine whether the information in this text was accurate or if following the advice in this book is safe?

Chapter Review

For study help, animations, and more quiz questions go to www.mybiology.com.

Summary

12.1 Biological Classification

- Bioprospectors seek to discover new drugs and other useful chemicals from the diversity of living organisms on Earth (p. 314).
- The number of known living species is estimated to be between 1.4 and 1.8 million, but the total number of species may be as high as 100 million (p. 314).
- The history of life on Earth has consisted of successive dynasties of organisms. All of these

Summary (Continued)

organisms can be grouped into a few, very broad groups, however (p. 315).

- Organisms are classified in domains according to evolutionary relationships and in kingdoms based on similarities in structure and lifestyle (p. 315).

web animation 12.1
The Tree of Life

12.2 The Diversity of Life

- Life on Earth began about 3.6 billion years ago with simple prokaryotes, but it would be 1.5 billion years before eukaryotes evolved (p. 320).
- Bacteria and Archaea are prokaryotes, simple single-celled organisms without a nucleus or other membrane-bound organelles. They are abundant, found in a variety of habitats, and rely on a variety of food sources. Prokaryotes may produce antibiotics or have chemicals that function in extreme conditions (p. 320).
- Eukaryotes, cells with nuclei and other membrane-bound organelles, probably evolved from symbioses among ancestral eukaryotes and prokaryotes (p. 321).
- The kingdom Protista is a hodgepodge of organisms that are typically unicellular eukaryotes. Algae are protists that are especially interesting to bioprospectors because they make defensive chemicals against predators and produce unique food-storage compounds, but fungus-like and animal-like protists also exist (p. 321).
- Multicellular organisms did not appear until approximately 600 million years ago, and this advance in form led to the diversity of species on Earth today (pp. 321–324).
- Animals are motile, multicellular eukaryotes that rely on other organisms for food. Animal groups evolved in a short period of time known as the Cambrian explosion. Bioprospectors are interested in animals that produce venom or defensive chemicals (pp. 325–328).
- Fungi are immobile, multicellular eukaryotes that rely on other organisms for food and are made up of thin, threadlike hyphae. Fungi often produce antibiotics that kill their competitors, and some can suppress their host's immune system response. (pp. 329–330).
- Plants are multicellular, photosynthetic eukaryotes. They have become increasingly adapted to land habitats over time. The diversity of flowering plants may be due partly to their production of defensive chemicals (pp. 331–332).
- Viruses are nonliving entities that reproduce by hijacking living cells (p. 333).

web animation 12.2
Endosymbiotic Theory

12.3 Learning About Species

- Some bioprospectors look for useful products by screening as many compounds as possible against a particular disease (p. 334).
- An understanding of the ecological relationships of organisms provides clues to the likelihood and nature of possible chemical compounds in organisms (pp. 334–335).
- Determining the evolutionary relationships among living organisms can help provide clues about an organism's traits. Phylogenies are created and tested by evaluating the shared traits of different species that indicate they shared a recent ancestor (pp. 336–337).
- Studying how indigenous healers called shamans use organisms can help bioprospectors identify species that may have useful chemicals (p. 338).
- Biopiracy occurs when a small group of people benefit from the knowledge of an indigenous culture; this practice may also undermine society's efforts to protect biodiversity (p. 338).

Roots to Remember

The following roots of words come mainly from Latin and Greek and will help you decipher terms:

anima- means breath or soul.
botan- means pertaining to plants.
eco- comes from the word for "home."
endo- means inside.
eu- means true.
-karyo- means kernel.
myco- means fungal.
pro- means before.
zoo- means pertaining to animals.

Learning the Basics

1. How did the nucleus of eukaryotic cells likely evolve?
2. How is knowledge of the ecology of an organism useful for predicting which types of valuable chemicals it may possess?
3. How are hypotheses about the evolutionary relationships among living organisms tested?
4. Which of the following kingdoms or domains is a hodgepodge of different evolutionary lineages?
A. Bacteria; **B.** Protista; **C.** Archaea; **D.** Plantae; **E.** Animalia
5. Comparisons of ribosomal RNA among many different modern species indicate that ________.
A. there are 2 very divergent groups of prokaryotes; **B.** the kingdom Protista represents a conglomeration of very unrelated forms; **C.** fungi are more closely related to animals than to plants; **D.** a and b are correct; **E.** a, b, and c are correct
6. On examining cells under a microscope, you notice that they occur singly and have no evidence of a nucleus. These cells must belong to:
A. Domain Eukarya; **B.** Domain Bacteria; **C.** Domain Archaea; **D.** Kingdom Protista; **E.** more than 1 of the above is correct.
7. The mitochondria in a eukaryotic cell ________.
A. serve as the cell's power plants; **B.** probably evolved from a prokaryotic ancestor; **C.** can live independently of the eukaryotic cell; **D.** a and b are correct; **E.** a, b, and c are correct
8. Fungi feed by________.
A. producing their own food with the help of sunlight; **B.** chasing and capturing other living organisms; **C.** growing on their food source and secreting chemicals to break it down; **D.** filtering bacteria out of their surroundings; **E.** producing spores
9. Which of the following is/are always true:?
A. Viruses cannot reproduce outside a host cell; **B.** viruses are not surrounded by a membrane; **C.** viruses are not made up of cells; **D.** a and c are correct; **E.** a, b, and c are correct.?
10. Phylogenies are created based on the principle that all species descending from a recent common ancestor ________.
A. should be identical; **B.** should share characteristics that evolved in that ancestor; **C.** should be found as fossils; **D.** should have identical DNA sequences; **E.** should be no more similar than species that are less closely related

Analyzing and Applying the Basics

1. Unless handled properly by living systems, oxygen can be quite damaging to cells. Imagine an ancient nucleated cell that ingests an oxygen-using bacterium. In an environment where oxygen levels are increasing, why might natural selection favor a eukaryotic cell that did not digest the bacterium but instead provided a "safe haven" for it?
2. Imagine you have found an organism that has never been described by science. The organism, made up of several hundred cells, feeds by anchoring itself to a submerged rock and straining single-celled algae out of pond water. What kingdom would this organism probably belong to, and why do you think so?
3. Vertebrate animals are much less likely to produce antibiotic compounds than bacteria and fungi. What features of vertebrate ecology makes antibiotic production unlikely? What features of vertebrate physiology makes antibiotic production unlikely?

Connecting the Science

1. Scientists initially rejected the endosymbiotic theory, which is the hypothesis that eukaryotic cells evolved from a set of cooperating independent cells. Most biologists still believe that competition for resources among organisms is the primary force for evolution. Do you think our modern culture, where competition is often valued over cooperation, might make scientists less likely to search for and see the role of cooperation in evolution, or do you think cooperation leading to evolution must be truly rare?
2. Do we have an obligation to future generations to preserve as much biodiversity as possible? Why or why not? Would simply preserving the information contained in an organism's genes (in a zoo or other collection) be good enough, or do we need to preserve organisms in their natural environments?

CHAPTER 13

Is the Human Population Too Large?

Population Ecology

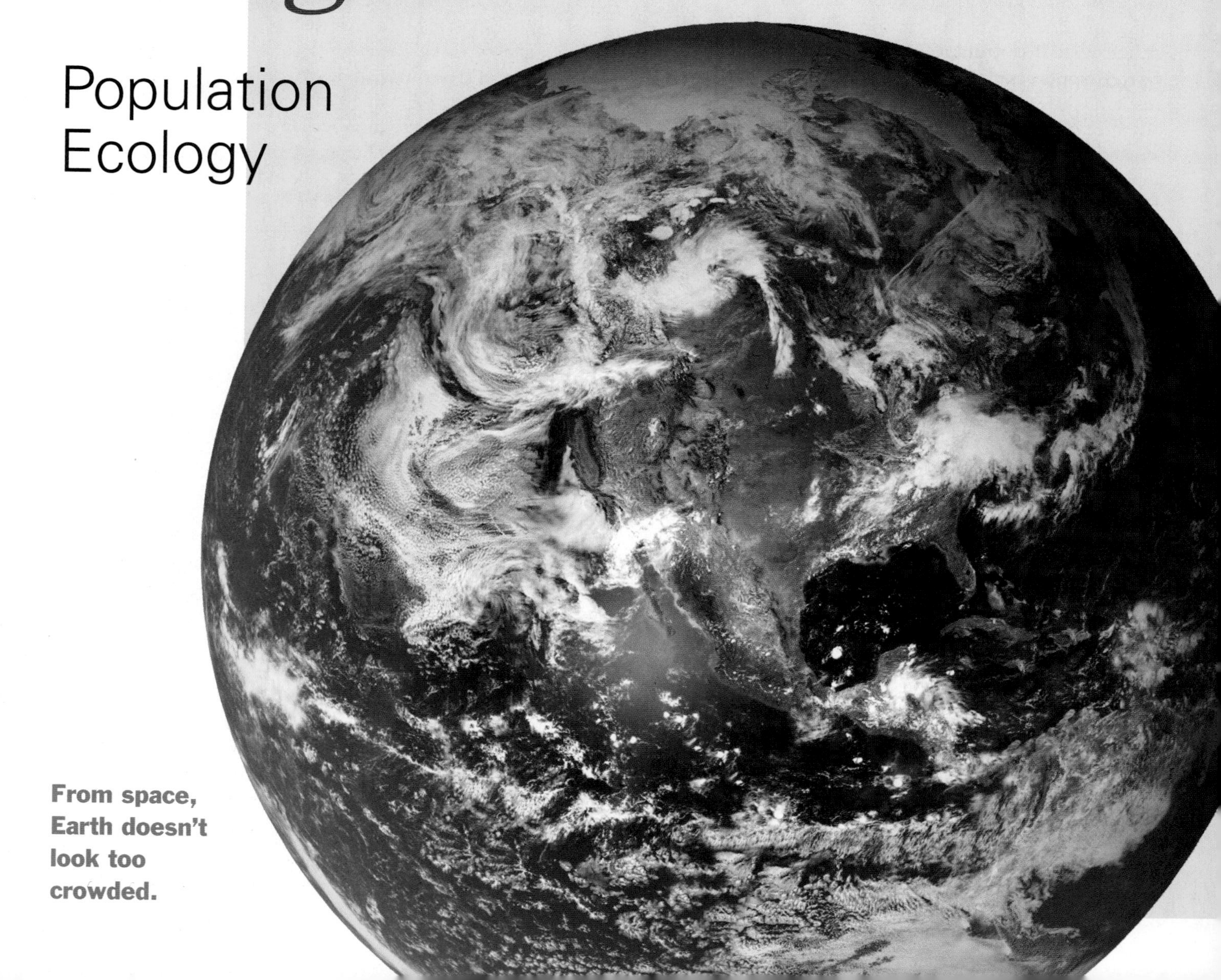

From space, Earth doesn't look too crowded.

But Earth's human population is 6.6 billion—and rising.

In its most recent estimate in 2005, the United Nations (UN) reported that the world's population is approximately 6.6 billion—double the population in 1960. The UN also predicted that the population would continue to grow for several more decades before stabilizing at as high as 10.3 billion by about 2050. As is usually the case, many observers greeted the report as another piece of bad news. While the UN's population projection is lower than past predictions (previous reports forecast a population of over 12 billion by 2050), many scientists and environmentalists wonder if our planet can support the current population for very long, let alone an additional 4 billion people.

Is this Ethiopian child hungry because the planet is overpopulated?

Other commentators, such as the late economist Julian Simon, a former senior fellow at the influential Cato Institute, are skeptical of environmentalists' statements about population growth. They point to predictions made in the best-selling book *The Population Bomb* (1968), in which author Paul Ehrlich forecast worldwide food and water shortages by the year 2000. In fact, most measures of human health have become more upbeat since 1970, including global declines in infant mortality rates, increases in life expectancy, and a 20% increase in per capita income—despite a near doubling in population since the publication of Ehrlich's book. By most measures, the average person is better off today than in 1970. Paul Ehrlich was clearly wrong in 1968; why should we believe his doom-and-gloom predictions about the future now?

Ehrlich and his colleagues counter that while they were wrong about how soon it would happen, there are some indications that the large human population is rapidly reaching a real limit to growth. For example, according to the UN, as of June 2008, thirty-three countries were facing food emergencies, meaning that famine could be imminent. Worldwide, 854 million people—including 150 million children under the age of 5—do not get enough food regularly for a healthy existence. A staggering 55% of the

Or can Earth support everyone at the same level as that of the average North American family?

nearly 12 million deaths each year among children under 5 in the developing world are associated with inadequate nutrition. Despite years of international attention and billions of dollars spent to address this problem, the situation has not improved dramatically—there are only 10% fewer children suffering from malnutrition today than there were in 1980. And rapidly rising food prices around the globe pose a serious threat to future human health and survival.

So what is the truth? Is the human population larger than Earth can support for much longer? Are we headed into a global food crisis and massive famine? Or are we gradually moving toward an era when all people on Earth will be as well-fed, long-lived, and affluent as the average North American?

web Tutorial 13.1
Population Growth

13.1 A Growing Human Population

Ecology is the field of biology that focuses on the interactions among organisms as well as between them and their environment. The relationship between organisms and their environments can be studied at many levels—from the individual, to populations of the same species, to communities of interacting species, and finally to the effects of biological activities on the nonbiological environment, such as the atmosphere. The three chapters in this unit present basic ecological principles obtained from the study of ecology at all the latter three of these levels.

In ecology, a **population** is defined as all of the individuals of a species within a given area. Populations exhibit a structure, which includes the spacing of individuals (that is, their distribution) and their density (abundance). Ecologists seek to explain the distribution and abundance of the individuals within populations to understand the factors that lead to a population's success or failure. The interactions among species described in Chapter 14 make up one set of influences on distribution and abundance, but another set is the dynamics of the population, including the relative numbers of individuals of different sexes and ages and the numbers that are born or die in a given time period.

Population Structure

The first task of a population ecologist is to estimate the size of the population of interest. Certain populations can be counted directly, as in a census of humans in an area or a survey identifying all individuals of a particular tree species in a forest tract.

The size of more mobile or inconspicuous species can be estimated by the **mark-recapture method** *(Figure 13.1)*. In this technique, researchers capture many individuals, mark them in some way (for instance, with an ear tag) and release them back into the environment. At some later time, the researchers capture another group of individuals and calculate the proportion of previously marked individuals in this group. This proportion can be used to estimate the size of the total population.

For example, imagine that a researcher captured, marked, and released 100 beetles. If the researcher returns a week later and finds that 10% of the beetles caught on the second round are marked, the researcher can assume that the 100 beetles originally marked represented 10% of the entire beetle population. According to this mark-recapture survey, the total population is approximately 1000 beetles.

① Researcher captures 100 beetles in a trap, marks each with a dab of paint.

② After one week, a trap is set again, resulting in a captured group of marked and unmarked individuals.

③ Total population is estimated as equivalent to the percentage of marked individuals in the second trap.

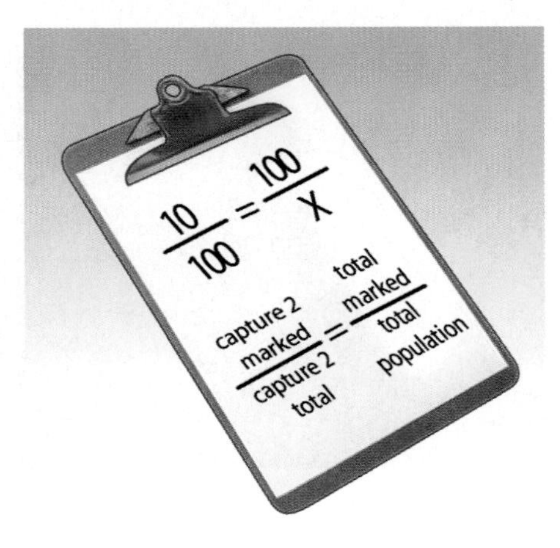

Figure 13.1 **The mark-recapture method.** (a) In the first step, a researcher captures, marks, and releases animals in a wild population. (b) Some time later, the researcher returns and determines what percentage of animals trapped again were previously marked. (c) In the second capture, the ratio of marked to total animals captured is approximately the ratio of the original number trapped to the whole population. **Visualize This:** Animals may become either "trap-happy," that is, attracted to traps, or "trap-shy," after one capture. How might each of these behaviors affect a population estimate?

Stop and Stretch 13.1: Imagine the same scenario described in the previous paragraph, but the researcher found that only 1% of the beetles were marked in the second round of captures. How large is the population likely to be in this case?

Another basic aspect of population structure is dispersion—that is, how organisms are distributed in space. Many species show a **clumped distribution**, with high densities of individuals in certain resource-rich areas and low densities elsewhere. Plants that require certain soil conditions and the animals that depend on these plants tend to be clumped *(Figure 13.2a)*. On a global scale, humans show a clumped distribution, with high densities found around transportation resources such as rivers and coastlines.

The clumped distribution of humans masks a more **uniform distribution** on a local scale; for instance, the spacing between houses in a subdivision or strangers in a classroom tends to equalize the distances among individual property owners or people. Species that show a uniform distribution are often territorial—they defend their own personal space from intruders. We can observe these same strong reactions among certain species of birds at a nesting site *(Figure 13.2b)*.

Nonsocial species with the ability to tolerate a wide range of conditions typically show a **random distribution**, in which no compelling factor is actively bringing individuals together or pushing them apart. The distribution of seedlings of trees with windblown seeds is often random *(Figure 13.2c)*.

A population's distribution and abundance provide a partial snapshot of its current situation. The dispersion of the human population—and recent changes in that pattern—profoundly affects the natural environment, as discussed in Chapter 15. However, to better understand how a population is responding to its environment, we need to determine how it is changing through time.

Population Growth: An Overview

Historians have been able to use archaeological evidence and written records to determine the size of the human population on Earth at various times during the past 10,000 years. This record, presented in *Figure 13.3*, dramatically illustrates the pattern of population growth.

(a) Clumped

(b) Uniform

(c) Random

Figure 13.2 **Patterns of population dispersion.** Individuals in a population may be (a) clumped, like these cattails growing in soil with the correct water content; (b) uniformly distributed, like these nesting penguins; or (c) randomly dispersed, like these seedlings in a forest.

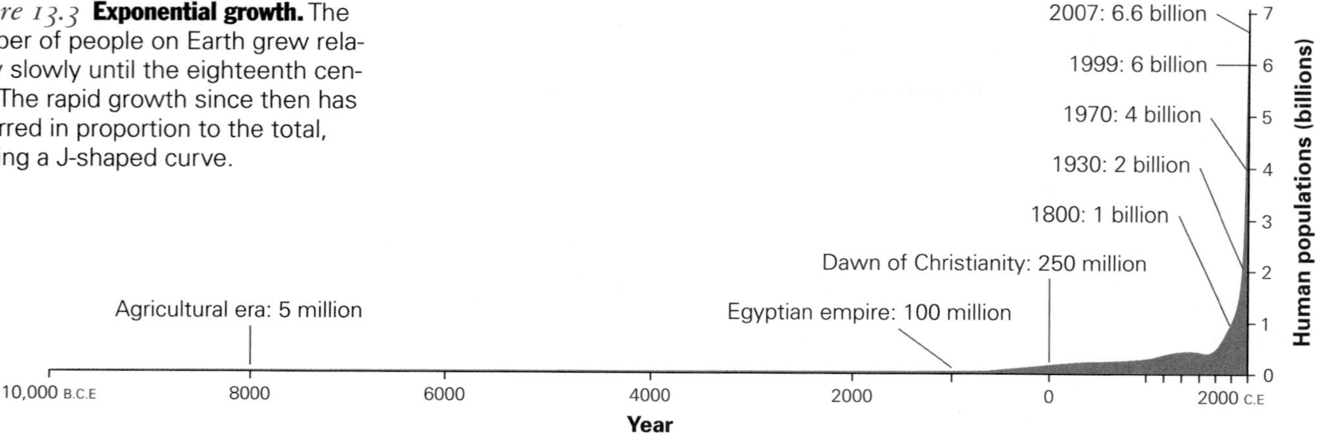

Figure 13.3 **Exponential growth.** The number of people on Earth grew relatively slowly until the eighteenth century. The rapid growth since then has occurred in proportion to the total, causing a J-shaped curve.

The graph of human population growth is a striking illustration of **exponential growth**—growth that occurs in proportion to the current total. In other words, populations growing exponentially do not add a fixed number of offspring every year; instead, the quantity of new offspring is an ever-growing number. Exponential growth results in the J-shaped growth curve seen in Figure 13.3.

A Closer Look: Population Growth

For most of our history, the human population has remained at very low levels. At the beginning of the agricultural era, about 10,000 years ago, there were approximately 5 million humans. There were 100 million people during the Egyptian Empire (7000 years later) and about 250 million at the dawn of the Christian religion in 1 A.D. The population was growing, but at a very slow rate—approximately 0.1% per year.

Beginning around 1750, the rate at which the human population was growing jumped to about 2% per year. The human population reached 1 billion in 1800, had doubled to 2 billion by 1930, and then doubled again to 4 billion by 1970. Although the current growth rate is slower, about 1.2% per year, the rapid increase in population looks quite dramatic on a graph of human population over time.

(A Closer Look continued)

The larger a population is, the more rapidly it grows because an increase in numbers depends on individuals reproducing in the population. So, while a growth rate of 1.2% per year may seem rather small, the number of individuals added to the 6.6-billion-strong human population every year at this rate of growth is a mind-boggling 77 million (approximately the entire population of Germany). Put another way, three people are added to the world population per second, and about a quarter of a million people are added every day.

What has fueled this enormous increase in human population? The annual **growth rate** of a population is the percent change in population size over a single year. Growth rate, which is mathematically represented as r, is a function of the birth rate of the population (the number of births averaged over the population as a whole) minus the death rate (the number of deaths averaged over the population as a whole). For example, 22 babies are born per year, on average, in a group of 1000 people—that is, the birth rate for the population is 2.2%:

$$\frac{22}{1000} = 0.022 = 2.2\%$$

In addition, each year 10 individuals die out of every 1000 people, resulting in a death rate of 1.0%:

$$\frac{10}{1000} = 0.01 = 1\%$$

This results in the current growth rate of 1.2%:

$$\text{Growth rate} = \text{Birth rate} - \text{Death rate}$$

$$1.2\% = 2.2\% - 1.0\%$$

Today's relatively high growth rate, compared to the historical average of 0.1%, is the result of a large difference between birth and death rates.

It can be easier to think of exponential growth in terms of the amount of time it takes for a population to double in size. At a growth rate of 0.1, it takes 693 years for the population to double. At a rate of 1.2%, it takes only about 58 years *(Figure 13.4)*.

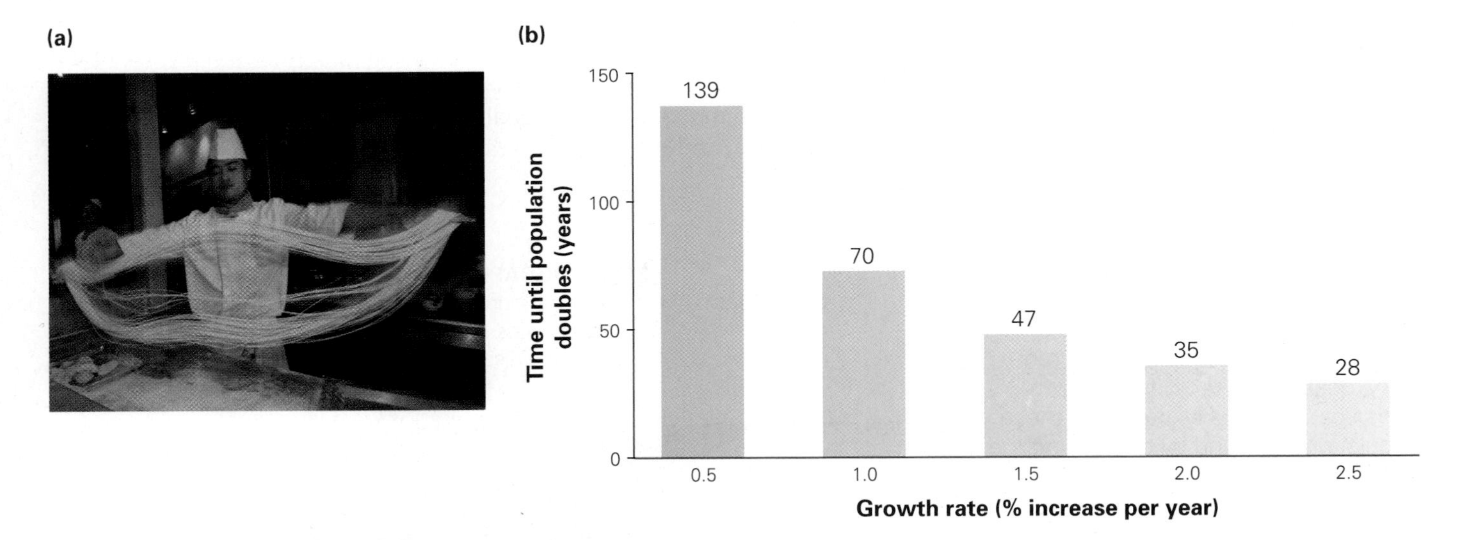

Figure 13.4 **Growth rate and doubling time.** (a) Hand-pulled noodles are made by folding and pulling dough repeatedly, doubling the number of strands in every fold. A single thick noodle folded only 12 times will produce 4096 fine noodles, illustrating the rapidity of growth when it is exponential. (b) The time it takes a population to double in size depends on its growth rate. A population growing at 2% per year doubles in 35 years, while one growing at 1% takes twice as long to double.

The Demographic Transition

In human populations, the tendency has been for decreases in death rate to be followed by decreases in birth rate. The speed of this adjustment helps to determine population growth in the future.

Before the Industrial Revolution, both birth and death rates were high in most human populations. Women gave birth to many children, but relatively few children lived to reach adulthood. Rapid population growth was triggered in the eighteenth century by a dramatic decline in infant mortality (the death rate of infants and children) in industrializing countries. In particular, advances in treating and preventing infectious disease has reduced the number of children who die from these illnesses.

Stop and Stretch 13.2: Life expectancy is the average age a newborn is expected to reach. Declines in infant mortality dramatically increase life expectancy in a country, even if adults are not living longer. Explain why this is the case.

Not long after death rates declined in these countries, birth rates followed suit, lowering growth rates again. Scientists who study human population growth refer to the period when birth rates are dropping toward lowered death rates as the **demographic transition** *(Figure 13.5)*. The length of time that a human population remains in the transition has an enormous effect on the size of that population. Countries that pass through the transition swiftly remain small, while those that take longer can become extremely large. Nearly all developed countries (those that have industrial economies and high individual incomes) have already passed through the demographic transition and have low population growth rates.

However, global human population growth rates have remained high because the least-developed countries (countries that are early in the process of industrial development and have low per capita incomes) remain in the demographic transition. In addition, several recent changes have decreased infant mortality even more dramatically. These changes include the use of pesticides to reduce rates of mosquito-borne malaria; immunization programs against cholera, diphtheria, and other fatal diseases; and the widespread availability of antibiotics.

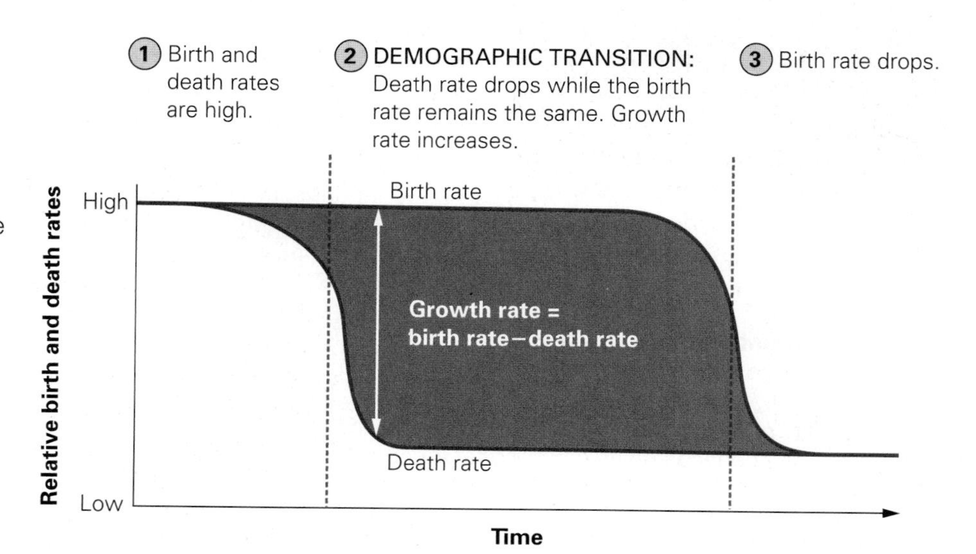

Figure 13.5 **The demographic transition.** As improvements in sanitation and medical care in human populations cause a decrease in infant mortality, death rates drop, and growth rates soar. Eventually, people in these populations respond by decreasing the number of children they have.
Visualize This: Where on this graph is growth rate of the population close to zero?

While birth rates are gradually declining in less-developed countries, they still remain high, contributing to high growth rates. This means that the vast majority of future population growth will occur within populations in the less-developed world, but these countries are also where the vast majority of food crises are occurring. Are the populations in these countries already too large to support themselves? Answering that question requires an understanding of the factors that limit population growth.

13.2 Limits to Population Growth

In their study of nonhuman species, ecologists see clear limits to the size of populations. They can also observe the sometimes-awful fates of individuals in populations that outgrow these limits. For this reason, many professional ecologists are gravely concerned about the rapidly growing human population.

You may know of several instances of nonhuman populations outgrowing their food supplies. The elk population in Yellowstone National Park suffered enormous mortality throughout the 1970s after it grew so large that it degraded its own rangeland. The massive migrations of Norway lemmings that occur every 5 to 7 years and lead to many lemming deaths result from population crowding. While these animals do not commit "mass suicide," as often assumed, the loss of high-quality food in an area as populations increase incites the lemmings to disperse, and they often meet their death in the process. Even yeast in brewing beer grow large populations that eventually use up their food source and die off during the fermenting process. Let us explore what ecology can tell us about the likelihood of human populations suffering the same fate as elk, lemmings, or yeast.

Carrying Capacity and Logistic Growth

The examples of elk in Yellowstone and Norway lemmings illustrate a basic biological principle. While populations have the capacity to grow exponentially, their growth is limited by the resources—food, water, shelter, and space—that individuals need to survive and reproduce. The maximum population that can be supported indefinitely in a given environment is known as the environment's **carrying capacity.**

A simplified graph of population size over time in resource-limited populations is S-shaped *(Figure 13.6)*. This model shows the growth rate of a population declining to zero as it approaches the carrying capacity. In other words, birth rate and death rate become equal, and the population stabilizes at its maximum size, which is mathematically represented as K. Not long after ecologists first predicted this pattern of growth, called logistic growth, populations of organisms as diverse as flour beetles, water fleas, and single-celled protists were shown in laboratory studies to conform to this projected growth curve.

The declining growth rate near a population's carrying capacity is caused by **density-dependent factors,** which are population-limiting factors that increase in intensity as a population increases in size. Density-dependent factors include limited food supplies, increased risk of infectious disease, and an increase in toxic waste levels. Density-dependent factors cause either declines in birth rate or increases in death rate. In organisms such as fruit flies growing in laboratory culture bottles, high populations lead to increased mortality of the flies as food supplies dwindle and wastes accumulate. Water fleas living in crowded aquariums do not have enough food to support egg production, and so birth rates drop. Females of white-tailed deer populations living in crowded natural habitats are less likely to have the energy reserves necessary to carry a pregnancy to term than deer in less-crowded environments.

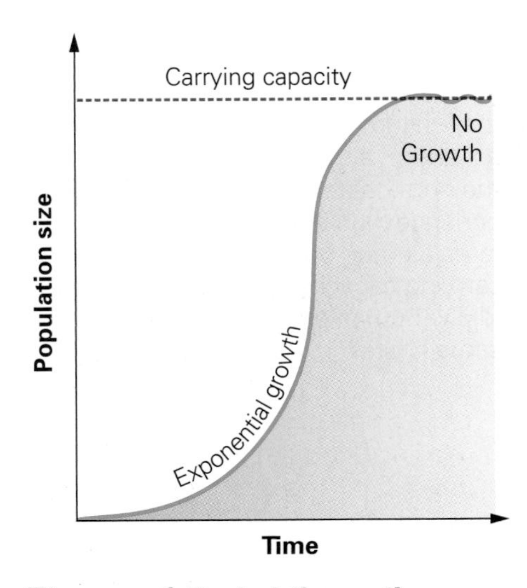

Figure 13.6 **The logistic growth curve.** The S-shaped curve is due to a gradual slowing of the population growth rate as it approaches the carrying capacity of the environment. **Visualize This:** Where on the graph is the number of individuals added to the population *per unit time* highest?

Density-dependent factors can be contrasted with **density-independent factors** that influence population growth rates—for instance, severe droughts that increase the death rate in plant populations regardless of their density or increased temperatures that increase the birth rate in cold-limited insects. However, density-independent factors do not occur in a vacuum; they can have more or less severe effects depending on the size of a population. For example, a density-independent factor such as an unusually cold winter can be deadly to individuals in a white-footed mouse population, but the likelihood of survival is also a function of how much food each individual has stored for the winter. How much food is stored depends on the density of mice competing for food sources during the autumn.

Stop and Stretch 13.3: Why are increases in infectious disease considered a density-dependent factor rather than density independent?

Are density-dependent factors beginning to reduce growth rates in a human population? That is, are humans nearing the carrying capacity of Earth for our population? If we are, will death rates increase as food resources dwindle and more people starve? Or will birth rates decline because fewer women will have enough food to support themselves and a developing baby *(Figure 13.7)*?

Earth's Carrying Capacity for Humans

One way to determine if the human population is reaching Earth's carrying capacity is to examine whether, and how rapidly, the growth rate is declining. As we saw in Figure 13.6, the S-shaped curve of population size over time results from a gradually declining growth rate as the population approaches carrying capacity.

Human population growth rates were at their highest in the early 1960s, about 2.1% per year, but they have since declined to the current rate of 1.2%.

(a) Fruit flies

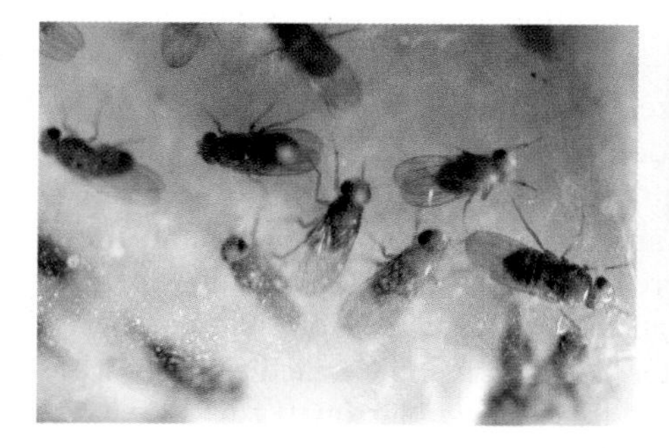

(b) Water fleas

(c) White-tailed deer

(d) Humans

Figure 13.7 **Limits to growth.** Populations of (a) fruit flies in a laboratory culture, (b) water fleas in an aquarium, and (c) white-tailed deer in the northeastern United States all experience high death rates or low birth rates as their populations approach the carrying capacity of the environment. (d) Do human populations face these same limits?

This steady decline is an indication that the population, though still currently growing, is nearing a stable number. Uncertainty about the future rate of growth has led the UN to produce differing estimates of this number and how soon population stability will be reached *(Figure 13.8)*. However, the unique characteristics of humanity make it difficult to determine exactly which population size represents Earth's carrying capacity for humans.

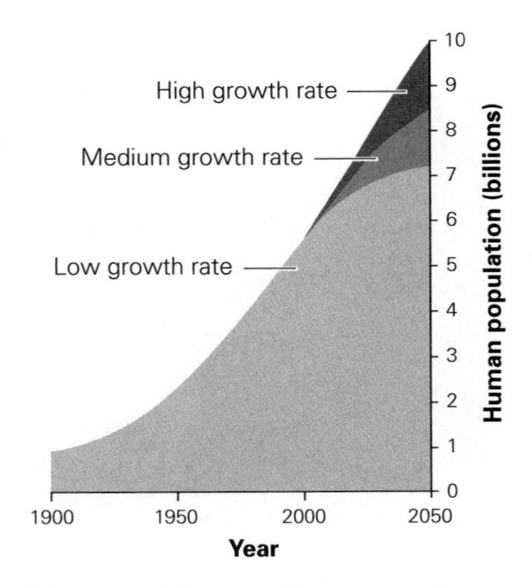

Figure 13.8 **Projected human population growth.** The United Nation's report is based on a number of uncertainties and returns 3 projections: a low-growth scenario of 7.2 billion people; medium growth resulting in 8.6 billion; and a high growth estimate of 10.3 billion.

Signs That the Population Is Not Near Carrying Capacity.

The rates of population increase of fruit flies and water fleas in the laboratory have slowed as these populations neared carrying capacity because their growth rates were forced down by density-dependent factors; lack of resources caused increased death rates or decreased birth rates. However, this is not the case in human populations. Even as the human population has rapidly increased, death rates continue to decline—an indication that people are not running out of food. Growth rates are declining because birth rates are falling faster than death rates. Unlike the water fleas and white-tailed deer, whose females are unable to have offspring when populations are near carrying capacity, birth rates in human populations are falling because women and families, especially those with adequate resources, are *choosing* to have fewer children.

Another way to determine if humans are near Earth's carrying capacity is to estimate the proportion of Earth's resources used by humans now. The amount of food energy available on the planet is referred to as the **net primary productivity (NPP).** NPP is a measure of plant growth, typically over the course of a single year *(Figure 13.9)*.

Several different analyses of the global extent of agriculture, forestry, and animal grazing estimate that humans use roughly one-third of the total land NPP. If we accept these rough estimates, we can approximate that the carrying capacity of Earth is three times the present population, or approximately 20 billion people. This theoretical maximum is the total number of humans that could be supported by all of the photosynthetic production of the planet—leaving no resources for millions of other species. Given the dependence of humans on natural systems (explored in Chapter 14), it is unlikely that our species could survive on a planet where no natural systems

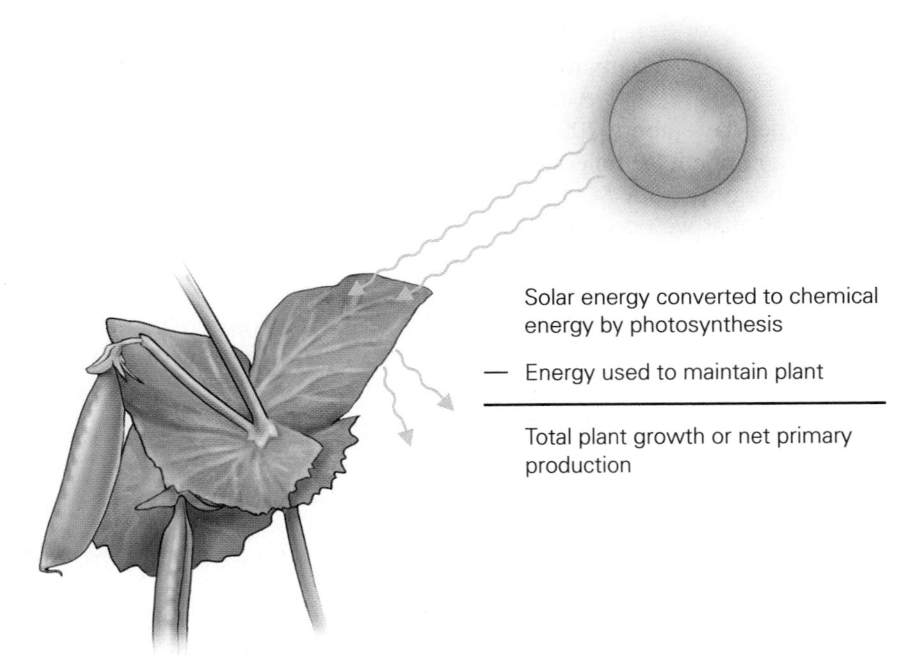

Figure 13.9 **Net primary production.** NPP is a measure of the total calories available on Earth's surface to support consumers.

American family

Malian (African) family

Figure 13.10 **Population plus consumption.** Americans consume more than people in the developing world and therefore have a greater impact on resources.

remained. However, even the largest population projection by the UN, 10.3 billion, falls well short of this theoretical maximum.

Signs That the Population Is Near Carrying Capacity. Ecologists caution that the resources required to sustain a population include more than simply food, and so the carrying capacity deduced from NPP estimates may be much too high. Humans also need a supply of clean water, clean air, and energy for essential tasks such as heating, food production, and food preservation.

The relationship between population size and the supply of these resources is not as straightforward as the relationship between population and food. For instance, every new person added to the population requires an equivalent amount of clean water, but every new person also introduces a certain amount of pollution to the water supply. We cannot simply divide the current supply of clean water by 10.3 billion to determine if enough will be available in the future since increased population leads to increased pollution and therefore less total clean water.

Furthermore, many essential supplies that sustain the current human population are **nonrenewable resources,** meaning that they are a one-time stock and cannot be easily replaced. The most prominent nonrenewable resource is fossil fuel, the buried remains of ancient plants transformed by heat and pressure into coal, oil, and natural gas.

The use of fossil fuel and other nonrenewable resources is a function not only of the number of people but also of average lifestyles, which vary widely around the globe. For example, Americans make up only 5% of Earth's population but are responsible for 24% of global energy consumption. The average American uses as many resources as 2 Japanese or Spaniards, 3 Italians, 6 Mexicans, 13 Chinese, 31 Indians, 128 Bangladeshis, 307 Tanzanians, or 370 Ethiopians *(Figure 13.10)*.

Americans also consume a total of 815 billion food calories per day—about 200 billion calories more than is required or enough to feed an additional 80 million people. Much of modern food production relies on the energy provided by fossil fuel. When these fuels begin to run out, we might find that we need far more of Earth's NPP than we do now to sustain abundant food production. In other words, the actual carrying capacity of our planet may be much lower than our approximations.

The question posed at the beginning of this section remains unanswered; there is no agreement among scientists concerning the carrying capacity of Earth for the human population. Given that uncertainty, what can ecologists tell us about the risks facing the human population that may result from massive, rapid population growth?

13.3 The Future of the Human Population

Unlike nearly all other species, human populations are not simply at the mercy of environmental conditions. With its ability to transform the natural world, human ingenuity has helped populations circumvent seemingly fixed natural limits. However, ingenuity has a dark side in that it can lull people

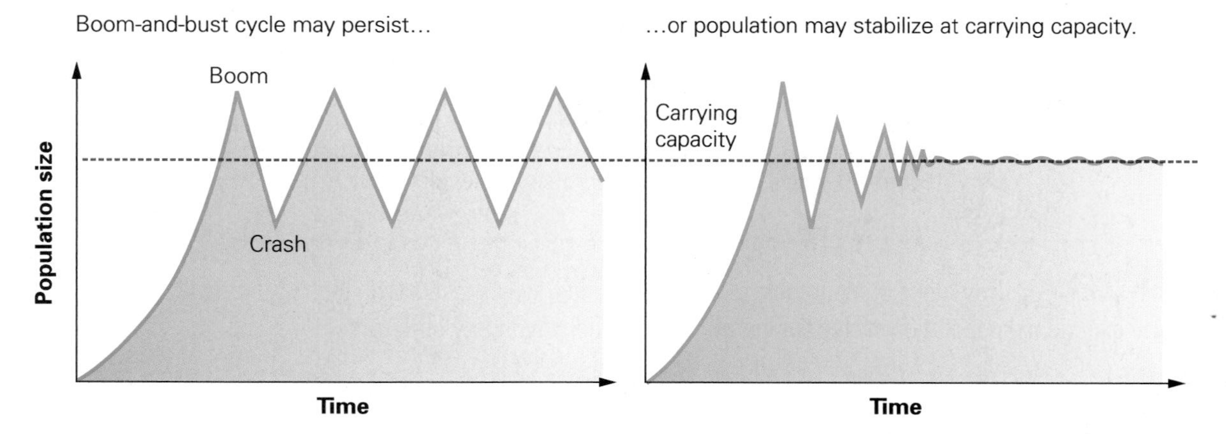

Figure 13.11 **Overshooting and crashing.** These graphs illustrate rapid population growth followed by a population crash. Over time, the population may stay in a "boom-and-bust" cycle, or it may stabilize at its carrying capacity.

into believing that nature has an almost infinite capacity to support their ever-growing needs. Managing the growth of human populations before even the most secure of us face environmental and economic disaster requires an understanding of the risks of continued rapid growth and the strategies that help reduce it.

A Possible Population Crash?

The use of nonrenewable resources creates a risk of the human population overshooting a still-unknown carrying capacity. Ecologists have long known that when populations have high growth rates, they may continue to add new members even as resources dwindle. This causes the population to grow larger than the carrying capacity of the environment. The members of this large population are then competing for far too few resources, and the death rate soars while the birth rate plummets. This results in a **population crash,** a steep decline in number *(Figure 13.11)*.

For instance, in some species of water flea, healthy offspring continue to be born for several days after the food supply becomes inadequate because females can use their fat stores to produce additional young. The size of the population continues to rise even when there is no food left for grazing; however, when the young water fleas run out of stored fat, most individuals die. For many species with high birth rates, rapid growth followed by dramatic crashes produce a **population cycle** of repeated "booms" and "busts" in number.

A population overshoot and subsequent crash affected the human population on the Pacific island of Rapa Nui (also known as Easter Island) during the eighteenth century *(Figure 13.12)*. This 150-square-mile island is separated from other landmasses by thousands of miles of ocean; therefore, its people were limited to using only the resources on or near their island. Archaeological evidence suggests that at one time, the human population on Rapa Nui was at least 7000—apparently a number far greater than the carrying capacity of the island. By 1775, the subsequent overuse and loss of Rapa Nui's formerly lush palm forest had resulted in a rapid decline to fewer than 700 people, a population likely much lower than the initial carrying capacity of the island.

Figure 13.12 **The crash of a human population.** On Rapa Nui, also known as Easter Island, the human inhabitants created these large statues. Soon after completely deforesting this small island, the large population suffered a severe crash.

Biological populations may also overshoot carrying capacity when there is a time lag between when the population approaches carrying capacity and when it actually responds to that environmental limit. Scientists who study human populations note a lag between the time when humans reduce birth rates and when population growth actually begins to slow. They call this lag

demographic momentum. The momentum occurs because while parents may be reducing their family size, their children will begin having children before the parents die, causing the population to continue growing. Even when families have an average of two children, just enough to replace the parents, demographic momentum causes the human population to grow for another 60 to 70 years before reaching a stable level.

Stop and Stretch 13.4: How does demographic momentum explain why species with high birth rates are more likely to experience population cycles compared to populations with low birth rates?

The demographic momentum of a human population can be estimated by looking at its **population pyramid,** a summary of the numbers and proportions of individuals of each sex and each age group. As *Figure 13.13* illustrates, the potential for high levels of demographic momentum occurs when the age structure most closely resembles a true pyramid, with a large proportion of young people. As this large group of young people age, they represent a very large group of potential parents, with the capacity to cause the population to swell. In more stable populations, the proportion who are young is not significantly larger than the proportion who are middle aged, and the pyramid looks more like a column. The number of "parents-in-waiting" is the same as the number of current parents, keeping the population stable.

Whether our reliance on stored resources and the potential demographic momentum in human populations will result in an overshoot of Earth's carrying capacity—followed by a severe crash, as on the island of Rapa Nui—remains to be seen. But human ecologists already know what factors help to slow population growth, so a crash may become less likely.

Figure 13.13 **Demographic momentum.** In a rapidly growing human population like that of (a) South Africa in 2000, most of the population is young, and the population will continue to grow as these children reach child-bearing age. In a slower-growing or stable human population, the ages are more evenly distributed, as in (b) the United States in 2000. **Visualize This:** How do the overall population sizes of the United States and South Africa compare?

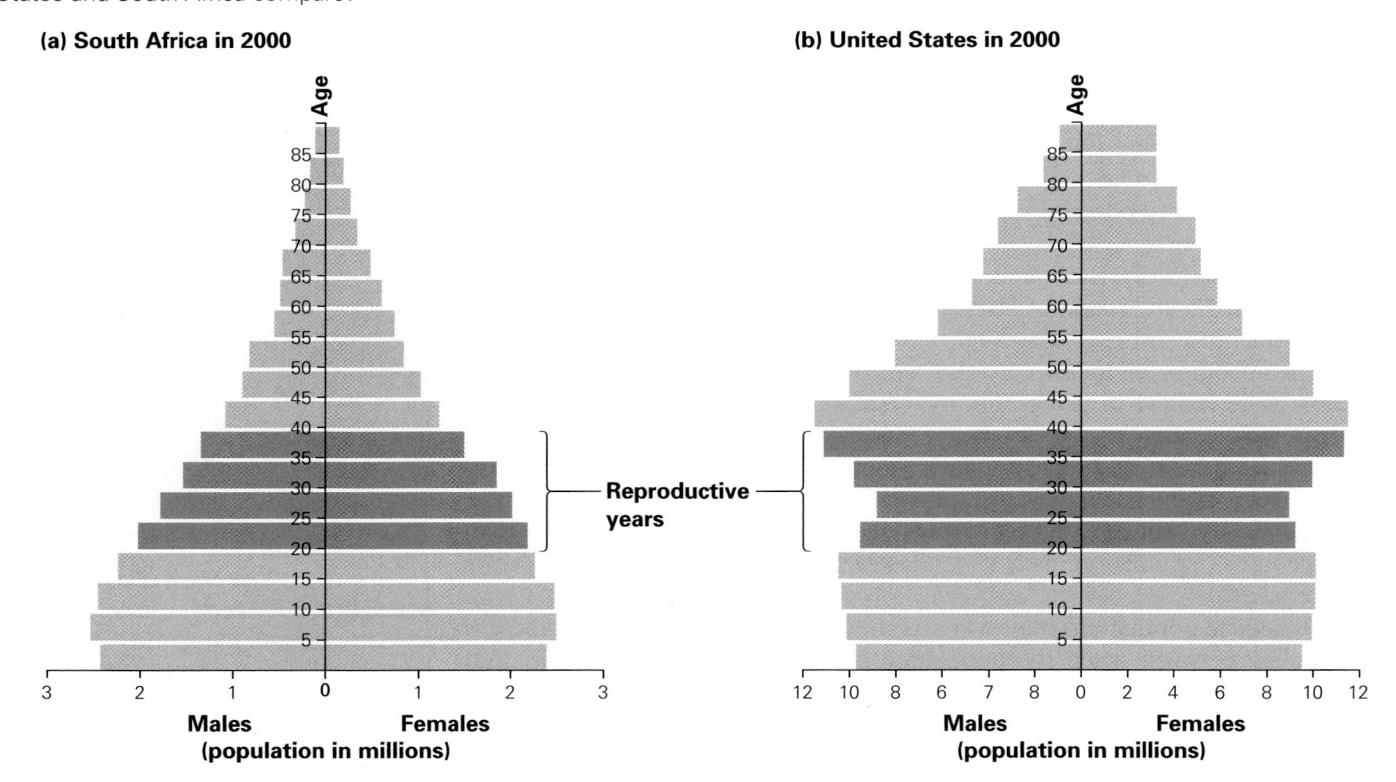

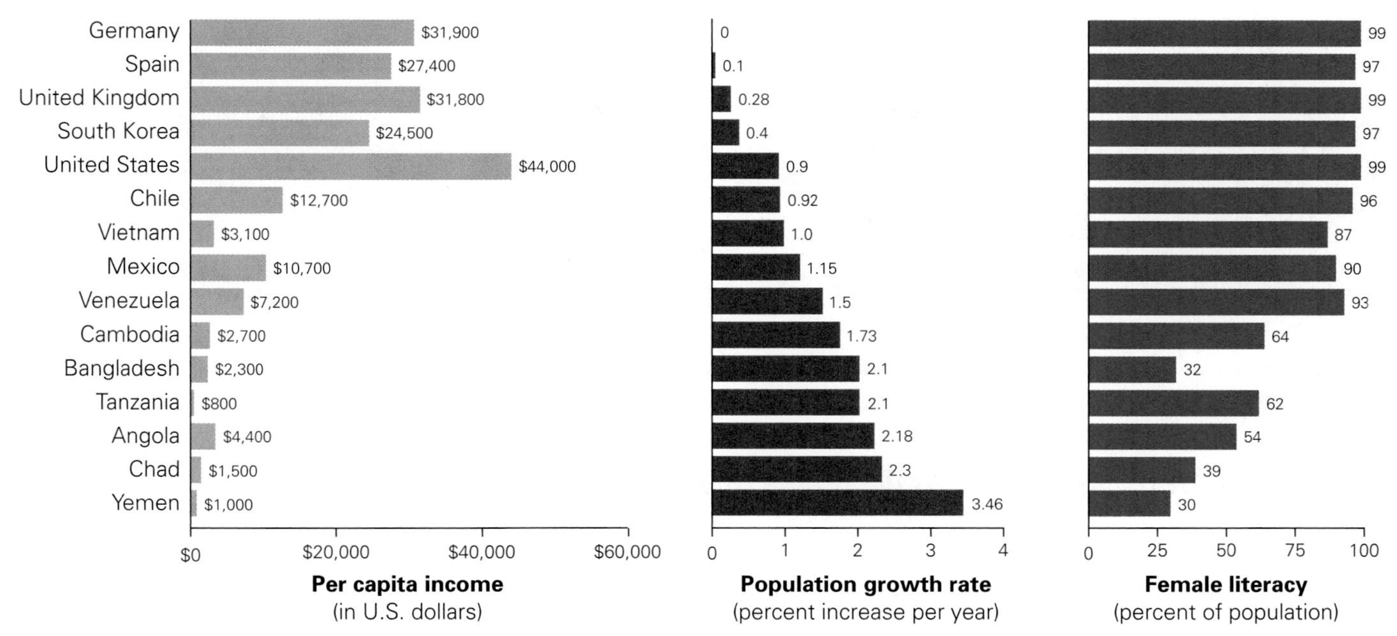

Figure 13.14 **Income, growth rate, and women's literacy.** These 3 graphs illustrate the relationships among income, population growth, and female literacy. Note that higher income and literacy are correlated with decreased birth rates and thus decreased population growth in most countries. **Visualize This:** Why do you suppose the United States has a relatively high population growth rate compared to other wealthy countries?

Avoiding Disaster

As discussed earlier in the chapter, when death rates drop in human population, birth rates eventually follow. Unlike any other species known to science, humans will voluntarily limit the number of babies they produce. When more opportunities become available outside of child rearing, most women delay motherhood and have fewer children. In fact, birth rates are lowest in countries where income is high and women are provided with education *(Figure 13.14)*.

This information provides a clear direction for public policies attempting to decrease population growth rates: improve conditions for women, including increasing access to education, health care, and the job market, and provide them with the information and tools that allow them to regulate their fertility *(Figure 13.15)*.

Slowing growth rates before the human population reaches some environmentally imposed limit has additional benefits. Determining Earth's carrying capacity for humans as simply a function of whether food and water will be available also ignores quality-of-life issues, or what some scientists call cultural carrying capacity. An Earth that was wholly given over to the production of food for the human population would lack wild, undisturbed places and the presence of species that nurture our sense of wonder and discovery. With human populations at the limits of growth, much of our creative energy would be used for survival, taking away our ability to make and enjoy music, art, and literature.

Limiting human population growth also leaves room for nonhuman species. As we discuss in Chapter 14, human activity is posing a direct threat to the survival of a significant percentage of Earth's biodiversity—a threat that increases in direct proportion to the size and affluence of the planet's human population.

What we have learned is that scientists cannot tell us exactly how many people Earth can support, partly because humans make unpredictable choices and partly because humans have the capacity to innovate and adjust seemingly fixed biological limits. Ultimately, the question of how many people Earth should support—and at what quality of life, or including support for nonhuman species—is a question not solely of science but also of values and ethics.

Figure 13.15 **Reducing population growth through opportunity.** The Grameen Bank in Bangladesh provides small loans to help raise people from poverty. Women make up 97% of the borrowers, and successful women borrowers are less likely to have large families.

What's Happening to Birds We Know and Love?

From The National Audubon Society "State of the Birds" Web Site

Audubon's unprecedented analysis of forty years of citizen-science bird population data from our own Christmas Bird Count plus the Breeding Bird Survey reveals the alarming decline of many of our most common and beloved birds.

Since 1967 the average population of the common birds in steepest decline has fallen by 68 percent; some individual species nose-dived as much as 80 percent. All 20 birds on the national Common Birds in Decline list [including northern bobwhite, whip-poor-will, and common grackle] lost at least half their populations in just four decades.

The findings point to serious problems with both local habitats and national environmental trends. Only citizen action can make a difference for the birds and the state of our future.

Which Species? Why?

The wide variety of birds affected is reason for concern. Populations of meadowlarks and other farmland birds are diving because of suburban sprawl, industrial development, and the intensification of farming over the past 50 years.

Greater scaup and other tundra-breeding birds are succumbing to dramatic changes to their breeding habitat as the permafrost melts earlier and more temperate predators move north in a likely response to global warming. Boreal forest birds like the boreal chickadee face deforestation from increased insect outbreaks and fire, as well as excessive logging, drilling, and mining.

The one distinction these common species share is the potential to become uncommon unless we all take action to protect them and their habitat.

1. According to this article, how does the National Audubon Society know that bird species are declining?
2. What information is missing from this article that could influence how you respond to this issue?

CHAPTER Review

For study help, animations, and more quiz questions go to www.mybiology.com.

Summary

13.1 A Growing Human Population

- A population is defined as a group of individuals of the same species living in a fixed area. The structure of a population can be described by the number of individuals and their dispersion (pp. 344–345).
- The human population has grown very rapidly over the last 150 years and exhibits a pattern of exponential growth, which is an increase in numbers as a function of the current population size (pp. 346–347).
- Human population growth is spurred by decreases in death rate caused by decreased infant mortality. In most populations, this decrease is followed by a decrease in birth rates. The gap between when death rates drop and birth rates follow is called the demographic transition (p. 348).

web Tutorial 13.1
Population Growth

13.2 Limits to Population Growth

- Nearly all populations eventually reach the carrying capacity of their environment. Near carrying capacity, density-dependent factors cause an increase in death rate (pp. 349–350).
- The growth rate of the human population is declining, but not as a result of density-dependent factors. Instead, birth rates are dropping because women are choosing to have fewer children (p. 351).
- Rough calculations indicate that the energy received from the sun each year could support a population of 20 billion people, well over the largest population projections (p. 351).
- Humans' reliance on nonrenewable resources may be temporarily inflating the actual carrying capacity of Earth (p. 352).

13.3 The Future of the Human Population

- Fast-growing populations that overshoot their environment's carrying capacity may experience a crash or go through periodic booms and busts (p. 353).
- It is possible that the human population will overshoot Earth's carrying capacity because of our reliance on nonrenewable resources and due to demographic momentum (pp. 353–354).
- Human population growth rates decline when women are empowered to seek an education and may choose to work outside the home (p. 355).

Roots to Remember

The following roots of words come mainly from Latin and Greek and will help you decipher terms:

demo- is from the word meaning people.

logo- means proportion or ratio.

Learning the Basics

1. What factors have led to the explosive increase in the human population over the past 150 years?

2. Explain why a decrease in population growth rate is expected as a nonhuman population approaches carrying capacity.

3. When individuals in a population are evenly spaced throughout their habitat, their dispersion is termed as _______________.

A. clumped; **B.** uniform; **C.** random; **D.** excessive; **E.** exponential

4. The growth of human populations over the past 150 years has increased primarily due to _______________.

A. increases in death rate; **B.** increases in birth rate; **C.** decreases in death rate; **D.** decreases in birth rate; **E.** increases in net primary production

5. According to *Figure 13.16*, the carrying capacity for fruit flies in the environment of the culture bottle is _______________.

A. 0 flies; **B.** 100 flies; **C.** 150 flies; **D.** between 100 and 150 flies; **E.** impossible to determine

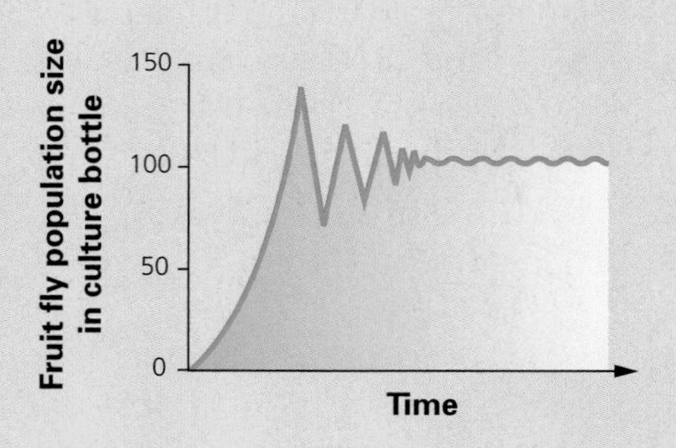

Figure 13.16

6. All of the following are density-dependent factors that can influence population size *except* _______________.

A. weather; **B.** food supply; **C.** waste concentration in the environment; **D.** infectious disease; **E.** supply of suitable habitat for survival

7. In contrast to nonhuman populations, human population growth rates have begun to decline due to _______________.

A. voluntarily increasing death rates; **B.** voluntarily decreasing birth rates; **C.** involuntary increases in death rates; **D.** involuntary decreases in birth rates; **E.** density-dependent factors

8. Populations that rely on stored resources are likely to overshoot the carrying capacity of the environment and consequently experience a _______________.

A. demographic momentum; **B.** cultural carrying capacity; **C.** decrease in death rates; **D.** population crash; **E.** exponential growth

9. The current carrying capacity of Earth for the human population may be inflated by _______________.

A. demographic momentum; **B.** the tendency for women to want to control family size; **C.** an artificially low number of density-independent factors; **D.** our use of fossil fuels; **E.** recent population crashes

10. Demographic momentum refers to the tendency for _______________.

A. low population growth rates to continue to decline; **B.** high population growth rates to continue to increase; **C.** populations to continue to grow in number even when growth rates reach zero; **D.** populations to continue to grow in number even when women are reducing the number of children they bear; **E.** women to continue to have children even though they no longer wish to

Analyzing and Applying the Basics

1. A researcher captures 50 penguins, marks them with a spot of paint on their bills, and releases them. One month later the researcher returns, captures another 50 penguins, and notes that only 1 has a previous mark. What is the likely size of the total penguin population in the researcher's study area?

2. Consider a population in which, for every thousand individuals in the population, 25 children are born each year and 27 individuals die. What is the growth rate of this population? Do you expect that it is near carrying capacity? Why or why not?

3. Review Figure 13.16 above. How would you expect the carrying capacity of the population to change if the flies are supplied with more food? What other factors might influence the carrying capacity in this environment?

Connecting the Science

1. Review your answer to Question 2 in "Analyzing and Applying the Basics." How are the factors that limit fruit fly populations in a culture bottle similar to the factors that limit human populations on Earth? How are they different?

2. Africa is the only continent where increases in food production have not outpaced human population growth. Many of the most severe food crises are in African countries. Should those of us in the more developed world assist African populations? How? What factors influence your thoughts on this question?

CHAPTER 14

Conserving Biodiversity

Community and Ecosystem Ecology

The Lost River sucker faces extinction ...

... but saving the fish has angered these farmers

They came with chainsaws, pry bars, blowtorches, and American flags to challenge the authorities. On their way to the facility, they passed throngs of cheering supporters holding handmade signs and carrying buckets to fill with purloined resources. Then the farmers from normally sleepy Klamath Falls, Oregon, set upon the fences and gates that were preventing them from working and threatening to destroy their way of life. It was July 2001, and events in the Klamath Basin had come to a head.

Who has the right to use Klamath Lake for their survival—farmers or fish?

The wrath of the people of Klamath Falls and surrounding communities was generated by the federal government's legal requirement to protect species that are recognized as in danger of extinction. In the case of the Klamath crisis, the species at risk are fish—the Lost River sucker and the shortnose sucker, which live in Upper Klamath Lake, and the coho salmon, which lives in the Klamath River. In the midst of a multiyear drought and dangerously low water levels in Upper Klamath Lake, the U.S. Fish and Wildlife Service stopped the pumps providing irrigation water from it. The canals that had provided water to barley, potato, and alfalfa fields in the high desert of the Klamath Basin since the early 1900s suddenly went dry. Without irrigation, thousands of farmers were unable to produce crops and faced the prospect of bankruptcy, foreclosure, and loss of their livelihood.

Lost River and shortnose suckers are dull-colored fish that feed on the mucky bottoms of lakes and streams in the region. These fish have not represented a viable economic resource for humans for several decades. In contrast, the crops produced annually by irrigated fields in the region produce millions of dollars in income. Ty Kliewer, a student at Oregon State University whose family farms in the basin, summarized the feelings of many when he told his senator in 2001 that he had learned

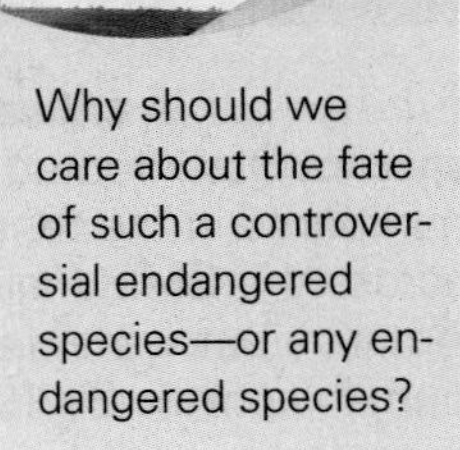

Why should we care about the fate of such a controversial endangered species—or any endangered species?

the importance of balancing mathematical and chemical equations in school. "It appears to me that the people who run the Bureau of Reclamation and the U.S. Fish and Wildlife Service slept through those classes," Kliewer said. "The solution lacks balance, and we've been left out of the equation." The Klamath crisis of 2001 was not unique; thousands of people all over the United States have had their jobs threatened or eliminated by the government's attempts to protect endangered species.

Why should the survival of one or a few species come before the needs of humans? Many biologists and environmentalists say that the Klamath Falls bumper sticker "Fish—or Farmers?" misstates the dilemma. Instead, they argue, humans depend on the web of life that creates and supports natural ecosystems, and they worry that disruptions to this web may become so severe that our own survival as a species will be threatened. In this view, protecting endangered species is not about pitting fish against farmers; it is about protecting fish to ensure the survival of farmers. In this chapter, we explore the causes and consequences of the loss of biological diversity.

web animation 14.1 *Tropical Deforestation and the Species-Area Curve*

web animation 14.2 *Habitat Destruction and Fragmentation*

14.1 The Sixth Extinction

The government agencies that stopped water delivery to the Klamath Basin farmers were acting under the authority of the **Endangered Species Act (ESA),** a law passed in 1973 with the purpose of protecting and encouraging the population growth of threatened and endangered species. Lost River and shortnose suckers were once among the most abundant fish in Upper Klamath Lake—at one time, they were harvested and canned for human consumption. Now, with populations of fewer than 500 and minimal reproduction, these fish are in danger of **extinction,** defined as the complete loss of a species. Critically imperiled species such as the Lost River and shortnose suckers are exactly the type of organisms that legislators had in mind when they enacted the ESA.

The ESA was passed because of the public's concern about the continuing erosion of **biodiversity,** the entire variety of living organisms. Gray wolves, passenger pigeons, black-footed ferrets *(Figure 14.1)*, and spotted skunks—once abundant species—are extinct or highly threatened in the United States as a result of human activity. The ESA was a response to the unprecedented and rapid rate of species loss at our hands.

Figure 14.1 **A critically endangered species.** Black-footed ferrets once numbered in the tens of thousands across the Great Plains of North America but were decimated by human activity. By 1986, they were completely gone from the wild, and only 18 remained alive in captivity.

Critics of the ESA argue that the goal of saving all species from extinction is unrealistic. After all, extinction is a natural process—the approximately 10 million species living today constitute less than 1% of the species that have ever existed. In the next section, we explore the scientific questions posed by ESA critics: How does the rate of extinction today compare to the rates in the past? Is the ESA just attempting to postpone the inevitable, natural process of extinction?

Measuring Extinction Rates

If ESA critics are correct in stating that the current rate of species extinctions is "natural," then the extinction rate today should be roughly equal to the rate in previous eras. The rate of extinction in the past can be estimated by examining the fossil record.

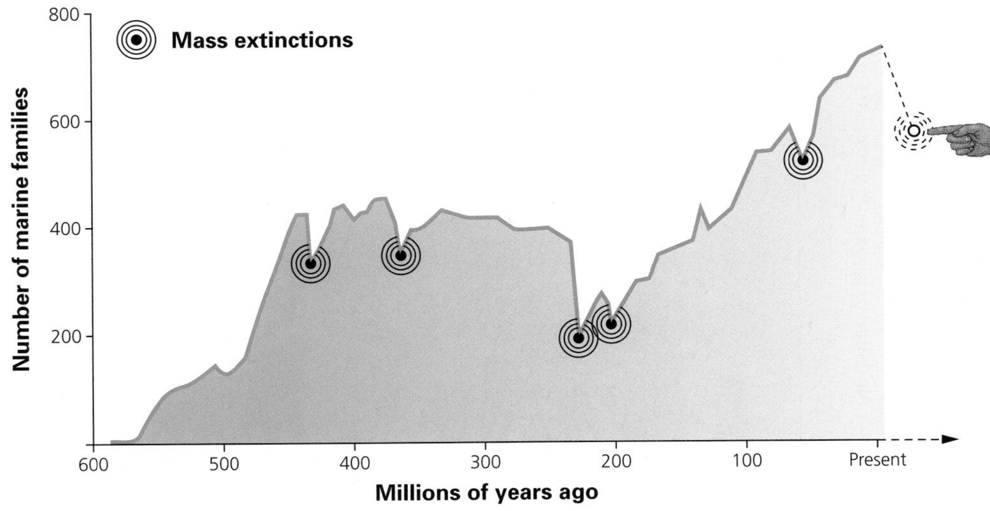

Figure 14.2 **Mass extinction.** This history of life on Earth is reflected by this graph plotting the change in the number of families of fossil marine organisms over time. The number of families has increased over the past 600 million years. However, this rise has been punctuated by five mass extinctions (marked here with black circles), each occurring during a global decline in biodiversity. **Visualize This:** According to the graph, about how long does it take after a mass extinction for diversity to return to preextinction levels?

Figure 14.2 illustrates what the fossil record tells scientists about the history of biodiversity. Since the rapid evolution of a wide variety of animal groups approximately 580 million years ago, the number of families of organisms has generally increased. However, this increase in biodiversity has not been smooth or steady.

The history of life has been punctuated by five **mass extinctions**—species losses that are global in scale, affect large numbers of species, and are dramatic in impact. Past mass extinctions were probably caused by massive global changes—for instance, climate fluctuations that change sea levels, continental drift that changed ocean and land forms, or asteroid impact causing widespread destruction and climate change. Many scientists argue that we are now seeing biodiversity's sixth mass extinction, this one caused by human activity.

Determining whether the current rate of extinction is unusually high requires knowledge of the **background extinction rate,** the rate at which species are lost through the normal evolutionary process. Normal extinctions occur when a species lacks the variability to adapt to environmental change or when a new species arises due to the evolution of the old species. The fossil record can provide clues about the background extinction rate that results from this continual process of species turnover.

The span of rock ages in which fossils of a species are found represents the life span of that species *(Figure 14.3)*. Biologists have thus estimated that the "average" life span of a species is around 1 million years (although there is tremendous variation). We would therefore expect that the background rate of extinction is about 1 species per million (0.0001%) per year. Some scientists have argued that this estimate is too low because it is based on observations of fossils, a record that may be biased toward long-lived species. However, it remains the best-supported approximation of background extinction rates.

The current rate of extinction is calculated from known species disappearances. Calculating current rates is a challenge because extinctions are surprisingly difficult to document. The only way to conclude that a species no longer exists is to exhaustively search all areas where it is likely to have survived. In the absence of a complete search, most conservation organizations

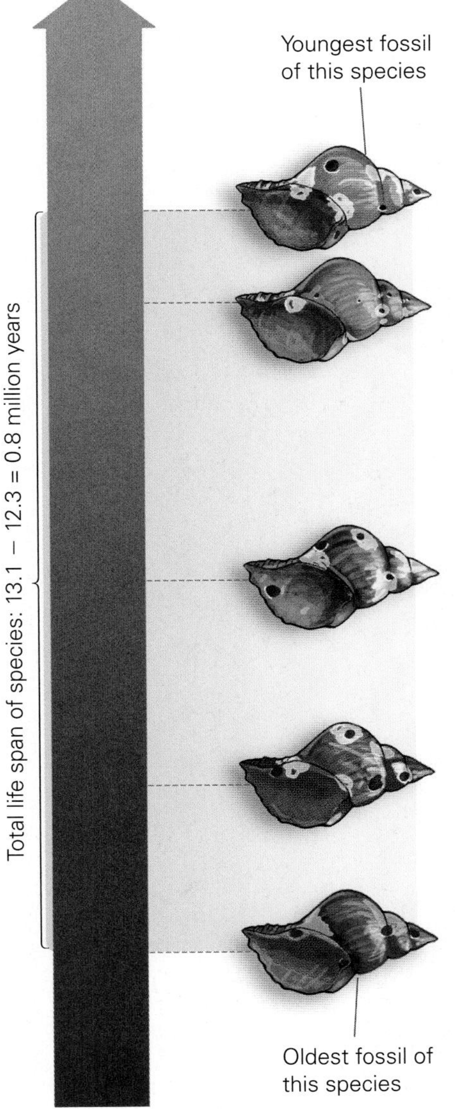

Figure 14.3 **Estimating the life span of a species.** Fossils of the same species are arranged on a timeline from oldest to youngest. The difference in age between the oldest and youngest fossil of a species is an estimate of the species' life span.

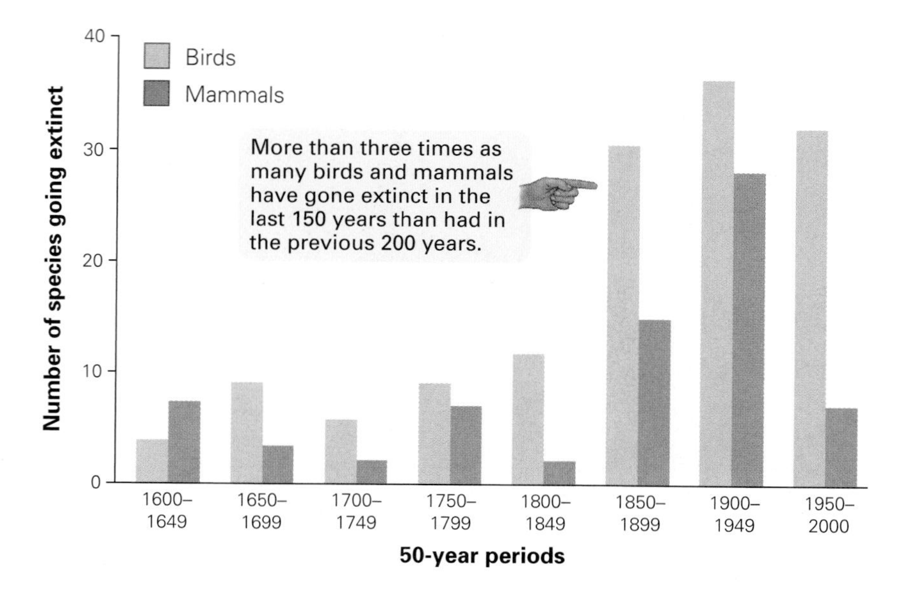

Figure 14.4 **Rate of extinction.** This graph illustrates the number of species of mammals and birds known to have become extinct since 1600.

have adopted this standard: To be considered extinct, no individuals of a species must have been seen in the wild for 50 years.

A few searches for specific species give hints to the recent extinction rate. In Malaysia, a 4-year search for 266 known species of freshwater fish turned up only 122. In Africa's Lake Victoria, 200 of 300 native fish species have not been seen for years. On the Hawaiian island of Oahu, half of 41 native tree snail species have not been found, and in the Tennessee River, 44 of the 68 shallow-water mussel species are missing. Despite these results, few of the missing species in any of these searches is officially considered extinct.

The most complete records of documented extinction occur in groups of highly visible organisms, primarily mammals and birds. Since 1600, of an approximate 4500 identified mammal species 83 have become extinct, while 113 of approximately 9000 known bird species have disappeared. The known extinctions of mammals and birds, spread out over the 400 years of these records, correspond to a rate of 0.005% per year. Compared to the background rate of extinctions calculated from the fossil record, the current rate of extinction is 50 times higher. If we examine the past 400 years more closely, we see that the extinction rate has actually accelerated over time *(Figure 14.4)* to about 0.01% per year, making the current rate 100 times higher than the background rate.

There are reasons to expect that the current elevated rate of extinction will continue into the future. The World Conservation Union (known by its French acronym, IUCN), a highly respected global organization composed of and funded by states, government agencies, and nongovernmental organizations from over 140 countries, collects and coordinates data on threats to biodiversity. According to the IUCN's most recent assessment, 11% of all plants, 12% of all bird species, and 24% of all mammal species are in danger of extinction.

Causes of Extinction

The IUCN attempts to identify all the threats that put a particular species at risk of extinction. For most, some type of human activity is to blame. The most severe threats belong to one of four general categories: loss or degradation of habitat, introduction of nonnative species, overexploitation, and effects of pollution *(Figure 14.5)*. Of these categories, the first poses the most serious threat; the IUCN estimates that 83% of endangered mammals, 89% of endangered birds, and 91% of endangered plants are directly threatened by damage to or destruction of the species' **habitat,** the place where they live and obtain their food, water, shelter, and space.

(a) Habitat destruction

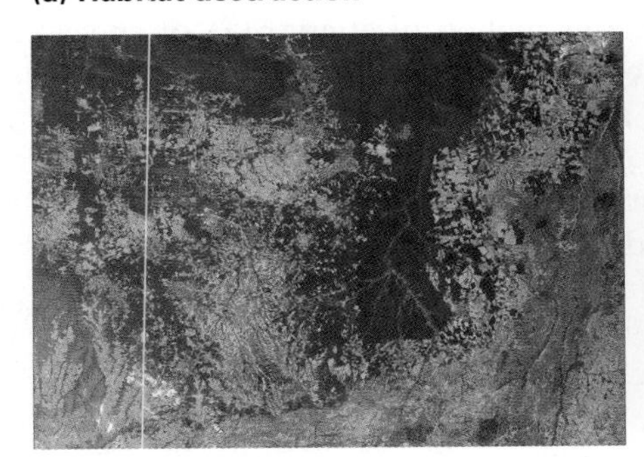

(b) Habitat fragmentation

(c) Introduced species

(d) Overexploitation of species

(e) Pollution

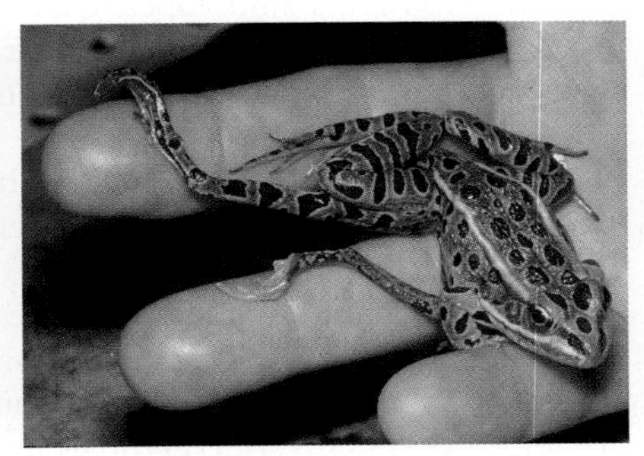

(f) Global warming

Figure 14.5 **The primary causes of extinction.** (a) This 1999 satellite photo illustrates the extent of destruction in an area of Brazilian rain forest that until 30 years ago contained no agricultural lands. The lighter parts of the photo are agricultural fields. The darker regions are intact forests. (b) This "island" of tropical forest was created when the surrounding forest was logged. Scientists have documented hundreds of localized extinctions within fragments such as this. (c) The introduced brown tree snake is responsible for the extinction of dozens of native bird species on the Pacific island of Guam. (d) These tiger skins represent a small fraction of the illegal harvest of tigers in Asia, primarily for the Chinese market. (e) Pollution from herbicides appears to be responsible for the increase of deformities in frogs in the midwestern United States. (f) Polar bears hunt for seals, their primary prey, from sea ice. The extent of sea ice in the Arctic Ocean has been steadily declining over the past 20 years, threatening the bears' survival.

Habitat Destruction. The dramatic reduction in numbers of shortnose and Lost River suckers in Upper Klamath Lake is almost entirely due to habitat destruction. At one time, 350,000 acres of wetlands regulated the overall quality and amount of water entering the lake. Most of these wetlands have been drained and converted to irrigated agricultural fields now. The disruption of natural water flows into and out of the lake has interfered with sucker reproduction and has reduced the number of offspring they produce by as much as 95%.

Habitat destruction is not limited to species in the more developed world (*Figure 14.6*). Rates of habitat destruction caused by agricultural, industrial, and residential development accelerated throughout the twentieth century as the human population swelled from less than 2 billion in 1900 to over 6 billion today. As the amount of natural landscape declines, the number of species supported by the habitats in these landscapes naturally also decreases.

The relationship between the size of a natural area and the number of species that it can support follows a general pattern called a **species-area curve.** A species-area curve for reptiles and amphibians on a West Indies archipelago is illustrated in *Figure 14.7a*. Similar graphs have been generated in studies of different groups of organisms in a variety of habitats. The general pattern in all these graphs is that the number of species in an area increases rapidly as the size of the area increases, but the rate of increase slows as the area becomes very large. This rule of thumb is shown in *Figure 14.7b*. From the graph, we can estimate that a 90% decrease in landscape area will cut the number of species living in the remaining area by half. We can use this curve to estimate rates of extinction in regions that are rapidly being modified by human activity but difficult to survey extensively.

Using images from satellites, scientists have estimated that approximately 20,000 square kilometers (about 7722 square miles, an area the size of Massachusetts) of rain forest are cut each year in South America's Amazon River basin. At this rate of habitat destruction, tropical rain forests will be reduced to

Figure 14.6 **Lost habitat = lost species.** Lemurs are found only on the island of Madagascar, first settled by humans 1500 years ago. Of the 48 species of lemur present on the island 2000 years ago, 16 have become extinct, and 15 are at risk of extinction.

10% of their original size within about 35 years. According to the species-area curve, the habitat loss translates into the extinction of about 50% of species living there. If this prediction proves accurate, the extinct species in the rain forest would include about 50,000 of all known 250,000 species of plants, 1800 species of birds, and 900 species of mammals.

Of course, habitat destruction is not limited to tropical rain forests. Freshwater lakes and streams, grasslands, and temperate forests are also experiencing high levels of modification. According to the IUCN, if habitat destruction around the world continues at its present rate, nearly one-fourth of *all* living species will be lost within the next 50 years.

Some critics have argued that these estimates of future extinction are too high because not all groups of species are as sensitive to habitat area as the curve in Figure 14.7b suggests. Many species may still survive and even thrive in human-modified landscapes. Other biologists contend that there are other threats to species, including habitat fragmentation, and therefore the rate of species loss is likely to be even higher than these estimates.

Habitat Fragmentation. Habitat destruction rarely results in the complete loss of a habitat type. Often what results from human activity is habitat fragmentation, in which large areas of intact natural habitat are subdivided. Habitat fragmentation is especially threatening to large predators, such as grizzly bears and tigers, because of their need for large hunting areas.

Large predators require large, intact hunting areas due to a basic rule of biological systems: Energy flows in one direction within an ecological system along a **food chain,** which typically runs from the sun to **producers**

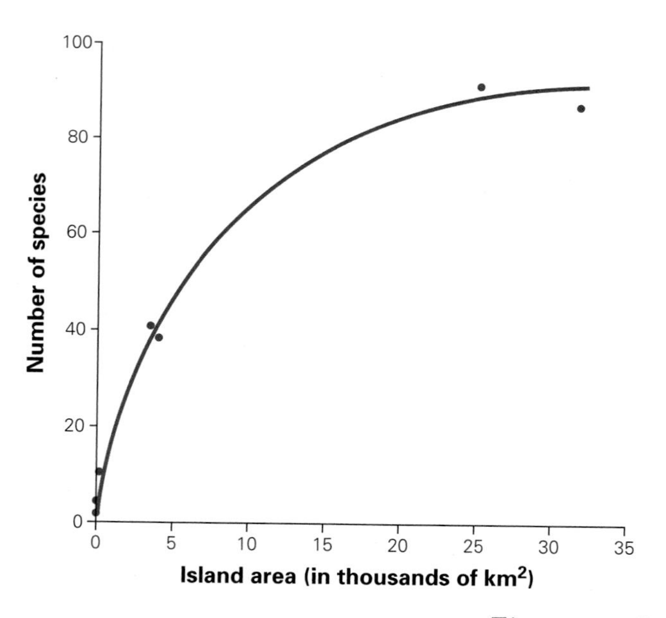

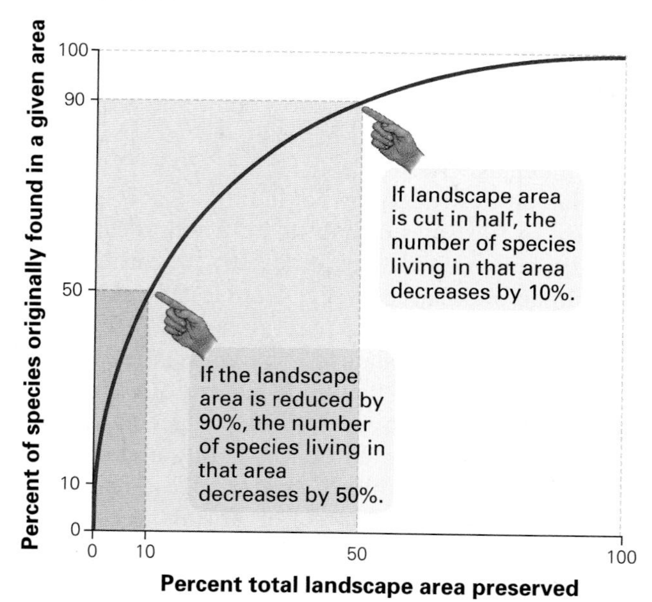

Figure 14.7 **Predicting extinction caused by habitat destruction.** (a) This curve demonstrates the relationship between the size of an island in the West Indies and the number of reptile and amphibian species living there. (b) We use a generalized species-area curve to roughly predict the number of extinctions in an area experiencing habitat loss. **Visualize This:** Using part (a) of this figure, calculate how many species of reptiles and amphibians you would expect to find on an island that is 15,000 square kilometers in area. Imagine that humans colonize this island and dramatically modify 10,000 square kilometers of the natural habitat. Using part (b), calculate the percentage of the species originally found on the island you would expect to become extinct.

(photosynthetic organisms), to the **primary consumers** that feed on them, to **secondary consumers** (predators that feed on the primary consumers), and so on. Along the way, most of the Calories taken in at one **trophic level** (that is, a level of the food chain) are used to support the activities of the individuals at that level. In other words, a substantial amount of the energy individuals take in is dissipated—that is, given off as heat. Each level of the food chain thus has significantly less energy available to it compared to the next lower level. You can see this in your own life; an average adult needs to consume between 1600 and 2400 kilocalories per day simply to maintain his or her current weight.

The flow of energy along a food chain leads to the principle of the **trophic pyramid,** the bottom-heavy relationship between the **biomass** (total weight) of populations at each level of the chain *(Figure 14.8)*. Habitat destruction and fragmentation can cause the lower levels of the pyramid to shrink, depriving the top predators of adequate calories for survival.

Habitat fragmentation also exposes wide-ranging predators to additional dangers. For example, grizzly bears need 200 to 2000 square kilometers of habitat to survive a Canadian winter, but the Canadian wilderness is increasingly crisscrossed by roads built for tree harvesting. Each road represents an increased chance of grizzly-human interaction. These encounters are often deadly—to the bears. For example, between 1970 and 1995, of the 136 grizzlies that died in Canada's national parks, 17 died of natural causes, and 119 were killed by humans.

The species that can remain in small fragments of habitat are also made more susceptible to extinction by isolation. Habitat fragmentation often makes it impossible for individuals to follow changes in food sources or available nesting or growing sites. Isolated populations are also subject to a lack of genetic diversity resulting from inbreeding, as we discuss in detail later in the chapter.

Although habitat destruction and fragmentation are the gravest threats to endangered species, the others are not insignificant. According to the IUCN, human activity unrelated to habitat modification plays a role in about 40% of all cases of endangerment. These activities include problems caused by species that do well in our presence.

Introduced Species. Introduced species include organisms brought by human activity, either accidentally or purposefully, to new environments. Introduced species are often dangerous to native species because they have not evolved with them; for instance, many birds on oceanic islands such as Hawaii and New Zealand are unable to defend themselves from introduced ground hunters. On Hawaii, the Pacific black rat, accidentally introduced from the holds of visiting ships, became very adept at raiding eggs from nests and

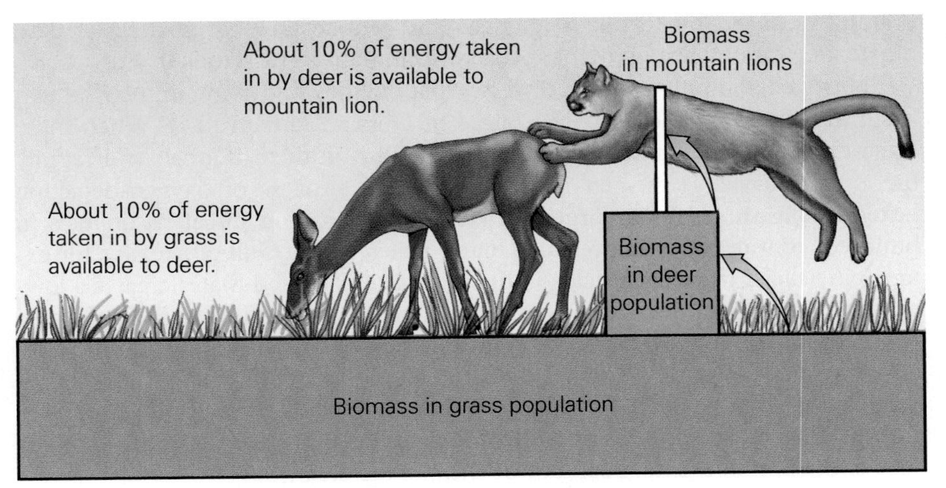

Figure 14.8 **A trophic pyramid.** Because most of the energy consumed by a trophic level is used within that level for maintenance, biomass decreases as position in the food chain increases.

Figure 14.9 **An introduced species.** Kudzu is the common name for the vine *Pueraria lobata,* introduced from Japan to the southeast United States as an ornamental plant and promoted as a solution to soil erosion. In the absence of its predators and parasites, Kudzu grows extremely rapidly, almost 50 centimeters per day, and effectively smothers other plants.

contributed to the extinction of dozens of species of honeycreepers, birds found nowhere else on Earth. Even domestic cats, deliberately introduced by people who sought to control rodents around their houses, can take an enormous toll on wildlife. A study in Wisconsin estimated that free-ranging cats in that state kill approximately 39 million birds every year.

Introduced species may also compete with native species for resources, causing populations of the native species to decline. For example, zebra mussels, accidentally introduced to the Great Lakes through the ballast water in European trading ships, crowd out native mussel species as well as other organisms that filter algae from water, and the introduced vine kudzu, deliberately brought to the United States from Japan, grows over native trees and vines in the Southeast *(Figure 14.9)*.

Humans continue to move species around the planet. As the global trade in agriculture and other goods continues to expand, the number of species introductions is likely to increase over the next century.

Overexploitation. When the rate of human use of a species outpaces its reproduction, the species experiences overexploitation. Overexploitation may occur when particular organisms are highly prized by humans, for example as exotic pets or houseplants *(Figure 14.10)* or for their medicinal value. For instance, 3 of the planet's 8 tiger species have become extinct, and the remaining species are gravely endangered, in part because their bones are believed to cure arthritis and because their genitals are believed to reverse male impotence. As a result, illegal hunting of these animals is very lucrative. Popular wild-grown herbal medicines—including echinacea, the common cold remedy discussed in Chapter 1—are also at risk of overexploitation.

Figure 14.10 **Imperiled oddities.** Dozens of animal and plant species made vulnerable by habitat loss are further endangered by the exotic wildlife trade. It is not just the beautiful that are endangered in this way; this sawfish, a type of ray, is at risk because of high demand for its unusual "nose."

Overexploitation is also likely when an animal competes directly with humans—as in the case of the gray wolf, which was nearly exterminated in the United States by human hunters determined to protect their livestock. Species that cross international boundaries during migration or live in the world's oceans are also susceptible to overexploitation since no single government regulates the total harvest. The near extinction of several whale species in the nineteenth century was the result of the unregulated harvest of these animals by many nations. Stocks of cod, swordfish, and tuna are now similarly threatened.

Pollution. The release of poisons, excess nutrients, and other wastes into the environment—a practice otherwise known as pollution—poses an additional threat to biodiversity. For example, the herbicide atrazine poisons frogs and salamanders in agricultural areas of the United States, and nitrogen pollution caused by fertilizer and car and smokestack exhaust has led to drastic declines of certain plant species within native grasslands in Europe.

Nitrogen and phosphorous fertilizer that pollutes waterways in the Klamath basin has increased the growth of algae in Upper Klamath Lake. When these algae explode in numbers, the bacteria that feed on them flourish and rapidly use up the available oxygen in the water. This process of oxygen-depleting **eutrophication** results in large fish kills. Eutrophication threatens animals in hundreds of waterways all over the United States. In the Gulf of Mexico, for example, fertilizer from farm fields in the midwestern United States creates a low-oxygen "dead zone" the size of New Jersey at the mouth of the Mississippi River.

Perhaps the most serious pollutant released by humans is carbon dioxide, also a principal cause of global climate change (see Chapter 4). Computer models that link predicted changes in climate to known ranges and requirements of over 1000 species of plants indicate that 15% to 37% of these plants face extinction in the next century as the climate changes.

Stop and Stretch 14.1: Many species of amphibians (frogs and toads) around the world are declining, and some are critically endangered. Describe some human activities that might be affecting these animals. What information would help you determine whether these widely dispersed species were put at risk by the same cause?

By nearly any measure, ESA critics who describe modern extinction rates as "natural" are incorrect. Over the past 400 years, humans have caused the extinction of species at a rate that far exceeds past rates. Human activities continue to threaten thousands of additional species around the world. Earth appears to be on the brink of a sixth mass extinction of biodiversity—and the massive global change responsible is human activity.

Many people feel a moral responsibility to minimize the human impact on other species and therefore support conservation. However, there is also a practical, human-centered reason to prevent the sixth extinction from occurring—the loss of biodiversity can hurt us as well.

14.2 The Consequences of Extinction

Concern over the loss of biodiversity is not simply a matter of an ethical interest in nonhuman life. Humans have evolved with and among the variety of species that exist on our planet, and the loss of these species often results in negative consequences for us.

Loss of Resources

The Lost River and shortnose suckers were once numerous enough to support fishing and canning industries on the shores of Upper Klamath Lake. Even before the arrival of European settlers, the native people of the area relied heavily on these fish as a mainstay of their diet. They still rely heavily on the salmon community in the Klamath River, of which endangered coho salmon is one species. The loss of the suckers and coho salmon species represents a tremendous impoverishment of wild food sources. And these are not the only wild species humans rely on directly.

The biological resources that are harvested directly from natural areas include wood for fuel and lumber, shellfish for protein, algae for gelatins, and herbs for medicines. The loss of any of these species affects human populations economically. One estimate places the value of wild species in the United States at $87 billion a year or about 4% of the gross domestic product.

Wild species also provide resources for humans in the form of unique biological chemicals. Chapter 12 detailed a number of valuable drugs, food additives, and industrial products that derive from wild species. One dramatic example is the rosy periwinkle (*Catharanthus roseus*), which evolved on the island of Madagascar, one of the most endangered regions on Earth *(Figure 14.11a)*. Two drugs extracted from this plant, vincristine and vinblastine, have dramatically reduced the death rate from leukemia and Hodgkin's disease. If species go extinct before they are well studied, we will never know which ones might have provided compounds that would improve human lives.

Wild relatives of domesticated plants and animals, such as agricultural crops and cattle, are also important resources for humans. Genes and alleles that have been "bred out" of domesticated species are often still found in their wild relatives. These genetic resources are a reservoir of traits that can be reintroduced into agricultural species through breeding or genetic engineering.

(a) Rosy periwinkle

(b) Teosinte

(c) Boll weevil wasp

Figure 14.11 **Resources from nature.** (a) Anticancer drugs vincristine and vinblastine were first isolated from rosy periwinkle (*Catharanthus roseus*), a species of flower native to Madagascar. (b) Teosinte is the ancestor of modern corn, first cultivated in Central America, and rediscovered in Mexico in 1968. (c) *Catolaccus grandis*, the boll weevil wasp, is a predator of one of the most damaging pests of cotton. *Catolaccus grandis* was discovered preying on boll weevils on wild cotton in southern Mexico.

Agricultural scientists attempting to produce better strains of wheat, rice, and corn look to the wild relatives of these crops for genes conferring pest resistance and improved yields. For example, the Mexican teosinte species *Zea diploperennis (Figure 14.11b)*, discovered in 1977, is an ancestor of modern corn. This species of teosinte is resistant to several viruses that plague cultivated corn; the genes that provide this resistance have been transferred to our domestic plants via hybridization.

By preserving wild relatives of domesticated crops in their natural habitats, scientists can also find resources that reduce pest damage and disease on the domestic crop. For example, the wasp *Catolaccus grandis* consumes boll weevils and is used to control infestations of these pests in cotton fields *(Figure 14.11c)*. *Catolaccus grandis* was discovered in the tropical forest of southern Mexico, where it parasitizes a similar pest in wild cotton populations.

Of course, introducing an insect such as *C. grandis* into a new environment carries risk, even if the introduction is meant to reduce environmental damage. We have already noted many examples of environmental disasters caused by introduced predators—another dramatic example is the cane toad, which was brought to Australia to control beetles in sugarcane crops but now threatens the survival of several native frog species. Often, a less-risky approach to reducing pest damage to crops is to preserve nearby habitats and the ecological interactions that persist there.

Predation, Mutualism, and Competition Derailed

Although humans receive direct benefits from thousands of species, most threatened and endangered species are probably of little or no use to people. Even the Lost River and shortnose suckers, as valuable as they once were to the native people of the Klamath Basin, are not especially missed as a food source; no one has starved simply because these fish have become less common.

In reality, most species are beneficial to humans because they are part of a biological **community,** consisting of all the organisms living together in a particular habitat area. Within a community, each species occupies a particular **ecological niche,** which can be thought of as the role or "job" of the species. The complex linkage among organisms inhabiting different niches in a community is often referred to as a **food web** *(Figure 14.12)*. As with a spider's web, any disruption in one strand of the web of life is felt by other

Figure 14.12 **The web of life.** Species are connected to each other and to their environment in various, complex ways. This drawing shows some of the important relationships among organisms in the Antarctic Ocean. Black arrows represent feeding relationships; for example, penguins eat fish and in turn are eaten by leopard seals.

portions of the web. Some tugs on the web cause only minor changes to the community, while others can cause the entire web to collapse. Most commonly, losses of strands in the web are felt by a small number of associated species. However, some disruptions caused by the loss of seemingly insignificant species have the potential to be felt by humans.

Mutualism: How Bees Feed the World. An interaction between two species that benefit each other is called mutualism. We find examples of mutualism in many environments. Cleaner fish that remove and consume parasites from the bodies of larger fish, fungi called mycorrhizae that increase the mineral absorption of plant roots while consuming the plant's sugars *(Figure 14.13)*, and ants that find homes in the thorns of acacia trees and defend the trees from other insects are all examples of mutualism. The mutualistic interaction between plants and bees is perhaps the most important to us.

Bees occupy a very important ecological niche as the primary pollinators of many species of flowering plants. In other words, they transfer sperm, in the form of pollen grains, from one flower to the female reproductive structures of another flower. The flowering plant benefits from this relationship because insect pollination increases the number of seeds that the plant produces. The bee benefits by collecting excess pollen and nectar to feed itself and its relatives in the hive *(Figure 14.14)*.

Wild bees pollinate at least 80% of all the agricultural crops in the United States, providing a net benefit of about $8 billion. In addition, populations of wild honeybees have a major and direct impact on many more billions of dollars of agricultural production around the globe.

Unfortunately, bees have suffered dramatic declines in recent years. Steady declines in wild and domesticated bee populations occurred from the 1970s through 2005; however, in 2006 and 2007, the declines reached crisis proportions, with 25% of surviving stocks disappearing as a result of "colony collapse disorder." The exact causes of these dramatic declines are not known but are believed to result from an increased level of bee **parasites** (infectious organisms that cause disease or drain energy from their hosts), competition with the invading Africanized honeybees ("killer bees"), and habitat destruction. The extinction of these inconspicuous mutualists of crop plants would be extremely costly to humans.

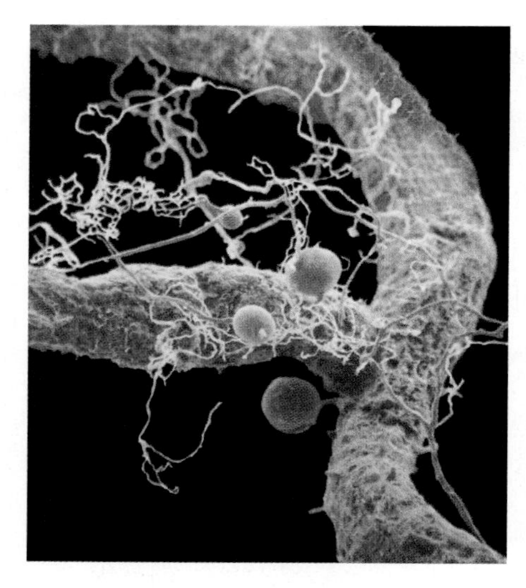

Figure 14.13 **A close partnership.** The relationship between plants and fungi in mycorrhizae is a clear example of mutualism. Here, thin fungal hyphae (colored yellow) surround and penetrate a plant root (colored orange).

Predation: How Songbirds May Save Forests. A species that survives by eating another species is typically referred to as a predator. The word conjures up images of some of the most dramatic animals on Earth: cheetahs, eagles, and killer whales. You might not picture wood warblers, a family of North American bird species characterized by their small size and colorful summer plumage, as predators; however, these beautiful songsters are voracious consumers of insects *(Figure 14.15a)*. The hundreds of millions of individual warblers in the forests of North America collectively remove tons of insects from forest trees and shrubs every summer.

Most of the insects that warblers eat prey on plants. By reducing the number of insects in forests, warblers reduce the damage that insects inflict on forest plants. Reducing the amount of damage likely increases the growth rate of the trees. Harvesting trees for paper and lumber production fuels an industry worth over $200 billion in the United States alone. At least some of this wood was produced because warblers were controlling insects in forests *(Figure 14.15b)*.

Many species of forest warblers appear to be experiencing declines in abundance. The loss of warbler species has several causes, including habitat destruction in their summer habitats in North America and their winter habitats in Central and South America. Warblers also face increased predation by human-associated animals such as raccoons and house cats. Although other, less-vulnerable birds may increase in number when warblers decline, these birds are typically less insect dependent. If smaller warbler populations correspond to lower forest growth rates and higher levels of forest disease, then these tiny, beautiful birds definitely have an important effect on the human economy.

Figure 14.14 **Mutualism.** Honeybees transfer pollen, allowing a plant to "mate" with another plant some distance away.

Benefit to bee:
It obtains food in the form of nectar and excess pollen.

Benefit to flower:
Its sperm (within the pollen) is carried to the female reproductive structures of another flower, enabling cross-pollination.

Competition: How a Deliberately Infected Chicken Could Save a Life. When two species of organisms both require the same resources for life, they will be in competition for the resources within a habitat. In general, competition limits the size of competing populations. To determine whether two species are competing, we remove one from an environment. If the population of the other species increases, then the two species are competitors.

We may imagine lions and hyenas fighting over a freshly killed antelope or weeds growing in our vegetable gardens as typical examples of competition, but most competitive interactions are invisible. The least-visible competition occurs among microorganisms. However, microbial competition is often essential to the health of both people and ecological communities.

Salmonella enteritidis is a leading cause of food-borne illness in the United States. Between 2 million and 4 million people in this country are infected by *S. enteritidis* every year, experiencing fever, intestinal cramps, and diarrhea as a result. In about 10% of cases, the infection results in severe illness requiring hospitalization. Four to six hundred Americans die as a result of *S. enteritidis* infection every year.

Most *S. enteritidis* infections result from consuming undercooked poultry products, especially eggs. The U.S. Centers for Disease Control and Prevention estimate that as many as 1 in 50 consumers are exposed to eggs contaminated with *S. enteritidis* every year. Surprisingly, most of these eggs look perfectly normal and intact. These pathogens contaminate the egg when it forms inside the hen. Thus, the only way to prevent *S. enteritidis* from contaminating eggs is to keep it out of hens.

A common way to control *S. enteritidis* is to feed hens antibiotics—chemicals that kill bacteria. However, like most microbes, *S. enteritidis* strains can evolve drug resistance that makes them more difficult to kill off. Another way to reduce *S. enteritidis* infection in poultry is to make sure another species is occupying its niche.

Most *S. enteritidis* infections originate in an animal's gut. If another bacterial species is already monopolizing the food and available space in a hen's digestive system, then *S. enteritidis* will have trouble colonizing there. Following this principle, some poultry producers now intentionally infect hens' digestive systems with harmless bacteria, via a practice called **competitive exclusion,** to reduce *S. enteritidis* levels in their flocks. This technique involves feeding cultures of benign bacteria to one-day-old birds. When the harmless bacteria become established in the niche in the chicks' intestines, the chicks will be less likely to host large *S. enteritidis* populations *(Figure 14.16)*. There is evidence that this practice is working; *S. enteritidis* infections in chickens have dropped by nearly 50% in the United Kingdom, where competitive exclusion is common practice.

The competitive exclusion of *S. enteritidis* in hens mirrors the role of some human-associated bacteria, such as those that normally live within our intestines and genital tracts. For instance, many women who take antibiotics for a bacterial infection will then develop vaginal yeast infections because the antibiotic kills noninfectious bacteria as well, including species that normally compete with yeast.

Maintaining competitive interactions between larger species can be important for humans as well. For instance, in temporary ponds, the main competitors for the algae food source are mosquitoes, tadpoles, and snails. In the absence of tadpoles and snails, mosquito populations can become quite large—potentially with severe consequences since these insects may carry deadly diseases such as malaria, West Nile virus, and yellow fever. With frogs and their tadpoles increasingly endangered, this risk is a real one.

(a) Black-throated blue warbler, predator of insects

(b) Forests suffer when insects are unchecked by predators.

Figure 14.15 **Predation.** (a) The black-throated blue warbler is one of many warbler species native to North American forests. These birds are active predators of plant-eating insects. (b) Insects can kill trees, as seen in this photo of a spruce budworm infestation. Warblers and other insect-eating birds probably reduce the number and severity of such insect outbreaks.

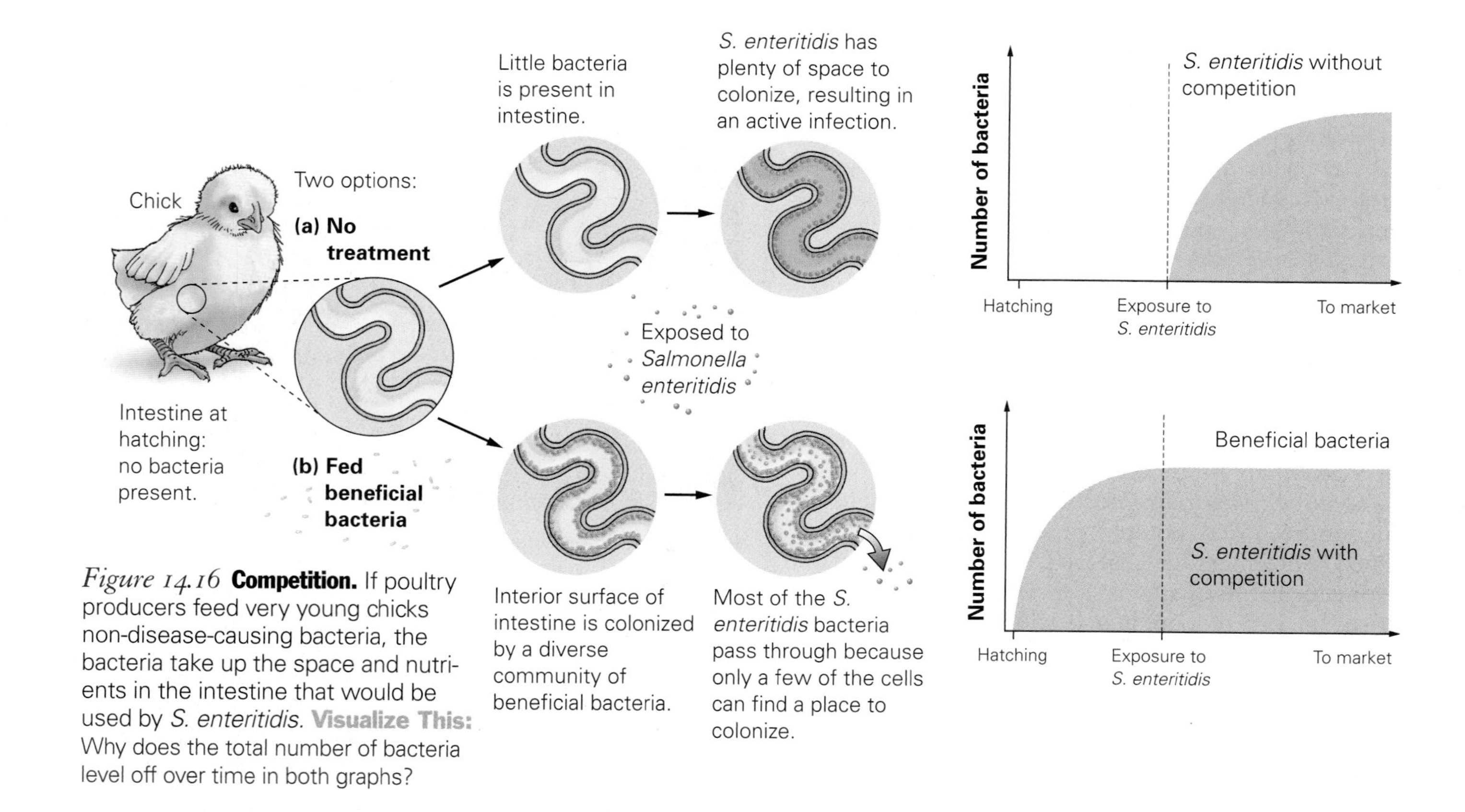

Figure 14.16 **Competition.** If poultry producers feed very young chicks non-disease-causing bacteria, the bacteria take up the space and nutrients in the intestine that would be used by *S. enteritidis.* **Visualize This:** Why does the total number of bacteria level off over time in both graphs?

Stop and Stretch 14.2: Examine the web of relationships among organisms depicted in Figure 14.12. Which of the following species pairs are likely competitors? In each case, describe what the competition involves. *A.* leopard seals, penguins; *B.* baleen whales, squid; *C.* toothed whales, leopard seals; *D.* sperm whale, elephant seals; *E.* fish, krill. How could you test your hypothesis that these animals are in competition with each other?

Keystone Species: How Wolves Feed Beavers. *Table 14.1* summarizes the major types of ecological interactions among organisms. However, this table emphasizes the effects of each interaction on the species directly involved; it does not illustrate that many of these interactions may have multiple indirect effects.

Look again at the food web pictured in Figure 14.12. None of the species in the Antarctic Ocean's biological community is connected to only one other species—they all eat something, and most of them are eaten by something else. You can imagine that penguins, by preying on squid, have a negative effect on elephant seals, which they compete with for these squid, and a more indirect positive effect on other sea birds, which compete with squid for krill. The existence of these **indirect effects** of varying importance has led ecologists to hypothesize that, in at least some communities, the activities of a single species can play a dramatic role in determining the composition of the system's food web. These organisms are called **keystone species** because their role in a community is analogous to the role of a keystone in an archway *(Figure 14.17a)*. Remove the keystone, and an archway collapses; remove the keystone species, and the web of life collapses. Biologists can point to several examples of keystone species, including the population of gray wolves in Yellowstone National Park.

TABLE 14.1 **Types of species interactions and their direct effects.**

Interaction	Example	Effect on Species 1	Effect on Species 2
Mutualism: Association increases the growth or population size of both species.	1. Ants 2. Acacia tree	+ The swollen thorns of the acacia provide shelter for the ants. The acacia leaves provide "protein bodies" that the ants harvest for food.	+ Ants kill herbivorous insects and destroy competing vegetation, benefiting the acacia.
Predation and Parasitism: Consumption of one organism by another.	1. Brown bear 2. Salmon	+ The brown bear catches the salmon and eats it, obtaining nourishment.	– The salmon does not survive.
Competition: Association causes a decrease or limitation in population size of both species.	1. Dandelion 2. Tomato plant	– The dandelion does not grow as well in the presence of the tomato plant. Dandelion produces fewer seeds and fewer offspring.	– The tomato plant does not grow optimally in the presence of the weed. Tomato plant produces fewer flowers and fruit.

Gray wolves were exterminated within Yellowstone National Park by the mid-1920s because of a systematic campaign to rid the American West of this occasional predator of livestock. However, by the 1980s, attitudes about the wolf had changed, leading to renewed interest in returning wolves to their historical homeland. In the mid-1990s, thirty-one wolves originally trapped in Canada were released into Yellowstone National Park. By the end of 2007, this number had grown to at least 1500 wolves, living both in the park and in surrounding public lands, and the animal was no longer considered endangered in the area.

During the time that wolves were extinct in the park, biologists noticed dramatic declines in populations of aspen, cottonwood, and willow trees. They attributed this decline to an increase in predation by elk, especially during winter when grasses become unavailable. However, just a few years after wolf reintroduction, aspen, cottonwood, and willow tree growth has rebounded in some areas of the park. Besides the regions near active wolf dens, these areas include places on the landscape where elk have limited ability to see approaching wolves or to escape. Thus, they will stay away from these areas to avoid wolf predation. Wolves, primarily by changing elk behavior, appear to be important to maintaining large populations of hardwood trees in Yellowstone Park *(Figure 14.17b)*.

(a)

(b)

Figure 14.17 **Keystone species.** (a) The keystone in an archway helps to stabilize and maintain the arch. (b) A keystone species, such as wolves in Yellowstone National Park, helps to stabilize and maintain other species in an ecosystem.

The rebound of aspen, cottonwood, and willow populations in Yellowstone has effects on other species as well. Beaver rely on these trees for food, and their populations appear to be growing in the park after decades of decline. Warblers, insects, and even fish that depend on shelter, food, and shade from these trees are increasing in abundance as well. Wolves in Yellowstone appear to fit the profile of a classic keystone species, one whose removal had numerous and surprising effects on biodiversity. Since biologists cannot usually predict which species, if any, will act as keystones in a community, it is often impossible to know whether the rippling effects of one extinction will change that community forever.

Disrupted Energy and Chemical Flows

As the examples in the previous section illustrate, the extinction of a single species can have sometimes surprising effects on other species in a habitat. What may be even less apparent is how the loss of seemingly insignificant species can change the environmental conditions on which the entire community depends.

Ecologists define an **ecosystem** as all of the organisms in a given area, along with their nonbiological environment. The function of an ecosystem is described in terms of the rate at which energy flows through it and the rate at which nutrients are recycled within it. The loss of some species can dramatically affect both of these ecosystem properties.

Energy Flow. In nearly all ecosystems, the primary energy source is the sun. Producers convert sun energy into biomass during the process of photosynthesis. The chemical energy stored in the biomass is passed through trophic levels in the ecosystem. Biomass is partitioned among the trophic levels, with most of it residing at the bottom of the pyramid (review Figure 14.8 on p. 373). As we move up the pyramid, only a portion of the energy available at one level can be converted into the biomass of the next level. The amount of biomass at the producer level effectively determines how large the population of organisms at the highest level can be.

The amount of sunlight reaching the surface of the Earth and the availability of water at any given location are the major determiners of both trophic pyramid structure and energy flow through it. (Chapter 15 provides a summary of how variance in sunlight and water availability leads to differences in Earth's ecosystem types.) However, the biodiversity found in an ecosystem can also have strong effects on energy flow within it.

Studies in grasslands throughout the world have provided convincing evidence that loss of species can affect energy flow. By comparing experimental prairie "gardens" planted with the same total number of individual plants but with different numbers of species, scientists at the University of Minnesota and elsewhere have discovered that the overall plant biomass tends to be greater in more diverse gardens. This research indicates that a decline in diversity, even without a decline in habitat, may lead to less energy being made available to organisms higher on the food chain, including people who depend on wild-caught food.

Nutrient Cycling. When essential mineral nutrients for plant growth pass through a food web, they are generally not lost from the environment—hence the term nutrient cycling. *Figure 14.18* illustrates the nitrogen nutrient cycle in a natural prairie. Nitrogen is a major component of protein, and abundant protein is essential for the proper growth and functioning of all living organisms nutrient cycling therefore often the nutrient that places an upper limit on production in most ecosystems—more nitrogen generally leads to greater production, while areas with less available nitrogen can support fewer plants (and therefore animals).

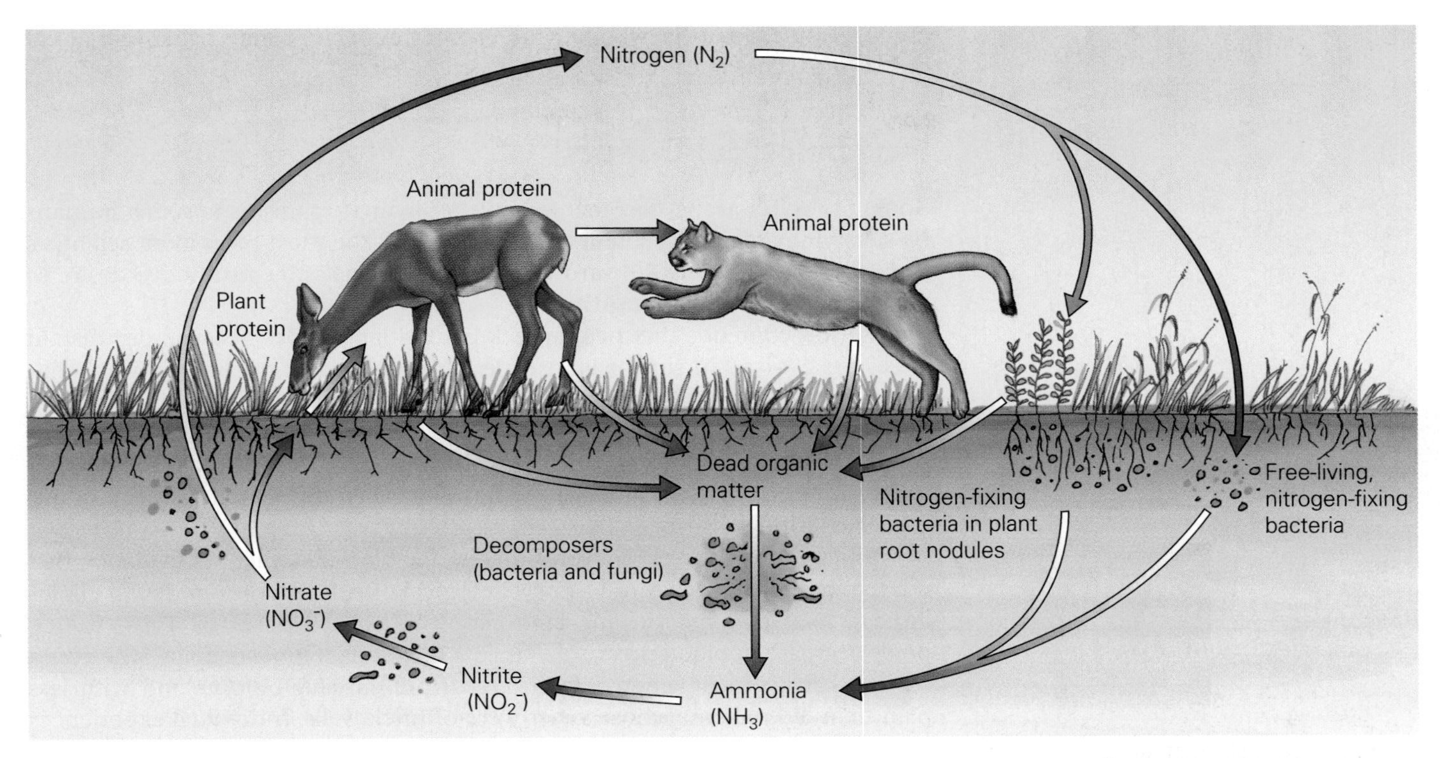

Figure 14.18 **Nutrient cycling.** Nutrients such as nitrogen are recycled in an ecosystem, flowing from soil to producers to consumers and then back into the soil, where complex nutrients are decomposed into simpler forms. **Visualize This:** Consider how people obtain nutrition and what happens to our waste and remains after death. How are nutrient flows in human systems different from natural cycles?

Stop and Stretch 14.3: Carbon is a nutrient that cycles on a global scale. Use Figure 14.18 as a resource to describe how the carbon cycle differs from the nitrogen cycle. How is human activity affecting the carbon cycle?

You should note as you review Figure 14.18 that plants absorb simple molecules from the soil and incorporate them into more complex molecules. These complex molecules move through the food web with relatively minor changes until they return to the soil. Here, complex molecules are broken down into simpler ones by the action of **decomposers,** typically bacteria and fungi.

Changes in the soil community can greatly affect nutrient cycling and thus the survival of certain species in ecosystems. Scientists investigating the effects of introduced earthworms, which have invaded forests throughout the northeastern United States, have observed dramatic reductions in the diversity and abundance of plants on the forest floor *(Figure 14.19)*. The introduced earthworms have apparently caused changes to the community of native soil organisms. As a result of these changes, the nutrient cycle is disrupted and the native plant community has suffered.

The loss of biodiversity can clearly have profound effects on the health of communities and ecosystems on which humans depend. However,

Figure 14.19 **Changes in ecosystem function.** Notice how barren of living vegetation the worm-infested forest floor appears. One reason for this dramatic change may be a disruption in the native nutrient cycle by this introduced species.

controversy exists over whether the current extinction may negatively affect our own psychological well-being.

Psychological Effects

Some scientists argue that the diversity of living organisms sustains humans by satisfying a deep psychological need. One of the most prominent scientists to promote this idea is Edward O. Wilson, who calls this instinctive desire to commune with nature **biophilia.**

Wilson contends that people seek natural landscapes because our distant ancestors evolved in similar landscapes *(Figure 14.20)*. According to this hypothesis, ancient humans who had a genetic predisposition driving them to find diverse natural landscapes were more successful than those without this predisposition, since diverse areas provide a wider variety of food, shelter, and tool resources. Wilson claims that we have inherited this genetic imprint of our preagricultural past.

While there is no evidence of a gene for biophilia, there is evidence that our experience with nature has psychological effects. Studies in dental clinics indicate that patients viewing landscape paintings experience a 10- to 15-point decrease in blood pressure. Patients in a Philadelphia hospital who could see trees from their windows recovered from surgery quicker and with less pain than did patients whose views were of brick walls. Individual experiences with pets and houseplants indicate that many people derive great pleasure from the presence of nonhuman organisms. Although not conclusive, these studies and experiences are intriguing since they suggest that a continued loss of biodiversity could make life in human society less pleasant overall.

The consequences of extinction are not confined to our generation. The fossil record illustrated in Figure 14.2 reveals that it takes 5 to 10 million years to recover the biological diversity lost during a mass extinction. The species that replaced those lost in previous mass extinctions were very different; for instance, after the mass extinction of the reptilian dinosaurs, mammals replaced them as the largest animals on Earth. We cannot predict what biodiversity will look like after another mass extinction. In other words, the mass extinction we may be witnessing today will have consequences felt by people in thousands of generations to come.

Figure 14.20 **Is our appreciation of nature innate?** Humans evolved in a landscape much like this one in East Africa. Some scientists argue that we have an instinctive need to immerse ourselves in the natural world.

14.3 Saving Species

So far in this chapter, we have established the possibility of a modern mass extinction occurring, and we have described the potentially serious costs of this loss of biodiversity to human populations. Since the sixth extinction is largely a result of human activity, reversing the trend of species loss requires political and economic, rather than scientific, decisions. But what can science tell us about how to stop the rapid erosion of biodiversity?

Protecting Habitat

Without knowing exactly which species are closest to extinction and where they are located, the most effective way to prevent loss of species is to preserve as many habitats as possible. The same species-area curve that is used to estimate the future rate of extinction also gives us hope for reducing this number. According to the curve in Figure 14.7b, species diversity declines rather slowly as habitat area declines. Thus, in theory we can lose 50% of a habitat but still retain 90% of its species. This estimate is optimistic because habitat destruction is not the only threat to biodiversity, but the species-area curve tells us that if the rate of habitat destruction is slowed or stopped, extinction rates will slow as well.

Protecting the Greatest Number of Species. Given the growing human population, it is difficult to imagine a complete halt to habitat destruction. However, biologist Norman Myers and his collaborators have concluded that 25 biodiversity "hot spots," making up less than 2% of Earth's surface, contain up to 50% of all mammal, bird, reptile, amphibian, and plant species *(Figure 14.21)*. Hot spots occur in areas of the globe where favorable climate conditions lead to high levels of plant production, such as rain forests, and where geological factors have resulted in the isolation of species groups, allowing them to diversify.

Stopping habitat destruction in biodiversity hot spots could greatly reduce the global extinction rate. By focusing conservation efforts on hot spot areas at the greatest risk, humans can very quickly prevent the loss of a large number of species. Of course, even with habitat protection, many species in these hot spots will likely become extinct anyway for other human-mediated reasons.

Figure 14.21 **Diversity "hot spots."** This map shows the locations of 25 identified biodiversity hot spots around the world. Notice how unevenly these regions of high biodiversity are distributed.

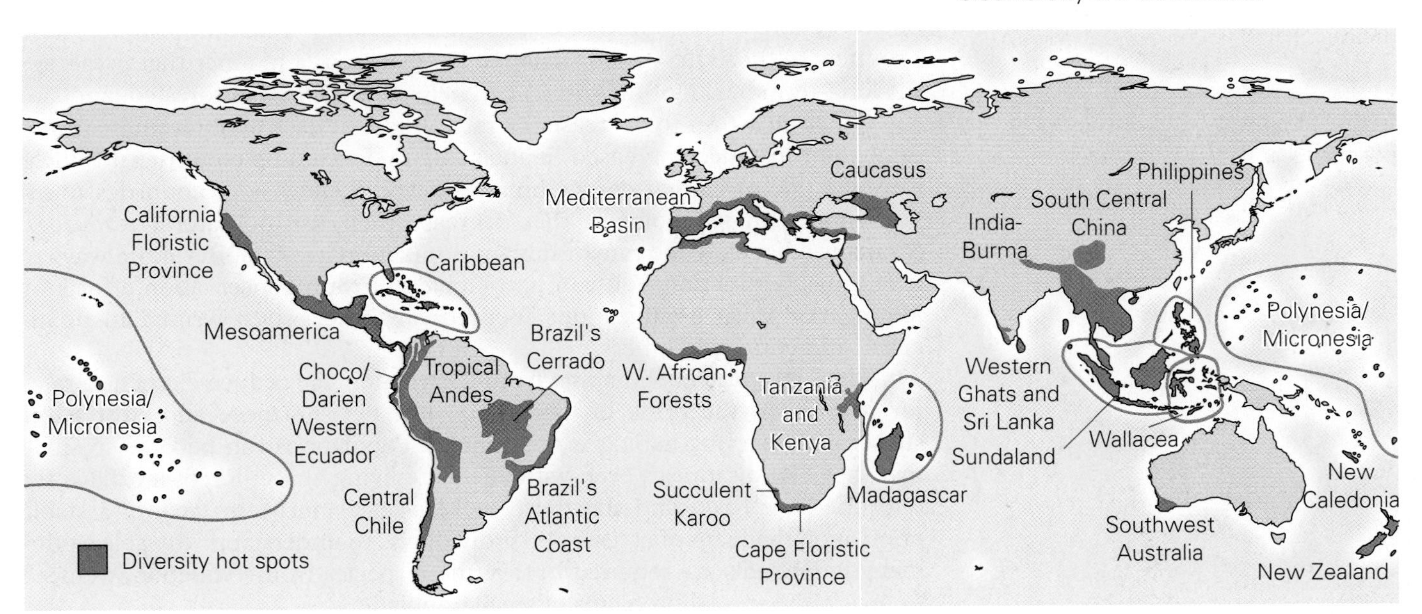

In the long term, we must find ways to preserve biodiversity while including human activity in the landscape. One option is **ecotourism,** which encourages travel to natural areas in ways that conserve the environment and improve the well-being of local people. Some hot spot countries, such as Costa Rica and Kenya, have used ecotourism to preserve natural areas and provide much-needed jobs; other countries have been less successful.

While preserving hot spots may greatly reduce the total number of extinctions, this approach has its critics, who say that by promoting a strategy that focuses intensely on small areas, we risk losing large amounts of biodiversity elsewhere. These critics promote an alternative approach—identifying and protecting a wide range of ecosystem *types*—designed to preserve the greatest diversity of biodiversity rather than just the largest number of species.

Protecting Habitat for Critically Endangered Species. Although preserving a variety of habitats ensures fewer extinctions, already endangered species require a more individualized approach. The ESA requires the U.S. Department of the Interior to designate critical habitats for endangered species, that is, areas in need of protection for the survival of the species. The amount of critical habitat that becomes designated depends on political as well as biological factors.

The biological part of a critical habitat designation includes conducting a study of habitat requirements for the endangered species and setting a population goal for it. The U.S. Department of the Interior's critical habitat designation has to include enough area to support the recovery population. However, federal designation of a critical habitat results in the restriction of human activities that can take place there. The U.S. Department of Interior has the ability to exclude some habitats from protection if there are "sufficient economic benefits" for doing so—a decision that is political in nature.

Decreasing the Rate of Habitat Destruction. Preserving habitat is not simply the job of national governments that designate protected areas or of private conservation organizations that purchase at-risk habitats. All of us can take actions to reduce habitat destruction and stem the rate of species extinction. Conversion of land to agricultural production is a major cause of habitat destruction, and so eating lower on the food chain and reducing your consumption of meat and dairy products from grain-fed animals is one of the most effective actions you can take. Reducing your use of wood and paper products and limiting your consumption of these products to those harvested sustainably (that is, in a manner that preserves the long-term health of the forest) can help slow the loss of forested land.

Other measures to decrease the rate of habitat destruction require group effort. For instance, increased financial aid to developing countries may help slow the rate of habitat destruction. This would allow poor countries to invest money in technologies that decrease their use of natural resources. Strategies that slow the rate of human population growth offer more ways to avoid mass extinction. You can participate in group conservation efforts by joining nonprofit organizations focused on these issues, writing to politicians, and educating others.

Although protecting habitat from destruction can reduce extinction rates for species on the brink of extinction—like the shortnose and Lost River suckers—preserving habitat is not enough. Populations can become so small that they can disappear, even with adequate living space. Recovery plans for both the Lost River and shortnose suckers set a short-term goal of a stable population made up of at least 500 individuals. To understand why at least this many individuals are required to save these species from extinction, we need to review the special problems of small populations.

Protection from Environmental Disasters

The growth rate of an endangered species influences how rapidly that species can attain a target population size. Shortnose and Lost River suckers have relatively high growth rates and will meet their population goals quickly if the environment is ideal *(Figure 14.22a)*. For slower-growing species, such as the California condor *(Figure 14.22b)*, populations may take decades to recover.

The rate of recovery is important because the longer a population remains small, the more it is at risk of experiencing a catastrophe that could eliminate it entirely. The story of the heath hen is a case study on just this point.

The heath hen was a small wild chicken that once ranged from Maine to Virginia. In the eighteenth century, the heath hen population numbered in the hundreds of thousands of individuals. Continued human settlement of the eastern seaboard of the United States resulted in the loss of habitat, causing a rapid and dramatic decline of the birds' population. By the end of the nineteenth century, the only remaining heath hens lived on Martha's Vineyard, a 100-square-mile island off the coast of Cape Cod. Farming and settlement on the island further reduced the habitat for heath hen breeding. By 1907, only 50 heath hens were present on the island.

In response to this precipitous decline, Massachusetts established a 2.5-square-mile reserve for the remaining birds on Martha's Vineyard in 1908. The response initially seemed effective; by 1915, the population had recovered to nearly 2000 individuals. However, beginning in 1916, a series of disasters struck. First, fire destroyed much of the remaining habitat on the island. The following winter was long and cold, and an invasion of starving predatory goshawks further reduced the heath hen population. Finally, a poultry disease introduced by imported domestic turkeys wiped out much of the remaining population. By 1927, only 14 heath hens remained—almost all males. The last surviving member of the species was spotted for the last time on March 11, 1932.

The final causes of heath hen extinction were natural events—fire, harsh weather, predation, and disease. But it was the population's small size that doomed it in the face of these relatively common challenges. A population of 100,000 individuals can weather a disaster that kills 90% of its members but leaves 10,000 survivors, but a population of 1000 individuals will be nearly eliminated by the same circumstances. Even when human-caused losses to the heath hen population were halted, the species' survival was still extremely precarious.

Figure 14.22 **The effect of growth rate on species recovery.** (a) This graph illustrates the rapid growth of a hypothetical population of quickly reproducing Lost River suckers. (b) The slow growth rate of the California condor has made the recovery of this species a long process. Today, nearly 30 years after recovery efforts began, the population of wild condors is still only in the dozens.

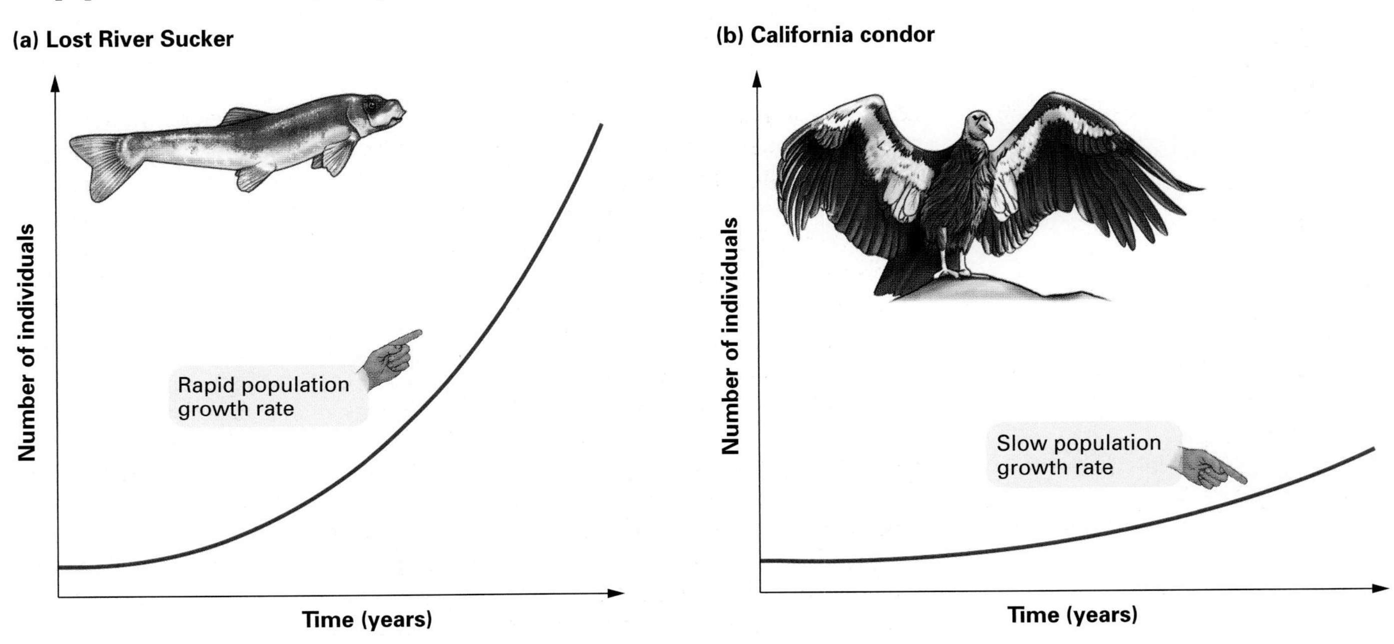

Small populations of endangered species can still be protected from the fate that befell the heath hen. Having additional populations of the species at sites other than Martha's Vineyard would have nearly eliminated the risk that *all* members of the population would be exposed to the same series of environmental disasters. This is the rationale behind placing captive populations of endangered species at several different sites. For instance, captive whooping cranes are located at the U.S. National Biological Service's Patuxent Wildlife Research Center in Maryland, the International Crane Foundation in Wisconsin, the Calgary Zoo in Canada, and the Audubon Center for Endangered Species Research in New Orleans. Even with multiple habitats, if populations of endangered species remain small in number, they are subject to a more subtle but potentially equally devastating situation—the loss of genetic variability.

An Overview of Conservation Genetics

A species' **genetic variability** is the sum of all of the alleles and their distribution within the species. For example, the gene that determines your ABO blood type comes in three different forms. A population containing all three alleles contains more genetic variability than does a population with only two alleles.

The loss of genetic variability in a population is a problem for two reasons. First, on an individual level, low genetic variability leads to low fitness because homozygotes are more likely to express mutant alleles and because they generally have narrower habitat requirements. Second, on a population level, rapid loss of genetic variability may lead to extinction due to the low fitness of individuals and because the population as a whole does not contain variants that will survive in changed environmental conditions, which are inevitable *(Table 14.2)*.

TABLE 14.2 **The costs of low genetic diversity.**

Subject of Risk	Outcomes of Low Diversity	Example
Individuals		
	Inbreeding depression: Deleterious recessive genetic mutations are expressed because low diversity leads to higher numbers of homozygous individuals.	Cheetahs exhibit poor sperm quality and low cub survival, probably because they lack genetic diversity and express more recessive alleles.
	Loss of heterozygote advantage: With only one type of protein produced for a homozygous functional gene, the range of environments in which an individual can be successful may be limited.	In humans, heterozygotes for the sickle-cell allele can have higher fitness than homozygotes for the normal allele because the allele provides protection from malaria.
Populations		
	Extinction vortex: Low population levels resulting in inbreeding depression make the population more vulnerable to the risks of small population size.	The heath hen population may not have grown rapidly enough due to inbreeding depression. As a result, the population could not weather a series of random environmental challenges.
	Inability to respond to environmental change: Population lacks variants that can survive in the face of environmental challenge.	Potato population in 1840s Ireland did not contain any variants that were resistant to potato blight, leading to the loss of nearly the entire crop.

A Closer Look: Conservation Genetics

The Importance of Individual Genetic Variability. As discussed in Chapter 10, fitness refers to an individual's ability to survive and reproduce in a given set of environmental conditions. Low individual genetic variability decreases fitness for two reasons. We can use an analogy to illustrate them.

First, imagine that you could own only two jackets *(Figure 14.23)*. If both are blazers, then you would be well prepared to meet a potential employer. However, if you had to walk across campus to your job interview in a snowstorm, you would be pretty uncomfortable. If you own two parkas, then you will always be protected from the cold, but you would look pretty silly at a dinner party. However, if you own one warm jacket and one blazer, you are ready for freezing weather as well as a nice date. In a way, individuals experience the same advantages when they carry two different functional alleles for a gene. In this case, since each allele codes for a functional protein, a heterozygous individual produces two slightly different proteins for the same function.

If the protein produced by each allele works best in different environments (for instance, if one works best at high temperatures and the other at moderate temperatures), then heterozygous individuals are able to function efficiently over a wider range of conditions than are homozygotes, who only make one version of the protein. The sickle-cell allele in humans (discussed in Chapter 11) is an example of this phenomenon, called *heterozygote advantage*. Individuals who carry one copy of the sickle-cell allele have greater fitness than those with no copies of the allele in areas where malaria is common. Their individual diversity in hemoglobin variants allows them to both effectively transport oxygen to their tissues and provides resistance to malaria infection.

The second reason high individual genetic variability increases fitness is that, in many cases, one allele for a gene is **deleterious**—that is, it produces a protein that is not very functional. In our jacket analogy, a nonfunctional, deleterious allele is equivalent to a badly torn jacket. If you have this jacket and an intact one, at least you have one warm covering *(Figure 14.24)*. In this case, heterozygosity is valuable because a heterozygote still carries one functional copy of the gene. Often these deleterious alleles are recessive, meaning that the activity of the functional allele in a heterozygote masks the fact that a deleterious allele is present (see Chapter 6). An individual who is homozygous (carries two identical copies of a gene) for the deleterious allele will have low fitness—in our analogy, two torn jackets and nothing else. For both of these reasons, when individuals are heterozygous for many genes (or, in our analogy, have two choices for all clothing items), the cumulative effect is greater fitness relative to individuals who are homozygous for many genes.

In a small population, in which mates are more likely to be related to each other simply because there are few mates to choose from, heterozygosity declines. When related individuals mate—known as **inbreeding**—the chance that their offspring will be homozygous for any allele is relatively high. The negative effect of homozygosity on fitness is known as **inbreeding depression.** In cheetahs,

Being heterozygous may confer higher fitness for responding to a changing environment.

Homozygote 1: Relatively low fitness
(only one type of jacket in wardrobe)

Homozygote 2: Relatively low fitness
(only one type of jacket in wardrobe)

Heterozygote: Relatively high fitness
(two types of jackets in wardrobe)

Figure 14.23 **Heterozygotes can inhabit a wider range of environments.** In this analogy, each jacket represents an allele. Just as having two different jackets prepares you for a wider range of situations than having only one type, two different proteins for the same gene may allow for optimal function over a wider range of conditions.

(continued on the next page)

(A Closer Look continued)

Being heterozygous may confer higher fitness by masking deleterious recessive alleles.

Homozygote 1: Relatively high fitness
(two functional jackets in wardrobe)

Homozygote 2: Relatively low fitness
(two nonfunctional jackets in wardrobe)

Heterozygote: Relatively high fitness
(one functional jacket in wardrobe)

Figure 14.24 **Heterozygotes avoid the deleterious effects of recessive mutations.** Again, each jacket represents an allele. If one type of jacket you can receive is nonfunctional, it is better to receive no more than one of them. Heterozygotes are likewise protected against the likelihood of having only nonfunctional alleles, as is present in homozygous recessive individuals. **Visualize This:** Explain why homozygotes for the normal allele also have higher fitness than homozygotes for the recessive allele.

high levels of inbreeding has led to poor sperm quality and low cub survival, both likely due to increased expression of deleterious alleles. The costs of inbreeding are seen in humans as well; the children of first cousins have higher rates of homozygosity and higher mortality rates (thus lower fitness) than children of unrelated parents. In a small population of an endangered species, inbreeding often causes low rates of survival and reproduction and can seriously hamper a species' recovery.

Stop and Stretch 14.4: What strategies could wildlife managers employ to try to reduce the risk of inbreeding in small animal populations?

How Variability Is Lost via Genetic Drift. Small populations lose genetic variability because of **genetic drift**, a change in the frequency of an allele within a population that occurs simply by chance. Although in Chapter 11 we discussed genetic drift as a process for causing evolutionary change, in a small population, genetic drift can have detrimental consequences also.

Imagine two human populations in which the frequency of blood-type allele *A* is 1%; that is, only 1 of every 100 blood-type genes in the population is the *A* form (we use the symbol I^A). In the first population of 20,000 individuals, there are 40,000 total blood group genes (that is, 2 copies per individual). At 1% frequency, this population contains 400 I^A alleles. If a few of the individuals who carry the I^A allele die accidentally before they reproduce, the number of copies of the allele will drop slightly in the next generation—let's say to 385. If the population size in this generation is still 20,000, the new allele frequency is 385 divided by 40,000, or 0.96%.

The change in allele frequency from 1% to 0.96% is the result of genetic drift. However, this 0.04% change in allele frequency is relatively minor in this large population; hundreds of individuals will still carry the I^A allele.

Now imagine the situation in a second, much smaller population. In a population of only 200 individuals and with an I^A frequency at 1%, only 4 copies of the allele are present. If 2 individuals carrying a copy of I^A fail to pass it on, its frequency will drop to 0.5%. Another chance occurrence in the following generation could completely eliminate the 2 remaining I^A alleles from the population. Genetic drift is more severe in small populations and is much more likely to result in the complete loss of alleles (*Figure 14.25*).

In most populations, alleles that are lost through genetic drift have relatively small effects on fitness at the time. However, many alleles that appear to be nearly neutral with respect to fitness in an environment may have positive fitness in another environment. When this is the case, their loss may spell disaster for the entire species.

The Consequences of Low Genetic Variability in a Population. Populations with low levels of genetic variability have an insecure future for two reasons. First, when alleles are lost, the level of inbreeding depression in a population increases, which means lower reproduction and

(A Closer Look continued)

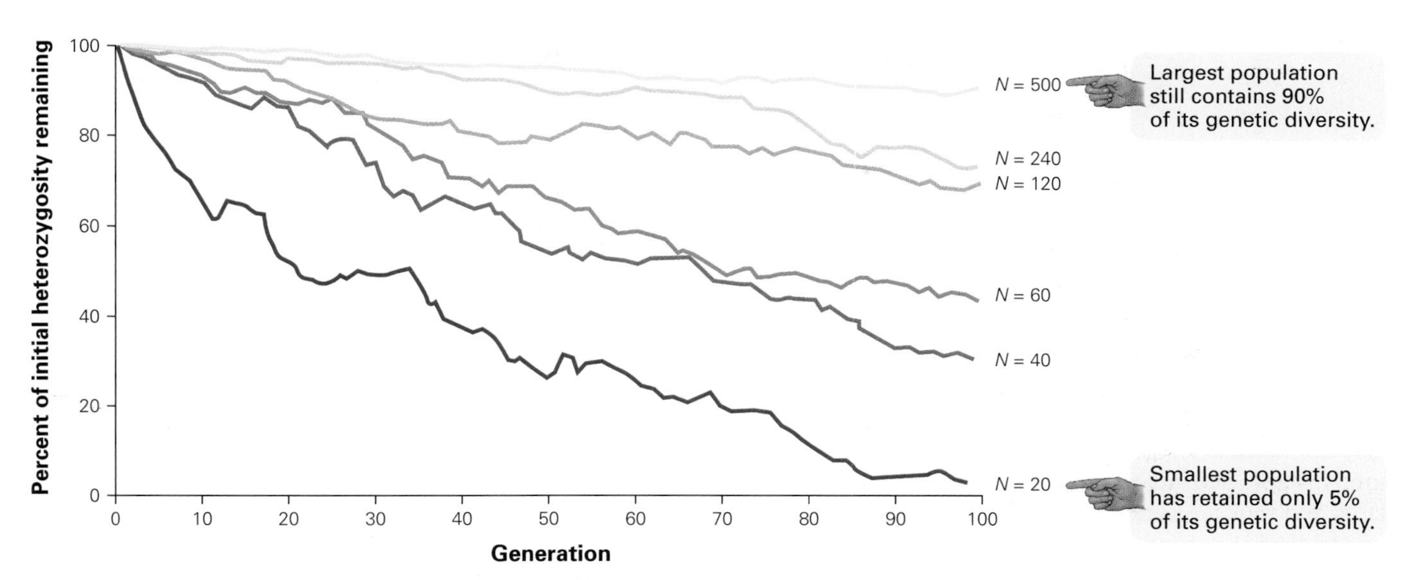

Figure 14.25 **Genetic drift affects small populations more than large populations.** In this graph, each line represents the average of 25 computer simulations of genetic drift for a given population size. After 100 generations, a population of 500 individuals still contains 90% of its genetic variability. In contrast, a population of 20 individuals has less than 5% of its original genetic variability.

higher death rates, leading to declining populations that are susceptible to all the other problems of small populations. This process is often referred to as the **extinction vortex** *(Figure 14.26)*. The heath hen discussed early in the chapter is an example of the extinction vortex; once the remnant population was reduced to small numbers by fire, disease, and predation, inbreeding depression prevented it from rebounding, thus dooming it to extinction.

Second, populations with low genetic variability may be at risk of extinction because they cannot evolve in response to changes in the environment. When few alleles are available for any given gene, it is possible that no individuals in a population possess an adaptation allowing them to survive an environmental challenge. For example, there is some evidence that people with type A blood are more resistant to cholera and bubonic plague than are people with type O or B blood. If a human population loses diversity and contains no individuals carrying the I^A allele, then it may become extinct on exposure to bubonic plague.

(continued on the next page)

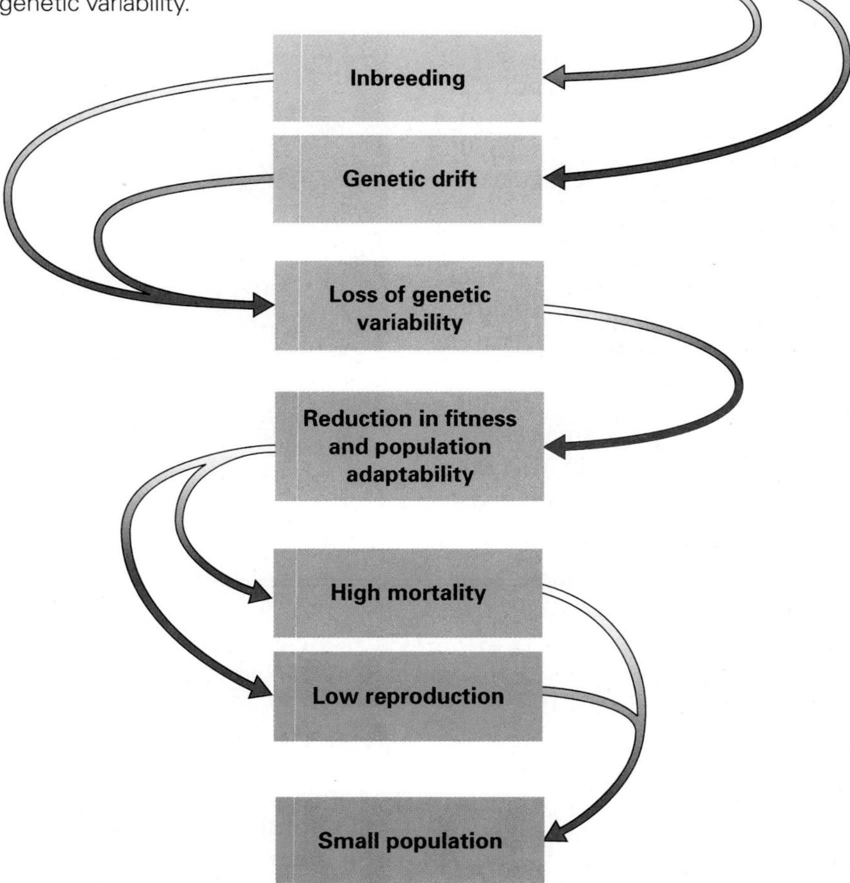

Figure 14.26 **The extinction vortex.** A small population can become trapped in a positive-feedback process that causes it to continue to shrink in size, eventually leading to extinction.

(A Closer Look continued)

Stop and Stretch 14.5: One counterweight to the loss of neutral or beneficial alleles due to genetic drift is that recessive deleterious alleles may be quickly lost from small populations due to natural selection. Explain why loss of recessive mutations would be faster in small populations compared to large ones.

As is often the case, there are some exceptions to the "rules" just described. For example, widespread hunting of northern elephant seals in the 1890s reduced its population to 20 individuals, thus wiping out much of the species' genetic variation. However, elephant seal populations have rebounded to include about 150,000 healthy individuals today. The dramatic recovery of elephant seals is apparently unusual. There are many more examples of populations that suffer because of low genetic variability. The Irish potato is perhaps the most dramatic example of this cost of low genetic diversity.

Potatoes were a staple crop of rural Irish populations until the 1850s. Although the population of Irish potatoes was high, it had remarkably low genetic variability. First, potatoes are not native to Ireland (in fact, they originated in South America), so the crop was limited to just a few varieties that were originally imported. Most of the potatoes grown on the island were of a single variety, called Lumper for its bumpy shape. Second, new potato plants are grown from potatoes harvested from the previous year's plants and thus are genetically identical to their parents. As a result, all of the potatoes in a given plot had identical alleles for every gene, and most potatoes on the whole island were extremely similar.

When the organism that causes potato blight arrived in Ireland in September 1845, nearly all of the planted potatoes became infected and rotted in the fields *(Figure 14.27)*. The few potatoes that by chance escaped the initial infection were used to plant the following year's crops. While some varieties of potatoes in South America carry alleles that allow them to resist potato blight, apparently very few or no Irish potatoes carried these alleles. As a result in 1846, the entire Irish potato crop failed. Because of this failure and another in 1848, along with the ruling British government's harsh policies that inhibited distribution of food relief, nearly 1 million Irish peasants died of starvation and disease, and another 1.5 million peasants emigrated to North America.

Because Irish potatoes lacked genetic diversity, even an enormous population of these plants could not escape the catastrophe caused by this disease. Similarly, preventing endangered species from declining to very small population levels, thus eroding genetic diversity, is critical to avoiding a similar genetic disaster even once populations recover.

(a) *Phytophthora infestans*

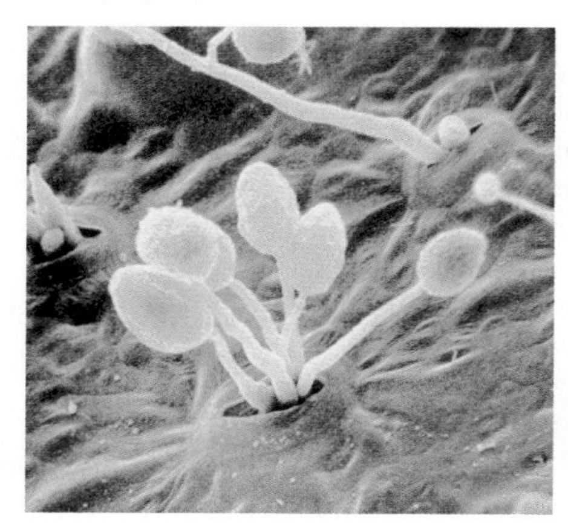

(b) Potatoes with blight

Figure 14.27 **Potato blight.** (a) The organism, *Phytophthora infestans,* that causes late blight in potatoes. (b) Infected potato tubers.

14.4 Protecting Biodiversity Versus Meeting Human Needs

Saving the Lost River and shortnose suckers from extinction requires protecting all of the remaining fish and restoring the habitat they need for reproduction. These actions cause economic and emotional suffering for humans who make their living in the Klamath Basin. In fact, many actions necessary to save endangered species result in immediate problems for people. As pointed out by the Oregon State University student, we need to balance the costs and benefits of preserving endangered species.

A provision of the Endangered Species Act allows members of a committee to weigh the relative costs and benefits of actions taken to protect endangered species. The Endangered Species Committee, which includes the U.S. secretaries of agriculture and the interior as well as the chairman of the Council of Economic Advisors, has convened a number of times for this purpose. This so-called God Squad decides if they should overrule a federal action protecting an endangered species to protect the livelihoods of people.

Farmers in the Klamath Basin advocated for a God Squad ruling on the diversion of water from Upper Klamath Lake but failed to receive one. History suggests that a decision is not likely to be in their favor. The Endangered Species Committee convened only four times from 1973 through 2001, and it has granted two exemptions, one of which was overturned by the subsequent presidential administration.

Instead of convening the God Squad, the George W. Bush administration took a different approach to the Klamath Crisis—they asked a committee of scientists from the prestigious National Academy of Science to examine the question. While some critics speculate that the committee was influenced by political concerns, their report indicated that the water-level requirements established in the fish management plan for the Lost River and shortnose suckers were not well supported by science. The committee recommended allowing water releases in subsequent years that could be adequate for the needs of both farmers and fish, at least in the short term. In 2002, water was released for irrigation—with apparently no harmful affect on the suckers. However, continued low water flows in the Klamath River have led to the collapse of the coho salmon population and an almost total ban on salmon fishing in the area. The federal government continues to purchase farmland from willing sellers in the Klamath Basin to protect and restore the watershed for all three species.

The ESA has been a successful tool for bringing species such as the peregrine falcon, American alligator, and bald eagle *(Figure 14.28)* back from the brink of extinction, but all of these successes have come with some cost to citizens. If the solution to these and other endangered species controversies is any guide, many Americans are willing to devote tax dollars to efforts that balance the needs of people and wildlife to protect our natural heritage.

As with any challenge that humans face, the best strategy is prevention. *Table 14.3* provides a list of actions that can help reduce the rate at which species become endangered. Meeting the challenge of preserving biodiversity requires some creativity, but it is possible to provide for the needs of people while making space for other organisms. However, it will take all of us to help keep the equation in balance.

Figure 14.28 **An endangered species success story.** The bald eagle was "delisted" from endangered status in 2007, at an event celebrated at this ceremony in Washington, DC. Its dramatic recovery from near extinction was thanks in part to government protections. However, delisting is still a relatively rare event.

TABLE 14.3 **Taking action to preserve biodiversity.**

Objective	Why Do It?	Actions
Reduce fossil fuel use 	• Mining, drilling, and transporting fossil fuels modifies habitat and leads to pollution. • Burning fossil fuels contributes to global climate change, further degrading natural habitats.	• Buy energy-efficient vehicles and appliances. • Walk, bike, carpool, or ride the bus whenever possible. • Choose a home near school, work, or easily accessible public transportation. • Buy "clean energy" from your electric provider, if offered.

(continued on the next page)

TABLE 14.3 **Taking action to preserve biodiversity.** *(continued)*

Objective	Why Do It?	Actions
Reduce the impact of meat consumption 	• The primary cause of habitat destruction and modification is agriculture. • Modern beef, pork, and chicken production relies on grains produced on farms. One pound of beef requires 4.8 pounds of grains or about 25 square meters of agricultural land.	• Eat one more meat-free meal per week. • Make meat a "side dish" instead of the main course. • Purchase grass-fed or free-range meat.
Reduce pollution 	• Pollution kills organisms directly or can reduce their ability to survive and reproduce in an environment.	• Do not use pesticides. • Buy products produced without the use of pesticides. • Replace toxic cleaners with biodegradable, less-harmful chemicals. • Consider the materials that make up the goods you purchase and choose the least-polluting option. • Reuse or recycle materials instead of throwing them out.
Educate yourself and others 	• Change happens most rapidly when many individuals are working for it.	• Ask manufacturers or store owners about the environmental costs of their goods. • Talk to family and friends about the choices you make. • Write to decision makers to urge action on effective measures to reduce human population growth and curb habitat destruction and species extinction.

Savvy Reader Did Wildlife Managers Make a Terrible Mistake?

Right Time, Wrong Fish

BY ERIK STOKSTAD | SCIENCENOW DAILY NEWS | 30 AUGUST 2007

The greenback cutthroat trout (*Oncorhynchus clarkii stomias*) once lived throughout thousands of kilometers of rivers and streams on the eastern slopes of the Colorado Rockies. Its life changed drastically, however, after mining, pollution, and competition from other species that were stocked in its habitat for recreational fishing. In 1937, the greenback was declared extinct.

But beginning in 1953, several populations were discovered in the headwaters of the South Platte River and Arkansas River in Colorado. The U.S. Fish and Wildlife Service added the subspecies to the federal list of

(continued on the next page)

Savvy Reader (continued)

endangered species in 1978. To help it recover, wildlife managers first removed nonnative fish from greenback habitat. Then they took eggs and sperm from surviving greenbacks and reared fish in hatcheries. Last year, the Colorado Division of Wildlife reached the target of 20 self-sustaining populations—or so they thought.

While surveying the genetics of the greenback trout, [researchers at the University of Colorado, Boulder] discovered that only four populations actually consisted of greenback trout. The rest were Colorado River cutthroat trout (*O. c. pleuriticus*), a species that looks very similar. That means managers had been accidentally stocking streams with Colorado River cutthroat, which are not listed as endangered.

Instead of being fully recovered, the greenback inhabits just a dozen kilometers of streams. Moreover, the genetic diversity of the greenback population in the eastern Rockies is low compared to other subspecies of cutthroat, perhaps enough to have led to inbreeding. "Now there's no way they are coming off the list" anytime soon, says lead author Jessica Metcalf.

1. What apparently was the original hypothesis studied by the UC-Boulder researchers?
2. How did the researchers determine that the fish in several of the populations were not greenback trout?
3. What information is missing from this article that would help you understand if the researcher's results truly support her concluding statement? What information in the article might make you more skeptical about her results?

CHAPTER Review

For study help, animations, and more quiz questions go to www.mybiology.com.

Summary

14.1 The Sixth Extinction

- The loss of biodiversity through species extinction is exceeding historical rates by 50 to 100 times (p. 363–364).
- Species-area curves help us predict how many species will become extinct due to human destruction of natural habitat (p. 365–366).
- Species at the top of the food chain are more susceptible to extinction because less energy is available for survival at higher trophic levels (p. 367).
- Additional threats of habitat fragmentation, introduced species, overexploitation, and pollution also contribute to species extinction (p. 366–369).

web animation 14.1
Tropical Deforestation and the Species-Area Curve

web animation 14.2
Habitat Destruction and Fragmentation

14.2 The Consequences of Extinction

- Species are important to us as resources, either directly as consumed products or indirectly as organisms used to provide potential medicines or genetic resources (p. 369).
- Species are members of communities; their loss as mutualists, predators, competitors, and keystone species may change a community, making it less valuable or even harmful to humans (p. 370–375).
- Species also play a role in ecosystem function, including effects on energy flow and nutrient cycling. Changes to the biological components of an ecosystem may change its nonbiological properties as well (p. 376–377).
- Biodiversity may fulfill a human need to experience natural landscapes (p. 377–378).

Summary (continued)

14.3 Saving Species

- If habitat protection is focused on a few well-defined biodiversity hot spots, then the number of organisms becoming extinct can be markedly reduced (p. 379).
- When species are already endangered, restoring larger populations is critical for preventing extinction (p. 380).
- Small populations are at higher risk for extinction due to environmental catastrophes (p. 381–382).
- Small populations are at risk when individuals have low fitness due to inbreeding and thus are less able to increase population size (p. 383).
- Genetic variability is lost in small populations because of genetic drift—the loss of alleles from a population due to chance events. Therefore, small populations may be less able to evolve in response to environmental change (p. 384–386).

14.4 Protecting Biodiversity Versus Meeting Human Needs

- The political process enables people to develop plans for helping endangered species recover from the brink of extinction while minimizing the negative effects of these actions on people (p. 387).

Roots to Remember

The following roots of words come mainly from Latin and Greek and will help you decipher terms:

eco- means home or habitation.
eu- means true or good.
-philia means affection or love.
-trophic means food, nourishment.

Learning the Basics

1. Describe how habitat fragmentation endangers certain species. Which types of species do you think are most threatened by habitat fragmentation?

2. Compare and contrast the species interactions of mutualism, predation, and competition.

3. A mass extinction ________.
A. is global in scale; **B.** affects many different groups of organisms; **C.** is caused only by human activity; **D.** A and B are correct; **E.** A, B, and C are correct

4. Current rates of species extinction appear to be approximately ________ historical rates of extinction.
A. equal to; **B.** 10 times lower than; **C.** 10 times higher than; **D.** 50 to 100 times higher than; **E.** 1000 to 10,000 times higher than

5. According to the generalized species-area curve, when habitat is reduced to 50% of its original size, approximately ________ of the species once present there will be lost.
A. 10%; **B.** 25%; **C.** 50%; **D.** 90%; **E.** it is impossible to estimate the percentage

6. Which cause of extinction results from humans' direct use of a species?
A. overexploitation; **B.** habitat fragmentation; **C.** pollution; **D.** introduction of competitors or predators; **E.** global warming

7. The web of life refers to the ________.
A. evolutionary relationships among living organisms;
B. connections between species in an ecosystem;
C. complicated nature of genetic variability;
D. flow of information from parent to child;
E. predatory effect of humans on the rest of the natural world

8. Which of the following is an example of a mutualistic relationship?
A. moles catching and eating earthworms from the moles' underground tunnels; **B.** cattails and reed canary grass growing together in wetland soils; **C.** cleaner fish removing parasites from the teeth of sharks; **D.** Colorado potato beetles consuming potato plant leaves; **E.** more than 1 of the above

9. The risks faced by small populations include ________.
A. erosion of genetic variability through genetic drift;
B. decreased fitness of individuals as a result of inbreeding;
C. increased risk of experiencing natural disasters; **D.** A and B are correct; **E.** A, B, and C are correct

10. One advantage of preserving more than one population and more than one location of an endangered species is ________.
A. a lower risk of extinction of the entire species if a catastrophe strikes 1 location;
B. higher levels of inbreeding in each population;
C. higher rates of genetic drift in each population;
D. lower numbers of heterozygotes in each population;
E. higher rates of habitat fragmentation in the different locations

Analyzing and Applying the Basics

1. Off the coast of the Pacific Northwest, areas dominated by large algae, called kelp, are common. These "kelp forests" provide homes for small plant-eating fishes, clams, and abalone. These animals in turn provide food for crabs and larger fishes. A major predator of kelp in these areas is sea urchins, which are preyed on by sea otters. When sea otters were hunted nearly to extinction in the early twentieth century, the kelp forest collapsed. The kelp were only found in low levels, while sea urchins proliferated on the sea floor. Use this information to construct a simple food web of the kelp forest. Using the food web, explain why sea otters are a keystone species in this system.

2. In which of the following situations is genetic drift more likely and why?
A. A population of 500 in which males compete heavily for harems of females. About 5% of the males father all of the offspring in a given generation. Females produce 1 offspring per season. **B.** A population of 250 in which males and females form bonded pairs that last throughout the mating season. Females produce 3 to 4 offspring per season.

3. The piping plover is a small shorebird that nests on beaches in North America. The plover population in the Great Lakes is endangered and consists of only about 30 breeding pairs. Imagine that you are developing a recovery plan for the piping plover in the Great Lakes. What sort of information about the bird and the risks to its survival would help you to determine the population goal for this species as well as how to reach this goal?

Connecting the Science

1. From your perspective, which of the following reasons for preserving biodiversity is most convincing? (A) Nonhuman species have roles in ecosystems and should be preserved to protect the ecosystems that support humans; or (B) nonhuman species have a fundamental right to existence. Explain your choice.

2. If a child asks you the following question 20 or 30 years from now, what will be your answer, and why? "When it became clear that humans were causing a mass extinction, what did you do about it?"

CHAPTER 15

Where Do You Live?

Climate and Biomes

What is your "biological address"?

Who are your neighbors?

How do you answer the question, "Where do you live?" Most of us would respond with a neighborhood or street address to someone from our community, a city or town name to someone from elsewhere in our state, or a state or country name to someone who lives far away. Not too many of us would give a reply such as "the Sonoran Desert" or "the boreal forest." However, descriptions of the natural environment in which we live can be thought of as our *bioregional* addresses. Our bioregional addresses include the native vegetation and the resident animals, fungi, and microbes that share, or once shared, our living space.

It can be difficult to identify a biological address in a human-designed landscape.

In this chapter, we will explore factors that help to determine the qualities of a particular bioregional address by guiding you to learn more about your own natural neighborhood. See if you can answer the following questions that test your bioregional awareness:

- Is the native vegetation of the place where you live forest, grassland, or desert?
- What are the seasonal weather changes in your area?
- Can you describe the physical characteristics of three plant species that are native to your area?
- What is the largest mammalian predator that lives, or once lived, in the habitat that is now your neighborhood?
- Can you describe three native bird species that breed in your region?

Many Americans would have a difficult time answering at least some of these questions. But is it important to know the answers to questions like these?

You may have experienced one consequence of poor bioregional awareness—local water shortages occurring when large numbers of homeowners attempt to keep their thirsty lawns green during summertime droughts. Other costs of lack of bioregional awareness

And it is sometimes difficult to determine where the resources on which we rely come from.

may be more severe, including the consequences of home construction in areas where periodic fires are common or on sandy coastlines, which are unstable and prone to storm damage.

Understanding one's own bioregion may allow people to build human settlements that are better for both humans and the natural environment. For example, in the southwest desert regions of the United States, environmentally sensitive housing developers use xeriscaping, a kind of landscaping that involves native, drought-tolerant plants. Xeriscaping not only prevents the overconsumption of water but also provides a habitat for resident wildlife.

Another aspect of bioregional awareness includes understanding how our activities fit into natural cycles. Human populations, like all natural populations, require resources and produce waste. Surprisingly few Americans have an understanding of where our resources come from and where our waste goes. See if you do by considering the following questions:

- What body of water serves as the source of your tap water?
- What primary agricultural crops are produced in your area?
- How is your electricity generated?
- Where does your garbage go?
- How is the waste handled when you flush the toilet or pour liquids down the drain?

Why is knowing the answers to these questions important? Humans remain dependent on the natural world for our resources and waste disposal. If we don't understand how our human communities fit into the surrounding biological community, the results can include air and water pollution that harms our biological neighbors—and ourselves. In this chapter, we explore how the ecology of a bioregion intersects with the biology of human habitats.

web animation 15.1
Tropical Atmospheric Circulation and Global Climate

15.1 Global and Regional Climate

Why is Buffalo, New York, so much snowier than frigid Winnipeg, Manitoba? Why is Miami, Florida, hot and humid while Tucson, Arizona, is hot and dry? Why does India experience monsoons? Why is the weather on tropical Pacific islands seemingly always beautiful?

The answers to all of these questions require an understanding of **climate,** the average conditions of a place as measured over many years. Climate is different from **weather,** which is the current conditions in terms of temperature, cloud cover, and **precipitation** (rain or snowfall). Simply put, weather information about a region will tell you if you have to shovel snow tomorrow morning, while climate information will tell you if you even need to own a snow shovel.

An Overview of Temperature and Precipitation Patterns

In general, temperatures are warmer in the tropics than in the Arctic and Antarctic and at elevations close to sea level compared to high elevations. Variation in temperature throughout the year correlates to geographic location as well—the closer one is to the Arctic or Antarctic, the more seasonality one experiences. On a local scale, the proximity of water influences temperatures by moderating them over the course of year, and small-scale elevation differences have an effect—valleys are colder than hilltops.

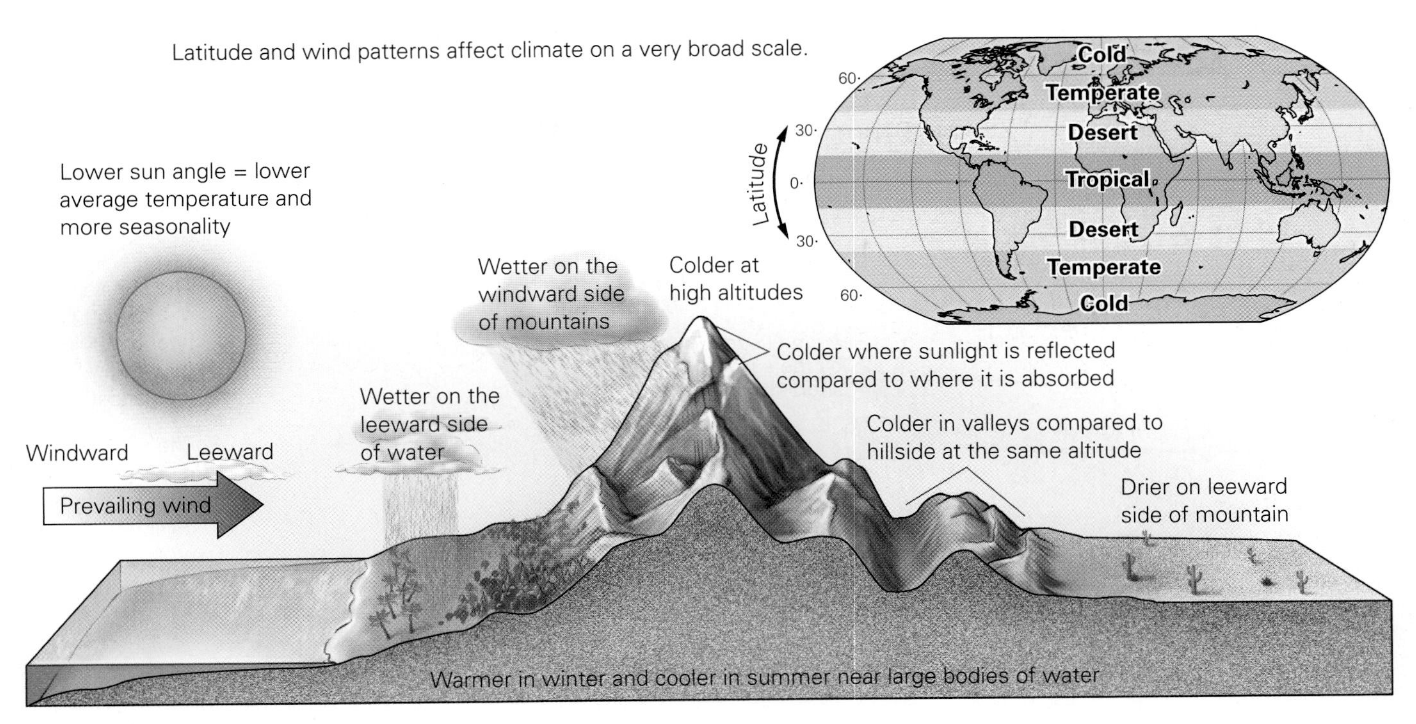

Figure 15.1 **Factors that determine climate.** Average temperature and precipitation depend on several aspects of geography.

Rain and snowfall patterns are more variable than temperature patterns on a global scale. There is a wide region of high rainfall in the tropics, while north and south of this point Earth's great deserts can be found. Locally, nearby oceans and large lakes cause increased precipitation on leeward (opposite the wind direction) land masses, while mountains cause dry conditions on their leeward sides and high precipitation on their windward sides. *Figure 15.1* summarizes many of the factors influencing global and regional climates.

A Closer Look: Temperature and Precipitation Patterns

Both global and regional factors influence climate in a particular bioregion.

Global Temperature Patterns. On a broad scale, the average temperature of any region is determined by the amount of **solar irradiance** it receives—that is, the amount of light energy per unit area. Locations that receive large amounts of solar irradiance in a year have a higher average temperature than places receiving small amounts.

Earth's axis is roughly perpendicular to the flow of energy from the sun. The extremes of this axis are called **poles,** while the circle around the planet that is equidistant to both poles is called the **equator.** The amount of solar irradiance varies between these regions because of the planet's spherical shape. *Figure 15.2* illustrates two identical streams of solar energy flowing from the sun. One strikes Earth's surface directly at the equator, while the other strikes the surface closer to the pole, where Earth's surface is "curving away" from the sun. The difference in geometry means that the surface area warmed at the equator is much smaller than the surface area warmed at the poles. In other words, the solar irradiance (energy per area) is greatest at the equator and lowest near the poles. This is why southern Florida has a warmer climate than northern Vermont.

Solar irradiance also varies in a particular location on an annual basis. This occurs because Earth's axis actually tilts approximately 23.5° from perpendicular to the sun's rays *(Figure 15.3)*. Due to this tilt, as Earth orbits the sun, the Northern Hemisphere (north of the equator) is tilted toward the sun during the northern summer and away from the sun during the northern winter. The tilt also helps explain why the position of the sunrise changes over the course of a year, moving from south to north as winter turns to summer *(Figure 15.4)*.

(continued on the next page)

(A Closer Look continued)

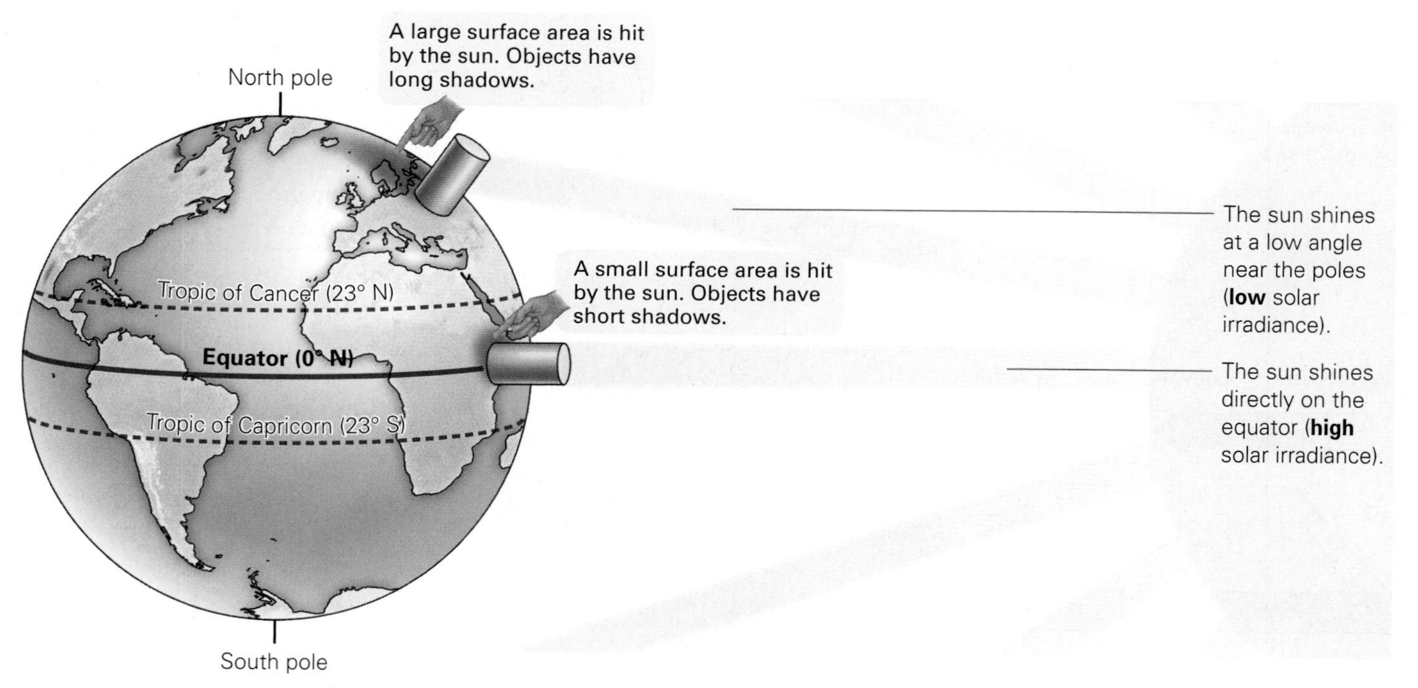

Figure 15.2 **Solar irradiance on Earth's surface.** The annual average temperature in a location on Earth's surface is most directly determined by its solar irradiance. Areas near the equator receive the greatest amount of solar energy, while areas near the poles receive the least.

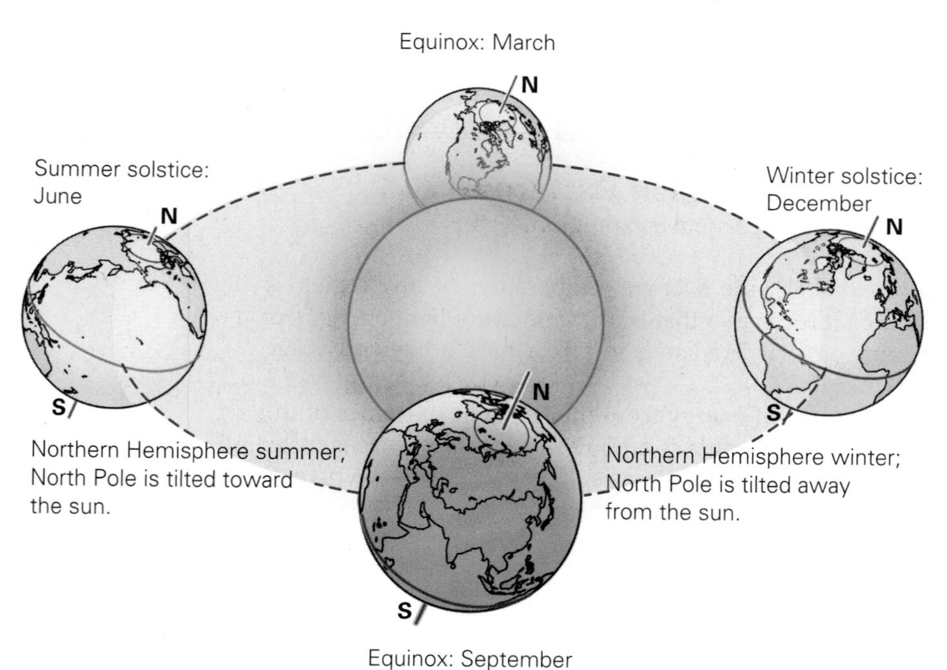

Figure 15.3 **Earth's tilt leads to seasonality.** Since Earth's axis is 23° from perpendicular to the sun's rays, as Earth orbits the sun during a year, the poles appear to move toward and away from the sun. Solar irradiance is increased at and near a pole when it tilts toward the sun and is decreased when that pole tilts away. **Visualize This:** Why is it summer in Australia when it is winter in North America?

Stop and Stretch 15.1: There is evidence that Earth has "wobbled" on its axis throughout its history, such that the axis was more perpendicular to the sun at some times and tilted at a greater angle at others. How would a reduction (or increase) in the tilt affect seasonality?

(A Closer Look continued)

Solar irradiance is at its annual maximum in the Northern Hemisphere during the summer **solstice,** when the sun reaches its northern maximum and the North Pole is tilted closest to the sun. The length of day is also greatest at the summer solstice and least at the winter solstice. For example, in Chicago, the time from sunrise to sunset is approximately 15 hours at the summer solstice and 9 hours at the winter solstice. The closer a region is to a pole, the greater the variance in day length over the course of a year. So while day length in Chicago varies by 6 hours from winter to summer, in Fairbanks, Alaska, it varies by 18 hours: from less than 4 hours at the winter solstice to nearly 22 hours at the summer maximum.

Figure 15.4 **The sun's travels.** This image was produced by taking a picture of the sun at the same location and time approximately once each week throughout a year.

Although the summer solstice is the day of highest solar irradiance in the Northern Hemisphere, it is not the warmest day of the year. Instead, the warmest days of the year are about one month after the solstice. This lag occurs because Earth converts the energy from the sun to heat and releases it gradually into the atmosphere. You can think of the solar heat stored in Earth as a bank account. Heat is always dissipating, much like money being withdrawn regularly from a bank account. If money is added to a bank account at the same rate as it is withdrawn, the balance remains the same. Similarly, if heat is added to a location on Earth's surface at the same rate that it is dissipating, the temperature there remains constant.

As day length increases, the amount of heat deposited is greater than the amount withdrawn, and the bank account gets larger—that is, the temperature starts to rise. On the longest day of the year, the amount deposited is greatest, but even after that day, the heat deposits are greater than the withdrawals, and the balance (temperature) continues to increase. Similarly, the coldest days of the year are about a month after the winter solstice—in late January or early February—because withdrawals remain larger than deposits for several weeks *(Figure 15.5)*.

The same analogy can help explain why the warmest part of the day is in the middle-to-late afternoon, even though the solar irradiance is greatest at noon when the sun is directly overhead. In the first few hours of the afternoon, heat gain remains greater than heat loss, increasing temperature.

Because day length changes more dramatically near the poles than near the equator, the bank account balance also changes more dramatically closer to the poles. The amount of balance change explains why seasonal temperature swings are more pronounced closer to the poles than near the equator. The average low temperature for January in Tampa, Florida, is 10°C (50°F), and the average high in July is 32°C (90°F), a difference of 22°C (40°F). On the contrary, the January low in Missoula, Montana, is −10°C (14°F), and the July high

(continued on the next page)

Figure 15.5 **Thermal momentum.** The winter carnival in St. Paul, Minnesota, is traditionally scheduled for the coldest weeks of the year—at the end of January—both to ensure the survival of the ice sculptures and to celebrate the year's turning point toward the warm summer.

(A Closer Look continued)

is 29°C (84°F), a difference of 39°C (70°F).

Local Factors That Influence Temperature. Three additional characteristics of a location's setting have an effect on its temperatures: (1) altitude, (2) the proximity of a large body of water, and (3) characteristics of the land's surface and vegetation.

Temperature drops as altitude—the height above sea level—increases. Temperature differences due to elevation are dramatic; the summit of Mt. Everest, 8.8 km (5.5 mi) above sea level, averages −27°C (−16°F), while nearby Kathmandu, Nepal, at 1.3 km (4385 ft), averages 18°C (65°F). However, smaller differences in elevation within a region have a converse effect on air temperature. Because cold air is denser than warm air, cold air tends to "drain" to the lowest point on a landscape. Thus valleys will often remain colder than nearby hilltops.

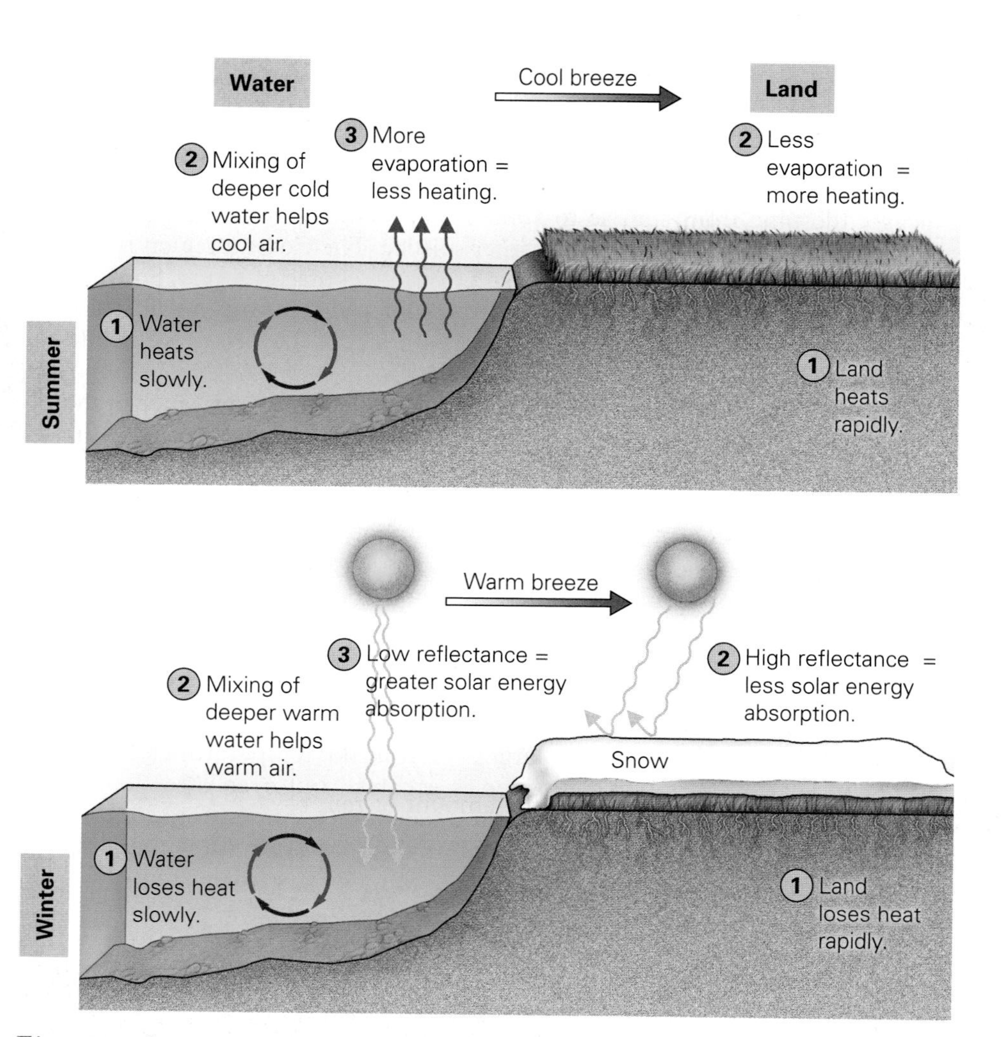

Figure 15.6 **The moderating influence of water.** Because water heats slowly, winds blowing across water in spring will cool nearby landmasses. Conversely, since water loses heat more slowly than land, it is a source of warmer breezes in fall and winter. **Visualize This:** Consider this figure along with Figure 15.5. Why is late July through mid-August traditionally the most popular time for people to visit the seashore?

Temperatures in areas near oceans, seas, and large lakes are influenced by the thermal properties of water. Water temperature rises and falls slowly in response to solar irradiation when compared to the rate of temperature change of land surfaces. Thus, air over a large body of water is comparatively cooler in summer and warmer in winter *(Figure 15.6)*, and nearby land areas experience more moderate temperatures than do regions further inland. Because of this phenomenon, the growing season on the shores of the Niagara peninsula in Canada, bordered by Lake Erie and Lake Ontario, is as much as 30 days longer than areas just a few miles inland from the lakes. Oceans also have an effect on local climates because heat is transferred through them via currents. The Gulf Stream is a current that carries water from the tropical Atlantic Ocean to the shores of northern Europe, producing a much milder climate there than in other areas at the same distance from the equator. The warmth of the Gulf Stream makes Dublin, Ireland, as warm as San Francisco, even though Dublin is 1600 km (1000 mi) closer to the North Pole.

The amount of light absorbed or reflected by the land's surface will also influence surrounding air temperature. A surface that reflects most light energy will have a lower nearby air temperature compared to a surface that absorbs most of that energy, heats up, and radiates that heat into the air. Snow reflects more light than a forest does, so air over a snowpack remains cold. The low reflectance of asphalt pavement and most building materials contributes to the urban heat island effect, which is the tendency for cities to be from 0.5° to 3°C (1° to 6°F) warmer than the surrounding areas *(Figure 15.7)*.

(A Closer Look continued)

Figure 15.7 **Urban heat island.** Urban areas are significantly warmer than nearby rural areas for a variety of reasons.

Stop and Stretch 15.2: One possible consequence of global warming is a decrease in the overall ice cover at the North Pole. Less ice is not only an effect of warming, it is likely to increase the rate of warming. Why?

Cities are also warmer because they contain little vegetation, which tends to reduce air temperature. Much of the solar energy absorbed by plants converts liquid water inside the plant to vapor, which also prevents the energy from being converted to heat.

As with temperature patterns, the amount of rain or snowfall experienced in any location on Earth is a product of both global and local factors.

Distribution of Precipitation on Earth's Surface. Energy from the sun is the primary driver of rain and snowfall patterns on Earth. To understand how sunlight causes rainfall, we must first understand some of the properties of water vapor. For water to remain as vapor, condensation must be less than evaporation; that is, the number of water molecules that clump together to form liquid droplets must be fewer than the number of water molecules that escape from droplets.

The rate of evaporation depends on temperature; at high temperatures, the evaporation rate is high, and the reverse is true at low temperatures. Thus when air cools and the rate of evaporation decreases, water molecules clump into larger and larger droplets. When the droplets are large enough, concentrations of them can be seen as clouds. As clouds grow even larger, droplets can become heavy enough to fall as rain. If the temperature inside the cloud is cold enough, droplets will freeze into ice crystals, which may fall as snow.

Rainfall patterns in tropical regions result from the air cycling near Earth's surface to high in the atmosphere and back down, as illustrated in *Figure 15.8*. Where solar irradiation is highest, at or near the equator, air temperatures rise quickly during the day. Because hot air is less dense than cold air, air at the equator rises. This leaves an area of low air pressure near Earth's surface, called the intertropical convergence zone, that is filled by breezes blowing from the north and south.

As the air rises, it cools; water vapor condenses to form clouds and falls to Earth as rain. The now-dry air flows in the upper atmosphere toward the poles, where pressure is lower because of the airflow in the lower atmosphere. The once-warm, wet air mass continues to cool and release water and finally drops back to Earth's surface at about 30° north and south latitudes. This very dry falling air displaces the ground-level air at these latitudes, and that ground-level air, having picked up moisture from the surface, flows toward the poles.

As the air masses drop, they are affected by Earth's rotation and thus deflect to the east or west. The movements of these vast air masses create the prevailing winds in various regions of the globe.

(continued on the next page)

(A Closer Look continued)

The pattern of air movement in the atmosphere helps explain the band of rain forests near the equator, the great deserts at 30° north and south of the equator, and the tendency for weather patterns in North America to come from the west.

Global rainfall patterns exhibit seasonality as well. The area of maximum solar irradiance travels from 23° north of the equator to 23° south over the course of a year due to the tilt in Earth's axis. Therefore, the intertropical convergence zone also moves, creating distinct rainy seasons wherever it is located. Rainy seasons occur in desert areas when the movement of the convergence zone results in changes in the prevailing winds. The monsoon seasons in India and southern Arizona are both associated with these wind shifts, as breezes move over long expanses of ocean, picking up water vapor that falls on land as rain.

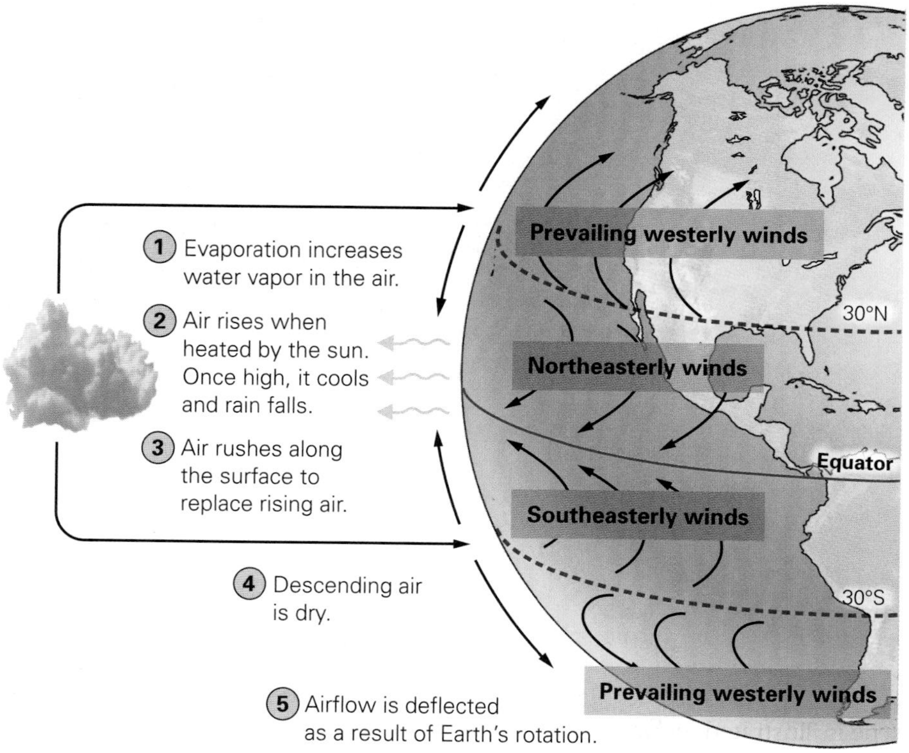

Figure 15.8 **Global precipitation and wind patterns.** High levels of solar irradiance at the solar equator lead to high levels of evaporation and rainfall. This phenomenon drives massive movements of air near the tropics into the temperate zones. **Visualize This:** The Doldrums and the Horse Latitudes are regions where there is little wind and thus where sailing ships are likely to stall. Based on these global wind patterns, where do you think these places are on the globe?

Local Precipitation Patterns. The amount of precipitation that falls in a given land region is highly dependent on the context of that area—in particular, the land's proximity to a large body of water. Wind blowing across warm water accumulates water vapor that condenses and falls when it reaches a cooler landmass. Communities surrounding the Great Lakes provide a dramatic example of this effect. Because the prevailing winds are from the west, areas immediately to the south and east of these lakes accumulate much more snow than do regions on the northern and western sides. For example, Toronto, on the northwest side of Lake Ontario, averages about 140 cm (55 in) of snow per year, while Syracuse, New York, on the southeast side averages almost twice that—274 cm (108 in).

Precipitation amounts are also affected by the presence of mountains or mountain ranges. When an air mass traveling horizontally approaches a mountain, it is forced upward. Cooling as it moves upward, the water vapor within it condenses to form clouds, rain, or snow, which falls on the windward side of the mountain. Warming again as it drops down the other side, the dry air mass causes water to evaporate from land on the sheltered or leeward side of the mountain. The dry area that results is often referred to as the mountain's "rain shadow." The Great Basin of North America, encompassing nearly all of Nevada and parts of Utah, Oregon, and California, is a desert because of the rain shadow cast by the Sierra Nevada.

The temperature and rainfall patterns in a geographical region play a major part in determining not only the primary vegetation in that area but also the human activities that are most successful there.

15.2 Terrestrial Biomes

Climate plays the greatest role in determining the physical appearance of the vegetation in a particular geographic area. Plants (and animals) native to a region are adapted to the water availability and temperatures experienced there.

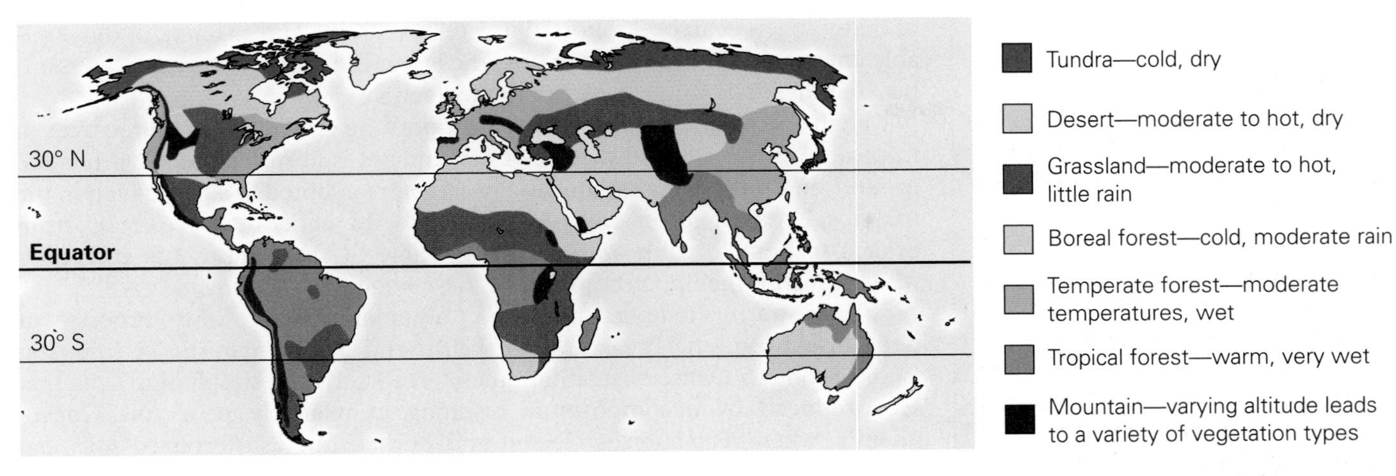

Figure 15.9 **The distribution of Earth's biomes.** The primary vegetation type in a given area is determined by the region's climate. **Visualize This:** Why is the east side of Australia tropical forest, while the west side is desert?

In general, the size of the vegetation is limited by water availability—large trees require large amounts of water. Water availability is a function of total precipitation, but it is also influenced by temperature; frozen water cannot be taken up by plants.

Four basic land **biome** categories, or primary vegetation types, are typically recognized: forest, grassland, desert, and tundra. Each of these biome categories may contain several biome types; for instance, a grassland may be either prairie, steppe, or savanna. The relationship between climate and biome type is illustrated in *Figure 15.9*.

Forests and Shrublands

Forests are vegetation communities dominated by trees and other woody plants. They occupy approximately one-third of Earth's land surface and, when all forest-associated organisms are included, contain about 70% of the biomass, the total weight of living organisms, found on land. Forests are generally categorized into three groups based on their distance from the equator: (1) tropical forests at or near the equator, (2) temperate forests from 23° to 50° north and south of the equator, and (3) boreal forests close to the poles *(Figure 15.10)*.

Tropical Forests. Extensive areas of tropical forest were once found throughout Earth's equatorial region—in Central and South America, central Africa, India, southeast Asia, and Indonesia. Tropical forests contain a large amount of biological diversity; 1 hectare (10,000 m^2) may contain as many as 750 tree species.

One hypothesis regarding why tropical forests are so diverse is the high solar irradiance they experience. Because the energy level is high, populations

Figure 15.10 **Forest biomes.** (a) Tropical rain forests are highly diverse and efficient at capturing light before it reaches the ground. (b) Temperate forests contain mostly deciduous trees that drop all of their leaves annually. (c) Boreal forests are often highly uniform, made up of one or two species of coniferous trees.

(a) Tropical rain forest

(b) Temperate forest

(c) Boreal forest

of many species can be supported in a relatively small area. Think of the available energy as analogous to a pizza—the larger the pizza, the greater the number of individuals who can be fed to satisfaction.

High energy and water levels also support the growth of very large trees in tropical forests. As a result, most of the sunlight is absorbed before it hits the ground, and most living organisms are therefore adapted to survive high in the treetops. Tropical forest animals are able to fly, glide, or move freely from branch to branch, while small plants are able to obtain nutrients and water while living on the upper branches of these huge trees *(Figure 15.11)*.

Figure 15.11 **Life in the canopy.** This bromeliad, often found more than 60 m from the forest floor, is vase shaped as an adaptation to acquire water and nutrients from rainfall.

With warm temperatures and abundant water, the process of **decomposition,** the breakdown of waste and dead organisms, is rapid in tropical forests. Dense vegetation quickly reabsorbs the simple nutrients that are produced by decomposition, resulting in relatively poor soil. Consequently, when vegetation is cleared and burned, the ash-fertilized soils can support crop growth for only 4 to 5 years. Once its soil is depleted of nutrients, a farm field carved out of tropical forest is abandoned, and a new field is cleared using the same method.

Among human populations in tropical forest areas, this slash-and-burn (or swidden) agricultural system is common. Tropical forests appear able to support swidden agriculture for many generations if the human population is small enough and abandoned plots have several decades to recover. However, increasing human population levels in tropical countries may be too large to allow adequate recovery time, and road building into interior forests has exposed untouched wilderness to swidden fields, endangering thousands of organisms *(Figure 15.12)*.

Figure 15.12 **Loss of tropical forest.** The amount of tropical forest land converted to agriculture is greater now than at any other time in human history.

Temperate Forests. Some areas of tropical forest may demonstrate seasonal changes between annual wet and dry periods. But most people associate major seasonal change with forests in temperate areas, where winter temperatures drop below the freezing point of water. Large areas of temperate forest appear in eastern North America, and remnants of these forests remain in western and central Europe and eastern China.

In temperate forests, water is abundant enough during the growing season to support the growth of large trees, but cold temperatures during the winter limit photosynthesis and freeze water in the soil. Trees with broad leaves can grow faster than narrow-leaved trees, so they have an advantage in the summer. However, broad leaves allow tremendous amounts of water to evaporate from a plant, leading to potentially fatal dehydration when water supplies are limited. To balance these two seasonal challenges, most broad-leaved trees in temperate forests have evolved a **deciduous** habit, meaning that they drop their leaves every autumn. In preparation for shedding its leaves, a deciduous tree reabsorbs their chlorophyll, the green pigment essential for photosynthesis. The colorful fall leaves that grace eastern North America in September and October result from secondary photosynthetic pigments and sugars that are left behind.

Stop and Stretch 15.3: Why do you suppose deciduous trees reabsorb chlorophyll, but not all pigments, from their leaves before shedding them?

The short time lag between the onset of warm temperatures and the releafing of deciduous trees provides an opportunity for plants on the temperate forest floor to receive full sunlight. Spring in these forests triggers the blooming of wildflowers that flower, fruit, and produce seeds quickly before losing the competition for light and water to their towering companions *(Figure 15.13)*. The lighter leaves of deciduous trees allow more sunlight through than do their tropical forest equivalents, permitting a shrub layer typically missing in the tropics. In addition, animals in temperate forests are more evenly distributed throughout the forest rather than concentrated in the treetops.

Nearly all of the forested lands in the eastern United States were converted to farmland within 100 years of the American Revolution. However, in the late nineteenth and early twentieth centuries, farms in the eastern United States were abandoned as production switched to the south and west. The regrowth of these forests provided a unique opportunity for ecologists to examine **succession**, the progressive replacement of different suites of species over time. Succession in most forests follows a predictable pattern. The first colonizers of vacant habitat are fast-growing species that produce abundant, easily dispersed seeds. These pioneers are replaced by plants that grow more slowly but are better competitors for light and nutrients. Eventually, a habitat patch is dominated by a set of species—called the climax community—that cannot be displaced by other species without another environmental disturbance *(Figure 15.14)*.

Today, these resurgent forests are once again threatened—this time by expanding urban development. The World Wildlife Fund estimates that worldwide, only 5% of temperate deciduous forests remain relatively untouched by humans.

Figure 15.13 **Springtime in a temperate forest.** Forest wildflowers are adapted to take advantage of a short period between snowmelt and tree leafing by emerging from the soil, flowering, and producing fruit rapidly.

Boreal Forests. The largest biome on Earth is the boreal forest, covering vast expanses of northern North America, northern Europe and Asia, and high-altitude areas in the western United States. Coniferous plants that produce seed cones instead of flowers and fruits dominate boreal forests *(Figure 15.15)*. In fact, boreal forests are the only land areas where flowering plants are not the dominant vegetation type.

Climate conditions for boreal forests include very cold, long, and often snowy winters and short, moist summers. The dominant conifers in these forests are evergreens, meaning that they maintain their leaves throughout the year. Coniferous tree leaves are needle shaped and coated with a thick, waxy coat—both adaptations to reduce water loss during winter. The evergreen habit likely explains conifers' dominance over flowering trees in boreal forests—the growing season is so short that the ability to begin photosynthesizing immediately after thawing gives conifers an advantage over faster-growing but slower-to-start deciduous trees.

Since they occupy regions with such cold winters and short summers, boreal forests tend to be lightly populated by humans. These areas represent some of the "wildest" landscapes on Earth—home to moose, wolves, bobcat, beaver, and a surprising diversity of summer-resident birds. Trees in these landscapes are valuable for both building materials and paper products, and logging in the boreal forest is extensive. There are increasing concerns that the boreal forest in North America is being cut at rates faster than it can be replaced.

Figure 15.14 **Forest succession in the eastern United States.** (a) Abandoned farms are quickly colonized by fast growing herbs. (b) The second stage of succession consists of a dense landscape of shrubs and quickly growing trees. (c) The climax community is made up of beech and maple trees in northeastern United States.

(a) (b)

(c)

Figure 15.15 **Coniferous trees.** Cone-bearing plants produce sperm and eggs on cones instead of in flowers. Their evergreen habit is an adaptation that allows them to be dominant in boreal forests.

Chaparral. One major biome is dominated by woody plants but is not a forest. This landscape is known as chaparral, and its vegetation consists mostly of spiny evergreen shrubs *(Figure 15.16)*. Chaparral is found extensively in areas surrounding the Mediterranean Sea and in smaller patches in southern California, South Africa, and southwest Australia.

Long, dry summers and frequent fires maintain the shrubby nature of chaparral. In fact, chaparral vegetation is uniquely adapted to fire. Several species have seeds that will germinate only after experiencing high temperatures. Many chaparral plants have extensive root systems that quickly resprout after aboveground parts are damaged. Chaparral will grow into temperate forest when fire is suppressed. As a result, natural selection has favored chaparral shrubs that actually encourage fire—such as rosemary, oregano, and thyme, which contain fragrant oils that are also highly flammable.

Fire can be an important contributor to the structure and function of many biomes. In southern California, the flammability of chaparral vegetation has come directly into conflict with rapid urbanization. In fall 2003, this area experienced its most extensive and expensive fire season ever as 750,000 acres of shrubland and surrounding forest burned—consuming over 3,200 buildings and killing 17 people in San Diego County alone. In response to this disaster, recommendations by a state task force have called for a policy of suppressing fires immediately rather than letting them burn and thus renewing the chaparral. In fact, the policy may contribute to the buildup of fuel and more intense fires that are destructive to both people and the wildlands. Fire policy in Southern California along with the long-term human modification of chaparral around the Mediterranean makes this biome one of the most threatened on the planet.

Grasslands

Grasslands are regions dominated by nonwoody grasses and containing few or no shrubs or trees. These biomes occupy geographic regions where precipitation is too limited to support woody plants. Grasslands can be further categorized into tropical and temperate categories.

Tropical grasslands are known as **savannas** and are characterized by the presence of scattered individual trees *(Figure 15.17)*. Savanna covers about half of the African continent as well as large areas of India, South America, and Australia. Savannas are maintained by periodic fires or clearing. In regions where wet and dry seasons are distinct, yearly grass fires during the dry season kill off woody plants; in regions where elephants and other large grazing animals are present, the damage these animals do to trees helps to maintain the grass expanse *(Figure 15.18)*.

Figure 15.16 **Chaparral.** Much of the scrubby landscape of southern California is made up of this biome type.

Figure 15.17 **Savanna.** Grasslands in tropical areas can support huge herds of grass-eating mammals—and their associated predators.

Figure 15.18 **Savanna maintenance.** Elephants severely prune acacia trees when feeding on them, keeping the trees' size relatively small; they actively destroy thorny myrrh bushes that they do not eat, which provides additional habitat for grass to grow. **Visualize This:** Why aren't domestic cattle as effective as elephants at maintaining savannah?

Because grazing animals eat the tops of plants, natural selection in these environments has favored plants, like grasses, that keep their growing tip at or below ground level. Because grass grows from its base, grazing (or mowing) is equivalent to a haircut—trimming back but not destroying. Woody plants grow from the tips, so intense grazing can destroy these plants.

Grass leaves are very fibrous and thus difficult to digest. The most common adaptation of animals that use grass as food is a mutualism with fiber-digesting bacteria. The bacteria live in a specialized chamber of the grazer's digestive system and partially digest the grass so that more nutrients are available to the animal. Although grazing mammals are characteristic of savannas, the introduction of large numbers of cattle and other grazers in these habitats has transformed the landscape from grassland to bare, sandy soil. In the Sahel region of central Africa, overgrazing has destroyed thousands of square kilometers of savanna.

Temperate grassland biomes include tallgrass **prairies** and shortgrass **steppes.** Generally, the height of the vegetation corresponds to the precipitation—greater precipitation can support taller grasses. Prairies and steppes are found in central North America, central Asia, parts of Australia, and southern South America. These landscapes are generally flat or slightly rolling and contain no trees.

In the cooler temperate regions, decomposition is relatively slow, and the soil of prairies and steppes is rich with the partially decayed plants *(Figure 15.19)*. These soils provide an excellent base for agriculture. As a result, most native prairies and steppes have been plowed and replanted with crops. In North America, less than 1% of native prairie remains.

Figure 15.19 **A midwestern prairie in late summer bloom.** Although the dominant plants on prairies are grasses, some of the less-numerous plants produce large, colorful flowers. **Visualize This:** Why might peak flowering period in a prairie be late summer, when it is early spring in temperate forests?

Desert

Where rainfall is less than 50 cm (20 in) per year, the biome is called **desert.** This biome can be found throughout the world, but the world's great deserts include the Sahara in northern Africa, the Gobi of central Asia, and the deserts of the Middle East, central Australia, and the southwestern United States. Most of these deserts are close to 30° north or 30° south of the equator, where the air masses that were "wrung" of water in the rain forest regions around the equator fall back to Earth's surface as hot and dry winds.

Although the image of a desert is often hot and sandy, some deserts can be quite cold; most deserts have vegetation, although it can be sparse. Plants and animals in desert regions have evolutionary adaptations to retain and conserve water. For example, plants are often thickly coated with waxes to reduce evaporation, contain photosynthetic adaptations that reduce water loss through leaf pores, and may be protected from predators of their stored water by spines and poisonous compounds *(Figure 15.20)*.

Figure 15.20 **Adaptations to constant drought.** Thick, whitish, and waxy coatings reflect light and reduce evaporation. Column-like forms reduce exposure to the high-intensity midday sun. Stems store water. The production of spines discourages predators from accessing their water or damaging their protective surface.

Deserts are also home to many flowering plants that complete their entire life cycle from seed to seed in a single season. In deserts, the wet season is quite short; many desert-adapted plants can germinate, flower, and produce seeds in a matter of 2 or 3 weeks. The seeds they produce are hardy and adapted to survive in the hot, dry soil for many years until the correct rain conditions return.

Some animals in these dry environments have physiological adaptations that allow them to survive with little water intake. The most amazing of these animals are the various species of kangaroo rat, which apparently never consume water directly. Instead, they conserve water produced during the chemical reactions of metabolism and have kidneys that produce urine four times more concentrated than our own.

The appeal of the sunny, warm, and dry climate of the deserts of Arizona and New Mexico has led to the greatest rate of human population growth in the United States. The increasing population is putting stress on water supplies, causing conflicts and depleting water sources for native animals. The Colorado River is so extensively used that often it cannot sustain a flow through its historic outlet at the Gulf of California in Mexico. As a result, up to 75% of the species that lived in the Colorado River delta are now gone.

Figure 15.21 **Tundra.** A marmot feeds on lush green tundra vegetation in Alaska. The short growing season and the year-round chance of below-freezing temperatures limit the vegetation of the tundra to ground-hugging plants.

Tundra

The biome type where temperatures are coldest—close to Earth's poles and at high altitudes—is known as **tundra** *(Figure 15.21)*. Here plant growth can be sustained for only 50 to 60 days during the year, when temperatures are high enough to melt ice in the soil. High temperatures are not sustained long enough to melt all of the soil's ice. Therefore, places like the arctic tundra near the North Pole are underlain by **permafrost,** icy blocks of gravel and finer soil material. The permafrost layer impedes water drainage, and soils above permafrost are often boggy and saturated.

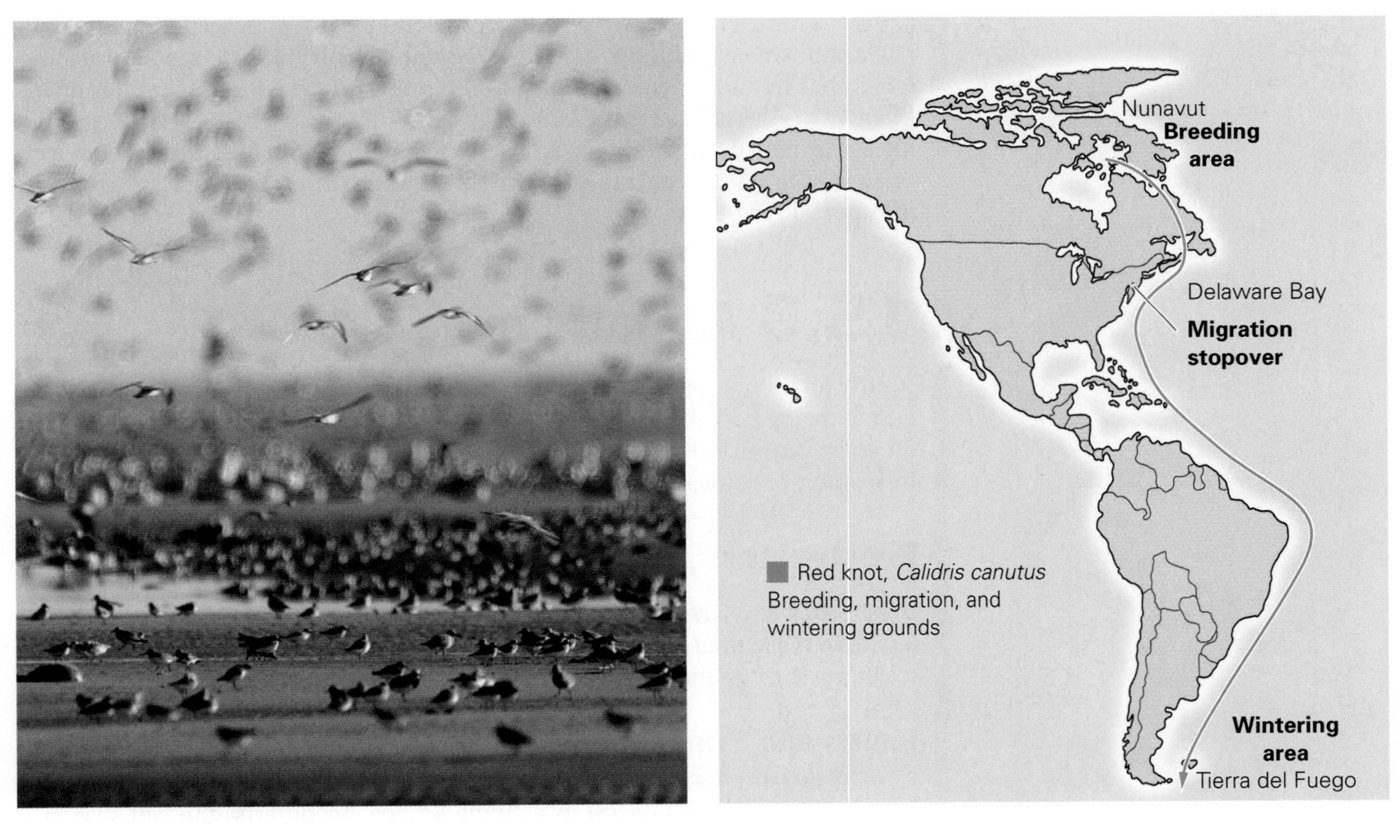

Figure 15.22 **Migration.** Birds that breed in the Arctic, such as these red knots massed on the beach at Delaware Bay, undertake a fatiguing and perilous journey to take advantage of the abundant daylight and insect life available during summer on the tundra.

Plants in tundra regions are adapted to windswept expanses and freezing temperatures, often growing in mutualistic multispecies "cushions" where all individuals are the same height and shelter each other. This low vegetation supports a remarkably large and diverse community of grazing mammals, such as caribou and musk oxen, and their predators, such as wolves.

Animals in tundra regions have evolved to survive long winters with structural adaptations, such as storing fat and producing extra fur or feathers. Other animals, such as ground squirrels, have adapted by evolving hibernation; they enter a sleeplike state and reduce their metabolism to maximize energy conservation. Grizzly bears and female polar bears also spend many of the coldest months in deep sleep; although this is not true hibernation, these bears are so lethargic that females give birth in this state without fully awakening. Other animals survive by migrating south to avoid the hardships of a long, frigid winter. Hundreds of millions of birds, migrating from the arctic tundra or other polar regions toward warmer climates and back again, take part in one of the great annual dramas of life on Earth *(Figure 15.22)*.

Tundra is very lightly settled by humans, but it is threatened by our dependence on fossil fuels—oil, natural gas, and coal that formed from the remains of ancient plants. Some of the largest remaining untapped oil deposits are found in tundra regions, including northern Alaska, Canada, and Siberia, and the proposed development of these resources threatens thousands of square kilometers of tundra with destruction caused by road building, oil spills, and water pollution. Fossil fuel use also contributes to global warming, which has had the greatest impact at the poles, where tundra is predominant. Winters in Alaska have warmed by 2° to 3°C (4°–6°F), whereas elsewhere they have warmed by about 1°C. As climate conditions change, areas that were once tundra have begun to support shrub and tree growth and are changing into boreal forest.

15.3 Aquatic Biomes

Nearly all human beings live on Earth's land surface, but most people also live near a major body of water and are both influenced by and have an influence on these **aquatic** systems. Aquatic biomes are typically classified as either freshwater or saltwater.

Freshwater

Freshwater is characterized as having a low concentration of salts—typically less than 0.1% total volume. Scientists usually describe three types of freshwater biomes: lakes and ponds, rivers and streams, and wetlands.

Lakes and Ponds. Bodies of water that are inland, meaning surrounded by land surface, are known as lakes or ponds. Typically, ponds are smaller than lakes, although there is no set guideline for the amount of surface area required for a body of water to reach lake status. Some ponds, however, are small enough that they dry up seasonally. These vernal (springtime) ponds are often crucial to the reproductive success of frogs, salamanders, and a variety of insects that spend part of their lives in water.

The aquatic environment of lakes and ponds can be divided into different zones: the surface and shore areas, which are typically warmer, brighter, and thus full of living organisms; and the deepwater areas, which are dark, low in oxygen, cold, and home to mostly decomposers. The biological productivity of lakes in temperate areas is increased by seasonal turnovers—times of the year when changes in air temperature and steady winds lead to water mixing, redistributing nutrients that had settled on the bottom of the lake and bringing fresh oxygen to deeper water *(Figure 15.23)*.

Nutrients applied to agricultural lands and residential lawns near lakes and ponds can increase their algae populations as the nutrients leach into the

Figure 15.23 **Nutrients in lakes.** (a) A lake is stable during the summer, and oxygen levels in deeper regions diminish as nutrient levels near the surface decline. (b) Nutrients and oxygen are mixed throughout a lake during fall and spring "turnover," providing the raw materials for algae growth.

(a) Lake in summer

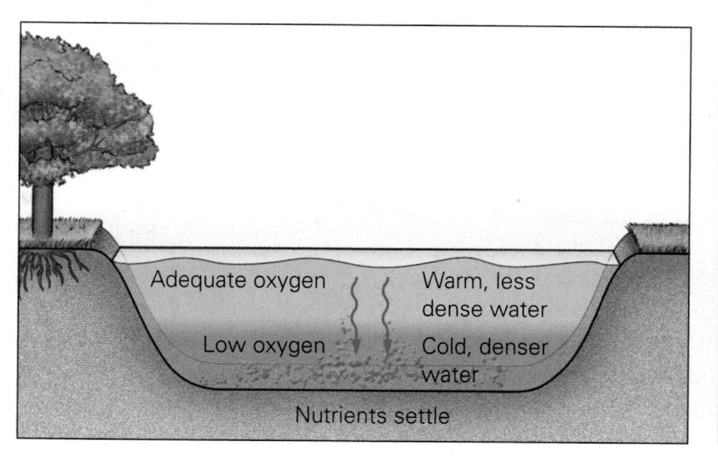

(b) Lake in fall and spring

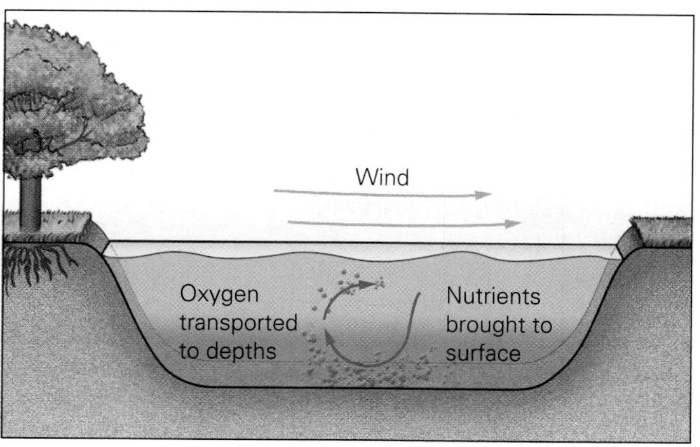

water. Ironically, too many nutrients, a process known as eutrophication, can lead to the "death" of these water bodies. Their degradation occurs because large populations of algae lead to large populations of microbial decomposers. These microbes consume a large fraction of oxygen in the water, suffocating the native fish *(Figure 15.24)*.

Figure 15.24 **Eutrophication.** High nutrient levels from fertilizer runoff lead to "blooms" of algae growth, severe oxygen depletion, and a risk of fish kills.

Rivers and Streams. Rivers and streams are flowing water moving in one direction. These waterways can be divided into zones along their lengths. At the headwaters of a river—often a lake, an underground spring, or near a melting snowpack—the water is clear, cold, and fast flowing; it is thus high in oxygen, providing an ideal habitat for cold-water fish such as trout *(Figure 15.25a)*. Near the middle reaches of a river, the width typically increases. Also increasing is the diversity of fish, reptile, amphibian, and insect species that the river supports because the warming water provides a better habitat for photosynthetic plants and algae.

At the mouth of a river, where it flows into another body of water, the speed of water flow is often slower, and the amount of sediment—soil and other particulates carried in the water—is high. High levels of sediment reduce the amount of light in the water and therefore the diversity of photosynthesizers that survive there. Oxygen levels are also typically lower near the mouth since the activities of decomposers increase relative to photosynthesis. Many of the fish found at the mouth of a river are bottom-feeders such as carp and catfish, which eat the dead organic matter that flowing water has picked up along its way *(Figure 15.25b)*.

Rivers and streams are threatened by the same pollutants that damage lakes. But their habitats also face wholesale destruction with the development of dams and channels—dams to provide hydropower or reservoirs for cooling fossil-fuel-powered, electricity-generating plants and channels to simplify and expedite boat traffic.

Wetlands. Areas of standing water that support emergent, or above-water, aquatic plants are called wetlands *(Figure 15.26)*. In numbers of species supported, wetlands are comparable to tropical rain forests. The high biological productivity of wetlands results from the high nutrient levels found at the interface between aquatic and land environments.

Besides their importance as biological factories, wetlands provide health and safety benefits by slowing the flow of water. Slower water flows reduce the likelihood of flooding and allow sediments and pollutants to settle before the water flows into lakes or rivers.

Figure 15.25 **River habitats.** (a) A river's headwaters are cold, fast flowing, and rich in oxygen. (b) With sluggish water flow, warmer temperatures, and high nutrient levels, the mouth of a river is a very different habitat.

(a) Headwaters

(b) River mouth

Figure 15.26 **Wetlands.** This dragonfly rests on a cattail emerging from a lakeside marsh.

Stop and Stretch 15.4: How might the loss of wetlands contribute to the death of lakes and rivers due to eutrophication?

Since the European settlement of the continental United States, over 50% of wetlands have been filled, drained, or otherwise degraded. Although legislation passed in the last few decades has greatly slowed the rate of wetland loss in the United States, about 58,000 acres are still destroyed every year.

Saltwater

About 75% of Earth's surface is covered with saltwater, or **marine,** biomes. Marine biomes can be categorized into three types: oceans, coral reefs, and estuaries.

Oceans. The open ocean covers about two-thirds of Earth's surface but is the least well-known biome of all. In truth, it can be thought of as a shifting mosaic of different environments that vary according to temperature, nutrient availability, and depth of the ocean floor. About 50% of the oxygen in Earth's atmosphere is generated by single-celled photosynthetic plankton in the open ocean. The open ocean also generates most of Earth's freshwater because water molecules evaporating from its surface condense and fall on adjacent landmasses as rain and snow.

Photosynthetic plankton serve as the base of the ocean's food chain, providing energy to microscopic animals called zooplankton, which in turn feed fish, sea turtles, and even large marine mammals such as blue whales. These predators provide a source of food for yet another group of predators, including sharks and other predatory fish, as well as ocean-dwelling birds such as albatross.

Unlike lakes, oceans experience tides—regular fluctuations in water level caused by the gravitational pull of the moon. As a result of tides, oceans contain unique habitats known as **intertidal zones,** which are underwater during high tide and exposed to air during low tide. Organisms in intertidal zones must be able to survive the daily fluctuations and rough wave action they experience. Adaptations such as strong anchoring structures, burrow building, and water-retaining, gelatinous outer coatings allow animals and seaweeds to take advantage of the high-nutrient environment found along the shore (*Figure 15.27*).

Figure 15.27 **Intertidal zones.** Pools of water that form in the intertidal zone contain animals and plants that are relatively sessile and thus relatively easy to observe, mark, and move. This intertidal area on Tatoosh Island off the coast of Washington State is the site of the longest-running ecological research program in the United States.

Oceans also contain a habitat known as the abyssal plain, the deep ocean floor. In these areas, sunlight never penetrates, temperatures can be quite cold, and the weight of the water above creates enormous pressure. Once thought to be lifeless because of these conditions, the abyssal plain is surprisingly rich in life, supported primarily by the nutrients that rain down from the upper layers of the ocean. In the 1970s, researchers studying the deep ocean discovered an entire ecosystem supported by bacteria that use hydrogen sulfide escaping from underwater volcanic vents as an energy source. Animals in this ecosystem either use the bacteria as a food source directly or have evolved a mutualistic relationship with them, providing living space for the bacteria while benefiting from the bacteria's metabolism. The abyssal plain represents the last major unexplored frontier on Earth.

The open ocean is heavily exploited by human fishing fleets—a research analysis published in the journal *Nature* in 2005 provided evidence that species diversity in the open ocean has declined by 50% over the last 50 years as a result of fishing pressure.

Figure 15.28 **Coral reefs.** These coral animals have their filtering appendages extended and are straining small pieces of organic matter from the water. Most of their calories are provided from the algae that live in their bodies and give the animals in this picture their orange color.

Coral Reefs. Coral reefs are unusual in that the structure of the habitat is not determined by a geological feature but composed of the skeletons of the dominant organism in the habitat: coral animals. Coral animals are very simple in structure but have a unique lifestyle—they filter dead organic material from the water but also feed on the photosynthetic algae they harbor inside their bodies. Up to 90% of a coral's nutrition is provided by the algae, which receives in return a protected site for photosynthesis and easy access to the coral's nutrients and carbon dioxide. Reef-building coral live in large colonies, and each individual coral secretes a limestone skeleton that protects it from other animals and from wave action *(Figure 15.28)*.

Coral reefs are found throughout the tropics in warm and well-lit water, providing ample resources for abundant plankton and algae growth. Their complex structure and high biological productivity make them the most diverse aquatic habitats, rivaling terrestrial tropical rain forests in species per area.

Coral reefs are sensitive to environmental conditions and prone to "bleaching," which occurs when host coral animals expel their algae companions. Without their algae mutualists, the animals may starve. Bleaching can occur for various reasons, but recent episodes seem to be associated with high ocean temperatures. Although coral can recover from bleaching, increased global temperatures due to global warming may lead to more frequent bleaching and the death of especially sensitive reef systems.

Estuaries. The zone where freshwater rivers drain into salty oceans is known as an **estuary** *(Figure 15.29)*. The mixing of fresh and saltwater combined with water-level fluctuations produced by tides create an extremely productive habitat. Some familiar and economically important estuaries are Tampa Bay, Puget Sound, and Chesapeake Bay.

Estuaries provide a habitat for up to 75% of commercial fish populations and 80% to 90% of recreational fish populations; they are sometimes called the "nurseries of the sea." Estuaries are also rich sources of shellfish—crabs, lobsters, and clams. Vegetation surrounding estuaries, including extensive salt marshes consisting of wetland plants that can withstand the elevated saltiness compared to freshwater, provides a buffer zone that stabilizes a shoreline and prevents erosion. Unfortunately, estuaries are threatened by human activity as well, including eutrophication from increasing fertilizer pollution and outright loss as a result of housing and resort development.

As is clear from our discussion of both aquatic and terrestrial biomes, no habitat on Earth has escaped the effects of humans. Consequently, to truly know our bioregion, we must learn how human populations use our environment.

Figure 15.29 **Estuary.** The estuary in Chesapeake Bay is the largest in North America; it supports an enormous number of organisms, some with significant commercial value, such as the blue crab. **Visualize This:** How is an estuary different from an intertidal zone?

Preserving our biological neighborhoods requires being able to meet human needs and at the same time respecting the needs of other species with which we coexist.

15.4 Human Habitats

According to the United Nations, humans have modified 50% of Earth's land surface. Most of this modification is for agriculture and forestry, but a surprisingly large amount—2% to 3% of Earth's land surface, or 3 to 4.5 million square kilometers, larger than the area of India—has been modified for human habitation *(Figure 15.30)*. And this is only direct conversion; human activities have environmental effects far beyond their geographic boundaries, including changes to Earth's atmosphere, which receives some of our waste. In fact, it is fair to say that no natural habitat has escaped the human fingerprint.

Today, half of the human population of 6.6 billion lives in cities. If current migrations from rural to urban areas continue, by the year 2050 two-thirds of the then population, about 6 billion people, will live in urban areas. Clearly, cities are going to become increasingly important in determining the human impact on native biomes.

Energy and Natural Resources

Consider the requirements of a forest. The only energy required for its growth is the solar irradiance striking the area, and most of the nutrients available to support this growth are already present in the soil. In addition, nearly all of the waste produced by the organisms in a forest is processed on site; that is, it becomes part of the soil or air and is recycled in the system.

In contrast, cities rely extensively on energy imported from elsewhere and tend to be the central link in a linear flow of materials—from natural or agricultural landscapes and into waste disposed of elsewhere.

Energy Use. In more developed countries, much of the energy required to power the activities within cities—from running buses and heating and cooling buildings to lighting traffic signals and energizing wireless phones—is

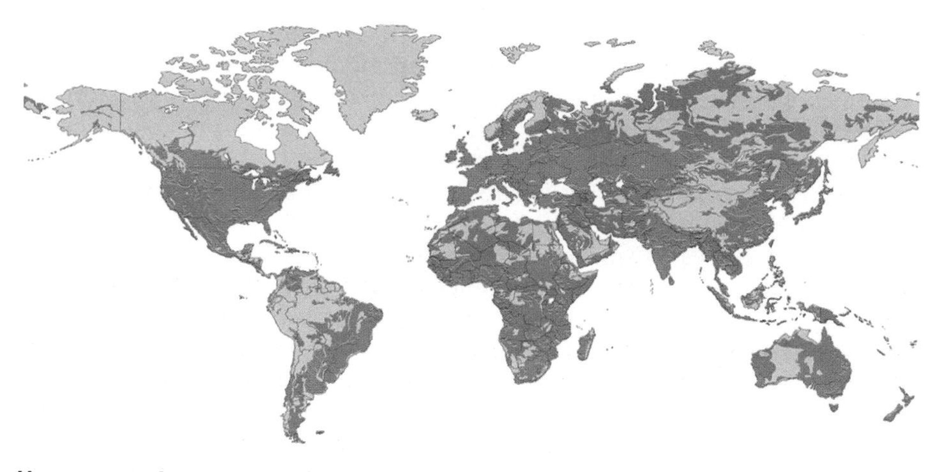

Figure 15.30 **Human modification of Earth's land surface.** Most of the land that is relatively untouched by humans lies in the vast boreal forest covering much of Canada and Siberia, the great deserts of Africa and Australia, and the dense Amazonian rain forest in north and central South America.

Figure 15.31 **The cost of fossil fuel extraction.** In the 1970s, mining companies developed a technique known as "mountaintop removal"—using explosives to blast off the rock above the coal deposit and dumping the waste rock into nearby valleys—to access this coal. Since 1981, more than 500 square miles of West Virginia have been destroyed by this method of mining.

derived from fossil fuels. These fuels, extracted from wells and mines around the globe, include the region surrounding the Persian Gulf, source of much of the world's oil, and the Appalachian Mountains, a rich source of coal. For most people in developed countries, the energy we use is not associated with the bioregion in which we live. The environmental impacts of acquiring and transporting fossil fuel can be substantial, ranging from the degradation of oceans and estuaries resulting from oil spills to the wholesale dismantling of forested mountains during coal mining *(Figure 15.31)*.

In less-developed countries, fossil fuel use is still relatively low, and energy sources are more directly tied to the surrounding natural environment. A primary source of energy for heating and cooking in these countries is wood and other plant-based materials (including animal dung). As urban areas in less-developed countries grow, surrounding forests are stripped of trees, often at a faster rate than they can regrow.

In both less-developed and more developed countries, an awareness of cleaner sources of energy within a given bioregion—such as abundant solar or wind resources—could help to reduce the environmental costs of energy consumption.

Stop and Stretch 15.5: Which historical and present factors do you think cause fossil fuels, rather than solar or wind power, to be the most common source of energy in the developed world?

Natural Resources. In addition to energy, human settlements require materials for survival—food from agricultural production and harvesting of natural resources, metals and salts from mines, freshwater for human and industrial consumption, petroleum for asphalt and manufactured goods, and trees for processing into paper and packaging. Developing and extracting all these raw materials changes biomes far from a city's center.

The amount of material needed to support a city can be tremendous. A year 2000 estimate calculated that the **ecological footprint** of the city of London—that is, the amount of land needed to support the human activity there—was 293 times the actual size of the city, equal to twice the entire land surface

Figure 15.32 **Treatment of human waste.** In more developed countries, human waste is treated in expensive and technologically advanced sewage treatment plants. This is a great improvement over the past, when untreated waste was dumped directly in rivers, lakes, and oceans.

of the United Kingdom. Clearly, the resources to support London must come from other countries around the world—from the boreal forests of Russia and Scandinavia that supply wood for building and paper production to the tropical forests of India and China that supply Londoners' daily spot of tea.

London is not exceptional; most cities in more developed countries have similarly sized footprints. A city's footprint can be reduced by its citizens, especially if they reduce the amount of energy used for transportation and make sensitive consumer choices, including buying locally produced foods and less meat.

Waste Production

The waste produced by a city presents a special problem. The density of the population is so great in cities that specialized systems are required to handle the enormous amounts of human waste generated in a small area. Human wastes can be liquid, in the form of wastewater; solid, in the form of garbage; or gaseous, in the form of air pollution.

Wastewater. In developed countries, cities have sewage treatment systems to handle the wastewater that drains from sinks, tubs, toilets, and industrial plants *(Figure 15.32)*. These systems typically treat water by removing semisolid wastes through settling, using chemical treatment to kill any disease-causing microorganisms, and eventually discharging the treated water to lakes, streams, or oceans. Unfortunately, older treatment systems can be overwhelmed by storm water, causing the discharge of large volumes of wastewater directly into waterways. Untreated wastewater can cause nutrient levels to spike and algae and bacteria growth to increase, thus leading to beaches being closed to swimming and other water-based recreation.

Disposal of semisolid wastes, called sludge, is a significant challenge. Sludge is often composted (allowed to decompose) and applied to land as fertilizer or trucked to landfills. Land application of composted sludge has its own problems; this material contains not only human waste—a valuable fertilizer if properly composted—but also industrial wastewater, which can contain a wide variety of toxic chemicals.

In less-developed countries, safe disposal of treated sewage is often a distant dream. In many rapidly industrialized regions, the emigration of people to urban areas has overwhelmed antiquated and inadequate sewer systems. Large numbers of people in these cities live in slums where running water is scarce, and untreated human waste flows in open gutters. The consequences of inadequate disposal of human waste can be severe—intestinal diseases due to contact with waste-contaminated drinking water cause the preventable deaths of over 2 million children under 5 years old every year.

Figure 15.33 **Disposal of garbage.** Solid waste contained in a sanitary landfill is locked within a thick liner. Garbage is compacted and "capped" with soil or other material.

Garbage and Recycling. In addition to dealing with wastewater, people in urban areas also must find ways to control their solid waste—that is, their garbage. In more developed countries, most of the solid waste finds its way into sanitary landfills, which are pits lined with resistant material such as plastic. Landfills have systems for collecting liquid that drains through the waste and exhaust pipes to vent dangerous gases that result from decomposition *(Figure 15.33)*.

In most areas, landfill space is becoming more and more limited and farther removed from cities. To stave off a looming "garbage crisis," states and communities mandate recycling of paper, glass, and metal, and many have or are considering composting programs to reduce the amount of food and yard waste. Unfortunately, even where recycling rates are relatively high, household garbage production continues to increase.

In less-developed countries, the problem of solid waste disposal is more severe. Many large cities in these areas have large, open dumps in which unstable, fire-prone piles of garbage provide living space for desperately poor immigrants in the city.

Air Pollution. Because of human dependence on fossil fuels, urban areas produce large amounts of gaseous waste. These air emissions include carbon dioxide as well as combustion by-products, including nitrogen and sulfur oxides, small airborne particulates, and fuel contaminants such as mercury. Exposure to sunlight and high temperatures can cause some of these by-products to react with oxygen in the air to form ground-level ozone or smog *(Figure 15.34)*. For individuals with asthma, heart disease, or reduced lung function, smog exposure can lead to severe illness, even death.

When gaseous pollution enters the upper reaches of Earth's atmosphere, it can be carried on air currents to less-settled areas throughout the globe. Air

Figure 15.34 **Smog.** This brownish haze over downtown Los Angeles, California, results from automobile emissions and other air pollutants causing the conversion of oxygen to ozone in sunny, still conditions.

Figure 15.35 **Fitting human needs into the bioregion.** Lake Erie was once so polluted that it was considered "dead" in the 1960s. Thanks to government and citizen efforts, the lake has now recovered and provides clean swimming beaches and a healthy fish population.

emissions from coal-fired power plants throughout the Midwest cause severe acid rain in lightly settled regions of the northeastern United States, and airborne toxins such as benzene and PCBs (polychlorinated biphenyls) have been found in high levels in animals in the Arctic.

Air pollution is a problem in both developed and less-developed countries. In less-developed countries, pollution control is weak or lacking altogether; in more developed countries, the sheer volume of fossil fuel use contributes to poor air quality. In the United States, the number of miles driven by car per household has nearly doubled in the last 25 years, partially due to an increase in the distance that individuals live from their workplaces. Development of suburban settlements outside the geographical limits of cities has been termed **urban sprawl.** Urban sprawl not only contributes to an increase in fossil fuel consumption but also affects wildlife through habitat destruction and fragmentation. Urban sprawl also impairs water quality via the destruction of wetlands and the increasing amount of paved surfaces, which funnel pollutants and warmed water into lakes, streams, and rivers.

The effect of human settlements on surrounding natural biomes can be significant and severe, as we have discovered in this chapter. However, many of these impacts can be mitigated with thoughtful planning and the use of improved technology. In the United States, laws such as the Clean Air Act and the Clean Water Act—passed in 1970 and 1972, respectively—have greatly reduced air and water pollution and have contributed to the recovery of once severely impaired habitats *(Figure 15.35)*.

Cities throughout the more developed world are supporting projects aimed at creating sustainable communities that are both economically vital and environmentally intelligent. *Table 15.1* poses a set of questions that will help you become more knowledgeable about your own bioregion. Perhaps by getting to know our biological neighbors and understanding how our choices affect these organisms, we can be inspired to help create communities that are safe and healthy for humans and other species.

TABLE 15.1 **Know your bioregion.**

Environmental Factor	Your Bioregion
Climate	Describe the climate of the place where you live, including the annual average temperature, average precipitation, average number of sunny days, and names of seasons. Which global and local factors influence the climate where you live?
Vegetation	Describe the native vegetation of the place where you live. How much native vegetation remains, and what has taken its place? Which ecological factors (climate, fire, etc.) have influenced the native vegetation type?
Aquatic habitats	What are the aquatic habitats nearest to you? Can you name some of the dominant species in these habitats? What threats do these habitats face in your area?
Energy resources	What is the primary source for the electricity you use in your home? Is your bioregion rich in any less environmentally damaging energy resources?
Food resources	What agricultural products are produced in your bioregion? How easy is it to buy local produce?
Natural resources	What natural resources does your bioregion supply?
Wastewater and sewage	Where does your wastewater go? Does your community have any problems handling wastewater and sewage?
Solid waste	Where does your garbage go? What is the rate of recycling in your community and can it be improved?
Air quality	How is the air quality in your community? What are the major air pollutants and their sources?

Savvy Reader Greenwashing?

The term *greenwashing* refers to the practice of covering up environmentally damaging actions or policies with public relations techniques that emphasize ecologically friendly values. Some environmentalists have accused the oil company BP (formerly British Petroleum) of greenwashing with its "beyond petroleum" campaign. Here is one of the print ads from that campaign:

Building a diverse portfolio is one way we're investing in the new energy future. Over the last 5 years, we've invested $30 billion in U.S. energy supplies, like cleaner burning natural gas. We're investing up to $8 billion over ten years in solar, wind, natural gas and hydrogen to provide low cabon electricity. And, we're expanding our longtime involvement in biofuels, investing $500 million to bring the next generation of biofuels to market.
It's a start.

1. In which technologies other than oil (petroleum) is BP investing?
2. Can you determine from this ad how much money BP invests in a yearly basis in "green" fuels? If so, approximately how much are they investing?
3. What would you need to know to determine if BP is making a significant commitment to alternative fuels compared with oil exploration and development?

CHAPTER Review

For study help, animations, and more quiz questions go to www. mybiology. com.

Summary

15.1 Global and Regional Climate

- Climate is the weather of a place as measured over many years. Climate in an area is determined by global temperature patterns, which are driven by solar irradiance and local factors such as proximity to large bodies of water (p. 394).
- Temperatures are warmer at the equator than at the poles because solar irradiance is greater at the equator (p. 395).
- Seasonal changes are caused by the tilt of Earth's axis. For example, during summer in the Northern Hemisphere, total solar irradiance during the day is much greater than it is during winter (pp. 395–397).
- Local temperatures are influenced by altitude, proximity to a large body of water, nearby ocean currents, the light reflectance of the land surface, and the amount of surrounding vegetation (p. 398).
- Global precipitation patterns are driven by solar energy, with areas of heavy rainfall near the solar equator and dry regions at latitudes of 30° north and south of the equator (p. 399).
- Local precipitation patterns are influenced by the presence of a large body of water, which tends to increase rainfall, and the presence of a mountain range, which tends to reduce rainfall on its leeward or sheltered side (p. 400).

web animation 15.1
Tropical Atmospheric Circulation and Global Climate

Summary (continued)

15.2 Terrestrial Biomes

- Categories of primary vegetation types found on Earth's land surface are called biomes (p. 401).
- Forests are dominated by trees and categorized by distance from the equator into tropical, temperate, and boreal types (pp. 401–403).
- The chaparral biome is dominated by woody shrubs (p. 404).
- The major vegetation type on grasslands is non-woody grasses. These biomes are categorized by distance from the equator into tropical and temperate types (p. 404).
- Deserts are found where precipitation is 50 cm (20 in) per year or less and contain mostly drought-resistant plants (p. 405).
- Tundra occurs in areas where the growing season is less than 60 days, both near the poles and at high altitudes (p. 406).

15.3 Aquatic Biomes

- Freshwater biomes include lakes, rivers, and wetlands (pp. 408–409).
- Marine biomes include oceans, coral reefs, and estuaries (pp. 410–411).

15.4 Human Habitats

- Humans have modified over 50% of Earth's land surface. Much of this modification now supports the activities of cities (p. 412).
- Cities must rely on imported energy and other resources to survive; extraction of these resources carries an environmental cost felt by other bioregions (pp. 412–413).
- Waste disposal is a significant challenge for large urban areas. Sewage must be treated to avoid contaminating drinking water; garbage must be disposed of effectively, and air emissions must be controlled. More developed countries have more resources to consume and handle the vast amounts of waste that they produce, while less-developed countries struggle with inadequate systems (pp. 414–415).
- Increasing the bioregional awareness of citizens and communities can help them devise methods for supporting human activities in a more sustainable manner (p. 416).

Roots to Remember

The following roots of words come mainly from Latin and Greek and will help you decipher terms.

aqua- means water.
deciduous- means that which falls off.
sol- means sun.
-stice means to stand still.

Learning the Basics

1. Explain why the northern United States experiences a cold season in winter and a warm season in summer.

2. Why does proximity to a large body of water moderate climate on a nearby landmass?

3. Areas of low solar radiation are ________.
A. closer to the equator than to the poles; **B.** closer to the poles than the equator; **C.** at high altitudes; **D.** close to large bodies of water; **E.** more than one of the above is correct.

4. The solar equator, the region of Earth where the sun is directly overhead, moves from 23° N to 23° S latitudes and back over the course of a year. Why?
A. Earth wobbles on its axis during the year; **B.** The position of the poles changes by this amount annually; **C.** Earth's axis is 23° from perpendicular to the sun's rays; **D.** Earth moves 23° toward the sun in summer and 23° away from the sun in winter; **E.** Ocean currents carry heat from the tropical ocean north in summer and south in winter.

5. Which of the following biomes is most common on Earth's terrestrial surface?
A. chaparral; **B.** desert; **C.** temperate forest; **D.** tundra; **E.** boreal forest

6. Tundra is found ________.
A. where average temperatures are low, and growing seasons are short; **B.** near the poles; **C.** at high altitudes; **D.** A and B are correct; **E.** A, B, and C are correct

7. Which statement best describes the desert biome?
A. It is found wherever temperatures are high; **B.** It contains the largest amount of biomass per unit area than any other biome; **C.** Its dominant vegetation is adapted to conserve water; **D.** Most are located at the equator; **E.** It is not suitable for human habitation.

8. Which of the following biomes has a structure made up primarily of the remains of its dominant organisms?
A. coral reefs; **B.** freshwater lakes; **C.** rivers; **D.** estuaries; **E.** oceans

9. An ecological footprint ________.
A. is the position an individual holds in the ecological food chain; **B.** estimates the total land area required to support a particular person or human population; **C.** is equal to the size of a human population; **D.** helps determine the most appropriate wastewater treatment plan for a community; **E.** is often smaller than the actual land footprint of residences in a city.

10. In more developed countries, wastewater containing human waste ________.
A. is dumped directly into waterways; **B.** is applied directly to farm fields; **C.** is piped into sanitary landfills; **D.** is treated to remove sludge and disease-causing organisms and released into nearby waterways; **E.** is a major source of infectious disease in human populations

Analyzing and Applying the Basics

1. Consider the following geographic factors and predict both the climate and biome type found in the location described. Explain the reasoning that you used to determine your answer. This small city is

- On the coast of the Pacific Ocean
- 20° north of the equator
- 20° m above sea level
- At the base of a mountain range

2. One prediction of global climate change models is that significant amounts of melting ice will change the salt content of the ocean, causing the Gulf Stream current in the Atlantic Ocean to stop altogether. How will this change likely affect Europe?

3. What can you infer about the geographical relationship among the cities in the following table and the Cascades, the primary mountain range that influences their climate?

City	Approximate Average Annual Rainfall
Bend, OR	12 inches
Eugene, OR	43 in
Portland, OR	36 in
Tacoma, WA	39 in
Walla Walla, WA	21 in
Yakima, WA	8 in

Connecting the Science

1. How many biomes do you rely on to supply your food? Many grocery stores label the origin of their produce. The next time you go to the grocery store, try to determine the number of different countries from which your groceries come. Could you easily change your diet and shopping habits to rely on locally produced food? Why or why not?

2. Consider the setting of your home. What kind of changes to the environment of your home would help it fit better into the bioregion? Are these changes feasible? Are they desirable?

Appendix

Metric System Conversions

To Convert Metric Units:	Multiply by:	To Get English Equivalent:
Length		
Centimeters (cm)	0.3937	Inches (in)
Meters (m)	3.2808	Feet (ft)
Meters (m)	1.0936	Yards (yd)
Kilometers (km)	0.6214	Miles (mi)
Area		
Square centimeters (cm^2)	0.155	Square inches (in^2)
Square meters (m^2)	10.7639	Square feet (ft^2)
Square meters (m^2)	1.1960	Square yards (yd^2)
Square kilometers (km^2)	0.3831	Square miles (mi^2)
Hectare (ha) (10,000 m^2)	2.4710	Acres (a)
Volume		
Cubic centimeters (cm^3)	0.06	Cubic inches (in^3)
Cubic meters (m^3)	35.30	Cubic feet (ft^3)
Cubic meters (m^3)	1.3079	Cubic yards (yd^3)
Cubic kilometers (km^3)	0.24	Cubic miles (mi^3)
Liters (L)	1.0567	Quarts (qt), U.S.
Liters (L)	0.26	Gallons (gal), U.S.
Mass		
Grams (g)	0.03527	Ounces (oz)
Kilograms (kg)	2.2046	Pounds (lb)
Metric ton (tonne) (t)	1.10	Ton (tn), U.S.
Speed		
Meters/second (mps)	2.24	Miles/hour (mph)
Kilometers/hour (kmph)	0.62	Miles/hour (mph)

To Convert English Units:	Multiply by:	To Get Metric Equivalent:
Length		
Inches (in)	2.54	Centimeters (cm)
Feet (ft)	0.3048	Meters (m)
Yards (yd)	0.9144	Meters (m)
Miles (mi)	1.6094	Kilometers (km)
Area		
Square inches (in^2)	6.45	Square centimeters (cm^2)
Square feet (ft^2)	0.0929	Square meters (m^2)
Square yards (yd^2)	0.8361	Square meters (m^2)
Square miles (mi^2)	2.5900	Square kilometers (km^2)
Acres (a)	0.4047	Hectare (ha) (10,000 m^2)
Volume		
Cubic inches (in^3)	16.39	Cubic centimeters (cm^3)
Cubic feet (ft^3)	0.028	Cubic meters (m^3)
Cubic yards (yd^3)	0.765	Cubic meters (m^3)
Cubic miles (mi^3)	4.17	Cubic kilometers (km^3)
Quarts (qt), U.S.	0.9463	Liters (L)
Gallons (gal), U.S.	3.8	Liters (L)
Mass		
Ounces (oz)	28.3495	Grams (g)
Pounds (lb)	0.4536	Kilograms (kg)
Ton (tn), U.S.	0.91	Metric ton (tonne) (t)
Speed		
Miles/hour (mph)	0.448	Meters/second (mps)
Miles/hour (mph)	1.6094	Kilometers/hour (kmph)

Metric Prefixes

Prefix		Meaning
giga-	G	$10^9 = 1{,}000{,}000{,}000$
mega-	M	$10^6 = 1{,}000{,}000$
kilo-	k	$10^3 = 1{,}000$
hecto-	h	$10^2 = 100$
deka-	da	$10^1 = 10$
		$10^0 = 1$
deci-	d	$10^{-1} = 0.1$
centi-	c	$10^{-2} = 0.01$
milli-	m	$10^{-3} = 0.001$
micro-	μ	$10^{-6} = 0.000001$

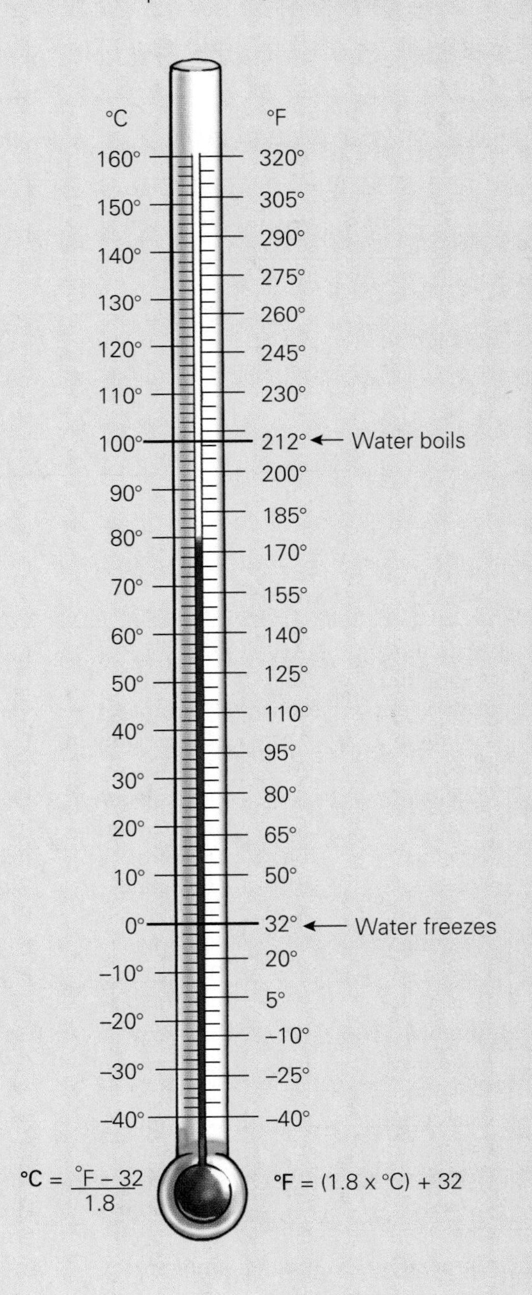

Periodic Table of the Elements

Metals | Metalloids | Nonmetals

(Key: Group number; Atomic number)

Period	1	2											3	4	5	6	7	8
1	1 H 1.008																	2 He 4.003
2	3 Li 6.941	4 Be 9.012											5 B 10.81	6 C 12.01	7 N 14.01	8 O 16.00	9 F 19.00	10 Ne 20.18
3	11 Na 22.99	12 Mg 24.31	Transition elements										13 Al 26.98	14 Si 28.09	15 P 30.97	16 S 32.07	17 Cl 35.45	18 Ar 39.95
4	19 K 39.10	20 Ca 40.08	21 Sc 44.96	22 Ti 47.87	23 V 50.94	24 Cr 52.00	25 Mn 54.94	26 Fe 55.85	27 Co 58.93	28 Ni 58.69	29 Cu 63.55	30 Zn 65.41	31 Ga 69.72	32 Ge 72.64	33 As 74.92	34 Se 78.96	35 Br 79.90	36 Kr 83.80
5	37 Rb 85.47	38 Sr 87.62	39 Y 88.91	40 Zr 91.22	41 Nb 92.91	42 Mo 95.94	43 Tc (98)	44 Ru 101.1	45 Rh 102.9	46 Pd 106.4	47 Ag 107.9	48 Cd 112.4	49 In 114.8	50 Sn 118.7	51 Sb 121.8	52 Te 127.6	53 I 126.9	54 Xe 131.3
6	55 Cs 132.9	56 Ba 137.3	57 La 138.9	72 Hf 178.5	73 Ta 180.9	74 W 183.8	75 Re 186.2	76 Os 190.2	77 Ir 192.2	78 Pt 195.1	79 Au 197.0	80 Hg 200.6	81 Tl 204.4	82 Pb 207.2	83 Bi 209.0	84 Po (209)	85 At (210)	86 Rn (222)
7	87 Fr (223)	88 Ra (226)	89 Ac (227)	104 Rf (261)	105 Db (262)	106 Sg (266)	107 Bh (264)	108 Hs (269)	109 Mt (268)	110 Ds (271)	111 — (272)	112 — (285)	113 — (284)	114 — (289)	115 — (288)			

Lanthanides	58 Ce 140.1	59 Pr 140.9	60 Nd 144.2	61 Pm (145)	62 Sm 150.4	63 Eu 152.0	64 Gd 157.3	65 Tb 158.9	66 Dy 162.5	67 Ho 164.9	68 Er 167.3	69 Tm 168.9	70 Yb 173.0	71 Lu 175.0
Actinides	90 Th 232.0	91 Pa 231.0	92 U 238.0	93 Np (237)	94 Pu (244)	95 Am (243)	96 Cm (247)	97 Bk (247)	98 Cf (251)	99 Es 252	100 Fm 257	101 Md 258	102 No 259	103 Lr 260

Name	Symbol
Actinium	Ac
Aluminum	Al
Americium	Am
Antimony	Sb
Argon	Ar
Arsenic	As
Astatine	At
Barium	Ba
Berkelium	Bk
Beryllium	Be
Bismuth	Bi
Bohrium	Bh
Boron	B
Bromine	Br
Cadmium	Cd
Calcium	Ca
Californium	Cf
Carbon	C
Cerium	Ce
Cesium	Cs
Chlorine	Cl
Chromium	Cr
Cobalt	Co
Copper	Cu
Curium	Cm
Darmstadtium	Ds
Dubnium	Db
Dysprosium	Dy
Einsteinium	Es
Erbium	Er
Europium	Eu
Fermium	Fm
Fluorine	F
Francium	Fr
Gadolinium	Gd
Gallium	Ga
Germanium	Ge
Gold	Au
Hafnium	Hf
Hassium	Hs
Helium	He
Holmium	Ho
Hydrogen	H
Indium	In
Iodine	I
Iridium	Ir
Iron	Fe
Krypton	Kr
Lanthanum	La
Lawrencium	Lr
Lead	Pb
Lithium	Li
Lutetium	Lu
Magnesium	Mg
Manganese	Mn
Meitnerium	Mt
Mendelevium	Md
Mercury	Hg
Molybdenum	Mo
Neodymium	Nd
Neon	Ne
Neptunium	Np
Nickel	Ni
Niobium	Nb
Nitrogen	N
Nobelium	No
Osmium	Os
Oxygen	O
Palladium	Pd
Phosphorus	P
Platinum	Pt
Plutonium	Pu
Polonium	Po
Potassium	K
Praseodymium	Pr
Promethium	Pm
Protactinium	Pa
Radium	Ra
Radon	Rn
Rhenium	Re
Rhodium	Rh
Rubidium	Rb
Ruthenium	Ru
Rutherfordium	Rf
Samarium	Sm
Scandium	Sc
Seaborgium	Sg
Selenium	Se
Silicon	Si
Silver	Ag
Sodium	Na
Strontium	Sr
Sulfur	S
Tantalum	Ta
Technetium	Tc
Tellurium	Te
Terbium	Tb
Thallium	Tl
Thorium	Th
Thulium	Tm
Tin	Sn
Titanium	Ti
Tungsten	W
Uranium	U
Vanadium	V
Xenon	Xe
Ytterbium	Yb
Yttrium	Y
Zinc	Zn
Zirconium	Zr

Answers

CHAPTER 1

Stop and Stretch

1.1. Given the information here, it is most likely that someone brought a box of doughnuts to share, and that Homer ate a large number of them. This hypothesis is based on inductive reasoning.

1.2. No, because even though Homer is a likely culprit, there are probably other individuals in your workplace who may have eaten some or all of the doughnuts.

1.3. Participants could have been exposed to different cold viruses; early arrivals may have been infected by a severe virus that was spreading in the hospital, while later arrivals may have been infected by a milder virus.

1.4. The independent variable is stress, and the dependent variable is susceptibility to developing a cold when exposed to a virus.

1.5. According to the poll, the actual percentage of the population who favor candidate A has a 95% probability of being between 44% and 50%, while the percentage who favor B is likely between 48% and 54%. Since these ranges overlap, it is currently unclear whether most of population favor A or B.

1.6. It is much easier to relate to individual experiences than to a more abstract piece of information. Anecdotes are the way humans have learned from each other throughout history; polling or statistical data are much more recent.

Visualize This

Figure 1.3. Even with a well-designed experiment, it is possible that an alternative factor can explain the results, and researchers need to consider if that might be the case in their experiment.

Figure 1.4. Preventing virus attachment and invasion of cells. Preventing activation of the immune system response. Other answers may be acceptable.

Figure 1.10. There may not have been enough participants at a particular stress level to allow effective analysis of the correlation. If individuals at stress level 3 have the same cold susceptibility as those at stress level 4, that should cause us to question the correlation since we'd expect those at stress level 4 to have greater susceptibility.

Figure 1.15. Hypothesis: true. Difference between control and experimental groups: large. Sample size: large. This result is very likely to be both statistically and practically significant.

Savvy Reader

1. Red flags for questions 1, 2, 9, 10, and 11. Most other questions from the checklist do not pertain to this particular excerpt.

Learning the Basics

1. A placebo allows members of the control group to be treated exactly like the members of the experimental group *except* for only their exposure to the independent variable.

2. b; **3.** b; **4.** d; **5.** a; **6.** d; **7.** e; **8.** c; **9.** b; **10.** c

Analyzing and Applying the Basics

1. No, another factor that causes type 2 diabetes could lead to obesity as well, or it is possible that type 2 diabetes causes obesity, not vice versa.

2. Neither the participants nor the researchers were blind, meaning that both subject expectation and observer bias might influence the results. If participants felt that vitamin C was likely to be effective, they may underreport their cold symptoms. If the clinic workers thought one or the other treatment was less effective, they might let that bias show when they were talking to the participants, influencing the results.

3. See the answer to number 1; BDNF may increase as a result of exercise, or excess BDNF might cause an increased interest in exercise.

CHAPTER 2

Stop and Stretch

2.1. The carbons should each be bonded to two hydrogens to complete their valence shells.

2.2. Yes, because ribosomes are not bounded by membranes and because prokaryotic cells need to synthesize proteins

Visualize This

Figure 2.4. Oxygen

Figure 2.7. 100×

Figure 2.16. Purines are larger and colored blue. Pyrimidines are smaller and yellow.

Figure 2.19. Central vacuole, cell wall, and chloroplast

Figure 2.21. Yes

Savvy Reader

1. No evidence presented to back up any of the claims made

2. Affiliation of scientist, funding source, data

Learning the Basics

1. Carbohydrates, lipids, proteins, nucleic acids

2. See Table 2.1

3. Phospholipid bilayer with embedded proteins

4. b; **5.** a; **6.** a; **7.** d; **8.** d; **9.** d; **10.** d

Analyzing and Applying the Basics

1. No. A virus is not capable of reproduction without help from the host cell.

2. Oxygen partial negative charge; carbon partial positive.

3. Silicon, like carbon, has four spaces in its valence shell and would also be tetravalent.

CHAPTER 3

Stop and Stretch

3.1. These food pairings together provide all the essential amino acids.

3.2. 200 calories per day

3.3. Active transport
3.4. The amount of the total cholesterol that is LDL

Visualize This

Figure 3.5. Increase
Figure 3.7. No
Figure 3.8. 348 Calories
Figure 3.18. Fat; eat less of this

Savvy Reader

1. Promotional material only
2. Only scientific journal articles undergo peer review.
3. Scientists hired by a company
4. Can't know for sure

Learning the Basics

1. All of the building up and breaking down chemical reactions that occur in a cell
2. Carbon dioxide, oxygen, and water
3. d; **4.** c; **5.** c; **6.** b; **7.** d; **8.** d; **9.** e; **10.** b

Analyzing and Applying the Basics

1. Not getting all the vitamins, minerals, and antioxidants present in whole foods
2. Metabolic rate, age, gender
3. If your BMI is in the normal range, you are probably at a healthful weight. If your BMI indicates that you are overweight but you are very muscular, you are probably at a healthful weight.

CHAPTER 4

Stop and Stretch

4.1. Greenhouse gases hold heat in and moderate temperature swings on Earth.
4.2. Carbon dioxide is sequestered in organic matter, soil, and the ocean.
4.3. More oxygen is available for the conversion of lactic acid to pyruvic acid.
4.4. The products of respiration are used during photosynthesis. The products of photosynthesis are reactants in respiration.
4.5. Nitrogen, sulfur, and phosphorous

Visualize This

Figure 4.1. Increased heat retention
Figure 4.3. Human activity
Figure 4.5. 2000s
Figure 4.17. *Glyco* means sugar, and *lysis* means cutting.
Figure 4.20. A double bond is broken to allow for the bonding of hydrogen
Figure 4.22. Oxygen exists in cells as molecular oxygen (O_2). Since only one oxygen atom is required to make water (H_2O), the 1/2 is placed before O_2.
Figure 4.31. Ultimately, yes because no NADPH would be produced.

Savvy Reader

1. Melting ice causes polar bears to have to swim more than normal, leading to increased drowning rates.
2. No
3. Continue to track polar bear drownings and rate of ice melt to see if correlation continues. Also look for alternate explanations for increased rate of polar bear drownings. May be that less body fat on polar bears makes them less buoyant, and so on.

Learning the Basics

1. Reactants of photosynthesis are carbon dioxide and water. Reactants of respiration are oxygen and glucose. Products of photosynthesis are oxygen and glucose. Products of respiration are carbon dioxide and water.
2. c; **3.** d; **4.** c; **5.** e; **6.** a; **7.** c; **8.** d; **9.** a; **10.** True

Analyzing and Applying the Basics

1. Fats and proteins also feed into cellular respiration at various points in the cycle.
2. C_3 plants function best in environments with high concentrations of carbon dioxide. C_4 and CAM plants function well in environments with lower carbon dioxide levels.
3. Respiration occurs in the mitochondrion. The citric acid cycle occurs in the matrix and electron transport in the inner mitochondrial membrane. Photosynthesis occurs in the thylakoid. The light reactions occur in the thylakoid membrane, and the Calvin cycle occurs in the stroma.

CHAPTER 5

Stop and Stretch

5.1. If a cell has many mutations, it is better that it is not allowed to continue to replicate.
5.2. DNA is semiconservatively replicated; that is, each strand serves as a template for the synthesis of a new strand.
5.3. Descent with modification explains the existence of similar processes in disparate organisms.
5.4. Individuals will die as cells reach their cell division limit.
5.5. Because it damages DNA
5.6. Both meiosis I and mitosis produce two daughter cells. Meiosis I differs from mitosis in that meiosis I separates homologous pairs of chromosomes from each other. Mitosis does not do this.

Visualize This

Figure 5.5. Four DNA molecules would be produced. Two of those would be all purple; two would be half purple, half red.
Figure 5.18. The X chromosome is the larger of the two.
Figure 5.25. Four

Savvy Reader

1. No
2. Yes; yes
3. No
4. Statistics are preferred.
5. It could be that he is making more money than the fine costs him, so it is worth it for him.

Learning the Basics

1. Mitosis occurs in somatic cells, meiosis in cells that give rise to gametes. Mitosis produces daughter cells that are genetically identical to parent cells, while meiosis produces daughter cells with novel combinations of chromosomes and half as many chromosomes as compared to parent cells.
2. Cell division
3. d; **4.** a; **5.** c; **6.** d; **7.** b; **8.** c; **9.** d; **10.** a, diploid; b, diploid; c, haploid; d, haploid

Analyzing and Applying the Basics

1. No. Only mutations in cells that produce gametes are passed to offspring.

2. More cell divisions means more opportunities for mutations to occur.
3. Cancers that are thought to be localized (not yet spread) are treated with radiation.

CHAPTER 6

Stop and Stretch

6.1. A mutation in a promoter portion of the DNA may affect whether the gene is copied or translated. A mutation in a structural portion may weaken the chromosome or change its ability to be transcribed. A mutation in "nonsense" DNA is likely to have no effect.

6.2. Eight different gametes. The relationship is 2n, where n is the number of chromosome pairs.

6.3. Its effects are not seen until later adulthood, after an individual is likely to have had children.

6.4. Immigration from countries where average height is shorter may be changing the genetics of height in the United States such that there are a larger number of "small" alleles today than in the past. Dietary changes may also be leading to reduced growth or early puberty, which stunts growth in height.

6.5. Even a trait that is highly heritable can be strongly influenced by the environment. If the current environment results in some individuals having very low IQ scores relative to others, it is certainly possible that a changed environment would lead to smaller differences among individuals. See the case of the "maze-dull" and "maze-bright" rats.

Visualize This

Figure 6.3. Many possibilities

Figure 6.6. Blood group gene from his dad eye color gene from his dad; Blood group gene from his mom, eye color gene from his mom.

Figure 6.14. 50%

Figure 6.16. No, the majority of men were *not* 5 feet 10 inches. There is a large range of heights in this population.

Figure 6.20. Dots would be more widely scattered, and the correlation line would be flatter.

Savvy Reader

1. It appears that individuals become more similar to their parents on various measures of religious practice as they age. The researchers argue that this indicates that the genetic influence on this trait increases as we age.
2. They do not provide any justification for why certain behaviors ("external") would be more subject to environmental influence. We are left to assume that if the behavior is visible to others, they assume a greater environmental influence. Without solid support for this distinction, the study is somewhat undermined because the researchers are deciding without evidence which traits "should" be genetic and which are not.
3. The headline is misleading. While the research seemed to indicate that there was a larger influence of genes on religious behavior as people age, the relative contribution of genes and environment is not discussed. The headline suggests that religious behavior is primarily genetic.
4. Clearly, religious behavior is influenced by the environment. This study supports the idea that there is a genetic basis for religious behaviors as well.

Learning the Basics

1. Genotype refers to the alleles you possess for a particular gene or set of genes. Phenotype is the physical trait itself, which may be influenced by genotype *and* environmental factors.
2. Multiple genes or multiple alleles influencing a trait, environmental effects that affect the expression of a trait, or a mixture of both factors
3. b; **4.** a; **5.** b; **6.** b; **7.** b; **8.** c; **9.** e; **10.** d

Genetics Problems

1. All have genotype *Tt* and are tall.
2. One-quarter are *TT* and are tall; one-half are *Tt* and are tall; and one-quarter are *tt* and are short.
3. Both are alleles for the same trait; *D* is dominant, and *d* is recessive.
4. Both parents must be heterozygous, *Aa*.
5. 50%
6. a. yellow; b. YY and yy
7. Yellow parent = Yy; green parent = yy
8. a. 50%; b. 50%; c. 25%; d. yes
9. a. dominant; b. the phenotype does not inevitably surface when the allele is present.

Analyzing and Applying the Basics

1. The trait could be recessive, and both parents could be heterozygotes. If the trait is coded for by a single gene with two alleles, one would expect 25% of the children (1 of 4) to have blue eyes. That 50% (2 of 4) have blue eyes does not refute Mendel because the probability of any genotype is independent for each child and is not dependent on the genotype of his or her siblings. (Consider that some families may have 4 boys and no girls; this does not refute the idea that the Y chromosome is carried by 50% of a man's sperm cells.)
2. No, most quantitative traits are influenced by the environment. Even if differences in genes account for most of the differences among individuals in a trait, if the environment changed, the trait would change as well.
3. No, heritability doesn't explain why any two individuals differ. For instance, John and Jerry could be identical twins, but Jerry might suffered a minor accidental head injury that reduced his IQ.

CHAPTER 7

Stop and Stretch

7.1. 8 gametes; 64 boxes in the Punnett square

7.2. $I^C I^C$ type C; $I^A I^C$ type AC; $I^B I^C$ type BC; I^C I type C; along with previously described AB types

7.3. Her father must have the disease, and her mother is a carrier or has the disease.

7.4. Crossing over; they may be infertile.

7.5. They may or may not carry different alleles of gene.

7.6. Because each parent only contributes half of their genetic information to a child

Visualize This

Figure 7.6. ½; ½

Figure 7.8. A cross involving a hemophilic female

Figure 7.11. This pedigree would have a square connected to a circle connected to a square in the first generation. The second generation would have a square from one set of parents and a circle from the second set of parents.

Savvy Reader

1. No; no
2. Answers will vary.
3. The author misstates the idea of natural selection. One's gametes do not know which traits to select.

Learning the Basics

1. More than two alleles are present in the population. Each individual can have (at most) two different alleles.
2. Females inherit one X chromosome from their mom in the egg cell and one from their dad in the sperm cell. Males inherit an X chromosome from their mom in the egg cell and a Y chromosome from their dad in the sperm cell.
3. DNA is isolated and cut with restriction enzymes. The fragments are loaded on a gel, and an electrical current is applied to separate them; DNA fragments are transferred to a filter and then probed. The probe binds in complementary regions. X-ray film is placed over filter to generate a fingerprint.
4. I^ai (dad), I^Bi (mom)
5. $\frac{1}{4}$A+; $\frac{1}{4}$A−; $\frac{1}{8}$AB−; $\frac{1}{8}$AB+; $\frac{1}{8}$B+; $\frac{1}{8}$B−;
6. c; **7.** b; **8.** c; **9.** e; **10.** d

Analyzing and Applying the Basics

1. Alleles that are codominant are equally expressed. An allele that is incompletely dominant does not completely mask the presence of the recessive allele.
2. Answers will vary. Just be certain that each band found in a daughter is present in one of the parents. About 50% of the bands in a given daughter will be present in each parent. The sisters should have around 50% of their bands in common.
3. Pedigrees will vary. However, the first cousins should have one set of common grandparents. If a grandparent is a carrier of a rare recessive allele, he or she could pass the same rare recessive allele to each of the cousins. If these cousins (both carriers of a rare recessive allele) mate, their child has a 1 in 4 chance of having the recessive disease.

CHAPTER 8

Stop and Stretch

8.1. No because with 2 base codons, only 16 different codons are possible
8.2. New alleles give natural selection something to act on.
8.3. Gene expression would no longer be negatively regulated. A gene may be on all the time.
8.4. There could be more growth hormone in the milk of treated cows.
8.5. Preembryonic cells are capable of becoming any cell type.
8.6. Dolly's chromosomes were that of an older sheep and therefore shortened. Compare the lengths of chromosomes from Dolly's cells with those of a newborn lamb.

Visualize This

Figure 8.1. The 2' carbon on the sugar is ribose and has an oxygen atom, while the same carbon on the deoxyribose does not. Uracil has a hydrogen instead of a methyl group.
Figure 8.3. Sample answer starting after the fork in the DNA: AGCTGGGCAGGTAC. From this particular DNA, the mRNA would be UCGACCCGUCCAUG.
Figure 8.5. As paired with U's or C's paired with G's
Figure 8.9. Bottom
Figure 8.13. It could reanneal to itself.

Savvy Reader

1. Answers vary
2. No. Cartoons are not peer reviewed.
3. No. This is not an issue that can be resolved by hypothesis testing.

Learning the Basics

1. GCUAAUGAAU
2. gln, arg, ile, leu
3. mRNA, ribosome, amino acids, tRNAs
4. b; **5.** c; **6.** b; **7.** c; **8.** c; **9.** b; **10.** d

Analyzing and Applying the Basics

1. The same amino acid is often coded for by codons that differ in the third position. For instance, UUU and UUC both code for phenylalanine.
2. Transcription would be slowed or stopped in that cell.
3. GAATTC
4. CTTAAG

CHAPTER 9

Stop and Stretch

9.1. The change in dress size reflects a change in the environment—in particular, today there is increased access to abundant, calorie-rich food compared to 50 years ago. This did not come about because of a change in inherited traits of individuals.
9.2. Because statements of faith are based on nonrepeatable, mostly unobserved phenomenon, they cannot be falsified. Thus, they are not subject to hypothesis testing.
9.3. It may be that the human diet was harder on teeth in the past, so that teeth were lost due to breakage, leaving "room" for the wisdom teeth to erupt. These teeth may have been beneficial, in fact, replacing the role of missing molars.
9.4. They are very close. The oldest human ancestor is slightly younger than the estimated date of divergence between chimpanzees and humans, as is expected.
9.5. Not necessarily, although conditions on the planet where life first evolved may have been different from what scientists have inferred about early Earth, perhaps making the emergence of life more likely elsewhere.

Visualize This

Figure 9.1. All lice would have been eliminated by the pesticide, and the population would not have evolved.
Figure 9.6. Perhaps the presence of plant-eating tortoises made those cacti with longer stems (thus out of reach of the tortoises) more likely to survive, leading to the evolution of longer stems. There

also may be a lack of other trees, meaning that taller cacti were able to capture more sunlight and thus reproduce more than smaller plants.

Figure 9.16. Hair-covered bodies, birth to live young, other characteristics of all mammals (also forward-facing eyes and opposable thumbs)

Figure 9.17. Squirrel would be the farthest left branch, originating above the mammal ancestor. The shared ancestor is the junction point.

Figure 9.23. 3 million years

Savvy Reader

1. The argument is essentially that the theory of evolution is suspect because several hundred scientists are skeptical of the theory.

2. Science is not a democratic process, but is based on evidence. The theory of evolution has support from many different subfields of biology, as well as chemistry and geology. A scientific theory that fits the evidence more effectively has not been proposed by these skeptics. In addition, the article provides little evidence that the skeptics are experts in evolutionary theory. The scientist quoted is an expert in human medicine.

3. Look at the "About" page of an Internet site to learn about the mission of the host organization. An organization with a particular agenda is more likely to cherry pick data that supports their agenda than to present a balanced view of an issue.

Learning the Basics

1. The theory of common descent describes that the similarities among living species can be explained as a result of their descent from a common ancestor.

2. b; **3.** d; **4.** d; **5.** b; **6.** e; **7.** c, b, a; **8.** d; **9.** e; **10.** d

Analyzing and Applying the Basics

1. Parentheses surround groups that share a common ancestor ((((pasture rose and multiflora rose) spring avens) live forever) spring vetch)

2. (((((((Humans and chimpanzee) mouse) donkey) carp) yeast) corn) green algae)

3. The comparison is between humans and these organisms in genetic sequence. There is no comparison among the organisms to base that assertion on. The differences between humans and yeast may be in totally different segments of DNA than the differences between humans and algae.

CHAPTER 10

Stop and Stretch

10.1. The TB bacteria can continue to multiply, doing damage to the lungs and resulting in a larger population. More cells means more likelihood of resistant variants, as well as a more difficult time clearing out the infection.

10.2. Humans have invented technologies and disease cures that decrease the differences in fitness among individuals with different genetics (for example, corrective lenses have increased the survival of individuals with poor eyesight). We are still subject to natural selection for traits that are less amenable to technological fixes and even in our ability to respond to technology.

10.3. The cystic fibrosis allele may provide protection from diseases that are common in more urban environments. Individuals carrying the allele in these environments were more likely to survive, thus causing the allele to become more common.

10.4. Mutations appear by chance and may not have much affect on fitness unless the environment changes. The resistance allele appeared randomly.

Visualize This

Figure 10.8. Because they are using most of their resources in generation 3, a percentage of generation 4 will die from starvation.

Figure 10.9. It decreased by about 75%.

Figure 10.11. The population would change in the direction of longer legs and coats.

Figure 10.13. The individual with the fast metabolism of alcohol would not have proportionally more offspring than those without the allele, so the ratio of fast to slow metabolizers would stay the same on each "shelf."

Savvy Reader

1. Airline air is polluted and leads to virus spreading. Andrew Speaker knowingly put his bride and others at risk of contracting a dangerous form of TB. There is no evidence for either statement.

2. Other postings by this author; evidence that the author linked to credible sites to back up statements of fact

3. Opinion question

Learning the Basics

1. Fitness is a term that describes the survival and reproductive output of an individual in a population relative to other members of the population.

2. Artificial selection is a process of selection of plants and animals by humans who control the survival and reproduction of members of a population to increase the frequency of human-preferred traits. Artificial selection is like natural selection in that it causes evolution; however, it differs because humans are directly choosing which organisms reproduce. In natural selection, environmental conditions cause one variant to have higher fitness than other variants.

3. Within a single patient, the *Mycobacterium tuberculosis* consists of a large number of genetic variants. These variants differ in a number of traits, including their susceptibility to a particular antibiotic. When the patient takes an antibiotic, the individual bacteria that are more resistant to the drug have higher fitness, and natural selection leads to the evolution of a bacteria population that is resistant to the drug.

4. c; **5.** d; **6.** e; **7.** e; **8.** b; **9.** c; **10.** d

Analyzing and Applying the Basics

1. Farmers who desired larger strawberries must have saved seeds from the plants that produced the largest berries. Over the course of many generations of the same type of selection, average berry size increased dramatically.

Tradeoffs include less energy to put into roots and leaves, a large requirement for supplied water and nutrients, and perhaps fewer chemical defenses against insects and other predators.

2. Individuals with some sort of striping must have had greater survival than those who were not striped. Over the course of many generations of this type of selection, individuals in the population evolved dramatic stripes.

3. No, not all features are adaptations; some might arise by chance. A trait is an adaptation if it increases fitness relative to others without the trait. One can look for ways that the trait increases fitness or examine individuals who lack or have a less-dramatic version of the trait to measure its fitness effects.

CHAPTER 11

Stop and Stretch

11.1. Individuals with fewer or lighter spots had higher fitness, perhaps because they blended better into the savannah background, allowing them to sneak up on prey.

11.2. No, because the varieties are the same species, so the number of chromosomes should be the same in both parents.

11.3. Morphological concept

11.4. Multiregional concept

11.5. Average height varies but is not typically used to identify similar groups. Adding this category might lead to identifying a larger number of races.

11.6. "Ma" or "mmmm" is one of the first sounds an infant can make. Since the mother is so associated with infant care, this sound became associated with her. ("Da" or "Pa" is very early as well.)

11.7. Females put more investment (energy) in offspring compared to males, so have a higher incentive for making sure that the offspring has the greatest chance of survival and later reproduction.

Visualize This

Figure 11.5. No, because the chromosomes from the two different parents still could not pair properly

Figure 11.8. One common factor likely led to speciation in these pairs. The most likely factor is the separation event.

Figure 11.13. According to the definition, yes

Figure 11.16. The change occurred after the emergence of the blue race and distinguishes two populations in the blue race from the third population. It does not necessarily support the red/blue hypothesis.

Figure 11.17. Heterozygotes can form via (p)males × (q)females AND by (p)females × (q)males.

Figure 11.24. Various possibilities

Savvy Reader

1. The drug BiDil is ineffective in whites, but effective in African-Americans. Asthma drugs are less well absorbed by African-Americans compared to whites.

2. In studies of hypertension, some populations identified as black have very low rates of the disease, despite the fact that high rates of hypertension are assumed to be more common in blacks. Race is also associated with lower quality health care, even taking into account poverty, age, and condition severity, indicating that racism is at play. Race is often a proxy for socioeconomic status.

3. The existence of studies that identify certain alleles or groups of alleles that are more common in one race; studies that divide people by genetic markers rather than self-identified racial groupings and show a correlation between group and disease outcome

Learning the Basics

1. The gene pools of populations must become isolated from each other, they must diverge (as a result of natural selection or genetic drift) while they are isolated, and reproductive incompatibility must evolve.

2. Yes, as long as the gene pools of the populations are isolated. An example is when the timing of reproduction in two populations of a species is different.

3. Genetic drift can occur as a result of founder's effect, the bottleneck effect, or by the chance loss of alleles from small, isolated gene pools. By changing allele frequencies, all of these events result in the evolution of a population.

4. d; **5.** b; **6.** a; **7.** c; **8.** d; **9.** d; **10.** c

Analyzing and Applying the Basics

1. It depends on the species definition one uses. The two populations are not different biological species. They are likely different genealogical species—the definition that corresponds to "race." To determine if they are different races, one would have to look for the standard evidence: unique alleles or unique allele frequencies.

2. In Ireland, 0.01; Britain 0.007; Scandinavia 0.005. The allele frequency could be different as a result of genetic drift, particularly the founder effect, in these different populations. It also could be different because the allele has some fitness advantage in the environment of Ireland relative to the environment of Britain or Scandinavia.

3. If the early mutation becomes widespread, the timing difference between the early and normal populations should cause reproductive isolation. The same is possible with the expanded mutation if certain pollinators become associated with the individuals that bloom earlier than expected.

CHAPTER 12

Stop and Stretch

12.1. Superficial similarities allow observers to group things quickly, by obviously visible characteristics, and without knowing much biology.

12.2. Mitochondria in plant and animal cells are identical, indicating that they evolved before plants and animals diverged.

12.3. It is highly visible and easy to catch. Predators would quickly pick these animals off if they were not poisonous.

12.4. Early human societies were probably more likely to test plants as possible food and drug sources; the drug use of other organisms would not be as obvious (and microorganisms would be unknown).

12.5. They cannot reproduce, do not use energy to maintain homeostasis, and do not have a cellular structure.

12.6. A beak that can tear meat, long wings, large size, and the ability to soar. This classification can help beginner birders learn to quickly identify all birds that share these easily visible characteristics.

Visualize This

Figure 12.5. The bases that are common to all groups provide a framework; then older species are more likely to have the ancestral bases than younger species.

Figure 12.12. To advertise its poisonous nature, "scaring" predators away and preventing injury to the frog caused by a predator "testing" it

Figure 12.15. To allow dispersal of spores away from the food source that is already in use

Figure 12.22. It probably had little head striping, like the junco, but a sparrow colored body (brown striped) like all of the sparrows.

Savvy Reader

1. He has authored another book and somehow has insider knowledge about pharmaceutical companies and government regulators. It is not clear what his training or other experience is.

2. That organizations that people often consider untrustworthy (big government, large corporations, the mainstream media) are withholding information that can make your life better inexpensively; that someone who risks prosecution to help you deserves to be trusted (and supported with your book purchase)

3. Look for the references Trudeau uses to support his statements—are they from peer-reviewed journals? Do they represent a review of all of the data, or are they single, perhaps unusual, studies? Is the book itself reviewed by qualified independent physicians and scientists?

Learning the Basics

1. From folds of the plasma membrane that segregated nuclear material from the rest of the cell

2. Competition between organisms and the risk of predation favor the evolution of defensive chemicals, some of which may be valuable to humans. Organisms that live on or in other living organisms must have some way to evade their host's immune system, another potentially valuable characteristic.

3. By comparing the DNA sequences of organisms to see if the pattern of similarity and difference matches what is predicted and by examination of the fossil record, looking for the record of evolutionary change in the group of organisms

4. b; **5.** e; **6.** e; **7.** d; **8.** c; **9.** d; **10.** b

Analyzing and Applying the Basics

1. An oxygen-sensitive cell that kept an oxygen-consuming prokaryote alive inside of it would be protected from the damaging effects of oxygen. This would be an advantage over cells that did not have this prokaryote.

2. Although it is not very motile, this is probably an animal since it consumes other organisms (is not photosynthetic) by ingesting.

3. Vertebrates are unlikely to compete with bacteria for food resources, so they are unlikely to need to "kill them off." Vertebrates also have an immune system for dealing with infections, so typically do not need to invest in manufacturing antibiotics.

CHAPTER 13

Stop and Stretch

13.1. 10,000 (solve for x: 1/100 = 100/x)

13.2. Average life span equals the sum of the ages at death divided by the number of deaths. Deaths of infants thus greatly bring down the average life span, and reducing these deaths will raise it dramatically, even if the overall population health has not changed.

13.3. Infectious diseases require a population of susceptible individuals to survive and spread. If the population is small, the infectious organism cannot maintain a population.

13.4. A large population with a high birth rate produces a large number of offspring, potentially overshooting the carrying capacity. A large population with a low birth rate produces many fewer young, allowing a gradual approach to the carrying capacity.

Visualize This

Figure 13.1. Trap-happiness would cause the population estimate to be too small since a greater proportion of returnees will be marked. Trap-shyness causes the opposite effect since fewer returnees will be marked.

Figure 13.5. Where birth rates and death rates are closest to equal; that is, before the transition and after

Figure 13.6. Where the line is most vertical; that is, at the point where the "S" turns from rising to flattening out

Figure 13.13. The United States has about 10 times more people than does South Africa.

Figure 13.14. Several reasons: larger immigration rate, younger population on average, religious traditions

Savvy Reader

1. The Christmas Bird Count and Breeding Bird Surveys, "citizen science" programs (i.e., birdwatchers participate in data collection)

2. Many answers possible, such as, evidence that the causes Audubon identifies are truly responsible, the types of actions that are needed, how effective these actions are in preserving bird populations, etc.

Learning the Basics

1. A decrease in death rate, especially a decrease in infant mortality, has led to increasing growth rates. In addition, the process of exponential growth lends itself to these population explosions as the number of people added in any year is a function of the population of the previous year.

2. In most populations, growth rate declines because death rate increases (or birth rate decreases) as resources are "used up" by the population.

3. b; **4.** c; **5.** b; **6.** a; **7.** b; **8.** d; **9.** d; **10.** e

Analyzing and Applying the Basics

1. 2500. Solve for x: 1/50 = 50/x.

2. −0.002 (r = Birth rate − Death rate = 25/1000 − 27/1000). This population is likely near carrying capacity since it is not growing, but in fact declining slightly, as would be expected if resources are limited.

3. The carrying capacity is expected to rise since more resources are available. However, because the flies are in a restricted environment (a culture bottle), the accumulation of waste may become a problem, and a large population of flies may not be able to survive this accumulation. Depending on fly behavior, there may also be an increase in aggressive interactions or a decrease in mating in more dense situations.

CHAPTER 14

Stop and Stretch

14.1. Habitat destruction, pollution, introduction of exotics, even overexploitation of colorful or dramatic species. It would help to know the environmental situation of many of the species at risk to determine if they are threatened by a common cause. It may be just that these animals are especially vulnerable to all threats because of their unique biology.

14.2. All species pairs are competitors; the organisms they compete for are a) fish b) krill c) penguins d) squid e) herbivorous zooplankton. The most straightforward way to test if competition is occuring is to remove one of the competing species and see if the remaining species increases in population size.

14.3. Unlike nitrogen, carbon can be sequestered for long periods in buried vegetation and fossil fuels. Human activity is affecting the carbon cycle by releasing this stored carbon in a short amount of time.

14.4. Moving animals around so that close relatives are less likely to mate; importing animals from other areas for the same reason

14.5. If homozygous individuals are more common in small populations, individuals who express the disease (homozygous recessives) are also more common. If these individuals do not survive, the allele is less likely to be passed on to the next generation, so change in allele frequency is more rapid.

Visualize This

Figure 14.2. 10 to 50 million years

Figure 14.7. About 78 species. If the island lost 67% of its habitat (10,000 of 15,000 square kilometers), you'd expect about 5% of the species to become extinct.

Figure 14.16. The total number of bacteria reaches the carrying capacity of the environment (that is, the intestines)

Figure 14.18. In modern human systems, nutrients typically flow in one direction (from crops to humans to waste water and cemeteries), so we can no longer refer to the process as a nutrient cycle.

Figure 14.24. Because the normal allele does not cause disease, which is the problem with the homozygous recessive

Savvy Reader

1. That the 20 fish populations designated as greenback trout in Colorado actually consisted of greenback trout

2. By looking at the DNA of the fish in these populations

3. Missing information: what the genetic profile of the original source population was; how much genetic diversity is necessary in these populations for their long-term survival; whether the greenback has colonized other streams not investigated by the researchers. Makes one skeptical: Subspecies are indistinguishable in appearance and habitat except (apparently) genetically. Should they be classified as different species?

Learning the Basics

1. Habitat fragmentation endangers species because it interferes with their ability to disperse from unsuitable to suitable habitat and because it increases their exposure to humans and human-modified environments. Species that have very specialized habitat requirements, those that cannot move rapidly, and those that are very susceptible to human disturbance are most negatively affected by fragmentation.

2. Mutualism is a relationship among species in which all partners benefit. Predation is a relationship among species in which one benefits and others are consumed. Competition is a relationship among species in which all partners are harmed by the presence of the others.

3. d; **4.** c; **5.** a; **6.** a; **7.** b; **8.** c; **9.** d; **10.** a

Analyzing and Applying the Basics

1. Kelp → sea urchins → sea otters. Kelp also provides habitat for fishes, clams, and abalones. When sea otters are removed, sea urchins devastate the kelp forest, leading to the loss of these associated species. Sea otter thus serves as a keystone.

2. Even though (A) has more individuals, with so few fathers contributing genes to the next generation, the likelihood of large changes in allele frequency due to drift (i.e., the founder effect) is much higher.

3. How uncertain the environmental conditions are—that is, do density-independent factors cause the death of large numbers of individuals when they occur? What the mating system is—are a few males mating with the majority of the females, or is the species more monogamous? What is the likely carrying capacity for piping plovers in this environment?

CHAPTER 15

Stop and Stretch

15.1. A reduction in tilt would reduce seasonality because areas would be receiving the same solar irradiance throughout the year. An increase in tilt would increase the yearly range of temperatures in seasonal areas.

15.2. The ocean is dark and thus absorbs light and heats up. Ice reflects the light and keeps areas around it cold.

15.3. Chlorophyll may be "expensive" to make in some way. There may be more competitors for some of the elements in chlorophyll, so it may be risky for the tree to drop these elements and try to rely on reabsorption through the roots the next summer. Other pigments may be mostly carbon, which is easy to acquire.

15.4. Wetlands slow water flow and thus can absorb extra nutrients before they accumulate in the open water body.

15.5. Fossil fuels were easy to access just by digging. The technology to convert the fuels to electricity (burning and generating steam) or to power an engine (controlled explosions) were a lot simpler than the technology needed to convert sunlight into electricity. Wind and solar power is not as reliable.

Visualize This

Figure 15.3. Because when the Northern Hemisphere is tilted away from the sun, the Southern Hemisphere (where Australia is located) is tilted toward it

Figure 15.6. This period is the hottest part of the summer, and areas near the sea will be cooler than areas inland.

Figure 15.8. Near 30° north and south and near the equator, where air is more likely to be moving vertically rather than horizontally.

Figure 15.9. The east side is windward, and the west side is leeward.

Figure 15.19. Cattle are not nearly as large or strong as elephants and cannot uproot these large trees.

Figure 15.20. Prairie plants can accumulate energy all summer without being shaded over, while in temperate forests, sunlight disappears on the forest floor by early summer.

Figure 15.30. In most estuaries, there is a mixture of salt and freshwater. Intertidal zones will be saltwater only. Estuaries are unlikely to fluctuate in water availability so dramatically compared to intertidal zones.

Savvy Reader

1. Cleaner burning natural gas, solar, wind, hydrogen, and biofuels

2. It's not totally clear. The $8 billion over 10 years in solar, wind, hydrogen, and natural gas is $800 million per year, but most could be invested in natural gas, which is not "green" (it is still a fossil fuel).

3. You'd have to know how much they were investing in oil development and determine what percentage of their investment in alternatives is really in renewable fuels.

Learning the Basics

1. The tilt of Earth's axis means that as the planet revolves around the sun, the Northern Hemisphere is tilted toward the sun during part of the year and away during part of the year. The temperature at a given point on Earth's surface is determined in large part by the solar irradiance (i.e., the strength of sunlight) striking the surface. Because the Northern Hemisphere is tilted away in the winter, solar irradiance is low, as are temperatures. The opposite occurs in the summer.

2. Water is slow to gain and lose heat, so when surrounding land areas cool as a result of declining solar irradiance, warmer air from over the water keeps the temperature high. The opposite is true as the surrounding land warms as a result of increasing solar irradiance.

3. b; **4.** c; **5.** e; **6.** e; **7.** c; **8.** a; **9.** b; **10.** d

Analyzing and Applying the Basics

1. Semitropical with seasonal wet and dry periods as a result of latitude. Relatively rainy due to location on windward side of mountain range.

2. The temperature will become less moderate because warm water will not be carried to the coasts of Europe, warming nearby landmasses. Seasonality will be more pronounced and winters colder and longer.

3. Bend, Yakima, and perhaps Walla Walla are on the leeward side of the Cascades, while the other cities are on the windward side.

Glossary

A

ABO blood system A system for categorizing human blood based on the presence or absence of carbohydrates on the surface of red blood cells. (Chapter 7)

acid A substance that increases the concentration of hydrogen ions in a solution. (Chapter 2)

activation energy The amount of energy that reactants in a chemical reaction must absorb before the reaction can start. (Chapter 3)

activator A protein that serves to enhance the transcription of a gene. (Chapter 8)

active site Substrate-binding region of an enzyme. (Chapter 3)

active transport The ATP-requiring movement of substances across a membrane against their concentration gradient. (Chapter 3)

adaptation Trait that is favored by natural selection and increases an individual's fitness in a particular environment. (Chapters 9, 10)

adaptive radiation Diversification of one or a few species into large and very diverse groups of descendant species. (Chapter 12)

adenine Nitrogenous base in DNA, a purine. (Chapters 2, 4, 5, 8)

adenosine diphosphate (ADP) A nucleotide composed of adenine, a sugar, and two phosphate groups. Produced by the hydrolysis of the terminal phosphate bond of ATP. (Chapter 4)

adenosine triphosphate (ATP) A nucleotide composed of adenine, the sugar ribose, and three phosphate groups that can be hydrolyzed to release energy. Form of energy that cells can use. (Chapters 2, 3, 4)

aerobic An organism, environment, or cellular process that requires oxygen. (Chapter 4)

aerobic respiration Cellular respiration that uses oxygen as the electron acceptor. (Chapter 4)

agarose gel A jelly-like slab used to separate molecules on the basis of molecular weight. (Chapter 7)

algae Photosynthetic protists. (Chapter 12)

allele Alternate versions of the same gene, produced by mutations. (Chapters 5, 6, 7, 8, 10, 14)

allele frequency The percentage of the gene copies in a population that are of a particular form, or allele. (Chapters 11, 14)

allopatric Geographic separation of a population of organisms from others of the same species. Usually in reference to speciation. (Chapter 11)

alternative hypothesis Factor other than the tested hypothesis that may explain observations. (Chapter 1)

amenorrhea abnormal cessation of menstrual cycle. (Chapter 3)

amino acid Monomer subunit of a protein. Contains an amino, a carboxyl, and a unique side group. (Chapters 2, 3, 8)

anaerobic An organism, environment, or cellular process that does not require oxygen. (Chapter 4)

anaerobic respiration A process of energy generation that uses molecules other than oxygen as electron acceptors. (Chapter 4)

anaphase Stage of mitosis during which microtubules contract and separate sister chromatids. (Chapter 5)

anchorage dependence Phenomenon that holds normal cells in place. Cancer cells can lose anchorage dependence and migrate into other tissues or metastasize. (Chapter 5)

anecdotal evidence Information based on one person's personal experience. (Chapter 1)

angiogenesis Formation of new blood vessels. (Chapter 5)

angiosperm Plant in the phyla Angiospermae, which produce seeds borne within fruit. (Chapter 12)

animal an organism that obtains energy and carbon by ingesting other organisms and is typically motile for part of its life cycle (Chapter 12)

Animalia Kingdom of Eukarya containing organisms that ingest others and are typically motile for at least part of their life cycle. (Chapters 9, 12)

annual Plant that completes its life cycle in a single growing season. (Chapter 15)

annual growth rate Proportional change in population size over a single year. Growth rate is a function of the birth rate minus the death rate of the population. (Chapter 13)

anorexia Self-starvation. (Chapter 3)

antibiotic A chemical that kills or disables bacteria. (Chapters 10, 12)

antibiotic resistant Characteristic of certain bacteria; a physiological characteristic that permits them to survive in the presence of particular antibiotics. (Chapter 10)

anticodon Region of tRNA that binds to a mRNA codon. (Chapter 8)

antioxidant Certain vitamins and other substances that protect the body from the damaging effects of free radicals. (Chapter 3)

antiparallel Feature of DNA double helix in which nucleotides face "up" on one side of the helix and "down" on the other. (Chapters 2, 5)

aquaporin A transport protein in the membrane of a plant or animal cell that facilitates the diffusion of water across the membrane (osmosis). (Chapter 3)

aquatic Of, or relating to, water. (Chapter 15)

Archaea Domain of prokaryotic organisms made up of species known from extreme environments. (Chapters 2, 12)

artificial selection Selective breeding of domesticated animals and plants to increase the frequency of desirable traits. (Chapters 6, 10)

asexual reproduction A type of reproduction in which one parent gives rise to genetically identical offspring. (Chapters 5, 11)

assortative mating Tendency for individuals to mate with someone who is like themselves. (Chapter 11)

atom The smallest unit of matter that retains the properties of an element. (Chapter 2)

atomic number The number of protons in the nucleus of an atom. Unique to each element, this number is designated by a subscript to the left of the symbol for the element. (Chapter 2)

ATP synthase Enzyme found in the mitochondrial membrane that helps synthesize ATP. (Chapter 4)

autosome Non-sex chromosome, of which there are 22 pairs in humans. (Chapters 5, 6, 7)

B

background extinction rate The rate of extinction resulting from the normal process of species turnover. (Chapter 14)

Bacteria Domain of prokaryotic organisms. (Chapters 2, 4, 8, 12, 14)

basal metabolic rate Resting energy use of an awake, alert person. (Chapter 3)
base A substance that reduces the concentration of hydrogen ions in a solution. (Chapter 2)
behavioral isolation Prevention of mating between individuals in two different populations based on differences in behavior. (Chapter 11)
benign Tumor that stays in one place and does not affect surrounding tissues. (Chapter 5)
bias Influence of research participants' opinions on experimental results. (Chapter 1)
biodiversity Variety within and among living organisms. (Chapters 12, 14)
biogeography The study of the geographic distribution of organisms. (Chapter 9)
biological classification Field of science attempting to organize biodiversity into discrete, logical categories. (Chapters 9, 12)
biological diversity Entire variety of living organisms. (Chapter 12)
biological evolution See evolution. (Chapter 9)
biological population Individuals of the same species that live and breed in the same geographic area. (Chapters 9, 11)
biological race Populations of a single species that have diverged from each other. Biologists do not agree on a definition of "race." See also subspecies. (Chapter 11)
biological species concept Definition of a species as a group of individuals that can interbreed and produce fertile offspring but typically cannot breed with members of another species. (Chapter 11)
biology The study of living organisms. (Chapter 1)
biomass The mass of all individuals of a species, or of all individuals on a level of a food web, within an ecosystem. (Chapter 14)
biome A broad ecological community defined by a particular vegetation type (e.g. temperate forest, prairie), which is typically determined by climate factors. (Chapter 15)
biophilia Humans' innate desire to be surrounded by natural landscapes and objects. (Chapter 14)
biopiracy Using the knowledge of the native people in developing countries to discover compounds for use in developed countries. (Chapter 12)
bioprospecting Hunting for new organisms and new uses of old organisms. (Chapter 12)
biopsy Surgical removal of some cells, tissue, or fluid to determine if cells are cancerous. (Chapter 5)
bipedal Walking upright on two limbs. (Chapter 9)
blind experiment Test in which subjects are not aware of exactly what they are predicted to experience. (Chapter 1)
body mass index (BMI) Calculation using height and weight to determine a number that correlates to an estimate of a person's amount of body fat with health risks. (Chapter 3)
boreal forest A biome type found in regions with long, cold winters and short, cool summers. Characterized by coniferous trees. (Chapter 15)
botanist Plant biologist. (Chapter 12)
bulimia Binge eating followed by purging. (Chapter 3)

C

C_3 plant Plant that uses the Calvin cycle of photosynthesis to incorporate carbon dioxide into a 3-carbon compound. (Chapter 4)
C_4 plant Plant that performs reactions incorporating carbon dioxide into a 4-carbon compound that ultimately provides carbon dioxide for the Calvin cycle. (Chapter 4)
calcium Nutrient required in plant cells for the production of cell walls and for bone strength and blood clotting in humans. (Chapter 3)
Calorie A kilocalorie or 1000 calories. (Chapter 3)
calorie Amount of energy required to raise the temperature of one gram of water by 1 °C. (Chapter 3)
Calvin cycle A series of reactions that occur in the stroma of plants during photosynthesis that utilize NADPH and ATP to reduce carbon dioxide and produce sugars. (Chapter 4)
CAM plant A plant that uses Crassulacean acid metabolism, a variant of photosynthesis during which carbon dioxide is stored in sugars at night (when stomata are open) and released during the day (when stomata are closed) to prevent water loss. (Chapter 4)
Cambrian explosion Relatively rapid evolution of the modern forms of multicellular life that occurred approximately 550 million years ago. (Chapter 12)
cancer A disease that occurs when cell division escapes regulatory controls. (Chapter 5)
capillary The smallest blood vessel of the cardiovascular system, connecting arteries to veins and allowing material exchange across their thin walls. (Chapter 5)
carbohydrate Energy-rich molecule that is the major source of energy for the cell. Consists of carbon, hydrogen, and oxygen in the ratio CH_2O. (Chapters 2, 3)
carcinogen Substance that causes cancer or increases the rate of its development. (Chapter 5)
carrier Individual who is heterozygous for a recessive allele. (Chapters 6, 7)
carrying capacity Maximum population that the environment can support. (Chapter 13)
catalyst A substance that lowers the activation energy of a chemical reaction, thereby speeding up the reaction. (Chapter 3)
catalyze To speed up the rate of a chemical reaction. Enzymes are biological catalysts. (Chapter 3)
cell Basic unit of life, an organism's fundamental building-block units. (Chapters 2, 3)
cell cycle An ordered sequence of events in the life cycle of a eukaryotic cell from its origin until its division to produce daughter cells. Consists of M, G_1, S, and G_2 phases. (Chapter 5)
cell division Process a cell undergoes when it makes copies of itself. Production of daughter cells from an original parent cell. (Chapters 5, 6)
cell plate A double layer of new cell membrane that appears in the middle of a dividing plant cell and divides the cytoplasm of the dividing cell. (Chapter 5).
cell wall Tough but elastic structure surrounding plant and bacterial cell membranes. (Chapters 2, 4, 5)
cellular respiration Metabolic reactions occurring in cells that result in the oxidation of macromolecules to produce ATP. (Chapter 4)
cellulose A structural polysaccharide found in cell walls and composed of glucose molecules. (Chapters 2, 4, 5)
central vacuole A membrane enclosed sac in a plant cell that functions to store many different substances. (Chapter 2)
centriole A structure in animal cells that helps anchor for microtubules during cell division. (Chapters 2, 5)
centromere Region of a chromosome where sister chromatids are attached and to which microtubules bind. (Chapter 5)

chaparral A biome characteristic of climates with hot, dry summers and mild, wet winters and a dominant vegetation of aromatic shrubs. (Chapter 15)
checkpoint Stoppage during cell division that occurs to verify that division is proceeding correctly. (Chapter 5)
chemical reaction A process by which one or more chemical substances is transformed into one or more different chemical substances. (Chapter 2)
chemotherapy Using chemicals to try to kill rapidly dividing (cancerous) cells. (Chapter 5)
chlorophyll Green pigment found in the chloroplast of plant cells. (Chapter 4)
chloroplast An organelle found in plant cells that absorbs sunlight and uses the energy derived to produce sugars. (Chapters 2, 4, 12)
cholesterol A steroid found in animal cell membranes that affects membrane fluidity. Serves as the precursor to estrogen and testosterone. (Chapters 2, 3)
chromosome Subcellular structure composed of a long single molecule of DNA and associated proteins, housed inside the nucleus. (Chapters 5, 6, 7, 8, 11)
circulatory system The vessels that transport blood, nutrients, and waste around the body. (Chapter 5)
citric acid cycle A chemical cycle occurring in the matrix of the mitochondria that breaks the remains of sugars down to produce carbon dioxide. (Chapter 4)
cladistic analysis A technique for determining the evolutionary relationships among organisms that relies on identification and comparison of newly evolved traits. (Chapter 12)
classification system Method for organizing biological diversity. (Chapters 9, 12)
cleavage In embryology, the period of rapid cell division that occurs during animal development. (Chapter 5)
climate The average temperature and precipitation as well as seasonality. (Chapter 15)
climax community The group of species that is stable over time in a particular set of environmental conditions. (Chapter 15)
clinical trial Controlled scientific experiment to determine the effectiveness of novel treatments. (Chapter 5)
cloning Producing copies of a gene or an organism that are genetically identical. (Chapter 8)
clumped distribution A spatial arrangement of individuals in a population where large numbers are concentrated in patches with intervening, sparsely populated areas separating them. (Chapter 13)
codominant Two different alleles of a gene that are equally expressed in the heterozygote. (Chapters 6, 7)
codon A triplet of mRNA nucleotides. Transfer RNA molecules bind to codons during protein synthesis. (Chapter 8)
coenzyme (or **cofactor**) Substances such as vitamins that help enzymes catalyze chemical reactions. (Chapter 3)
cohesion The tendency for molecules of the same material to stick together. (Chapter 2)
common descent The theory that all living organisms on Earth descended from a single common ancestor that appeared in the distant past. (Chapter 9)
community A group of interacting species in the same geographic area. (Chapter 14)
competition Interaction that occurs when two species of organisms both require the same resources within a habitat; competition tends to limit the size of populations. (Chapter 14)
competitive exclusion Reduction or elimination of one species in an environment resulting from the presence of another species that requires the same or similar resources. (Chapter 14)
complementary base pair Nitrogenous bases that hydrogen bond to each other. In DNA, adenine is complementary to thymine, and cytosine is complementary to guanine. In RNA, adenine is complementary to uracil and guanine to cytosine. (Chapters 2, 5, 7, 8)
complete protein Dietary protein that contains all the essential amino acids. (Chapter 3)
complex carbohydrate Carbohydrate consisting of two or more monosaccharides. (Chapter 3)
compound A substance consisting of two or more elements in a fixed ratio. (Chapter 2)
confidence interval In statistics, a range of values calculated to have a given probability (usually 95%) of containing the true population mean. (Chapter 1)
consilience The unity of knowledge. Used to describe a scientific theory that has multiple lines of evidence to support it. (Chapter 9)
contact inhibition Property of cells that prevents them from invading surrounding tissues. Cancer cells may lose this property. (Chapter 5)
continuous variation A range of slightly different values for a trait in a population. (Chapter 6)
control Subject for an experiment who is similar to experimental subject except is not exposed to the experimental treatment. Used as baseline values for comparison. (Chapter 1)
convergent evolution Evolution of same trait or set of traits in different populations as a result of shared environmental conditions rather than shared ancestry. (Chapters 9, 11)
coral reef Highly diverse biome found in warm, shallow salt water, dominated by the limestone structures created by coral animals. (Chapters 12, 15)
correlation Describes a relationship between two factors. (Chapters 1, 6, 11)
covalent bond A type of strong chemical bond in which two atoms share electrons. (Chapter 2)
cross In genetics, the mating of two organisms. (Chapter 6)
crossing over Gene for gene exchange of genetic information between members of a homologous pair of chromosomes. (Chapters 5, 6)
cyst Noncancerous, fluid-filled growth. (Chapter 5)
cytokinesis Part of the cell cycle during which two daughter cells are formed by the cytoplasm splitting. (Chapter 5)
cytoplasm The entire contents of the cell (except the nucleus) surrounded by the plasma membrane. (Chapters 2, 8)
cytosine Nitrogenous base, a pyrimidine. (Chapters 2, 4, 5, 8)
cytoskeleton A network of tubules and fibers that branch throughout the cytoplasm. (Chapter 2)
cytosol The semifluid portion of the cytoplasm. (Chapters 2, 4)

D

data Information collected by scientists during hypothesis testing. (Chapter 1)
daughter cells The offspring cells that are produced by the process of cell division. (Chapter 5)
death rate Number of deaths averaged over the population as a whole. (Chapter 13)
deciduous Pertaining to woody plants that drop their leaves at the end of a growing season. (Chapter 15)
decomposer An organism, typically bacteria and fungi in the soil, whose action breaks down complex molecules into simpler ones. (Chapter 14)

decomposition The breakdown of organic material into smaller molecules. (Chapter 15)
deductive reasoning Making a prediction about the outcome of a test; "if / then" statements. (Chapter 1)
deforestation The removal of forest lands, often to enable the development of agriculture. (Chapters 4, 14)
degenerative disease Disease characterized by progressive deterioration. (Chapter 8)
dehydration Loss of water. (Chapter 3)
demographic momentum Lag between the time that humans reduce birth rates and the time that population numbers respond. (Chapter 13)
demographic transition The period of time between when death rates in a human population fall (as a result of improved technology) and when birth rates fall (as a result of voluntary limitation of pregnancy). (Chapter 13)
denature (1) In proteins, the process where proteins unravel and change their native shape, thus losing their biological activity. (Chapter 2) (2) For DNA, the breaking of hydrogen bonds between the two strands of the double-stranded DNA helix, resulting in single-stranded DNA. (Chapters 4, 7)
density-dependent factor Any of the factors related to a population's size that influence the current growth rate of a population—for example, communicable disease or starvation. (Chapter 13)
density-independent factor Any of the factors unrelated to a population's size that influence the current growth rate of a population—for example, natural disasters or poor weather conditions. (Chapter 13)
deoxyribonucleic acid (DNA) Molecule of heredity that stores the information required for making all of the proteins required by the cell. (Chapters 2, 5, 6, 7, 8, 10, 12)
deoxyribose The five-carbon sugar in DNA. (Chapters 2, 5, 6, 8)
dependent variable The variable in a study that is expected to change in response to changes in the independent variable. (Chapter 1)
desert The biome found in areas of minimal rainfall. Characterized by sparse vegetation. (Chapter 15)
development All of the progressive changes that produce an organism's body. (Chapter 12)
diabetes Disorder of carbohydrate metabolism characterized by impaired ability to produce or respond to the hormone insulin. (Chapter 3)
diastolic blood pressure The lowest blood pressure in the arteries, occurring during diastole of the cardiac cycle. (Chapter 3)
diffusion The spontaneous movement of substances from a region of their own high concentration to a region of their own low concentration. (Chapter 3)
dihybrid cross A genetic cross involving the alleles of two different genes. For example AaBb x AaBb. (Chapter 7)
diploid cell A cell containing homologous pairs of chromosomes (2n). (Chapters 5, 6)
directional selection Natural selection for individuals at one end of a range of phenotypes. (Chapter 10)
disaccharide A double sugar consisiting of two monosaccharides joined together by a glycosidic linkage. (Chapter 2)
diverge See divergence (Chapter 11)
divergence Occurs when gene flow is eliminated between two populations. Over time, traits found in one population begin to differ from traits found in the other population. (Chapters 9, 11, 12)
diversifying selection Natural selection for individuals at both ends of a range of phenotypes but against the "average" phenotype. (Chapter 10)
dizygotic twins Fraternal twins (nonidentical) that develop when two different sperm fertilize two different egg cells. (Chapter 6)
DNA See deoxyribonucleic acid. (Chapters 2, 5, 6)
DNA fingerprinting Powerful genetic identification technique that takes advantage of differences in DNA sequences between all people other than identical twins. (Chapter 7)
DNA polymerase Enzyme that facilitates base pairing during DNA synthesis. (Chapter 5)
DNA replication The synthesis of two daughter DNA molecules from one original parent molecule. Takes place during the S phase of interphase. (Chapters 5, 7)
domain Most inclusive biological category. Biologists group life into three major domains. (Chapters 9, 12)
dominant Applies to an allele with an effect that is visible in a heterozygote. (Chapter 6)
double blind Experimental design protocol when both research subjects and scientists performing the measurements are unaware of either the experimental hypothesis or who is in the control or experimental group. (Chapter 1)

E

ecological footprint A measure of the natural resources used by a human population or society. (Chapter 15)
ecological niche The functional role of a species within a community or ecosystem, including its resource use and interactions with other species. (Chapter 14)
ecology Field of biology that focuses on the interactions between organisms and their environment. (Chapters 12, 13)
ecosystem All of the organisms and natural features in a given area. (Chapter 14)
ecotourism The visitation of specific geographical sites by tourists interested in natural attractions, especially animals and plants. (Chapter 14)
egg cell Gamete produced by a female organism. (Chapters 5, 6, 7, 11, 12)
electron A negatively charged subatomic particle. (Chapters 2, 4)
electron shell An energy level representing the distance of an electron from the nucleus of an atom. (Chapter 2)
electron transport chain A series of proteins in the mitochondrial and chloroplast membranes that move electrons during the redox reactions that release energy to produce ATP. (Chapter 4)
electronegative The tendency to attract electrons to form a chemical bond. (Chapter 2)
element A substance that cannot be broken down into any other substance. (Chapter 2)
embryo The developmental stage commencing after the first mitotic divisions of the zygote and ending when body structures begin to appear; from about the second week after fertilization to about the ninth week (Chapters 6, 7, 8, 12)
Endangered Species Act (ESA) U.S. law intended to protect and encourage the population growth of threatened and endangered species enacted in 1973. (Chapter 14)
endocytosis The uptake of substances into cells by a pinching inward of the plasma membrane. (Chapter 3)
endoplasmic reticulum (ER) A network of membranes in eukaryotic cells.

When rough, or studded with ribosomes, functions as a workbench for protein synthesis. When devoid of ribosomes, or smooth, it functions in phospholipid and steroid synthesis and detoxification. (Chapter 2)

endosymbiotic theory Theory that organelles such as mitochondria and chloroplasts in eukaryotic cells evolved from prokaryotic cells that took up residence inside ancestral eukaryotes. (Chapter 12)

enzyme Protein that catalyzes and regulates the rate of metabolic reactions. (Chapters 2, 3, 4, 10)

equator The circle around Earth that is equidistant to both poles. (Chapter 15)

essential amino acid Any of the amino acids that humans cannot synthesize and thus must be obtained from the diet. (Chapter 3)

essential fatty acid Any of the fatty acids that animals cannot synthesize and must be obtained from the diet. (Chapter 3)

estuary An aquatic biome that forms at the outlet of a river into a larger body of water such as a lake or ocean. (Chapter 15)

eukaryote Cell that has a nucleus and membrane-bounded organelles. (Chapters 2, 8, 12)

eutrophication Process resulting in periods of dangerously low oxygen levels in water, sometimes caused by high levels of nitrogen and phosphorus from fertilizer runoff that result in increased growth of algae in waterways. (Chapter 14)

evolution Changes in the features (traits) of individuals in a biological population that occur over the course of generations. See also theory of evolution. (Chapters 1, 2, 9, 10)

evolutionary classification System of organizing biodiversity according to the evolutionary relationships among living organisms. (Chapter 12)

exocytosis The secretion of molecules from a cell via fusion of membrane-bounded vesicles with the plasma membrane. (Chapter 3)

experiment Contrived situation designed to test specific hypotheses. (Chapter 1)

exponential growth Growth that occurs in proportion to the current total. (Chapter 13)

extinction Complete loss of a species. (Chapter 14)

extinction vortex A process by which an endangered population is driven toward extinction via loss of genetic diversity, increased vulnerability to the effects of density–independent factors, and inbreeding depression. (Chapter 14)

F

facilitated diffusion The spontaneous passage of molecules, through membrane proteins, down their concentration gradient. (Chapter 3)

falsifiable Able to be proved false. (Chapter 1)

fat Energy-rich, hydrophobic lipid molecule composed of a three-carbon glycerol skeleton bonded to three fatty acids. (Chapters 2, 3, 4)

fatty acid A long acidic chain of hydrocarbons bonded to glycerol. Fatty acids vary on the basis of their length and on the number and placement of double bonds. (Chapters 2, 3)

fermentation A process that makes a small amount of ATP from glucose without using an electron transport chain. Ethyl alcohol and lactic acid are produced by this process. (Chapter 4)

fertilization The fusion of haploid gametes (in humans, egg and sperm) to produce a diploid zygote. (Chapters 5, 6, 7)

fitness Relative survival and reproduction of one variant compared to others in the same population. (Chapters 10, 14)

flowering plant Members of the kingdom Plantae which produce flowers and fruit. See also angiosperm. (Chapter 12)

fluid mosaic model The accepted model for how membranes are structured with proteins bobbing in a sea of phospholipids. (Chapter 2)

food chain The linear relationship between trophic levels from producers to primary consumers, and so on. (Chapter 14)

food web The feeding connections between and among organisms in an environment. (Chapter 14)

foramen magnum Hole in the skull that allows passage of the spinal cord. (Chapter 9)

forest Terrestrial community characterized by the presence of trees. (Chapter 15)

fossil Remains of plants or animals that once existed, left in soil or rock. (Chapters 9, 12, 14)

fossil fuel Nonrenewable resource consisting of the buried remains of ancient plants that have been transformed by heat and pressure into coal and oil. (Chapters 4, 14)

fossil record Physical evidence left by organisms that existed in the past. (Chapter 9)

founder effect Type of sampling error that occurs when a small subset of individuals emigrates from the main population to found a new population and results in differences in the gene pools of both. (Chapter 11)

founder hypothesis The hypothesis that the diversity of unique species in isolated habitats results from divergence from a single founding population. (Chapter 11)

frameshift mutation A mutation that occurs when the number of nucleotides inserted or deleted from a DNA sequence is not a multiple of three. (Chapter 8)

free radical A substance containing an unpaired electron that is therefore unstable and highly reactive causing damage to cells. (Chapter 3)

Fungi Kingdom of eukaryotes made up of members that are immobile, rely on other organisms as their food source, and are made up of hyphae that secrete digestive enzymes into the environment and that absorb the digested materials. (Chapters 9, 12, 14)

G

galls Tumor growths on a plant. (Chapter 8)

gamete Specialized sex cell (sperm and egg in humans) that contain half as many chromosomes as other body cells and are therefore haploid. (Chapters 5, 6)

gamete incompatibility An isolating mechanism between species in which sperm from one cannot fertilize eggs from another. (Chapter 11)

gel electrophoresis The separation of biological molecules on the basis of their size and charge by measuring their rate of movement through an electric field. (Chapter 7)

gene Discrete unit of heritable information about genetic traits. Consists of a sequence of DNA that codes for a specific polypeptide—a protein or part of a protein. (Chapters 5, 6)

gene expression Turning a gene on or off. A gene is expressed when the protein it encodes is synthesized. (Chapter 8)

gene flow Spread of an allele throughout a species' gene pool. (Chapter 11)

gene gun Device used to shoot DNA-coated pellets into plant cells. (Chapter 8)

gene pool All of the alleles found in all of the individuals of a species. (Chapter 11)
gene therapy Replacing defective genes (or their protein products) with functional ones. (Chapter 8)
genealogical species concept A scheme that identifies as separate species all populations with a unique lineage. (Chapter 11)
genetic code Table showing which mRNA codons code for which amino acids. (Chapters 8, 9)
genetic drift Change in allele frequency that occurs as a result of chance. (Chapters 11, 14)
genetic variability All of the forms of genes, and the distribution of these forms, found within a species. (Chapter 14)
genetic variation Differences in alleles that exist among individuals in a population. (Chapter 6)
genetically modified organisms (GMOs) Organisms whose genome incorporates genes from another organism; also called transgenic or genetically engineered organisms. (Chapter 8)
genome Entire suite of genes present in an organism. (Chapter 8)
genotype Genetic composition of an individual. (Chapters 6, 7)
genus Broader biological category to which several similar species may belong. (Chapters 9, 11)
geologic period A unit of time defined according to the rocks and fossils characteristic of that period. (Chapter 12)
germ line gene therapy Gene therapy that changes genes in a zygote or early embryo, thus the embryo will pass on the engineered genes to their offspring. (Chapter 8)
germ theory The scientific theory that all infectious diseases are caused by microorganisms. (Chapter 1)
global warming Increases in average temperatures as a result of the release of increased amounts of carbon dioxide and other greenhouse gases into the atmosphere. (Chapters 4, 14)
glycolysis The splitting of glucose into pyruvate which helps drive the synthesis of a small amount of ATP. (Chapter 4)
Golgi apparatus An organelle in eukaryotic cells consisting of flattened membranous sacs that modify and sort proteins and other substances. (Chapter 2)
gonads The male and female sex organs; testicles in human males or ovaries in human females. (Chapter 5)
gradualism The hypothesis that evolutionary change occurs in tiny increments over long periods of time. (Chapter 11)
grana Stacks of thylakoids in the chloroplast. (Chapters 2, 4)
grassland Biome characterized by the dominance of grasses, usually found in regions of lower precipitation. (Chapter 15)
greenhouse effect The retention of heat in the atmosphere by carbon dioxide and other greenhouse gases. (Chapter 4)
growth factor Protein that stimulates cell division. (Chapter 5)
growth rate Annual death rate in a population subtracted from the annual birth rate. (Chapter 13)
guanine Nitrogenous base in DNA, a purine. (Chapters 2, 4, 5, 6, 8)
guard cell Either of the paired cells encircling stomata that serve to regulate the size of the stomatal pore, helping to minimize water loss under dry conditions and maximize carbon dioxide uptake under wet conditions. (Chapter 4)

H

habitat Place where an organism lives. (Chapter 14)
habitat destruction Modification and degradation of natural forests, grasslands, wetlands, and waterways by people; primary cause of species loss. (Chapter 14)
habitat fragmentation Threat to biodiversity caused by humans that occurs when large areas of intact natural habitat are subdivided by human activities. (Chapter 14)
half-life Amount of time required for one-half the amount of a radioactive element that is originally present to decay into the daughter product. (Chapter 9)
haploid Describes cells containing only one member of each homologous pair of chromosomes (n); in humans, these cells are eggs and sperm. (Chapter 5)
Hardy-Weinberg theorem Theorem that holds that allele frequencies remain stable in populations that are large in size, randomly mating, and experiencing no migration or natural selection. Used as a baseline to predict how allele frequencies would change if any of its assumptions were violated. (Chapter 11)
heart attack An acute condition, during which blood flow is blocked to a portion of the heart muscle, causing part of the muscle to be damaged or die. (Chapter 3)
heat The total amount of energy associated with the movement of atoms and molecules in a substance. (Chapter 4)
hemophilia Rare genetic disorder caused by a sex-linked recessive allele that prevents normal blood clotting. (Chapter 7)
heritability The amount of variation for a trait in a population that can be explained by differences in genes among individuals. (Chapter 6)
heterozygote Individual carrying two different alleles of a particular gene. (Chapters 6, 7, 11)
heterozygous Genotype containing two different alleles of a gene. (Chapter 6)
high-density lipoprotein (HDL) A cholesterol-carrying particle in the blood that is high in protein and low in cholesterol. (Chapter 3)
homeostasis The steady state condition an organism works to maintain. (Chapter 2)
hominin Referring to humans and human ancestors. (Chapters 9, 11)
homologous pair Set of two chromosomes of the same size and shape with centromeres in the same position. Homologous pairs of chromosomes carry the same genes in the same locations but may carry different alleles. (Chapters 5, 6)
homology Similarity in characteristics as a result of common ancestry. (Chapter 9)
homozygous Having two copies of the same allele of a gene. (Chapters 6, 14)
Human Genome Project Effort to determine the nucleotide base sequences and chromosomal locations of all human genes. (Chapter 8)
hybrid Offspring of two different strains of an agricultural crop (see also interspecies hybrid). (Chapter 11)
hydrocarbon A compound consisting of carbons and hydrogens. (Chapters 2, 3)
hydrogen atom One negatively charged electron and one positively charged proton. (Chapters 2, 4)
hydrogen bond A type of weak chemical bond in which a hydrogen atom of one molecule is attracted to an electronegative atom of another molecule. (Chapters 2, 4)
hydrogen ion The positively charged ion of hydrogen (H^+) formed by removal of the electron from a hydrogen atom. (Chapters 2, 4)
hydrogenation Adding hydrogen gas under pressure to make liquid oils more solid. (Chapter 3)

hydrophilic Readily dissolving in water. (Chapter 2)
hydrophobic Not able to dissolve in water. (Chapter 2)
hypertension High blood pressure. (Chapter 3)
hyphae Thin, stringy fungal material that grows over and within a food source. (Chapter 12)
hypothesis Tentative explanation for an observation that requires testing to validate. (Chapters 1, 9, 11)

I

immortal Property of cancer cells that allows them to divide more times than normal cells. (Chapter 5)
in vitro fertilization Fertilization that takes place when sperm and egg are combined in glass or a test tube. (Chapter 8)
inbreeding Mating between related individuals. (Chapter 14)
inbreeding depression Negative effect of homozygosity on the fitness of members of a population. (Chapter 14)
incomplete dominance A type of inheritance where the heterozygote has a phenotype intermediate to both homozygotes. (Chapter 7)
independent assortment The separation of homologous pairs of chromosomes into gametes independently of one another during meiosis. (Chapters 6, 7)
independent variable A factor whose value influences the value of the dependent variable, but is not influenced by it. In experiments, the variable that is manipulated. (Chapter 1)
indirect effect In ecology, a condition where one species affects another indirectly through it intervening species. (Chapter 14)
induced fit A change in shape of the active site of an enzyme so that it binds tightly to a substrate. (Chapter 3)
inductive reasoning A logical process that argues from specific instances to a general conclusion. (Chapter 1)
infant mortality Death rate of infants and children under the age of 5. (Chapter 13)
insulin A hormone secreted by the pancreas that lowers blood glucose levels by promoting the uptake of glucose by cells and the storage of glucose as glycogen in the liver. (Chapter 3)
insulin-dependent diabetes mellitus (IDDM) Type 1 diabetes, which results from inability to produce insulin. (Chapter 3)
interphase Part of the cell cycle when a cell is preparing for division and the DNA is duplicated. Consists of G_1, S, and G_2. (Chapter 5)
interspecies hybrid Organism with parents from two different species. (Chapter 11)
intertidal zone The biome that forms on ocean shorelines between the high tide elevation and the low tide elevation. (Chapter 15)
introduced species A nonnative species that was intentionally or unintentionally brought to a new environment by humans. (Chapter 14)
invertebrate Animal without backbones. (Chapter 12)
ion electrically charged atom. (Chapter 2)
ionic bond A chemical bond resulting from the attraction of oppositely charged ions. (Chapter 2)

K

karyotype The chromosomes of a cell, displayed with chromosomes arranged in homologous pairs and according to size. (Chapter 7)
keystone species A species that has an unusually strong effect on the structure of the community it inhabits. (Chapter 14)
kingdom In some classifications, the most inclusive group of organisms; usually five or six. In other classification systems, the level below domain on the hierarchy. (Chapters 9, 12)

L

lactase The enzyme that cleaves the disaccharide lactose into glucose and galactose. Missing or deficient in people with lactose intolerance. (Chapter 3)
lactose intolerance Inability to digest lactose resulting in bloating, gas, and diarrhea. (Chapter 3)
lake An aquatic biome that is completely landlocked. (Chapter 15)
laparoscope A thin tubular instrument inserted through an abdominal incision and used to view organs in the pelvic cavity and abdomen. (Chapter 5)
leptin A hormone by fat cells that may be involved in the regulation of appetite. (Chapter 3)
life cycle Description of the growth and reproduction of an individual. (Chapter 6)
light reaction A series of reactions that occur on thylakoid membranes during photosynthesis and serve to convert energy from the sun into the energy stored in the bonds of ATP and evolve oxygen. (Chapter 4)
linked gene Genes located on the same chromosome. (Chapter 5)
lipid Hydrophobic molecule including fats, phospholipids, and steroids. (Chapters 2, 3)
logistic growth Pattern of growth seen in populations that are limited by resources available in the environment. A graph of logistic growth over time typically takes the form of an S-shaped curve. (Chapter 13)
low-density lipoprotein (LDL) Cholesterol carrying subsance in the blood that is high in cholesterol and low in protein. (Chapter 3)
lymph node Organ located along lymph vessels that filter lymph and help defend against bacteria and viruses. (Chapter 5)
lymphatic system A system of vessels and nodes that return fluid and protein to the blood. (Chapter 5)
lysosome A membrane-bounded sac of hydrolytic enzymes found in the cytoplasm of many cells. (Chapter 2)

M

macroevolution Large-scale evolutionary change, usually referring to the origin of new species. (Chapter 9)
macromolecule Any of the large molecules including polysaccharides, proteins, and nucleic acids, composed of subunits joined by dehydration synthesis. (Chapters 2, 4)
macronutrient Nutrient required in large quantities. (Chapter 3)
malignant Describes a tumor that is cancerous, whether it is invasive or metastatic. (Chapter 5)
marine Of, or pertaining to, salt water. (Chapter 15)
mark-recapture method A technique for estimating population size, consisting of capturing and marking a number of individuals, releasing them, and recapturing more individuals to determine what proportion are marked. (Chapter 13)
mass extinction Loss of species that is rapid, global in scale, and affects a wide variety of organisms. (Chapter 14)
matrix In a mitochondrion, the semifluid substance inside the inner mitochondrial membrane, which houses the enzymes of the citric acid cycle. (Chapters 2, 4)
mean Average value of a group of measurements. (Chapter 1)
mechanical isolation A form of reproductive isolation between species that

depends on the incompatibility of the genitalia of individuals of different species. (Chapter 11)

meiosis Process that diploid sex cells undergo in order to produce haploid daughter cells. Occurs during gametogenesis. (Chapters 5, 7, 11)

messenger RNA (mRNA) Complementary RNA copy of a DNA gene, produced during transcription. The mRNA undergoes translation to synthesize a protein. (Chapters 8, 10)

metabolic rate Measure of an individual's energy use. (Chapter 3)

metabolism All of the physical and chemical reactions that produce and use energy. (Chapters 2, 3, 4)

metaphase Stage of mitosis during which duplicated chromosomes align across the middle of the cell. (Chapter 5)

metastasis When cells from a tumor break away and start new cancers at distant locations. (Chapter 5)

microbe Microscopic organism, especially Bacteria and Archaea. (Chapter 12)

microbiologists Sientists who study microscopic organisms, especially referring to those who study prokaryotes. (Chapter 12)

microevolution Changes that occur in the characteristics of a population. (Chapter 9)

micronutrient Nutrient needed in small quantities. (Chapter 3)

microorganism See microbe. (Chapter 12)

microtubule Protein structure that moves chromosomes around during mitosis and meiosis. (Chapters 2, 5, 6)

mineral Inorganic nutrient essential to many cell functions. (Chapter 3)

mitochondria Organelles in which products of the digestive system are converted to ATP. (Chapters 2, 4)

mitosis The division of the nucleus that produces daughter cells that are genetically identical to the parent cell. Also, portion of the cell cycle in which DNA is apportioned into two daughter cells. (Chapter 5)

model organism Any nonhuman organism used in genetic studies to help scientists understand human genes because they share genes with humans. (Chapters 1, 8)

mold A fungal form characterized by rapid, asexual reproduction. (Chapters 9, 12)

molecular clock Principle that DNA mutations accumulate in the genome of a species at a constant rate, permitting estimates of when the common ancestor of two species existed. (Chapter 9)

molecule Two or more atoms held together by covalent bonds. (Chapter 2)

monosaccharide Simple sugar. (Chapter 2)

monosomy A chromosomal condition in which only one member of a homologous pair is present. (Chapter 7)

monozygotic twins Identical twins that developed from one zygote. (Chapter 6)

morphological species concept Definition of species that relies on differences in physical characteristics among them. (Chapter 11)

morphology Appearance or outward physical characteristics. (Chapter 11)

multicellular The condition of being composed of many coordinated cells. (Chapter 12)

multiple allelism A gene for which there are more than 2 alleles segregating in the population. (Chapter 7)

multiple hit model The notion that many different genetic mutations are required for a cancer to develop. (Chapter 5)

mutagen Substance that increases the likelihood of mutation occurring; increases the likelihood of cancer. (Chapter 5)

mutation Change to a DNA sequence that may result in the production of altered proteins. (Chapters 5, 6, 8, 10)

mutualism Interaction between two species that provides benefits to both species. (Chapter 14)

mycologist Scientists who specialize in the study of fungi. (Chapter 12)

N

natural experiment Situation where unique circumstances allow a hypothesis test without prior intervention by researchers. (Chapter 6)

natural selection Process by which individuals with certain traits have greater survival and reproduction than individuals who lack these traits, resulting in an increase in the frequency of successful alleles and a decrease in the frequency of unsuccessful ones. (Chapters 9, 10, 11)

net primary production (NPP) Amount of solar energy converted to chemical energy by plants, minus the amount of this chemical energy plants need to support themselves. A measure of plant growth, typically over the course of a single year. (Chapter 13)

neutral mutation A genetic mutation that confers no selective advantage or disadvantage. (Chapter 8)

neutron An electrically neutral particle found in the nucleus of an atom. (Chapter 2)

nutrients Atoms other than carbon, hydrogen, and oxygen that must be obtained from a plant's environment for photosynthesis to occur. (Chapter 3)

nicotinamide adenine dinucleotide Intracellular electron carrier. Oxidized form is NAD^+; reduced form is NADH. (Chapter 4)

nitrogen-fixing bacteria Organisms that convert nitrogen gas from the atmosphere into a form that can be taken up by plant roots; some species live in the root nodules of legumes. (Chapter 14)

nitrogenous base Nitrogen-containing base found in DNA: A, C, G, and T and in RNA : U. (Chapters 2, 4, 5, 6, 8)

nondisjunction The failure of members of a homologous pair of chromosomes to separate from each other during meiosis. (Chapter 7)

non-insulin-dependent diabetes mellitus (NIDDM) Type of diabetes that does not require insulin injections, Type II. (Chapter 3)

nonpolar Won't dissolve in water. Hydrophobic. (Chapter 2)

nonrenewable resource Resource that is a one-time supply and cannot be easily replaced. (Chapter 13)

normal distribution Bell-shaped curve, as for the distribution of quantitative traits in a population. (Chapter 6)

nuclear envelope The double membrane enclosing the nucleus in eukaryotes. (Chapters 2, 5)

nuclear transfer Transfer of a nucleus from one cell to another cell that has had its nucleus removed. (Chapter 8)

nucleic acids Polymers of nucleotides that comprise DNA and RNA. (Chapters 2, 4, 5)

nucleotides Building blocks of nucleic acids that include a sugar, a phosphate, and a nitrogenous base. (Chapters 2, 4, 5, 6, 7, 8, 10)

nucleus Cell structure that houses DNA; found in eukaryotes. (Chapters 2, 5, 8, 10)

nutrient cycling Process by which nutrients become available to plants. Nutrient cycling in a natural environment relies upon a healthy community of decomposers within the soil. (Chapter 14)

O

obesity Condition of having a BMI of 30 or greater. (Chapter 3)
objective Without bias. (Chapter 1)
observation Measurement of nature. (Chapters 1, 10)
ocean A biome consisting of open stretches of salt water. (Chapter 15)
oncogene Mutant version of a cell cycle controlling proto-oncogene. (Chapter 5)
organelle Subcellular structure found in the cytoplasm of eukaryotic cells that performs a specific job. (Chapters 2, 4)
organic chemistry The chemistry of carbon-containing substances. (Chapter 2)
osmosis The diffusion of water across a selectively permeable membrane. (Chapter 3)
osteoporosis A condition resulting in an elevated risk of bone breakage from weakened bones. (Chapter 3)
ovary In animals, the paired abdominal structures that produce egg cells and secrete female hormones. (Chapter 5)
overexploitation Threat to biodiversity caused by humans that encompasses overhunting and overharvesting. (Chapter 14)
oviduct Egg-carrying duct that brings egg cells from ovaries to uterus. (Chapter 5)
ovulation Release of an egg cell from the ovary. (Chapter 5)

P

paleontologist Scientist who searches for, describes, and studies ancient organisms. (Chapter 9)
parasite An organism that benefits from an association with another organism which is harmed by the association. (Chapter 14)
passive transport The diffusion of substances across a membrane with their concentration gradient and not requiring an input of ATP. (Chapter 3)
pedigree Family tree that follows the inheritance of a genetic trait for many generations. (Chapter 7)
peer review The process by which reports of scientific research are examined and critiqued by other researchers before they are published in scholarly journals. (Chapter 1)
peptide bond Covalent bond that joins the amino group and carboxyl group of adjacent amino acids. (Chapter 2)
perennial Plant that lives for many years. (Chapter 15)
permafrost Permanently frozen soil. (Chapter 15)
pH A logarithmic measure of the hydrogen ion concentration ranging from 0-14. Lower numbers indicate higher hydrogen ion concentrations. (Chapter 2)
phenotype Physical and physiological traits of an individual. (Chapters 6, 7)
phenotypic ratio Proportion of individuals produced by a genetic cross that possess each of the various phenotypes that cross can generate. (Chapter 7)
phosphodiester bond Covalent bond that joins adjacent nucleotides in DNA. (Chapter 2)
phospholipid One of three types of lipids, phospholipids are components of cell membranes. (Chapter 2)
phospholipid bilayer The membrane that surrounds cells and organelles and is composed of two layers of phospholipids. (Chapter 2)
phosphorylation To introduce a phosphoryl group into an organic compound. (Chapter 4)
photorespiration A series of reactions triggered by the closing of stomatal openings to prevent water loss. (Chapter 4)
photosynthesis Process by which plants, along with algae and some bacteria, transform light energy to chemical energy. (Chapter 4)
phyla (singular: **phylum**)The taxonomic category below kingdom and above class. (Chapters 9, 12)
phylogeny Evolutionary history of a group of organisms. (Chapter 12)
pituitary gland Small gland attached by a stalk to the base of the brain that secretes growth hormone, reproductive hormones, and other hormones. (Chapter 8)
placebo Sham treatments in experiments. (Chapter 1)
Plantae Multicellular photosynthetic eukaryotes, excluding algae. (Chapter 12)
plasma membrane Structure that encloses a cell, defining the cell's outer boundary. (Chapters 2, 3)
plasmid Circular piece of bacterial DNA that normally exists separate from the bacterial chromosome and can make copies of itself. (Chapter 8)
pleiotropy The ability of one gene to affect many different functions. (Chapter 7)
polar Describes a molecule with regions having different charges; capable of ionizing. (Chapter 2)
poles Opposite ends of a sphere, such as of a cell (Chapter 5) or of a planet such as Earth (Chapter 15).
pollen The male gametophyte of seed plants. (Chapters 11, 12, 14)
pollinator An organism that transfers sperm (pollen grains) from one flower to the female reproductive structures of another flower. (Chapter 14)
pollution Human-caused threat to biodiversity involving the release of poisons, excess nutrients, and other wastes into the environment. (Chapter 14)
polygenic trait A trait influenced by many genes. (Chapters 6, 7)
polymer General term for a macromolecule composed of many chemically bonded monomers. (Chapter 2)
polymerase An enzymee that catalyzes phosphodiester bond formation between nucleotides. (Chapter 7)
polymerase chain reaction (PCR) A laboratory technique that allows the production of many identical DNA molecules. (Chapter 7)
polyploidy A chromosomal condition involving more than two sets of chromosomes. (Chapter 11)
polysaccharide A carbohydrate composed of three or more monosaccharides. (Chapter 2)
polyunsaturated relating to fats consisting of carbon chains with many double bonds unsaturated by hydrogen atoms. (Chapter 3)
pond An aquatic biome that is completely landlocked. (Chapter 15)
population Subgroup of a species that is somewhat independent from other groups. (Chapters 11, 13)
population bottleneck Dramatic but short-lived reduction in population size followed by an increase in population. (Chapter 11)
population crash Steep decline in number that may occur when a population grows larger than the carrying capacity of its environment. (Chapter 13)
population cycle In some populations, the tendency to increase in number above the environment's carrying capacity, resulting in a crash, following by an overshoot of the carrying capacity and another crash, continuing indefinitely. (Chapter 13)
population genetics Study of the factors in a population that determine allele frequencies and their change over time. (Chapter 11)
population pyramid A visual representation of the number of individuals in different age categories in a population. (Chapter 13)

prairie A grassland biome. (Chapter 15)
precipitation When water vapor in the atmosphere turns to liquid or solid form and falls to Earth's surface. (Chapter 15)
predation Act of capturing and consuming an individual of another species. (Chapter 14)
predator Organism that eats other organisms. (Chapter 14)
prediction Result expected from a particular test of a hypothesis if the hypothesis were true. (Chapter 1)
primary consumers Organism that eats plants. (Chapter 14)
primary source Article reporting research results, written by researchers, and reviewed by the scientific community. (Chapter 1)
probability Likelihood that something is the case or will happen. (Chapter 1)
probe Single-stranded nucleic acid that has been radioactively labeled. (Chapter 7)
producer Organism that produces carbohydrates from inorganic carbon; typically via photosynthesis. (Chapter 14)
products The modified chemical that results from a chemical or enzymatic reaction. (Chapter 2)
prokaryote Type of cell that does not have a nucleus or membrane-bounded organelles. (Chapters 2, 8, 12)
promoter Sequence of nucleotides to which the polymerase binds to start transcription. (Chapter 8)
prophase Stage of mitosis during which duplicated chromosomes condense. (Chapter 5)
protein Cellular constituents made of amino acids coded for by genes. Proteins can have structural, transport, or enzymatic roles. (Chapters 2, 3, 4, 8, 10, 14)
protein synthesis Joining amino acids together, in an order dictated by a gene, to produce a protein. (Chapter 8)
Protista Kingdom in the domain Eukarya containing a diversity of eukaryotic organisms, most of which are unicellular. (Chapter 12)
proton A positively charged subatomic particle. (Chapters 2, 4)
proto-oncogenes Genes that encode proteins that regulate the cell cycle. Mutated proto-oncogenes (oncogenes) can lead to cancer. (Chapter 5)
punctuated equilibrium The hypothesis that evolutionary changes occur rapidly, and in short bursts followed by long periods of little change. (Chapter 11)
Punnett square Table that lists the different kinds of sperm or eggs parents can produce relative to the gene or genes in question and predicts the possible outcomes of a cross between these parents. (Chapter 6)
purine Nitrogenous base (A or G) with a two-ring structure. (Chapter 2)
pyrimidine Nitrogenous base (C, T or U) with a single-ring structure. (Chapter 2)
pyruvic acid The 3-carbon molecule produced by glycolysis. (Chapter 4)

Q

qualitative trait Trait that produces phenotypes in distinct categories. (Chapter 6)
quantitative trait Trait that has many possible values. (Chapter 6)

R

race See biological race. (Chapter 11)
racism Idea that some groups of people are naturally superior to others. (Chapter 11)
radiation therapy Focusing beams of reactive particles at a tumor to kill the dividing cells. (Chapter 5)
radioactive decay Natural, spontaneous breakdown of radioactive elements into different elements, or "daughter products." (Chapter 9)
radioimmunotherapy Experimental cancer treatment with the goal of delivering radioactive substances directly to tumors without affecting other tissues. (Chapter 5)
radiometric dating Technique that relies on radioactive decay to estimate a fossil's age (Chapters 9, 14)
random alignment When members of a homologous pair line up randomly with respect to maternal or paternal origin during metaphase I of meiosis, thus increasing the genetic diversity of offspring. (Chapters 5, 7)
random assignment Placing individuals into experimental and control groups randomly to eliminate systematic differences between the groups. (Chapter 1)
random distribution The dispersion of individuals in a population without pattern. (Chapter 13)
random fertilization The unpredictability of exactly which gametes will fuse during the process of sexual reproduction. (Chapter 6)
reactant Any starting material in a chemical reaction. (Chapter 2)
reading frame The grouping of mRNAs into 3 base codons for translation. (Chapter 8)
receptor Protein on the surface of a cell that recognizes and binds to a specific chemical signal. (Chapters 5, 9)
recessive Applies to an allele with an effect that is not visible in a heterozygote. (Chapter 6)
recombinant Produced by manipulating a DNA sequence. (Chapter 8)
recombinant bovine growth hormone (rBGH) Growth hormone produced in a laboratory and injected into cows to increase their size and ability to produce milk. (Chapter 8)
red blood cell Primary cellular component of blood, responsible for ferrying oxygen throughout the body. (Chapter 5)
remission The period during which the symptoms of a disease subside. (Chapter 5)
repressor A protein that suppresses the expression of a gene. (Chapter 8)
reproductive cloning Transferring the nucleus from a donor adult cell to an egg cell without a nucleus in order to clone the adult. (Chapter 8)
reproductive isolation Prevention of gene flow between different biological species due to failure to produce fertile offspring; can include premating barriers and postmating barriers. (Chapter 11)
restriction enzyme An enzyme that cleaves DNA at specific nucleotide sequences. (Chapters 7, 8)
restriction fragment length polymorphisms (RFLP) Differences among members of a population in the number and size of DNA fragments generated by cutting DNA with restriction enzymes. (Chapter 7)
Rh factor Surface molecule found on some red blood cells. (Chapter 7)
ribose The five-carbon sugar in RNA. (Chapters 5, 8)
ribosomal RNA (rRNA) RNA that makes up part of the structure of ribosomes. (Chapter 2)
ribosome Subcellular structure that helps translate genetic material into proteins by anchoring and exposing small sequences of mRNA. (Chapters 2, 8, 12)
risk factor Any exposure or behavior that increases the likelihood of disease. (Chapter 5)
river Aquatic biome characterized by flowing water. (Chapter 15)

RNA (ribonucleic acid) Information-carrying molecule composed of nucleotides. (Chapters 2, 8)
RNA polymerase Enzyme that synthesizes mRNA from a DNA template during transcription. (Chapter 8)
rough endoplasmic reticulum Ribosome studded subcellular membranes found in the cytoplasm and responsible for some protein synthesis. (Chapter 2)
rubisco Abbreviation for ribulose bisphosphate carboxylase oxygenase, the enzyme that catalyzes the first step in the Calvin cycle of photosynthesis. (Chapter 4)

S

salt A charged substance that ionizes in solution (Chapter 2)
sample Small subgroup of a population used in an experimental test. (Chapter 1)
sample size Number of individuals in both the experimental and control groups. (Chapter 1)
sampling error Effect of chance on experimental results. (Chapter 1)
saturated fat Type of lipid rich in single bonds. Found in butter and other fats that are solids at room temperature. This type of fat is associated with higher blood cholesterol levels. (Chapter 3)
savanna Grassland biome containing scattered trees. (Chapter 15)
scientific method A systematic method of research consisting of putting a hypothesis to a test designed to disprove it, if it is in fact false. (Chapter 1)
scientific theory Body of scientifically accepted general principles that explain natural phenomena. (Chapters 1, 9)
secondary compounds Chemicals produced by plants and some other organisms as side reactions to normal metabolic pathways and that typically have an antipredator or antibiotic function. (Chapter 12)
secondary consumers Animals that eat primary consumers; predators. (Chapter 14)
secondary sources Books, news media, and advertisements as sources of scientific information. (Chapter 1)
seed A plant embryo packaged with a food source and surrounded by a seed coat. (Chapter 12)
segregation Separation of pairs of alleles during the production of gametes. Results in a 50% probability that a given gamete contains one allele rather than the other. (Chapter 6)
semipermeable In biological membranes, a membrane that allows some substances to pass but prohibits the passage of others. (Chapter 2)
severe combined immunodeficiency disorder (SCID) Illness caused by a genetic mutation that results in the absence of an enzyme, and a severely weakened immune system. (Chapter 8)
sex chromosome Any of the sex-determining chromosomes (X and Y in humans). (Chapters 5, 7)
sex determination Determining the biological sex of an offspring. Humans have a chromosomal mechanism of sex determination in which two X chromosomes produce a female and an X and a Y chromosome produce a male. (Chapter 7)
sex-linked gene Any of the genes found on the X or Y sex chromosomes. (Chapter 7)
sexual reproduction Reproduction involving two parents that give rise to offspring that have unique combinations of genes. (Chapter 5)
sexual selection Form of natural selection that occurs when a trait influences the likelihood of mating. (Chapter 11)
sister chromatid Either of the two duplicated, identical copies of a chromosome formed after DNA synthesis. (Chapter 5)
smog Products of fossil fuel combustion in combination with sunlight, producing a brownish haze in still air. (Chapter 15)
smooth endoplasmic reticulum The subcellular, cytoplasmic membrane system responsible for lipid and steroid biosynthesis. (Chapter 2)
solar irradiance The amount of solar energy hitting Earth's surface at any given point. (Chapter 15)
solid waste Garbage. (Chapter 15)
solstice When the sun reaches its maximum and minimum elevation in the sky. (Chapter 15)
solute The substance that is dissolved in a solution. (Chapter 2)
solution A mixture of two or more substances. (Chapter 2)
solvent A substance, such as water, that a solute is dissolved in to make a solution. (Chapter 2)
somatic cell Any of the body cells in an organism. Any cell that is not a gamete. (Chapter 5)
somatic cell gene therapy Changes to malfunctioning genes in somatic or body cells. These changes will not be passed to offspring. (Chapter 8)
spatial isolation A mechanism for reproductive isolation that depends on the geographic separation of populations. (Chapter 11)
special creation The hypothesis that all organisms on Earth arose as a result of the actions of a supernatural creator. (Chapter 9)
speciation Evolution of one or more species from an ancestral form; macroevolution. (Chapter 11)
species A group of individuals that regularly breed together and are generally distinct from other species in appearance or behavior. (Chapters 2, 9, 11, 12, 14)
species–area curve Graph describing the relationship between the size of a natural landscape and the relative number of species it contains. (Chapter 14)
specificity Phenomenon of enzyme shape determining the reaction the enzyme catalyzes. (Chapter 3)
sperm Gametes produced by males. (Chapters 5, 6, 7, 12)
spore Reproductive cell in plants and fungi that is capable of developing into an adult without fusing with another cell. (Chapter 12)
stabilizing selection Natural selection that favors the average phenotype and selects against the extremes in the population. (Chapter 10)
standard error A measure of the variance of a sample; essentially the average distance a single data point is from the mean value for the sample. (Chapter 1)
statistical test Mathematical formulation that helps scientists evaluate whether the results of a single experiment demonstrate the effect of treatment. (Chapter 1)
statistically significant Low probability that experimental groups differ simply by chance. (Chapter 1)
statistics Specialized branch of mathematics used in the evaluation of experimental data. (Chapter 1)
stem cell Cells that can divide indefinitely and can differentiate into other cell types. (Chapter 8)
steppe Biome characterized by short grasses, found in regions with relatively little annual precipitation. (Chapter 15)
steroid naturally occurring or synthetic organic fat-soluble substance that produces physiologic effects. (Chapter 2)
stomata Pores on the photosynthetic surfaces of plants that allow air into the

internal structure of leaves and green stems. Stomata also provide portals through which water can escape. (Chapter 4)

stop codon An mRNA codon that does not code for an amino acid and causes the amino acid chain to be released into the cytoplasm. (Chapter 8)

stream Biome characterized by flowing water, sometimes seasonal. Typically smaller than rivers. (Chapter 15)

stroke Acute condition caused by a blood clot that blocks blood flow to an organ or other region of the body. (Chapter 3)

stroma The semi-fluid matrix inside a chloroplast where the Calvin cycle of photosynthesis occurs. (Chapters 2, 4)

subspecies Subdivision of a species that is not reproductively isolated but represents a population or set of populations with a unique evolutionary history. See also biological race, variety. (Chapter 11)

substrate The substance upon which an enzyme reacts. (Chapter 3)

succession Replacement of ecological communities over time since a disturbance, until finally reaching a stable state. (Chapter 15)

sugar-phosphate backbone Series of alternating sugars and phosphates along the length of the DNA helix. (Chapters 2, 5, 6)

supernatural Not constrained by the laws of nature. (Chapter 1)

symbiosis A relationship between two species. (Chapter 12)

sympatric In the same geographic region. (Chapter 11)

systematist Biologist who specializes in describing and categorizing a particular group of organisms. (Chapter 12)

systolic blood pressure Force of blood on artery walls when heart is contracting. (Chapter 3)

T

telomerase An enzyme that helps prevent the degradation of the tips of chromosomes, active during development and sometimes reactivated in cancer cells. (Chapter 5)

telophase Stage of mitosis during which the nuclear envelope forms around the newly produced daughter nucleus, and chromosomes decondense. (Chapter 5)

temperate forest Biome dominated by deciduous trees. (Chapter 15)

temperature A measure of the intensity of heat or kinetic energy. (Chapters 4, 15)

temporal isolation Reproductive isolation between populations maintained by differences in the timing of mating or emergence. (Chapter 11)

testable Possible to evaluate through observations of the measurable universe. (Chapter 1)

theory See scientific theory. (Chapters 1, 9)

theory of evolution Theory that all organisms on Earth today are descendants of a single ancestor that arose in the distant past. See also evolution. (Chapters 1, 2, 9)

therapeutic cloning Using early embryos as donors of stem cells for the replacement of damaged tissues and organs in another individual. (Chapter 8)

thylakoid Flattened membranous sac located in the chloroplast stroma. Function in photosynthesis. (Chapters 2, 4)

thymine Nitrogenous base in DNA, a pyrimidine. (Chapters 2, 4, 5, 8)

Ti plasmid Tumor-inducing plasmid used to genetically modify crop plants. (Chapter 8)

totipotent Describes a cell able to specialize into any cell type of its species, including embryonic membrane. Compare with multipotent, pluripotent. (Chapter 8)

trans fat Contains unsaturated fatty acids that have been hydrogenated, which changes the fat from a liquid to a solid at room temperature. (Chapter 3)

transcription Production of an RNA copy of the protein coding DNA gene sequence. (Chapters 8, 10)

transfer RNA (tRNA) Amino-acid-carrying RNA structure with an anticodon that binds to a mRNA codon. (Chapter 8)

transgenic organism Organism whose genome incorporates genes from another organism; also called genetically modified organism (GMO). (Chapter 8)

translation Process by which an mRNA sequence is used to produce a protein. (Chapters 8, 10)

transpiration Movement of water from the roots to the leaves of a plant, powered by evaporation of water at the leaves and the cohesive and adhesive properties of water. (Chapter 4)

trisomy A chromosomal condition in which three copies of a chromosome exist instead of the two copies of a chromosome normally present in a diploid organism. (Chapter 5)

trophic level Feeding level or position on a food chain; e.g. producers, primary consumers, etc. (Chapter 14)

trophic pyramid Relationship among the mass of populations at each level of a food web. (Chapter 14)

tropical forest Biome dominated by broad-leaved, evergreen trees; found in areas where temperatures never drop below the freezing point of water. (Chapter 15)

tuberculosis (TB) Degenerative lung disease caused by infection with the bacterium *Mycobacterium tuberculosis*. (Chapter 10)

tumor Mass of tissue that has no apparent function in the body. (Chapter 5)

tumor suppressor Cellular protein that stops tumor formation by suppressing cell division. When mutated leads to increased likelihood of cancer. (Chapter 5)

tundra Biome that forms under very low temperature conditions. Characterized by low-growing plants. (Chapter 15)

U

undifferentiated A cell that is not specialized. (Chapter 8)

unicellular Made up of a single cell. (Chapter 12)

uniform distribution Occurs when individuals in a population are disbursed in a uniform manner across a habitat. (Chapter 13)

unsaturated fat Type of lipid containing many carbon-to-carbon double bonds; liquid at room temperature. (Chapter 3)

uracil Nitrogenous base in RNA, a pyrimidine. (Chapters 2, 6, 8)

urban sprawl The tendency for the boundaries of urban areas to grow over time as people build housing and commercial districts farther and farther from an urban core. (Chapter 15)

V

valence shell The outermost energy shell of an atom containing the valence electrons which are most involved in the chemical reactions of the atom. (Chapter 2)

variable A factor that varies in a population or over time. (Chapter 1)

variable number tandem repeat (VNTR) A DNA sequence that varies in number between individuals. Used during the process of DNA fingerprinting. (Chapter 7)

variance Mathematical term for the amount of variation in a population. (Chapter 6)

variant An individual in a population that differs genetically from other individuals in the population. (Chapter 10)
vascular tissue Cells that transport water and other materials within a plant. (Chapter 12)
vertebrate Animal with a backbone. (Chapters 7, 12)
vestigial trait Modified with no, or relatively minor function compared to the function in other descendants of the same ancestor. (Chapter 9)
virus Infectious intracellular parasite composed of a strand of genetic material and a protein or fatty coating that can only reproduce by forcing its host to make copies of it. (Chapters 1, 8)
vitamin Organic nutrient needed in small amounts. Most vitamins function as coenzymes. (Chapter 3)

W

wastewater Liquid wastes produced by humans. (Chapter 15)
water One molecule of water consists of one oxygen and two hydrogen atoms. (Chapters 2, 3, 4, 15)
weather Current temperature and precipitation conditions. (Chapter 15)
wetland Biome characterized by standing water, shallow enough to permit plant rooting. (Chapter 15)
whole food Any food that has not undergone processing. Includes grains, beans, nuts, seeds, fruits, and vegetables. (Chapter 3)

X

X inactivation The inactivation of one of two chromosomes in the XX female. (Chapter 7)
x-axis The horizontal axis of a graph. Typically describes the independent variable. (Chapter 1)
X-linked gene Any of the genes located on the X chromosome. (Chapter 7)

Y

y-axis The vertical axis of a graph. Typically describes the dependent variable. (Chapter 1)
yeast Single-celled eukaryotic organisms found in bread dough. Often used as model organisms and in genetic engineering. (Chapters 4, 9, 12)
Y-linked gene Any of the genes located on the Y chromosome. (Chapter 7)

Z

zoologist Scientist who specializes in the study of animals. (Chapter 12)
zygote Single cell resulting from the fusion of gametes (egg and sperm). (Chapters 5, 7)

Credits

CHAPTER 1

Opener (L) Heidi Velten/Photolibrary.com. **Opener (R) Top** Indexopen.com. **Middle** Edyta Pawlowska/Shutterstock. **Bottom** Sinclair Stammers/Photo Researchers. **1-2a** Eye of Science/Photo Researchers. **1-2b** Tony Ashby/Getty Images. **1-4** Custom Medical Stock Photo. **1-5** The Natural History Museum, London. **1-6** Zina Seletskaya/Shutterstock. **1-17** Graca Victoria/Shutterstock.

CHAPTER 2

Opener (L) NASA. **Opener (R) Top** Fox/Photofest. **Middle** NASA. **Bottom** NASA. **2-1** Photos.com. **2-2a** NASA. **2-2b** DLR/FU Berlin/European Space Agency. **2-8** Charles D. Winters/Photo Researchers. **2-11** Richard Megna/ Fundamental Photographs. **2-16** National Cancer Institute. **2-17a** Visuals Unlimited. **2-19(1)** Gary Gaugler/Photo Researchers. **2-19(2)** B. Boonyaratanakornkit & D. S. Clark, G. Vrdoljak/EM Lab, UCBerk/ Visuals Unlimited. **2-19(3)** Brittany Carter Courville/iStockphoto. **2-19(4)** Jan Krejci/Shutterstock. **2-19(5)** webphotographer/iStockphoto. **2-19(6)** Michael Abbey/Visuals Unlimited. **2-19(7)** Dusan Zidar/iStockphoto. **2-19(8)** Vicki Beaver/iStockphoto. **2-20** Dept. of Natural Resources & Parks Water & Land Resources.

CHAPTER 3

Opener (L) Reuters/Guang Niu/ Landov. **Opener (R) Top** Bubbles Photolibrary/Alamy. **Middle** Ian Waldie/Getty Images. **Bottom** Christopher LaMarca/Redux Pictures. **3-2a** Olga Shelego/Shutterstock. **3-2b** indexopen.com. **3-3a** fonats/Shutterstock. **3-3b** Craig Veltri/iStockphoto. **3-15a(1)** Redux Pictures. **3-15a(2)** Creative Digital Visions, Pearson Science. **3-15b(1)** Michael Ochs Archives/Getty Images. **3-15b(2)** Madison/x17online.

CHAPTER 4

Opener (L) Rick Wilking/Reuters/Corbis. **Opener (R) Top** Milos Peric/ iStockphoto. **Middle** Jan Martin Will/Shutterstock. **Bottom** Frank Siteman/Mira.com/Digital Railroad. **4-4** Reuters/Corbis. **4-6** British Antarctic Survey/SPL/Photo Researchers. **4-18a** P. Motta & T. Naguro/SPL/Photo Researchers. **4-24a** Suza Scalora/ Getty Images. **4-24b** SciMAT/Photo Researchers. **4-25a** Jacques Regad, Department de la Sante des Forets– France, www.forestryimages.org. **4-25b** Beat Forster/Swiss Federal Research Institute Snow and Landscape Research (WSL). **4-28** Richard Kessel & Gene Shih/Visuals Unlimited/Getty Images. **4-29** George Chapman/Visuals Unlimited/Getty Images. **Table 4.1(1)** TongRo Image Stock/ Jupiter Images. **Table 4.1(2)** Bronwyn8/iStockphoto. **Table 4.1(3)** Biosphoto/Hazan Muriel/Peter Arnold.

CHAPTER 5

Opener (L) SPL/Photo Researchers. **Opener (R)** Jim Whitmer Photography. **Middle** Jim Whitmer Photography. **Bottom** Jim Whitmer Photography. **5-1a** James Cavallini/Photo Researchers. **5-1b** Arthur C. Fleischer, MD Professor Chief of Diagnostic Sonography Departments of Radiology and Obstetrics/Gynecology, Vanderbilt University Medical Center. **5-3a** Biophoto Associates/Photo Researchers. **5-3b** rossco/Shutterstock. **5-4a** Biophoto Associates/Photo Researchers. **5-4b** Biophoto Associates/Photo Researchers. **5-9a** Ed Reschke/Peter Arnold. **5-9b** David Phillips/Visuals Unlimited/Getty Images. **5-11(1**) Science, September 17, 1999 in the article "Antiangiogenic Activity of the Cleaved Conformation of Serpin Antithrombin" written by Judah Folkman, Michael S. O'Reilly, Steven Pirie-Shepard and William S. Lane. **5-11(2)** Robert Calentine/Visuals Unlimited. **5-11(3)** EnviroWatch, Inc. **5-16a** Biodisc/Visuals Unlimited. **5-16c** Ida Wyman/Phototake. **5-17** Jim Whitmer Photography. **5-18(1)** CNRI/SPL/Photo Researchers. **5-18(2)** CNRI/SPL/ Photo Researchers. **5-18(3)** CNRI/SPL/ Photo Researchers.

CHAPTER 6

Opener (L) Tony Brain/SPL/Photo Researchers. **Opener (R) Top** Melissa Schalke/Shutterstock. **Middle** Digital Vision/Getty Images. **Bottom** Thinkstock/SuperStock. **6-2a** Photo Courtesy Gerald McCormack. **6-2b** NINAN, C. A. 1958. Studies on the cytology and phylogeny of the pteridophytes VI. Observations on the Ophioglossaceae. Cytologia 23: 291-316. **6-8** Getty Images. **6-10** Linda Gordon. **6-11** Simon Fraser/Royal Victoria Infirmary/Photo Researchers. **6-12a** Dr. Kathy Lovell, Michigan State University. **6-16a** Darrick E. Antell MD, F.A.C.S. **6-16b** Darrick E. Antell MD, F.A.C.S. **6-18a** Photodisc/Getty Images. **6-18b** peace!/A. Collection/Getty Images. **6-21** Angel Franco/The New York Times. **6-23** UpperCut Images/SuperStock.

CHAPTER 7

Opener (L) The Granger Collection, NY. **Opener (R) Top** V. Velengurin/R.P.G./ CORBIS SYGMA. **Middle** Dr. Sergey Nikitin, Search Foundation Inc. **Bottom** Robert Voets/CBS-TV/ The Kobal Collection. **7-4(1)** Tiago Estima/ iStockPhoto. **7-4(2)** luri/Shutterstock. **7-4(3)** Heather Wall/iStockPhoto. **7-7** Andrew Syred/Photo Researchers. **7-10(1)** Joellen L Armstrong/Shutterstock. **7-10(2)** Oleg Kozlov, Sophy Kozlova/Shutterstock. **7-10(3)** Justyna Furmanczyk/Shutterstock. **7-12(1)** SPL/Photo Researchers. **7-12(2)** Spencer Grant/Photo Edit. **7-12(3)** Ed Kashi/Corbis. **7-17** David Parker/Photo Researchers. **7-19a** Getty Images. **7-19b** Popperfoto/Getty Images. **Table 7.3(1)** Johan Pienaar/Shutterstock. **Table 7.3(2)** Bruce Davidson/Nature Picture Library. **Table 7.3(3)** Vasilkin/ Shutterstock. **Table 7.3(4)** Arnon Ayal/Shutterstock. **Table 7.3(5)** Dennis Kunkel/Visuals Unlimited.

CHAPTER 8

Opener (L) Sion Touhig/Getty Images. **Opener (R) Top** brue/iStockPhoto. **Middle** Liangxue Lai et al. "Generation of cloned transgenic pigs rich in omega-3 fatty acids". Nature Biotechnology, 2006. Photograph courtesy of University of Missouri-Columbia. Reprinted by permissioni from Macmillan Publishers Ltd. **Bottom** H. Kuboto, R. Binrster and J E. Hayden, RBP, School of Veterinary Medicine, University of Pennsylvania.Proc. Natl. Acad. Sci. USA 101:16489-16494, 2004. **8-11a** Biology Media/Photo

Researchers. **8-11b** David McCarthy/SPL/Photo Researchers. **8-14** Pearson Science. **8-15** From Doebley, J. Plant Cell, 2005 Nov; 17(11): 2859-72. Courtesy of John Doebley/University of Wisconsin. **8-16** John Christensen, Pearson Education. **8-17** Brad Mogen/Visuals Unlimited. **8-19** Corbis/SuperStock. **8-20** Pascal Goetgheluck/SPL/Photo Researchers. **8-22** Van De Silva. 8-23 Photo Courtesy of Rosolin Institute. **Savvy Reader (L)** Jimmy Margulies, North American Syndicate. **Savvy Reader (R)** Rob Rogers © The Pittsburgh Post-Gazette/Dist. by United Feature Syndicate, Inc.

CHAPTER 9

Opener (L) James Balog/Image Bank/Getty Images Inc. **Opener (R) Top** Michelangelo, "Creation of Adam" (Detail of Sistine Chapel Ceiling) © Scala/Art Resource. **Middle** David Gifford/Photo Researchers. **Bottom** William Campbell/Corbis/Sygma. **9-3** ARCHIV/Photo Researchers. **9-5a** Schafer & Hill/Getty Images. **9-5b** Tui De Roy/The Roving Tortoise Nature Photography. **9-6a** D.C. Lowe/Superstock. **9-6b** Paul Sterry/Alamy. **9-10a** Stan Elems/Visuals Unlimited. **9-10b** B.Runk/S. Schoenberger/Grant Heilman Photography, Inc. **9-11** Richardson, M. K., et al. Science. Vol 280: Pg 983c, Issue # 5366, May 15, 1998. Embryo from Professor R. O'Rahilly. National Museum of Health and Medicine/Armed Forces Institute of Pathology. **9-12(1)** Lori Skelton/Shutterstock. **9-12(2)** Steve Kaufman/CORBIS. **9-12(3)** Hans Dieter Brandl/INTERFOTO Pressebildagentur/Alamy. **9-12(4)** Tom Vezo/Minden Pictures. **09-13a** Tui DeRoy/Minden Pictures. **9-13b** Frans Lanting/Minden Pictures. **9-13c** Robert Scanlon. **9-15** Anup Shah/Minden Pictures. **9-18(1)** Mitsuaki Iwago/Minden Pictures. **9-18(2)** Michael Najjar/Getty Images. **9-20** Javier Trueba/Madrid Scientific Films/Photo Researchers. **9-22a** The Natural History Museum, London. **9-22b** Maxwell Museum of Anthropology. **9-25** SPL/Photo Researchers. **9-28(1)** Photos.com. **9-28(2)** Dean Turner/iStockPhoto. **9-28(3)** Stanko Mravljak/iStockPhoto. **9-28(4)** R. Calentine/Visuals Unlimited. **9-30a** Associated Press. **9-30b** Steven Kazlowski/Peter Arnold Inc.

CHAPTER 10

Opener (L) Mike Cassese/Reuters/CORBIS. **Opener (R) Top** American Broadcasting Companies, Inc. **Middle** mediacolor's/Alamy. **Bottom** Michael Coddington/Shutterstock. **10-1** Kwangshin Kim/Photo Researchers. **10-2a** ISM/Phototake NYC. **10-2b** Carolina Biological Supply Company/Newscom. **10-3** Bart's Medical Library/Phototake NYC. **10-4** Nick Gregory/Alamy. **10-5** Courtesy of Caufield & Shook Collection, Special Collections, University of Louisville. **10-6a** Lynn Stone/Photolibrary. **10-6b** sil63/Shutterstock. **10-7** Tom McHugh/Photo Researchers. **10-10** Archana Bhartia/Shutterstock. **10-14** Katherine Feng/Globio/Minden Pictures. **10-15** B. Runk/S. Schoenberger/Grant Heilman. **10-19** PhototakeMedical/Newscom.

CHAPTER 11

Opener (L) Anderson-Ross/Brand X Pictures/Alamy. **Opener (R) Top** Jim Whitmer Photography. **Middle** David Butow/CORBIS SABA. **Bottom** SW productions/Getty Images. **11-1a** Keith Levit/Shutterstock. **11-1b** Keith Levit/Shutterstock. **11-2a** Tui De Roy/Minden Pictures. **11-2b** Tui De Roy/Minden Pictures. **11-3a** Steven J. Kazlowski/Danita Delimont Photography. **11-3b** Steven J. Kazlowski/Danita Delimont Photography. **11-4** Visuals Unlimited/Corbis. **11-5(1)** Eric Isselée/Shutterstock. **11-5(2)** Eric Isselée/Shutterstock. **11-5(3)** Bob Langrish/Dorling Kindersley Media Library. **11-10(1)** Suzannah Skelton/iStockphoto. **11-10(2)** Corbis RF. **11-10(3)** Peter Grosch/Shutterstock. **11-11** Getty Images. **11-13** McDonald Wildlife Photography/Animals Animals/Earth Scenes. **11-22a** Peter Turnley/CORBIS. **11-22b** Roger De LaHarpe/Gallo Images/Corbis. **11-23(1)** Corbis RF. **11-23(2)** Hiroshi Sato/Shutterstock. **11-27(1)** Brad Mogen/Visuals Unlimited. **11-27(2)** Derek Hall/Dorling Kindersley. **11-28a** Dave Logan/iStockphoto. **11-28b** PurestockX. **11-28c** Ricco Smith/iStockphoto. **11-29** Eric Gay/AP Photo. **11-30** AP Photo.

CHAPTER 12

Opener (L) David Strong, Penn State University. **Opener (R) Top** Robert Plowes/Beth Plowes. **Middle** Rabbit Creek, Yellowstone National Park. Image courtesy of Dr. Ken Stedman. **Bottom** IndexOpen. **12-1** WHOI/Edmund/Visuals Unlimited. **12-2** D. Cavagnaro/DRK Photo. **12-3a** Michel Viard/Peter Arnold. **12-3b** Mauro Fermariello/SPL/Photo Researchers. **12-6** J. William Schopf University of California at Los Angeles. **12-7a** Dennis Kunkel/Visuals Unlimited/Getty Images. **12-7b** Carolina Biological Supply/Visuals Unlimited. **12-7c** Marli Miller/Visuals Unlimited **12-9a** Jeffrey Lord. **12-9b** J.L Kerwin. **12-9c** David M. Phillips/Visuals Unlimited. **12-9d** Walter H. Hodge/Peter Arnold. **12-10** National Musem of Natural History (c) 2003 Smithsonian Institution. **12-11** Wayne Lynch/DRK Photo**12-12a** Michael Fogden/DRK. **12-12b** pomortzeff/iStockphoto. **12-12c** J Hindman/Shutterstock. **12-13** Gregory Ochoki/Photo Researchers. **12-15** Photo Researchers. **12-16a** Zoran Ivanovic/iStockphoto. **12-16b** Frans Lanting/Minden Pictures. **12-16c** Patti Murray/Animals Animals/Earth Scenes. **12-18** Dr. Linda Stannard, UCT/Photo Researchers. **12-19** Ray Pfortner/Peter Arnold, Inc. **12-20** Mayer/Le Scanff/Garden Picture Library/Photolibrary.com. **12-21a** Johan Swanepoel/Shutterstock. **12-21b** Cay-Uwe Kulzer/iStockphoto. **12-21c** Bill Dyer/Photo Researchers. **12-23** Photo by Peter S. Goltra for the National Tropical Botanical Garden. Used with permission. **12-24** Alison Wright/Corbis. **Table 12-3(1)** M. I. Walker/Photo Researchers. **Table 12-3(2)** Dennis Kunkel/Visuals Unlimited. **Table 12-3(3)** Wim van Egmond/Visuals Unlimited/Getty Images. **Table 12-3(4)** Bill Beatty/Visuals Unlimited. **Table 12-3(5)** Dee Breger/Photo Researchers. **Table 12-3(6)** Jeffrey Waibel/iStockphoto. **Table 12-3(7)** Roland Birke/Peter Arnold, Inc. **Table 12-4(1)** photos.com. **Table 12-4(2)** xxx. **Table 12-4(3)** Robert Calentine/Visuals Unlimited. **Table 12-4(4)** Dr. Richard Kessel & Dr. Gene Shih/Visuals Unlimited/Getty Images. **Table 12-4(5)** Lavigne Herve/Shutterstock. **Table 12-4(6)** E.R. Degginger/Photo Researchers. **Table 12-4(7)** Tom McHugh/Photo Researchers. **Table 12-4(8)** Randy Morse/Animals Animals. **Table 12-4(9)** Tom McHugh/Photo Researchers. **Table 12-5(1)** Dr. M.F. Brown/Visuals Unlimited/Getty Images. **Table 12-5(2)**

Shutterstock. **Table 12-5(3)** Henk Bentlage/Shutterstock. **Table 12-6(1)** Lucie Zapletalova/Shutterstock. **Table 12-6(2)** Clay Perry/Corbis. **Table 12-6(3)** Prenzel Photo/Animals Animals Earth Scenes. **Table 12-6(4)** Digital Vision/Getty Images.

CHAPTER 13

Opener (L) NASA. **Opener (R) Top** Digital Vision/Getty Images. **Middle** Pat Hamilton/Reuters New Media, Inc./Corbis. **Bottom** Larry Dale Gordon/The Image Bank/Getty Images. **13-2a** Peter Haigh/Alamy Images. **13-2b** Doug Allen/NPL/Minden Pictures. **13-2c** Carr Clifton/Minden Pictures. **13-4a** Pete Oxford/Danita Delimont/Alamy. **13-7a** Dung Vo Trung/Corbis/Sygma. **13-7b** Robert Pickett/Corbis. **13-7c** Mr. Jamsey/iStockPhoto. **13-7d** Steve Wolper/DRK Photo. **13-10(1)** Peter Menzel. **13-10(2)** Peter Menzel. **13-12** Happy Alex/Shutterstock. **13-15** Shehzad Noorani/Peter Arnold, Inc.

CHAPTER 14

Opener (L) Tupper Ansell Blake/USFWS. **Opener (R) Top** AP Photos. **Middle** Peter Essick/Aurora/Getty Images. **Bottom** AP Photos. **14-1** Laura Romin & Larry Dalton/Alamy. **14-5a** NASA/MODIS Rapid Response Team. **14-5b** Buddy Mays/Corbis. **14-5c** David Dennis/Animals Animals Earth Scenes. **14-5d** Belinda Wright/DRK Photo. **14-5e** Allen Blake Sheldon/Animals Animals. **14-5f** Thomas Pickard/iStockphoto. **14-6** Petr Maaek/iStockphoto. **14-9** Chuck Pratt/Bruce Coleman. **14-10** Ariel Bravy/Shutterstock. **14-11a** David Muench/Corbis. **14-11b** Professor John Doebley, University of Wisconsin. **14-11c** J. Morales-Ramos, USDA-ARS. **14-13** M.F.Brown/Visuals Unlimited/Getty. **14-14** Michael Durham/Minden Pictures. **14-15a** Ron Austing/Photo Researchers. **14-15b** W.M. Ciesla, FHMI. **14-17a** h46it/iStockphoto. **14-17b** Peter Weinmann/Animals Animals Earth Scenes. **14-19-01** Dave Hansen, University of Minnesota Agricultural Experiment Station. **14-19-02** Dave Hansen, University of Minnesota Agricultural Experiment Station. **14-20** franck camhi/Shutterstock. **14-27a** Andrew Syred/SPL/Photo Researchers. **14-27b** Astrid & Hanns-Frieder Michler/SPL/Photo Researchers. **14-28** Defene Dept. photo by Petty Officer 2nd Class Molly A. Burgess, USN.

CHAPTER 15

Opener (L) Loretta Hostettler/iStockphoto. **Opener (R) Top** IndexOpen. **Middle** Jim Kruger/iStockPhoto. **Bottom** Don Farrall/Getty Images. **15-4** Frank Zullo/Photo Researchers. **15-5** Jim Mone/AP Photos. **15-7** Getty Images. **15-10a** Les Cunliffe/iStockphoto. **15-10b** R.S. Ryan/Shutterstock. **15-10c** Bierchen/Shutterstock. **15-11** Geoff Bryant/Photo Researchers. **15-12** Mark Edwards/Peter Arnold. **15-13** Hans Dieter Brand/Frank Lane Picture Agency/Corbis. **15-14a** Barnpix.com/Alamy. **15-14b** James P. Jackson/Photo Researchers. **15-14c** Mares Lucian/Shutterstock. **15-15** B. Runk/S. Schoenberger/Grant Heilman Photography, Inc. **15-17** Walt Anderson/Visuals Unlimited. **15-18** Markus Divis/iStockphoto. **15-19** Chris Johns/National Geographic Society/Getty Images. **15-20** George Burba/Shutterstock. **15-21** Comstock Complete. **15-22** Tatarszkij/Shutterstock. **15-23** Steve Winter/National Geographic Society/Getty Images. **15-25** Theo Heimann/Getty Images. **15-26a** Linda Mirro/iStockphoto. **15-26b** Ernest Manewal/Photolibrary.com. **15-27** Cathy Keifer/Shutterstock. **15-28** David Muench/Getty Images. **15-29** Specta/Shutterstock. **15-30** Millard H. Sharp/Photo Researchers. **15-32** Steven Wayne Rotsch. **15-33** David R. Frazier Photolibrary/Alamy. **15-34** James Leynse/Corbis. **15-35** Daniel Stein/iStockphoto. **15-36** Virginia Borden.

Credits

CHAPTER 2, SAVVY READER
Caroline Stacey, "Detox Drinks a Clear Winner" *The Independent*, 1/3/04. Copyright © 2004. Used with permission.

CHAPTER 4, SAVVY READER
Jim Carlton, "Is global warming killing the polar bears?" *Wall Street Journal*, 12/14/05. Copyright © 2005 Dow Jones & Co., Inc. Used with permission.

CHAPTER 5, SAVVY READER
"Shark Cartilage Therapy Evaluated" by Dr. William Lane, *Dynamic Chiropractic*, 2/25/94. Copyright © 1994. Reprinted by MPA Media.

CHAPTER 6, SAVVY READER
"Nature helps create religious adults," *ScienceDaily*, 4/5/05. Copyright © 2005. Used by permission. www.sciencedaily.com

CHAPTER 7, SAVVY READER
Alan S. Miller and Satoshi Kara Zawa, "10 Politically Incorrect Turths about Human Nature," *Psychology Today*, July/August 2007, p. 93. Copyright © 2007 Sussex Publishers, LLC. Used with permission.

CHAPTER 9, SAVVY READER
© Discovery Institute

CHAPTER 10, SAVVY READER
Fwtandy, 6/8/07

CHAPTER 11, SAVVY READER
Gregory M. Lamb, "A place for race in medicine?" *Christian Science Monitor*, March 3, 2005. Copyright © 2005 The Christian Science Monitor. All rights reserved. www.csmonitor.com

CHAPTER 12, SAVVY READER
From *More Nature Cures Revealed* by Kevin Trudeau. Copyright © 2006.

CHAPTER 13, SAVVY READER
Copyright © 2008 Audubon. Used with permission. www.audubon.org

CHAPTER 14, SAVVY READER
Erik Stokstad, *ScienceNow* 8/30/07
Figure 14-4 From R.B. Primack, *Essentials of Conservation Biology*. Copyright © 1993 Sinauer Associates, Inc. Used with permission. **Figure 14-16** Adapted from N. Myers et al, *Biodiversity Hotspots for Conservation Priorities*. Copyright © 2000 Nature Publishing. Used with permission.

Index

Page references that refer to figures are in **bold**.
Page references that refer to tables are in *italics*.

D

H

T

U

V

W

X

Y

Z